CANCER ETIOLOGY, DIAGNOSIS AND TREATMENTS SERIES

HANDBOOK OF PROSTATE CANCER CELL RESEARCH: GROWTH, SIGNALLING AND SURVIVAL

Cancer Etiology, Diagnosis and Treatments Series

Cell Apoptosis and Cancer
Albina W. Taylor (Editor)
2007. ISBN: 1-60021-506-8

Chronic Lymphocytic Leukemia Research Focus
Chadi Nabhan (Editor)
2007. ISBN: 1-60021-526-2

Cervical Cancer Research Trends
Eleanor P. Bankes (Editor)
2007. ISBN: 1-60021-648-x

Lung Cancer in Women
Varetta N. Torres (Editor)
2008. ISBN: 1-60021-659-5

Lung Cancer in Women
Varetta N. Torres (Editor)
2008. ISBN: 978-1-60692-765-6 (Online book)

Cancer Research at the Leading Edge
Ignatius K. Martakis (Editor)
2008. ISBN: 1-60021-728-1

Chronic Lymphocytic Leukemia: New Research
Inès B. Moreau (Editor)
2008. ISBN: 978-1-60456-081-7

Cancer and Stem Cells
Thomas Dittmar and Kurt S. Zander (Editors)
2008. ISBN: 978-1-60456-478-5

Cancer Prevention Research Trends
Louis Braun and Maximilian Lange (Editor)
2008. ISBN: 978-1-60456-639-0

Clinical, Genetic and Molecular Precursor Features in Colorectal Neoplasia
Kjetil Søreide and Håvard Søiland (Editors)
2008. ISBN: 978-1-60456-714-4

Human Polyomaviruses: Molecular Mechanisms for Transformation and their Association with Cancers
Ugo Moens, Marijke Van Gheule and Mona Johannessen
2009. ISBN: 978-1-60692-812-7

Molecular Therapy of Breast Cancer: Classicism Meets Modernity
Marc Lacroix
2009. ISBN: 978-1-60741-593-0

Molecular Therapy of Breast Cancer: Classicism Meets Modernity
Marc Lacroix
2009. ISBN: 978-1-60876-726-7 (Online Book)

Aromatase Inhibitors: Types, Mode of Action and Indications
Jean R. Lamonte (Editor)
2009. ISBN: 978-1-60741-711-8

Anticancer Drugs: Design, Delivery and Pharmacology
Peter Spencer and Walter Holt (Editors)
2009. ISBN: 978-1-60741-004-1

Anticancer Drugs: Design, Delivery and Pharmacology
Peter Spencer and Walter Holt (Editors)
2009. ISBN: 978-1-60876-629-1 (Online Book)

Cancer Biology: An Updated Global Overview
Tarek H. EL-Metwally
2009. ISBN: 978-1-60876-193-7

Drug Resisant Neoplasms
Ethan G. Verrite (Editor)
2009. ISBN: 978-1-60741-255-7

Handbook of Prostate Cancer Cell Research: Growth, Signalling and Survival
Alan T. Meridith (Editor)
2009. ISBN: 978-1-60741-954-9

CANCER ETIOLOGY, DIAGNOSIS AND TREATMENTS SERIES

HANDBOOK OF PROSTATE CANCER CELL RESEARCH: GROWTH, SIGNALLING AND SURVIVAL

ALAN T. MERIDITH
EDITOR

Nova Biomedical Books
New York

For permission to use material from this book please contact us:
Telephone 631-231-7269; Fax 631-231-8175
Web Site: http://www.novapublishers.com

Library of Congress Cataloging-in-Publication Data

Handbook of prostate cancer cell research : growth, signalling, and survival / [edited by] Alan T. Meridith. p. ; cm.
Includes bibliographical references and index.
ISBN 978-1-60741-954-9 (hardcover : alk. paper)
1. Prostate--Cancer--Molecular aspects--Handbooks, manuals, etc. 2. Cancer cells--Proliferation--Handbooks, manuals, etc. 3. Cell transformation. I. Meridith, Alan T.
[DNLM: 1. Prostatic Neoplasms--pathology. 2. Cell Proliferation. 3. Cell Survival. 4. Cell Transformation, Neoplastic. 5. Intracellular Signaling Peptides and Proteins. 6. Signal Transduction. WJ 752 H2358 2009]
RC280.P7H36 2009
616.99'463071--dc22 2009024628

Published by Nova Science Publishers, Inc. ✦ New York

Contents

Preface

Prostate cancer is a disease in which cancer develops in the prostate, a gland in the male reproductive system. It occurs when cells of the prostate mutate and begin to multiply out of control. These cells may spread (metastasize) from the prostate to other parts of the body, especially the bones and lymph nodes. Prostate cancer may cause pain, difficulty in urinating, problems during sexual intercourse, erectile dysfunction and other symptoms. However those symptoms are present only in an advanced stage of the disease. Prostate cancer is the most common malignancy diagnosed in men and is the second leading cause of cancer deaths among males. Androgens play a major role in the development of prostate cancer since males castrated during childhood do not develop this disease. Another factor known to be involved in prostate cancer development is age, since the majority of prostate cancer patients are aging men. Rates of detection of prostate cancers vary widely across the world with South and East Asia detecting less frequently than in Europe, and especially the United States. This new important handbook gathers the latest research from around the globe in this field.

Chapter 1 - Since the pioneering studies of Huggins in 1941, it has been known that prostate cancer (PC) cells, like certain normal epithelial cells, chronically depend on a critical level of androgenic stimulation for their continuous growth and survival. Development of resistance to androgen deprivation therapy (ADT) is an inevitable evolution of prostate carcinogenesis thought to occur because a portion of cells can survive without androgen stimulation due to takeover of the tumor mass by androgen-independent (AI) tumor cells, as well as to emergence of apoptosis-resistant clones as a result of further genetic alterations. Although advances in terms of symptoms palliation and quality of life improvement have been addressed with current treatment options, innovative approaches are needed to improve survival rates. In this regard, knowledge of the mechanisms underlying resistance is of extreme importance for drug development.

In the past century, natural products have been invaluable as platforms for developing front-line active molecules. Evidence is provided by the fact that natural products-based drugs (parent compounds, derivatives, analogues, and mimics) are major entities among the Food and Drug Administration (FDA)-approved drugs (over 50%), at present. Moreover, recent literature has suggested a role for certain dietary agents in cancer chemoprevention. Optimized bioassay-guided fractionation of extracts of various plants, marine organisms, and microorganisms has made reasonable amounts of pure compounds available for unequivocal characterization and *in vitro* and *in vivo* testing, in order to establish biological activities and mechanisms of action.

This chapter describes the action of selected natural products and their synthetic analogues on PC cell growth, signaling and survival focusing on chemical structures, molecule source and targets. This discussion is intended to provide the reader with a multidisciplinary overview of natural products chemistry with potential for PC treatment and/or prevention as well as with a valuable tool for lead compound selection in future drug development.

Chapter 2 - In recent years, positron emission tomography/computed tomography (PET/CT) with [11C]Choline has been increasingly used for the evaluation of prostate cancer. In vitro and in vivo magnetic resonance spectroscopy studies have shown increased choline compounds and over-expression of choline kinase in prostate cancer. Liver and kidney are the major sites for choline oxidation, while negligible activity is present in the bladder, representing a significant advantage for prostate imaging.

In patients with prostate cancer, [11C]Choline has limited sensitivity and specificity for distinguishing malignant vs. benign prostate cells. There is no significant association between prostate [11C]Choline uptake and PSA levels, Gleason score and pathological stage, while anti-androgenic therapy interferes with the uptake of the tracer. The reduction of prostate [11C]Choline uptake following anti-androgenic hormonal therapy indicates that PET/CT with [11C]Choline could be used to monitor the response to treatment. The accuracy of the technique for lymph node staging has to be further investigated.

There is good agreement among several centres on the utility of PET/CT with [11C]Choline for the detection of recurrence of disease after radical prostatectomy in patients with biochemical failure. The percent of positive scans increased with increasing PSA values, being this technique clinically valuable for restaging prostate cancer patients even with low PSA levels. Clinical guidelines to proper indicate a PET/CT with [11C]Choline study in prostatectomized patients will be defined.

Chapter 3 - Prostate cancer (PCa) is initially regarded as an androgen-dependent disease, as therapeutic androgen ablation induces massive apoptosis of malignant cells and clinically leads to tumor regression. However, after a remission period of up to several years, mutations occur and tumors usually progress despite low circulating androgen levels. Currently, no therapeutic options are available for the treatment of androgen-independent PCa.

One relatively new paradigm in explaining the biology of the Prostate cancer is the possible existence of cancer stem cells (CSC). The mechanisms of progression of prostate cancer have been extensively studied, yet are poorly understood. To explain these mechanisms and the heterogeneity in prostate cancer, one of the concepts that has evolved is that cancer arises from the neoplastic transformation of normal prostatic epithelial stem cells or transit amplifying cells. Based on this stem cell model, understanding the stages of cell differentiation in normal prostate epithelium is essential for the identification of cells that are involved in prostate carcinogenesis and ablation-resistant prostate cancer. Normal tissues arise from stem cells that grow and differentiate to create progenitor and mature cell populations. For normal stem cells, the most important standard is that the cell can engage a multilineage repopulation of a tissue or organ. Both somatic stem cells and CSCs have the capacity for organogenesis, but somatic stem cells generate normal tissues and tumor stem cells generate tumor tissues. The most crucial standard for a CSC is its ability to re-initiate a serially transplantable tumor that recapitulates the original heterogeneity and tumor histology. The CSCs theory postulates that CSCs are rather a functional definition, inferring that the expression of a putative stem cell marker from specific cell population does not mean at all that these cells are stem cells. Using this gold standard, many so-called prostate CSCs reported so far are based on correlative studies rather than being true CSCs.

Recent evidence has shown that cancer stem cells (CSCs) are a subset of the tumor cells that are responsible for initiating and maintaining the disease. Understanding normal stem cells and CSCs may provide insight into the origin of and new therapeutics for prostate cancer. Normal stem cells and CSCs have been identified in prostate tissue by the use of several markers or techniques. Although research on stem cells has been limited by the lack of suitable in vitro systems, recent studies show that not only primary cells but also several established cell lines may exhibit stem cell properties.

Various in vitro culture systems to propagate normal prostate stem cells and prostate CSCs together with molecular markers are reviewed together with transgenic models of prostate cancer in conjunction with meta-analyses based on genomics and proteomics data supporting a novel hypothesis regarding the pathways and genes involved in the transformation of normal prostate cells into neoplastic and their subsequent loss of androgen-receptor (AR) sensitivity.

Based on CSC theory, CSCs are biologically distinct from non-CSCs in the overall population and are thought to be resistant to conventional chemotherapy or radiation therapy. Most prostate cancer cells express AR and PSA, and this phenotype mimicking luminal cells has been reported. AR signaling in luminal cells not only regulates PSA expression, but it is also required for survival, and this requirement is maintained in early cancers.

Therefore, AR has a crucial role in the progression and treatment of prostate cancer. Although prostate cancer cells usually express AR, AR expression is heterogeneous in most primary prostate cancers and their metastases.

The signal transduction pathways in androgen-dependent and -independent prostate cancer cell proliferation were also being reviewed. Some of them include Akt, extracellular-signal-regulated kinase (ERK), and mitogen-activated protein kinase members' family (MAPK) and are involved in poor clinical outcome of prostate cancers. Akt activation, but not ERK activation, is correlated with proliferation in human prostate tumors. In connection with these pathways, a role has been established of the mTOR, whose specific inhibition has a growth-inhibitory effect on androgen-independent prostate carcinoma cell lines. The latter is involved in the metabolic stress pathways in tumor cells, thus giving rise to the hypothesis regarding the "shift" to metabolic and regulatory pathways supporting the overall survival in prostate cancer cells.

Chapter 4 - Prostate cancer is the second most prevalent type of cancer diagnosed in men in the western world and the leading cause of cancer mortality in older men. Since early-stage development of prostate cancer depends on androgens for growth and survival, androgen deprivation by surgical or chemical castration has been used as the main therapeutic intervention in the treatment of the disease. Despite initial responsiveness to androgen ablation therapy, tumor cells invariably progress to an androgen-independent and anti-androgen resistance that is untreatable, ultimately leading to mortality. It is well established that androgen receptor mediated signaling is critical not only in normal prostate functioning but also in the development and progression of prostate cancer. Given its pivotal role in prostate cancer progression, it is important to delineate its interaction with other signaling pathways that may contribute to prostate cancer growth and development. This is particularly necessary since the signaling mechanisms involved in the transition from normal to androgen-dependent and then to androgen-independent prostate cancer are poorly understood. It has become increasingly evident that crosstalk exists between the androgen receptor and other signaling pathways. This suggests the existence of a coordinated mechanism where there is a converging point for generated signals, thereby driving prostate cancer growth and aggressiveness, downstream. This chapter therefore focuses on the role that the androgen receptor plays in the major signaling pathways implicated in prostate cancer development and progression.

Chapter 5 - Normal prostate gland development requires many coordinated cellular processes through prostate stem cells, including epithelial proliferation, mesenchymal-epithelial interaction, ductal branching morphogenesis, and ductal canalization.

Stem cells are defined functionally as cells that have the capacity to self-renew as well as the ability to generate differentiated cells. Stem cells generate daughter cells identical to their mother (self-renewal) as well as produce progeny with more restricted potential (differentiation). Self-renewal is achieved by symmetrical cell division while maintaining pluripotency; and this can be modulated by extrinsic factors, transcriptional regulator, and effectors.

Even though current research is focused on the identification, characterization, and isolation of stem cells from the adult, to date, most adult stem cells have restricted potential, and achieving indefinite proliferation and expansion of the stem cells in culture is still not routine.

One of the critical questions in the adult stem cell field concerns the mechanisms that regulate the decision between self-renewal and differentiation. During embryogenesis, cells are initially proliferative and pluripotent; they only gradually become restricted to a different cell fate. Adult stem cells have two fundamental properties: a long-term capacity to divide and the ability to produce daughter cells that either retain stem cell identity or initiate differentiation along the appropriate lineage(s). The balance between self-renewal and initiation of differentiation is crucial.

In regard to prostate cells, although signals from the luminal layer specify self-renewal or stem cells identity, other signals from the microenvironment also play roles in regulating the fate of daughters of stem cell decisions. Interactions between progenitor cells (basal layer) and mature cells (luminal layer) appear to restrict expression of stem cell characteristics and ensure that stem cell daughters displaced away from the basal cells for initiating differentiation. Thus, the luminal cells play an important guardian role, preventing accumulation of basal cells with ambiguous identity.

The signal that controls which daughter cell of an adult stem cell remains as a stem cell and which begins the process of differentiation may be mediated through a number of signaling pathways including, Wnt, Hedgehog (Hh), Notch, Oct4, Bone Morphogenetic Proteins (BMPs), JAK/Stat, or other signaling pathways.

Chapter 6 - Prostate cancer is the most frequently diagnosed cancer and a leading cause of cancer deaths. Epidemiological and clinical intervention studies support the protective role of selenium against the development of prostate cancer. Earlier studies have shown that inorganic forms of selenium significantly decrease the expression of the androgen receptor (AR) and prostate-specific antigen (PSA), which are widely used clinically for prostate cancer diagnostics and as indicators of therapeutic efficacy and recurrence. More recently, organoselenium compounds were found to exhibit differential effects on growth, PSA secretion, oxidative stress and selective gene responses in human prostate cancer. Furthermore, literature reports have consistently shown that the dose and chemical forms of selenium are important factors in anticancer properties. However, the mechanisms responsible for the inhibitory activity by this micronutrient remain elusive and could involve separate pathways. One hypothesis concerns selenium metabolites, especially methylselenol, which are implicated as active metabolites for the anticancer effect. Other possible mechanisms involve changes in enzyme activities and/or mRNA levels of glutathione transferase (GST), apoptosis induction, caspase-3 activation, protein kinase regulation and cell cycle modulation. The information outlined above supports the idea that organoselenium compounds could be important new agents in prostate cancer treatment. In recent years the authors' team has designed and synthesised several organoselenium derivatives that have shown interesting antiproliferative activity on the PC-3 cell line. The authors present here a review of significant organic structures that contain selenium and have marked prostate cancer antitumoural activity.

Chapter 7 - Prostate cancer is the second leading cause of cancer-related mortality in the human male population. Conventional external beam doses below 68 Gy have a limited curative potential for the locally advanced stages of prostate cancer (LAPC) because two-thirds are hypothesized to experience disease relapse every 5 years after a treatment. Since it is well-established that ionizing radiation (IR) induces cellular toxicity through mechanisms involving generation of intracellular reactive oxygen species (ROS), the fate of irradiated cells is believed to be controlled by networks of signaling elements leading to different modes of cell death or cell survival. In this light, molecular mechanisms for the LAPC radiation resistance are not quite clear. In this chapter, a combined approach, based on experimental cell biology studies and systems biology strategies, has been shown to be of value for identifying signaling molecules with therapeutic relevancy, thus making more localized and efficient treatments of various ROS-mediated pathologies substantially facilitated. Since morphological, physiological and molecular studies on death of cells in tissues are usually carried out on isolated cell populations due to known difficulties manifested by interactions with surrounding cells, systems biology is herein considered and established to be a means of overcoming these difficulties. The practical usefulness of a systems biology approach in rationalizing the ability of cells to interpret the same TNF-induced signal action in distinct fashions, survival vs. death signal transduction, has been confirmed. Taking into account the fact that the low-LET-IR effects are primarily reflected by ROS generation, in the context of the radioresistance of PC-3 cells, the IR-induced expression both of antioxidant enzymes (MnSOD, CuZnSOD, catalase) and of the transcriptional nuclear factor-kappa B (NF-κB) has been discussed. Besides, a systems biology model, aimed to shed more light on how MnSOD affects the biological state of

PC-3 cells depending on the production of H_2O_2, has been described. Due to an excellent agreement between the experimental results and the systems biology insights, the radioresistance of PC-3 cells has been suggested to be a consequence of the positive, feed-forward vicious circle established between the H_2O_2-mediated elevation of MnSOD expression. The inclusion of some other important pathways, such as the hydroxyl radical (·OH)-mediated activation of NF-κB, has been demonstrated to bring oscillations in the model, thus indicating a broad range of possible instabilities that might have undesirable consequences on the development of successful therapies. Although the NF-κB signaling pathway in the model consists of 26 molecular components linked by 64 unidirectional reactions, a phase plane interpretation of the model has indicated that a much lower dimensional representation of the system of 64 kinetic equations could narrow down the range of possible instabilities that the oscillatory signaling may exhibit.

Chapter 8 - Transcription is currently one of the most prominent fields in cell biology, endocrinology and cancer biology, with the ubiquitin-proteasome system in particular being an area of intensive study. The androgen receptor (AR) is a transcription factor belonging to the family of nuclear receptors that mediate the action of androgen. The AR plays an important role in normal development of the prostate as well as in the progression of prostate cancer. The AR is regulated by several post-translational modifications, including phosphorylation, acetylation and ubiquitination. Activation of the AR affects the cell growth of prostate cancer and the expression and secretion of prostate-specific antigen (PSA), one of the most reliable diagnostic markers for prostate cancer. Furthermore, the AR has been reported to regulate several cell cycle regulators such as cyclins, retinoblastoma tumor suppressor (Rb) and cyclin-dependent kinase inhibitor (CKI), suggesting that AR is directly involved in cell cycle regulation, especially in prostate cancer cells. Recently, several enzymes for modifications of the AR have been identified and the molecular mechanism for AR activation is gradually being elucidated. It is important to clarify the mechanism of AR-dependent cell proliferation for identifying a novel diagnostic tool and a potentially therapeutic target for prostate cancer.

Chapter 9 - The drug, cisplatin (*cis*-diamminedichloroplatinum(II)) is routinely used worldwide to cure testicular, ovarian and many other cancers. Cisplatin and its analogs DWA-2114R {2-aminomethylpyrrolidine(1,1-cyclobutanedicarboxylato) platinum(II) and JM216 (cis-(ammine)(cyclohexylamine)dichloro-*trans*-(diacetato) platinum(IV)) are also found to be active against prostate cancer. Unfortunately, these and other chemotherapeutic drugs are toxic. Recently, some organorhenium(I) complexes have been reported to be cytotoxic against various cancer cell lines including prostate cancer cell lines. Surprisingly, these complexes are not toxic to prostate and other cancer cell lines according to these researchers.

This prompted the authors to synthesize a wide variety of mono-, di-, and tetrarhenium (diphosphine) and (diimine) carbonyl complexes. They were characterized by spectroscopic techniques and X-ray crystal structure determinations. As expected, the rhenium(I) (α-diimine) tricarbonyl complexes exhibit fluorescence. Both rhenium(I) (diphosphine) and (diimine) carbonyl complexes are cytotoxic against prostate cancer cell lines. However, the rhenium(I) (α-diimine) tricarbonyl complexes are preferable to diphosphines due to their fluorescence properties.

Chapter 10 - Eicosanoids have received considerable research attention as mediating agents in inflammatory and physiological processes, and there is some evidence for a role in neoplasia.

Eicosanoids′ well-recognized ability to modulate DNA/RNA synthesis, cell membrane activity, cell communication and immune response may mediate neoplastic processes including: tumor cell proliferation, tissue invasiveness, the spread of metastasis, immune surveillance and hormone responsiveness.

COX-2 induction is rapid but transient and there can be rapid induction in response to a variety of stimuli occurring over a similar time course to other immediate-early genes, like c-fos and c-myc.

Experimental studies have suggested that biosynthesis of arachidonic acid-derived eicosanoids such as prostaglandin E_2 via the cyclooxygenase pathway may play a significant role in supporting cell proliferation in human prostate cancer.

However, the exact role of COX-2 expression and eicosanoid production in prostate carcinogenesis and cancer progression remains unclear since precise mechanisms at the cellular and molecular level have not been established.

Chapter 11 - Delineating signaling pathways active in prostate epithelial cells will lead to a better understanding of those processes essential for prostate cancer progression, such as uncontrolled proliferation, survival as a consequence of resistance to cell death, invasion and metastasis, and for development of resistance to therapies such as androgen ablation and radiation therapy. Additionally, once mechanistic information is acquired, identified gene products can be used as determinants of high-grade tumor development leading to acquisition of prognostic biomarkers or therapeutic targets for intervention. The authors' approach involves identifying gene expression pathways that regulate these essential functions, especially focusing on those genes related to proliferation and genotoxic stress response, examining interactions of transcription factors (TFs) such as androgen receptor (AR), p53, retinoblastoma (Rb), E2F, and WT1. Although many gene targets regulated by these TFs are known, there is relatively little information available about common regulatory pathways shared between these genes. Thus, the authors have sought evidence of overlapping target genes common between these TFs. A mechanistic understanding of how these key TFs regulate their gene targets, such as identification of essential cofactors and other cooperating or competing TFs will lead to additional therapeutic and diagnostic applications. Here, they describe their global gene expression studies applied to prostate cancer epithelial cells using LNCaP derivative C4-2 cells, with altered androgen responsiveness. Irradiation used as a prototypic genotoxic stress response model system has revealed many induced and repressed genes. For those genes repressed by genotoxic stress that are candidate radiation sensitive genes they seek a better understanding of their mechanisms of coordinate expression. A list of genes repressed at two-fold level or higher among irradiated C4-2 cells was identified, and functional groups of genes were designated using Gene Ontology annotations. Additionally, promoter sequences of genes from the top two functional clusters were examined to identify the presence of TF binding sites (TFBS) of interest. Genes with detected TFBS were further considered as candidates for functional studies as potential target genes. Here the authors report the co-occurrence of predicted TFBS for E2F, p53, and WT1 within the promoters of radiation suppressed mitotic genes. Examining relationships between TFs and their potential target genes will provide important insight into pathways essential for prostate cancer progression that could lead to the development of the next generation of diagnostics and therapeutics.

Chapter 12 - Due to intensive investigations for the last ten years it has become increasingly clear that tumor microenvironment plays a critical role in prostate carcinogenesis. Cancer progression depends on either molecular alterations of epithelial cells and tumor-associated stroma, which provides required changes for growth, invasion and distant metastasis of tumor epithelium. Tumor microenvironment undergoes significant modification such as protein expression alterations and various genetic changes of stromal cells. Accumulation of multiplicity genetic alterations such as tumor suppressor genes hypermethylation, allelic imbalance, loss of heterozygosity, mutations, telomere shortening is typical not only for cancer epithelial cells but for tumor-associated fibroblasts and endothelial cells as well. Furthermore, these stromal cell alterations are not tumor epithelium specific.

The authors investigated epigenetic changes and allelic imbalance (LOH/AI) in prostate cancer epithelia, tumor-associated stroma and prostate intraepithelial neoplasia (PIN) in prostatectomy specimens from patients with pT1–T4 stage prostate cancer by using laser capture microdissection. Microsatellite allelotyping was evaluated using 4 highly polymorphic markers for chromosomal regions 8p22, 16q23-24 and 13q14. The methylation status of four genes significant in prostate carcinogenesis *P16*, *HIC1*, *N33* and *GSTP1* were investigated in tumor epithelia, tumor-associated stroma, normal epithelia, foci of PIN and benign prostate hyperplasia, and stroma adjacent to tumor tissues.

Their results show that genetic and epigenetic aberrations of stromal cells are the frequent event present on all stages of prostate cancer development. First, frequencies of genetic aberrations are as high in stromal cells as in tumor cells. Second, molecular aberration spectrum in stroma and tumor are discordant. Third, stromal AI frequencies for initial prostate cancer stages demonstrate at least a 2-fold excess over epithelium ones. Fourth, some methylation in both hyperplastic epithelium and stromal cells were observed in normal-appearing tissues located adjacent to tumors.

Thus numerous stroma investigations testify that cancer is not an epithelium cell disease but one of the whole tumor microenvironment. "Which genetic alterations are primary?" That is the question for elucidation of prostate cancer's main mechanisms. Further research of the molecular mechanisms underlying tumor-stromal interaction may yield improved medical treatments for prostate cancer growth and metastasis and open potentially novel compartments for targeted therapy and prevention.

Chapter 13 - Prostate cancer pathogenesis is the result of loss of balance in a number of cellular functions including cell proliferation, growth, death and survival. It has been established that misregulation of cell cycle progression can lead to an accelerated cell division, loss of appropriate checkpoints, or increased genetic instability. Accumulating data showed that the number of signal transduction pathways that control these events leading to carcinogenesis is not unlimited in the cell. Therefore, interrupting these pathways should conceivably lead to the attenuation of cancer. However, the clinical resolution in prostate cancer cure is very limited. Despite the discovery of many oncogenes and tumor suppressors, their specific roles in prostate cancer development remain a considerable dichotomy. Researches have provided increasing evidences that oncogenes can promote apoptosis while the expression of tumor suppressors can be correlated with tumor progression. To understand the molecular and cellular mechanisms of prostate cancer, signaling pathways dictated by the anti- and pro-oncogenic pathways need to be integrated in the context-dependent manner, rather than as a generalization. This chapter will review some recent progresses and problems facing prostate cancer research. For future directions, defining the specific roles of pro- and anti-oncogenic elements in their respective signaling pathways at the different stages of prostate cancer growth and survival presents unique opportunities as well as challenges for drug designs and therapeutic intervention.

Chapter 14 - Nowadays, cancer is one of the major leading causes of death worldwide. Because of the improvement of primary cancer site control by significant advances in treatment, cancer bone metastases have become a major problem for patients with cancers. Indeed, prostate cancer metastases to bone are observed in around 80% of prostate cancer patients and represent the most critical complication of advanced prostate cancer, frequently resulting in significant morbidity and mortality.

Bone is a well-known target organ of cancer metastasis, as well as the lungs and the liver, and represents a good microenvironment for the colonization and progression of a tumor mass. The vicious cycle formed between cancer cells and bone cells in the bone microenvironment has been proposed to explain the high trend of tumor development in bone.

Recently, the biological mechanisms that govern prostate cancer metastases to bone at the molecular level were disclosed. Accumulating findings have strongly suggested the important involvement of the key regulators of bone metabolism, Receptor Activator of NF-κB ligand (RANKL)/RANK/Osteoprotegerin (OPG), in prostate cancer bone metastasis development, prostate cancer cell migration and its survival. It has been reported that blocking of RANKL-RANK binding significantly diminished prostate cancer progression in bone both in indirect and direct fashion.

The present chapter summarizes the current knowledge on the development of bone metastases derived from prostate cancer. Recent data on cancer cell migration and molecular pathways involved in cancer cell survival in the bone microenvironment identifying novel therapeutic targets of the prostate cancer bone metastases are described.

Chapter 15 - Prostate cancer has become the major public health issue for men in the past decade. Prostate cancer incidences have increased in the past few years, although rates vary considerably across the world. More advanced technology to increase early detection rates must also be taken into consideration and have somewhat countered the increased rates of prostate cancer. Nevertheless, prostate cancer is the second most leading cause of death after lung cancer in both the UK and USA. Risk factors, include eating habits (e.g., fats vs. vegetables and antioxidants, vitamin D deficiency), race, sexual habits, and inherented genetic susceptibilities all contribute to prostate cancer rates. The exact causes of prostate cancer are unclear and promise to be complex. As in all other cancers, the development of prostate cancer is a multi-step process involving genes that govern cell growth and differentiation and their interaction with environmental and micro-environmental changes during life. Major host factors affecting the progression of prostate cancer include hormonal levels, androgen receptor (AR) and polymorphisms in AR, intracellular heat shock proteins, the extracellular heat shock factor, clusterin, and cell cycle regulators (e.g., cyclin A1). A prognosis of prostate cancer depends greatly on patient responses to androgen withdrawal therapy. Treatment options have varied greatly in the past decade and have matched our knowledge of the disease. For example, antisense therapies against proposed cancer stem cell targets, bcl-2, clusterin, or telomerase have been proposed. Treatments targeting tumour angiogenesis and metastasis, as well as more conventional chemotherapeutic or radiotherapeutic methods, are described. We focus on prostate cancer risk factors, causes, and progression and shed more light into some of the newer treatment modalities available.

Chapter 16 - In the United States, prostate cancer ranks as the second leading cause of cancer-related mortality for men, behind only lung cancer. Prostate cancer is also the most frequently diagnosed cancer overall, affecting almost 200,000 each year. While the 5-year survival rates are excellent, especially in comparison to all other cancers, metastasis remains a problem for clinicians and is a leading contributor to mortality. In addition, the molecular mechanisms governing the development of prostate disease need to be elucidated. Lysophosphatidic acid (LPA) signaling induces significant molecular pathways that may contribute and/or exacerbate cellular effects in prostate cancer. Changes in extracellular protein expression, motility, cross-communication with the epidermal growth factor signaling pathway and enhanced ERK signaling are effects mediated by LPA in prostate cell lines. In addition, intracellular enzymatic production of LPA may provide a survival signal in prostate. Herein we revisit and summarize these LPA-mediated molecular mechanisms and receptor signaling which contribute to prostate cancer cell differentiation, migration, survival and growth and discuss LPA signaling as an early inducer of abnormalities that occur in prostate disease. Future drug targeting of LPA signaling pathways may have clinical benefit against prostate cancer.

Chapter 17 - Cellular or tissue reduction-oxidation (redox) state is the result of the net balance of molecular reducing and oxidizing equivalents, and can be subdivided into intra-or extracellular redox state (inside or outside the cell), respectively. Under physiological conditions, redox state regulates cell metabolism, survival, and the cell cycle. Prostate cancer is the most common malignancy diagnosed in men and is the second leading cause of cancer deaths in males in the U.S.A. Many studies have documented that prostate cancer cell lines have altered levels of intracellular reactive oxygen species/reactive nitrogen species and/or antioxidant levels in relation to nonmalignant cells. This abnormal intracellular redox state may play an important role in prostate cancer cell cycle progression and cell survival through regulation of redox- modulated transcription factors. The authors' laboratory recently reported that prostate cancer cells in the cell cultures exhibited an increased glutathione/glutathione disulfide ratio in the media in comparison to immortalized but nonmalignant prostate epithelial cells, suggesting an alteration of extracellular redox state of prostate cancer cells. The authors also demonstrated that modulation of extracellular redox state by overexpression of extracellular superoxide dismutase resulted in inhibition of prostate cancer cell invasiveness through a mechanism responsive to superoxide radical and/or nitric oxide. They propose that both intra- and extracellular redox states play complementary role(s) in regulation of prostate cancer cell proliferation and/or invasion. This possible relationship between redox states may lead to the development of anti-cancer therapeutics utilizing redox-modulating compounds that are specifically designed to target prostate cancer cells and/or the prostate cancer microenvironment.

Chapter 18 – Purpose: To investigate human telomerase mRNA in the serum of prostate cancer patients.

Experimental Design: The aim of this study was to investigate the telomerase mRNA detection in the serum of patients with a prostate tumour by using Real-Time reverse transcription PCR. The results were compared with biological samples obtained by age-matched normal donors.

Results: The data demonstrated that telomerase mRNA is detectable in the serum of patients with different prostate tumours whereas it is not amplifiable in normal donors.

Conclusions: This marker, revealed with the molecular method of quantitative PCR in serum, may be useful for diagnosing and monitoring prostate cancer patients.

Chapter 19 - Electromagnetic fields (EMF) may contribute to the pathogenesis of prostate cancer, but the mechanism is obscure. Numerous calcium/phosphorus crystalloids with various morphologies have been found in the normal prostate gland and in both benign and malignant prostate lesions. Their distribution within the gland is heterogeneous and diffuse. The authors propose that an environmental EMF may exert a piezoelectric effect on these crystalloids, resulting in simultaneous, multidirectional and diffuse compression or expansion and consequently a slight mechanical distortion of the prostate. Such distortion is known to alter gene expression in many kinds of cells and therefore cell phenotype; such alterations can suppress apoptosis. In this paper we explore the scientific and clinical evidence favoring this hypothesis and present a mathematical model of the proposed cell-mechanical effect.

Chapter 20 - In men, prostate cancer (PCa) is the most common malignancy and a leading cause of cancer-related deaths in the western world. So far, measurements of prostate specific antigen (PSA) levels in the blood has proved fruitful diagnostically, increasing the number of prostate cancer patients diagnosed with the disease at its early-stage, when it is still confined to the prostate gland. This allows for early-detection and treatment of the disease at the early stages resulting in a significant decline in mortality rates. However, current treatments have limited efficacies. Following androgen ablation therapy, there is the onset of the hormone-refractory or androgen-independent prostate cancer. In general, a major problem faced in the management of prostate cancer is the lack of a reliable prognostic biomarker that will predict the clinical course of the disease.

In: Handbook of Prostate Cancer Cell Research...
Editor: Alan T. Meridith

ISBN: 978-1-60741-954-9

Chapter 1

Natural Products Action on Prostate Cancer Cell Growth, Signaling and Survival

Jorge A. R. Salvador[a,b,*] ***and Vânia M. Moreira***[a]

[a]Laboratório de Química Farmacêutica, Faculdade de Farmácia, Universidade de Coimbra, Pólo das Ciências da Saúde, Azinhaga de Santa Comba, 3000-548 Coimbra, Portugal

[b]Instituto Pedro Nunes-Labpharm, Rua Pedro Nunes, 3030-199 Coimbra, Portugal

Abstract

Since the pioneering studies of Huggins in 1941, it has been known that prostate cancer (PC) cells, like certain normal epithelial cells, chronically depend on a critical level of androgenic stimulation for their continuous growth and survival. Development of resistance to androgen deprivation therapy (ADT) is an inevitable evolution of prostate carcinogenesis thought to occur because a portion of cells can survive without androgen stimulation due to takeover of the tumor mass by androgen-independent (AI) tumor cells, as well as to emergence of apoptosis-resistant clones as a result of further genetic alterations. Although advances in terms of symptoms palliation and quality of life improvement have been addressed with current treatment options, innovative approaches are needed to improve survival rates. In this regard, knowledge of the mechanisms underlying resistance is of extreme importance for drug development.

In the past century, natural products have been invaluable as platforms for developing front-line active molecules. Evidence is provided by the fact that natural products-based drugs (parent compounds, derivatives, analogues, and mimics) are major entities among the Food and Drug Administration (FDA)-approved drugs (over 50%), at present. Moreover, recent literature has suggested a role for certain dietary agents in cancer chemoprevention. Optimized bioassay-guided fractionation of extracts of various plants, marine organisms, and microorganisms has made reasonable amounts of pure

* E-mail address : salvador@ci.uc.pt

compounds available for unequivocal characterization and *in vitro* and *in vivo* testing, in order to establish biological activities and mechanisms of action.

This chapter describes the action of selected natural products and their synthetic analogues on PC cell growth, signaling and survival focusing on chemical structures, molecule source and targets. This discussion is intended to provide the reader with a multidisciplinary overview of natural products chemistry with potential for PC treatment and/or prevention as well as with a valuable tool for lead compound selection in future drug development.

1. Introduction

PC is a common disease worldwide. Apart from skin cancer, PC is one of the most common cancers in men in developed countries such as the USA and Australia, as well as in the European Union [1-4]. It is also a leading cause of mortality and the second most common cause of cancer-related death in both the USA and Australia, and the third most common cause of cancer-related death in the European Union [1-4].

Androgens have been known to play a compulsory role in the embryonic development and adult function of the prostate [5]. Prostatic secretory epithelial cells express the androgen receptor (AR) and require chronic androgenic stimulation for survival and functional integrity. If androgen levels drop below a threshold as is the case in castration, the secretory cells undergo apoptosis, causing glandular involution [5]. Apart from male physiology, androgens are also involved in the development, growth and progression of PC. Huggins *et al.* introduced ADT for advanced and metastatic PC in 1941 [6, 7] and ever since this strategy has been applied for advanced PC treatment.

The progression from androgen-dependent (AD) to hormone-refractory PC (HRPC) has been the subject of intensive study. This transition typically occurs in less than 2 years following the onset of ADT [8]. Mechanisms that may mediate PC adaptation to decreased androgen levels include AR amplification [9], AR gene mutations [10, 11], alterations in the balance between transcriptional coactivators and corepressors and activation of other signal transduction pathways that by-pass the AR [8, 12, 13]. Moreover, a major reason why AI PC cells are resistant to the currently available therapies is their extremely low rate of proliferation. Less than 10% of such cells proliferate during a given day [14]. In contrast, high numbers of normal host cells proliferate daily (around 10-20% per day for gut, skin and blood cells). Therefore, there is an extremely small therapeutic index for anti-proliferation drugs, targeted at the limited number of proliferating AI PC cells, and clinical development of agents that can delay the onset and/or progression of human PC could have significant impact on disease-related cost, morbidity and mortality for a large fraction of the population [15]. In addition, the fact that it is now possible to diagnose PC at an early stage of the patient's life opens a large window of opportunity for intervention to prevent or slow the progression of the disease.

Primary hormonal therapy to perform androgen ablation includes castration either by orchiectomy or using luteinizing hormone-releasing hormone (LHRH) agonists. Hormonal resistance is found with almost no exception after an average of 18 to 24 months of treatment and when this occurs, the prognosis in patients with metastatic, hormone-refractory disease is dismal with a median survival of 12 to 18 months [16, 17]. The goal of therapy becomes palliation of the symptoms of the tumor and metastases in an attempt to improve quality of life. Patients can be offered secondary hormonal therapy as a means to delay the time to initiate systemic chemotherapy. In this regard, ketoconazole, a nonsteroidal antifungal compound that inhibits cytochrome P450 17α-hydroxylase-$C_{17,20}$-lyase (CYP17), one of the enzymes responsible for the production of androgen precursors in the body, and 11β-hydroxylase, remains one of the most active agents used for secondary hormonal therapy due to its efficacy and relative ease of administration [17-22]. Following ketoconazole, several steroidal and nonsteroidal CYP17 inhibitors have been designed and tested, in the hope of achieving higher potency and a better side effects profile [15,23,24]. Recently, docetaxel **1** (Figure 16) and prednisone chemotherapy has become the first-line standard of care for men with metastatic HRPC [25].

Natural products have been extensively used in traditional medicine to treat a myriad of diseases including various types of cancers [26-28]. Over the past 20 years there has been only one FDA-approved drug (sunitinib for renal carcinoma treatment in 2005) resulting from high-throughput screening of combinatorial chemistry libraries followed by the optimization of hits whereas natural products-based drugs have traditionally been major entities among the FDA-approved drugs [29]. Moreover, recent research suggests that antioxidant levels may have an inverse relationship with cancer incidence, namely PC, with phytochemicals being one of the main sources of antioxidants in the diet [30].

The most common study models for the assessment of the effects of novel compounds as potential agents for PC treatment *in vitro* are LAPC4, LNCaP, PC-3 and DU145 cells. LAPC4 and LNCaP cells are derived from human lymph node metastatis of PC [31-33] and differ in the type of AR that they express, however both have AD behavior. For LNCaP cells this behavior was found to be concentration-dependent [32]. LAPC4 cells bear the original wild-type AR [31] whereas LNCaP cells bear a mutated AR which acts promiscuously and can be bound by ligands other than dihydrotestosterone (DHT) due to a mutation in the androgen binding region [33-35]. This mutation consists of a switch from threonine (Thr877) to alanine (T877A). PC-3 cells derive from bone metastasis of PC and do not express the AR, thus representing a model of study for PC in the more advanced stages of the disease together with DU145 cells which derive from brain metastasis, also do not express the AR and bear the same AI behavior [36]. Both LAPC4 and DU145 tested positive for wild-type TP53 and p53 expression (90%), whereas on LNCaP cells, a mutated TP53 is present however expression of p53 still occurs (10%), and on PC-3 cells p53 expression is absent, probably because gene mutations resulted in truncated translation products [33]. The apparent absence of a p53-dependent pathway in prostatic epithelial cells, even in those bearing wild-type p53, together with the fact that PC cells with mutant p53 occur fairly frequently, are thought to account for ineffectiveness of many standard chemotherapeutic drugs on HRPC [25, 37, 38].

In this chapter the reader will find a description of selected natural products that have been tested on PC models *in vitro* and *in vivo*. Although a large number of molecules have been identified and their ability to arrest PC cell growth has been reported, focus will be directed towards the ones with established mechanisms of action. An integrated approach with highlights on chemical structures, molecule source and targets is provided. Hormones and vitamins have been left out of this discussion. The selected natural products have been divided into sections according to their origin (marine, microbial, plant, and animal), and subsections according to their chemical nature. Some success examples of synthetic analogues obtained from natural products have been included throughout the text.

2. Natural Products Action on Prostate Cancer Cells

2.1. Marine Compounds

2.1.1. Peptidic Compounds

Kahalalide F **2** (Figure 1) belongs to the family of dehydroaminobutyric acid-containing peptides that have been isolated from the herbivorous marine species of mollusc *Elysia rufescens*, an organism living in the seas near Hawaii, and later from its green algal diet *Bryopsis pennata* (Bryopsidaceae) [39]. The mechanism by which kahalalide F **2** acts on human PC cells was found to be induction of oncosis, a term describing the progression of cellular events leading to necrotic cell death [40] and lysosomes have been reported to be intracellular targets of kahalalide F **2** [41]. Dose-dependent inhibition of DNA synthesis was found for LNCaP and DU145 cell lines with an IC_{50} of 0.2 μM whereas for the AI PC-3 cell line the IC_{50} was 0.07 μM [40]. Also, kahalalide F **2**-induced cytotoxicity did not require gene expression or caspase activity and did not cause cell cycle arrest or DNA degradation. Loss of mitochondrial membrane potential, alteration of lysosomal function and profound alterations of cell architecture were features of kahalalide F **2**-induced cell death [40]. Two Phase I clinical trials in adult patients bearing advanced PC or pretreated solid tumors have identified the recommended dose for two different schedules (once a week and 5 consecutive days a week every 3 weeks). Dose-limiting toxicity was elevation of aminotransferases in both trials [42-44].

Recently, the results of a phase II clinical study of kahalalide F **2** in patients with advanced non-small cell lung cancer (NSCL), hepatocarcinoma, and advanced malignant melanoma have been reported at the 31st European Society for Medical Oncology Congress, revealing an excellent tolerability profile with no serious adverse events [45]. A positive response and stable disease were observed in a number of patients.

Dolastatin 10 **3**, jasplakinolide **4** (Figure 2), pachymatismin and HTI-286 **6** (Figure 3), bear a common target: the cytoskeleton. Dolastatin 10 **3** is an anticancer peptide isolated from the shell-less mollusc *Dolabella auricularia* [46, 47]. Structurally, it consists of four aminoacids (dolavaline, valine, dolaisoleucine and dolaproine) three of which are unique, linked to an unsual primary amine (dolaphenine) at its COOH terminus [47]. The absolute configuration of its nine asymmetric centers has been established and total synthesis of the compound has been achieved [46]. Mechanistically, dolastatin 10 **3** is a highly lipophylic pentapeptide that interacts with tubulin [48]. On the DU145 human prostate cell line dolastatin 10 **3** has been found to inhibit growth at 1 nM [49]. Cell cycle arrest at the G2/M phase and α-tubulin depolymerization correlated with the growth inhibition. Similar results were observed on LNCaP and PC-3 cells. *In vivo* efficacy was demonstrated at a 5 μg dosage of dolastatin 10 **3**, after intra-peritoneal (i.p.) administration every 4 days, in athymic mice [49]. This agent reduced the number and size of tumors on the diaphragms of mice and prevented invasion of the musculature, as compared to the prostate tumors implanted in mice used as controls [49]. Toxicity studies of dolastatin 10 **3** in mice, rats and dogs showed that granulocytopenia was dose-limiting and the recommended clinical starting dose was established as 65 μg/m^2 [50]. These results were confirmed in a phase I clinical trial of patients with advanced solid tumors using dolastatin 10 **3** [51]. However, phase II clinical trials in patients with hormone-refractory metastatic prostate adenocarcinoma have shown lack of significant clinical activity for dolastatin 10 **3** as a single agent [52].

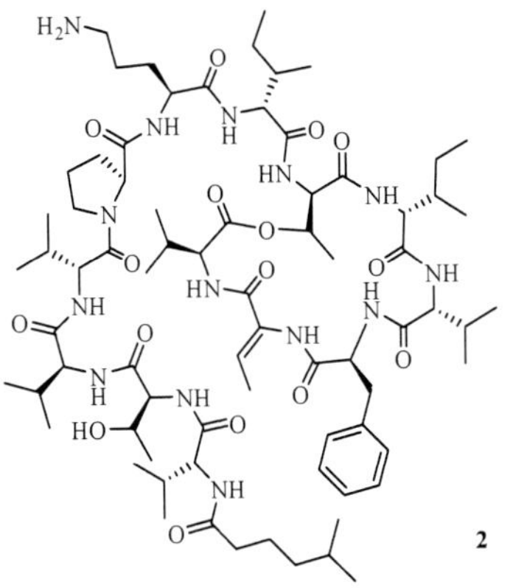

2

Source: *Elysia refescens, Bryopsis pennata* (Bryopsidaceae)

Target: lysosomes

Excelent tolerability profile with no side effects on a Phase II clinical trial for NSCL carcinoma, hepatocarcinoma and malignant melanoma.

Figure 1. Kahalalide F **2**.

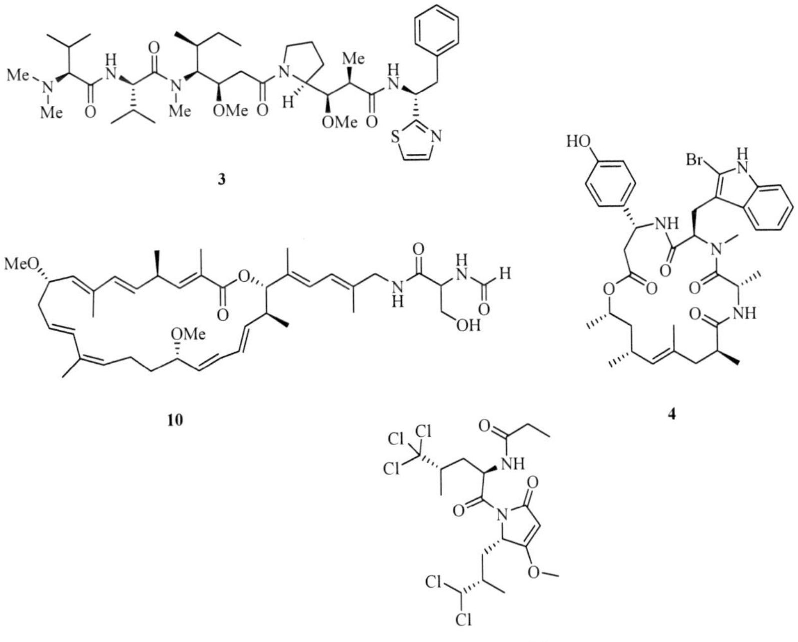

3 **10** **4** **11**

Figure 2. Peptidic compounds of marine origin.

Jasplakinolide **4**, a cyclodepsipeptide isolated from *Jaspis johnstonii* [53] has been shown to inhibit the proliferation of human PC cells [54]. Exposure of PC-3 cells to jasplakinolide **4** resulted in the formation of multinucleated cells and disruption of the actin cytoskeleton [55]. At 1 μM, jasplakinolide **4** inhibited the survival of DU145, PC-3, and LNCaP cells, after 48 hours of treatment [55]. The effect of pachymatismin, a glycoprotein extracted from the marine sponge *Pachymatisma johnstonii* Bowerbank 1842 [56], has also been reported to involve interference with microtubules [57]. On the DU145 cell line, the cytotoxicity of pachymatismin has been linked to its capacity to interact with microtubules causing cytoskeletal dysfunction and microtubule depolymerization (which was complete at 16 μM), in an analogous fashion to estramustine. On E4 cells, a variant of DU145 with poor sensitivity to estramustine, however, even at the high concentration of 16 μM, pachymatismin failed to induce microtubule depolymerization [57]. Resistance to pachymatismin was suggested to be associated with changes in microtubule composition that may alter both the stability of microtubules and their ability to bind to the compound [57].

HTI-286 **6**, a fully synthetic analogue of the marine sponge (genus *Auletta*) tripeptide product hemiasterlin **5** (Figure 3), has been reported to inhibit tubulin polymerization and circumvent transport-based resistance to taxanes [58]. HTI-286 **6** binds to the α-tubulin heterodimer at a unique site that appears to be in the interface of the tubulin subunits and is close to but distinct from *Vinca* alkaloids, at high concentrations. HTI-286 **6** depolymerizes, rather than stabilizes, existing microtubules. More importantly, however, at low concentrations both *Vinca* alkaloids and HTI-286 **6** interfere with spindle-microtubule dynamics [58]. The IC_{50} values for inhibition of LNCaP, C4-2, PC-3 and PC-3dR (docetaxel **1**-refractory cells) were 0.65, 1.7, 1.9, and 4.6 nM, respectively, with HTI-286 **6**. *In vivo*, the compound significantly inhibited the growth of PC-3 and LNCaP xenografts and retained potency in PC-3dR tumors, when after intra-venous (i.v.) administration at 1.5 mg/kg on days 1, 5, and 9 of treatment. Simultaneous castration plus HTI-286 **6** (1.2 mg/kg) was superior to sequential treatment in the LNCaP model [58].

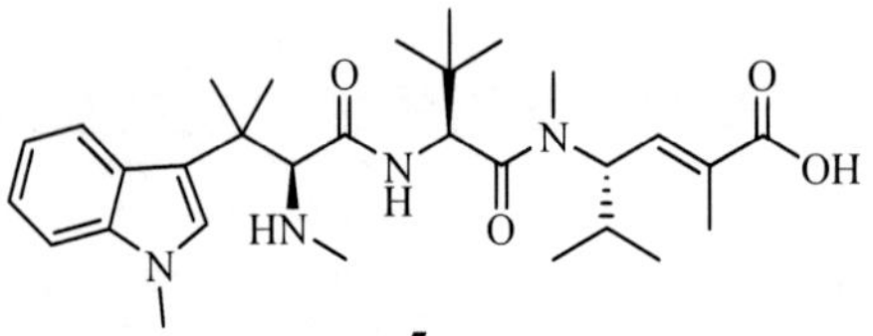

5

Source: marine sponge genus *Auletta*

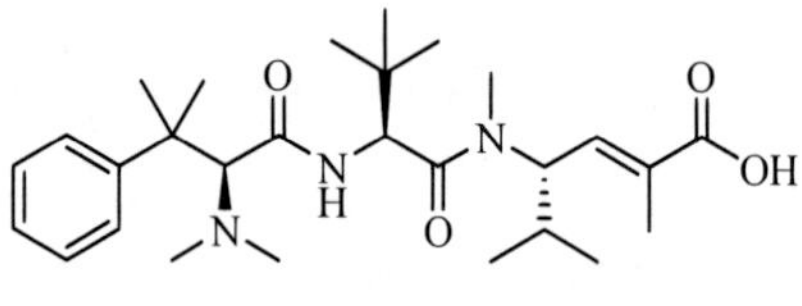

6

Cytotoxicity (IC_{50})
LNCaP = 0.65 nM
C4-2 = 1.7 nM
PC-3 = 1.9 nM
PC-dR = 4.6 nM

Target: microtubules
circumvents transport-based resistance to taxanes

Figure 3. Hemiasterlin **5** and its synthetic analogue HTI-286 **6**.

A family of naturally occurring cyclodepsipeptides called didemnins (Figure 4) has been isolated from marine tunicates and identified as a class of highly potent anticancer agents [59]. A related depsipeptide dehydrodidemnin B (aplidine **7**) was later isolated from the Mediterranean tunicate *Aplidium albicans* [60]. Aplidine **7**, didemnin B **8** and isohomohalichondrin B **9**, another marine compound isolated from the sponge *Lissodendoryux spp.* [61], have been evaluated on the human PC cell lines DU145, PC-3 and LNCaP-FGC [62]. The compounds were found to potently inhibit the growth of the human PC cell lines at levels between 5 and 50 pmol/ml. However, severe neurotoxicity accompanied this effect [62]. Didemnin B **8** was evaluated in a phase II trial for the treatment of metastatic, hormone-refractory adenocarcinoma of the prostate [63]. Thirteen patients were treated with didemnin B **8** (3.5 mg/m^2) whereas twenty patients were treated with 6.3 mg/m^2 i.v. of compound every 28 days. Of 32 evaluable patients there was one partial response for an overall response rate of 3%. The most common toxicities were nausea, vomiting, and diarrhea. Serious cardiac and pulmonary toxicities were also noted [63]. Thus, further evaluation of didemin B **8** for the treatment of this disease, as a single agent, has been discouraged.

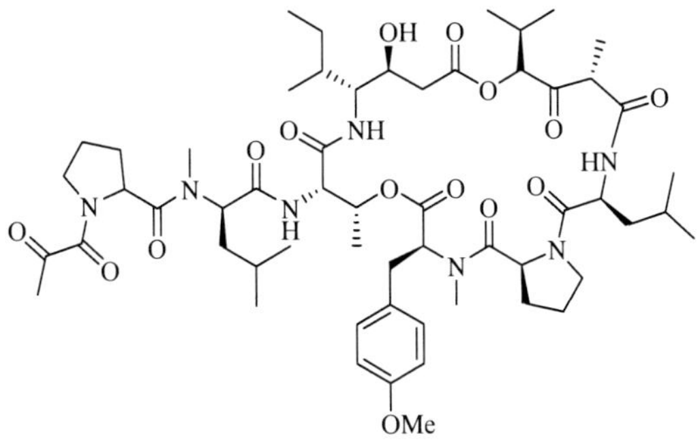

7

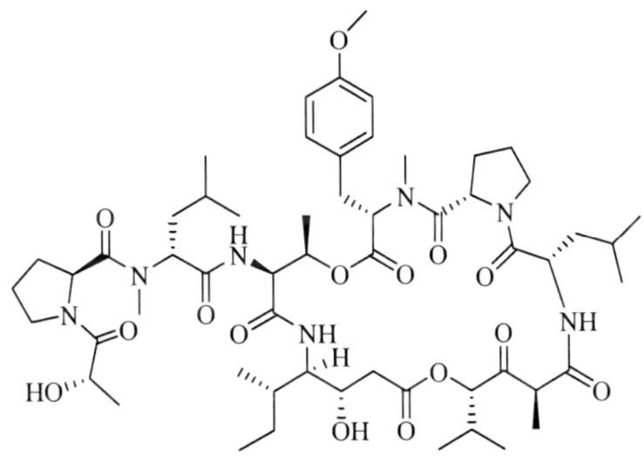

8

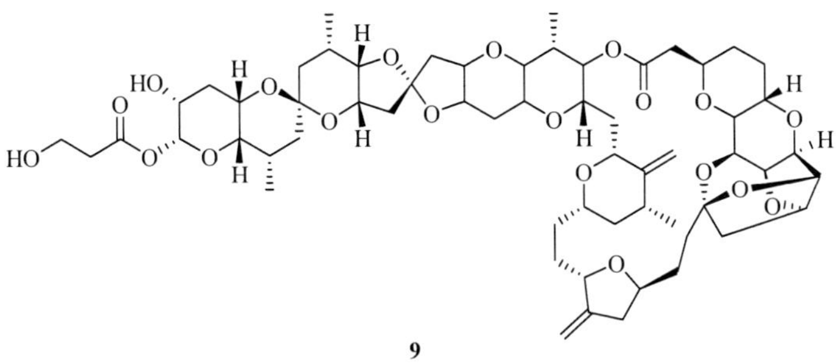

9

Figure 4. Peptidic compounds of marine origin.

Iejimalide B **10** (Figure 2), a marine macrolide extracted from the tunicate *Eudistoma cf. rigida*, has been found to have a dose and time-dependent effect on LNCaP and PC-3 cell proliferation [64]. In LNCaP cells, concentrations of 30 or 50 nM of the compound induced a dose-dependent G0/G1 arrest and apoptosis, after 48 hours of treatment. Iejimalide B **10** was found to modulate the steady-state levels of many gene products associated with the cell cycle and death. In contrast, the same concentrations of compound initially induced a G0/G1 arrest followed by S phase arrest on PC-3 cells, without triggering apoptosis [64]. Increased expression of the cyclin kinase inhibitor $p21^{WAF1/CIP1}$ and downregulation of cyclin A expression were also observed with no modulation of the genes associated with cell death. Comparison of the effects of iejimalide B **10** on the two cell lines suggested that the compound induced cell cycle arrest by two different mechanisms and that the induction of apoptosis on LNCaP cells was p53-dependent [64].

Sintokamide A **11** (Figure 2) is a chlorinated peptide isolated from the marine sponge *Dysidea* sp. collected in Indonesia [65]. This compound has been recently reported to be a selective inhibitor of *N*-terminus transactivation of the AR in PC cells. At a concentration of 5 μg/ml, the compound blocked AR activity and was found to be as effective as bicalutamide in blocking androgen-induced proliferation of LNCaP cells. The same effect on proliferation was not found for non-AR-expressing PC-3 cells [65].

2.1.2. Others

Other compounds of marine origin with reported effects on PC cells are depicted on figure 5. Spisulosine **12** (PharmaMar, ES-285) is a novel anti-cancer agent recently isolated from the sea mollusc *Spisula polynyma* [66] which has been reported to inhibit the growth of human PC-3 and LNCaP cells with an IC_{50} of 1-10 μM, through intracellular accumulation of ceramide and activation of the atypical protein kinase isoform PKCζ (Table 1) [66-68]. Thus, intracellular ceramide levels were increased after 48 hours of treatment with 1 μM spisulosine **12**. Both fumonisin B1, a mycotoxin produced by *Fusarium verticillioides* and a reported inhibitor of ceramide synthase, and myriocin, a potent fungal inhibitor of serine palmitoyltransferase, were able to prevent this effect pointing to a *de novo* synthesis of ceramide in the mechanism of action of spisulosine **12** on PC cells. The same effect was not observed with specific inhibitors of the stress-related mitogen activated protein (MAP) kinases as well as of the peroxisome proliferator-activated receptor PPARγ, phosphatidylinositol 3-kinase (PI3K)/protein kinase B (Akt), and classical protein kinases C (PKCs) pathways, rulling out interference of these pathways in the mechanism of action of spisulosine **12** [66].

Table 1.

Target: *De novo* synthesis of ceramide
Ceramide generation is a nearly universal feature of programmed cell death. PCKζ directly interacts with ceramide and modulates the cellular response to tumor necrosis factor α (TNFα). Moreover, ceramide increases the affinity of PCKζ towards its inhibitor protein proapoptotic prostate apoptosis response-4 (Par-4), which is a mediator of apoptosis with special relevance in prostate cells. Therefore, stimulation of the *de novo* synthesis of ceramide could be of great interest for PC drug development.

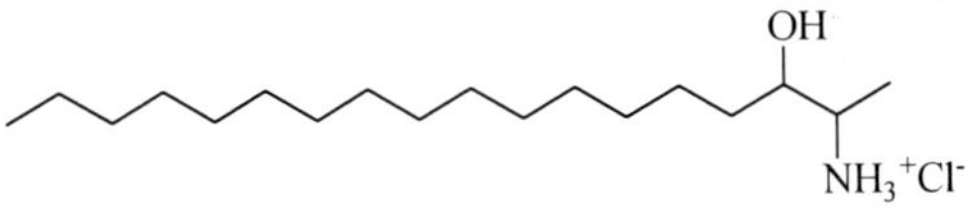

12

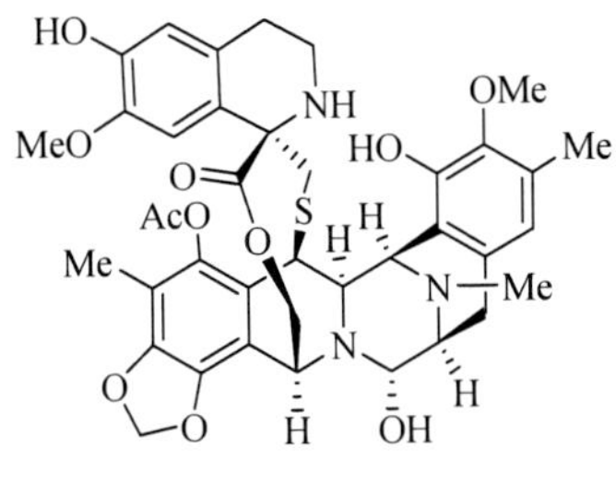

13

Source: *Ecteinascidia turbinata*

Phase II/III for HRPC

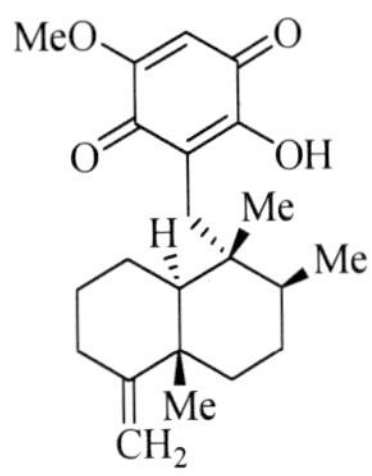

14

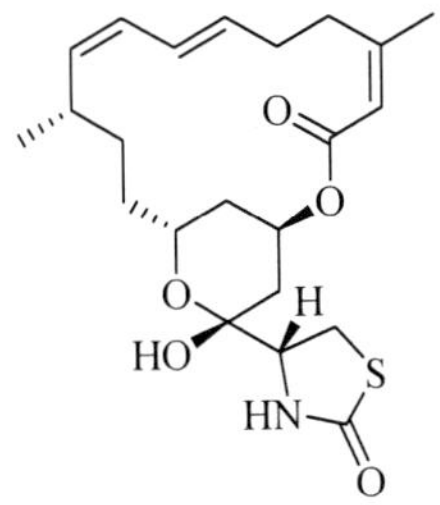

15

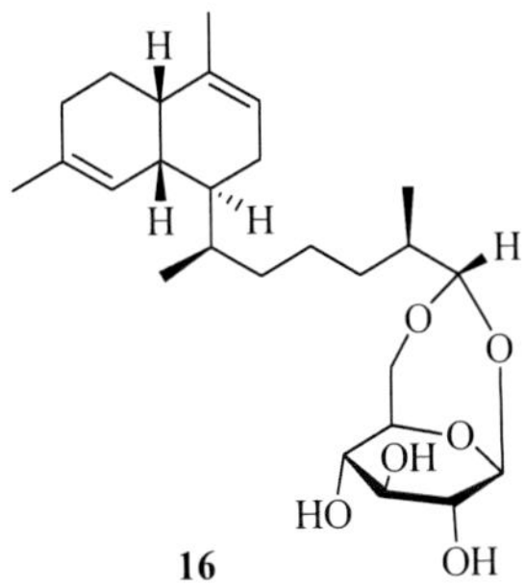

16

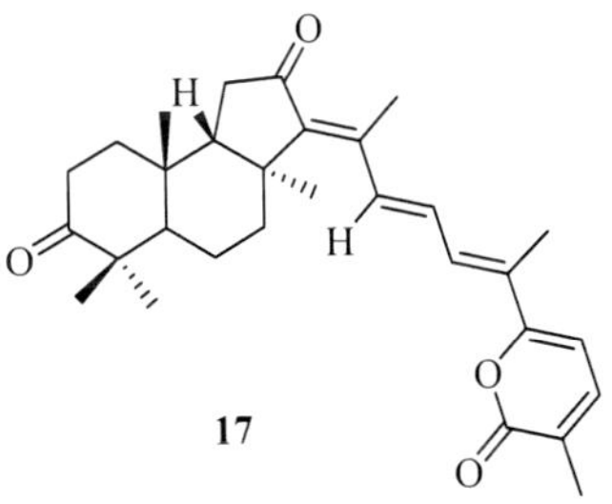

17

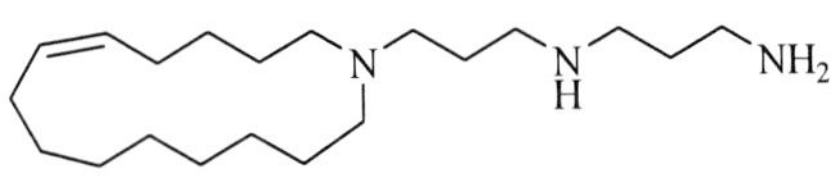

18

Figure 5. Other compounds of marine origin.

Trabectedin **13** (Yondelis™, ET-743, PharmaMar, Figure 5) is a novel marine-derived anticancer compound found in the ascidian *Ecteinascidia turbinata* [69]. It is currently under phase II/III development in breast cancer [69], HRPC [70], sarcomas, and ovarian cancer [71].

Ilimaquinone **14** (Figure 5), a marine sponge metabolite, was first isolated from *Hippospongia metachromia* [72]. Ilimaquinone **14** inhibited PC-3 cell proliferation with an IC_{50} value of 2.6 μM [73]. Degradation of cyclin E was prevented by ilimaquinone **14** and cell cycle arrest at G1 phase was observed. Both the nuclear translocation and the DNA binding ability of translocation nuclear factor-κB (NF-κB) were inhibited on PC-3 cells after treatment with 10 μM of ilimaquinone **14** [73]. However, cell proliferation inhibition was also seen on LNCaP cells, in which NF-κB is not ubiquitously activated, with an IC_{50} value of 2.4 μM, suggesting that NF-κB activity may play a minor role on the ilimaquinone **14**-induced antitumor effect. An increase in the expression and nuclear translocation of the transcription factor CHOP/GADD153 was then suggested to account for the observed effects of ilimaquinone **14** on the cell cycle [73]. Thus, the ilimaquinone **14**-mediated anti-proliferative effect was significantly reduced in the antisense CHOP/GADD153-overexpressing cells.

Latrunculins are macrolides from the red sea sponge *Latrunculia magnifica* [74]. The most important biological effect of latrunculins is their ability to disrupt microfilament organization thus inhibiting microfilament-mediated processes, without affecting the organization of the microtubular system [75]. They bind reversibly to the cytoskeleton actin monomers as 1:1 complexes with G-actin and disrupt polymerization. Latrunculin A **15** (Figure 5) remarkably inhibited baseline calcitonin (CT)-stimulated invasion of highly metastatic human PC-3M-CT^+ cells, at 100 nM, an effect most likely mediated by disruption of actin cytoskeleton remodeling [76]. Cytotoxicity was observed only at 500 nM. At this concentration, latrunculin A **15** decreased the disaggregation and cell migration of PC-3M-CT^+ spheroids by 3-fold [76].

Induction of apoptotic pathways has been reported to be mediated by marine compounds such as stellettins and lemnabourside **16** (Figure 5). Stellettins belong to a group of isomalabaricane triterpenes present in several marine sponge genera such as *Jaspis*, *Stelletta* and *Rhabdastrella*. They have also been found in *Geodia japonica* [77]. Stellettin A **17** (Figure 5) was reported to induce oxidative stress on LNCaP cells at 120 μg/ml by virtue of reactive oxygen species (ROS) generation, leading to Fas ligand (FasL)-caspase 3-mediated apoptosis [78]. Lemnabourside **16**, a component of the soft coral *Nephthea chabroli*, was found to be a weak inhibitor of rat 5α-reductase [79]. When tested on LNCaP, PC-3 and DU145 cells, it inhibited the growth of all three cell lines in a similar fashion with IC_{50} values ranging from 37.5 to 40 μM. Reduction of the expression of genes encoding all prostate biomarkers, decrease of antiapoptotic bcl-2 gene and increase of caspase-3 gene were found after treatment of LNCaP cells with a 12.5 μM concentration of lemnabourside **16** [79].

Belonging to the family of marine sponge macrocyclic alkaloids containing a spermidine-like structure (the motuporamines), motuporamine C **18** (Figure 5) has been found to inhibit invasion of basement membrane gels by PC-3 cells, at 5 μM [80]. The inhibition of migration occurred with little effect on cell proliferation for motuporamine C **18**. The compound was also shown to inhibit angiogenesis of human endothelial cells both *in vitro* and *in vivo* and this effect did not occur through toxic or antiproliferative effects either [80]. It was suggested that motuporamine C **18** acts to dampen "inside-out" signaling mechanisms that regulate the affinity of cell surface integrin receptors for their extracellular matrix ligands, due to a subtle decrease in the affinity state of β1-integrins.

2.2. Microbial Compounds

Microbial compounds with known action on PC cells are depicted on figure 6. Epothilones, 16-membered macrolides isolated from the myxobacterium *Sorangium cellulosum* [81, 82], bind to tubulin and prevent the depolymerization of microtubules although no structural similarity to taxanes such as paclitaxel **19** (Figure 16) exists [83]. Epothilones bind to tubulin and cause the hyperstabilization of microtubules with subsequent mitotic arrest and apoptotic cell death. Both epothilones and taxanes bind to α,β-tubulin heterodimers in a 1:1 stoichiometric ratio and epothilones are capable of displacing bound, labeled paclitaxel **19** at concentrations and with kinetics similar to those displayed by paclitaxel **19** itself [83]. The main difference resides in the fact that taxanes are substrates for P-glycoprotein and epothilones are not and this ability together with ease of formulation have been considered advantages for overcoming taxane resistance of tumor cells [83]. Moreover, total synthesis of epothilones A **20** and B **22** (Patupilone, EPO 906) has been accomplished [81, 84-86].

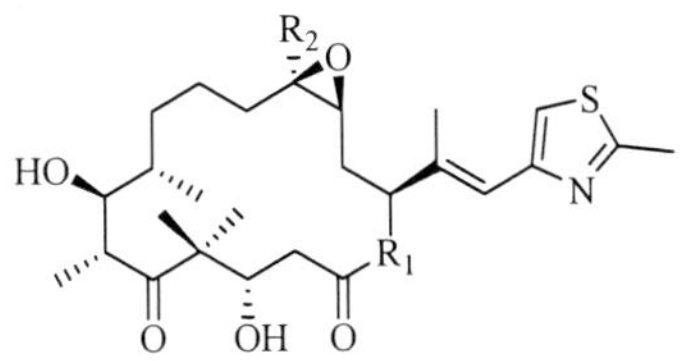

R_1 = O; R_2 = H **20**
R_1 = O; R_2 = Me **22**
R_1 = NH; R_2 = Me **24**

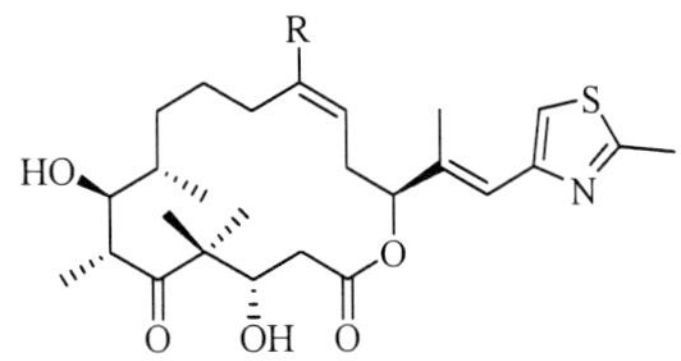

R = H **21**
R = Me **23**

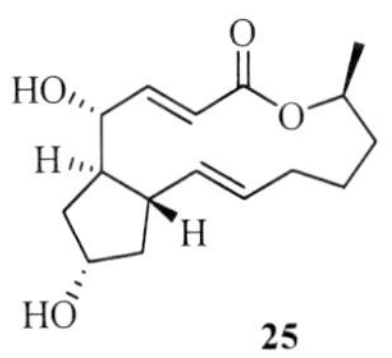

25

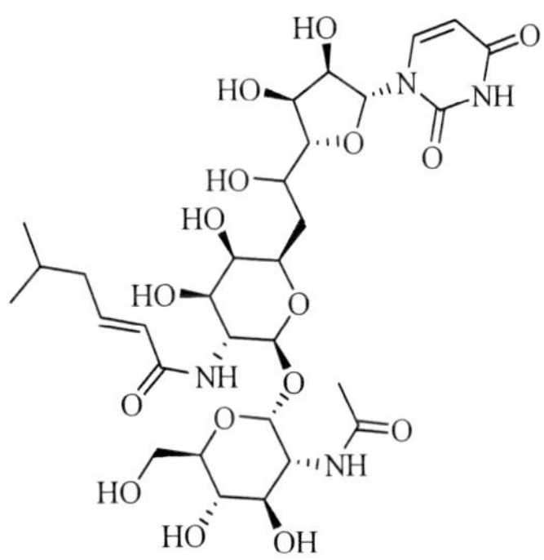

26

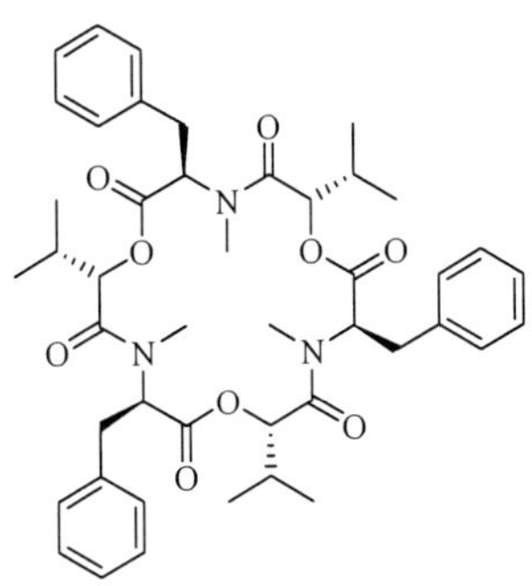

27

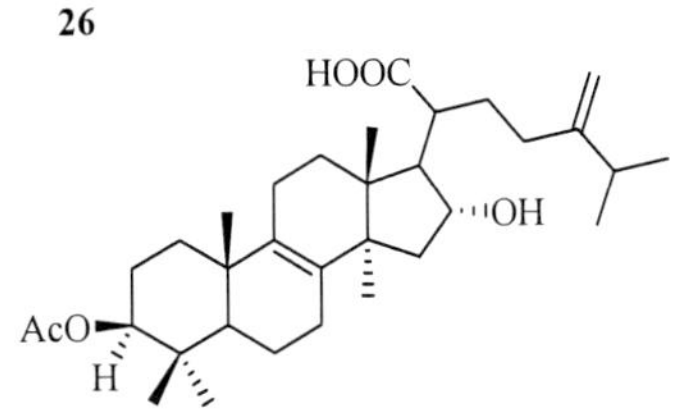

30

Figure 6. Microbial compounds.

Epothilones and desoxyepothilones A **21** and B **23** (Figure 6) have been evaluated on DU145 and TSU-Pr1 PC cell lines (Table 2) and treatment with the compounds caused hyperpolymerization of microtubules, formation of microtubular fibers in G1, and mitotic arrest resulting in apoptosis [87]. Patupilone **22** was found to be the most potent compound in inhibiting cell growth and inducing mitotic arrest (Table 2, Entry 4). Moreover, pretreatment of DU145 cells with epothilones A **20** and desoxyepothilones A **21** and B **23** (1 to 100 nM), as well as with patupilone **22** (0.1 to 3 nM), sensitized cells to the farnesyl transferase inhibitor L-744,832, in an analogous fashion to paclitaxel **19** at 10-333 nM [87].

Table 2. *In vitro* assay data for epothilones.

Entry	Compound	PC cell line IC_{50}		
		DU145	TSU-Pr1	
		Growth inhibition	Growth inhibition	Mitotic arrest
1	**19**	2.2×10^{-10} M	6×10^{-10} M	200-700 pM
2	**20**	1.3×10^{-9} M	1.2×10^{-9} M	2 nM
3	**21**	2×10^{-8} M	2×10^{-8} M	20 nM
4	**22**	6×10^{-11} M	5×10^{-11} M	50-75 pM
5	**23**	1.7×10^{-10} M	2.5×10^{-10} M	250 pM

The effect of patupilone **22** on the proliferation and cell cycle of both LNCaP and DU145 cells as well as of a transformed rat prostate epithelial cell line with a temperature sensitive mutant p53, was found to be dependent on the p53 status of the cell model [88]. Cells bearing mutant p53 were more sensitive to patupilone **22**. The compound was capable of dephosphorylating Cdc2 kinase in both wild-type and p53 mutant cells, which was associated with G2/M arrest [88]. *In vivo* studies with patupilone **22** have shown activity against human PC xenografts either when transplanted s.c. or orthotopically [89]. Patupilone **22** produced transient to long-lasting tumor regressions, including apparent cures, in s.c. transplanted DU145 and PC-3M PC xenografts in mice, with the 2.5 mg/kg regimen twice every 7 days being superior to a single dose of 4 mg/kg patupilone **22** [89]. The anti-tumor activity was superior to that of paclitaxel **19**, which failed to produce tumor regressions, although displaying a better tolerability profile. The effect of a single administration of patupilone **22** was equivalent to paclitaxel **19** on the s.c. PC-3M tumor model. However, two administrations within a 7 day interval produced marked suppression of primary tumor growth as well as metastasis to mesenteric lymph nodes, in an orthotopic model of PC-3M PC [89]. Despite the enthusiastic *in vivo* results, an objective tumor response rate of only 20% has been reported for patupilone **22** on HRPC in which 78% of patients had received one (unspecified) prior line of chemotherapy, a number comparable to the 12% objective tumor response obtained with docetaxel **1** and prednisone combined treatment, in the phase III TAX 327 study [84]. Randomized studies are recruiting in order to further define the efficacy and safety of patupilone **22** in comparison to standard treatment for HRPC [84]. Nonetheless, patupilone **22** seems to hold promise for HRPC treatment due to a better side effects profile than other epothilones or taxanes, inasmuch as diarrhea is the dose-limiting toxicity rather than neutropenia and neuropathy, and also due to paucity of other active agents [84]. A clinical trial with patupilone **22** is currently recruiting to assess the response of combinations of patupilone **22** with prednisone as compared to docetaxel **1** and prednisone, on prostate specific antigen (PSA), in patients with metastatic HRPC, whereas another study has been designed to determine the efficacy of patupilone **22** and to find out what its beneficial and/or deleterious effects on patients with PC that has progressed following hormone treatment and docetaxel **1** chemotherapy are [71].

Aza-epothilone B **24** (BMS-247550, Ixabepilone, Bristol-Myers-Squibb Co., Figure 6) is a semi-synthetic analogue of patupilone **22** [90]. Ixabepilone **24** has demonstrated activity in patients with chemotherapy-naive metastatic HRPC [91, 92]. Six clinical trials are currently recruiting for evaluation of the effect of ixabepilone **24** on advanced PC [71]. Phase I clinical trials of ixabepilone **24** administered i.v. as a 1 hour infusion daily, for 5 consecutive days every 21 days, at doses of 1.5 mg/m^2/day, demonstrated antitumor responses in solid tumors and in patients with prior taxane treatment [25]. Major toxicities were neutropenia and neuropathy. Single agent ixabepilone **24** was evaluated in two first line studies [25]. PSA response was observed in 30-35% of cases with an objective clinical response rate of over 15% in patients with measurable disease. The estimated median progression-free survival was about 6 months and median survival ranged between 16 and 18 months. Again, neuropathy was found to be dose-limiting [25]. When combined with estramustine phosphate (ixabepilone **24** 35 mg/m^2 i.v. on day 2 with a 280 mg oral dose of estramustine, 3 times/day, on days 1-5, every 3 weeks), it was found to be well tolerated and have antitumor activity in patients with castrate metastatic PC [93]. Incomplete clinical cross-resistance between taxanes and epothilones has been reported [94]. Second-line taxane chemotherapy after ixabepilone **24** resulted in a substantial frequency of PSA declines [94]. Although patients with ixabepilone **24**-refractory disease were less likely to respond to second-line taxane chemotherapy, 36% did achieve a PSA response [94]. For patients with taxane-resistant HRPC, a randomized phase II study compared ixabepilone **24** to mitoxantrone/prednisone [25]. For the 8 patients enrolled, a PSA decline of over 50% was seen in 20% and median survival of 13 months was found for patients on the mitoxantrone arm, as compared to a PSA decline of over 50% in 17% and median survival of 12 months for ixabepilone **24** treatment [25]. Ixabepilone **24** has recently been approved by the FDA for the treatment of metastatic breast cancer [90].

Brefeldin A **25** (Figure 6), a fungal macrolytic lactone [95], acts to inhibit intracellular transport by disrupting the vesicular coating process, thereby preventing transport of proteins from the endoplasmic reticulum (ER) to the Golgi complex, and causing its disintegration [96, 97]. This compound has been reported to inhibit the growth of both androgen-responsive and non-responsive human PC cells [98, 99]. Suppression of LNCaP growth was accompanied by a drastic reduction (over 90%) of AR expression suggesting that this effect was mainly due to AR downregulation [98]. The growth of non-AR expressing PC-3 cells was also reduced by 70% with brefeldin A **25** (30 ng/ml) treatment without induction of apoptosis, and this effect was found to be mediated by dephosphorylation of retinoblastoma protein (pRB) [99]. Studies conducted on DU145 cells, which are known to express defective or mutated pRB and p53, showed inhibition of cell growth with brefeldin A **25** at concentrations of over 30 ng/ml, after 72 hours of treatment [100]. An accumulation of cells at the G1 phase with downregulation of the cell cycle regulators cyclin-dependent kinases (CDKs) 2 and 4, and cyclin D1 occurred concomitantly. This growth arrest ultimately resulted in apoptosis, presumably through a p53-independent pathway [100]. The effect of this compound on the growth of primary cultures of human prostatic epithelial cells revealed an IC_{50} value of 5 ng/ml (18 nM) [101]. Cells derived from low-grade as well as high grade cancers responded similarly to treatment. Morphological manifestations of apoptosis, a decline in cell viability and DNA fragmentation were visible, with unaltered levels of p53 and the presence of hypophosphorylated pRB [101].

Brefeldin A **25** is capable of inhibiting the growth of all PC cell lines via a set of diverse mechanisms which would make it a promising candidate for treatment of the disease. However, the fact that the compound is poorly soluble in aqueous medium and that it possesses a very narrow range of concentrations that can be safely used are current drawbacks in drug development [101]. Brefeldin A **25** has been shown to upregulate the transcription factor CHOP and the death receptor 5 (DR5) in PC-3 cells sensitizing them to TNF-related apoptosis-inducing ligand (TRAIL)-induced apoptosis [102]. A similar effect has been demonstrated for the naturally occurring antibiotic tunicamycin **26** (Figure 6), highlighting the relevance of the use of natural products, which are commonly very potent, in combination regimens [102]. Thus, the side effects of both tunicamycin **26** and TRAIL might be reduced because a lower dose of each of the compounds would be enough to achieve the desired effect [95]. Tunicamycin **26** blocks the first step in the biosynthesis of *N*-linked oligosaccharides in cells. It has been found to induce apoptosis on PC-3 cells *per se* only moderately [102]. However, it potently enhanced TRAIL-induced apoptosis through upregulation of DR5 expression. The presence of both compounds cooperatively activated caspases 3, 8, 9, and 10 as well as cleavage of the proapoptotic Bcl-2 family member Bid. Tunicamycin **26**-mediated induction of CHOP and DR5 protein expression was not observed in normal human peripheral blood mononuclear cells nor was enhanced sensitivity to apoptosis [102].

Beauvericin **27** (Figure 6) has been encountered in a number of fungal strains including *Beauveria bassiana* [103], *Paecilomyces fumoso-roseus* [104], and *Polyporus sulphureus* [105], among others [106]. This compound has been found in fruit tissues of two cultivars of melon fruits artificially infected with a strain of *F. oxysporum* f. sp. *Melonis* [107]. More recently, wound-healing assay-guided fractionation of an extract of a solid culture of *F. oxysporum* strain $EPH2R_{AA}$ also led to the isolation of beauvericin **27** [106]. Beauvericin **27** has been found to inhibit PC-3M cell proliferation and migration with IC_{50} values of 3.8 and 3.0 μM, respectively [106]. Beauvericin **27** clearly showed potent antiangiogenic activity at sublethal concentrations on endothelial HUVEC-2 cells. Complete disruption of HUVEC-2 network formation was observed at 3.0 μM, below the IC_{50} of 7.5 μM for this cell type [106].

Tyropeptin A **28** (Figure 7) is produced by *Kitasatospora* sp. MK993-dF2 and has been found to specifically inhibit the chymotrysin-like activity of the 20S proteasome [108]. TP-110 **29** (Figure 7), a synthetic analogue, was also found to inhibit the chymotrypsin-like activity of the 20S proteasome and the growth of various cell lines *in vitro*. The proteasome is a molecular target with growing relevance on the cancer setting since the demonstration that cancer cells are more sensitive to inhibition of protein degradation than normal cells, and therefore proteasome inhibitors might act as novel specific anticancer agents (Table 3) [109, 110]. The IC_{50} value for inhibition of PC-3 cell proliferation was 0.05 μM [108]. G2/M arrest was also found to occur with accumulation of p21 and p27 proteins. The apoptotic process triggered by TP-110 **29** was caspase-dependent. The compound inhibited not only the degradation of the NF-κB inhibitor IκBα and the nuclear translocation of NF-κB, but also the DNA binding activity of NF-κB on PC-3 cells. However, only weak tumor activity was observed with TP-110 **29** *in vivo* due to poor metabolic stability [108].

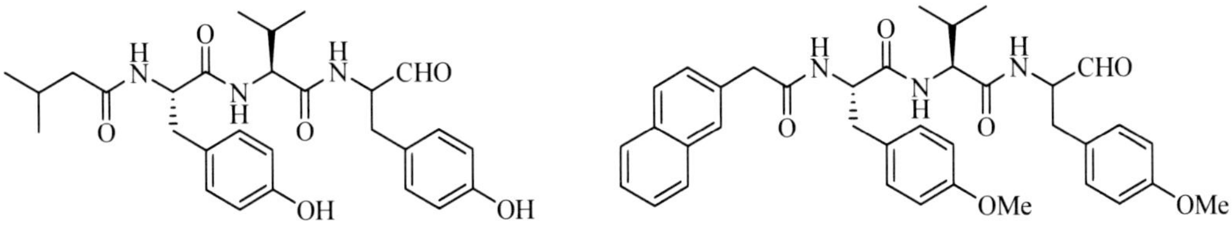

28 **29**

Figure 7. Tyropeptin A **28** from Kitasatospora sp. MK993-dF2 and TP-110 **29**, a synthetic derivative.

Table 3

Target: The proteasome
The 26 S proteasome is the major proteolitic complex responsible for the non-lisosomal degradation of cellular proteins. It is comprised of a catalytic complex (20S) and two regulatory units (19S), and exists in the cytoplasm, nucleus and smooth ER. The regulatory subunits bear α subunits that function as binding sites for ubiquitinated proteins and allow the release of ubiquitin as they are degraded by the 20S complex. The catalytic subunits of the 20S complex (β1, β2 and β5) possess *N*-Thr terminal residues acting as nucleophiles in the active site that hydrolyze peptidic bonds. The proteolytic activities of the β1, β2, and β5 subunits are designated caspase-like, trypsin-like and chymotrypsin-like, respectively. The proteasome is biologically relevant from the point of view of the organism's protein turnover and quality, cell cycle and apoptosis, transcription and cellular signaling, immune response and antigen presentation, and its inhibition causes apoptosis in tumor cells. Inhibition of the proteasome can help improve the efficacy of the available therapeutic regimens for a number on oncologic diseases by a variety of mechanisms yet to determine.

Pachymic acid **30** (Figure 6) is a natural triterpenoid isolated from the fungal species *Poria cocos*, also known as Fuling, which is known to inhibit the phospholipase A2 family of arachidonic acid producing enzymes [111]. The proliferation of both LNCaP and DU145 cells was inhibited by 40 μg/ml of this compound and at this concentration apoptosis was triggered on both cell lines. Pachymic acid **30**-induced apoptosis was suggested to occur by mitochondrial dysfunction and reduction of prostaglandin synthesis and Akt activity [111].

2.3. Plant Compounds

2.3.1. Polyphenols

Tea consumption in the world is high and ranks second to water consumption [112]. In this regard, green tea consumption accounts for approximately 20% of total tea consumption whereas black tea accounts for the remaining 80%. Green and black teas are derived from *Camellia sinensis* (Theaceae family) and green tea is the aqueous infusion of dried unfermented leaves. The naturally occurring antioxidants present in tea have received growing attention as possible agents for chemoprevention of several types of cancer among which is PC [30, 113-115]. Tea is an ideal candidate for chemoprevention studies because little or no adverse effects are known from tea consumption, extracts have been reported to be effective on several cancer types at achievable dose levels, it is orally consumed, has a low cost, and a history of use by the human population [112]. Thus, the biological effects of green and black tea extracts on PC both *in vitro* [116-119] and *in vivo* [120] have been well documented. At present, clinical trials are recruiting for chemoprevention of PC with green and black teas [71].

The major green tea phenols are (-)-epicatechin **31**, (-)-epigallocatechin **32**, (-)-epicatechin-3-gallate **33**, (-)-epigallocatechin-3-gallate **34**, (Figure 8) (+)- gallocatechin and (+)-catechin [113]. Through fermentation, a large percentage of the catechins are converted to oligomeric theaflavins and polymeric thearubigins in black tea. The resulting brewed black tea contains 3–10% catechins, 2–6% theaflavins and 20% thearubigins [113]. Thus, black tea theaflavins are theaflavin **35**, theaflavin-3-gallate **36**, theaflavin-3'-gallate **37** and theaflavin-3,3'-digallate **38** (Figure 8).

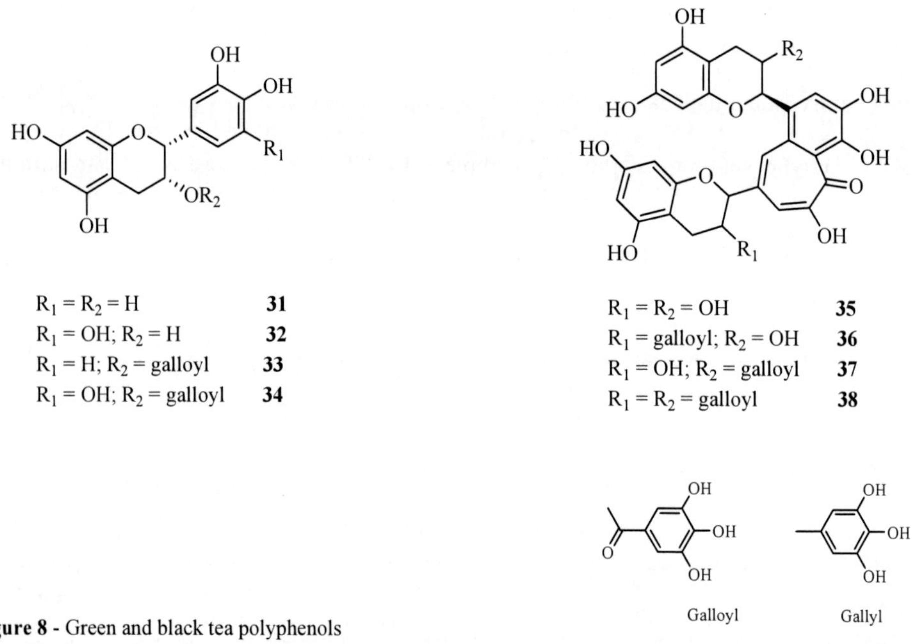

Figure 8 - Green and black tea polyphenols

Figure 8. Green and black tea polyphenols

(-)-Epigallocatechin-3-gallate **34** accounts for about 50% of the solid matter (about 5-10 mM) in a hot water extract of green tea beverage and has been found to be the most potent of the compounds. Treatment of DU145 cells with 40 μg/ml of compound **34** resulted in decreased levels of PI3K and phosphorylated Akt after 24 hours [121]. A similar effect was found for theaflavins from black tea at 50 μg/ml [121]. Induction of apoptosis and 5α-reductase inhibition (the enzyme responsible for the conversion of testosterone to its more potent metabolite DHT) on a cell-free assay which did not, however, correlate to the results obtained on a whole-cell assay, have been reported with compound **34** on PC cells (Table 4, Entry 1) [112, 122-124]. The compound strongly inhibited the growth of LNCaP cells at 1 μM (90%) and was also active on PC-3 cells with an IC_{50} value of 1 μM [122].

Table 4. Inhibitory potential of several polyphenolic and flavonoid compounds on 5α-reductase. Values in brackets are percent inhibition of 5α-reductase activity in the presence of 100 μM of the indicated compound.

Entry	Compound	5α-Reductase Inhibition			
		Cell-free assay IC_{50} (μM)		Whole-cell assay IC_{50} (μM)	
		Type 1	Type 2	Type 1	Type 2
1	**34**	15 (99)	74 (74)	> 100 (6)	> 100 (0)
2	**39**	> 100 (7)	> 100 (9)	ND	ND
3	**41**	> 100 (13)	> 100 (8)	ND	ND
4	**48**	7 (99)	21 (99)	7 (100)	6 (99)
5	**51**	19 (99)	50 (80)	19 (99)	22 (99)
6	**59**	> 100 (16)	23 (76)	> 100 (22)	20 (89)
7	**60**	23 (96)	> 400 (14)	> 100 (15)	> 100 (29)
8	**61**	> 100 (3)	29 (69)	> 100 (13)	7 (89)
9	**64**	100 (50)	17 (74)	64 (64)	5 (93)
10	**81**	29 (79)	99 (51)	> 100 (24)	> 100 (4)
11	**156**	25 (97)	> 100 (36)	8 (99)	7 (98)

A more modest effect was seen on DU145 cells (IC_{50} of 25 μM). Treatment of LNCaP and DU145 cells with 200 μM of compound **34** resulted in nuclear morphology changes and DNA fragmentation [122]. Despite being a poor inhibitor of 5α-reductase, compound **34** was found to inhibit cytochrome P450 17α-hydroxylase,$C_{17,20}$-lyase (CYP17) more potently than ketoconazole, a reference compound used in the clinic for advanced PC treatment [119]. The IC_{50} for enzyme inhibition was 24.5 μM for the compound and 35 μM for ketoconazole (Table 5, Entries 1 and 6). All theaflavins tested were more potent than ketoconazole with IC_{50} values ranging from 11.5 to 25.0 μM (Table 5, Entries 2-6).

Table 5. Inhibitory potential of compound 34 and black tea theaflavins 35-38 on rat testis CYP17.

Entry	Compound	CYP17 Inhibition IC_{50} (μM)
1	**34**	24.5
2	**35**	25.0
3	**36**	11.5
4	**37**	12.5
5	**38**	15.0
6	ketoconazole	35.0

Daily i.p. injection of 1 mg of compound **34** starting 2 weeks after the injection of PC-3 tumor cells into nude mice halted tumor growth, and in some cases actually reduced the initial tumor size by 20-30% within 1-2 weeks, when compared to control [125]. Treatment suspension resulted in tumor regrowth to 330-470% of the initial tumor size in 7-14 days. A similar effect was seen on LNCaP 104-R tumors of castrated male nude mice. However, daily i.p. injection of compounds **31-33** for 14 days did not inhibit tumor growth in mice [125]. It was pointed out that the galloyl group of compound **34** appears to be necessary for activity since compound **32** was not active.

Compound **34** was tested alone and in combination with specific cyclooxygenase-2 (COX-2) inhibitors on the growth of human PC cells *in vitro* and *in vivo*. LNCaP, PC-3 and CWR22R1 cells were treated with compound **34** (10-40 μM) and NS398 alone (10 μM) or in combination and the combination regimen was found to be advantageous in cell growth inhibition and apoptosis induction [126]. Numerous health benefits have been proposed for the consumption of tea, however, the effectiveness of tea as a cancer preventive agent in humans remains unclear [113]. One of the major issues is the fact that the concentrations of catechins or theaflavins used in cell culture systems exceed the plasma concentrations obtained in animal studies by 10- to 100-fold. Therefore, mechanisms based on the use of such high concentrations may be relevant for cancers of the gastrointestinal tract but most likely not for sites such as the lung, prostate and breast, which depend on systemic bioavailability.

The fruit of the *Punica granatum* (pomegranate) contains hundreds of phytochemicals and pomegranate extracts have been shown to exhibit antioxidant properties thought to be due to the action of ellagic acid **39** (Figure 9), the main active polyphenol in pomegranate [127]. Pomegranate is thus a rich source of polyphenolic compouns such as anthocyanins, which give red colour to its juice and seeds, and hydrolysable tannins such as ellagic acid **39** and punicalagin **40** (a polymeric ellagitannin which produces ellagic acid by spontaneous lactonisation on hydrolysis) (Figure 9) [127].

The biological effects of crude fruit extracts both *in vitro* on PC cells and *in vivo* on animal models have been reported [128-130]. More recently, isolation of pure compounds from the crude extract has made it possible to evaluate their effects on PC cell lines such as RWPE-1 (immortalized prostate epithelial cells), 22Rv1 (metastatic PC cells) and PC-3 [128, 131]. However, neither ellagic acid **39** nor punicalagin **40** at concentrations of 12.5 to 100 μg/ml were found to be more potent than pomegranate juice in inhibiting the proliferation of both RWPE-1 and 22Rv1 cells [128]. The ability to inhibit lipid peroxidation at 10 μg/ml was also higher for pomegranate juice followed by punicalagin **40** and ellagic acid **39** [128]. Thus, pomegranate juice is a more potent antioxidant than its separated and individual polyphenols which suggests synergistic and/or additive effects from other phytochemicals present in the juice. Four pure compounds from pomegranate ellagic acid **39**, caffeic acid **41**, punicic acid **42** and luteolin **43** (Figure 9), significantly inhibited invasion of PC-3 cells in an assay employing Matrigel™ artificial membranes [131]. When caffeic acid **41**, punicic acid **42** and luteolin **43** were equally combined at the same gross dosage of 4 μg/ml, suppradditive inhibition of invasion was observed. Ellagic acid **39** and caffeic acid **41** do not significantly inhibit the activity of 5α-reductase, as determined on a cell-free assay (Table 4, Entries 2 and 3) [124]. Currently, clinical trials are recruiting to establish a possible role for pomegranate on PC chemotherapy and/or chemoprevention [71].

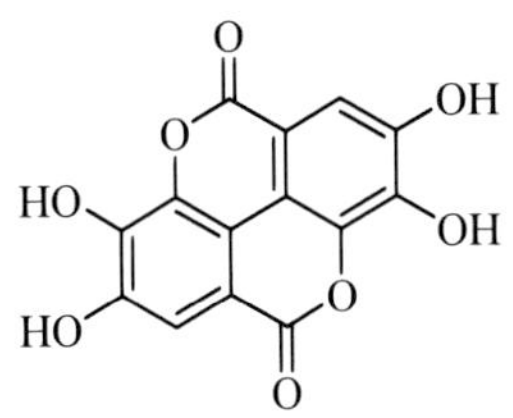

39

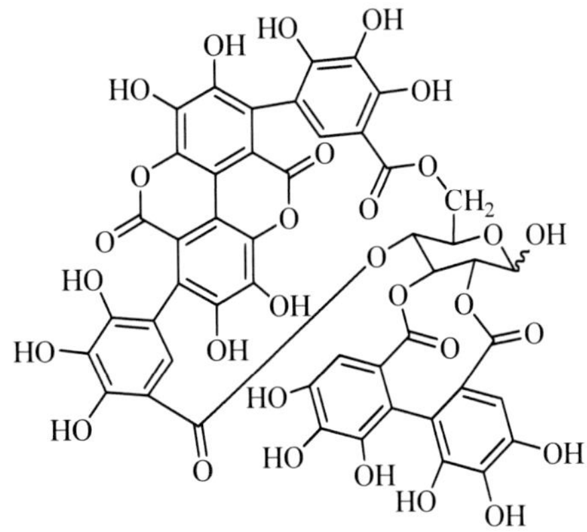

40

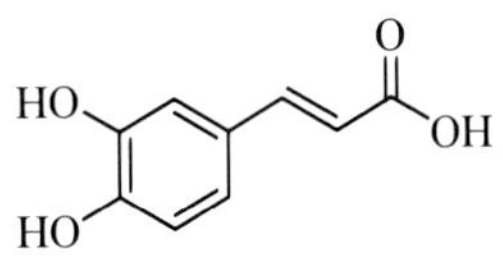

41

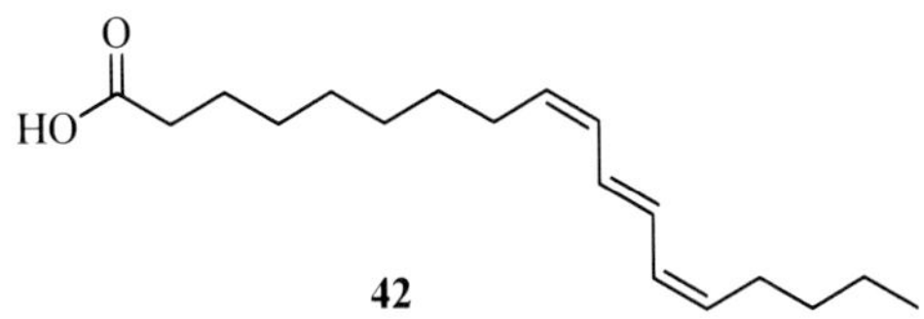

42

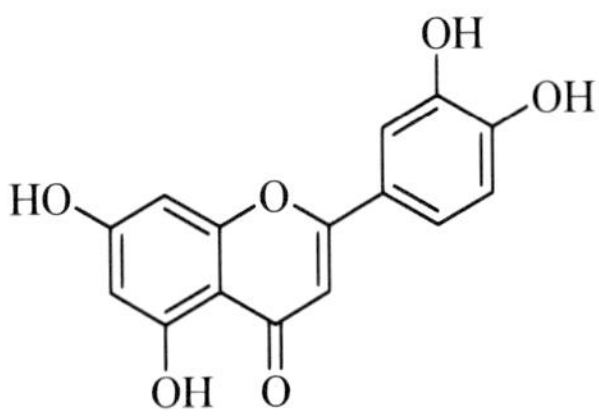

43

Figure 9. Pomegranate polyphenopls

Resveratrol **44** (Figure 10) is a phytoalexin present in the skin of red grapes and berries [30]. Grape fruit and wine are good sources of resveratrol **44**. The effect of resveratrol **44** on carcinogenesis has been extensively studied. The reader is encouraged to consider a very recent review on the subject [132]. It was concluded that no clear answers can be derived from epidemiologic considerations, so the possible efficacy of resveratrol **44** remains an open question, despite the incredible amount of time, money, and intellectual capital that has been invested in the study of this compound.

Curcumin **45** (turmeric, Figure **10**) can be extracted from the rhizome of the plant *Curcuma longa L.* [30]. Treatment of both AD and AI PC cell lines with curcumin **45** decreased their proliferation and increased their apoptotic potential [133], an effect largely attributable to the downregulation of antiapoptotic Bcl-2 and Bcl-xL levels, with proapoptotic Bax levels remaining unaffected. Curcumin **45** (5-50 μM) inhibited the tyrosine kinase activity of the epidermal growth factor (EGF) receptor as well as its expression, suggesting that it interferes with EGF-mediated signaling pathways that are critical for the progression of cancer [134]. Significant suppression of tumor growth (LNCaP tumors in nude mice) in animals fed with an experimental diet containing 2% of curcumin **45** was observed [135]. Curcumin **45** was found to suppress both constitutive and inducible NF-κB activation on DU145 (at 75 μM) and LNCaP (at 25 μM) cells, respectively (Table **6**), as well as constitutive activation of the transcription factor AP-1 on DU145 cells at 50 μM, and to potentiate TNF-induced apoptosis [136]. Apoptosis induction with 50 μM curcumin **45** on both LNCaP and DU145 cells correlated with the activation of caspases 3 and 8 and downregulation of Bcl-2 and Bcl-xL protein levels. At this concentration, complete inhibition of growth of the two cell lines occurred [136]. The effect of curcumin **45** on TRAIL-induced apoptosis was also investigated. LNCaP cells do not undergo apoptosis in the presence of 5-10 μM of neither curcumin **45** or resveratrol **44** nor in the presence of 5-50 ng/ml of TNFα or TRAIL [137]. Nonetheless, a combination regimen of 10 μM curcumin and 20 ng/ml TRAIL (but not TNFα) enhanced LNCaP cell death by 2-3 fold. Increased cleavage of procaspases 3, 8 and 9, truncation of Bid and release of cytochrome c from the mitochondria were triggered [137]. The mechanism of TRAIL-induced apoptosis enhancement by curcumin **45** was found to be the inhibition of NF-κB activation by blockage of IκBα phosphorylation which eventually prevents its degradation [138].

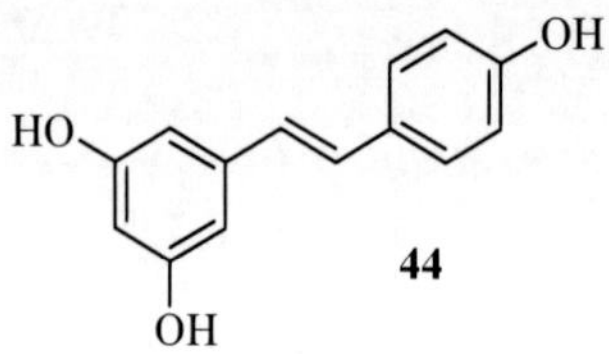

44

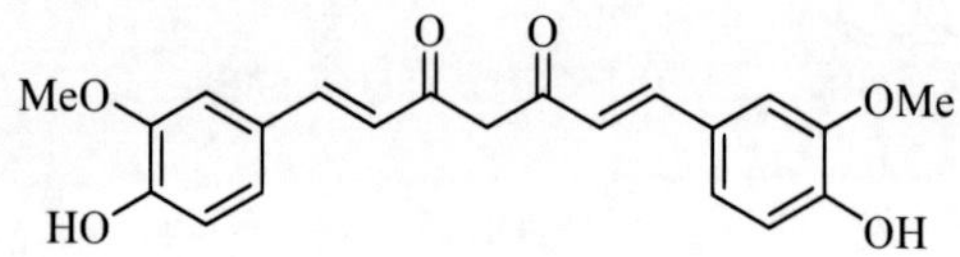

45

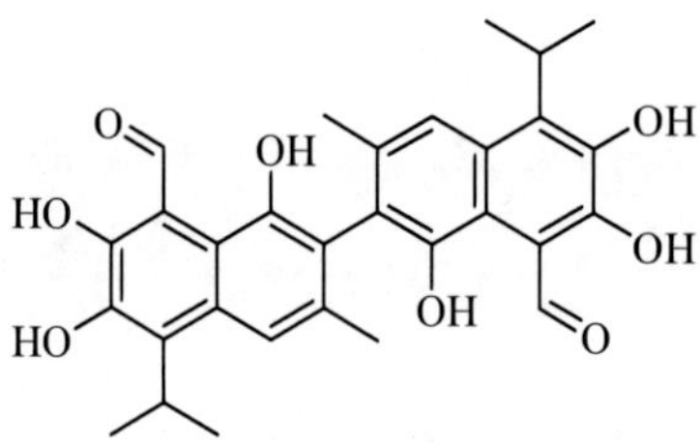

48

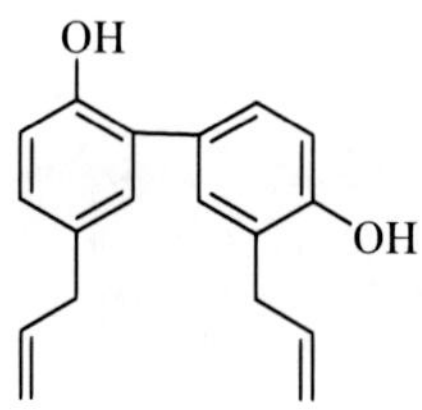

49

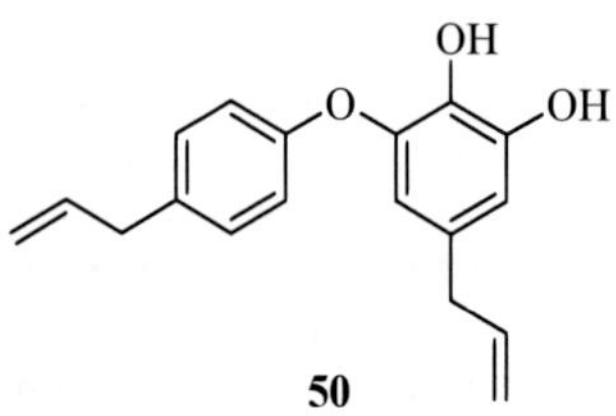

50

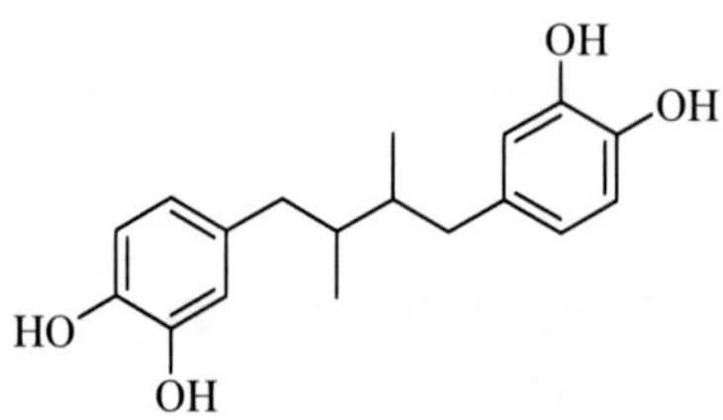

51

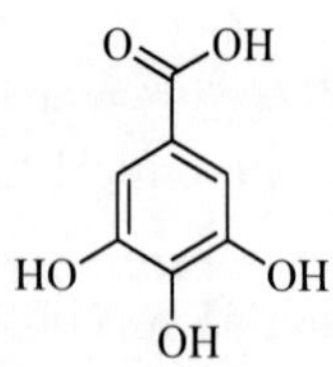

52

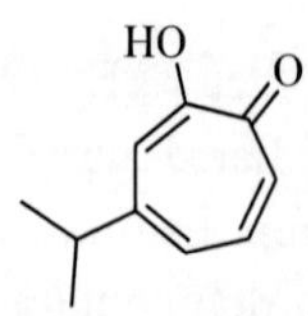

57

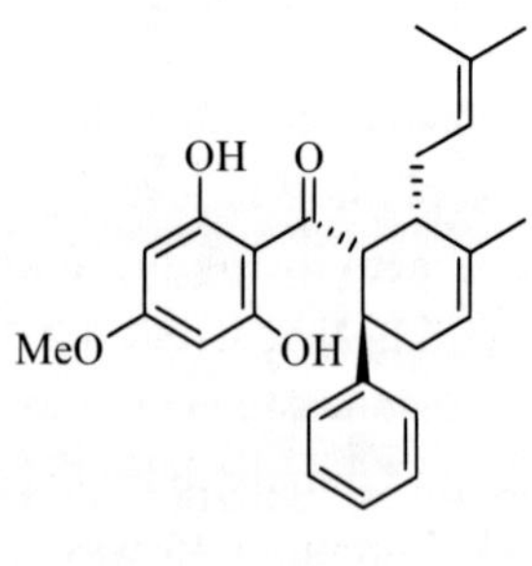

58

Figure 10. Plant polyphenols.

Curcumin **45** has been evaluated in combination with capsaicin **46** (Figure 25) and the nutraceutical Bioflavone α on LNCaP cells [139]. Although no inhibitory activity of cell growth was found for the isolated compounds, cocktails of Bioflavone α with capsaicin or curcumin **45** inhibited cell growth by 73 and 47%, respectively, as compared to control. It has also been reported that combination of curcumin **45** (25 μM) with β-phenethyl isothiocyanate **47** (Figure 27) at 10 μM exerts additive inhibitory activity on PC-3 cells through targeting epidermal growth factor receptor (EGFR), Akt and NF-κB, leading to apoptosis [140, 141]. A common drawback to the use of curcumin **45** for further studies on cancer chemotherapy and chemoprevention is its poor oral bioavailability [142] which has been minimized by coformulation with piperine, a compound that inhibits the hepatic and intestinal degradation of curcumin **45** [143].

Table 6.

Target: Suppression of NF-κB
Studies have shown that androgen-refractory DU145 and PC-3 cells are resistant to TNFα treatment. This resistance may be due to constitutive activation of NF-κB, a characteristic that has been observed in these cell lines and that may help cells survive the damage induced by proapoptotic stimuli. Activation of NF-κB renders TNFα ineffective in inducing apoptotic cell death of cancer cells and may cause the induction of a toxic TNFα environment on PC patients. Because the NF-κB pathway plays such an important role in PC cell survival, proliferation and resistance to chemotherapy and radiation, this signaling pathway represents a key molecular target for anticancer strategies.

Gossypol **48** (Figure 10), a polyphenolic compound naturally occurring in cottonseed [144], was originally identified as a male antifertility agent. Gossypol **48** treatment resulted in a marked reduction of the prostate weight as well as induction of structural and functional changes of the gland somewhat similar to the ones observed after castration [145, 146]. Antiproliferative and antimetastatic activities in AI Dunning R3327-MAT-LyLu PC cell-bearing Copenhagen rats have been reported with gossypol **48** [147]. Gossypol **48** has also been reported to potently inhibit both isoforms of 5α-reductase (Table 4, Entry 4) [124]. The compound was found to inhibit the proliferation without affecting the viability of AI PC-3 cells with an IC_{50} value of less than 1 μM, and to induce a dose- and time-dependent accumulation of cells at the G0/G1 phase of the cell cycle [148]. This cytostatic effect occurred by a mechanism that included amplification of transforming growth factor (TGF)-β1 mRNA expression on PC-3 cells. On DU145 cells, at a 5-10 μM dose-level, gossypol **48** significantly enhanced apoptosis through downregulation of Bcl-2 and Bcl-xL and upregulation of Bax at the mRNA and protein levels [144]. Activation of caspases 3, 8, and 9 was also found to occur as well as increased PARP cleavage. *R*-(–)-Gossypol acetic acid (AT-101, Ascenta Therapeutics, Inc.) is currently being studied on phase II clinical trials in combination with docetaxel **1** and prednisone for the treatment of HRPC [71]. AT-101 was found to inhibit the proliferation of both VCaP and PC-3 cells with IC_{50} values of 5 and 10 μM, respectively [149]. Activation of caspase-mediated apoptosis on VCaP cells was accompanied by downregulation of Bcl-2, and antiapoptotic Mcl-1 mRNA and protein expression was also observed. AT-101 (15 mg/kg, orally, 5 days/week) in combination with surgical castration, delayed the onset of male SCID mice VCaP xenograft tumors growth *in vivo* [149].

One of the bioactive constituents of *Magnolia officinalis*, honokiol **49** (Figure 10) [150], has attracted a great deal of interest due to its very diverse set of biological effects [151-153]. The desirable effects of honokiol **49** on osteoclastogenesis and angiogenesis inhibition [154, 155] together with its ability to inhibit NF-κB activation (Table 6) [155, 156] made it a good candidate for study on marrow and prostate stromal cells, known to be closely associated with tumor cells in the primary and their subsequent metastasis to bone in patients [151]. Honokiol **49** inhibited the growth of LNCaP, C4-2 and PC-3 cells with IC_{50} values of 18.75 to 37.5 μM (5 to 10 μg/ml, a concentration achievable *in vivo* with 100 mg/kg of honokiol **49** treatment) as well as both normal/benign and cancer-associated prostate stromal cells established from patients, at 10 to 20 μg/ml. Induction of apoptosis on AI C4-2 cells was found to occur with activation of caspases 3, 8 and 9 and increased cleavage of PARP [151]. Honokiol **49** (10 to 20 μg/ml) exerted additive growth inhibitory effects when combined with docetaxel **1** (1 nM) on C4-2 cells in culture, and in mice harbouring C4-2 tumors. The combination regimen of 100 mg/kg honokiol **49** and 5 mg/kg docetaxel **1** significantly inhibited human PC xenografts grown in mouse skeleton as evaluated by mouse serum levels of PSA, X-ray of mouse bone and histomorphometric analysis of the mouse skeleton [151].

The decrease in the viability of LNCaP and PC-3 cells after treatment with 20-60 μM honokiol **49** has been found to be associated with G0/G1 arrest [152]. ROS generation was also found with honokiol **49** treatment [152]. Apoptosis induction by honokiol **49** on LNCaP and PC-3 cells occurred with overexpression of Bax, Bak and Bad and dowregulation of Bcl-xL and Mcl-1 protein levels [153]. Oral gavage of 2 mg honokiol **49**/mouse, 3 times a week, significantly retarded the growth of PC-3 xenografts without causing weight loss. Tumors from honokiol **49**-treated mice exhibited markedly higher counts of apoptotic bodies and reduced proliferation indexes as well as neovascularization when compared to control tumors [153].

Another constituent of *Magnolia officinalis*, obovatol **50** (Figure 10), has been recently reported to dose-dependently inhibit the growth of LNCaP and PC-3 cells with IC_{50} values of 18 and 23 μM, respectively [157]. TNFα and tetradecanoyl phorbol acetate (TPA)-induced growth was also inhibited by obovatol **50** with IC_{50} concentrations of 13.5 and 18 μM for LNCaP cells, and 18 and 18 μM for PC-3 cells, respectively. Induction of apoptosis in which inhibition of NF-κB activation may have a significant role was suggested as the mechanism of cell-mediated death for obovatol **50** (Table 6) [157].

Androgen ablation on AI PC cells does not induce the changes in the intracellular concentrations of Ca^{2+} that would be required to induce apoptosis [158]. However, it has been demonstrated that both AD normal prostatic glandular cells and AD PC cells undergo apoptosis following androgen deprivation, which is associated with enhanced expression of a series of genes, that ultimately mediate activation of Ca^{2+}/Mg^{2+}-dependent endonucleases present in the nucleus to fragment dsDNA leading to cell death [159-162]. Agents that can cause a sustainable elevation in the intracellular concentrations of Ca^{2+} would therefore be of interest for HRPC treatment. One of such agents is nordihydroguaiaretic acid **51** (Figure 10), found in the long-lived creosote bush, and commonly used as an inhibitor of lipoxygenases

[163]. Compound **51** is also an inhibitor of 5α-reductase types 1 and 2 with similar potency (Table 4, Entry 5) [124]. Inhibition of PC-3 cell growth with compound **51** was found to occur with an IC_{50} value of 17 μM [164]. The compound was reported to induce concentration-dependent increases of intracellular Ca^{2+} levels both in the presence and absence of extracellular Ca^{2+}, in PC-3 cells, with an EC_{50} of 30 μM [165]. The rise in intracellular Ca^{2+} level was suggested to activate K^+ and Cl^- currents and trigger the secretion of proteases. Moreover, the source of Ca^{2+} involved on compound **51**-mediated action on PC-3 cells was found to be the ER [165].

Gallic acid **52** (Figure 10) one of the major active constituents of grape seed extract, has been reported to cause massive DU145 cell death after 72 hours of treatment, at 100 μg/ml [166]. Mechanistic studies demonstrated a strong caspase 9 and caspase 3 activation as well as PARP cleavage by compound **52** on DU145 cells [166]. The 3,3'-di-*O*-gallate ester of procyanidin dimer B2 (Epi-Epi), also present in grape seed extract, exhibited dose-dependent effects on DU145 cells over the range of 25-100 μM concentration, whereas compound **52** exhibited comparable activity at lower doses but was highly lethal at 100 μM [167]. Further studies on DU145 cells have shown that gallic acid **52** causes strong cell growth inhibition, cell cycle arrest, and apoptotic death in a dose- and time-dependent manner, together with a decrease in CDKs and cyclins levels, however with strong increase in p21 expression [168].

The activity of four lignans **53-56** (Figure 11) isolated from *Campylotropis hirtella* (Franch.) Schindl. on PSA production and AR expression on LNCaP cells has recently been reported [169]. The IC_{50} values for PSA secretion inhibition were found to be 19, 45, 110, and 128 μM, for compounds **53-56**, respectively. These compounds decreased AR expression on LNCaP cells and triggered caspase-mediated apoptosis [169].

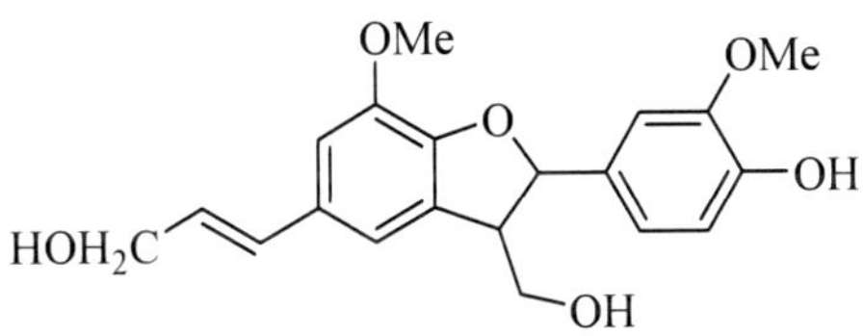

53

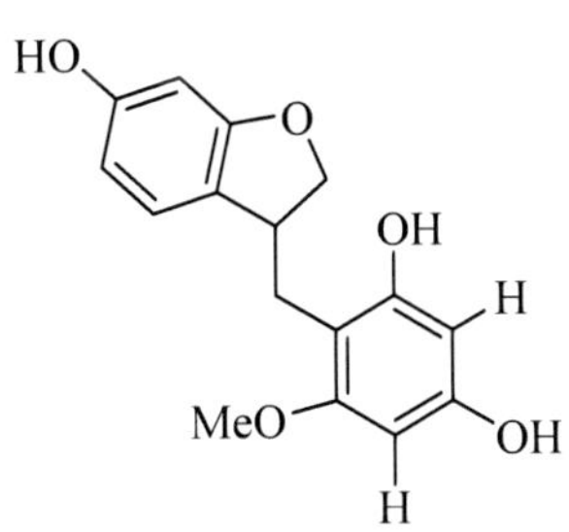

54

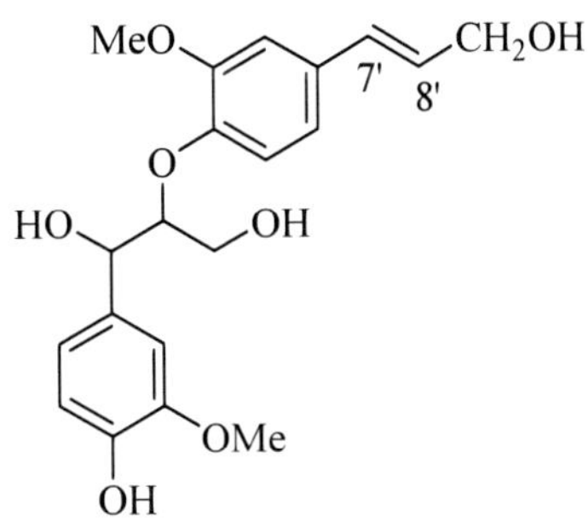

7',8'-*erythro* **55**
7',8'-*threo* **56**

Figure 11. Lignans from *Campylotropis hirtells* (Franch) Schindl.

Hinokitiol **57** (β-thujaplicin, Figure 10), a metal chelator found in the heartwood of cupressaceous plants, has been found to suppress PC cell growth and disrupt AR signaling [170]. Half-maximal LNCaP proliferation inhibition was found with 50 μM of compound whereas on PC-3 cells, even at the high concentration of 100 μM no growth inhibition was found. On LNCaP cells, a decrease in PSA levels and AR mRNA and protein expression was observed with hinokitiol **57** [170].

Panduratin A **58** (Figure 10), a cyclohexenyl chalcone derivative isolated from the methanolic extract of *K. pandurata*, has been found to inhibit the proliferation of PC-3 and DU145 cells with IC_{50} values of 13.5 and 14 μM, respectively [171]. Panduratin A **58** treatment triggered caspase-mediated apoptosis on both cells lines and cell cycle arrest at the G2/M phase.

2.3.1.1. Flavonoids

The beneficial effects that have been attributed to soy as a dietary chemopreventive agent are due to its isoflavone content out of which genistein **59** (Figure 12) is the major constituent, also found in other legumes such as peas, lentils or beans, together with quercetin **60** (Figure 12), a main representative of the flavonol class and a polyphenolic antioxidant also present in onions, broccoli, apples, grapes (red wine) and a component of *ginko biloba* [172]. A very thorough set of *in vitro* and *in vivo* activities have been reported for genistein **59** on PC cells. The *in vitro* effects include inhibition of tumor cell growth (LNCaP, PC-3 and DU145 cells below 20 μM, as well as PNT-1, PNT-2, LNCaP in SCID mice cells), induction of apoptosis, inhibition of PSA expression, suppression of DNA synthesis, downregulation of cyclin B, upregulation of both $p21^{WAF1/CIP1}$ and $p27^{KIP1}$ proteins, inhibition of NF-κB activation, modulation of the Akt signaling pathway, induction of G1 cell cycle blockage and G2/M arrest, inhibition of mitogenic signaling pathways, reduction of 25-hydroxyvitamin D-1α-hydroxylase (CYP27B1) and 25(OH)D_3-24-hydroxylase (CYP24) protein levels, and inhibition of 5α-reductase type 2 (Table 4, entry 6) [172]. *In vivo*, genistein **59** has been found to induce DNA fragmentation, reduce tumor vessel density and serum insulin-like growth factor (IGF) level, inhibit tumor growth and trigger apoptosis, reduce AR and estrogen receptor expression and to downregulate the EGF and ErbB2/Neu receptors in the rat prostate [172].

A series of phase I and II clinical trials have also been carried out with genistein **59** [172]. Analysis of the results from these clinical trials has however been inconclusive as to the role of this compound on PC chemotherapy/prevention. A common difficulty in clinical studies with dietary agents is the fact that the doses and concentrations of the compounds used in the studies are empirically derived, and the manufacturing and preparation of the products containing the desired molecules is not standardized. Thus, the choice of protocols for phase I and II clinical trials that will result in comparable sets of data for different studies is usually a very hard task.

Quercetin **60** (10-100 μM) has been shown to have an inhibitory effect on the expression and function of the AR in LNCaP and LAPC4 cells, and consequently to inhibit PSA and human glandular kallikrein (hK2) secretion as well as on other androgen regulated genes [173]. Optimal inhibitory effects on AR function were seen with 100 μM of quercetin **60**. It was postulated that the compound increased the binding affinity of Sp1 transcription factor to the AR and therefore reduced its function on PC cells [174]. It has recently been demonstrated that on LNCaP cells, quercetin **60** induces apoptosis and that this process is mediated by the dissociation of Bax from Bcl-xL and activation of caspases [175]. On PC-3 cells, quercetin **60** was found to block G2/M transition, with significant induction of apoptosis [176]. Apoptosis markers levels like Bcl-2 and Bcl-xL were significantly decreased and Bax and caspase-3 levels were increased. Addition of 100 μM of quercetin **60** to PC-3 cells led to a substantial decrease in the expression of Cdc2/CDK1, cyclin B1 and phosphorlyated pRb and to an increase in the expression of p21. Quercetin **60** has also been found to enhance TRAIL-induced apoptosis in human PC DU145 cells via Akt dephosphorylation [175]. Quercetin **60** has been found not to inhibit the activity of 5α-reductase significantly (Table 4, Entry 7) [124].

Daidzein **61** (Figure 12) is another major constituent of total soybean isoflavone content. It is speculated that some humans possess the necessary intestinal flora to convert daidzein **61** to equol, a molecule similar to estrogen [177]. Daidzein **61** inhibits the growth of PC cells *in vitro* and PC tumors in *vivo*, with less potent antiprostate cancer activity than genistein **59** [164, 178]. Both genistein **59** and daidzein **61** have been found to downregulate prostate androgen regulated transcript-1 (PART-1) gene expression induced by DHT in LNCaP cells with IC_{50} values of 25 and 50 μM, respectively [178]. A recent phase II trial of isoflavone (including a mixture of genistein **59** and daidzein **61**) in PC patients using soy milk (containing 47 mg of isoflavone *per* 8 oz serving, 3 times daily, for 12 months) has shown that isoflavone may slow the rate of increase of serum PSA in a subset of patients with biochemical progression after radiation therapy or radical prostatectomy [179]. Daidzein **61** was found to inhibit 5α-reductase type 2 (Table 4, Entry 8) [124].

Apigenin **62** (Figure 12) is yet another abundant flavone present in fruits and vegetables [30], that has been shown to promote cell cycle arrest and apoptosis in various malignant cell lines. Half-maximal cell growth inhibition of 22Rv1 cells was observed after 48 hours of treatment with 20 μM of compound [180]. Apigenin **62**-mediated apoptosis was studied on 22Rv1 and isogenic PC-3 ($p53^{-/-}$) and PC-3 ($p53^{+/+}$) cells and found to be initiated by a ROS-dependent disruption of the mitochondrial membrane potential through transcriptional-dependent and –independent pathways. *In vivo*, prior intake of apigenin **62** delayed the establishment of 22Rv1 tumor xenografts in athymic nude mice, after oral administration at 20 and 50 μg/mouse/day for 2 weeks, by 44 and 59%, respectively. Apoptosis induction through a p53-dependent pathway was also observed during this experiment [180]. Apigenin **62** caused G0/G1 arrest of the cell cycle with sub-G1 accumulation on both LNCaP and PC-3 cells, and xenografts [181, 182]. The observed dose- and time-dependent inhibition of Rb-phosphorylation, as well as inhibition of p38-MAPK, PI3K-Akt kinase phosphorylation and CDKs suggested that apigenin **62** exerts its effects at various phases of the cell cycle.

Acacetin **63** (Figure 12) is a flavonoid present in *Cirisium rhinoceros* Nakai (Compositae), a herbaceous perennial native to Korea, used in folk medicine [183]. LNCaP and DU145 cell proliferation has been found to be inhibited by this compound at 50 μM, after 48 hours of treatment. The compound induced either G1 or G2/M arrest on both cell lines, depending on the concentrations used and exposure times.

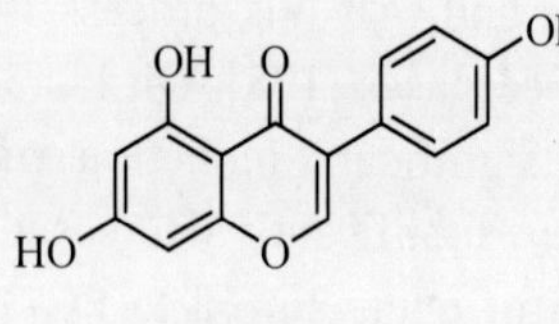

59

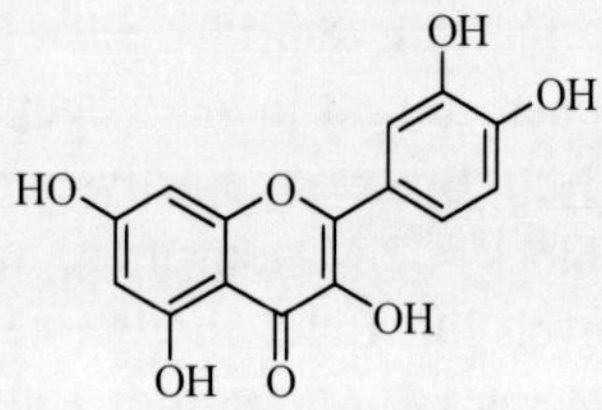

60

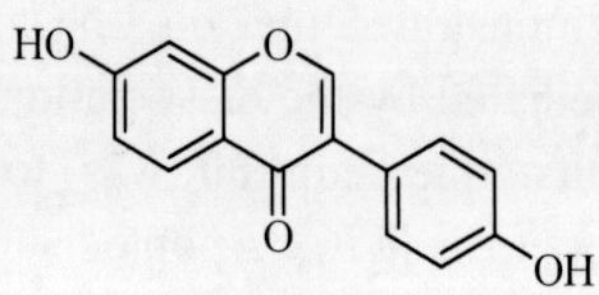

61

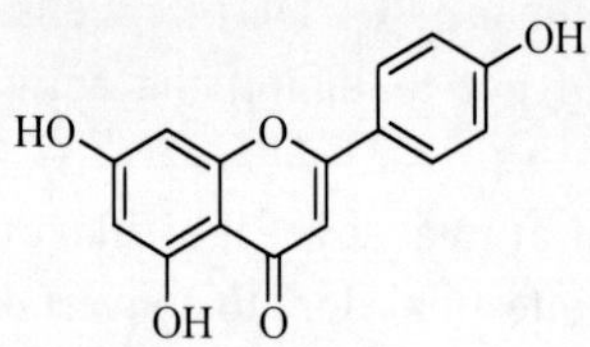

62

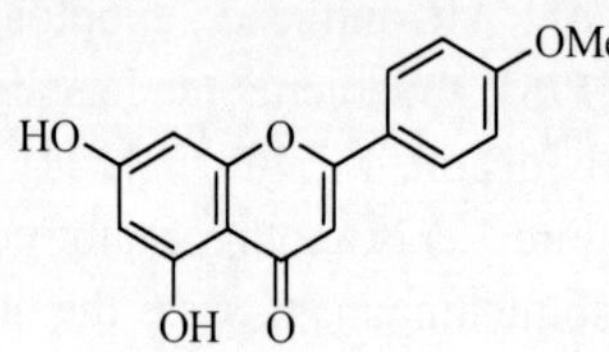

63

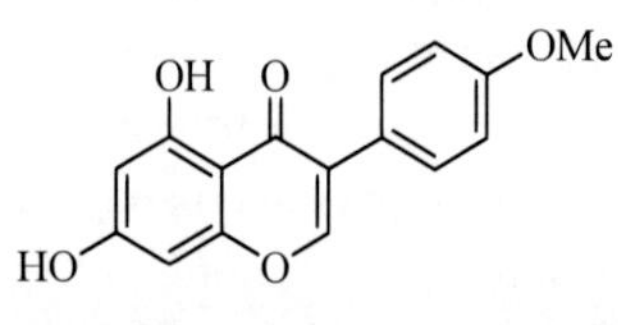

64

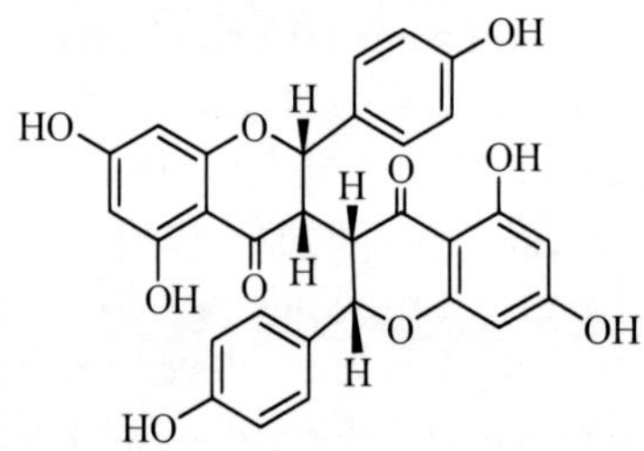

83

Figure 12. Plant flavonoids.

Biochanin A **64** (Figure 12), the major isoflavone present in red clover but not soy-derived products [184] has been reported to induce upregulation of UDP-glucuronosyl transferase, an enzyme that conjugates steroids, on LNCaP cells thereby causing inactivation and elimination of the compounds from the cells [185]. At 63 μM, biochanin A **64** inhibited 50% of LNCaP viability, after 48 hours of treatment [186]. The compound was found to induce G1 arrest and apoptosis, and cDNA microarray analysis identified 29 downregulated genes with 6 reduced below assay detection limits. I.p. administration of 4 μg/μl of compound to mice bearing LNCaP xenografts resulted in lower tumor incidence at 3 weeks, as compared to control, whereas at 6 weeks tumor incidence was virtually the same for the two groups suggesting a cytostatic effect for biochanin A **64** [186]. Growth inhibition of PC-3 cells with an IC_{50} value of 29 μM has also been reported for this compound [164] as well as 5α-reductase inhibition (Table 4, Entry 9) [124].

Silymarin is a crude extract of flavolignans obtained from milk thistle (*Silybum marianum* L. Gaertn.) out of which 7 compounds have been isolated and purified that include silybin A **65**, silybin B **66**, isosilybin A **67**, isosilybin B **68**, silydianin **69**, silychristin **70**, isosilychristin **71** and one flavonoid taxifolin **72** (Figure 13) [187]. Results from a survival assay with the compounds on LNCaP, DU145 and PC-3 cells are depicted on Table 7 [188].

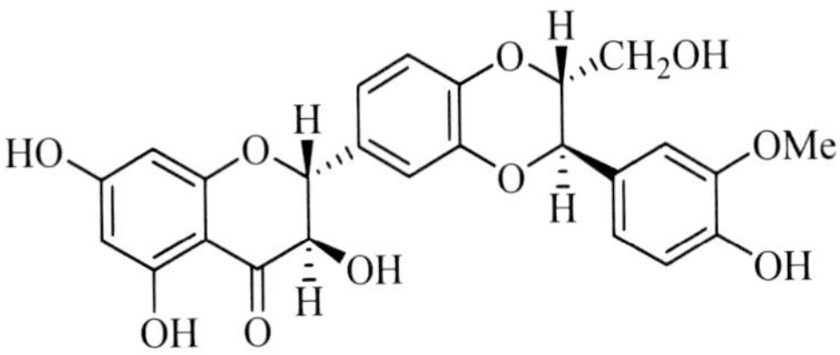

65

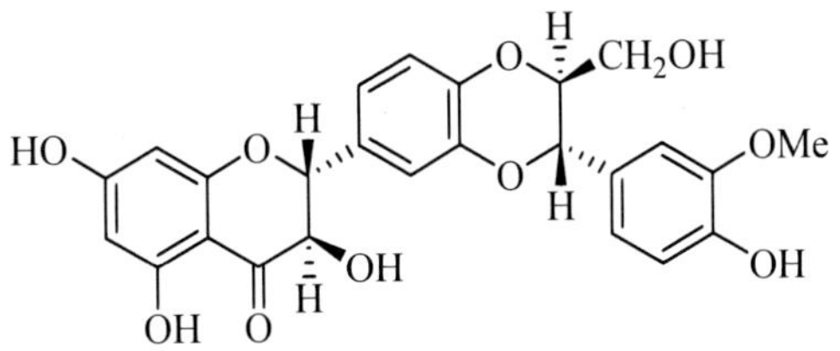

66

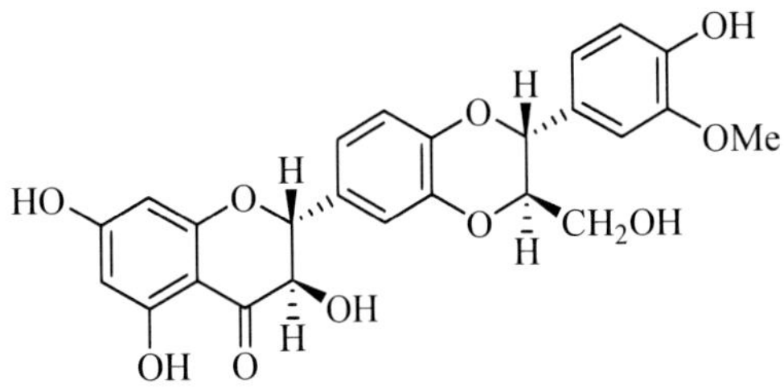

67

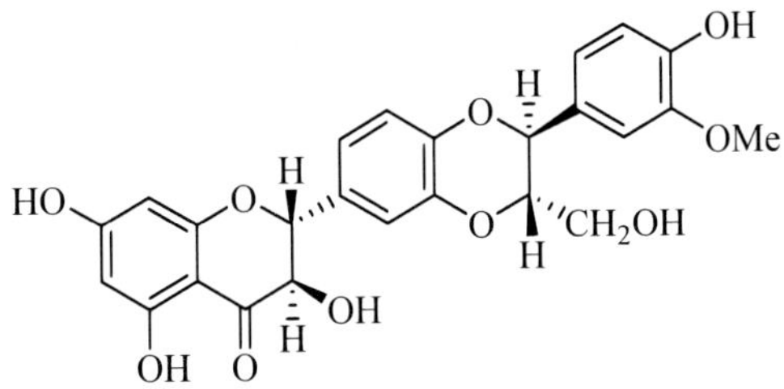

68

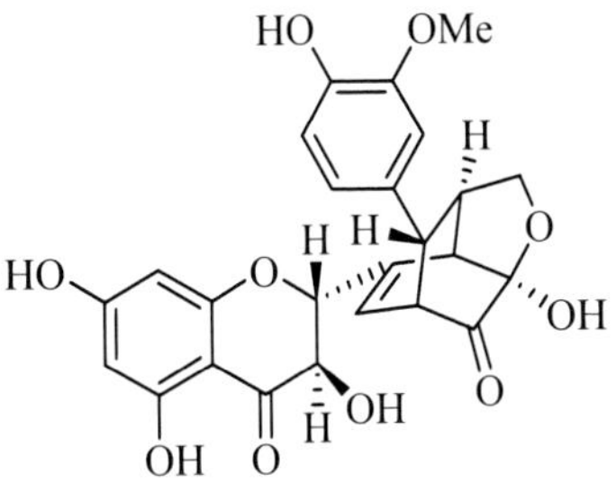

69

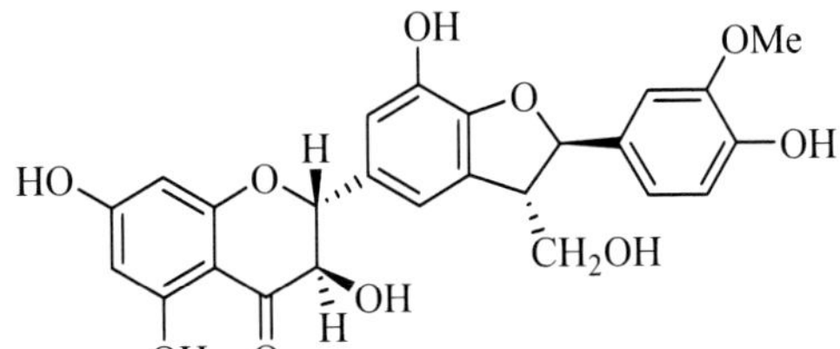

70

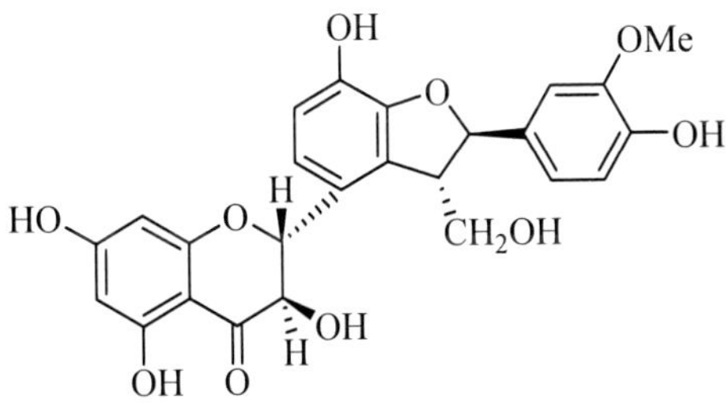

71

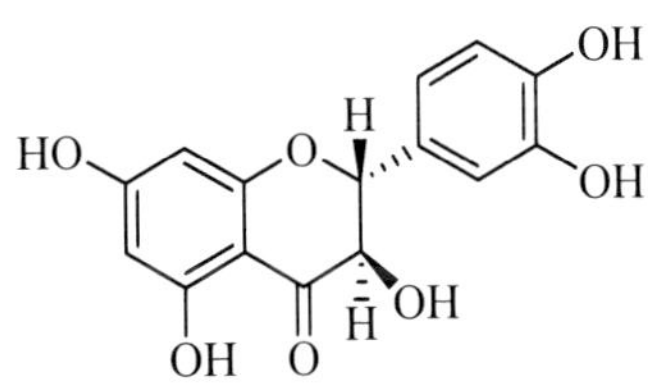

72

Figure 13. Silymarin compounds.

Thus, isosilybin B **68** ranked the most potent flavolignan for inhibiting PC cell survival (Table 7, Entry 4). Further study showed that silybin A **65**, silybin B **66**, isosilybin A **67** and isosilybin B **68** were effective in modulating the expression of PC-3 cell cycle regulators and inducing cell cycle arrest, while there was marginal or no effect of silydianin **69**, silychristin **70** or isosilychristin **71** on cell cycle and regulation [189]. At 90 μM, isosilybin B **68** was the most potent compound in inhibiting colony formation, cell growth, causing cell cycle arrest, and altering the expression of cell cycle regulators in PC-3 cells. All 7 compounds bear the same chemical formula and molecular mass as well as the common flavonoid nucleus, and differ only in the conformation of the lignan moiety. In the plant, these conformations are catalyzed by a peroxidase reaction between the flavonoid dihydroquercetin (taxifolin) and coniferyl alcohol. Silybin A **65**, silybin B **66**, isosilybin A **67** and isosilybin B **68** contain the lignan moiety with identical functional groups but differ with respect to the stereochemical positions of the terminal aryl(3-methoxy-4-hydroxy-phenyl) and hydroxymethyl groups at the α and β carbons. Therefore, the overall inactivity of silydianin **69**, silychristin **70** and isosilychristin **71** was suggested to be related to their different condensation products as compared to the previously mentioned compounds [189]. A comparative study of isosilybin A **67** and B **68** showed that both compounds inhibited the proliferation and induced G1 phase arrest and apoptosis in human PC cells [190]. Isosilybin B **68** was found to be more effective in reducing the levels of cyclin D1 and cyclin D3, in LNCaP cells. Similarly, isosilybin B **68** was more potent in reducing the levels of cyclins (D1, D3, E, and A) as well as of CDKs (4 and 2) and Cdc25A in 22Rv1 cells. A higher apoptotic index was however found with isosilybin A **67** on both cell types [190]. Isosilybin B **68** was also reported to decrease AR and PSA levels in LNCaP, 22 Rv1, and LAPC4 cells [187]. Its mechanism of action was suggested to occur through enhancement of the formation of a complex between Akt, MDM2 (a ring-finger protein with E3-ligase activity bearing a crucial role in ubiquitination and degradation of proteins), and AR, which promoted phosphorylation-dependent AR ubiquitination and its degradation by the proteasome.

Table 7. Cell survival assay for Silymarin compounds.

Entry	Compound	Cell line % survival (90 μM)		
		LNCaP	DU145	PC-3
1	**65**	55.9 ± 3.1	15.2 ± 0.9	21.9 ± 7.9
2	**66**	29.2 ± 6.3	33.9 ± 3.6	12.7 ± 1.6
3	**67**	54.4 ± 10.5	19.9 ± 0.5	22.9 ± 4.0
4	**68**	26.2 ± 4.3	16.0 ± 0.3	18.1 ± 3.1
5	**69**	99.1 ± 2.4	76.7 ± 3.0	51.3 ± 4.3
6	**70**	100.4 ± 3.0	16.6 ± 2.5	21.4 ± 0.9
7	**71**	94.8 ± 1.5	99.3 ± 1.5	76.9 ± 3.2

Hop (*Humulus lupulus L.*), an essential raw material for beer brewing, contains numerous phenolic and polyphenolic compounds. Phenolic acids, catechins, proanthocyanidins, and flavonols represent the major classes and their relative proportions vary widely depending on the hop variety and the growing conditions [191]. Prenylflavonoids from Hop are depicted on figure 14. Xanthohumol **73** is the most abundant prenylchalcone present in hop (up to 1 % w/w) and is accompanied by its analogue desmethylxanthohumol **74** albeit at lower concentrations. Both prenylchalcones are readily isomerized by intramolecular Michael-type addition to the corresponding prenylflavanones with compound **73** giving rise to isoxanthohumol **75**, and compound **74** being converted into a mixture of 8-prenylnaringenin

76 and 6-prenylnaringenin **77** [191]. All hop compounds reduced PC-3 and DU145 cell viability by at least 50% after 2 hours of incubation when added to the cells at 200 μM. No specific apoptotic features were seen with prenylflavonoids nor was activation of the caspase pathway. Formation of vacuoles was however observed in DU145 cells treated with isoxanthohumol **75** and 6-prenylnaringenin **77**, and in PC-3 cells treated with isoxanthohumol **75**, 8-prenylnaringenin **76** and 6-prenylnaringenin **77**, which was suggested to be due to autophagy induction and consequent cell death [191]. The effects of xanthohumol **73** on BPH-1 cells have been reported as inhibition of cell viability at 5 μM and apoptosis induction with activation of proapoptotic Bax, and p53 as well as decreased activation of NF-κB [192].

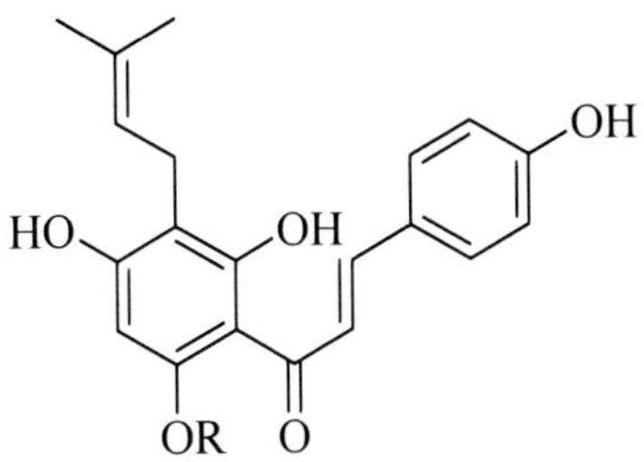

R = Me **73**
R = H **74**

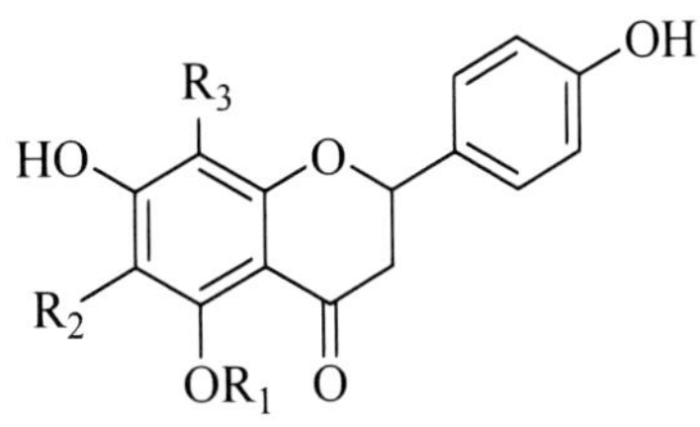

R_1 = Me; R_2 = H; R_3 = prenyl **75**
R_1 = R_2 = H; R_3 = prenyl **76**
R_1 = R_3 = H; R_2 = prenyl **77**

Figure 14. Prenylflavonoids from Hop (*Humulus luphulus L.*).

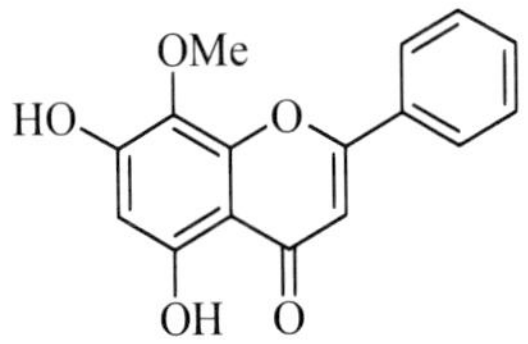

78

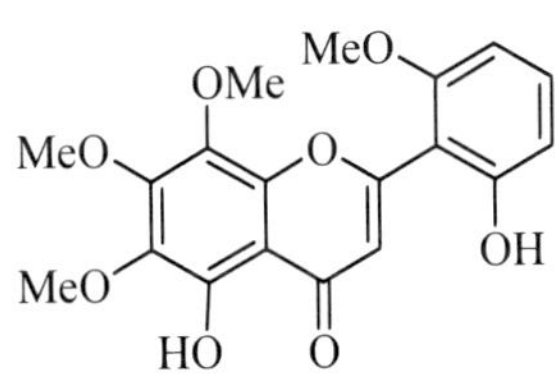

79

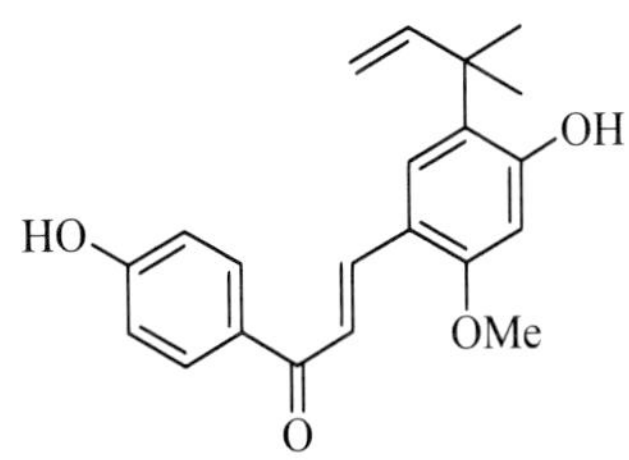

82

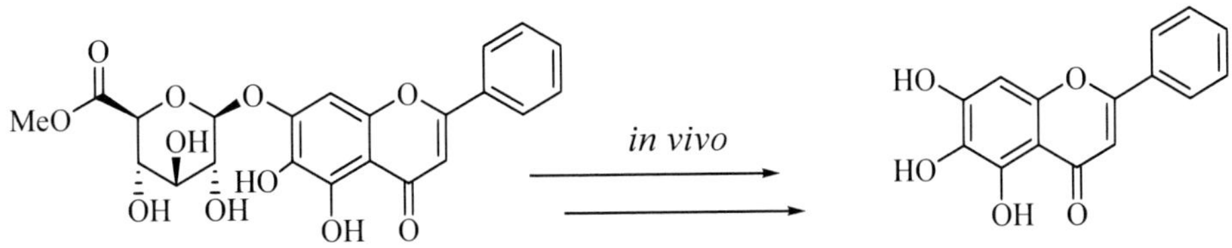

80 **81**

Figure 15. Flavonoids from the herbal mixture PC-SPES

Table 8. Cell proliferation assay for *S. baicalensis* flavonoids.

Entry	Compound	Cell line IC_{50} (μM)	
		LNCaP	PC-3
1	**78**	42	50
3	**79**	22	35
2	**81**	13	25

Wogonin **78**, neobaicalein **79**, and baicalein **81** are components of *Scutellaria baicalensis*, a botanical with a long history of medicinal use in China and a constituent of the herbal mixture PC-SPES, a product that inhibits PC cell growth both *in vitro* and *in vivo* (Figure 15) [30, 193]. The individual compounds were found to inhibit the proliferation of LNCaP and PC-3 cells (Table 8) inducing a G1 arrest in LNCaP cells, and to possess antiandrogenic activities.

Reduced expression of the PSA gene with AR protein levels remaining unchanged was also observed [193]. *In vivo*, baicalein **81** (20 mg/kg/day, orally) reduced the growth of PC xenografts in nude mice by 55% at 2 weeks compared to placebo and delayed the average time for tumor progression. Baicalein **81** is the most abundant active compound in the *S. baicalensis* extract and most of the activity seen with the complete extract was found to be due to this compound. Baicalein **81** is obtained by cleavage of the glycoside moiety of baicalin **80** *in vivo* (Figure 15) [194]. Baicalein **81** inhibits 12-lipoxygenase at nanomolar concentrations, a finding that may be significant for PC treatment seeing that elevated levels of 12-lipoxygenase and 12(*S*)-hydroxyeicosatetraenoic acid have been associated with advanced stage, poorly differentiated, metastatic PC [195]. This compound has been found to inhibit 5α-reductase on a cell-free assay, however the results did not correlate to the ones obtained in a whole-cell assay (Table 4, Entry 10) [124]. When tested on PC-3 and DU145 cells, baicalein **81** (25 μM) caused cell cycle arrest at G0/G1 [194, 195] which was associated with suppression of cyclins D1 and D3 levels [195]. A strong decrease in the expression of the pRB protein was also seen on PC-3 cells whereas the expression of other retinoblastoma proteins (p107 and p130) was inhibited on DU145 cells. Treatment with the compound triggered apoptosis in both cell lines which occurred by decreased activation of Akt, loss of survivin and subsequent activation of caspases 3 and 7 in each cell line. A decrease on Bcl-2 and Bcl-xL expression was found in DU145 cells, and a shift in Bcl-2/Bax levels favored apoptosis in PC-3 cells [195]. Suppression of AR expression on LNCaP cells has been reported with both baicalin **80** and baicalein **81** however the latter being twice as potent for this effect. More recently, it has been shown that p53-dependent transcriptional induction of p53 upregulated modulator of apoptosis (PUMA) and oligomerization of Bax play an important role in the sensitivity of LNCaP cells to apoptosis induced by caspase activation triggered by wogonin **78** at 50-100 μM [196].

Another flavonoid that has been isolated from PC-SPES composition licorice root *G. uralensis*, licochalcone A **82** (Figure 15), has been shown to elicit time and dose-dependent suppression of proliferation of PC-3 cells, with an IC_{50} value of 12.5 μM, which correlated with downregulation of the expression of regulatory proteins involved in control of cell cycle checkpoints. At the IC_{50} concentration, the compound caused G2/M arrest after 3 days of treatment [197].

Neochamaejasmin A **83** (Figure 12) is a biflavonoid isolated from the roots of a traditional Chinese medicine, *Stellera chamaejasme L.* which has shown ability to inhibit the proliferation of LNCaP cells with an IC_{50} value of 23 μM [198]. Treatment of LNCaP cells with 6.25 μg/ml of compound suppressed the DNA-binding activity of both NF-κB and AP-1 to the promoter of cyclin D and also the expression of the cell regulatory protein cyclin D, proliferating cell nuclear antigen, and nucleolin, therefore arresting cells at the G1 phase [198]. At the higher dose of 12.5 μg/ml, production of ROS has been found to occur with dissipation of mitochondrial membrane potential, upregulation of CDK inhibitor p21 and induction of apoptosis. Treatment with neochamaejasmin A **83** ultimately resulted in promotion of the Fas-caspase-8-caspase 3 apoptotic machinery [198].

2.3.2. Terpenes and steroids

2.3.2.1. Monoterpenes

Perillyl alcohol **84** (Figure 16) is a hydroxylated monocyclic monoterpene produced by plants and found in many commonly consumed fruits and vegetables including citrus and food flavorings (peppermint, spearmint, sage, cherries, orange and tangerine) [199]. The effects of perillyl alcohol **84** on LNCaP cells were evaluated in the presence of mibolerone (Mib), a non-metabolizable synthetic androgen [199]. Thus, in the presence of 1 nM Mib, concentrations of 0.1 to 1 μM of perillyl alcohol **84** significantly reduced the expression of PSA and hk2 proteins. Both AR gene expression and AR-mediated transcription were inhibited in the presence of perillyl alcohol **84** whereas c-Jun protein expression was increased [199]. A phase II multicenter trial of perillyl alcohol **84** in patients with advanced HRPC that failed at least one prior chemotherapeutic or experimental regimen concluded however, that a 1200 mg/m^2/dose of perillyl alcohol **84** given orally 4 times daily had no significant role in the treatment of the disease due to poor drug tolerability and lack of objective benefit [200].

2.3.2.2. Diterpenes

Paclitaxel **19** (Taxol™, Figure 16) was first isolated from the bark of the Western yew tree in 1971 and found to possess unique pharmacological actions as an inhibitor of mitosis, differing from the *Vinca* alkaloids and colchicine derivatives in that it promotes rather than inhibits microtubule formation [201]. The compound binds specifically to the β-tubulin subunit of microtubules and antagonizes the disassembly of this key cytoskeletal protein, so that bundles of microtubules and aberrant structures derived from microtubules appear in paclitaxel **19** treated cells, thus causing mitotic arrest [201]. Resistance to paclitaxel **19** is associated to high levels of P-glycoprotein expressed by tumor cells. Paclitaxel **19** is a diterpenoid compound that contains a complex taxane ring as its nucleus and the side chain at carbon 13 has been found to be essential for its antitumor activity. Modification of the side chain resulted in a more potent analogue docetaxel **1** (Taxotere™, Figure 16) which has clinical activity against breast and ovarian cancers [201]. A recently published review focuses on the present status and perspectives with taxane-based treatments, which the reader is advised to consider [25]. The authors found that the results of two independent research teams have revealed that a taxane-based chemotherapy regimen can lead to a survival benefit in men with HRPC. Whereas previous chemotherapy studies supported only a palliative role for cytotoxic chemotherapy, these studies demonstrate a clear survival advantage that justifies their use in asymptomatic patients. Although the results may appear to be somewhat modest, a 20–24% decrease in the risk of death from PC compares favorably with the benefits seen with chemotherapy in other solid tumors. These trials have also established every 3-week docetaxel **1** as the standard platform upon which targeted therapies may be combined, and support further investigation of docetaxel **1** based regimens in earlier clinical stages of disease [25]. Regarding paclitaxel **19**, results from large-scale phase III studies with paclitaxel **19**-based regimens are needed in order to establish an effective role in HRPC treatment. Thus, the major conclusion drawn from the analysis of literature concerning paclitaxel **19** and docetaxel **1** is that docetaxel **1** and prednisone chemotherapy has become the first-line standard of care for men with metastatic HRPC. Future studies evaluating the use of this regimen in earlier stages of disease, and in combination with targeted agents, have been considered promising and will hopefully result in greater clinical benefit [25].

84

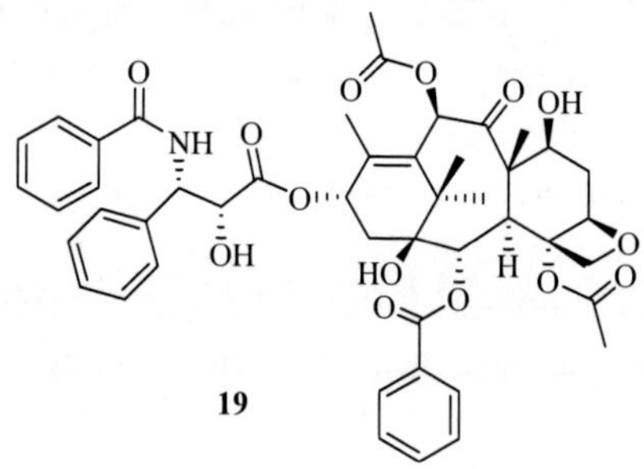

19

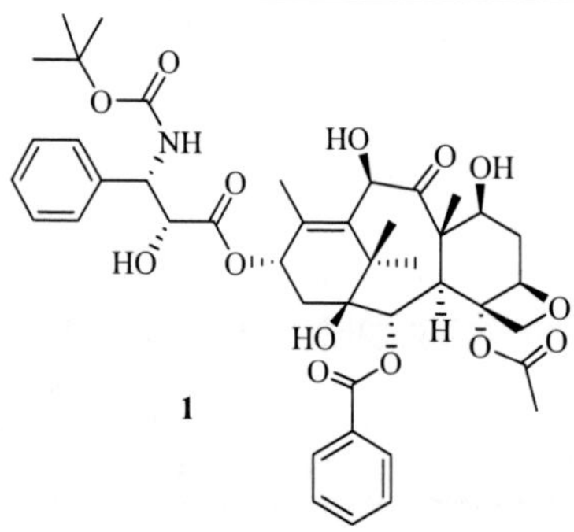

1

Target: the microtubules

Docetaxel and prednisone: recently been established as first-line standard of care for men with metastatic HRPC

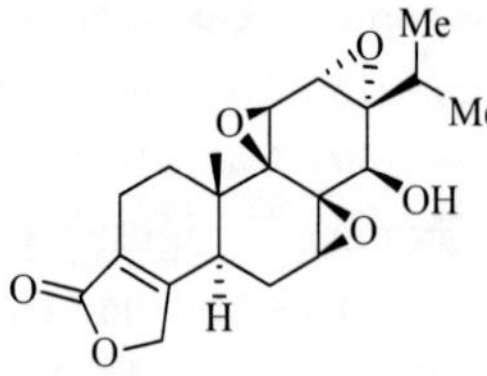

85

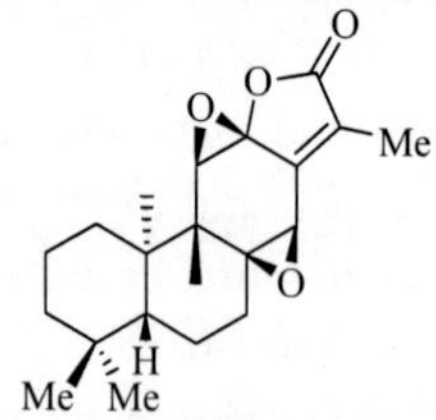

86

Figure 16. Plant derived mono and diterpenes.

Triptolide **85** (PG 490, Figure 16), a diterpene triepoxide, is purified from *Tripterygium wilfordii* Hook F [202]. Interest in this compound for PC research began with the observation that activation of p53 in response to γ-irradiation or other stresses such as hypoxia or DNA-damaging drugs was shown to be attenuated by triptolide **85** in primary cultures of normal and malignant PC epithelial cells, despite the presence of the wild-type p53 gene [203]. Moreover, triptolide **85** was found to inhibit LNCaP growth with an IC_{50} value of 10 ng/ml [204]. Half-maximal growth inhibition of five PC cell strains, E-PZ-10 (derived from histologically normal tissue of the periferal zone of the prostate), and four cell lines derived from prostatic adenocarcinomas of Gleason grade 3/3 (E-CA-11), 30% intraductal carcinoma / 70% Gleason grade 4 (E-CA-12) and 3/4 (E-CA-13 and E-CA-14), was observed with 0.1 ng/ml triptolide **85** [205]. Whereas low concentrations of triptolide **85** induced senescence of cells with G1 arrest, higher concentrations (50-100 ng/ml) triggered apoptosis. However, in contrast with the previous findings, p53 levels were significantly increased and predominantly accumulated in the nuclei of prostatic epithelial cells [205]. Nonetheless, downregulation of p53 target gene products such as Hdm-2 and $p21^{WAF1/CIP1}$ and Bcl-2 was also found to occur. It was suggested that triptolide **85** may use multiple signaling pathways, some of which involve p53 in cells with wild-type p53, growth arrest, and induction of apoptosis [205]. PG490-88Na (Pharmagenesis Inc.), a water-soluble prodrug of triptolide **85** (14-succinyl sodium salt) that is converted into triptolide **85** in the serum, has been approved for phase I clinical trials [206]. It has been shown that the compound suppresses tumor cell growth *in vivo* without toxicity and that it acts in synergy with chemotherapy. Thus, this compound has been pointed out as a promising novel antineoplastic agent for the treatment of patients with solid tumors [206].

Jolkinolide B **86** (Figure 16) is a diterpenic abietane-type lactone found in *Euphorbia fischeriana* extracts [207]. LNCaP, PC-3 and DU145 cell proliferation was totally inhibited by jolkinolide B **86** with IC_{50} values of 12.5, 75 and 45 μg/ml (40, 244 and 145 μM), respectively [207]. Further studies on LNCaP cells revealed apoptotic changes after 48 hours of exposure to jolkinolide B **86** at 25 μg/ml and G1 arrest was found to occur. Moreover, immunoblotting analysis of jolkinolide B **86**-treated LNCaP cells showed a dose-dependent increase of neuron specific hydrolase (NSE) protein and keratins 8 and 18, suggesting that the compound induced neuroendocrine differentiation of the cells [207].

Out of the six clerodane-type diterpenoids isolated from *Casearia membranacea* **87-92** (Figure 17), caseamembrin C **89** was the most effective compound in inhibiting the growth of PC-3 cells and causing apoptosis [208]. The IC_{50} values for caseamembrins **87-92** were found to be 1.2, 24.2, 0.5, 1.4, 1.9 and 2.6 μM, respectively. Caseamembrin C **89**-induced apoptosis occurred predominantly through activation of intrinsic apoptotic pathways. Kaurene diterpenes are widely distributed in nature, and have been found in the *Flacourtiaceae* genera [209]. Three *ent*-kaurene diterpenes **93-95** (Figure 17) isolated from *Laetia* genus and one semi-synthetic analogue **96** have been found to inhibit the growth of 22Rv1 (IC_{50} values of 15, 21.8, 65, and 6.8 μg/ml) and LNCaP cells (IC_{50} values of 28, 12.8, 17.6 and 45.5 μg/ml).

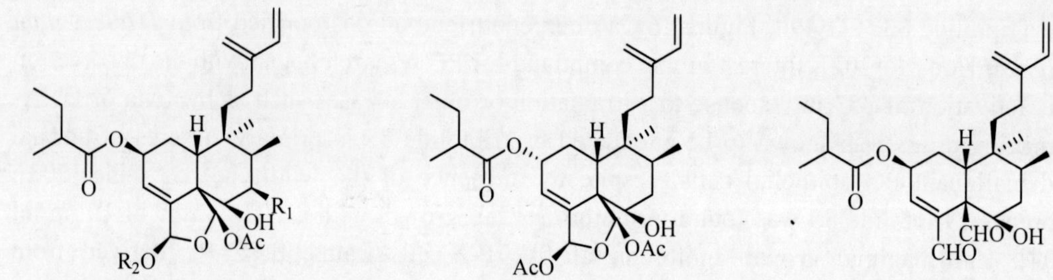

$R_1 = H; R_2 = CO(CH_2)_2CH_3$ **87**
$R_1 = H; R_2 = Me$ **88**
$R_1 = OH; R_2 = CO(CH_2)CH_3$ **89**
$R_1 = OAc; R_2 = CO(CH_2)_2CH_3$ **90**

91

92

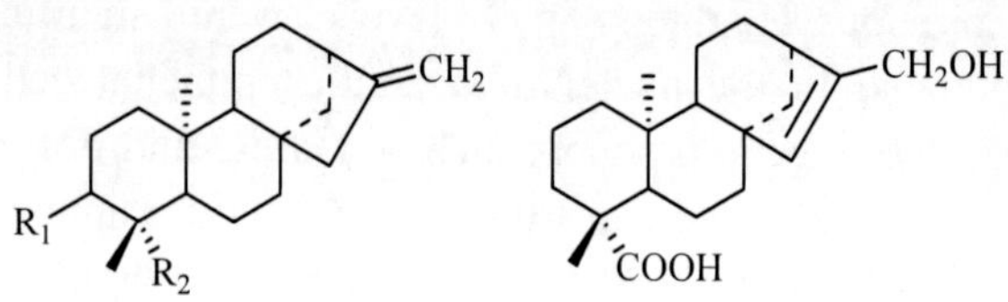

$R_1 = H; R_2 = COOH$ **93**
$R_1 = OH; R_2 = Me$ **94**
$R_1 = H; R_2 = COOCH_3$ **96**

95

Figure 17. Plant derived diterpenes.

2.3.2.3. Sesquiterpenes

Thapsigargin **97** (Figure 18), a sesquiterpene γ-lactone isolated from the root of the umbeliferous plant *Thapsia garganica*, is highly lipophilic and irreversibly inhibits the sarcoplasmic reticulum and ER (SERCA) enzymes with an IC_{50} of 30 nM (ki ca 2 pM) [210]. This compound has become an indispensable tool in cell physiology and in the study of calcium homeostasis. Its intracellular targets are the Ca^{2+}-dependent ATPases present in the ER. Thapsigargin **97** has been found to block entrance of AI PC cell lines into the S phase of the cell cycle thus triggering apoptosis [211]. Treatment of rat AT-3 and human TSU-Pr1, DU145 and PC-3 AI PC cells with 30 nM of thapsigargin **97** resulted in sustained intracellular Ca^{2+} level elevation [212]. It was found that the mechanism for this event was inhibition of ER Ca^{2+}-ATPase which prevented requestration of intracellular Ca^{2+} as it leaked out of the ER. This depletion of Ca^{2+} levels from the ER impaired its function that includes changes in protein synthesis, assembly and processing. Thapsigargin **97** treatment originated G0 arrest and loss of clonogenic activity of PC cells. DNA fragmentation and formation of apoptotic bodies was also documented [212].

Parthenolide **98** (Figure 18), a sesquiterpene lactone obtained from *Tanacetum parthenum*, is the purported active ingredient of the herbal remedy feverfew and is used to prevent migraines and treat arthritis [30]. Parthenolide **98** has been shown to inhibit NF-κB binding to DNA in PC cell lines, downregulate multiple genes under NF-κB control [213] including intracellular adhesion molecules (ICAM) 2 and 5, interleukin 1α (IL-1), interferon 1β fibroblast (IFN), interleukin-1 receptor-associated kinase (IRAK), and TNF receptor associated factors (TRAF) 1 and 5 [214]. Inhibition of LNCaP and PC-3 proliferation was also reported to occur in a dose-dependent fashion with parthenolide **98** at 0.5-5 μM [214]. Parthenolide **98** was shown to inhibit the proliferation of AI PC CWR22Rv1 cells with an IC_{50} of 2.5 μM and endothelial HUVEC cells with an IC_{50} of 7.5 μM [213]. NF-κB binding to DNA was blocked on both cell lines, with a decrease in the expression of TRAF1 and 2, and sustained activation of Jun-*N*- terminal kinase (JNK) [213]. *In vitro*, combination of 5 nM docetaxel **1** with 1 μM of parthenolide **98** resulted in a 75% reduction in clone formation of CWR22Rv1 cells. *In vivo*, despite its poor oral bioavailability, both tumor growth and tumor volume change rates were significantly lowered (55%) by the combination treatment of 40 mg/kg of parthenolide **98** (oral gavage) and 10 mg/kg docetaxel **1** administered i.p., in athymic nude mice bearing CWR22Rv1 tumors [213]. A combination regimen of 40 mg/kg by oral gavage of parthenolide **98** and 50 mg/kg of the antiandrogen bicalutamide caused a 45% reduction in tumor volume and significantly slowed down the rate of tumor growth, as compared to bicalutamide or parthenolide **98** alone treated groups, thus proving the ability of parthenolide **98** to restore CWR22Rv1 cell sensitivity to bicalutamide *in vivo* [213].

2.3.2.4. Triterpenes

Betulinic acid **99** (Figure 19) is a natural product identified in various bark extracts and is readily synthesized from betulin, a major constituent of the bark of birch trees [30]. It is currently undergoing development as a therapeutic agent against melanoma [71]. Betulinic acid **99** inhibited LNCaP cell proliferation with an IC_{50} of 5 μM, and at 10 μM it triggered apoptosis probably due to AR downregulation that led to decreased expression of vascular endothelial growth factor (VEGF) and survivin, and caspase-dependent PARP cleavage [215]. Betulinic acid **99** was shown to induce proteasome-dependent degradation of Sp proteins in LNCaP cells [215-217] (Tables 3 and 9). *In vivo*, compound **99** (10 and 20 mg/kg/day) inhibited tumor growth in athymic nude mice bearing LNCaP cell xenografts, which was accompanied by decreased expression of Sp1, Sp3 and Sp4 proteins and VEGF, and increased apoptosis [215].

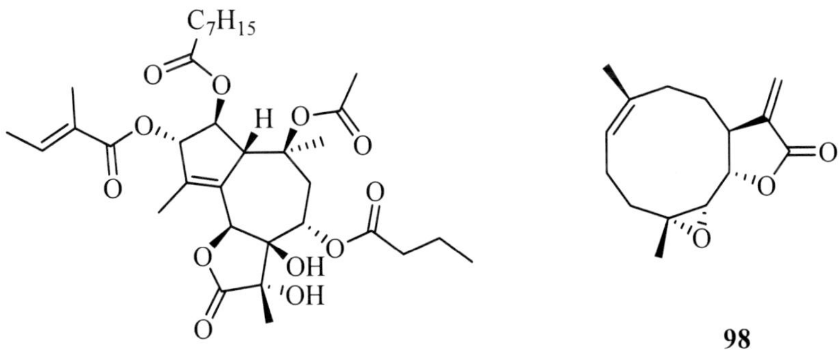

98

97

Figure 18. Plant derived sesquiterpenes.

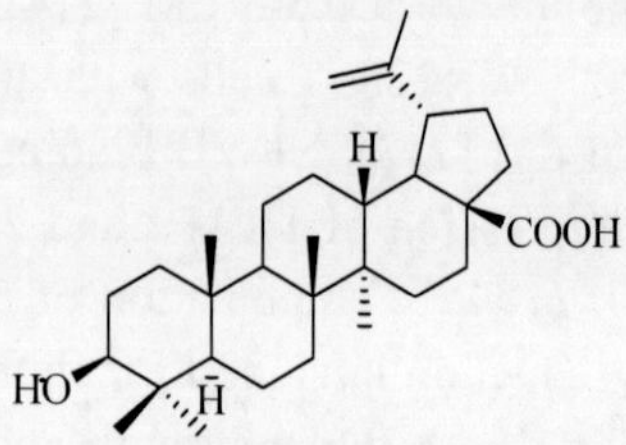

99

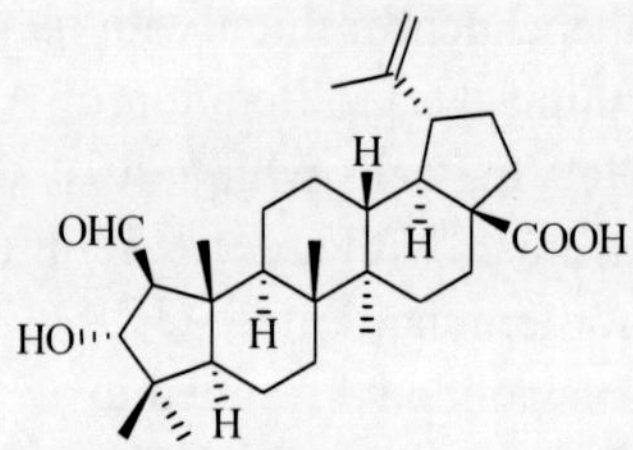

101

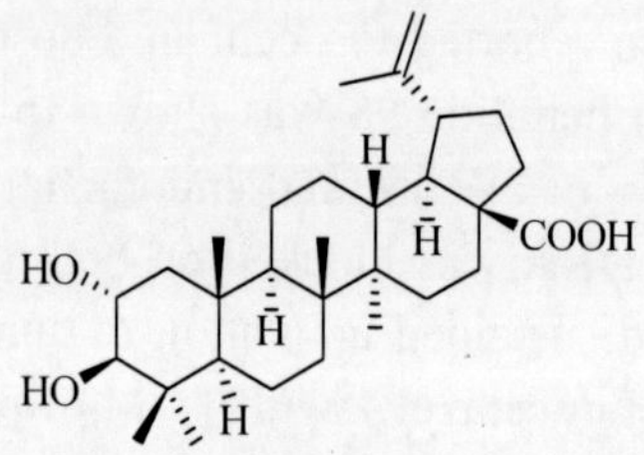

102

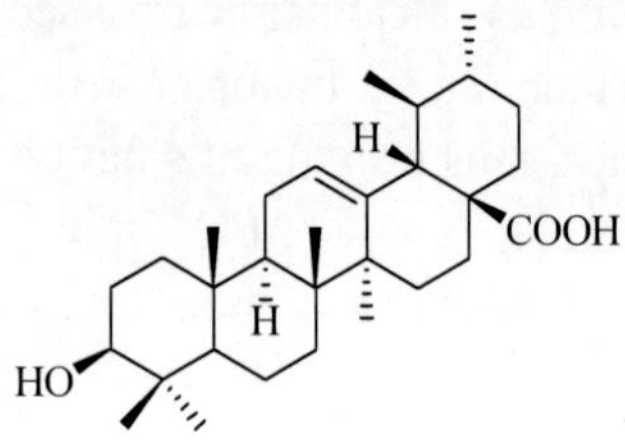

103

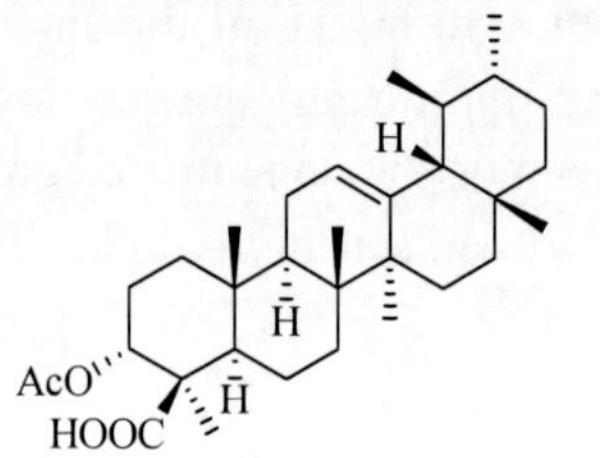

104

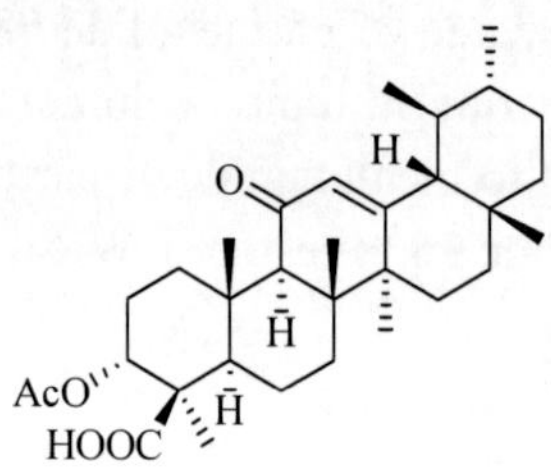

105

Figure 19. Plant derived triterpenes.

Table 9.

Target: Sp proteins
Sp proteins are overexpressed in many tumors and targeted degradation of Sp transcription factors results in activation of apoptotic and antiangiogenic pathways in tumors but not in non-target tissues that exhibit low Sp expression, thus making this molecular target appealing for drug development.

In androgen-refractory PC-3 cells, betulinic acid **99** has been found to effectively inhibit NF-κB binding to DNA and its translocation to the nucleus, IκBα phosphorylation, and IκB kinase complex (IKK) activation, and in addition to sensitize cells to TNFα-induced apoptosis through suppression of NF-κB [218] (Table **6**). A major advantage in using betulinic acid **99** for cancer therapy is its low toxicity. Doses as high as 500 mg/kg, administered every 4th day for 6 weeks to athymic mice carrying human melanomas have been shown to have no detectable toxic side effects [219]. However, a study demonstrated that betulinic acid **99** cannot be used to potentiate cell death in combination with most anticancer drugs that induce apoptosis via formation of topoisomerase I-DNA cleavable complexes such as camptothecin **100** (Figure 25), staurosporine and etoposide [220]. Thus, the compound inhibits all types of topoisomerase-I DNA cleavable complexes formation irrespective of the process of initiation, therefore disrupting the series of events that leads to cell death. Recently, the extract of *Zizyphus cambodiana* was found to contain betulinic acid **99** and another two pentacyclic triterpenes, colubrinic **101** and alphitolic **102** acids which inhibited the proliferation of DU145 cells with IC_{50} values of 78 and 70 μM, respectively (Figure 19) [221].

Ursolic acid **103** (Figure 19) is a pentacyclic triterpene that occurs in many plants, such as *Salvia officinalis* and Basil (*Ocimum sanctum*), and a constituent of several herbal medicines marketed in Asia and worlwide for inflammatory conditions [30, 222, 223]. Ursolic acid **103** inhibited the proliferation of PC-3 and LNCaP cells with IC_{50} values of 32.6 and 15.7 μM, respectively [223]. At 55 μM, induction of apoptosis was seen on PC-3 cells with a decrease on bcl-2 gene expression. Ursolic acid **103**-induced apoptosis on human prostate epithelial cells has been suggested to involve caspase activation through downregulation of cellular inhibitor of apoptosis proteins (c-IAPs) family proteins and without mitochondrial dysfunction [224].

Boswellic acids, particularly acetyl-boswellic acids such as acetyl-β-boswellic acid **104**, and acetyl-11-keto-β-boswellic acid **105** (Figure 19), can be isolated to chemical homogeneity from the gum resins of various *Boswellia* species and are well-known for their anti-inflammatory activity [30]. It has been reported that acetyl-boswellic acids promote apoptosis of AI PC-3 cells [225]. PC-3 cell proliferation was inhibited, with an IC_{50} value of 6.5 μM, after a 48 hours treatment with either compound **104** or **105** [225]. Exposure to 10 μM of either compound resulted in expression of phosphatidylserine on the outer membrane leaflet, reduced or hyploid DNA content, cytochrome c release from the mitochondria and DNA laddering. Moreover, inhibition of constitutively activated NF-κB was seen on PC-3 cells with both compounds, without direct interference with NF-κB binding to DNA (Table 6). Instead, inhibition of NF-κB signaling was found to occur by interception of IKK activity which finally led to diminished expression of the antiapoptotic proteins Bcl-2 and Bcl-xL as well as inhibiton of cyclin D1 levels [225]. In all of the described effects, acetyl-11-keto-β-boswellic acid **105** was more potent than acetyl-β-boswellic acid **104**.

In vivo, a water-soluble acetyl-11-keto-β-boswellic acid **105**-γ-cyclodextrin complex had a concentration-dependent inhibitory effect on PC-3 xenografts of chick chorioallantoic membrane tumor size and proliferation. Daily i.p. injections of 100 μmol/kg of the acetyl-11-keto-β-boswellic acid **105**-γ-cyclodextrin complex to nude mice bearing PC-3 xenografts for a period of 3 weeks reduced tumor volume in about 50% as compared to control [225]. Pharmacokinetic analysis of the acetyl-11-keto-β-boswellic acid **105**-γ-cyclodextrin complex showed plasma levels of 100 μM twenty minutes after i.p. injection.

The effect of boswellic acid **105** was also studied on LNCaP cells [226]. Acetyl-11-keto-β-boswellic acid **105** caused inhibition of LNCaP proliferation with an EC_{50} value of 20 μM [226]. At a concentration of 30 μM of acetyl-11-keto-β-boswellic acid **105**, G1 arrest occurred which was associated with upregulation of $p21^{WAF1/CIP1}$ expression and downregulation of cellular cyclin D1 expression [226]. Significant downregulation of AR mRNA and protein expression as well as supression of AR-dependent PSA and hk2 expression was seen with acetyl-11-keto-β-boswellic acid **105** treatment. Inhibition of Sp1 binding to DNA was suggested to be mechanism by which downregulation of the AR promoter activity occurred [226].

Pristimerin **106** (Figure 20) is a natural compound found in the *Celastraceae* and *Hippocrateaceae* families that are rich in quinone-methide triterpenes [227]. This compound targets the proteasome to induce apoptosis in human PC-3 and AR-positive C4-2B cells (Table 3) [228]. Proliferation of PC-3 cells was inhibited by 0.5-1 μM concentrations of pristimerin **106** and at these concentrations, apoptotic features were revealed (caspase-3-

activation and PARP cleavage) which correlated with inhibition of cellular proteasome activity by 30%, with accumulation of polyubiquitinated proteins [228]. The IC_{50} value for the inhibition of purified rabbit 20S proteasome with pristimerin **106** was 2.2 μM. Similarly, pristimerin **106** also potently inhibited 26S proteasome chymotrypsin-like activity in a extract prepared from exponentially grown PC-3 cells, with an IC_{50} of 3.0 μM [228]. About 40% inhibition of cellular proteasome activity occurred in C4-2B cells with 5 μM of pristimerin **106** and was accompanied by AR suppression with consequent apoptosis induction. Computational modeling showed that pristimerin **106** interacts with the catalytically active amino acid *N*-Thr of the proteasome β5 subunit [228].

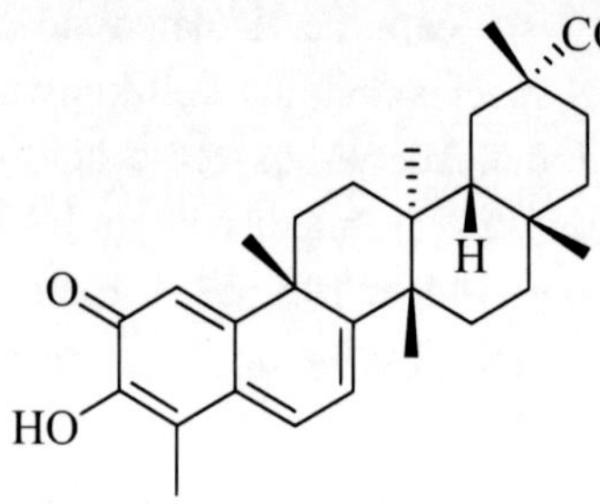

106

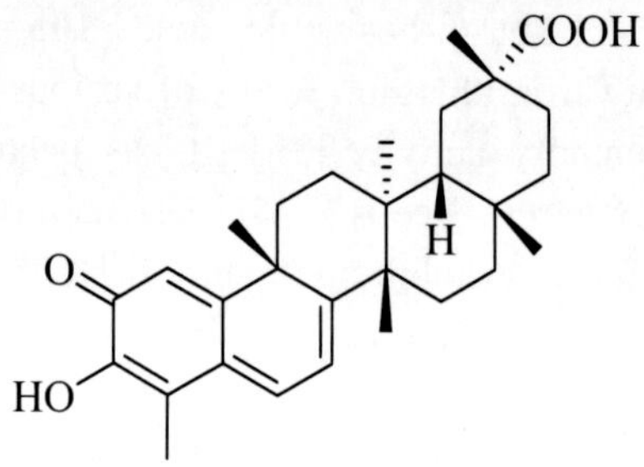

107

Figure 20. Plant derived triterpenes.

Celastrol **107** (Figure 20), an active compound extracted from the root bark of the Chinese medicine ‘‘Thunder of God Vine’’ (*Tripterygium wilfordii Hook F.*), has been used for years as a natural remedy for inflammatory conditions [229]. Celastrol **107** has been found to potently and preferentially inhibit the chymotrypsin-like activity of human PC cellular 26S proteasome at 1-5 μM (Table 3) [229]. Inhibition of the proteasome activity by celastrol **107** in both PC-3 and LNCaP cells resulted in the accumulation of ubiquitinated proteins and three natural proteasome substrates (IκB-α, Bax, and p27), accompanied by suppression of AR protein expression (in LNCaP cells) and induction of apoptosis. Treatment of PC-3 tumor–bearing nude mice with celastrol **107** (1-3 mg/kg/d, i.p., 1-31 days) resulted in a 65-93% inhibition of tumor growth, as compared to control. Multiple assays using the animal tumor tissue samples from both early and end time points showed *in vivo* inhibition of the proteasomal activity and induction of apoptosis after celastrol **107** treatment [229]. The enthusiastic results seen with celastrol **107** in the cancer setting show that this molecule has great potential for cancer prevention and treatment.

Ginseng is used in many cultures, especially in China and other Asian countries, for the treatment and prevention of various diseases, including cancer [230]. People who consume ginseng preparations are at lower risk of cancers of the oral cavity, stomach, lung, liver, pancreas, ovaries, and colon [230]. *Panax ginseng* (Korean ginseng), *Panax quinquefolius* (American ginseng), and other related plants, including *Panax notoginseng/pseudoginseng* (*P. notoginseng*, Buck FH Chen), are frequently used for medicinal purposes. In the US, ginseng is one of the most widely used medicinal plants [230]. The compounds mainly responsible for the activities of ginseng are the ginsenosides (saponin triterpenic glycosides). The activity of 25-hydroxyprotopanaxadiol **108** and 25-hydroxyprotopanaxatriol **109**, isolated from *Panax ginseng* [231], as well as of 20(*S*)-25-methoxyl-dammarane-3β,12β,20-triol **110** (Figure 21), isolated from *Panax notoginseng* [232], on PC cells has recently been reported [230, 231]. The IC_{50} values for LNCaP and PC-3 cell growth inhibition are depicted on table 10.

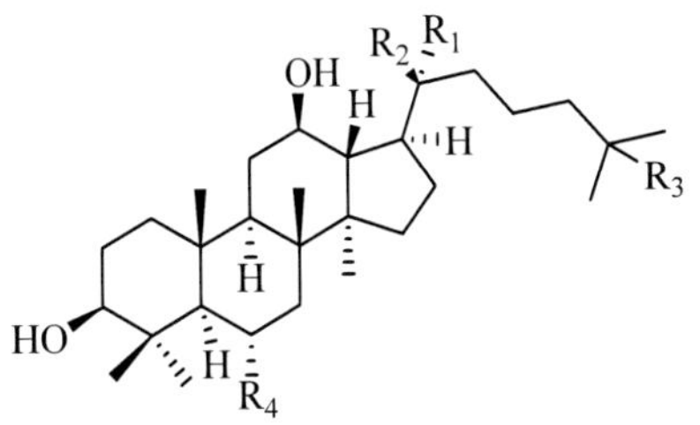

$R_1 = R_3 = OH$; $R_2 = Me$; $R_4 = H$ **108**
$R_1 = R_3 = R_4 = OH$; $R_2 = Me$ **109**
$R_1 = R_4 = H$; $R_2 = OH$; $R_3 = OMe$ **110**

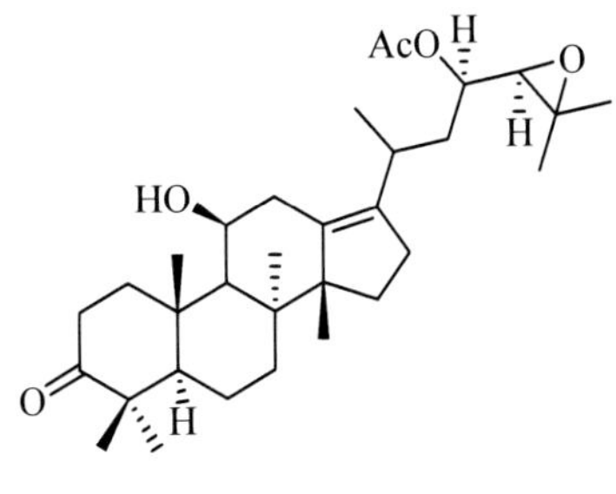

111

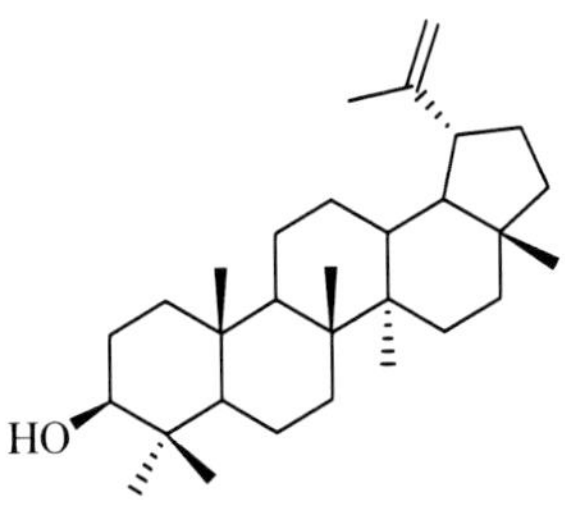

112

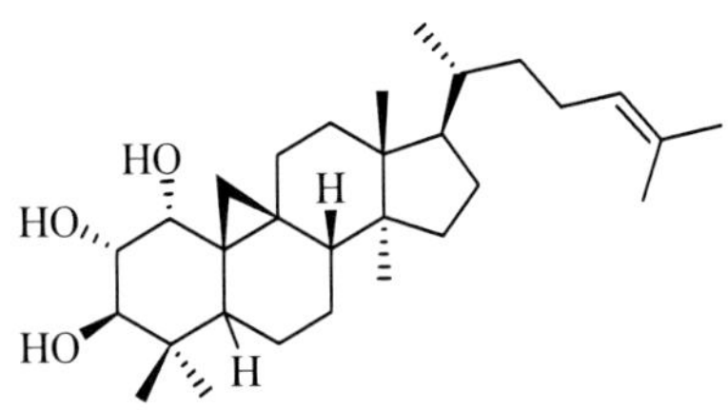

113

Figure 21. Plant derived triterpenes.

Table 10. Cell proliferation assay data for ginsenosides.

Entry	Compound	Cell line IC_{50} (μM)			
		LNCaP	PC-3	IMR90-EEA	IMR90-EIA
1	**108**	37	67	> 100	> 100
2	**109**	> 100	> 100	> 100	> 100
3	**110**	12.0	5.6	> 100	> 100

Under the same assay conditions no significant disturbance of the growth of fibroblasts (IMR90-EEA and IMR90-EIA cells) was observed [230, 231]. Both 25-hydroxyprotopanaxadiol **108** and 20(*S*)-25-methoxyl-dammarane-3β,12β,20-triol **110** triggered apoptosis on the tested cell lines with PARP cleavage, decreased levels of Bax and Bcl-2 as well as activation of caspases 3 and 9. G1 arrest occurred and the levels of p21 and p27 were consistently increased whereas CDKs 2, 4, and 6, MDM2, E2F1, and cyclin D1 levels were decreased by exposure to the compounds [230, 231]. The levels of p53 remained unaffected. Administration of 25-hydroxyprotopanaxadiol **108** to nude mice bearing PC3 xenografts at a regimen of 5 mg/kg/day for 5 days, aiming to fight the short half-lives and low bioavailabilities of ginsenosides, resulted in a 50% inhibition of tumor growth as compared to control whereas a dosage of 10 mg/kg/day inhibited tumor growth by more than 84% [231]. Moreover, combination regimens of this compound (10 mg/kg/day, 5 days a week) with docetaxel **1** (15 mg/kg) and gemcitabine (160 mg/kg) led to almost complete inhibition of tumor growth as compared to control [231]. 20(*S*)-25-Methoxyl-dammarane-3β,12β,20-triol **110** was found to decrease the expression of AR and PSA as effectively as resveratrol **44** (Figure 10), at concentrations ranging from 4 to 50 μM [230]. When administered to mice bearing PC-3 xenografts, 10 mg/kg/daily of compound (for a 5 days period) caused inhibition of tumor growth by 70% as compared to control. A synergistic effect with docetaxel **1** and gemcitabine in combination regimens was also found for 20(*S*)-25-methoxyl-dammarane-3β,12β,20-triol **110** [230].

Alisol B acetate **111** (Figure 21), a triterpene from *Alismatis rhizoma* was shown to induce apoptosis in human AI PC-3 cells in a time and dose-dependent fashion [233]. The IC_{50} value for the cell proliferation assay was 13.5 μM. Alisol B acetate **111**, at a concentration of 30 μM, induced a time-dependent decrease of the mitochondrial membrane potential of PC-3 cells which was accompanied by an increase in Bax levels and its translocation to the nucleus. Caspase activation was also found to occur [233].

Lupeol **112** (Figure 21), a triterpene found in fruits such as olive, mango, strawberry, grapes, and figs, as well as in vegetables and several medicinal plants is used worldwide by native people [234]. Lupeol **112** has been found to activate the Fas receptor-mediated apoptotic machinery of LNCaP cells and to inhibit the growth of tumors in athymic mice implanted with CWR22Rv1 cells, with concomitant reduction in PSA secretion, when administered i.p. at 1 mg/animal, in corn oil [234]. The IC_{50} values for lupeol **112** were estimated to be 21 and 18.5 μM for LNCaP and CWR22Rv1 cells, respectively. Lupeol **112** has been reported to possess protective effects against testosterone induced alterations in the mouse prostate and to exhibit its antiproliferative effect by ROS mediated apoptosis in an *in vitro* experimental model [235]. Treatment of LNCaP cells with lupeol **112** (75 μM) resulted in dose-dependent inhibition of cell proliferation, causing cell death by apoptosis with a lower Bcl-2:Bax ratio and activation of caspase 3 [235]. Exposure to lupeol **112** (IC_{50} value of 500 μM) resulted in dose-dependent inhibition of proliferation in PC-3 cells [236]. G2/M arrest was found to occur and apoptosis was found to be mediated by ROS generation with downregulation of the antiapoptotic gene bcl-2 and upregulation the apoptotic genes bax and apaf1 as well as with activation of caspases 9 and 3 [236].

Recently, a cycloartane-type triterpenoid **113** (Figure 21) isolated from *Commiphora opobalsamum* has been shown to have moderate activity towards PC-3 and LNCaP cells with IC_{50} values of 5.7 and 22.1 μM, respectively [237]. AR protein levels were decreased with 10 μM of compound, in the presence of Mib, and PSA expression was also decreased at this concentration.

2.3.2.5. Tetraterpenes

Lycopene **114** (Ψ,Ψ-carotene, Figure 22) is the most abundant carotenoid in tomatoes and is the red pigment of not only tomatoes but also rosehips, watermelon, papaya, pink grapefruit, and guava [238]. A very recent review has been published on the multitargeted action of this compound in the cancer setting, including PC, which the reader is advised to consider [238]. Unlike β-carotene, lycopene **114** lacks a β-ionone ring and therefore has no pro-vitamin A activity. However, the 11 conjugated and two non-conjugated double bonds in lycopene **114** make it highly reactive towards oxygen and free radicals, and this anti-oxidant activity probably contributes to its efficacy as a chemoprevention agent [238].

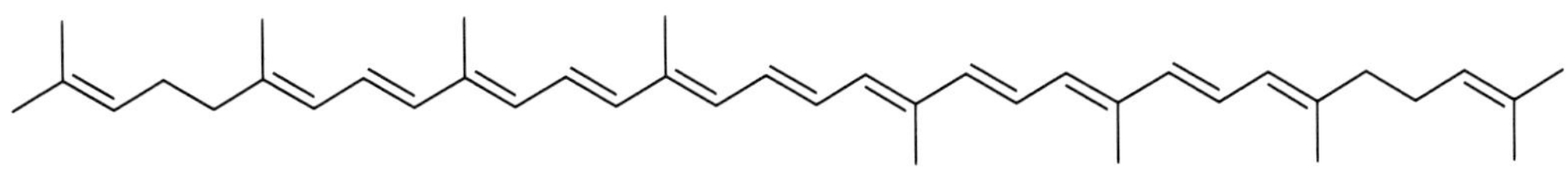

114

Figure 22. Lycopene **114**

Prospective epidemiological studies provide strong support for the hypothesis that the consumption of tomato products and probably lycopene **114** reduce the risk of PC. Epidemiological studies indicate an inverse relationship between lycopene **114** intake and PC risk and are supported by *in vitro* and *in vivo* experiments showing that oral lycopene **114** is bioavailable, accumulates in prostate tissue and is localized in the nucleus of prostate epithelial cells [238]. Among the many actions of lycopene **114** on PC cells is upregulation of the antioxidant response element inducing the synthesis of enzymes that protect against electrophilic metabolites and oxidative stress. In addition, lycopene **114** can induce apoptosis in cancer cells and inhibit their proliferation by promoting cell cycle arrest. The antimetastatic activity of lycopene **114** has also been shown in models of PC [238].

2.3.2.6. Steroids

Guggulsterone is a plant sterol derived from the gum resin of the Indian Ayurvedic medicinal plant *Commiphora mukul*, that has been safely used for thousands of years for the treatment of different disorders including bone fracture, arthritis, and inflammation [239]. The (*Z*)-isomer of guggulsterone **115** (Figure 23) was reported to inhibit the growth of PC-3, DU145, and LNCaP cells by causing apoptosis, at concentrations ranging from 20 to 40 μM [240, 241]. The effect of the compound on PC-3 cell death was not influenced by Bcl-2 protein levels but correlated with the induction of Bax and Bak [241]. The induction of apoptosis was later found to be initiated by ROS-mediated activation of JNK [240]. Compound **115** is also a potent suppressor of *in vitro* angiogenesis at 10 μM, as was evidenced by the inhibition of HUVEC cells tube formation and inhibition of HUVEC and DU145 cell migration [242]. In addition, oral gavage of compound **115** (40 mg/kg, 5 times/week) to male nude mice implanted with DU145-Matrigel™ plugs significantly inhibited *in vivo* angiogenesis [242].

Cardiac glycosides are a class of natural products that are used to increase cardiac contractile force in patients with congestive heart failure and cardiac arrhythmias [201]. Their mechanism of action is well established and involves inhibition of plasma membrane Na^+/K^+-APTase, leading to alterations in intracellular K^+ and Ca^{2+} levels [201]. The most common are digoxin **116** and digitoxin **117** (Figure 24), from the plant genera *Digitalis*, and ouabain **118** from *Strophanthus gratus* [243]. Oleandrin **119** and oleandrigenin **120** (its less water-soluble aglycone) (Figure 24) are cardiac glycosides derived from the ornamental evergreen plant *Nerium oleander* [244].

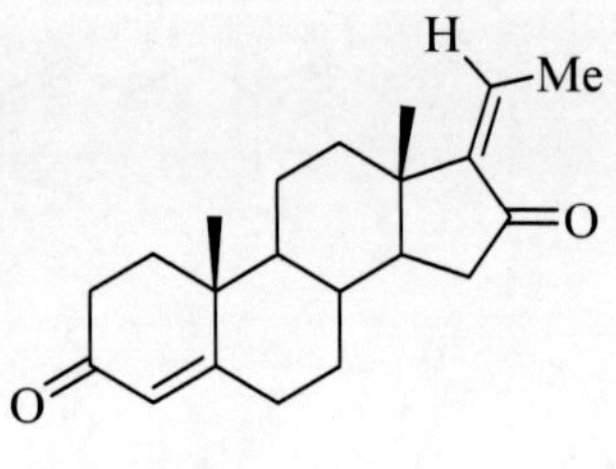

115

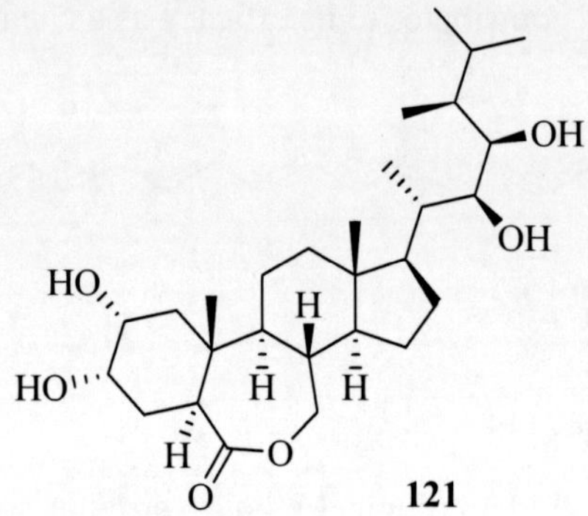

121

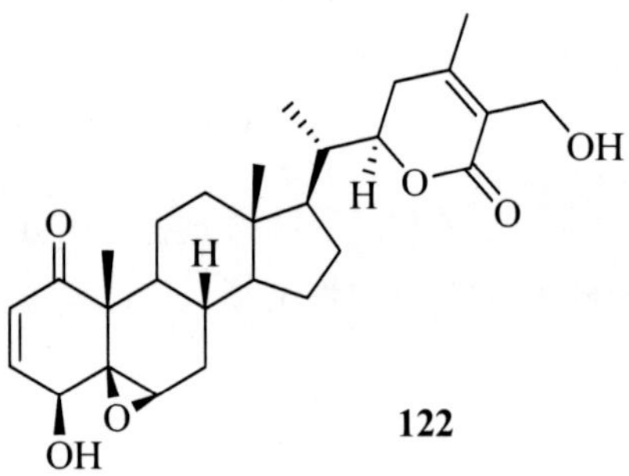

122

Source: *Withania somnifera*

Target: proteasome, Par-4

Suppresses growth of metastatic cells and can be exploited in combination regimens with antiandrogens to target resistant clones

Figure 23. Plant steroids.

Due to the fact that thapsigargin **97** (Figure 18) as well as other Ca^{2+} active agents exert potent cytotoxic effects in both normal prostate epithelial cells and cell lines derived from advanced cancer [212], and because Ca^{2+}-induced apoptosis is independent of cell cycle progression, it has been suggested that cardiac glycosides might be useful for PC treatment. Oleandrin **119** was reported to induce apoptosis of poorly metastatic (PC-3 M-Pro4) and highly metastatic (PC-3 M-LN4) PC cells, at 50 nM, by means of an early sustained Ca^{2+} level increase that preceded the release of cytochrome c from mitochondria and caspase activation [245]. More importantly, it was confirmed that PC cells express the Na^{+}/K^{+}-ATPase that serves as the molecular target for oleandrin **119** and related cardiac glycosides. Oleandrin **119** treatment also caused arrest of the cell cycle at the G2/M phase [245]. Ouabain **118** and digoxin **116** were also reported to induce apoptosis of the same cell types however less potently than oleandrin **119**. Thus, the relative potency of oleandrin **119** compared with the Anvirzel crude oleander extract (about 50 fold higher) strongly suggested that much of the proapoptotic activity of Anvirzel is directly attributable to this compound which makes up about 0.25% of the extract [245].

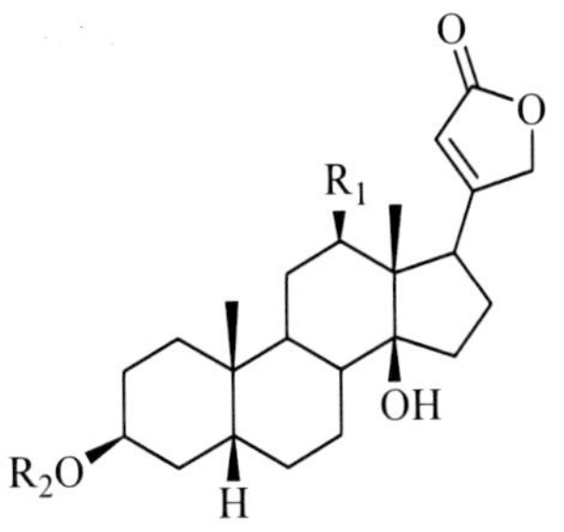

R_1 = OH; R_2 = Dox-β1-4-Dox-β1-4-Dox-β1 **116**

R_1 = H; R_2 = Dox-β1-4-Dox-β1-4-Dox-β1 **117**

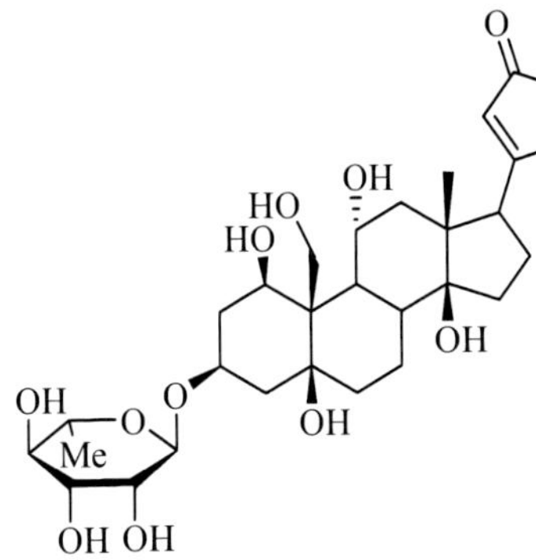

118

Dox = digitoxose

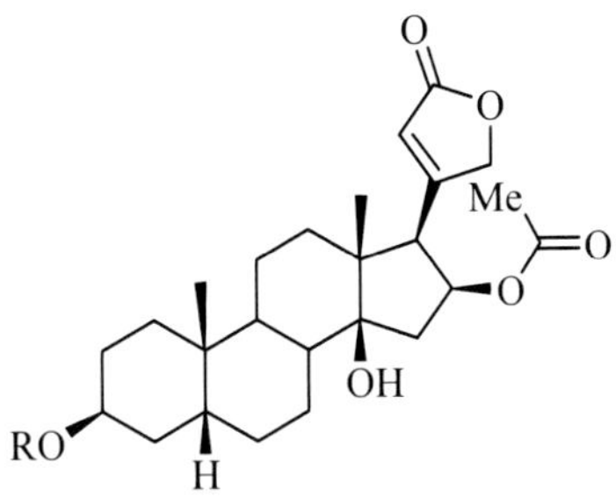

R = dideoxy arabinose **119**

R = H **120**

Figure 24. Plant derived cardiac steroids.

The IC_{50} values for PC-3 and DU145 cell proliferation inhibition with oleandrin **119** and oleandrigenin **120** were found to be 0.001 and 0.015 μg/ml and 0.002 and 0.018 μg/ml, respectively [244]. Inhibition of fibroblast growth factor FGF-2 export *in vitro* from both cell lines in a concentration and time-dependent fashion has been reported for oleandrin **119** and Anvirzel, and also for ouabain **118** (however at cardiotoxic doses) [244]. Digoxin **116**, digitoxin **117** and ouabain **118** were found to inhibit the proliferation of LNCaP, PC-3, and DU145 cell lines at 1 to 10 μM [243].

Brassinolide **121** (Figure 23) was first isolated from pollen of rape (*Brasica napus* L.) in 1979 [246]. It is a naturally occurring plant hormone that promotes growth, increases yields for grain and fruit crops, and makes plants more resistant to drought and cold weather. At a concentration of 40 μM, brassinolide **121** was found to inhibit 50% of PC-3 cell growth, after 24 hours of treatment. The compound caused a G2/M arrest of the cell cycle and after treatment, PC-3 cells revealed typical apoptotic features. Caspase 3 levels were found to be increased whereas Bcl-2 levels were decreased [246].

Withaferin A **122** (Figure 23) is a steroidal lactone purified from the medicinal plant "Indian Winter Cherry" or *Withania somnifera*, that is widely researched for its variety of properties, including antitumor effects [247]. Withaferin A **122** was found to inhibit the chymotrypsin-like activity of purified rabbit 20S proteasome (IC_{50} of 4.5 μM) and of 26S proteasome in cultured PC-3 cells (IC_{50} of 10-20 μM) and tumors (Table 3) [248]. Inhibition of PC tumor cellular proteasome both *in vitro* and *in vivo* by compound **122** was accompanied by accumulation of the proteasome target proteins Bax, IκB-α and p27, and induction of apoptosis [248]. Proteasome inhibition occurred before the apoptotic events of caspase-3 activation with PARP cleavage and the expected morphological changes (rounding and condensation), on PC-3 cells, at 10 μM of withaferin A **122**. On LNCaP cells, withaferin A **122** was found to decrease AR protein levels at 5 to 10 μM and to trigger apoptosis by the same mechanisms seen on PC-3 cells [248]. Treatment of human PC-3 xenografts with withaferin A **122** at 4.0 and 8.0 mg/kg/day for 24 days caused 54 to 70% inhibition of tumor growth, as compared to control, that was associated with a 28 to 56% inhibition of proteasomal chymotrypsin-like activity. Interestingly, neither the reduced form of withaferin A **122**, nor dihydrocelastrol, the reduced form of celastrol **107** (Figure 20), were as effective as their parent structures when used *in vitro*, a finding that highlights the need for the conjugated ketone structure for proteasome inhibition [248]. Induction of apoptosis by withaferin A **122** on PC cells bears additional interest because it has been shown to be mediated by the cancer specific target protein Par-4 (Table 11) [249].

Table 11

Target: Par-4
Par-4 selectively induces apoptosis in cancer cells which often lack p53 or phosphatase and tensin homolog gene (PTEN) function, but not in normal cells, and this cancer-selective function is mediated by its central core domain designated as SAC domain.

Out of a panel of several natural products including plumbagin (5 μM), diosgenin (10 nM), genistein **59** (30 μM), rutin (60 μM), psoralidin (15 μM), gingerol (40 μM), curcumin **45** (4 μM), resveratrol **44** (1 μM), lycopene **114** (1 μM), and (-)-epigallocatechin-3-gallate **34** (35 μM), withaferin A **122** (4 μM) was the only compound found to induce Par-4 expression on AI PC-3 cells transfected with the Par-4 promoter-luciferase reporter construct [250]. Moreover, treatment of androgen-responsive PC cells with antiandrogens such as flutamide or bicalutamide which do not induce apoptosis on their own, rendered these cells amenable to apoptosis triggered by withaferin A **122**. Apoptosis by withaferin A **122** required upregulation of Par-4 expression for inhibition of NF-κB activity and activation of the caspase cascade [250]. Therefore, withaferin A **122** shows great promise for HRPC treatment seeing that it not only seems to suppress growth of metastatic cells but can also be exploited in combination regimens with antiandrogens to target resistant clones.

2.3.3. Alkaloids

Evodiamine **123** (Figure 25), one of the major bioactive compounds isolated from the Chinese herb Wu-Chu-Yu, was reported to decrease testosterone production via reduction of 17β-hydroxysteroid dehydrogenase activity in testicular intersticial cells [251]. Thus, evodiamine **123** (100 μM) effectively reduced both basal and human chorionic gonadotropin-stimulated testosterone secretion in rat collagenase-dispersed testicular interstitial cells after 1, 2, or 4 hours of incubation [251]. The effect of evodiamine **123** on the proliferation of both AD LNCaP cells and AI PC-3 and DU145 cells was found to be caused by a G2/M arrest of the cell cycle progression and induction of apoptosis via the caspase pathway [252, 253]. The proliferation of all three cell lines was significantly blocked after 4 days of treatment with evodiamine **123** at concentrations ranging from 1 to 100 μM.

Mahanine **124** (Figure 25) is a carbazole alkaloid isolated from the leaves of a small shrub, widely available in East Asia, *Murraya koenigii* [30]. In a 72 hours assay, mahanine **124** was found to reduce the viability of both PC-3 and LNCaP cells at a concentration of 2 μg/ml (5,8 μM), and in addition was non-toxic to hepatocytes, cardiomyocytes or skeletal muscle cells in primary cultures, at 3 μg/ml [254]. The compound inhibited the phosphorylation of phosphoinositide-dependent kinase-1 (Pdk1) thereby inactivating Akt, and decreased Bcl-xL expression on PC-3 cells, which later resulted in cytochrome c release from the mitochondria and apoptotic death. Caspase activation was also seen with mahanine **124** treatment of PC-3 cells, as well as chromatin condensation resulting in oligonucleosomal DNA fragmentation and cleavage of PARP [254]. Moreover, mahanine **124** was recently shown to inhibit DNA methyltransferase activity and to induce the expression of an epigenetically silenced tumor supressor gene, RASSF1A, in both PC-3 and LNCaP cells, which in turn resulted in dowregulation of cyclin D1 to inhibit cell growth [255].

Berberine **125** (Figure 25), a naturally occurring isoquinoline alkaloid, is present in the roots, rhizome, and stem bark of a number of important medicinal plants such as *Berberis vulgaris* (barberry), *Berberis aquifolium* (Oregon grape), *Berberis aristata* (tree turmeric) and *Tinospora cordifolia* [256]. The compound was tested on LNCaP, PC-3, and DU145 PC cells and on a normal epithelial cell line (PWR-1E) at concentrations ranging from 10 to 100 μM [256]. A reduction of cell viability by 50% was obtained at 25 μM for both DU145 and PC-3 cells and at 50 μM for LNCaP cells. PWR-1E cells remained unaffected by the concentrations of berberine **125** tested. The effect of berberine **125** on DU145 cell viability was accompanied by G1 arrest of the cell cycle, with significant decrease on the levels of cyclins D1, D2 and E and CDKs 2, 4 and 6 [256]. Treatment of DU145 cells with berberine resulted in an increased ratio of Bax/Bcl-2, mitochondrial membrane potential disruption and release of cytochrome c from the mitochondria to the cytosol. Finally, activation of caspases 3 and 9 was also seen during berberine **125**-induced apoptosis of DU145 cells [256].

Indole-3-carbinol **126** (Figure 25) is an autolysis product of glucobrassicin, a compound found in vegetables of the *Brassica* genus, such as cabbage, broccoli, and Brussels sprouts [30]. Indole-3-carbinol **126** has been demonstrated to inhibit the growth of both AR-positive cell lines such as LNCaP and MDA-PCa-3b, as well as AR-negative DU145 cells [257]. Half-maximal response for DNA synthesis inhibition on LNCaP cells was 60 μM. This growth arrest was further studied with indole-3-carbinol **126** at 100 μM and found to be accompanied by regulation of expression or activity of specific G1 cell cycle CDKs, such as 2, 4, and 6 and CDK inhibitor $p21^{WAF1/CIP1}$ [257]. Furthermore, downregulation of PSA expression also occurred both with treatment of LNCaP cells with indole-3-carbinol **126** alone (100 μM) or in combination with antiandrogen flutamide (100 μM indole-3-carbinol **126** and 5 μM antiandrogen) [257]. Apoptosis induction on LNCaP, DU145 and PC-3 cells by indole-3-carbinol **126** was characterized by cell shrinkage, chromatin condensation, DNA fragmentation, and PARP cleavage, and was p53, Bax, Bcl-2 and FasL-independent [258]. Indole-3-carbinol **126** was also found to upregulate BRCA genes expression in PC cells, which are mediators of the ER stress response [259]. Finally, indole-3-carbinol **126** has also been found to inhibit AR transcription and androgen responsiveness in human LNCaP cells, at a concentration of 200 μM [260]. Currently, a clinical trial is ongoing to compare the effects of *Brassica* vegetables and indole-3-carbinol **126** in treating patients with PSA recurrence after surgery for PC [71]. Indole-3-carbinol **126** originates condensed polymeric products during metabolism among which is 3,3'-diindolylmethane **127** (Figure 25). This metabolite has been found to induce apoptosis on PC-3 cells at 75 μM, through the mitochondrial pathway [261].

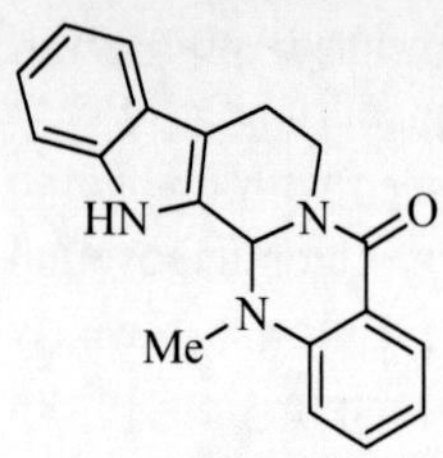

123

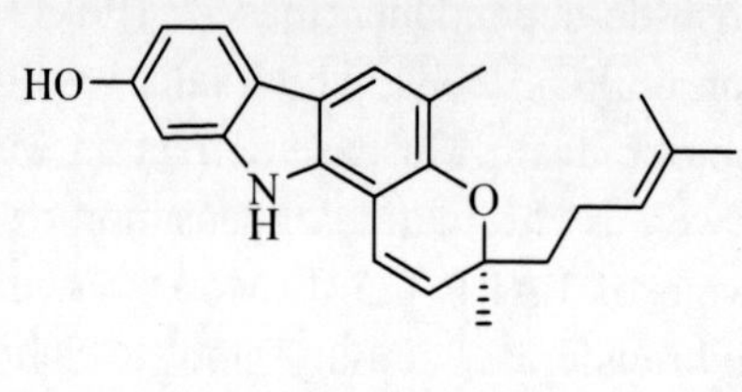

124

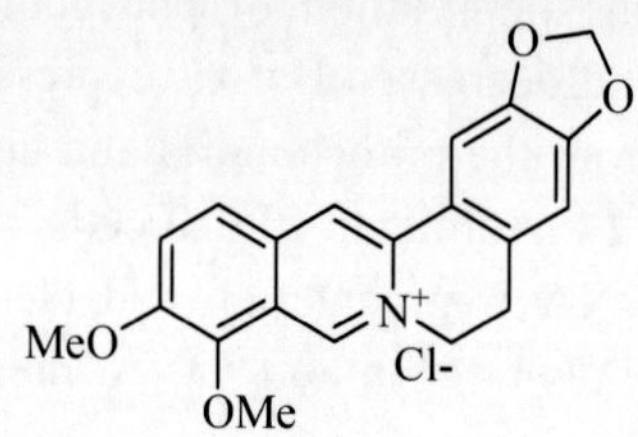

125

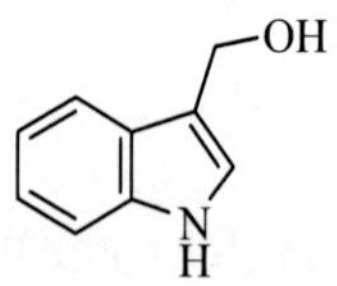

126

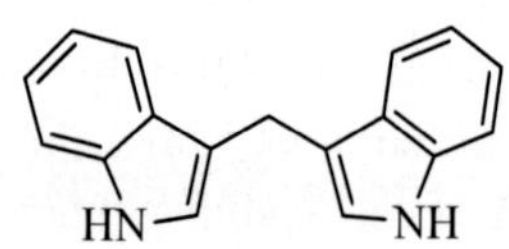

127

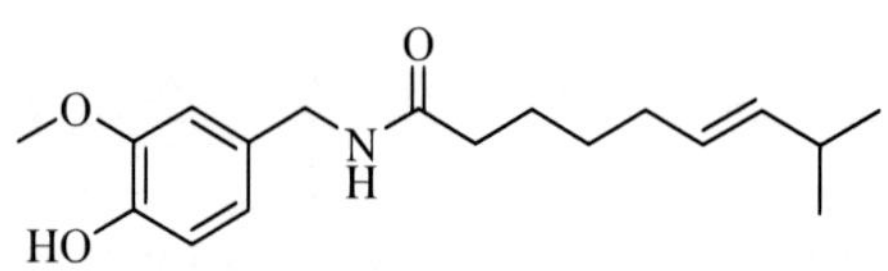

46

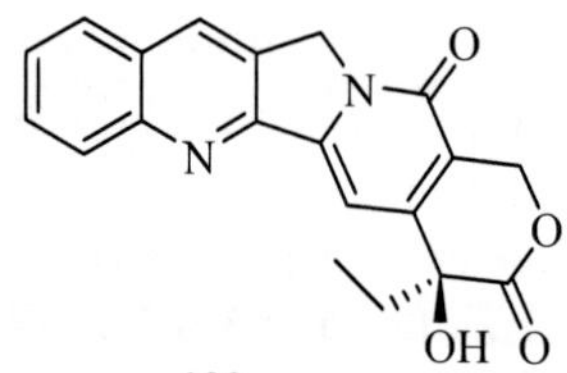

100

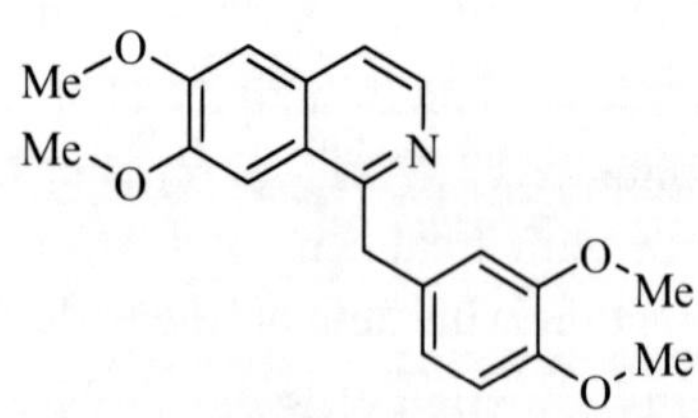

128

Figure 25. Plant alkaloids.

Capsaicin **46** (Figure 25) is the major pungent ingredient in red peppers [262]. It has been shown to inhibit the proliferation of PC cells, inducing the apoptosis of LNCaP, PC-3 and DU145 cells (at 500 μM the growth of each cell type was completely inhibited), which was associated with an increase in p53, p21 and Bax levels [262]. Capsaicin **46** (100-200 μM) down-regulated the expression of both PSA and AR. The compound was found to inhibit proteasome activity which suppressed the degradation of IκBα, preventing the activation of NF-κB (Table **6**). When given orally to nude mice bearing PC-3 xenografts, capsaicin **46** (5 mg/kg/day) significantly slowed tumor growth when compared to vehicle treated mice, as measured by size and weight [262]. The IC_{50} value for inhibition of PC-3 cell proliferation with capsaicin **46** was reported to be 20 μM. At this concentration, capsaicin **46**-induced apoptosis in PC-3 cells was found to occur by a mechanism involving ROS generation, dissipation of the mitochondrial inner membrane potential and activation of caspase 3 [263]. Further studies on the signaling pathways involved in the antiproliferative effect of the compound, revealed that capsaicin **46**-induced ROS-mediated apoptosis on PC-3 cells activates JNK which originates ceramide accumulation by sphingomyelin hydrolysis, which in turn induces extracellular signal-regulated protein kinase (ERK) activation and the apoptotic response [264]. Recently, the effect of capsaicin **46** in the gene expression profile of PC-3 cells has been analyzed using a microarray approach [265]. The most remarkable change was the upregulation of GADD153/CHOP, an ER stress-regulated gene.

Camptothecins, as a promising new class of anticancer drugs, have advanced to the forefront of therapeutic and developmental chemotherapy, when either used alone or in combination [266]. The demonstration that topoisomerase I is the molecular target of camptothecins, and the development of water-soluble derivatives with less side effects, have translated this class of drugs into agents in the treatment of cancer [266]. In the presence of camptothecin **100** (Figure 25), significant increases in both p53 and p21 proteins were detected in human wild-type p53 gene (Adwtp53)-infected PC3 cells [267]. While Adwtp53 and camptothecin **100**, as single agents, caused apoptosis and cell death, combinations of camptothecin and Adwtp53 were better in inducing apoptosis and cell death in PC3 cells [267]. A combinatory treatment of AI DU145 cells with sublethal concentrations of camptothecin **100** (100 ng/ml) and very low concentrations of anti-Fas IgM (200 ng/ml) resulted in 75-95% cell death, in 24 hours [268]. The mechanism of this cooperation was found to be independent of ROS generation. Two water-soluble camptothecin **100** analogs have been approved by the FDA for clinical application: topotecan, as a second-line therapy for ovarian cancer or small-cell lung cancer, and irinotecan, for the treatment of colorectal carcinoma refractory to 5-fluorouracil or as initial therapy in combination with 5-fluorouracil for the treatment of metastatic colorectal cancer [266]. NSC606985 (NSC), a highly water-soluble camptothecin analog, caused more than 50% reduction of viable cell numbers after 72 hours of treatment as compared to the original cell numbers, at 50 nM, in DU-145 cells [266]. Similar NSC effects were observed in PC-3 cells. However, LNCaP cells were less sensitive to NSC, and the decrease in viability after 72 hours was not so obvious when compared to the original cell number. NSC induced cell apoptosis and growth inhibition was suggested to occur by interaction with topoisomerase I, cell cycle arrest and activation of the mitochondrial apoptotic pathway [266].

Papaverine **128** (Figure 25), an opium alkaloid, has been shown to induce terminal differentiation in LNCaP cells, which is correlated with an intracellular cyclic adenosine monophosphate (cAMP)-mediated pathway [269]. Thus, at 10 μM, papaverine **128** effectively induced morphologic changes and raised cAMP levels in LNCaP cells, an effect which was not seen on PC-3 and DU145 cells. Additionally, the proliferation and invasive potential of LNCaP cells was also significantly inhibited by papaverine **128**.

2.3.4. Quinones

Emodin **129** (Figure 26), a natural compound extracted from *Aloe vera* [30] and *Rheum palmatum*, has more potent activity on PC than genistein **59** (Figure 12) or curcumin **45** (Figure 10) [270]. The proliferation of AR expressing LNCaP and PC3-AR cells was inhibited by emodin **129** at 10 and 40 μM, respectively. DU145 and PC-3 cells, which lack the AR, were more resistant to this effect [270]. Emodin **129** inhibited AR transcriptional activity by preventing AR nuclear translocation. Treatment with the compound resulted in decreased association of AR and heat shock protein 90 (hsp90) but increased association of AR and MDM2, thus inducing AR degradation in a ligand-independent manner. Through targeting of the AR, emodin **129** suppressed PC cell growth *in vitro* and prolonged the survival of PC-producing C3(1)/SV40 transgenic mice *in vivo* (at 40 mg/kg) [270].

β-Lapachone **130** (Figure 26) is a simple plant product obtained by sulfuric acid treatment of the naturally occurring lapachol, which is readily isolated from *Tabebuia avellanedae*, that is native to South America [30] and inhibits the catalytic activity of DNA topoisomerase I by a mechanism other than that of camptothecin **100** (Figure 25) [271].The compound was first reported to induce apoptotic cell death in human PC cells in a p53-independent fashion [272, 273]. The IC_{50} values for the cell proliferation assays on DU145 and PC-3 cells were 5 and 8 μM, respectively. LNCaP cells were found to be equally sensitive to β-lapachone **130** [272]. The effect of β-lapachone on PC-3 cells involved a significant decrease in the phosphorylated forms of tumor suppressor pRB and consequent G1 arrest [274]. It was later reported that damage to genomic DNA was the trigger for the apoptotic-induced process seen on PC-3 and DU145 cells with β-lapachone **130** [275]. On DU145 cells, concentrations of up to 5 μM of β-lapachone **130** induced apoptosis with an increase on the Bax/Bcl-2 ratio and activation of caspase 3 [276]. The activity of β-lapachone **130** on PC cells has been found to be dependent on NAPDH quinone reductase (NQO1), a flavoenzyme that catalyzes the two-electron reduction of quinones into their hydroquinone form, bypassing the often mutagenic semiquinone intermediate and the formation of free radicals [277]. It has been shown that combination regimens of β-lapachone **130** (2.5 μM) and genistein **59** (Figure 12), at 10-70 μg/ml offer advantage over the use of the single agents for inhibition of both PC-3 and LNCaP cell proliferation [278].

Chemotherapeutic and chemopreventive effects of black seed extracts from *Nigella sativa*, an herb used for culinary and medicinal purposes in Asian and Middle Eastern countries, are attributed to quinones, mainly thymoquinone **131** (Figure 26), which is present in the volatile oil of the seed [279]. Thymoquinone **131** was found to inhibit DNA synthesis, proliferation and viability of cancerous (LNCaP, C4-2B, DU145, and PC-3) however not of noncancerous (BPH-1) prostate epithelial cells by downregulating the AR and E2F1 [280]. The IC_{50} values for inhibition of cell proliferation were approximately 60, 70 and 80 μM for PC-3, DU145 and LNCaP cells, respectively. In LNCaP cells, at 50 μM, a dramatic increase in $p21^{WAF1/CIP1}$, $p27^{KIP1}$ and Bax was observed. Thymoquinone **131** blunted progression of the cell cycle from the G1 to the S phase, with a concomitant decrease in AR and E2F-1 as well as the E2F-1-regulated proteins necessary for cell cycle progression. In a xenograft prostate tumor model, thymoquinone (20 mg/kg administered s.c. daily for 31 days) inhibited the growth of C4-2B-derived tumors in nude mice, which was associated with a decrease in AR, E2F-1 and cyclin A levels [280].

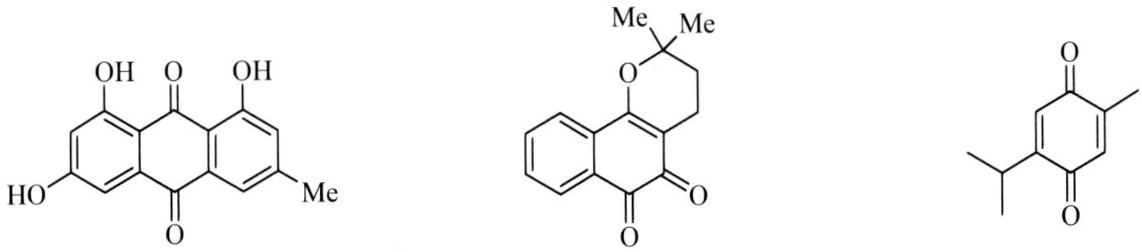

129 **130** **131**

Figure 26. Plant quinones

2.3.5. Organosulfur compounds

The consumption of cruciferous vegetables has long been associated with a reduced risk in the occurrence of cancer at various sites, including the prostate, lung, breast and colon [281]. This protective effect is attributed to isothiocyanates present in these vegetables, and to sulforaphane **132** (Figure 27), present in broccoli, that is by far the most extensively studied to uncover the mechanisms behind this chemoprotection [281]. Among the many reported effects for sulforaphane **132** on PC cells are retardation of the growth of human PC-3 xenografts in nude mice, significant induction of phase 2 enzyme expression and activity in LNCaP, MDA, PCa 2a, MDA PCa 2b, PC-3 and TSU-Pr1 PC cell lines at 0.1-15 μM (sulforaphane **132** produced robust and sustained transcriptional induction of NQO1 gene expression that was accompanied by similar increases in NQO1 enzymatic activity in human PC cells *in vitro*), as well as induction of apoptosis through activation of caspases 8 and 9 on both DU145 and PC-3 cells, with 20 μM sulforaphane **132**, and with an increased Bax:Bcl-2 ratio on PC-3 cells. An involvement of the JNK/MAPK on sulforaphane **132**-mediated apoptosis was found on PC-3 and DU145 cells [281]. Decrease in the mitochondrial potential in response to sulforaphane **132** treatment was detected in the prostate. On LNCaP cells, p53 was found not to be involved in the apoptotic process.

The effects of sulforaphane **132** on the cell cycle of PC cells have also been reported. Arrest at the G1 phase was found for DU145 and LNCaP cells whereas arrest at the G2/M was seen for PC-3 cells [281]. The G1 arrest in response to sulforaphane **132** treatment was associated with reduced phosphorylation pRb that activates the transition from G1 to the S phase. The G2/M arrest in PC-3 cells activated the DNA damage checkpoint pathway through activation of checkpoint kinase 2 (Chk2) and subsequent phosphorylation of cell division cycle 25C [281]. The levels of $p21^{WAF1/CIP1}$ were raised by 9-50 μM treatment with sulforaphane **132** on DU145 cells. Finally, histone deacetylase inhibition was found with 15 μM sulforaphane **132** on LNCaP and PC-3 cells along with increased levels of histones 3 and 4 [281].

β-Phenethyl isothiocyanate **47** (Figure 27) is present in rather large quantities in cruciferous vegetables including broccoli, cabbage or watercress [282]. Treatment of PC-3 cells with compound **47** indicated that p53 was not essential for apoptosis induction by this compound and that the apoptosis induction was mediated by ERKs [282]. Thus, treatment of

PC-3 cells with 10 μM of β-phenethyl isothiocyanate **47** resulted in rapid and sustained activation of ERK1/2 that was evident as early as 1 hour after treatment and persisted throughout the experiment, for 24 hours. The IC_{50} value for the survival assay on PC-3 cells was found to be 6 μM [282]. A study on PC-3 C4 cells showed that the inhibition caused by sulforaphane **132** (7.5 μM) and β-phenethyl isothiocyanate **47** (10 μM) on NF-κB transcriptional activity as well as NF-κB-regulated VEGF, cyclin D1, and Bcl-xL expression is mainly mediated through the inhibition of IKKs phosphorylation, particularly IKKβ, and the inhibition of IκBα phosphorylation and degradation, as well as the decrease of nuclear translocation of p65 [283].

Epidemiological studies have documented an inverse correlation between dietary intake of *Allium* vegetables and the risk for certain types of cancer, including PC [284]. Laboratory findings indicate that the anticarcinogenic activity of *Allium* vegetables is due to organosulfur compounds (OSCs) that are released upon processing (cutting or chewing) of these vegetables [284]. Garlic-derived OSCs include diallyl sulfide **133**, diallyl disulfide **134** and diallyl trisulfide **135** (Figure 27). These compounds were tested for PC-3 and DU145 cell proliferation inhibition and induction of apoptosis [285, 286]. The IC_{50} value for PC-3 cell proliferation inhibition with diallyl trisulfide **135** was 22 μM and with diallyl disulfide **134** it rounded 35 μM, and both compounds triggered apoptosis, whereas no effect was seen with diallyl sulfide **133** [286]. A similar activity profile was seen on DU145 cells suggesting that the presence of allyl groups and the oligosulfide chain length affects the activity of OSCs on the proliferation of these cells. Diallyl trisulfide **135**-induced apoptosis was found to be associated with phosphorylation of Bcl-2 which was mediated by JNK1 and to some extent by ERK1/2. It was suggested that diallyl trisulfide **135**-induced JNK activation involved hydrogen peroxide as a second messenger [286].

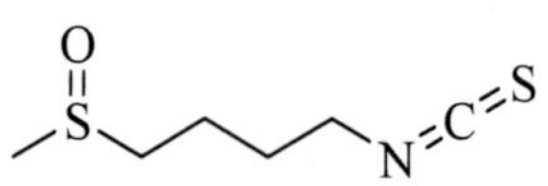

132

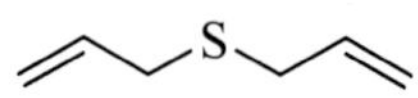

133

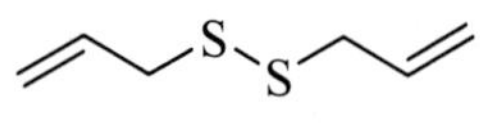

134

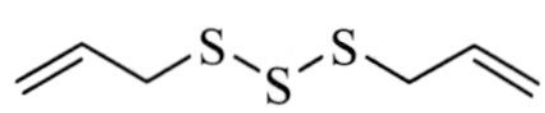

135

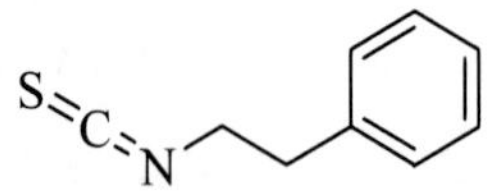

47

Figure 27. Plant OSC compounds.

2.3.6. Others

Recently, an aliphatic alcohol glycoside **136** (Figure 28) isolated from *Commiphora opobalsamum* has been shown to have moderate activity towards PC-3 and LNCaP cells with IC_{50} values of 7.1 and 23.6 μM, respectively [237]. AR protein levels were decreased with 10 μM of compound, in the presence of Mib, and PSA expression was also decreased at this concentration.

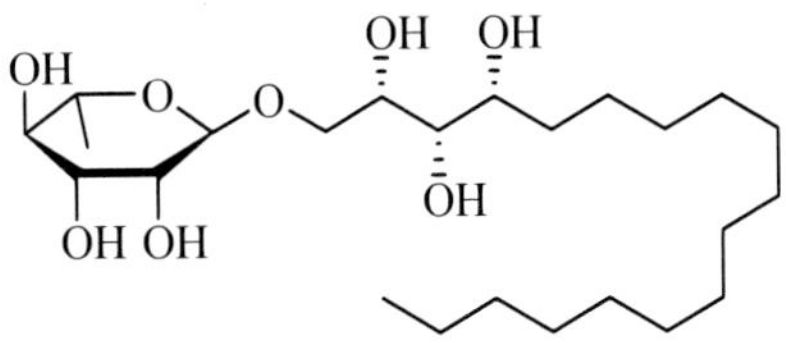

136

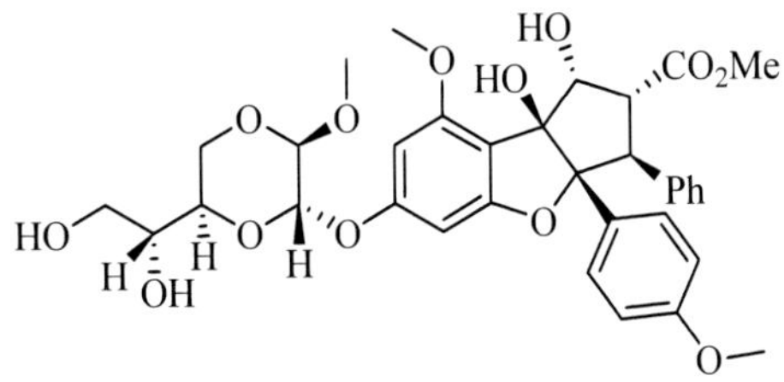

137

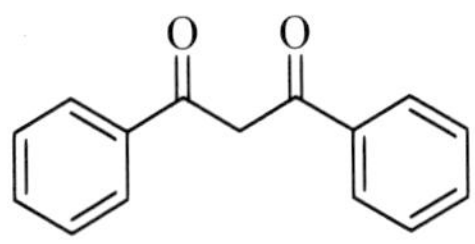

138

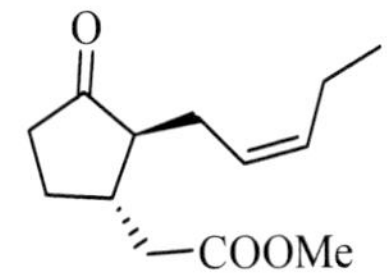

139

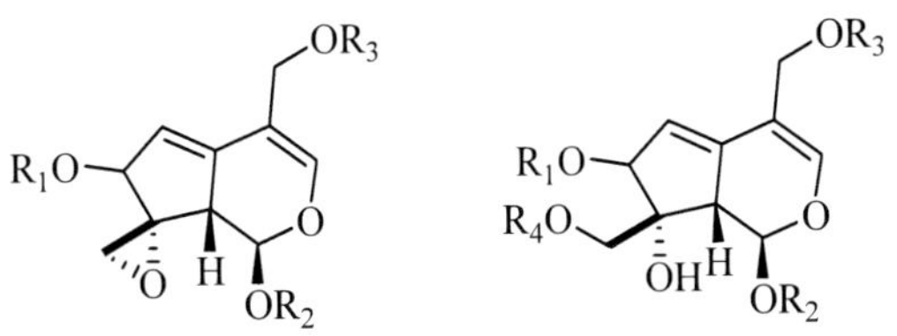

$R_1 = R_3 = Iv; R_2 = Cr$	**140**
$R_1 = Mv; R_2 = Cr; R_3 = Ac$	**141**
$R_1 = R_3 = R_4 = Iv; R_2 = Hiv$	**142**
$R_1 = Mv; R_2 = Cr; R_3 = Ac; R_4 = Iv$	**143**
$R_1 = Ac; R_2 = R_3 = Iv$	**144**
$R_1 = R_2 = Iv; R_3 = Ac$	**145**
$R_1 = Iv; R_2 = Cr; R_3 = Ac$	**146**
$R_1 = R_4 = Iv; R_2 = Hiv; R_3 = Ac$	**147**
$R_1 = R_4 = Iv; R_2 = Cr; R_3 = Ac$	**148**

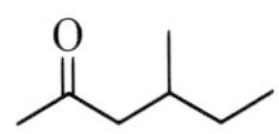

3-methylvaleryl (Mv)

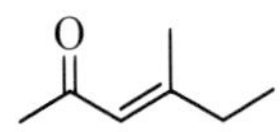

3-methylcrotonyl (Cr)

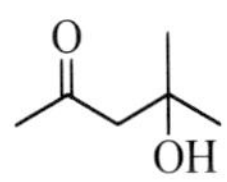

3-hydroxyisovaleryl (Hiv)

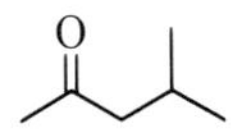

isovaleryl (Iv)

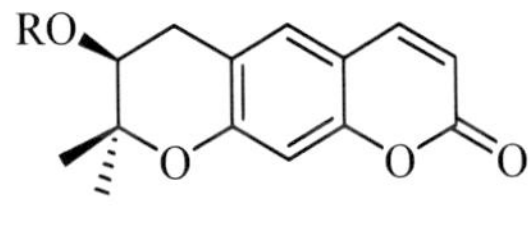

$R = COCHC(CH_3)_2$	**149**
$R = COC(CH_3)CHCH_3$	**150**
R = H	**151**

Figure 28. Other plant derived compounds.

The cyclopenta[*b*]benzofuran core, found only in plants of the gents *Aglaia* (Meliaceae), has afforded interesting lead structures due to its unique carbon skeleton and the potent biological activity of some members of this compound class, in which silvestrol **137** (Figure 28), a cyclopenta[*b*]benzofuran bearing a dioxanyloxy group at the C-6 position, is included

[287]. The cytotoxicity of silvestrol **137** in LNCaP cells was shown to occur through apoptosis [288]. The number of LNCaP colonies was significantly inhibited by 30 nM of compound, after 48 hours of treatment, revealing a cytotoxic rather than a cytostatic effect for silvestrol **137**. Activation of caspases-2, -9, and -10 was detected in silvestrol **137**-induced apoptosis, as well as upregulation of apoptotic protease activating factor 1 (Apaf-1). Together with the apoptosome, involvement of mitochondria was also indicated in silvestrol **137**-induced apoptosis. Although the mitochondrial membranes did not undergo substantial depolarization in response to silvestrol **137** exposure when compared to untreated cells, silvestrol **137** treatment did induce cytochrome *c* release from the mitochondria to the cytoplasm, upregulation of the proapoptotic protein Bak, and phosphorylation of the antiapoptotic protein, Bcl-xL. Activation of caspases-3 or -7 was however not detected in silvestrol **137**-induced apoptosis, suggesting that an alternative programmed cell death program may also be engaged along with the classic apoptotic mechanism [288]. *In vivo*, silvestrol **137** was found to reduce tumor weight by 60% in PC-3 xenografts implanted on the dorsal flank region of athymic mice, when administered i.p. at 3 mg/kg, 3 times a week for 29 days [287, 288].

Dibenzoylmethane **138** (Figure 28) is a small β-diketone compound and a minor constituent of licorice and sunscreens [30]. Dibenzoylmethane **138** has been shown to inhibit LNCaP and DU145 cell growth in a dose-dependent fashion, with IC_{50} values of approximately 25 and 40 μM, respectively, and in conjunction with cell cycle arrest at the G1 phase [20]. The action of dibenzoylmethane **138** on PC-3 and LNCaP cells, at concentrations of up to 100 μM, was found to be related to stabilization of hypoxia-inducible factor 1 subunit α (HIF1α) at a step prior to ubiquitination thus preventing its degradation and causing an increase in its transcriptional activation, as well as induction of the secretion of VEGF under normoxic conditions [289]. Dibenzoylmethane **138** was also reported do downregulate AR protein levels and diminish PSA secretion in LNCaP cells [290].

Methyl jasmonate **139** (Figure 28), a plant stress hormone with striking resemblance to lipoxygenase products, has been reported to inhibit the proliferation of human DU145 and PC-3 cells *in vitro* and to interact specifically with the 5-lipoxygenase enzyme pathway [291]. IC_{50} values for the cell proliferation assay were 1.1 and 2.3 mM, in PC-3 and DU145 cells, respectively, after 48 hours of treatment. Cell cycle arrest at the G_o/G_1 phase was also observed with 2 mM of methyl jasmonate **139** [291]. Recently the use of this compound on irradiated PC-3 cells has been found to suppress the radiation-induced Bcl-2 expression and to enhance the radiation sensitivity of human PC cells [292].

Sorbifolivaltrates **140-148** (Figure 28) are diene valepotriates from *Valeriana sorbifolia*. This class of compounds has recently been shown to inhibit PC-3M cell growth with IC_{50} values in the μM range (Table 12) [293].

The pyranocoumarin compound decursin **149** (Figure 28) from the herb *Angelica gigas* possesses potent anti-AR activity [294]. The activity of this compound has been recently compared with two natural analogues decursinol angelate **150** and decursinol **151**, isolated from *Angelica gigas* Nakai (Figure 28). Decursin **149** and its isomer decursinol angelate **150** decreased PSA expression, with an IC_{50} value of 1 μM. Both compounds inhibited androgen-stimulated AR-nuclear translocation and transactivation, decreased AR protein abundance through proteasomal degradation, and induced G0/G1 arrest and morphological differentiation [294]. The compounds also induced caspase-mediated apoptosis and ROS generation at higher concentrations. Decursinol **151**, which does not contain a side chain, lacked the ability to induce apoptosis and ROS generation [294].

Table 12. Inhibitory potential of sorbifolivaltrates from *Valeriana sorbifolia.*

Entry	Compound	PC-3M proliferation inhibition IC_{50} (μM)
1	**140**	2.6
2	**141**	12.2
3	**142**	8.9
4	**143**	7.3
5	**144**	3.1
6	**145**	1.7
7	**146**	20.0
8	**147**	11.1
9	**148**	8.9

Irofulven **153** (MGI Pharma, Inc., Figure 29), an anticancer agent derived from the mushroom natural product illudin S **152** (Figure 29) has been found to inhibit DU145, PC-3, and LNCaP cell proliferation with IC_{50} values of 25, 50 and 180 ng/ml (100, 200, and 700 nM) [295]. The cytotoxicity of irofulven **153** was not affected by loss of p53 or mismatch repair function, and the drug was found not to be a substrate for multidrug transporters such as P-glycoprotein and multidrug resistance protein 1.

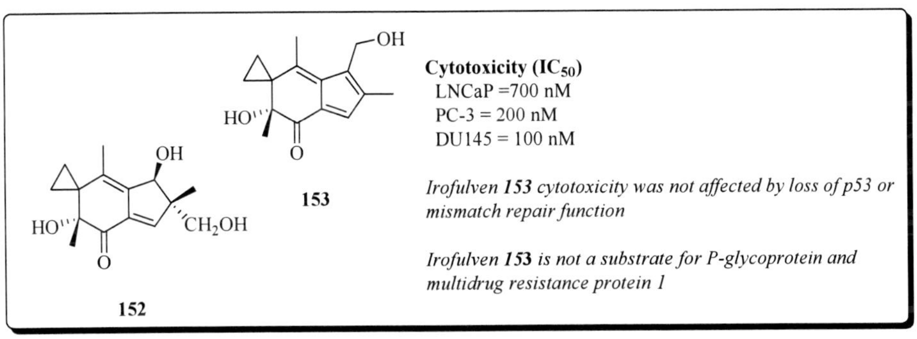

Figure 29. Illudin S **152** from mushroom and its synthetic derivative irofulven **153**.

3. Animal Derived Compounds

Bufalin **154** and cinobufagin **155** (Figure 30) are cardiotonic steroids isolated from Chansu, a galenical preparation of the dried white venom of Chinese *Bufo bufo gargarizans* [296]. The compounds were found to be cytotoxic towards LNCaP, DU145, and PC-3 cells at concentrations ranging from 0.1 to 10 μM [296]. This effect was associated with an elevation of intracellular Ca^{2+} and caspase-mediated apoptosis.

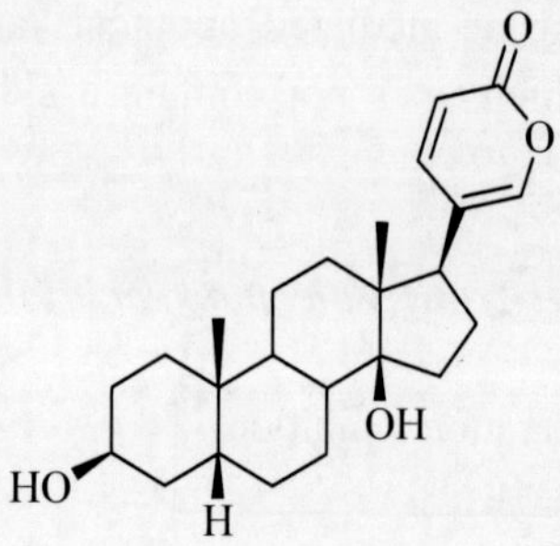

154

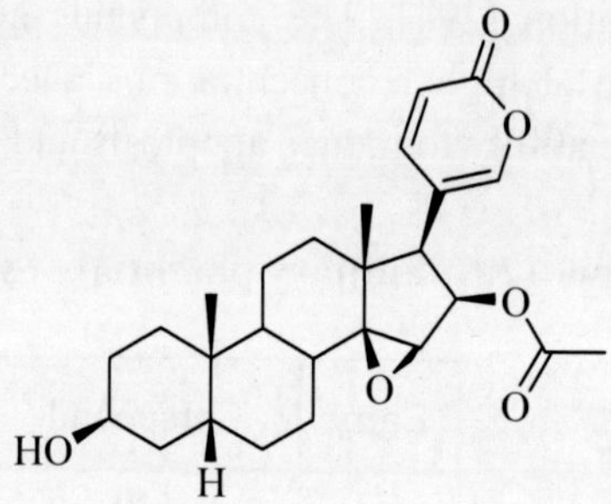

155

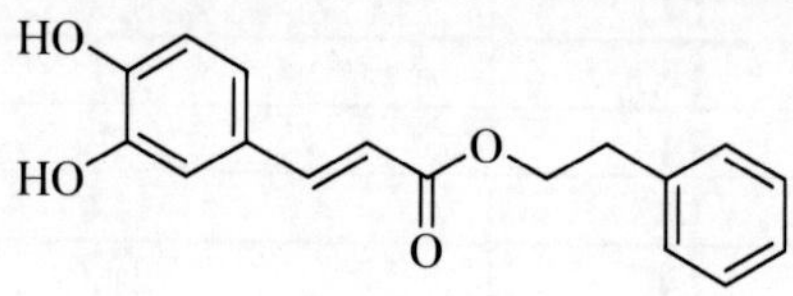

156

Figure 30. Animal derived compounds.

Caffeic acid phenethyl ester **156** (Figure 30) is an active component of propolis (a honey bee hive extract) [30] and has antiviral, anti-inflammatory and immunomodulatory properties [297]. Compound **156** is a potent inhibitor of NF-κB activation (maximal effect was seen at 25 μg/ml) (Table 6) [298] and is a 5α-reductase inhibitor (Table 4, Entry 11), indicating a potential role in the treatment of PC [124]. Caffeic acid phenethyl ester **156** has been found to be able to prevent paclitaxel **19** and TNFα-mediated NF-κB activation in PC-3 cells, at 25 μg/ml [297]. Its ability to induce apoptosis in a dose-dependent manner was associated with the loss of cIAP-1, cIAP-2 and X-linked mammalian inhibitor of apoptosis protein (XIAP) expression. The apoptotic mechanism as well as the corresponding loss of IAP expression were found to be caspase-dependent [297].

Conclusion

Developments in the extraction and characterization techniques of compounds from natural sources have supplied scientists with reasonable quantities of pure compounds, making it now possible to unequivocally establish a correlation between a naturally occurring compound and an *in vitro* or *in vivo* observed effect. Awareness of these correlations thus becomes a major step in understanding molecular targets that are disease specific. In the particular case of PC, throughout this discussion targets such as the AR, the proteasome, ceramide synthesis, Sp proteins, Par-4 and NF-κB have been found to be of extreme importance.

The most relevant issue on PC treatment strategies development is the ability to target both AD and AI cells in order to avoid recurrence phenomena. From this point of view the fact that most of the compounds described in this chapter revealed a promiscuous action on PC cells is rather appealing. Moreover, an intense search for synergistic effects between these compounds and the ones widely used in cancer chemotherapy such as paclitaxel **19**, docetaxel **1**, etoposide, 5-fluorouracil and campthotecin **100** with the hope of minimizing the adverse effects profile and enhance potency, is currently underway. Thus, a great number of compounds isolated from natural sources have interesting effects on PC cells both *in vitro* and *in vivo*. Recent research has brought into light compounds such as iejimalide B, spisulosine **12**, pristimerin **106**, celastrol **107** as well as withaferin A **122**, which show promise for future drug development. Both kahalalide F **2** and trabectedin **13** are currently undergoing clinical trials. Success examples of synthetic compounds based on the naturally occurring parent structures include the synthetic epothilone ixabepilone **24**, which is FDA approved for breast cancer treatment, the gossypol **48** derived AT-101, which is currently in phase II clinical trials for HRPC treatment, the triptolide **85** derivative PG490-88Na, approved for phase I clinical trials in patients with solid tumors, docetaxel **1**, which is now first-line therapy for HRPC combined with prednisone, and irofulven **153**, which is currently being tested on a number of solid tumors among which is PC.

Chemoprevention with dietary agents for PC has been extensively studied but controversy still surrounds this subject. Oriental countries with a soy-products based diet such as China and Japan have the lowest PC death rates (1 and 5.7 per 100,000 inhabitants, respectively) [4]. It has been demonstrated that immigrant populations coming to countries such as the USA (with fat-rich based diets) from countries with low risk of PC development, gradually undertake the PC incidence and death rates of the American population [299]. Therefore, a correlation between consumption of certain dietary agents and PC risk seems to exist. However, despite the enormous volume of publications available concerning dietary agents and PC, correlation between these two variables has not yet been unequivocally established. Thus, it has very recently been reported that increased serum levels of 25-hydroxyvitamin D did not correlate with improved outcomes in men screened for PC [300]. Nonetheless, clinical studies that are PC related using plant extracts are still currently underway. Dietary agents are appealing because their widely spread use on the human population makes the question of adverse effects not so relevant when compared to marine or microbial derived compounds, which are generally highly active. However, it is difficult to establish whether the observed effects are due to the compound itself or to a mixture of substances present on the dietary agents. Moreover, the question as to whether the concentrations used *in vitro* will be achievable *in vivo* remains.

References

[1] Black, R. J., Bray, F., Ferlay, J. & Parkin, D. M. (1997). Cancer Incidence and Mortality in the European Union: Cancer Registry Data and Estimates of National Incidence for 1990. *Eur. J. Cancer*, 33, 1075-1107.

[2] Greenlee, R. T., Murray, T., Bolden, S. & Wingo, P. A. (2000). Cancer Statistics, 2000. *CA Cancer J Clin*, 50, 7-33.

[3] Jemal, A., Siegel, R., Ward, E., Murray, T., Xu, J., Smigal, C. & Thun, M. J. (2006). Cancer Statistics, 2006. *CA Cancer J Clin*, 56, 106-130.

[4] Jemal, A., Siegel, R., Ward, E., Murray, T., Xu, J. & Thun, M. J. (2007). Cancer Statistics, 2007. *CA Cancer J Clin*, 57, 43-66.

[5] Chatterjee, B. (2003). The Role of the Androgen Receptor in the Development of Prostatic Hyperplasia and Prostate Cancer. *Mol. Cell Biochem.*, 253, 89-101.

[6] Huggins, C. & Hodges, C. V. (1941). Studies on Prostatic Cancer. I. The Effect of Castration, of Estrogen and of Androgen Injection on Serum Phosphatases in Metastatic Carcinoma of the Prostate. *Cancer Res.*, 1, 293-297.

[7] Huggins, C., Stevens, R. E. & Hodges, C. V. (1941). Studies on Prostatic Cancer. II. The Effect of Castration on Clinical Patients with Carcinoma of the Prostate. *Arch. Surg.*, 43, 209-223.

[8] Richterz, E., Srivastava, S. & Dobi, A. (2007). Androgen Receptor and Prostate Cancer. *Prostate Cancer Prostatic Dis.*, 10, 114-118.

[9] Li, R., Wheeler, T., Dai, H., Frolov, A., Thompson, T. & Ayala, G. (2004). High Level of Androgen Receptor is Associated With Aggressive Clinicopathologic Features and Decreased Biochemical Recurrence-free Survival in Prostate Cancer Patients Treated With Radical Prostatectomy. *Am. J. Surg. Pathol.*, 28, 928-934.

[10] Tilley, W. D., Buchanan, G., Hickey, T. E. & Bentel, J. M. (1996). Mutations in the Androgen Receptor Gene Are Associated With Progression of Human Prostate Cancer to Androgen Independence. *Clin. Cancer Res.*, 2, 277-285.

[11] Njar, V. C. O., Kato, K., Nnane, I. P., Grigoryev, D. N., Long, B. J. & Brodie, A. M. H. (1998). Novel 17-azolyl steroids, potent inhibitors of human cytochrome 17α-hydroxylase-C-17,C-20-lyase (P45017 α): Potential agents for the treatment of prostate cancer. *J. Med. Chem.*, 41, 902-912.

[12] Chen, C. D., Welsbie, D. S., Tran, C., Baek, S. H., Chen, R., Vessella, R., Rosenfeld, M. G. & Sawyers, C. L. (2004). Molecular Determinants of Resistance to Antiandrogen Therapy. *Nat. Med.*, 10, 33-39.

[13] Debes, J. D. & Tindall, D. J. (2004). Mechanisms of Androgen-refractory Prostate Cancer. *N. Engl. J. Med.*, 351, 1488-1490.

[14] Nemoto, R., Hattori, K., Uchida, K., Shimazui, T., Nishijima, Y., Koiso, K. & Harada, M. (1990). S-phase fraction of human prostate adenocarcinoma studied with *in vivo* bromodeoxyuridine labeling. *Cancer*, 66, 509-514.

[15] Moreira, V. M., Salvador, J. A. R., Vasaitis, T. S. & Njar, V. C. (2008). CYP17 Inhibitors for Prostate Cancer Treatment - An Update. *Curr. Med. Chem.*, 15, 868-899 and references cited therein.

[16] Gleave, M. E. & Small, E. J. (2004). Report to the Nation on Prostate Cancer. *Medscape*, 27-36.

[17] Lam, J. S., Leppert, J. T., Vemulapalli, S. N., Shvarts, O. & Belldegrun, A. S. (2006). Secondary Hormonal Therapy for Advanced Prostate Cancer. *J. Urol.*, 175, 27-34.

[18] Small, E. J., Baron, A. D. & Bok, R. (1997). Simultaneous Antiandrogen Withdrawal and Treatment with Ketoconazole and Hydrocortisone in Patients With Advanced Prostate Carcinoma. *Cancer*, 80, 1755-1759.

[19] Small, E. J., Baron, A. D., Fippin, L. & Apodaca, D. (1997). Ketoconazole Retains Activity in Advanced Prostate Cancer Patients With Progression Despite Flutamide Withdrawal. *J. Urol.*, 157, 1204-1207.

[20] Jackson, K. M., DeLeon, M., Verret, C. R. & Harris, W. B. (2002). Dibenzoylmethane induces cell cycle deregulation in human prostate cancer cells. *Cancer Lett.*, 178, 161-165.

[21] Small, E. J., Halabi, S., Dawson, N. A., Stadler, W. M., Rini, B. I., Picus, J., Gable, P., Torti, F. M., Kaplan, E. & Vogelzang, N. J. (2004). Antiandrogen Withdrawal Alone or in Combination With Ketoconazole in Androgen-independent Prostate Cancer Patients: a Phase III Trial (CALGB 9583). *J. Clin. Oncol.*, 22, 1025-1033.

[22] Available from: http://www.nccn.org/professionals/physician_gls/PDF/prostate.pdf

[23] Moreira, V. M., Vasaitis, T. S., Njar, V. C. & Salvador, J. A. R. (2007). Synthesis and evaluation of novel 17-indazole androstene derivatives designed as CYP17 inhibitors. *Steroids*, 72, 939-948.

[24] Moreira, V. M., Vasaitis, T. S., Guo, Z., Njar, V. C. & Salvador, J. A. R. (2008). Synthesis of Novel C17 Steroidal Carbamates. Studies on CYP17 Action, Androgen Receptor Binding and Function, and Prostate Cancer Cell Growth. *Steroids*, 73, 1217-1227.

[25] Mancuso, A., Oudard, S. & Sternberg, C. N. (2007). Effective chemotherapy for hormone-refractory prostate cancer (HRPC): Present status and perspectives with taxane-based treatments. *Critical Rev.Oncol. Hematol.*, 61, 176-185 and references cited therein.

[26] Newman, D. J., Cragg, G. M. & Snader, K. M. (2003). Natural products as sources of new drugs over the period 1981-2002. *J. Nat. Prod.*, 66, 1022-1037.

[27] Butler, M. S. (2004). The role of natural product chemistry in drug discovery. *J. Nat. Prod.*, 67, 2141-2153.

[28] Koehn, F. E. & Carter, G. T. (2005). The evolving role of natural products in drug discovery. *Nat. Rev. Drug Discov.*, 4, 206-220.

[29] Ojima, I. (2008). Miniperspectives: Natural Products in Drug Discovery. *J. Med. Chem.*, 51, 2587-2588.

[30] Aggarwal, B. B. & Shishodia, S. (2006). Molecular targets of dietary agents for prevention and therapy of cancer. *Biochem. Pharmacol.*, 71, 1397-1421.

[31] Klein, K. A., Reiter, R. E., Redula, J., Morad, H., Zhu, X. L., Brothman, A. R., Lamb, D. J., Marcelli, M., Belldegrun, A. S., Witte, O. N. & Sawyers, C. L. (1997). Progression of metastatic human prostate cancer to androgen independence in immunodeficient SCID mice. *Nat. Med.*, 3, 402-408.

[32] Horoszewicz, J. S., Leong, S. S., Kawinski, E., Karr, J. P., Rosenthal, H., Chu, T. M., Mirand, E. A. & Murphy, G. P. (1983). LNCaP model of human prostatic carcinoma. *Cancer Res.*, 43, 1809-1818.

[33] Van Bokhoven, A., Varella-Garcia, M., Korch, C., Johannes, W. U., Smith, E. E., Miller, H. L., Nordeen, S. K., Miller, G. J. & Lucia, M. S. (2003). Molecular characterization of human prostate carcinoma cell lines. *The Prostate*, 57, 205-225.

[34] Culig, Z., Klocker, H., Eberle, J., Kaspar, F., Hobish, A., Cronauer, M. V. & Bartsch, G. (1993). DNA Sequence of androgen receptor in prostatic tumor cells lines and tissue specimens assessed by means of polymerase chain reaction. *Prostate*, 22, 11-22.

[35] Veldscholte, J., Berrevoets, C. A. & Mulder, E. (1994). Studies on the human prostatic cancer cell line LNCaP. *J. Steroid Biochem. Mol. Biol.*, 49, 341-346.

[36] Kaighn, M. E., Narayan, K. S., Ohnuki, Y., Lechner, J. F. & Jones, L. W. (1979). Establishment and characterization of a human prostatic carcinoma cell line PC-3. *Invest. Urol.*, 17, 16-23.

[37] Lowe, S. W., Ruley, H. E., Jacks, T. & Housman, D. E. (1993). p53-Dependent apoptosis modulates the cytotoxicity of anticancer agents. *Cell*, 74, 957-967.

[38] Hannun, Y. A. (1997). Apoptosis and the dilemma of cancer chemotherapy. *Blood*, 89, 1845-1853.

[39] Hamann, M. T. & Scheuer, P. J. (1993). Kahalalide F: a bioactive depsipeptide from the Sacoglossan Mollusk *Elysia rufescens* and the green alga *Bryopsis sp. J. Am. Chem. Soc.*, 115, 5825-5826.

[40] Suarez, Y., Gonzalez, L., Cuadrado, A., Berciano, M., Lafarga, M. & Munoz, A. (2003). Kahalalide F, a new marine-derived compound, induces oncosis in human prostate and breast cancer cells. *Mol. Cancer Ther.*, 2, 863-872.

[41] Garcia-Rocha, M., Bonay, P. & Avila, J. (1996). The antitumoral compound Kahalalide F acts on cell lysosomes. *Cancer Lett.*, 99, 43-50.

[42] Rademaker-Lakhai, J. M., Horenblas, S., Meinhardt, W., Stokvis, E., de Reijke, T. M., Jimeno, J. M., Lopez-Lazaro, L., Martin, J. A. L., Beijnen, J. H. & Schellens, J. H. M. (2005). Phase I clinical and pharmacokinetic study of Kahalalide F in patients with advanced androgen refractory prostate cancer. *Clin. Cancer Res.*, 11, 1854-1862.

[43] Brown, A. P., Morrissey, R. L., Faircloth, G. T. & Levine, B. S. (2002). Preclinical toxicity studies of kahalalide F, a new anticancer agent: single and multiple dosing regimens in the rat. *Cancer Chemother. Pharmacol.*, 50, 333-340.

[44] Ciruelos, E., Trigo, T., Pardo, J., Paz-Ares, L., Estaun, N., Cuadra, C., Dominguez, M., Marin, A., Jimeno, J. & Izquierdo, M. (2002). A phase I clinical and pharmacokinetic (PK) study with Kahalalide F (KF) in patients (pts) with advanced solid tumors (AST) with a continuos weekly (W) 1-hour iv infusion schedule. *Eur. J. Cancer*, 38, S33.

[45] PharmaMar Biopharmaceutical Co. (2006). 31st European Society for Medical Oncology Congress (ESMO), Istanbul, Turkey.

[46] Petit, G. R., Singh, S. B., Hogan, F., Loloyd-Williams, P., Herald, D. L., Burkett, D. D. & Clewlow, P. J. (1989). The absolute configuration and synthesis of natural (-)-Dolastatin 10. *J. Am. Chem. Soc.*, 111, 5463-5465.

[47] Petit, G. R., Kamano, Y., Herald, C. L., Tuinman, A. A., Boettner, F. E., Kizu, H., Schmidt, J. M., Baczynskyj, L., Tomer, K. B. & Bontems, R. (1987). The isolation and structure of a remarkable marine animal antineoplastic constituent: Dolastatin 10. *J. Am. Chem. Soc.*, 109, 6883-6885.

[48] Bai, R., Pettit, G. R. & Hamel, E. (1990). Dolastatin-10, a Powerful Cytostatic Peptide Derived from a Marine Animal - Inhibition of Tubulin Polymerization Mediated through the Vinca Alkaloid Binding Domain. *Biochem. Pharmacol.*, 39, 1941-1949.

[49] Turner, T., Jackson, W. H., Pettit, G. R., Wells, A. & Kraft, A. S. (1998). Treatment of human prostate cancer cells with dolastatin 10, a peptide isolated from a marine shell-less mollusc. *Prostate*, 34, 175-181.

[50] Mirsalis, J. C., Schindler-Horvat, J., Hill, J. R., Tomaszewski, J. E., Donohue, S. J. & Tyson, C. A. (1999). Toxicity of dolastatin 10 in mice, rats and dogs and its clinical relevance. *Cancer Chemother. Pharmacol.*, 44, 395-402.

[51] Madden, T., Tran, H. T., Beck, D., Huie, R., Newman, R. A., Pusztai, L., Wright, J. J. & Abbruzzese, J. L. (2000). Novel marine-derived anticancer agents: A phase I clinical, pharmacological, and pharmacodynamic study of dolastatin 10 (NSC 376128) in patients with advanced solid tumors. *Clin. Cancer Res.*, 6, 1293-1301.

[52] Vaishampayan, U., Glode, M., Du, W., Kraft, A., Hudes, G., Wright, J. & Hussain, M. (2000). Phase II study of dolastatin-10 in patients with hormone-refractory metastatic prostate adenocarcinoma. *Clin. Cancer Res.*, 6, 4205-4208.

[53] Crews, P., Manes, L. V. & Boehler, M. (1986). Jasplakinolide, a cyclodepsipeptide from the marine sponge *Jaspis* sp. *Tetrahedron Lett.*, 27, 2797-2800.

[54] Takeuchi, H., Ara, G., Sausville, E. A., Teicher, B., Jasplakinolide: interaction with radiation and hyperthermia in human prostate carcinoma and Lewis lung carcinoma. *Cancer Chemother. Pharmacol.*, 1998, 42, 491-496.

[55] Senderowicz, A. M. J., Kaur, G., Sainz, E., Laing, C., Inman, W. D., Rodriguez, J., Crews, P., Malspeis, L., Grever, M. R., Sausville, E. A. & Duncan, K. L. K. (1995). Jasplakinolide Inhibition of the Growth of Prostate Carcinoma Cells *in Vitro* with Disruption of the Actin Cytoskeleton. *J. Natl. Cancer Inst.*, 87, 46-51.

[56] Zidane, M., Pondaven, P., Roussakis, C., Quemener, B. & More, M. T. (1996). Pachymatismin: a novel cytotoxic factor from the marine sponge (*Pachymatisma johnstonii*). *Comp. Biochem. Physiol.*, 115C, 47-53.

[57] Sangrajrang, S., Zidane, M., Berda, P., More, M. T., Calvo, F. & Fellous, A. (2000). Different microtubule network alterations induced by pachymatismin, a new marine glycoprotein, on two prostatic cell lines. *Cancer Chemother. Pharmacol.*, 45, 120-126.

[58] Hadaschik, B. A., Ettinger, S., Sowery, R. D., Zoubeidi, A., Andersen, R. J., Roberge, M. & Gleave, M. E. (2008). Targeting prostate cancer with HTI-286, a synthetic analog of the marine sponge product hemiasterlin. *Int. J. Cancer*, 122, 2368-2376 and references cited therein.

[59] Ribehart, K. L., Gloer, J. B., Hughes, R. G., Renis, H. E., McGovren, J. P., Swynenberg, E. B., Stringfellow, D. A., Kuentzel, S. L. & Li, L. H. (1981). Didemnins: antiviral and antitumor depsipeptides from a caribbean tunicate. *Science*, 212, 933-935.

[60] Urdiales, J. L., Morata, P., DeCastro, I. N. & Sanchez-Jimenez, F. (1996). Antiproliferative effect of dehydrodidemnin B (DDB), a depsipeptide isolated from Mediterranean tunicates. *Cancer Lett.*, 102, 31-37.

[61] Hirata, Y. & Uemura, D. (1986). Halichondrins - antitumor polyether macrolides from a marine sponge. *Pure Appl. Chem.*, 58, 701-710.

[62] Geldof, A. A., Mastbergen, S. C., Henrar, R. E. C. & Faircloth, G. T. (1999). Cytotoxicity and neurocytotoxicity of new marine anticancer agents evaluated using *in vitro* assays. *Cancer Chemother. Pharmacol.*, 44, 312-318.

[63] Williamson, S. K., Wolf, M. K., Eisenberger, M. A., O'Rourke, M., Brannon, W. & Crawford, E. D. (1995). Phase-II Evaluation of Didemnin-B in Hormonally Refractory Metastatic Prostate Cancer. A Southwest Oncology Group Study. *Invest. New Drugs*, 13, 167-170.

[64] Wang, W. L., McHenry, P., Jeffrey, R., Schweitzer, D., Helquist, P. & Tenniswood, M. (2008). Effects of Iejimalide B, a marine macrolide, on growth and apoptosis in prostate cancer cell lines. *J. Cell. Biochem.*, doi:10.1002/jcb.21898.

[65] Sadar, M. D., Williams, D. E., Mawji, N. R., Patrick, B. O., Wikanta, T., Chasanah, E., Irianto, H. E., Soest, R. V. & Andersen, R. J. (2008). Sintokamides A to E, chlorinated peptides from the sponge *Dysidea* sp. that inhibit transactivation of the *N*-terminus of the androgen receptor in prostate cancer cells. *Org. Lett.*, , doi:10.1021/ol802021w.

[66] Sanchez, A. M., Malagarie-Cazenave, S., Olea, N., Vara, D., Cuevas, C. & Diaz-Laviada, I. (2008). Spisulosine (ES-285) induces prostate tumor PC-3 and LNCaP cell death by *de novo* synthesis of ceramide and PKCζ activation. *Eur. J. Pharmacol.*, 584, 237-245.

[67] Wang, G., Silva, J., Krishnamurthy, K., Tran, E., Condie, B. G. & Bieberich, E. (2005). Direct binding to ceramide activates protein kinase Cζ before the formation of a pro-apoptotic complex with PAR-4 in differentiating stem cells. *J. Biol. Chem.*, 28, 26415-26424.

[68] Ruvolo, P. (2003). Intracellular signal transduction pathways activated by ceramide and its metabolites. *Pharmacol. Res.*, 47, 383-392.

[69] Zelek, L., Yovine, A., Brain, E., Turpin, F., Taamma, A., Riofrio, M., Spielmann, M., Jimeno, J. & Misset, J. L. (2006). A phase II study of Yondelis (trabectedin, ET-743) as a 24-h continuous intravenous infusion in pretreated advanced breast cancer. *Br. J. Cancer*, 94, 1610-1614.

[70] Michaelson, M. D., Gilligan, T., Oh, W., Kantoff, P., Taplin, M. E., Izquierdo, M. A., Flores, L. & Smith, M. R. (2005). Phase II study of three hour, weekly infusion of trabectedin (ET-743) in men with metastatic, androgen-independent prostate carcinoma (AIPC). *J. Clin. Oncol.*, 23, 382S-382S.

[71] Available from: http://clinicaltrials.gov/ct2/search

[72] Kwak, J. H., Schmitz, F. J. & Kelly, M. (2000). Sesquiterpene quinols/quinones from the Micronesian sponge *Petrosaspongia metachromia*. *J. Nat. Prod.*, 63, 1153-1156.

[73] Lu, P. H., Chueh, S. C., Kung, F. L., Pan, S. L., Shen, Y. C. & Guh, J. H. (2007). Ilimaquinone, a marine sponge metabolite, displays anticancer activity via GADD 153-mediated pathway. *Eur. J. Pharmacol.*, 556, 45-54.

[74] Kashman, Y., Groweiss, A. & Shmueli, U. (1980). Latrunculin, a new 2-thiazolidinone macrolide from the marine sponge *Latrunculia magnifica*. *Tetrahedron Lett.*, 21, 3629-3632.

[75] Spector, I., Shochet, N. R., Blasberger, D. & Kashman, Y. (1989). Latrunculins - novel marine macrolides that disrupt microfilament organization and affect cell growth: I. Comparison with cytochalasin D. *Cell Motil. Cytoskel.*, 13, 127-144.

[76] El Sayed, K. A., Khanfar, M. A., Shallal, H. M., Muralidharan, A., Awate, B., Youssef, D. T. A., Liu, Y., Zhoo, Y. D., Nagle, D. G. & Shah, G. (2008). Latrunculin A and its C-17-O-carbamates inhibit prostate tumor cell invasion and HIF-1 activation in breast tumor cells. *J. Nat. Prod.*, 71, 396-402.

[77] Tasdemir, D., Mangalindan, G. C., Concepcion, G. P., Verbitski, S. M., Rabindran, S., Miranda, M., Greenstein, M., Hooper, J. N. A., Harper, M. K. & Ireland, C. M. (2002). Bioactive Isomalabaricane Triterpenes from the Marine Sponge *Rhabdastrella globostellata*. *J. Nat. Prod.*, 65, 210-214.

[78] Liu, W. K., Cheung, F. W. K. & Che, C. T. (2006). Stellettin A induces oxidative stress and apoptosis in HL-60 human leukemia and LNCaP prostate cancer cell lines. *J. Nat. Prod.*, 69, 934-937.

[79] Liu, W. K., Wong, N. L. Y., Huang, H. M., Ho, J. K. C., Zhang, W. H. & Che, C. T. (2002). Growth inhibitory activity of lemnabourside on human prostate cancer cells. *Life Sci.*, 69, 843-853.

[80] Roskelley, C. D., Williams, D. E., McHardy, L. M., Leong, K. G., Troussard, A., Karsan, A., Andersen, R. J., Dedhar, S. & Roberge, M. (2001). Inhibition of tumor cell invasion and angiogenesis by motuporamines. *Cancer Res.*, 61, 6788-6794.

[81] Hofle, G., Bedorf, N., Steinmetz, H., Schomburg, D., Gerth, K. & Reichenbach, H. (1996). Epothilone A and B - novel 16-membered macrolides with cytotoxic activity: isolation, crystal structure, and conformation in solution. *Angew. Chem. Int. Ed. Engl.*, 35, 1567-1569.

[82] Gerth, K., Bedorf, N., Hofle, G., Irschik, H. & Reichenbach, H. (1996). Epothilons A and B : Antifungal and Cytotoxic Compounds from *Sorangium cellulosum* (Myxobacteria). Production, Physico-chemical and Biological Properties. *J. Antibiotics*, 49, 560-563.

[83] Wartmann, M. & Altmann, K. H. (2002). The biology and medicinal chemistry of epothilones. *Curr. Med. Chem. Anti Cancer Agents*, 2, 123-148 and references cited therein.

[84] Larkin, J. M. G. (2007). Patupilone. *Drugs Future*, 32, 323-336.

[85] Su, D. S., Meng, D., Bertinato, P., Balog, A., Sorensen, E. J., Danishefsky, S. J., Zheng, Y. H., Chou, T. C., He, L. & Horwitz, S. B. (1997). Total synthesis of (-)-Epothilone B: an extension of the Suzuki coupling method and insights into structure-activity relationships of the epothilones. *Angew. Chem. Int. Ed. Engl.*, 36, 757-759.

[86] Balog, A., Meng, D., Kamenecka, T., Bertinato, P., Su, D. S., Sorensen, E. J. & Danishefsky, S. J. (1996). Total synthesis of (-)-Epothilone A. *Angew. Chem. Int. Ed. Engl.*, 35, 2801-2803.

[87] Sepp-Lorenzino, L., Balog, A., Su, D. S., Meng, D., Timaul, N., Scher, H. I., Danishefsky, S. J. & Rosen, N. (1999). The microtubule-stabilizing agents epothilones A and B and their desoxy-derivatives induce mitotic arrest and apoptosis in human prostate cancer cells. *Prostate Cancer Prostatic Dis.*, 2, 41-52.

[88] Ioffe, M. L., White, E., Nelson, D. A., Dvorzhinski, D. & DiPaola, R. S. (2004). Epothilone induced cytotoxicity is dependent on p53 status in prostate cells. *Prostate*, 61, 243-247.

[89] O'Reilly, T., McSheehy, P. M., Wenger, F., Hattenberger, M., Muller, M., Vaxelaire, J., Altmann, K. H. & Wartmann, M. (2005). Patupilone (epothilone B, EPO906) inhibits growth and metastasis of experimental prostate tumors *in vivo*. *Prostate*, 65, 231-240.

[90] Goodin, S., Novel cytotoxic agents: Epothilones. *Am. J. Health Syst. Pharm.*, 2008, 65, S10-5.

[91] Hussain, M., Tangen, C. M., Lara, P. N., Vaishampayan, U. N., Petrylak, D. P., Colevas, A. D., Sakr, W. A. & Crawford, E. D. (2005). Ixabepilone (epothilone B analogue BMS-247550) is active in chemotherapy-naive patients with hormone-

refractory prostate cancer: A southwest oncology group trial S0111. *J. Clin. Oncol.*, 23, 8724-8729.

[92] Lee, D. (2004). Activity of epothilone B analogues ixabepilone and patupilone in hormone-refractory prostate cancer. *Clin. Prostate Cancer*, 3, 80-82.

[93] Galsky, M. D., Small, E. J., Oh, W. K., Chen, I., Smith, D. C., Colevas, A. D., Martone, L., Curley, T., DeLaCruz, A., Scher, H. I. & Kelly, W. K. (2005). Multi-institutional randomized phase II trial of the epothilone B analog ixabepilone (BMS-247550) with or without estramustine phosphate in patients with progressive castrate metastatic prostate cancer. *J. Clin. Oncol.*, 23, 1439-1446.

[94] Rosenberg, J. E., Galsky, M. D., Rohs, N. C., Weinberg, V. K., Oh, W. K., Kelly, W. K. & Small, E. J. (2006). A retrospective evaluation of second-line chemotherapy response in hormone-refractory prostate carcinoma - Second-line taxane-based therapy after first-line epothilone-B analog ixabepilone (BMS-247550) therapy. *Cancer*, 106, 58-62.

[95] Gunatilaka, A. A. L. (2006). Natural products from plant-associated microorganisms: Distribution, structural diversity, bioactivity, and implications of their occurrence. *J. Nat. Prod.*, 69, 509-526.

[96] Pelham, H. R. B. (1991). Multiple targets for Brefeldin A. *Cell*, 67, 449-451.

[97] Schwartz-Lippincott, J., Yuan, L. C., Bonifacino, J. S. & Klausner, R. D. (1989). Rapid redistribution of Golgi proteins into the ER in cells with Brefeldin A: evidence for membrane cycling from Golgi to ER. *Cell*, 56, 801-813.

[98] Konno, S., Mordente, J. A., Chen, Y. P., Wu, J. M., Tazaki, H. & Mallouh, C. (1998). Effects of brefeldin A on androgen receptor-mediated cellular responses in human prostatic carcinoma LNCaP cells. *Mol. Urol.*, 2, 7-11.

[99] Mordente, J. A., Konno, S., Chen, Y. P., Wu, J. M., Tazaki, H. & Mallouh, C. (1998). The effects of brefeldin A (BFA) on cell cycle progression involving the modulation of the retinoblastoma protein (pRB) in PC-3 prostate cancer cells. *J. Urol.*, 159, 275-279.

[100] Chapman, J. R., Tazaki, H., Mallouh, C. & Konno, S. (1999). Brefeldin A induced apoptosis in prostatic cancer DU-145 cells: a possible p53-independent death pathway. *BJU Int.*, 83, 703-708.

[101] Wallen, E., Sellers, R. G. & Peehl, D. M. (2000). Brefeldin A induces p53-independent apoptosis in primary cultures of human prostatic cancer cells. *J. Urol.*, 164, 836-841.

[102] Shiraishi, T., Yoshida, T., Nakata, S., Horinaka, M., Wakada, M., Mizutani, Y., Miki, T. & Sakai, T. (2005). Tunicamycin enhances tumor necrosis factor-related apoptosis-inducing ligand-induced apoptosis in human prostate cancer cells. *Cancer Res.*, 65, 6364-6370.

[103] Hamill, R. L., Higgens, C. E., Boaz, H. E. & Gorman, M. (1969). The structure of beauvericin, a new depsipeptide antibiotic to *Artemia salina*. *Tetrahedron Lett.*, 49, 4255-4258.

[104] Bernardini, M., Carilli, A., Pacioni, G. & Santurbano, B. (1975). Isolation of beauvericin from *Paecilomyces-fumoso-roseus*. *Phytochemistry*, 14, 1865-186.

[105] Deol, B. S., Ridley, D. D. & Singh, P. (1978). Isolation of Cyclodepsipeptides from Plant Pathogenic Fungi. *Aust. J. Chem.*, 31, 1397-1399.

[106] Zhan, J. X., Burns, A. M., Liu, M. P. X., Faeth, S. H. & Gunatilaka, A. A. L. (2007). Search for cell motility and angiogenesis inhibitors with potential anticancer activity: Beauvericin and other constituents of two endophytic strains of *Fusarium oxysporum*. *J. Nat. Prod.*, 70, 227-232.

[107] Moretti, A., Belisario, A., Tafuri, A., Ritieni, L., Corazza, L. & Logrieco, A. (2002). Production of beauvericin by different races of *Fusarium oxysporum f. sp. melonis*, the Fusarium wilt agent of muskmelon. *Eur. J. Plant Pathol.*, 108, 661-666.

[108] Momose, I., Iijima, M., Kawada, M. & Ikeda, D. (2007). A new proteasome inhibitor, TP-110, induces apoptosis in human prostate cancer PC-3 cells. *Biosci. Biotechnol. Biochem.*, 71, 1036-1043 and references cited therein.

[109] Meiners, S., Ludwig, A., Stangl, V. & Stangl, K. (2008). Proteasome inhibitors: poisons and remedies. *Med. Res. Rev.*, 28, 309-327.

[110] Lee, D. H., Goldberg, A. L., Proteasome inhibitors: valuable new tools for cell biologists. *Trends Cell. Biol.*, 1998, 8, 397-403.

[111] Gapter, L., Wang, Z. S., Glinski, J. & Ng, K. Y. (2005). Induction of apoptosis in prostate cancer cells by pachymic acid from *Poria cocos*. *Biochem. Biophys. Res. Commun.*, 332, 1153-1161.

[112] Ahmad, N., Feyes, D. K., Nieminen, A. L., Agarwal, R. & Mukhtar, H. (1997). Green tea constituent epigallocatechin-3-gallate and induction of apotposis and cell cycle arrest in human carcinoma cells. *J. Natl. Cancer Inst.*, 89, 1881-1886.

[113] Lambert, J. D. & Yang, C. S. (2003). Mechanisms of cancer prevention by tea constituents. *J. Nutr.*, 133, 3262S-3267S.

[114] Moon, C., Park, J. C., Chae, Y. K., Yun, J. H. & Kim, S. (2008). Current status of experimental therapeutics for prostate cancer. *Cancer Lett.*, 266, 116-134.

[115] Siddiqui, I. A., Raisuddin, S. & Shukla, Y. (2005). Protective effects of black tea extract on testosterone induced oxidative damage in prostate. *Cancer Lett.*, 227, 125-132.

[116] Prasad, S., Kaur, J., Roy, P., Kalra, N. & Shukla, Y. (2007). Theaflavins induce G2/M arrest by modulating expression of p21 (waf1/cip1), cdc25C and cyclin B in human prostate carcinoma PC-3 cells. *Life Sci.*, 81, 1323-1331.

[117] Kalra, N., Seth, K., Prasad, S., Singh, M., Pant, A. B. & Shukla, Y. (2007). Theaflavins induced apoptosis of LNCaP cells is mediated through induction of p53, down-regulation of NF-kB and mitogen-activated protein kinases pathways. *Life Sci.*, 80, 2137-2146.

[118] Ren, F. G., Zhang, S. B., Mitchell, S. H., Butler, R. & Young, C. Y. F. (2000). Tea polyphenols down-regulate the expression of the androgen receptor in LNCaP prostate cancer cells. *Oncogene*, 19, 1924-1932.

[119] Kimura, K. I., Itakura, Y., Goto, R., Tojima, M., Egawa, N. & Yoshihama, M. (2007). Inhibition of 17α-Hydroxylase/C-17,(20)-Lyase (CYP17) from rat testis by green tea catechins and black tea theaflavins. *Biosci. Biotechnol. Biochem.*, 71, 2325-2328.

[120] Siddiqui, I. A., Zaman, N., Aziz, M. H., Reagan-Shaw, S. R., Sarfaraz, S., Adhami, V. M., Ahmad, N., Raisuddin, S. & Mukhtar, H. (2006). Inhibition of CWR22R nu 1 tumor growth and PSA secretion in athymic nude mice by green and black teas. *Carcinogenesis*, 27, 833-839.

[121] Siddiqui, I. A., Adhami, V. M., Afaq, F., Ahmad, N. & Mukhtar, H. (2004). Modulation of phosphatidylinositol-3-kinase/protein kinase B- and mitogen-activated protein kinase-pathways by tea polyphenols in human prostate cancer cells. *J. Cell. Biochem.*, 91, 232-242.

[122] Paschka, A. G., Butler, R. & Young, C. Y. F. (1998). Induction of apoptosis in prostate cancer cell lines by the green tea component, (-)-epigallocatechin-3-gallate. *Cancer Lett.*, 130, 1-7.

[123] Liao, S. & Hiipakka, R. A. (1995). Selective inhibition of steroid 5α-reductase isoenzymes by tea epicatechin-3-gallate and epigallocatechin-3-gallate. *Biochem. Biophys. Res. Commun.*, 214, 833-838.

[124] Hiipakka, R. A., Zhang, H. Z., Dai, W., Dai, Q. & Liao, S. T. (2002). Structure-activity relationships for inhibition of human 5α-reductases by polyphenols. *Biochem. Pharmacol.*, 63, 1165-1176.

[125] Liao, S., Umekita, Y., Guo, J., Kokontis, J. M. & Hiipakka, R. A. (1995). Growth inhibition and regression of human prostate and breast tumors in athymic mice by tea epigallocatechin gallate. *Cancer Lett.*, 96, 239-243.

[126] Mustafa, V. M., Malik, A., Zaman, N., Sarfaraz, S., Siddiqui, I. A., Syed, D. N., Afaq, F., Pasha, F. S., Sallem, M. & Mukhtar, H. (2007). Combined inhibitory effects of green tea polyphenols and selective cyclooxygenase-2 inhibitors on the growth of human prostate cancer cells both *in vitro* and *in vivo*. *Clin. Cancer Res.*, 13, 1611-1619.

[127] Bell, C. & Hawthorne, S. (2008). Ellagic acid, pomegranate and prostate cancer - a mini review. *J. Pharm. Pharmacol.*, 60, 139-144 and references cited therein.

[128] Seeram, N. P., Adams, L. S., Henning, S. M., Niu, Y. T., Zhang, Y. J., Nair, M. G. & Heber, D. (2005). *In vitro* antiproliferative, apoptotic and antioxidant activities of punicalagin, ellagic acid and a total pomegranate tannin extract are enhanced in combination with other polyphenols as found in pomegranate juice. *J. Nutr. Biochem.*, 16, 360-367.

[129] Malik, A., Afaq, F., Sarfaraz, S., Adhami, V. M., Syed, D. N. & Mukhtar, H. (2005). Pomegranate fruit juice for chemoprevention and chemotherapy of prostate cancer. *Proc. Natl. Acad. Sci. USA*, 102, 14813-14818.

[130] Albrecht, M., Jiang, W. G., Kumi-Diaka, J., Lansky, E. P., Gommersall, L. M., Patel, A., Mansel, R. E., Neeman, I., Geldof, A. A. & Campbell, M. J. (2004). Pomegranate extracts potently suppress proliferation, xenograft growth, and invasion of human prostate cancer cells. *J. Med. Food*, 7, 274-283.

[131] Lansky, E. P., Harrison, G., Froom, P. & Jiang, W. G. (2005). Pomegranate (*Punica granatum*) pure chemicals show possible synergistic inhibition of human PC-3 prostate cancer cell invasion across Matrigel™. *Invest. New Drugs*, 23, 121-122.

[132] Pezzuto, J. M. (2008). Resveratrol as an inhibitor of carcinogenesis. *Pharm. Biol.*, 46, 443-573 and references cited therein.

[133] Dorai, T., Gehani, N. & Katz, A. (2000). Therapeutic potential of curcumin in human prostate cancer. I. Curcumin induces apoptosis in both androgen-dependent and androgen-independent prostate cancer cells. *Prostate Cancer Prostatic Dis.*, 3, 84-93.

[134] Dorai, T., Gehani, N. & Katz, A. (2000). Therapeutic potential of curcumin in human prostate cancer. II. Curcumin inhibits tyrosine kinase activity of epidermal growth factor receptor and depletes the protein. *Mol. Urol.*, 4, 1-6.

[135] Dorai, T., Cao, Y. C., Dorai, B., Buttyan, R. & Katz, A. (2001). Therapeutic potential of curcumin in human prostate cancer. III. Curcumin inhibits proliferation, induces apoptosis, and inhibits angiogenesis of LNCaP prostate cancer cells *in vivo*. *Prostate*, 47, 293-303.

[136] Mukhopadhayay, A., Bueso-Ramos, C., Chatterjee, D., Pantazis, P. & Aggarwal, B. B. (2001). Curcumin downregulates cell survival mechanisms in human prostate cancer cell lines. *Oncogene*, 20, 7597-7609.

[137] Deeb, D., Xu, Y. X., Jiang, H., Gao, X. H., Janakiraman, N., Chapman, R. A. & Gautam, S. C. (2003). Curcumin (diferuloyl-methane) enhances tumor necrosis factor-related apoptosis-inducing ligand-induced apoptosis in LNCaP prostate cancer cells. *Mol. Cancer Ther.*, 2, 95-103.

[138] Deeb, D., Jiang, H., Gao, X. H., Hafner, M. S., Wong, H., Divine, G., Chapman, R. A., Dulchavsky, S. A. & Gautam, S. C. (2004). Curcumin sensitizes prostate cancer cells to tumor necrosis factor-related apoptosis-inducing ligand/Apo2L by inhibiting nuclear factor-kB through suppression of IkBα phosphorylation. *Mol. Cancer Ther.*, 3, 803-812.

[139] Somasundaram, S. G., Arrambide, D. & Gunasekera, R. S. (2004). Enhancement of the chemotherapeutic activity of curcumin and capsaicin by Bioflavone α on human prostate cancer cells. *J. Nutr.*, 134, 3537S-3537S.

[140] Kim, J. H., Xu, C. J., Keum, Y. S., Reddy, B., Conney, A. & Kong, A. N. T. (2006). Inhibition of EGFR signaling in human prostate cancer PC-3 cells by combination treatment with β-phenethyl isothiocyanate and curcumin. *Carcinogenesis*, 27, 475-482.

[141] Sumitomo, M., Tachibana, M., Nakashima, J., Murai, M., Miyajima, A., Kimura, F., Hayakawa, M. & Nakamura, H. (1999). An essential role for nuclear factor kB in preventing TNFα-induced cell death in prostate cancer cells. *J. Urol.*, 161, 674-679.

[142] Sharma, R. A., McLelland, H. R., Hill, K. A., Ireson, C. R., Euden, S. A., Manson, M. M., Pirmohamed, M., Marnett, L. J., Gescher, A. J. & Steward, W. P. (2001). Pharmacodynamic and pharmacokinetic study of oral *Curcuma* extract in patients with colorectal cancer. *Clin. Cancer Res.*, 7, 1894-1900.

[143] Shoba, G., Joy, D., Joseph, T., Majeed, M., Rajendran, R. & Srinivas, P. (1998). Influence of piperine on the pharmacokinetics of curcumin in animals and human volunteers. *Planta Medica*, 64, 353-356.

[144] Huang, Y. W., Wang, L. S., Chang, H. L., Ye, W. P., Dowd, M. K., Wan, P. J. & Lin, Y. C. (2006). Molecular mechanisms of (-)-gossypol-induced apoptosis in human prostate cancer cells. *Anticancer Res.*, 26, 1925-1933.

[145] Wong, Y. C. & Tam, C. C. (1989). Structural-Changes of the Guinea-Pig Prostatic Epithelial-Cells after Gossypol Treatment. *Acta Anat.*, 134, 18-25.

[146] Moh, P. P., Chang, C. J. G., Brueggemeier, R. W. & Lin, Y. C. (1992). Effect of Gossypol (Gp) on 5α-Reductase and 3α-Hydroxysteroid Dehydrogenase (3α-Hsd) in Adult-Rat Testes. *Faseb J.*, 6, A2031-A2031.

[147] Chang, C. J. G., Ghosh, P. K., Hu, Y. F., Brueggemeier, R. W. & Lin, Y. C. (1993). Antiproliferative and Antimetastatic Effects of Gossypol on Dunning Prostate Cell-Bearing Copenhagen Rats. *Res. Commun. Chem. Pathol. Pharmacol.*, 79, 293-312.

[148] Shidaifat, F., Canatan, H., Kulp, S. K., Sugimoto, Y., Chang, W. Y., Zhang, Y., Brueggemeier, R. W., Somers, W. J. & Lin, Y. C. (1996). Inhibition of human prostate cancer cells growth by gossypol is associated with stimulation of transforming growth factor β. *Cancer Lett.*, 107, 37-44.

[149] Loberg, R. D., McGregor, N., Ying, C., Sargent, E. & Pienta, K. J. (2007). *In Vivo* Evaluation of AT-101 (R-(–)-Gossypol Acetic Acid) in Androgen-Independent Growth of VCaP Prostate Cancer Cells in Combination with Surgical Castration. *Neoplasia*, 9, 1030-1037.

[150] Li, T. S. C. (2002). Chinese and related North American herbs: phytopharmacology and therapeutic values. Boca Raton, FL: CRC Press.

[151] Shigemura, K., Arbiser, J. L., Sun, S. Y., Zayzafoon, M., Johnstone, P. A. S., Fujisawa, M., Gotoh, A., Weksler, B., Zhau, H. E. & Chung, L. W. K. (2007). Honokiol, a natural plant product, inhibits the bone metastatic growth of human prostate cancer cells. *Cancer*, 109, 1279-1289.

[152] Hahm, E. R. & Singh, S. V. (2007). Honokiol causes G(0)-G(1) phase cell cycle arrest in human prostate cancer cells in association with suppression of retinoblastoma protein level/phosphorylation and inhibition of E2F1 transcriptional activity. *Mol. Cancer Ther.*, 6, 2686-2695.

[153] Hahm, E. R., Arlotti, J. A., Marynowski, S. W. & Singh, S. V. (2008). Honokiol, a constituent of oriental medicinal herb *Magnolia officinalis*, inhibits growth of PC-3 xenografts *in vivo* in association with apoptosis induction. *Clin. Cancer Res.*, 14, 1248-1257.

[154] Bai, X., Cerimele, F., Ushio-Fukai, M., Waqas, M., Campbell, P. M., Govindarajan, B., Der, C. J., Battle, T., Frank, D. A., Ye, K., Murad, E., Dubiel, W., Soff, G. & Arbiser, J. L. (2003). Honokiol, a small molecular weight natural product, inhibits angiogenesis *in vitro* and tumor growth *in vivo*. *J. Biol. Chem.*, 278, 35501-35507.

[155] Ahn, K. S., Sethi, G., Shishodia, S., Sung, B., Arbiser, J. L. & Aggarwal, B. B. (2006). Honokiol potentiates apoptosis, suppresses osteoclastogenesis, and inhibits invasion through modulation of nuclear factor-kB activation pathway. *Mol. Cancer Res.*, 4, 621-633.

[156] Tse, A. K. W., Wan, C. K., Shen, X. L., Yang, M. & Fong, W. F. (2005). Honokiol inhibits TNFα-stimulated NF-kB activation and NF-kB-regulated gene expression through suppression of IKK activation. *Biochem. Pharmacol.*, 70, 1443-1457.

[157] Lee, S. Y., Yuk, D. Y., Song, H. S., Yoon, D. Y., Jung, J. K., Moon, D. C., Lee, B. S. & Hong, J. T. (2008). Growth inhibitory effects of obovatol through induction of apoptotic cell death in prostate and colon cancer by blocking of NF-kB. *Eur. J. Pharmacol.*, 582, 17-25.

[158] Martikainen, P., Kyprianou, N., Tucker, R. W. & Isaacs, J. T. (1991). Programmed Death of Nonproliferating Androgen-independent Prostatic Cancer Cells. *Cancer Res.*, 51, 4693-4700.

[159] Furuya, Y. & Isaacs, J. T. (1993). Differential gene regulation during programmed death (apoptosis) *versus* proliferation of prostatic glandular cells induced by androgen manipulation. *Endocrinology*, 133, 2660-2666.

[160] Kyprianou, N., English, H. F. & Isaacs, J. T. (1988). Activation of a Ca^{2+}-Mg^{2+}-dependent endonuclease as an early event in castration-induced prostatic cell death. *Prostate*, 13, 103-117.

[161] Martikainen, P. & Isaacs, J. T. (1990). Role of calcium in the programmed death of rat prostatic glandular cells. *Prostate*, 17, 175-187.

[162] English, H. F., Kyprianou, N. & Isaacs, J. T. (1989). Relationship Between DNA Fragmentation and Apoptosis in the Programmed Cell Death in the Rat Prostate Following Castration. *Prostate*, 15, 233-250.

[163] Trang, T., Sutak, M., Quirion, R. & Jhamandas, K. (2003). Spinal administration of lipoxygenase inhibitors suppresses behavioural and neurochemical manifestations of naloxone-precipitated opioid withdrawal. *Br. J. Pharmacol.*, 140, 295-304.

[164] Hempstock, J., Kavanagh, J. P. & George, N. J. R. (1998). Growth inhibition of prostate cell lines *in vitro* by phyto-oestrogens. *Br. J. Urol.*, 82, 560-563.

[165] Huang, J. K., Chen, W. C., Huang, C. J., Hsu, S. S., Chen, J. S., Cheng, H. H., Chang, H. T., Jiann, B. P. & Jan, C. R. (2004). Nordihydroguaiaretic acid-induced Ca^{2+} handling and cytotoxicity in human prostate cancer cells. *Life Sci.*, 75, 2341-2351.

[166] Veluri, R., Singh, R. P., Liu, Z. J., Thompson, J. A., Agarwal, R. & Agarwal, C. (2006). Fractionation of grape seed extract and identification of gallic acid as one of the major active constituents causing growth inhibition and apoptotic death of DU145 human prostate carcinoma cells. *Carcinogenesis*, 27, 1445-1453.

[167] Agarwal, C., Veluri, R., Kaur, M., Chou, S. C., Thompson, J. A. & Agarwal, R. (2007). Fractionation of high molecular weight tannins in grape seed extract and identification of procyanidin B2-3,3'-di-O-gallate as a major active constituent causing growth inhibition and apoptotic death of DU145 human prostate carcinoma cells. *Carcinogenesis*, 28, 1478-1484.

[168] Agarwal, C., Tyagi, A. & Agarwal, R. (2006). Gallic acid causes inactivating phosphorylation of cdc25A/cdc25C-cdc2 via ATM-Chk2 activation, leading to cell cycle arrest, and induces apoptosis in human prostate carcinoma DU145 cells. *Mol. Cancer Ther.*, 5, 3294-3302.

[169] Han, H. Y., Wang, X. H., Wang, N. L., Ling, M. T., Wong, Y. C. & Yao, X. S. (2008). Lignans isolated from *Campylotropis hirtella* (Franch.) Schindl. decreased prostate specific antigen and androgen receptor expression in LNCaP cells. *J. Agric. Food Chem.*, 56, 6928-6935.

[170] Liu, S. & Yamauchi, H. (2006). Hinokitiol, a metal chelator from natural plants, suppresses cell growth and disrupts androgen receptor signaling in prostate carcinoma cell lines. *Biochem. Biophys. Res. Commun.*, 351, 26-32.

[171] Yun, J. M., Kweon, M. H., Kwon, H., Hwang, J. K. & Mukhtar, H. (2006). Induction of apoptosis and cell cycle arrest by a chalcone panduratin A isolated from *Kaempferia pandurata* in androgen-independent human prostate cancer cells PC3 and DU145. *Carcinogenesis*, 27, 1454-1464.

[172] Perabo, F. G. E., Von Low, E. C., Ellinger, J., Von Rucker, A., Muller, S. C. & Bastian, P. J. (2008). Soy isoflavone genistein in prevention and treatment of prostate cancer. *Prostate Cancer Prostatic Dis.*, 11, 6-12 and references cited therein.

[173] Xing, N. Z., Chen, Y., Mitchell, S. H. & Young, C. Y. F. (2001). Quercetin inhibits the expression and function of the androgen receptor in LNCaP prostate cancer cells. *Carcinogenesis*, 22, 409-414.

[174] Yuan, H. Q., Gong, A. Y. & Young, C. Y. F. (2005). Involvement of transcription factor Sp1 in quercetin-mediated inhibitory effect on the androgen receptor in human prostate cancer cells. *Carcinogenesis*, 26, 793-801.

[175] Lee, D. H., Szczepanski, M. & Lee, Y. J. (2008). Role of Bax in quercetin-induced apoptosis in human prostate cancer cells. *Biochem. Pharmacol.*, 75, 2345-2355.

[176] Vijayababu, M. R., Kanagaraj, P., Arunkumar, A., Ilangovan, R., Aruldhas, M. M. & Arunakaran, J. (2005). Quercetin-induced growth inhibition and cell death in prostatic carcinoma cells (PC-3) are associated with increase in p21 and hypophosphorylated retinoblastoma proteins expression. *J. Cancer Res. Clin. Oncol.*, 131, 765-771.

[177] Magee, P. J., Rasche, M., Steiner, C., Duffin, J. G., Pool-Zobel, B. L., Jokela, T., Wahala, K. & Rowland, I. R. (2006). Equol: a comparison of the effects of the racemic compound with that of the purified *S*-enantiomer on the growth, invasion, and DNA integrity of breast and prostate cells *in vitro*. *Nutr. Cancer*, 54, 232-242.

[178] Yu, L. Y., Blackburn, G. L. & Zhou, J. R. (2003). Genistein and daidzein downregulate prostate androgen-regulated transcript-1 (PART-1) gene expression induced by dihydrotestosterone in human prostate LNCaP cancer cells. *J. Nutr.*, 133, 389-392 and references cited therein.

[179] Pendleton, J. M., Tan, W. W., Anai, S., Chang, M., Hou, W., Shiverick, K. T. & Rosser, C. J. (2008). Phase II trial of isoflavone in prostate-specific antigen recurrent prostate cancer after previous local therapy. *BMC Cancer*, 8, doi:10.1186/1471-2407-8-132.

[180] Shukla, S. & Gupta, S. (2008). Apigenin-induced prostate cancer cell death is initiated by reactive oxygen species and p53 activation. *Free Rad. Biol. Med.*, 44, 1833-1845.

[181] Shukla, S. & Gupta, S. (2007). Apigenin-induced cell cycle arrest is mediated by modulation of MAPK, PI3K-Akt, and loss of cyclin D1 associated retinoblastoma dephosphorylation in human prostate cancer cells. *Cell Cycle*, 6, 1102-1114.

[182] Shukla, S. & Gupta, S. (2006). Molecular targets for apigenin-induced cell cycle arrest and apoptosis in prostate cancer cell xenograft. *Mol. Cancer Ther.*, 5, 843-852.

[183] Singh, R. P., Agrawal, P., Yim, D., Agarwal, C. & Agarwal, R. (2005). Acacetin inhibits cell growth and cell cycle progression, and induces apoptosis in human prostate cancer cells: structure-activity relationship with linarin and linarin acetate. *Carcinogenesis*, 26, 845-854.

[184] Gikas, E., Alesta, A., Economou, G., Karamanos, A. & Tsarbopoulos, A. (2008). Determination of isoflavones in the aerial part of red clover by HPLC-diode array detection. *J. Liq. Chromatogr. Relat. Technol.*, 31, 1181-1194.

[185] Sun, X. Y., Plouzek, C. A., Henry, J. P., Wang, T. T. Y. & Phang, J. M. (1998). Increased UDP-glucuronosyltransferase activity and decreased prostate specific antigen production by biochanin A in prostate cancer cells. *Cancer Res.*, 58, 2379-2384.

[186] Rice, L., Samedi, V. G., Medrano, T. A., Sweeney, C. A., Baker, H. V., Stenstrom, A., Furman, J. & Shiverick, K. T. (2002). Mechanisms of the growth inhibitory effects of the isoflavonoid biochanin A on LNCaP cells and xenografts. *Prostate*, 52, 201-212.

[187] Deep, G., Oberlies, N. H., Kroll, D. J. & Agarwal, R. (2008). Isosilybin B causes androgen receptor degradation in human prostate carcinoma cells via PI3K-Akt-Mdm2-mediated pathway. *Oncogene*, 27, 3986-3998.

[188] Davis-Searles, P. R., Nakanishi, Y., Kim, N. C., Graf, T. N., Oberlies, N. H., Wani, M. C., Wall, M. E., Agarwal, R. & Kroll, D. J. (2005). Milk thistle and prostate cancer: Differential effects of pure flavonolignans from *Silybum marianum* on anti proliferative end points in human prostate carcinoma cells. *Cancer Res.*, 65, 4448-4457.

[189] Deep, G., Oberlies, N. H., Kroll, D. J. & Agarwal, R. (2008). Identifying the differential effects of silymarin constituents on cell growth and cell cycle regulatory molecules in human prostate cancer cells. *Int. J. Cancer*, 123, 41-50.

[190] Deep, G., Oberlies, N. H., Kroll, D. J. & Agarwal, R. (2007). Isosilybin B and isosilybin A inhibit growth, induce G(1) arrest and cause apoptosis in human prostate cancer LNCaP and 22Rv1 cells. *Carcinogenesis*, 28, 1533-1542.

[191] Delmulle, L., Berghe, T. V., De Keukeleire, D. & Vandenabeele, P. (2008). Treatment of PC-3 and DU145 prostate cancer cells by prenylflavonoids from hop (*Humulus lupulus* L.) induces a caspase-independent form of cell death. *Phytother. Res.*, 22, 197-203.

[192] Colgate, E. C., Miranda, C. L., Stevens, J. F., Bray, T. M. & Ho, E. (2007). Xanthohumol, a prenylflavonoid derived from hops induces apoptosis and inhibits NF-kB activation in prostate epithelial cells. *Cancer Lett.*, 246, 201-209.

[193] Bonham, M., Posakony, J., Coleman, I., Montgomery, B., Simon, J. & Nelson, P. S. (2005). Characterization of chemical constituents in *Scutellaria baicalensis* with antiandrogenic and growth-inhibitory activities toward prostate carcinoma. *Clin. Cancer Res.*, 11, 3905-3914.

[194] Chen, S., Ruan, Q., Bedner, E., Deptala, A., Wang, X., Hsieh, T. C., Traganos, F. & Darzynkiewicz, Z. (2001). Effects of the flavonoid baicalin and its metabolite baicalein on androgen receptor expression, cell cycle progression and apoptosis of prostate cancer cell lines. *Cell Prolif.*, 34, 293-304.

[195] Pidgeon, G. P., Kandouz, M., Meram, A. & Honn, K. V. (2002). Mechanisms controlling cell cycle arrest and induction of apoptosis after 12-lipoxygenase inhibition in prostate cancer cells. *Cancer Res.*, 62, 2721-2727.

[196] Lee, D. H., Kim, C., Zhang, L. & Lee, Y. J. (2008). Role of p53, PUMA, and Bax in wogonin-induced apoptosis in human cancer cells. *Biochem. Pharmacol.*, 75, 2020-2033.

[197] Fu, Y., Hsieh, T. C., Guo, J. Q., Kunicki, J., Lee, M., Darzynkiewicz, Z. & Wu, J. M. (2004). Licochalcone-A, a novel flavonoid isolated from licorice root (*Glycyrrhiza glabra*), causes G2 and late-G1 arrests in androgen-independent PC-3 prostate cancer cells. *Biochem. Biophys. Res. Commun.*, 322, 263-270.

[198] Liu, W. K., Cheung, F. W. K., Liu, B. P. L., Li, C., Ye, W. & Cho, C. T. (2008). Involvement of p21 and FasL in induction of cell cycle arrest and apoptosis by neochamaejasmin A in human prostate LNCaP cancer cells. *J. Nat. Prod.*, 71, 842-846.

[199] Chung, B. H., Lee, H. Y., Lee, J. S. & Young, C. Y. F. (2006). Perillyl alcohol inhibits the expression and function of the androgen receptor in human prostate cancer cells. *Cancer Lett.*, 236, 222-228.

[200] Liu, G., Oettel, K., Bailey, H., Van Ummersen, L., Tutsch, K., Staab, M. J., Horvath, D., Alberti, D., Arzoomanian, R., Rezazadeh, H., McGovern, J., Robinson, E., DeMets, D. & Wilding, G. (2003). Phase II trial of perillyl alcohol (NSC 641066) administered daily in patients with metastatic androgen independent prostate cancer. *Invest. New Drugs*, 21, 367-372.

[201] Goodman, A. G. (1996). The pharmacological basis of therapeutics *(ninth edition)*. New York-St. Louis-San Francisco-Auckland-Bogota-Caracas-Lisbon-Madrid-Mexico City-Milan-Montreal-New Delhi-San Juan-Singapore-Sydney-Tokyo-Toronto: McGraw-Hill.

[202] Kupchan, S. M., Coutr, W. A., Dailey, R. G., Gilmore, C. J. & Bryan, R. F. (1972). Triptolide and tripdiolide, novel antileukemic diterpenoid triepoxides from *Tripterygium wilfordii*. *J. Am. Chem. Soc.*, 94, 7194-7195.

[203] Girinsky, T., C, K., Graeber, T. G., Peehl, D. M. & Giaccia, A. J. (1995). Attenuated Response of p53 and p21 in Primary Cultures of Human Prostatic Epithelial Cells Exposed to DNA-damaging Agents. *Cancer Res.*, 55, 3726-3731.

[204] Shamona, L. A., Pezzuto, J. M., Graves, J. M., Metha, R. R., Wangcharoentrakul, S., Sangsuwand, R., Chaichanad, S., Tuchindad, T., Cleasond, P. & Reutrakuld, V. (1997). Evaluation of the mutagenic, cytotoxic, and antitumor potential of triptolide, a highly oxygenated diterpene isolated from *Tripterygium wilfovdii*. *Cancer Lett.*, 112, 113-117.

[205] Kiviharju, T. M., Lecane, P. S., Sellers, R. G. & Peehl, D. M. (2002). Antiproliferative and proapoptotic activities of triptolide (PG490), a natural product entering clinical trials, on primary cultures of human prostatic epithelial cells. *Clin. Cancer Res.*, 8, 2666-2674.

[206] Fidler, J. M., Li, K., Chung, C., Wei, K., Ross, J. A., Gao, M. & Rosen, G. D. (2003). PG490-88, a derivative of triptolide, causes tumor regression and sensitizes tumors to chemotherapy. *Mol. Cancer Ther.*, 2, 855-862.

[207] Liu, W. K., Ho, J. C. K., Qin, G. W. & Che, C. T. (2002). Jolkinolide B induces neuroendocrine differentiation of human prostate LNCaP cancer cell line. *Biochem. Pharmacol.*, 63, 951-957.

[208] Huang, D. M., Shen, Y. C., Wu, C., Huang, Y. T., Kung, F. L., Teng, C. M. & Guh, J. H. (2004). Investigation of extrinsic and intrinsic apoptosis pathways of new clerodane diterpenoids in human prostate cancer PC-3 cells. *Eur. J. Pharmacol.*, 503, 17-24.

[209] Henry, G. E., Adams, L. S., Rosales, J. C., Jacobs, H., Heber, D. & Seeram, N. P. (2006). Kaurene diterpenes from *Laetia thamnia* inhibit the growth of human cancer cells *in vitro*. *Cancer Lett.*, 244, 190-194.

[210] Thastrup, O., Cullen, P. J., Drobak, B. K., Hanley, M. R. & Dawson, A. P. (1990). Thapsigargin, a tumor promoter, discharges intracellular Ca^{2+} stores by specific inhibition of the endoplasmic reticulum Ca^{2+}-ATPase. *Proc. Natl. Acad. Sci. USA*, 87, 2466-2470.

[211] Furuya, Y. & Isaacs, J. T. (1994). Proliferation-dependent vs. independent programmed cell death of prostatic cancer cells involves distinct gene regulation. *Prostate*, 25, 301-309.

[212] Furuya, Y., Lundmo, P., Short, A. D., Gill, D. L. & Isaacs, J. T. (1994). The role of calcium, pH and cell proliferation in the programmed (Apoptotic) death of androgen-independent prostatic cancer cells induced by thapsigargin. *Cancer Res.*, 54, 6167-6175.

[213] Shanmugam, R., Jayaprakasan, J., Gokmen-Polar, Y., Kelich, S., Miller, K. D., Yip-Schneider, M., Cheng, L., Bhat-Nakshatri, P., Sledge, G. W. J., Nakshatri, H., Zheng, Q. H., Miller, M. A., DeGrado, T., Hutchins, G. D. & Sweeney, C. J. (2006). Restoring chemotherapy and hormone therapy sensitivity by parthenolide in a xenograft hormone refractory prostate cancer model. *Prostate*, 66, 1498-1511.

[214] Sweeney, C., Li, L., Shanmugam, R., Bhat-Natshtri, P., Jayaprakasan, V., Baldridge, L. A., Gardner, T., Smith, M., Nakshatri, H. & Cheng, L. (2004). Nucler factor-kB is constitutively activated in prostate cancer *in vitro* and is overexpressed in prostatic intraepithelial neoplasia and adenocarcinoma of the prostate. *Clin. Cancer Res.*, 10, 5501-5507.

[215] Chintharlapalli, S., Papineni, S., Ramaiah, S. K. & Safe, S. (2007). Betulinic acid inhibits prostate cancer growth through inhibition of specificity protein transcription factors. *Cancer Res.*, 67, 2816-2823.

[216] Yao, J. C., Wang, L., Wei, D., Gong, W. R., Hassan, M., Wu, T. T., Mansfield, P. & Xie, K. (2004). Association between expression of transcription factor Sp1 and increased vascular endothelial growth factor expression, advanced stage, and poor survival in patients with resected gastric cancer. *Clin. Cancer Res.*, 10, 4109-4117.

[217] Wang, L., Wei, D., Huang, S., Peng, Z., Le, X., Wu, T. T., Yao, J., Ajani, J. & Xie, K. (2003). Transcription factor Sp1 expression is a significant predictor of survival in human gastric cancer. *Clin. Cancer Res.*, 9, 6371-6380.

[218] Rabi, T., Shukla, S. & Gupta, S. (2008). Betulinic acid suppresses constitutive and TNFα-induced NF-kB activation and induces apoptosis in human prostate carcinoma PC-3 cells. *Mol. Carcinogen.*

[219] Pisha, E., Chan, H., Lee, I. S., Chagwedera, T. E., Farnsworth, N. R., Cordell, G. A., Beecher, C. W. W., Fong, H. H. S., Kinghorn, A. D., Brown, D. M., Wani, M. C., Wall, M. E., Hieken, T. J., Gupta, T. K. & Pezzuto, J. M. (1995). Discovery of betulinic acid as a selective inhibitor of human melanoma that functions by induction of apoptosis. *Nat. Med.*, 1, 1046-1051.

[220] Ganguly, A., Das, B., Roy, A., Sen, N., Dasgupta, S. B., Mukhopadhayay, S. & Majumder, H. K. (2007). Betulinic acid, a catalytic inhibitor of topoisomerase I, inhibits reactive oxygen species-mediated apoptotic topoisomerase I-DNA cleavable complex formation in prostate cancer cells but does not affect the process of cell death. *Cancer Res.*, 67, 11848-58.

[221] Arai, M. A., Tateno, C., Hosoya, T., Kouyano, T., Kowithayakorn, T. & Ishibashi, M. (2008). Hedgehog/GLI-mediated transcriptional inhibitors from *Zizyphus cambodiana*. *Bioorg. Med. Chem.*, doi:10.1016/j.bmc.2008.09.053.

[222] Neto, C. C. (2007). Cranberry and its phytochemicals: A review of *in vitro* anticancer studies. *J. Nutr.*, 137, 186S-193S.

[223] Kassi, E., Papoutsi, Z., Pratsinis, H., Aligiannis, N., Manoussakis, M. & Moutsatsou, P. (2007). Ursolic acid, a naturally occurring triterpenoid, demonstrates anticancer activity on human prostate cancer cells. *J. Cancer Res. Clin. Oncol.*, 133, 493-500.

[224] Choi, Y. H., Baek, J. H., Yoo, M. A., Chung, H. Y., Kim, N. D. & Kim, K. W. (2000). Induction of apoptosis by ursolic acid through activation of caspases and down-regulation of c-IAPs in human prostate epithelial cells. *Int. J. Oncol.*, 17, 565-571.

[225] Syrovets, T., Gschwend, J. E., Buchele, B., Laumonnier, Y., Zugmaier, W., Genze, F. & Simmet, T. (2005). Inhibition of IkB kinase activity by acetyl-boswellic acids promotes apoptosis in androgen-independent PC-3 prostate cancer cells *in vitro* and *in vivo*. *J. Biol. Chem.*, 280, 6170-6180.

[226] Yuan, H. Q., Feng, K., Wang, X. L., Young, C. Y. F., Hu, X. Y. & Lou, H. X. (2008). Inhibitory effect of acetyl-11-keto-β-boswellic acid on androgen receptor by interference of Sp1 binding activity in prostate cancer cells. *Biochem. Pharmacol.*, 75, 2112-2121.

[227] Filho, W. B., Corsino, J., Bolzani, V. S., Furlan, M., Pereira, A. M. S. & França, S. C. (2002). Quantitative determination of cytotoxic friedo-*nor*-oleanane derivatives from five morphological types of *Maytenus ilicifolia* (Celastraceae) by reverse-phase high performance liquid chromatography. *Phytochemical Anal.*, 13, 75-78.

[228] Yang, H. J., Landis-Piwowar, K. R., Lu, D., Yuan, P., Li, L. H., Reddy, G. P. V., Yuan, X. & Dou, Q. P. (2008). Pristimerin induces apoptosis by targeting the proteasome in prostate cancer cells. *J. Cell. Biochem.*, 103, 234-244.

[229] Yang, H., Chen, D., Cui, D. C., Yuan, X. & Dou, Q. P. (2006). Celastrol, a Triterpene Extracted from the Chinese ''Thunder of God Vine,'' Is a Potent Proteasome Inhibitor and Suppresses Human Prostate Cancer Growth in Nude Mice. *Cancer Res.*, 66, 4758-4765.

[230] Wang, W., Wang, H., Rayburn, E. R., Zhao, Y., Hill, D. L. & Zhang, R. (2008). 20(S)-25-methoxyl-dammarane-3β,2β,20-triol, a novel natural product for prostate cancer therapy: activity *in vitro* and *in vivo* and mechanisms of action. *Br. J. Cancer*, 98, 792-802 and references cited therein.

[231] Wang, W., Rayburn, E. R., Hao, M., Zhao, Y., Hill, D. L., Zhang, R. W. & Wang, H. (2008). Experimental therapy of prostate cancer with novel natural product anti-cancer ginsenosides. *Prostate*, 68, 809-819.

[232] Zhao, Y., Wang, W., Han, L., Rayburn, E. R., Hill, D. L., Wang, H. & Zhang, R. (2007). Isolation, structural determination, and evaluation of the biological activity of 20(*S*)-25-methoxyl-dammarane-3β,12β,20-triol [20(S)-25-OCH_3-PPD], a novel natural product from *Panax notoginseng*. *Med. Chem.*, 3, 51-60.

[233] Huang, Y. T., Huang, D. M., Chueh, S. C., Teng, C. M. & Guh, J. H. (2006). Alisol B acetate, a triterpene from *Alismatis rhizoma*, induces Bax nuclear translocation and apoptosis in human hormone-resistant prostate cancer PC-3 cells. *Cancer Lett.*, 231, 270-278.

[234] Saleem, M., Kweon, M. H., Yun, J. M., Adhami, V. M., Khan, N., Syed, D. N. & Mukhtar, H. (1995). A novel dietary triterpene lupeol induces Fas-mediated apoptotic death of androgen-sensitive prostate cancer cells and inhibits tumor growth in a xenograft model. *Cancer Res.*, 65, 11203-11213 and references cited therein.

[235] Prasad, S., Kalra, N. & Shukla, Y. (2008). Induction of Apoptosis by Lupeol and Mango Extract in Mouse Prostate and LNCaP Cells. *Nutr. Cancer*, 60, 120-130.

[236] Prasad, S., Nigam, N., Kalra, N. & Shukla, Y. (2008). Regulation of signaling pathways involved in lupeol induced inhibition of proliferation and induction of apoptosis in human prostate cancer cells. *Mol. Carcinogen.*, doi: 10.1002/mc.20442.

[237] Shen, T., Wan, W. Z., Yuan, H. Q., Kong, F., Guo, H. F., Fan, P. H. & Lou, H. X. (2007). Secondary metabolites from *Commiphora opobalsamum* and their antiproliferative effect on human prostate cancer cells. *Phytochemistry*, 68, 1331-1337.

[238] van Breemen, R. B. & Pajkovic, N. (2008). Multitargeted therapy of cancer by lycopene. *Cancer Lett.*, 269, 339–351 and references cited therein.

[239] Shen, T. & Lou, H. X. (2008). Bioactive constituents of myrrh and frankincense, two simultaneously prescribed gum resins in Chinese traditional medicine. *Chem. Biodiver.*, 5, 540-553.

[240] Singh, S. V., Choi, S., Zeng, Y., Hahm, E. R. & Xiao, D. (2007). Guggulsterone-induced apoptosis in human prostate cancer cells is caused by reactive oxygen intermediate-dependent activation of c-Jun NH_2-terminal kinase. *Cancer Res.*, 67, 7439-7449.

[241] Singh, S. V., Zeng, Y., Xiao, D., Vogel, V. G., Nelson, J. B., Dhir, R. & Tripathi, Y. B. (2005). Caspase-dependent apoptosis induction by guggulsterone, a constituent of Ayurvedic medicinal plant *Commiphora mukul*, in PC-3 human prostate cancer cells is mediated by Bax and Bak. *Mol. Cancer Ther.*, 4, 1747-1754.

[242] Xiao, D. & Singh, S. V. (2008). z-guggulsterone, a constituent of Ayurvedic medicinal plant Commiphora mukul, inhibits angiogenesis in vitro and in vivo. *Mol. Cancer Ther.*, 7, 171-180.

[243] Yeh, J. Y., Huang, W. J., Kan, S. F. & Wang, P. S. (2001). Inhibitory effects of digitalis on the proliferation of androgen dependent and independent prostate cancer cells. *J. Urol.*, 166, 1937-1942.

[244] Smith, J. A., Madden, T., Vijjeswarapu, M. & Newman, R. A. (2001). Inhibition of export of fibroblast growth factor-2 (FGF-2) from the prostate cancer cell lines PC3 and DU145 by Anvirzel and its cardiac glycoside component, oleandrin. *Biochem. Pharmacol.*, 62, 469-472.

[245] McConkey, D. J., Lin, Y., Nutt, L. K., Ozel, H. Z. & Newman, R. A. (2000). Cardiac glycosides stimulate Ca^{2+} increases and apoptosis in androgen-independent, metastatic human prostate adenocarcinoma cells. *Cancer Res.*, 60, 3807-3812.

[246] Wu, Y. D. & Lou, Y. J. (2007). Brassinolide, a plant sterol from pollen of *Brassica napus* L., induces apoptosis in human prostate cancer PC-3 cells. *Pharmazie*, 62, 392-395.

[247] Glotter, E., Waitman, R. & Lavie, D. (1966). Constituents of *Withania Somnifera Dun* . 8. A New Steroidal Lactone 27-Deoxy-14-Hydroxy-Withaferin A. *J. Chem. Soc. C*, 1765-1767.

[248] Yang, H. J., Shi, G. Q. & Dou, Q. P. (2007). The tumor proteasome is a primary target for the natural anticancer compound withaferin A isolated from "Indian Winter Cherry". *Mol. Pharmacol.*, 71, 426-437.

[249] El-Guendy, N., Zhao, Y., Gurumurthy, S., Burikhanov, R. & Rangnekar, V. M. (2003). Identification of a unique core domain of Par-4 sufficient for selective apoptosis induction in cancer cells. *Mol. Cell. Biol.*, 23, 5516-5525.

[250] Srinivasan, S., Ranga, R. S., Burikhanov, R., Han, S. S., Chendil, D. (2007). Par-4-dependent apoptosis by the dietary compound withaferin A in prostate cancer cells. *Cancer Res.*, 67, 246-253.

[251] Lin, H. M., Tsai, S. C., Chen, J. J., Chiao, Y. C., Wang, S. W., Wang, G. J., Chen, C. F. & Wang, P. S. (1999). Effects of evodiamine on the secretion of testosterone in rat testicular interstitial cells. *Metabolism*, 48, 1532-1535.

[252] Kan, S. F., Huang, W. J., Lin, L. C. & Wang, P. S. (2004). Inhibitory effects of evodiamine on the growth of human prostate cancer cell line LNCaP. *Int. J. Cancer*, 110, 641-651.

[253] Kan, S. F., Yu, C. H., Pu, H. F., Hsu, J. M., Chen, M. J. & Wang, P. S. (2007). Anti-proliferative effects of evodiamine on human prostate cancer cell lines DU145 and PC3. *J. Cell Biochem.*, 101, 44-56.

[254] Sinha, S., Pal, B. C., Jagadeesh, S., Banerjee, P. P., Bandyopadhaya, A. & Bhattacharya, S. (2006). Mahanine inhibits growth and induces apoptosis in prostate cancer cells through the deactivation of Akt and activation of caspases. *Prostate*, 66, 1257-1265.

[255] Jagadeesh, S., Sinha, S., Pal, B. C., Bhattacharya, S. & Banerjee, P. P. (2007). Mahanine reverses an epigenetically silenced tumor suppressor gene RASSF1A in human prostate cancer cells. *Biochem. Biophys. Res. Commun.*, 362, 212-217.

[256] Mantena, S. K., Sharma, S. D. & Katiyar, S. K. (2006). Berberine, a natural product, induces G(1)-phase cell cycle arrest and caspase-3-dependent apoptosis in human prostate carcinoma cells. *Mol. Cancer Ther.*, 5, 296-308.

[257] Zhang, J., Hsu, J. C., Kinseth, M. A., Bjeldanes, L. F. & Firestone, G. L. (2003). Indole-3-carbinol induces a G1 cell cycle arrest and inhibits prostate-specific antigen production in human LNCaP prostate carcinoma cells. *Cancer*, 98, 2511-2520.

[258] Nachshon-Kedmi, M., Yannai, S., Haj, A. & Fares, F. A. (2003). Indole-3-carbinol and 3,3'-diindolylmethane induce apoptosis in human prostate cancer cells. *Food Chem. Toxicol.*, 41, 745-752.

[259] Fan, S., Meng, Q., Auborn, K., Carter, T. & Rosen, E. M. (2006). BRCA1 and BRCA2 as molecular targets for phytochemicals indole-3-carbinol and genistein in breast and prostate cancer cells. *Br. J. Cancer*, 94, 407-426.

[260] Hsu, J. C., Zhang, J., Dev, A., Wing, A., Bjeldanes, L. F. & Firestone, G. L. (2005). Indole-3-carbinol inhibition of androgen receptor expression and downregulation of androgen responsiveness in human prostate cancer cells. *Carcinogenesis*, 26, 1896-1904.

[261] Nachshon-Kedmi, M., Yannai, S. & Fares, F. A. (2004). Induction of apoptosis in human prostate cancer cell line, PC3, by 3,3'-diindolylmethane through the mitochondrial pathway. *Br. J. Cancer*, 91, 1358-1363.

[262] Mori, A., Lehmann, S., O'Kelly, J., Kumagai, T., Desmond, J. C., Pervan, M., McBride, W. H., Kizaki, M. & Koeffler, H. P. (2006). Capsaicin, a component of red peppers, inhibits the growth of androgen-independent, p53 mutant prostate cancer cells. *Cancer Res.*, 66, 3222-3229.

[263] Sanchez, A. M., Sanchez, M. G., Malagarie-Cazenave, S., Olea, N. & Diaz-Laviada, I. (2006). Induction of apoptosis in prostate tumor PC-3 cells and inhibition of xenograft prostate tumor growth by the vanilloid capsaicin. *Apoptosis*, 11, 89-99.

[264] Sanchez, A. M., Malagarie-Cazenave, S., Olea, N., Vara, D., Chiloeches, A. & Diaz-Laviada, I. (2007). Apoptosis induced by capsaicin in prostate PC-3 cells involves ceramide accumulation, neutral sphingomyelinase, and JNK activation. *Apoptosis*, 12, 2013-2024.

[265] Sanchez, A. M., Martinez-Botas, J., Malagarie-Cazenave, S., Olea, N., Vara, D., Lasuncion, M. A. & Diaz-Laviada, I. (2008). Induction of the endoplasmic reticulum stress protein GADD153/CHOP by capsaicin in prostate PC-3 cells: A microarray study. *Biochem. Biophys. Res. Commun.*, 372, 785-791.

[266] Tan, C., Cai, L. Q., Wu, W., Qiao, Y., Imperato-McGinley, J., Chen, G. Q. & Zhu, Y. S. (2008). NSC606985, a novel camptothecin analog, induces apoptosis and growth arrest in prostate tumor cells. *Cancer Chemother. Pharmacol.*, doi:10.1007/s00280-008-0740-8 and references cited therein.

[267] Arah, I. N., Song, K. M., Seth, P., Cowan, K. H. & Sinha, B. K. (1998). Role of wild-type p53 in the enhancement of camptothecin cytotoxicity against human prostate tumor cells. *Anticancer Res.*, 18, 1845-1849.

[268] Costa-Pereira, A. P. & Cotter, T. G. (1999). Camptothecin sensitizes androgen-independent prostate cancer cells to anti Fas-induced apoptosis. *Br. J. Cancer*, 80, 371-378.

[269] Goto, T., Matsushima, H., Kasuya, Y., Hosaka, Y., Kitamura, T., Kawabe, K., Hida, A., Ohta, Y., Simizu, T. & Takeda, K. (1999). The effect of papaverine on morphological differentiation, proliferation and invasive potential of human prostatic cancer LNCaP cells. *Int. J. Urol.*, 6, 314-319.

[270] Cha, T. L., Qiu, L., Chen, C. T., Wen, Y. & Hung, M. C. (2005). Emodin down-regulates androgen receptor and inhibits prostate cancer cell growth. *Cancer Res.*, 65, 2287-2295.

[271] Li, C. J., Averboukh, L. & Pardee, A. B. (1993). β-Lapachone, a novel DNA topoisomerase I inhibitor with a mode of action different from campothecin. *J. Biol. Chem.*, 268, 22463-22468.

[272] Li, C. J., Wang, C. L. & Pardee, A. B. (1995). Induction of Apoptosis by β-Lapachone in Human Prostate-Cancer Cells. *Cancer Res.*, 55, 3712-3715.

[273] Planchon, S. M., Wuerzberger, S., Frydman, B., Witiak, D. T., Hutson, P., Church, D. R., Wilding, G. & Boothman, D. A. (1995). β-Lapachone-Mediated Apoptosis in Human Promyelocytic Leukemia (Hl-60) and Human Prostate Cancer Cells - a P53-Independent Response. *Cancer Res.*, 55, 3706-3711.

[274] Choi, Y. H., Kang, H. S. & Yoo, M. A. (2003). Suppression of human prostate cancer cell growth by β-lapachone via down-regulation of pRB phosphorylation and induction of Cdk inhibitor p21(WAF1/CIP1). *J. Biochem. Mol. Biol.*, 36, 223-229.

[275] Don, M. J., Chang, Y. H., Chen, K. K., Ho, L. K. & Chau, Y. P. (2001). Induction of CDK inhibitors (p21WAF1 and p27Kip) and Bak in the β-lapachone-induced apoptosis of human prostate cancer cells. *Mol. Pharmacol.*, 59, 784-794.

[276] Lee, J. H., Cheong, J. H., Park, Y. M. & Choi, Y. H. (2005). Down-regulation of cyclooxygenase-2 and telomerase activity by β-lapachone in human prostate carcinoma cells. *Pharmacol. Res.*, 51, 553-560.

[277] Planchon, S. M., Pink, J. J., Tagliarino, C., Bornmann, W. G., Varnes, M. E., Boothman, D. A. (2001). β-Lapachone-induced apoptosis in human prostate cancer cells: Involvement of NQO1/xip3. *Exp. Cell Res.*, 267, 95-106.

[278] Kumi-Diaka, J. (2002). Chemo sensitivity of human prostate cancer cells PC3 and LNCaP to genistein isoflavone and β-lapachone. *Biol. Cell*, 94, 37-44.

[279] Ghosheh, O. A., Houdi, A. A. & Crooks, P. A. (1999). High performance liquid chromatography analysis of the pharmacologically active quinones and related compounds in the oil of the black seed (*Nigella sativa L.*). *J. Pharm. Biomed Anal.*, 19, 757-762.

[280] Kaseb, A. O., Chinnakannu, K., Chen, D., Sivanandam, A., Tejwani, S., Menon, M., Dou, Q. P. & Reddy, G. P. V. (2007). Androgen receptor- and E2F-1-targeted thymoquinone therapy for hormone-refractory prostate cancer. *Cancer Res.*, 67, 7782-7788.

[281] Juge, N., Mithen, R. F. & Traka, M. (2007). Molecular basis for chemoprevention by sulforaphane: a comprehensive review. *Cell. Mol. Life Sci.*, 64, 1105-1127 and references cited therein.

[282] Xiao, D. & Singh, S. V. (2002). Phenethyl isothiocyanate-induced apoptosis in p53-deficient PC-3 human prostate cancer cell line is mediated by extracellular signal-regulated kinases. *Cancer Res.*, 62, 3615-3619.

[283] Xu, C. J., Shen, G. X., Chen, C., Gelinas, C. & Kong, A. N. T. (2005). Suppression of NF-kB and NF-kB-regulated gene expression by sulforaphane and PEITC through IkBα, IKK pathway in human prostate cancer PC-3 cells. *Oncogene*, 24, 4486-4495.

[284] Hsing, A. W., Chokkalingam, A. P., Gao, Y. T., Madigan, P., Deng, J., Gridley, G. & Fraumeni, J. F. (2002). *Allium* vegetables and risk of prostate cancer: a population-based study. *J. Natl. Cancer Inst.*, 94, 1648-1651 and references cited therein.

[285] Arunkumar, A., Vijayababu, M. R., Kanagaraj, P., Balasubramanian, K., Aruldhas, M. M. & Arunakaran, J. (2005). Growth suppressing effect of garlic compound diallyl disulfide on prostate cancer cell line (PC-3) *in vitro*. *Biol. Pharm. Bull.*, 28, 740-743.

[286] Xiao, D., Choi, S., Johnson, D. E., Vogel, V. G., Johnson, C. S., Trump, D. L., Lee, Y. J. & Singh, S. V. (2004). Diallyl trisulfide-induced apoptosis in human prostate cancer cells involves c-Jun N-terminal kinase and extracellular-signal regulated kinase-mediated phosphorylation of Bcl-2. *Oncogene*, 23, 5594-5606.

[287] Kim, S., Salim, A. A., Swanson, S. M. & Kinghorn, A. D. (2006). Potential of cyclopenta[b]benzofurans from *Aglaia* species in cancer chemotherapy. *Anti Cancer Agents Med. Chem.*, 6, 319-345 and references cited therein.

[288] Kim, S., Hwang, B. Y., Su, B. N., Cha, H. B., Mi, Q. W., Kinghorn, A. D., Wild, R. & Swanson, S. M. (2007). Silvestrol, a potential anticancer rocaglate derivative from *Aglaia foveolata*, induces apoptosis in LNCaP cells through the mitochondrial/apoptosome pathway without activation of executioner caspase-3 or -7. *Anticancer Res.*, 27, 2175-2183.

[289] Mabjeesh, N. J., Willard, M. T., Harris, W. B., Sun, H. Y., Wang, R. X., Zhong, H., Umbreit, J. N. & Simons, J. W. (2003). Dibenzoylmethane, a natural dietary compound, induces HIF-1α and increases expression of VEGF. *Biochem. Biophys. Res. Commun.*, 303, 279-286.

[290] Jackson, K. M., Frazier, M. C. & Harris, W. B. (2007). Suppression of androgen receptor expression by dibenzoylmethane as a therapeutic objective in advanced prostate cancer. *Anticancer Res.*, 27, 1483-1488.

[291] Ezekwudo, D. E., Wang, R. C. & Elegbede, J. A. (2007). Methyl jasmonate induced apoptosis in human prostate carcinoma cells via 5-lipoxygenazse dependent pathway. *J. Exp. Ther. Oncol.*, 6, 267-277.

[292] Ezekwudo, D. E., Shashidharamurthy, R., Devineni, D., Bozeman, E., Palaniappan, R. & Selvaraj, P. (2008). Inhibition of expression of anti-apoptotic Bcl-2 and induction of cell death in radioresistant human prostate adenocarcinoma cell line (PC-3) by methyl jasmonate. *Cancer Lett.*, 270, 277-285.

[293] Xu, Y. M., McLaughlin, S. P. & Gunatilaka, A. A. L. (2007). Sorbifolivaltrates A-D, diene valepotriates from *Valeriana sorbifolia*. *J. Nat. Prod.*, 70, 2045-2048.

[294] Guo, J. M., Jiang, C., Wang, Z., Lee, H. J., Hu, H. B., Malewicz, B., Lee, H. J., Lee, J. H., Baek, N. I., Jeong, J. H., Kim, D. K., Kang, K. S., Kim, S. H. & Lu, J. X. (2007). A novel class of pyranocoumarin anti-androgen receptor signaling compounds. *Mol. Cancer Ther.*, 6, 907-917 and references cited therein.

[295] Poindessous, V., Koeppel, F., Raymond, E., Comisso, M., Waters, S. J. & Larsen, A. K. (2003). Marked activity of irofulven toward human carcinoma cells: Comparison with cisplatin and ecteinascidin. *Clin. Cancer Res.*, 9, 2817-2825.

[296] Yeh, J. Y., Huang, W. J., Kan, S. F. & Wang, P. S. (2003). Effects of bufalin and cinobufagin on the proliferation of androgen dependent and independent prostate cancer cells. *Prostate*, 54, 112-124.

[297] McEleny, K., Coffey, R., Morrissey, C., Fitzpatrick, J. M. & Watson, R. W. G. (2004). Caffeic acid phenethyl ester-induced PC-3 cell apoptosis is caspase-dependent and mediated through the loss of inhibitors of apoptosis proteins. *BJU Int.*, 94, 402-406 and references cited therein.

[298] Natarajan, K., Singh, S., Burke, E. R., Grunberger, D. & Aggarwal, B. B. (1996). Caffeic acid phenethyl ester is a potent and specific inhibitor of activation of nuclear transcription factor NF-kB. *Proc. Natl. Acad. Sci. USA*, 93, 9090-9095.

[299] Pienta, K. J. & Lehr, J. E. (1993). Inhibition of Prostate-Cancer Growth by Estramustine and Etoposide - Evidence for Interaction at the Nuclear Matrix. *J. Urol.*, 149, 1622-1625.

[300] Ahn, J. (2008). Vitamin D concentration does not affect the risk of prostate cancer. *Nat. Clin. Pract. Oncol.*, doi:10.1038/ncpuro1159.

In: Handbook of Prostate Cancer Cell Research...
Editor: Alan T. Meridith
ISBN: 978-1-60741-954-9

Chapter 2

Molecular Imaging in Prostate Cancer: [11C]Choline PET/CT

M. Picchio[1], G. Giovacchini[2], R. Garcia Parra[2], C. Crivellaro[2], C. Landoni[1,2], L. Gianolli[1] and C. Messa[2,3,4]

[1] Department of Nuclear Medicine, Scientific Institute San Raffaele, Milan, Italy
[2] Molecular Bioimaging Center, University of Milano-Bicocca, Milan, Italy
[3] IBFM, National Research Council, Milan, Italy
[4] Department of Nuclear Medicine, San Gerardo Hospital, Monza, Italy

Abstract

In recent years, positron emission tomography/computed tomography (PET/CT) with [11C]Choline has been increasingly used for the evaluation of prostate cancer. In vitro and in vivo magnetic resonance spectroscopy studies have shown increased choline compounds and over-expression of choline kinase in prostate cancer. Liver and kidney are the major sites for choline oxidation, while negligible activity is present in the bladder, representing a significant advantage for prostate imaging.

In patients with prostate cancer, [11C]Choline has limited sensitivity and specificity for distinguishing malignant vs. benign prostate cells. There is no significant association between prostate [11C]Choline uptake and PSA levels, Gleason score and pathological stage, while anti-androgenic therapy interferes with the uptake of the tracer. The reduction of prostate [11C]Choline uptake following anti-androgenic hormonal therapy indicates that PET/CT with [11C]Choline could be used to monitor the response to treatment. The accuracy of the technique for lymph node staging has to be further investigated.

There is good agreement among several centres on the utility of PET/CT with [11C]Choline for the detection of recurrence of disease after radical prostatectomy in patients with biochemical failure. The percent of positive scans increased with increasing PSA values, being this technique clinically valuable for restaging prostate cancer patients even with low PSA levels. Clinical guidelines to proper indicate a PET/CT with [11C]Choline study in prostatectomized patients will be defined.

Introduction

Positron emission tomography (PET) is a diagnostic examination involving the development of functional images based on the detection of subatomic particles. PET shows molecular function and activity, physiological data not available with other techniques, and can often detect abnormalities before structural changes have occurred. Computed tomography (CT) and magnetic resonance imaging (MRI) provide morphological and structural information only. This is the primary difference between the CT and MRI scans, as compared to PET that provides molecular imaging of biological processes [1].

PET is an analytical imaging technology developed to use compounds labeled with positron emitting radioisotopes as molecular probes to image and measure biochemical processes of mammalian biology in vivo [2].

There are many radioactive isotopes that can be produced in the cyclotron. In order to use them in a PET apparatus for imaging the body, they must be capable of emitting positrons when they decay radioactively, i.e, when they transform from an unstable isotope into another, stable one. The positron is the anti-particle of an electron. The isotopes used in PET imaging generally have a short period of unstability, i.e, it will take a few minutes for any given atom of this isotope to emit a positron and become a stable element. The average measure of this instability is called half-life, which is the time it takes to decrease by half the number of radioactive isotopes of that element. Bombarding target material with protons that have been accelerated in a cyclotron produces positron-emitting radionuclides, which are then used to synthesize radiopharmaceuticals that are part of biochemical pathways in the human body. The most frequently used radioisotopes in PET are: Carbon-11, with a half-life of 20 min, Nitrogen-13, with a half-life of 10 min, Oxygen-15, with a half-life of 2 min, Fluorine-18, with a half-life of 110 min. Positrons are readily incorporated into an radiopharmaceutical of physiological interest, by chemical synthesis. Since carbon, nitrogen and oxygen exist in all naturally occurring substances of the body, many different molecules can be synthesized and used for specific imaging tasks [3].

The radioisotopes of oxygen [15O], nitrogen [13N], and carbon [11C] that can be administered to a subject and detected externally are all positron emitters. There is no positron emitter of hydrogen, so fluorine-18 [18F] is used as a hydrogen substitute. Positron emitters of Cu, Br, Rb, I, Ga and others are also used.

Molecular probes for PET are developed by first identifying a target process to be studied and then synthesizing a positron labeled molecule through which an assay can be performed. Because PET cannot provide direct chemical analysis of reaction products in tissue, labeled molecules are used to trace a small number of steps of a biochemical process. Drugs are designed to have limited interactions because the goal is to modify the function of a key step in a biological process with minimal interaction with other processes. In a typical molecular assay, a positron-labeled probe is injected intravenously, and PET scans provide measures of the tissue distribution of the probe and labeled products over time. The primary radiopharmaceutical used in clinical PET for oncologic imaging is 2-deoxy-2-[F18]fluoro-D-glucose, [18F]FDG. Images obtained with this tracer provide both qualitative and quantitative information about the regional distribution of glucose metabolism. [18F]FDG is injected intravenously, distributes throughout the body via bloodstream, and enters into organs, where it traces transport and phosphorylation of glucose. Positrons emitted from the nucleus of [18F] are positrons that travel a short distance and combine with an electron, and annihilation occurs with their masses converted into their energy equivalent ($E=mc^2$) through emission of two 511-keV photons 180° apart. The two 511-keV photons are electronically detected as a coincidence event when they strike opposing detectors simultaneously. Special crystals, called photomultiplier-scintillator detectors, within the PET scanner detect emitted gamma rays. The camera records the millions of gamma rays being emitted, and a connected computer uses the information and algorithms to map an image of the area where the radioactive substance has accumulated. After correction for photon attenuation, tomographic images of tissue concentration are reconstructed different colours or degrees of brightness on a PET image represent different levels of tissue or organ function [1].

PET is employed especially in oncology because the most malignancies have greater glucose utilization than normal tissues and accordingly exhibit greater [18F]FDG uptake [4]. [18F]FDG enters the cell by the same membrane transport mechanism as glucose. After penetration of the cellular membrane via glucose transporters, both [18F]FDG and glucose are phosphorylated by hexokinase. Unlike glucose-6-phosphate, [18F]FDG-6-phosphate is not a substrate of glucose-6-phosphate isomerase and does not undergo further metabolism in the glucose pathway. [18F]FDG is therefore trapped within cells. Three phenomena have been recognized to cause increased glucose utilization in tumor cells: overexpression of membrane glucose transporters, especially Glut-1 and Glut-3; increased hexokinase activity and decreased levels of glucose-6-phosphatase [5-7].

There are different methods for assessment of radiotracer uptake by normal and pathologic tissues, such as visual inspection, the standardized uptake value (SUV) and glucose metabolic rate calculation. The most frequently used method is the visual inspection. Images are first assessed visually, using transaxial, sagittal, and coronal displays: any abnormal focus of increased uptake that could not be related to the normal physiologic biodistribution of the radiopharmaceutical is considered pathological. SUVs are used for semiquantification of [18F]FDG uptake. The SUV is a semiquantitative assessment of the radiotracer uptake from a static (single point in time) PET image. The SUV is calculated with the following formula: tracer activity in tissue / (injected radiotracer dose/patient weight) where tissue tracer activity is in microcuries per gram, injected radiotracer dose is in millicuries, and patient weight is in kilograms. The SUV of a tissue can be depicted as the minimum, maximum, or mean in the region of interest. The mean SUV is the mathematical mean of all the pixels in the region of interest, whereas the minimum and maximum SUV are values of the pixel with the lowest and highest SUV, respectively. The nuclear physicians rely on visual inspection and use SUV in assessing questionable lesions, or for example in the follow-up of [18F]FDG-avid masses. Typically, malignant tumors have an SUV of greater than 2.5–3.0, whereas normal tissues such as the liver, lung, and marrow have SUVs ranging from 0.5 to 2.5. Variations in the SUV have been described such as the glucose-corrected SUV and SUV normalized by surface area or lean body mass. It is useful to know the tumor SUV before initiation of therapy to assess tumor grade and evaluate treatment response following radiation therapy or chemotherapy [8]. It is important to standardize the time interval between injection of the radiotracer and the PET study because SUV variability with time has been well documented [9]. Another method of quantification of dynamic [18F]FDG PET results is the more complex glucose metabolic rate calculation.

The number of clinical applications for [18F]FDG and PET continues to increase, especially in the field of oncology because it has been known for many years that increased glycolysis is a distinctive feature of malignant tumors compared with normal tissues.

The development of PET whole body imaging provided an accurate way to examine all organ systems of the body for the presence of primary and metastatic disease in a single study. Metabolic information from PET provides advantages over anatomic information from x-ray films, ultrasound, CT, MRI because PET can: 1) Differentiate benign from malignant lesions. For example, 60–80% of breast and 20–40% of lung biopsies are because x-ray mammography, CT, and MRI have limited accuracy in separating benign from malignant processes. 2) Identify the biological changes of cancer early in its course. 3) Image all organ systems of the body to detect the primary tumor and determine whether metastases are present so, their extent throughout the body. Because the whole body is imaged with PET, symptoms to direct an imaging procedure to a particular organ system are not required. Thus, asymptomatic disease is routinely detected. 4) Differentiate malignant tissue from necrosis, edema swelling, and scarring, for accurate, early assessment therapeutic response to continue, stop, or change therapy. In multicenter trials for lung and colorectal cancers, melanoma and lymphoma, PET has been shown to have an higher accuracy for detection of primary and metastatic disease vs. conventional workups in head-on comparisons and to the choice of therapy selected by conventional evaluations 20–40% of cases, depending on the specific clinical question [1]. About 65% of therapy changes occur because PET showed there was more extensive disease than originally determined conventional tests whereas, in about 35% of patients, showed that previously diagnosed metastases were benign lesions, resulting in changes to simpler therapies, improved prognosis, and lower costs. Similar results are reported for cancers such as breast, ovarian, head, neck, and renal [1].

Over the course of their disease, each patient and their providers face the challenge of having to make a myriad of decisions that range from the selection of a primary treatment modality, the specific approach used, the choice of adjuvant treatment, the intensity of follow-up and, should the disease recur, the choice and timing of any secondary interventions. PET is a valid tool in cancer diagnosis because the results may assist in avoiding an invasive diagnostic procedure may assist in determining the optimal anatomical location to perform an invasive diagnostic procedure. [18F]FDG PET allows disease staging at diagnosis, and restaging in the case of suspected recurrence in a single step. The role of PET is complementary to standard diagnostic workup including conventional imaging and the clinical management of the patient will differ depending on the stage of the cancer identified. This technique is used also after completion of treatment for the purpose of detecting residual, to detect suspected recurrence or metastasis, to determine the extent of known recurrence or could potentially replace one or more conventional imaging studies when it is expected that conventional study information is insufficient for the clinical management of the patient. PET can help to avoid unnecessary surgeries or invasive procedure and has an impact on patient management. It can be successfully used in radiotherapy at different steps, including patients selection, target volume and treatment planning definition by evaluating Biological Target Volumes (BTV) and evaluation of radiotherapy biological effect by monitoring the disease. The role of PET with [18F]FDG in the assessment of response at the end of treatment is established. However, the possibility to predict response during the treatment itself is a major challenge in oncology, in particular for the recent clinical introduction of individualized treatment regimens that require improved early risk stratification [10].

Recently, thanks to technological development, PET/CT was introduced: it is dual-modality PET/CT tomograph combining both PET and CT scanning in one device consisting in a PET tomograph and CT scanner mounted to the same rotating support. PET/CT imaging is changing the care of cancer patients in several ways: a) metabolic and anatomic whole-body staging of patients can be performed in one examination and much reduced scan times, thus, increasing patient comfort, b) because of limited patient motion, near ideal fusion of metabolic and anatomic images can be achieved, c) anatomic landmarks provided by CT will greatly facilitate the assignment of functional abnormalities to anatomic structures, d) images of regions of the body such as the head and neck, mediastinum, and postsurgical abdomen will be evaluated with a high diagnostic accuracy, e) fused images can be used to target radiation treatment more accurately and monitor the effects of chemotherapy, surgery, and radiation treatment [11].

Hybrid PET/CT improves the diagnostic interpretation of [18F]FDG PET and CT in cancer patients and has an impact on both diagnostic and therapeutic aspects of patient management [12].

The functional and anatomic information offered by PET/CT is being recognized as crucial in the care of oncology patients. PET and PET/CT are playing an ever-increasing role in the management of oncologic disease [8].

The use of [18F]FDG is based upon increased dependence of cancer for energy from glycolitic metabolism and a concomitant increase in glucose transporters in cell membranes of malignant cells. This process has been demonstrated in prostate cancer cells culture where uptake of [18F]FDG correlates with clinical measures of the severity of malignancy assessed by Gleason score and prostate specific antigen (PSA) levels. Despite these promising findings, [18F]FDG imaging of prostate cancer has been disappointed showing a limited efficacy for detection of primary and metastatic prostate cancer. Lower [18F]FDG uptake in prostate tumors, as compared to other cancers, has been attributed to a relatively slower metabolic rate of glucose transport proteins. This decreased metabolic dependence of prostate cancer upon glycolisis and glucose uptake coupled with the metabolism of [18F]FDG that includes renal excretion with activity throughout ureters and bladder obscures target organs (prostate) and closely related adjacent tissue (seminal vescicles) and it harbours local and regional metastases [13].

[11C]Choline PET in Imaging Cancer

Because of poor results of [18F]FDG in prostate cancer imaging, new radiopharmaceuticals were developed and [11C]Choline is one of the most promising.

In the human body, choline is needed for the synthesis of phospholipids in cell membranes, methyl metabolism, transmembrane signaling, and lipid-cholesterol transport and metabolism [14].

The background of the increased uptake of choline in prostate cancer is not clearly understood. Intracellular choline is rapidly metabolized to phosphorylcholine: it may also undergo acetylation to form acetylcholine or oxidation to form betaine (mainly in liver and kidney). The phosphorylation is catalyzed by the enzyme choline kinase. Once phosphorylated, the polar phosphorylcholine molecule is trapped within the cell. So far two mechanisms have been suggested. The first explanation is increased cell proliferation in tumours. Choline is one of the components of phosphatidylcholine, an essential element of phospholipids in the cell membrane. As a malignancy is characterised by increased proliferative activity, it is likely that choline uptake is a marker for cell proliferation. The second mechanism that may account for the increased uptake of choline in cancer cells is up-regulation of choline kinase. Choline kinase catalyses the phosphorylation of choline, the first step in the cascade of three enzymatic steps that make up the Kennedy pathway after choline is transported into the cell [15]. Overexpression of choline kinase has been found in cancer cell lines, including human derived prostate cancer [16].

Various studies have revealed an increased choline uptake as well as an upregulated activity of choline kinase and elevated levels of phosphorylcholine in cancer cells [17, 18]. These observations gave rise to the development and clinical evaluation of MRS imaging, which revealed a high content of phosphorylcholine in prostate cancer, whereas in normal tissue this choline metabolite was found in low concentrations or was undetectable. Based on these observations, [11C]Choline was introduced for imaging of malignancies including prostate carcinoma [19, 20]. [11C]Choline blood clearance is very rapid, the major amount of tracer remains trapped within cells. This allows for imaging as early as 3 to 5 min after tracer injection and provides images of good diagnostic quality [19, 21]. Physiologically increased tracer uptake is noted in liver, kidneys, salivary and adrenal glands, bowel and pancreas [22]. Intense bowel activity can be observed with all choline compounds and can be a reason for false-positive findings [23, 24]. Imaging of the abdomen and pelvis may benefit from fasting, as pancreatic juice and bile both contain phosphorylcholine [23, 25]. Currently, it is unclear whether choline uptake in prostate cancer lesions can serve as an indicator of biologic aggressiveness.

Breeuwsma et al. [26] evaluated the correlation between the uptake of [11C]Choline with PET and cell proliferation as assessed by tumour histology and Ki-67 expression, in human prostate cancer. The Ki-67 antigen is a nuclear protein complex that is expressed during cell replication. It is generally used as a marker of cellular proliferative activity, because expression of this antigen gives an approximation of the number of cells undergoing active division (growth fraction). It is not expressed in the resting cell (G0). Prostate cancer tissue from 18 patients who had undergone a radical prostatectomy for histologically proven disease was studied. [11C]Choline PET scan was performed prior to surgery. Post-prostatectomy specimens were prepared and stained with the antibody MIB-1 for Ki-67. Statistical analysis showed no significant correlation between [11C]Choline uptake and Ki-67 staining: the authors were unable to demonstrate a relation between in vivo uptake of [11C]Choline and cell proliferation. No statically significant relationships were found between the uptake of

[11C]Choline (SUV) and either preoperative PSA or Gleason sum score. In contrast, the relation between SUV and pathological stage (pT) reached statistical significance.

It is possible that some other mechanisms of chemical trapping could be responsible for the uptake of [11C]Choline in prostate cancer. One mechanism that has been proposed is up-regulation of choline kinase by overexpression in human prostate carcinoma cell lines. However, no evidence has been presented that overexpression of choline kinase is responsible for the uptake of [11C]Choline in prostate cancer. It is also unknown whether this overexpression is involved in the carcinogenic process [16].

[11C]Choline was proven to play a significant role in malignancies other than prostate cancer. The typical biodistribution of [11C]Choline (very high in the liver, pancreas, salivary glands, and kidneys but very low in the others organs) allows the evaluation of the other anatomical sites beside the pelvis, in particular, the brain, the torax, and the bones. So many researchers tested it for the evaluation of [18F]FDG-negative cancers or for the evaluation of physiologically increased [18F]FDG uptake areas.

For example, as [18F]FDG distributes to the brain, and the interpretation of the scan can be difficult in case of low grade brain malignancies.

Ohtani et al. [27] tested [11C]Choline in 22 patients affected by different kinds of brain tumours (high grade and low grade) and benign diseases. The authors found that [11C]Choline PET has the potential to easily enable differentiation between low-grade and high-grade gliomas, but not to differentiate low-grade gliomas from non-neoplastic lesions, and that the combination of [11C]Choline PET and magnetic resonance imaging may provide an accurate means by which to identify high-grade gliomas.

Another example is the evaluation of bladder cancer. Urothelial cancer, in fact, is known to be [18F]FDG -negative in the majority of cases. Furthermore, the bladder is one of those sites in wich unspecific presence of [18F]FDG is found due to the renal excretion of the tracer. As choline is only excreted late by the kidneys, it may allow the evaluation of bladder cancer patients.

Picchio et al. [28] analyzed 27 patients affected by urothelial bladder cancer. Their preliminary results suggested a possible role of [11C]Choline. In fact, [11C]Choline PET turned out to be comparable to computed tomography for detecting residual bladder cancer after TURB but appeared to be superior to CT for the evaluation of potential additional lymph node metastases.

Another big chapter on the use of [11C]Choline in non-prostatic cancer includes thoracic tumors. Several low-grade lung cancer like bronchoalveolar carcinoma, in fact, mildly bind [18F]FDG, which, by the way, can lead to an overestimation of the stage of the disease (especially in case of non-small-cell lung cancer) due to the high incidence of associated inflammation.

Khan et al. [29] analyzed both [11C]Choline and [18F]FDG of 17 patients affected by lung cancer. The results were compared to CT and histopathology in all cases. Despite few limitations, in particular the small number of patients and the absence of patients affected by benign diseases, this study concludes that both techniques have clinical value for the non-invasive detection of primary lung cancer that is 2 cm or greater in size. However, [18F]FDG PET may be superior to [11C]Choline PET in the detection of lung cancer that is less than 2 cm in diameter and in mediastinal lymph nodes metastases.

Nanni et al. [30] used [11C]Choline in a preliminary study to evaluate multiple myeloma patients, as despite its sensitivity, the uptake of [18F]FDG assessed with the maximum SUV can be very low, sometimes even comparable to the SUV of a benign lesion. The distinction between a benign lesion and a low-metabolic myeloma patients lesion can therefore be difficult to achieve. Ten patients with bone lesions were studied both with [11C]Choline and [18F]FDG PET. According to these preliminary data, [11C]Choline PET/CT appeared to be more sensitive than [18F]FDG PET/CT for the detection of bony myelomatous lesions, and surprisingly, the SUV turned out higher with [11C]Choline compared to [18F]FDG. If these data are confirmed in larger series of patients, [11C]Choline may be considered a more appropriate functional imaging in association with MRI for multiple myeloma bone staging.

Tian et al. [31] compared the results of [11C]Choline PET and [18F]FDG PET in 125 patients affected by brain tumors, head and neck tumors, bone tumors, lung tumors, and soft tissue tumors, concluding that in brain, head and neck, bone, and soft tissue tumors, [11C]Choline is characterized by higher contrast than [18F]FDG and that it can be used for differentiation between malignant and benign tumors. However, attention needs to be drawn to the high uptake of [11C]Choline in some benign tumors and tumor-like lesions, as this will be of significance in clinical practice.

As it is clear from these briefly described data, [11C]Choline seems to have a high diagnostic potential not only for prostate cancer, but also for a large series of malignancies. The main limitation of this tracer compared to the more common [18F]FDG, however, is the short half-life (20 min) that prevents from distribution in PET centers not provided with a cyclotron-based radiopharmacy. In conclusion, if this will be further demonstrated, [11C]Choline could be considered a promising positron-emitter tracer to be routinely used in the clinical practice for the functional evaluation of [18F]FDG -negative malignancies or for the evaluation of physiologically increased [18F]FDG -uptake areas.

[11C]Choline PET: Detection of Primary Prostate Cancer

The main diagnostic tools used to look for evidence of prostate cancer include digital rectal examination, serum concentration of PSA, and transrectal ultrasound (TRUS) guided biopsies [32]. During the last decade, there has been a significant advancement in imaging of urologic diseases.

TRUS, computerized tomography (CT), magnetic resonance imaging (MRI), magnetic resonance spectroscopy (MRS), and positron emission tomography (PET) are still experiencing new developments in urology. Despite these many technological advances, the initial diagnostic procedure for a patient with suspected prostate cancer is multiple site prostate biopsies. There is a need for a noninvasive metabolic imaging modality to direct the site of biopsy to decrease the sampling error. However, a number of problems have prevented success in this area. The small size and location deep of the prostate in the pelvis interferes with the accurate detection of diseases with noninvasive imaging at an early stage. The other problem for nuclear imaging of the prostate is the proximity of the urinary bladder. Most radiopharmaceuticals are significantly excreted by the kidneys, resulting in bladder activity that significantly obscures the prostate [33]. TRUS is the most commonly used imaging modality to evaluate the prostate gland. TRUS uses a high-frequency (5-7.5 MHz) endorectal transducer in close proximity to the prostate and produces high-resolution images (gray-scale imaging) [34]. Prostate cancer most commonly appears hypoechoic (60-70%) compared with the normal peripheral zone, but up to 40% of lesions are isoechoic and are not detected. Benign processes like prostatitis also frequently appear hypoechoic, therefore, hypoechogenicity is not specific for cancer. The positive predictive value of gray-scale is low (18-52%) and is not appropriate to make diagnosis of prostate cancer. Use of color Doppler, power Doppler, and contrast agents may increase the sensitivity and specificity. A positive predictive value as high as 77% has been reported [35, 36].

Sonoelasticity imaging is another way to increase the accuracy of the TRUS. With the exception of gray-scale imaging, all other forms of TRUS are at an experimental stage. Currently, TRUS is not used during biopsy to guide the needle to any specific abnormal site; rather it is used to localize the usual sites for biopsy. Therefore, although it is called TRUS-guided prostate biopsy, it is basically blind, ie, irrespective of the presence of ultrasound abnormalities [35, 36]. The yield from standard TRUS-mediated 6- to 12-zone prostate biopsy is found to be higher than biopsy only of lesions detected by TRUS [34]. CT lacks soft tissue contrast resolution for the detection of cancer within the prostate and offers no advantages over TRUS in biopsy guidance [34]. MRI using a combination of pelvic phased array and an endorectal coil provides the highest resolution images of the prostate and it is the best imaging modality for demonstrating normal zonal anatomy but it is of limited value in detecting primary prostate cancer. Prostate cancer usually appears as abnormal areas of low signal intensity within the normal homogeneous high signal intensity background in T2-weighted images [34]. Cancer may not be detected if it does not demonstrate low signal intensity in T2, if it is located primarily within the central gland (does not demonstrate uniform high intensity like the peripheral zone), or if the peripheral zone does not possess a uniform high signal intensity. Low intensity in the peripheral zone is sensitive but not specific for cancer. Benign conditions including prostatitis, hemorrhage, or dystrophic changes related to radiation or androgen-deprivation therapy produce similar findings [34, 36]. High cost and lack of a sufficiently high positive predictive value for cancer detection make MRI inappropriate for routine screening of primary tumor.

MRS imaging provides metabolic information about the prostate gland in vivo [37]. Currently, it is possible to obtain a 3D metabolic map of the entire prostate with a resolution of 0.24 cm. [38, 39]. Normal prostate tissue is rich in citrate, and prostate cancer contains high levels of choline and often diminished citrate. This is due to conversion from citrate-producing to citrateoxidating metabolism (low citrate) and a high phospholipids cell membrane turnover (high choline) in the cancer [15, 39]. Therefore, prostate cancer is distinguished from the normal tissue by an increased (choline + creatine)/citrate ratio. 31P MRS in vivo and in vitro has revealed an elevated level of phosphatidylcholine in tumors, which is the most abundant phospholipid in the cell membranes of all eukaryotic cells [40], it is thought that this elevation was the result of increased uptake of choline, a precursor of the biosynthesis of phosphatidylcholine [41]. Though promising, the major limitations of MRS are cost and availability, and lack of supporting clinical data.

Several studies have been conducted using [18F]FDG PET in localized prostate cancer. Most of the results were disappointing [42]. Effert et al. [43] studied 48 prostate cancer patients and 16 patients with benign prostatic hyperplasia. They reported a low grade FDG uptake in 81% of these tumors and there was a significant overlap between benign prostatic hyperplasia and prostate cancer in uptake values. Hofer et al. [44] reported similar findings. In a study of 24 patients Liu et al. [45] found a sensitivity of 4% for FDG PET in the diagnosis of primary prostate cancer. Oyama et al. [46] examined 44 consecutive patients with histologically proven adenocarcinoma of the prostate. In visual analysis, 5 patients with benign prostatic hyperplasia (control group) showed low [18F]FDG uptake. On the other hand, 28 of 44 (sensitivity 64%) patients with cancer were visually positive, showing intermediate and high FDG uptake. There was a trend for FDG uptake in prostate cancer to correlate Gleason score. The possible explanations for disappointing results as suggested by Schöder and Larson [42] include: 1) a relatively low metabolic rate in the majority of prostate cancers, 2) older image reconstruction techniques (ie, use of filtered back projection instead of iterative reconstruction), 3) location of the prostate adjacent to the urinary bladder, and 4) the lack of appropriate patient selection (FDG may be more useful in patients with clinical suspicion of a high-grade cancer).

In conclusion in uro-oncology the use of [18F]FDG PET has been limited because urinary excretion of [18F]FDG may mask pathological uptake. Furthermore, prostate tumors often have low glucose uptake. At present, PET tracers with more favourable properties have been developed: one of the most promising tracer is [11C]Choline.

Controversies exist about the accuracy of [11C]Choline for the detection of primary prostate cancer and several reports on this issue have been contradictory.

Hara et al. [21] in 1998 introduced [11C]Choline as a tracer for the visualisation of primary prostate cancer. Because of negligible urinary excretion, the tracer had the potential characteristics for in vivo imaging prostate cancer. Ten patients with known prostate cancer underwent both [11C]Choline PET and [18F]FDG PET studies before the beginning of treatment, and results of the two studies were compared. Imaging of prostate cancer and its local recurrence was difficult when [18F]FDG was used because, within the pelvis, the areas of high uptake were concealed by the overwhelmingly abundant radioactivity in urine (in ureters and bladder). By contrast, it was easy when [11C]Choline was used because the urinary activity was negligible and tumor uptake was marked. The radioactivity concentration of [11C]Choline in prostate cancer and metastatic sites has a SUV > 3 in most cases and the high uptake of [11C]Choline was found not only in already proved metastatic sites but also in other sites. Hara et al. concluded that prostate cancer and its local recurrence were visualized clearly in PET using [11C]Choline and the advantage of' [11C]Choline PET over [18F]FDG PET in the imaging of prostate cancer involved not only the absence of [11C]Choline activity in the urine but also the high uptake of in the prostate cancer itself.

De Jong et al. [25], to define the potential role of [11C]Choline PET in visualization of prostate cancer, prospectively studied 25 patients with histologically proven prostate cancer (patients with T4 tumors and/or known distant metastases at conventional imaging were excluded) and five patients with a benign prostate. The normal prostate was visualized with a mean SUV of 2.3 (range 1.3–3.2). The primary tumor could be visualized with a mean SUV of 5.0 (range 2.4–9.5) as an hot spot within the prostate gland in 24 of the 25 patients and as an uptake of in the entire prostate in one patient. Five patients had proven lymph node metastases after pelvic lymph node dissection: in four of these five patients, uptake of [11C]Choline in pelvic lymph nodes with metastases of 5 mm to 3 cm was seen with a mean SUV of 4.7 (range 2.9–9.1) and in one patient, with a micro-metastasis [11C]Choline PET failed to detect a metastatic site. One false positive PET was seen in a lymph node with inflammatory changes. In 19 patients without lymph node metastases, [11C]Choline PET was negative. The results of this study showed that [11C]Choline was avidly taken up in prostate cancer, both primary tumor and lymph node metastases, in the virtual absence of urinary radioactivity. This allowed further research on the value of [11C]Choline PET in clinical management of patients with prostate cancer because a metabolic imaging technique like [11C]Choline PET could especially be of value in those areas where conventional imaging have restrictions and limited sensitivity.

Sutinen et al. [47] evaluated the kinetics of the uptake of [11C]Choline in prostate cancer and benign prostatic hyperplasia. The value of [11C]Choline PET in assessing the malignant potential of prostate tumours was analysed by correlating tracer uptake with tumour grade, Gleason score, volume of the prostate and PSA. Fourteen patients with histologically confirmed prostate cancer and five patients with benign prostatic hyperplasia were studied with [11C]Choline PET. A mean dose of 430±31 MBq of [11C]Choline was injected intravenously and a dynamic emission acquisition of prostate was performed for 30 min. The uptake of [11C]Choline was measured as SUV and as a kinetic influx constant (Ki) obtained from graphical analysis. Both cancerous and hyperplastic prostate were well visualised with [11C]Choline against low or moderate tracer accumulation in the bladder and rectal wall. At visual inspection all 14 carcinomas showed increased accumulation of [11C]Choline on PET imaging. Two cancer patients who had started adjuvant hormonal treatment 2 months before PET exhibited the lowest accumulation of [11C]Choline in prostate of all studied patients, including those who presented with benign prostatic hyperplasia. This limited experience suggested that [11C]Choline PET may be applicable in trials where metabolic activity serves as a surrogate endpoint for drug effect on prostate carcinoma growth. Indeed, the prostate was well seen in hyperplasia without malignant transformation and visually it would have been difficult to evaluate the malignancy grade in most cases in which increased uptake was seen in the prostate only. One of the two histologically confirmed lymph node metastases (measuring <0.5 cm) was not detected with [11C]Choline PET, but the bone metastasis seen on scintigraphy in this patient was also detected on PET images. The [11C]Choline activity in plasma decreased rapidly after injection and the initial uptake of [11C]Choline in prostate during the first 5 min was rapid as well. Thereafter, the dynamic curve showed level activity in the prostate with no apparent difference in pattern between patients with cancerous and those with hyperplastic tissue. The mean Ki of untreated prostate carcinoma was 0.205±0.089 min−1 (range 0.128–0.351, including both primaries and metastases), and the mean SUV was 5.6±3.2 (range 1.9–15.5). There was a close correlation between the [11C]Choline influx constant and the SUV of tumours (r=0.964, P=0.0005; n=7). The SUV of the prostate in two hormonally treated patients was low, at 1.8 and 2.8. The mean SUV of benign hyperplastic prostate was 3.5±1.0 (range 2.0–4.5; n=4) and the mean Ki of hyperplastic prostate in two patients with metabolite-corrected measurement of plasma activity was 0.119± 0.076 min−1 (range 0.065–0.173). The difference between the SUVs of the prostate cancer and benign prostatic hyperplasia was not statistically significant (P=0.067), although the highest SUVs were found in cancers. The dynamic curve of [11C]Choline uptake showed no apparent difference in pattern between patients with cancerous or hyperplastic tissue. For that reason the diagnostic power of quantitative [11C]Choline PET for discrimination between benign and malignant prostatic tumours remained to be clarified in a larger population. The mean free PSA of cancer patients was 2.0±1.9 (range 0.35–6.2; n=12) and that of hyperplasia patients was 0.7±0.7 (range 0.1–1.6; n=4). The mean ratio of free PSA to total PSA was 14.7%±8.8% (range 7.6–39.4%; n=12) in cancer patients and 19.0%±11.1% (range 8.1–32.7%; n=4) in hyperplasia patients, respectively. The mean volume of the prostate in cancer patients was 43±21 cm3 (range 19–89; n=12). There was no correlation between the SUVs of prostate carcinoma and total PSA, free PSA, PSA ratio, histological grade, Gleason score or prostate volume, although the small sample size may have masked true biological differences. One clear explanation for the overlap and variability of SUVs is the low proliferative activity in the majority of primary prostatic adenocarcinomas and the different fractions of tissue components containing cancer, hyperplasia, and normal prostate in individual patients. In conclusion, a high uptake of [11C]Choline characterised not only neoplastic but also hyperplastic prostatic tissue. Dynamic imaging of the uptake of [11C]Choline in the prostate showed a good applicability of the graphical analysis model with an irreversible compartment, but a close correlation between the Ki values and semiquantitative SUVs of tumours supports the use of the simpler SUV in the clinical setting [47].

Yamaguchi et al. [48] conducted a prospective comparative study of MRI combined with proton MRS and [11C]Choline PET in patients with proven prostate cancer to assess the clinical utility of [11C]Choline PET for localizing and evaluating cancer lesions. PET and MR imaging combined with proton MRS were performed in 20 patients with prostate cancer. [11C]Choline PET uptake was visualised in the prostate gland in all patients. Correlations among the metabolite ratio of choline + creatine to citrate (Cho+Cr/Ci) on MRS, serum PSA and SUV of [11C]Choline were assessed. The location of the primary lesion was assessed by the site of highest tracer uptake and the laterality of the highest Cho+Cr/Ci ratio and confirmed by examination of surgical pathology specimens (n=16). Asymmetrical uptake in the prostate that exceeded the surrounding background activity was visualised in all patients; [11C]Choline PET revealed primary lesions in all 20 patients. In eight patients, the main primary cancer lesions could not be identified on MR images. MRS revealed the cancerous lesions (with a Cho+Cr/Ci ratio of more than 0.85) in 13 of 20 (65%) patients. No statistically significant difference was evident between the diagnostic sensitivity of MRI for prostate cancer and that of MRS. However, the diagnostic sensitivity of choline PET based on visual interpretation was superior to that of MRI ($p<0.05$). Significant linear correlations were observed between SUV and serum PSA ($r=0.52$, $p<0.05$), and between SUV and the ratio of Cho+Cr/Ci ($r=0.49$, $p<0.05$). On the other hand, no significant correlations were apparent between SUV and Gleason score, or between SUV and tumour grade. In 16 patients who underwent total prostatectomy, the authors investigated the accuracy of primary lesion localization obtained with choline PET and MRS. PET results were in agreement with pathological findings in 13 patients (81%), while MRS results were in accordance with pathological findings in 8 patients (50%). This preliminary study concluded that [11C]Choline PET may provide more accurate information regarding the localization of main primary prostate cancer lesions than MRI/MRS. The authors suggested that when prostate cancer is highly suspected due to a high serum PSA level but the results of biopsies are negative, obtaining information regarding localization of the main primary lesion by [11C]Choline PET may increase the success rate of biopsy and avoid unnecessary repeated biopsies, which would in turn reduce discomfort for patients.

Up to this point previous studies considered only PET, but thanks to development of technologies PET/CT has been performed in abdominal and pelvic cancer patients. PET/CT is advantageous for localizing tumour sites that may be difficult to localize by PET alone and in identifying normal sites of uptake that may be difficult to discriminate from tumours. The role of [11C]Choline PET/CT in localization of prostate cancer was investigated in the following studies.

Farsad et al. [49] evaluated the potential usefulness of [11C]Choline PET/CT for detection and localization of tumors within the prostate. They first correlated PET/CT findings with histopathologic analysis of the whole prostate on a sexant basis. They reviewed the results of the [11C]Choline PET/CT scans of 36 patients with prostate cancer and of 5 control subjects with bladder cancer. All patients underwent [11C]Choline PET/CT and, subsequently, radical prostatectomy with lymph node dissection within 1 month. [11C]Choline PET/CT scans were obtained 5–10 min after intravenous injection of 370–555 MBq of [11C]Choline. Images were reviewed visually and semiquantitatively using maximum SUV and tumor-to-background ratio (background activity was considered the uptake of 4 small regions of interest in the sextants without pathologic uptake or in areas showing the lowest activity). Sextant-per-sextant step-section histology confirmed multifocal stage T2 or T3 prostate cancer in all 36 patients, with a mean Gleason score of 5.9 (range, 2–7). On a sextant basis, histopathologic analysis identified cancer foci in 143 of 216 sextants, HGPIN (high-grade prostate intraepithelial neoplasm) foci in 89 (in 59 sextants in association with carcinoma, in 30 sextants alone), acute prostatitis foci in 7 (in 3 sextants in association with carcinoma, in 4 sextants alone), and normal tissue or benign hyperplasia in 39. At least 1 primary prostatic tumor focus could correctly be visualized through PET/CT in a total of 35 of 36 patients. On a sextant basis, PET/CT demonstrated focal [11C]Choline uptake in 108 sextants (50%): histologic examination showed that 94 of 108 were affected by tumor, 10 by HGPIN, 2 by HGPIN and acute prostatitis, and 2 by benign prostate hyperplasia or no pathologic findings. In sextants with areas of abnormal [11C]Choline uptake, the mean SUV was 5.3 ± 2.2 (range, 2.2–12) and the mean T/B ratio was 2.0 ± 0.5 (range, 1–3.4). PET/CT found 108 sextants without evidence of abnormal [11C]Choline uptake. These findings were true negative in 59 of the sextants and false negative in 49; in those 49, histologic examination proved the presence of cancer foci. PET/CT had a sensitivity of 66%, a specificity of 81%, an accuracy of 71%, a positive predictive value of 87%, and a negative predictive value of 55%. No statistically significant difference between tumors and HGPIN was found with either SUV or T/B ratio. They found no statistically significant linear correlation between SUV and tumor grade (R2=0.007), SUV and Gleason score (R2=0.05), or SUV and PSA (R2=0.009). In 3 patients the initial prostate biopsy failed to detect prostate cancer because the tumor was located anteriorly in the transition zone: biopsy repeated on the basis of PET/CT results revealed prostate cancer in the transition zone, and step-section histopathologic examination after radical prostatectomy confirmed the presence of cancer in only this region. Four patients had lymph node metastasis at pathologic evaluation; in 2 of these 4, PET/CT revealed lymph node involvement with clear focal accumulation of

[11C]Choline. In 2 patients, PET/CT failed to detect lymphatic involvement, whereas histologic examination showed micrometastases (4 and 7 mm). In the 5 control subjects, high-grade prostate intraepithelial neoplasm was detected at histologic examination in 16 of 30 sextants. PET/CT showed increased [11C]Choline uptake in 5 of 16 sextants. In summary, this study showed the feasibility of using [11C]Choline to identify cancer foci but the high rate of false negative results do not support the routine use of this technique. Prostate cancer was characterized by multiple foci (143 foci in 36 patients), which often are small. The limited spatial resolution of PET/CT scanners is well known, with the scanners unable to detect lesions smaller than 5 mm. In addition, the partial-volume effect could be another cause for failure to detect small lesions. Furthermore, prostate cancer may show a low [11C]Choline uptake; in this study, 13 of 93 cancer foci detected with [11C]Choline PET/CT had an SUV inferior to 3. A faint increase of tracer uptake (i.e., SUV < 2.5) was hardly detectable. The wide SUV range measured for cancer foci reflects the heterogeneity of prostate cancer, and one could hypothesize that [11C]Choline PET/CT failed to detect all cancer foci because of their differences in metabolic state. These two conditions (small dimension and low uptake) could explain the high rate of false-negative results with [11C]Choline PET/CT. Although more than half of histologically detected HGPIN foci showed no abnormal [11C]Choline uptake (18 of 30 sextants), this study demonstrated that HGPIN can accumulate [11C]Choline (5 sextants of the control group and 10 sextants of the patient population with proven HGPIN showed high uptake of [11C]Choline). Because HGPIN and carcinoma have a strong tendency to be present simultaneously and exhibit the same "zonal" origin and anatomic proximity, a possible explanation for pathologic uptake of [11C]Choline in only some HGPIN foci could be that some of these regions harbored a small focus of cancer undetected by pathologists. The complete overlap of SUV and T/B ratio between cancer foci and HGPIN foci detected by PET/CT supported this hypothesis and pointed out that no single SUV or T/B ratio cut-off can differentiate between cancer and HGPIN. Another reason for the high [11C]Choline uptake of some HGPIN foci could be the different metabolic states of HGPIN lesions. Farsad et al. concluded that in a selected patient population with negative findings on the first biopsy but at a high risk of cancer, the use of [11C]Choline PET/CT could be helpful as a second-line imaging method. In fact, additional biopsies should be directed at the region of abnormal uptake on PET/CT and at the immediately adjacent areas [49].

Reske et al. [50] investigated the ability of [11C]Choline and multimodality fusion imaging with integrated PET and contrast-enhanced CT (PET/CT) to delineate prostate carcinoma within the prostate and to differentiate cancer tissue from normal prostate, benign prostate hyperplasia, and focal chronic prostatitis. Twenty-six patients with clinical stage T1, T2, or T3 and biopsy-proven prostate cancer underwent [11C]Choline PET/CT after intravenous injection of about 1110 MBq of [11C]Choline, radical retropubic prostatovesiculectomy, and standardized prostate tissue sampling. Maximal [11C]Choline SUVs within 36 segments of the prostate were determined. PET/CT results were correlated with histopathologic results, prostate-specific antigen (PSA), Gleason score, and pT stage. Two experienced observers visually assessed PET/CT images and judged 26 of 26 patients positive for prostate cancer. Histopathology showed 66, 147, 33, 241, and 449 segments with normal prostate tissue, benign prostatic hyperplasia (BPH), prostatitis, BPH mixed with focal prostatitis, or prostate cancer, respectively. [11C]Choline uptake in carcinoma (SUV, 3.5 ± 1.3; mean ± SD) was significantly higher (P <0.001) than that in normal prostate tissue (1.7 ± 0.5), BPH (2.0 ± 0.6), focal prostatitis (2.1 ± 0.6), BPH mixed with focal prostatitis (2.1 ± 0.7), and all segments with benign histopathologic lesions together (2.0 ± 0.6). The reliability of segmental [11C]Choline uptake for detecting prostate cancer was analyzed using ROC analysis. The area under the ROC curve for diagnosing prostate cancer using segmental [11C]Choline SUV was 0.89 ± 0.01. For a SUV threshold of 2.65 chosen to yield the maximal accuracy, they obtained a sensitivity of 0.81 (363/449), a specificity of 0.87 (426/487), a positive predictive value of 0.86 (363/424), a negative predictive value of 0.83 (426/512), and an accuracy of 0.84 (789/936). The SUV of small tumors was significantly lower than that of larger tumors: SUV=3.8 ± 0.3 vs. 5.3 ± 1.7 (P= 0.004). The maximal segmental [11C]Choline SUV in all patients was not related to PSA or Gleason score. However, they observed a weak, but significant, correlation between tumoral [11C]Choline SUV and pT stage. The results of this study indicated that [11C]Choline PET/CT identified tumor territories within the prostate gland in virtually all patients examined. Visual assessment and quantitative segmental analysis of [11C]Choline uptake detected prostate cancer in 26 of 26 patients and 25 of 26 patients, respectively. In addition, a segmental SUV>2.65 had an accuracy of 84% for correctly detecting segments with prostate cancer. In agreement with Breeuwsma et al. [26], Reske et al. [50] observed a weak correlation of maximal segmental SUV with tumor stage. This might suggest an association between the extent of neoplasm within the prostate and upregulation of choline metabolism in tumor tissue. In agreement with other authors, they did not observe a correlation of tumoral [11C]Choline SUV with serum PSA concentration, Gleason score, or Ki67 labeling index, indicating that biologic aggressiveness may not be a dominant influencing factor of choline metabolism in prostate cancer [26, 47, 49]. The weak correlation of [11C]Choline SUV with serum PSA reported by Yamaguchi et al. [48] might be influenced by a relatively large BPH-related component of the serum PSA concentration. In conclusion, according to Reske et al. [50], [11C]Choline PET/CT can accurately detect and locate major areas with prostate cancer and differentiate segments with neoplasm from those with benign hyperplasia, chronic prostatitis, or normal prostate tissue.

Martorana et al. [51] assessed the sensitivity of PET/CT for intraprostatic localization of primary prostate cancer on a nodule-by-nodule basis, and compared its performance with 12-core transrectal biopsy. In 43 patients with known prostate cancer who had received [11C]Choline PET/CT before initial biopsy, they evaluated sensitivity of PET/CT for localization of nodules 5 mm or greater using radical prostatectomy histopathology as the reference standard. Comparison with transrectal ultrasound guided biopsy was based on sextant assessment of all cancer foci following sextant-by-sextant matching and reconstruction.

All 43 patients showed 1 or more focal uptake (total 70); focal uptake was bilateral in 32 (74%) patients and unilateral in 11 (26%). In 8 patients focal uptake was recorded in the presence only of high grade PIN. At nodule assessment 57 (of the 70) instances of focal uptake were in correspondence with one of the 69 histopathologically identified nodules 5 mm or greater in diameter, corresponding to a sensitivity of 83% for localization of tumor nodules. Of note, 58 instances of focal uptake corresponded to 1 of the 92 cancer foci (sensitivity of 63% for detection of any focus of cancer). PET/CT detected only 1 of 23 (4%) cancer focus 5 mm or less.

At logistic regression analysis only nodule size appeared to influence sensitivity; encouragingly neither nodule grade, type of location, prostate volume, PSA, nor tumor stage seemed to affect sensitivity. The authors also looked for correlations between SUV (measured in each of the 57 detected tumor nodules) and clinical/pathological variables. SUV correlated positively with PSA (p=0.002) and age (p=0.032), and negatively with nodule grade (p=0.0027). This finding is in line with the concept that [11C]Choline uptake seems to be influenced by metabolic factors that are distinct from tumor proliferation. No relationship was observed for nodule diameter (p=0.9) or location (p=0.5), tumor staging (p=0.9) or prostate volume (p=0.08). It is also noteworthy that no difference in SUV values was apparent in the false-positives and true positives, suggesting that this criterion cannot be used to distinguish between primary prostate cancer and high grade PIN or benign prostatic hyperplasia. At sextant PET/CT had slightly better sensitivity than transrectal ultrasound guided biopsy (66% vs 61%, p=0.434) but was less specific (84% vs 97%, p=0.008). The limited specificity of PET/CT in these patients with known prostate cancer could be explained by the likely confounding influence of the metabolic activity of high grade PIN. The restricted sensitivity of PET/CT can be attributed to several factors. Prostate tumors are characteristically multifocal and the limited spatial resolution of PET (approximately 5 mm) is likely to hinder visualization of the smaller foci that are readily detectable at radical prostatectomy histopathology. The recorded sensitivity could also have been affected by the difficulties inherent in sextant matching and reconstruction. The only fair concordance between PET/CT and TRUS biopsy results is doubtless partially due to the intrinsic pitfalls in sextant-by-sextant localization and matching process. The discrepancy could also be related to the lack of ability of TRUS biopsy (especially at the first sampling) in detecting anterior/transitional zone prostate cancers, whereas PET/CT does not appear to have this particular technical limitation. In conclusion [11C]Choline PET/CT had good sensitivity for intraprostatic localization of primary prostate cancer nodules 5 mm or greater and similar sensitivity compared with transrectal ultrasound guided biopsy for localization of any cancer focus.

Scher et al. [52] investigated the diagnostic efficacy of [11C]Choline PET and PET/CT in 58 patients with clinical suspicion of prostate cancer. Qualitative image analysis as well as semiquantitative SUV measurement was carried out and the reference standard was histopathological examination of resection specimens or biopsy. Prevalence of prostate cancer in this selected patient population was 63.8% (37/58). [11C]Choline PET and PET/CT on the basis of a predominantly qualitative image analysis, showed a sensitivity of 86.5% (32/37) and a specificity of 61.9% (13/21) in the detection of the primary malignancy. The positive predictive value was 80.0% (32/40), the negative predictive value was 72.2% (13/18) and the accuracy was 77.6%. The mean SUV of the primary malignancies was 4.3±1.7 with a range of 2.2–9.8. SUV of the prostate gland in patients without malignancy was 3.3±0.9 with a range of 1.4–4.7. The SUV of primary malignancies was found to be significantly higher than the SUV of prostate glands without malignant changes (p=0.027). Five cases were false negative: the primary malignancies missed by PET in these five cases exhibited a SUV ranging from 2.2 to 3.3. The main reason for interpreting these examinations as non-malignant was the fact that the pattern of tracer uptake did not have an inhomogeneous character. In eight of 21 cases, PET was false positive; SUV levels ranged from 3.3 to 4.7. Acute prostatitis was present in one of these cases and chronic prostatitis was present in another four. Three of these patients also had BPH. The pattern of [11C]Choline uptake was of inhomogeneous character, possibly due to inflammatory changes and BPH. On the basis of an ROC analysis of the assessed SUV levels of the prostate gland, a cut-off value of 3.3 for differentiation between benign and malignant processes would have provided the best results. This approach would have yielded a sensitivity of 70.3% with a specificity of 57.1% in the detection of the primary cancer. In comparison, an ROC analysis of the PSA level suggested a cut-off value of 7.2 to achieve the same sensitivity level of 70.3%, with a lower specificity of 52.4%. However, these values were too poor to warrant the definition of clear SUV or PSA cut-off values for differentiation between malignant and benign processes. Not surprisingly, a predominantly qualitative image interpretation taking into account the actual uptake pattern, and using the SUV primarily as guidance, yielded superior results. The high sensitivity of 86.5% was, however, achieved at the expense of a moderate specificity of 61.9%. An overlap of the [11C]Choline uptake in benign and malignant prostatic changes has to be considered responsible for the comparatively high rate of eight false positives.

Although it was beyond the scope of this study to compare stand-alone PET versus PET/CT, it has to be noted that readers found bladder activity as well as unspecific intestinal activity only to be an issue in stand-alone PET studies, and not in PET/CT studies. Because of the possibility of exact anatomical localization of foci seen on PET, combined PET/CT systems may eventually be favoured. The limitations of stand-alone PET may be further substantiated by the fact that radiolabelled choline in general provides poorer anatomical information than does [18F]FDG. From a clinical point of view, combined PET/CT scanners may also be favoured in scenarios where additional morphological information is indispensable, for example for planning treatment [52].

Testa et al. [53] retrospectively compared sensitivity and specificity of MRI imaging, three-dimensional (3D) MRS, combined MR imaging and 3D MRS, and [11C]Choline PET/CT for intraprostatic tumor sextant localization, with histologic findings as gold standard. Patient population was composed by 26 men with biopsy-proved prostate cancer who underwent radical prostatectomy. Sensitivity, specificity, and accuracy were, respectively, 55%, 86%, and 67% at PET/CT; 54%, 75%, and 61% at MR imaging; and 81%, 67%, and 76% at 3D MRS. When PET/CT analysis was performed quantitatively by using the maximum SUV for each sextant with a SUV threshold of 2.9 or higher, sensitivity was substantially increased (72%) at the expense of specificity (65%). The overall accuracy of the more quantitative approach (72%) was similar to that obtained with the qualitative interpretation (67%). There was a large overlap of maximum SUVs between sextants with tumors and those without tumors. It was proposed that high uptake of [11C]Choline is not specific for prostate cancer and that areas of prostate cancer may have SUVs that overlap those obtained from benign hyperplastic prostatic tissue. Compared with MR imaging and PET/CT results, 3D MR spectroscopic imaging results showed the highest sensitivity, while PET/CT results showed the highest specificity, for sextant localization of prostate cancer. The best results for localizing cancer within the prostate were obtained by combining MR imaging with 3D MR spectroscopic imaging. While the overall accuracy of combined MR imaging and 3D MR spectroscopic imaging for sextant localization of prostate cancer was similar to that of PET/CT, both good sensitivity (88%) and good specificity (90%) could be achieved with the appropriate combination of anatomic and metabolic information. All imaging techniques demonstrated a relatively large number of false-positive findings, which is possibly related to the presence of HGPIN. This could, in principle, stem from the fact that HGPIN is a precursor lesion leading to adenocarcinoma of the prostate and could give rise to the same metabolic and [11C]Choline uptake behavior as prostate cancer at both combined 3D MR spectroscopic imaging and MR imaging and PET/CT. On the other hand, not all voxels containing HGPIN demonstrated abnormal metabolic patterns and increased [11C]Choline uptake suggestive of prostate cancer. In fact, only 36% of sextants with HGPIN alone had positive findings at 3D MRS, and only 29% had positive findings at PET/CT. The slight concordance between PET/CT and 3D MR spectroscopic imaging depiction of prostate cancer could be explained by several factors. The comparison of cancer localization abilities could be affected by the intrinsic difference in spatial resolution between the techniques and by difficulties inherent in sextant correlation of imaging data. The authors showed that PET/CT correctly depicted transition zone tumors in all cases. This finding suggests a role for PET/CT in the identification of transition zone tumors, particularly those in the anterior portion of the central gland, which are often missed at TRUS guided biopsy and at combined MR imaging and 3D MR spectroscopic imaging. In conclusion, comparable specificity was obtained by using either combined MR imaging and 3D MR spectroscopic imaging or [11C]Choline PET/CT for localizing cancer to a sextant of the prostate; however, [11C]Choline PET/CT demonstrated much lower sensitivity relative to 3D MR spectroscopic imaging alone or to combined MR imaging and 3D MR spectroscopic imaging.

Giovacchini et al. [54] assessed the dependence of [11C]Choline SUV in the prostate gland on cell malignancy, PSA levels, Gleason score, tumour stage and anti-androgenic hormonal therapy. The effect of anti-androgenic hormonal therapy on [11C]Choline prostate uptake was an interesting issue not yet explored. There are currently no guidelines addressing the question as to whether hormonal therapy should be withdrawn in staging or restaging PET/ CT studies. If a significant inhibitory effect of hormonal therapy on [11C]Choline uptake is exerted, the capacity of the technique to detect the disease might be limited. In this prospective study, PET/CT with [11C]Choline was performed in 19 prostate cancer patients who subsequently underwent prostatectomy with histologic sextant analysis (group A) and in six prostate cancer patients before and after anti-androgenic hormonal therapy (bicalutamide 150 mg/day; median treatment of 4 months; group B). Histologic sextant-by-sextant analysis of the surgical specimen revealed cancer foci in 67 of 114 (58.8%) sextants. On a patient basis, the median number of sextants with evidence of cancer foci was 4. A SUV of 2.5 was the best cutoff to differentiate between malignant and benign histopathology. Among the 67 sextants with histologic evidence of cancer, 48 (71.6%) displayed [11C]Choline SUV greater than 2.5. Among the 47 sextants without histologic evidence of cancer, 20 (42.6%) did not show significant [11C]Choline uptake. True-positive (TP), true-negative (TN), false-negative (FN) and false-positive (FP) sextants were 48 (42.1%), 20 (17.5%), 19 (16.7%) and 27 (23.7%) of 114, respectively. Thus, [11C]Choline PET/CT sensitivity, specificity, positive predictive value, negative predictive value and accuracy in localizing tumour within the prostate gland were 72%, 43%, 64%, 51% and 60%, respectively (Fig. 1). To assess the impact of anti-androgenic therapy on the imaging technique, the sextant analysis was performed separately for the two sub-groups. In patients that were not receiving hormonal therapy sensitivity, specificity and accuracy were, respectively, 93%, 25% and 63%, while in patients that were receiving hormonal therapy sensitivity, specificity and accuracy were, 41%, 80% and 55%, respectively. Focal uptake was more frequently encountered in malignant disease, while diffuse uptake was more frequent in benign conditions and during antiandrogenic therapy [49, 50, 52]. Univariate analysis showed no significant correlation between [11C]Choline SUV and PSA, Gleason score or stage. A statistically significant negative correlation was detected between SUV and anti-androgenic therapy, indicating a potential interfering effect of hormonal therapy on prostate [11C]Choline uptake. Comparing baseline and post-treatment bicalutamide SUV values for each patient of group B, a decline in SUV after bicalutamide treatment was evident in all patients, despite sizeable inter-subject differences. In the whole group, SUV values after therapy were significantly ($P<0.05$) reduced compared to baseline (6.4±4.6 and 11.8±5.3, respectively), with a percent decrease of 45±32%. The decrease in PSA levels was of greater magnitude than SUV [54]. These results indicated that a clinically effective anti-androgenic therapy is associated with a substantial inhibition of prostate [11C]Choline uptake. Antiandrogenic therapy induced atrophy of glandular cells, with increased matrix and thinning of both malignant and normal epithelial cells [55]. A substantial time-dependent loss of the prostatic metabolites choline, creatine and citrate was observed during hormone deprivation therapy [56]. However there was no significant relationship between length of bicalutamide therapy and percent changes in [11C]Choline uptake (the effect of longer treatment was not assessed). Anti-androgenic therapy was associated with down-regulation of the expression of several genes, including genes involved in lipid metabolism [57]. Thus, it is conceivable that reduced activity of the choline transporter or choline kinase activity may occur after treatment with bicalutamide. However, the relation between [11C]Choline prostate uptake and cell cycle is uncertain. In vitro, a positive correlation was found between [14C]choline radioactivity and incorporation of [3H]methyl-thymidine into deoxyribonucleic acid [58]. Conversely , in vivo [11C]Choline SUV did not correlate with the expression of Ki-67 [26]. In conclusion the low accuracy of the technique indicated that [11C]Choline PET/CT should not be used as a first-line technique for the initial diagnosis of prostate cancer. Giovacchini et al. demonstrated that hormonal therapy exerts a sizable inhibition on prostate [11C]Choline accumulation and in patients with either local or systemic disease treated with hormonal therapy, PET/CT with [11C]Choline could be used, in combination with routine PSA level measurements, to monitor the response [54].

Li et al. [59] in a recent study on 49 pts investigated the potential role of [11C]Choline PET/CT for differentiating prostate cancer from benign prostate hyperplasia: PET data were analyzed visually and semiquantitatively by measuring SUV of the prostate lesions (target) and the muscles and calculating their ratios (P/M). Twenty-one prostate cancer and 28 benign prostatic hyperplasia (BPH) lesions were proven histologically. The mean values of the SUV of prostate cancer and BPH were 7.87 ± 5.74 and 4.95 ± 5.14, respectively without a significant statistically difference between the 2 groups. On the other hand, the group difference in P/M values was significant. The mean values of P/M of the prostate cancer and BPH were 4.21 ± 1.61 and 1.87 ± 0.98, respectively. Using 2.3 as cut-off, [11C]Choline PET/CT showed a sensitivity of 90.5%, a specificity of 85.7%, and a negative predictive value of 92.3%. They also observed that PET/CT precise localization of the hot spot in different parts of the prostate could contribute to the diagnosis.

Table 1: Principal [11C]Choline PET and PET/CT studies on diagnosis of prostate cancer (modified from Bouchelouche and Oehr [124])

Authors	N. Pts	% Sn	% Sp	% PPV	% NPV	Tomograph	Comments
Hara et al. [21]	10	-	-	-	-	PET	Choline vs FDG PET comparison. SUVchol>3 in neoplastic tissue
de Jong et al. [25]	30	100	-	-	-	PET	Moderate Choline uptake also in BPH. 1 pt FN (micrometastasis) and 1FP (inflammation) at LN
Sutinen et al. [47]	19	100	-	-	-	PET	No significant differences between SUV in cancer and BPH. Lower uptake in hormonal therapy
Breeuwsma et al. [26]	18	100	-	-	-	PET	No relation between uptake and cell proliferation (Ki67)
Yamaguchi et al. [48]	20	100	-	-	-	PET	Comparison with MRI/MRS. Sn 60% for MRI and 65% for MRS
Farsad et al. [49]	41	66	81	87	55	PET/CT	Sexant-based analysis. No SUV differences between cancer and HGPIN. PET not recommended as first line screening. Possibly helpful to guide biopsy in selected high-risk pts. Accuracy 71%.
Reske et al. [50]	26	81	87	86	83	PET/CT	Sexant-based analysis. SUV theresold of 2.65. No correlation between SUV and GS, PSA; weak relation between SUV and pT. Accuracy 84%.
Martorana et al. [51]	43	66	83	-	-	PET/CT	Sexant-based analysis, comparison with TRUS. PET/CT detected 4% of cancer foci <5mm. SUV correlated positively with PSA and age, negatively with grading, no correlation with pT. TRUS: Sn 61%, Sp 97%.
Scher et al. [52]	58	86	62	80	72	PET/CT	SUV theresold of 3.3: Sn 70%, Sp 57%. Overlap of Choline uptake in benign and malignant disease. Accuracy 78%.
Testa et al. [53]	26	55	86	-	-	PET/CT	Comparison with MRI. PET/CT accuracy was 67% (qualitative) and 72% (quantitative with SUV>2.9. Overlap SUV in benign and malignant disease. MRI: Sn 54%, Sp 75%, accuracy 61%. 3D MRS: Sn 81%, Sp 67%, accuracy 76%.
Giovacchini et al. [54]	25	72	43	64	51	PET/CT	Sexant-based analysis, SUV cut-off : 2.5. Overall accuracy 60%. For pts not receiving hormonal therapy: Sn 93%, Sp 25%, accuracy 63%. For pts receiving hormonal therapy: Sn 41%, Sp 80%, accuracy: 55%. SUV values after therapy significantly reduced.
Li et al. [59]	49	90	86	83	92	PET/CT	No significant SUV differences between cancer and BPH. Accuracy 88%. Values obtained using a SUV lesion/SUV muscle cut off of 2.3.

In summary, several studies confirmed the capability of [11C]Choline PET/CT in detection of prostate cancer but, actually, the low accuracy of the technique indicates that it should not be used as a first-line technique for the initial diagnosis of prostate cancer.

[11C]Choline PET and PET/CT studies on the detection of primary prostate cancer (diagnosis) are summarized in Table 1.

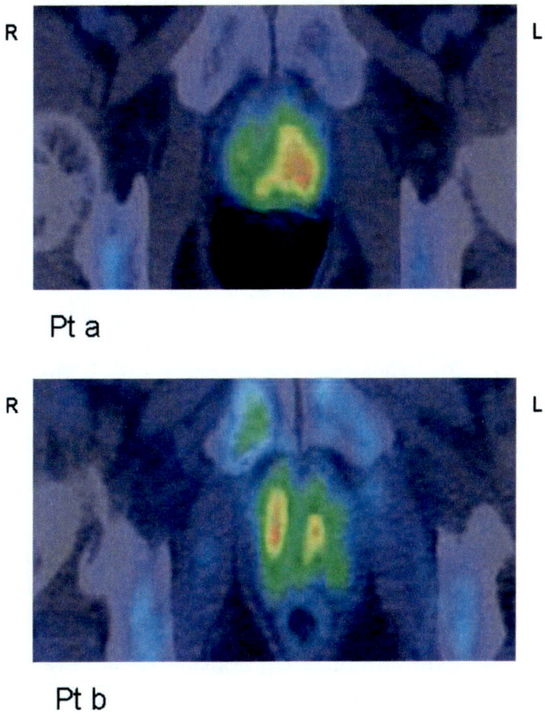

Figure 1. [11C]Choline PET/CT findings in two patients (Pt a and Pt b) with trans-rectal ultrasonography (TRUS)-guided biopsies indicating prostate cancer. Patients were studied with [11C]Choline PET/CT for primary staging before being treated with prostatectomy and pelvic lymphadenectomy. In both cases [11C]Choline PET/CT resulted bilaterally positive at prostate level. However, while in patient a there was an agreement between [11C]Choline uptake and histological analysis, reporting a prostate cancer of large size predominantly present in the left part, in patient b, a significant [11C]Choline bilateral uptake was present with a very small cancer documented at histology.

R: Right; L: Left

[11C]Choline PET: Staging of Primary Prostate Cancer

The role of [11C]Choline PET in staging prostate cancer has been largely investigated. An accurate primary staging is, in fact, foundamental in therapeutical management [32]. There are well-established therapeutic options in the management of localized cancer patients according to European Association of Urology. For stage N+/M0 and M+ standard hormonal therapy is recommended.. A theoretical prediction of pathological stage can be made on a patient-by-patient basis by using nomograms with various clinical variables (PSA level, Gleason-score, clinical state). However, the use of these nomograms is limited because they only estimate the likelihood for organ-confined disease, but they do not provide exact staging information in the individual patient. Transrectal ultrasound (TRUS) is the most commonly used modality for local staging of the prostate. However, only 60% of tumors are visualized by TRUS, and the method often is used simply to localize the prostate, as a guide to biopsy. Nonetheless, it has been reported that, in experienced hands, TRUS detects extracapsular extension with accuracy between 58% and 86%. However, a recent multicenter trial casts doubt on this statement because TRUS was no more accurate at predicting organ-confined disease than was the digital rectal examination [60, 61]. The role of CT in the local staging of prostate cancer is limited because of its low accuracy (24% for extracapsular extension and 69% for seminal vesicle invasion) [62]. In experienced hands, the most accurate imaging test for primary staging is MRI using an endorectal coil in conjunction with a pelvic phased-array coil. Although not perfect, this test can identify and localize the primary tumor and assess for potential extracapsular extension (sensitivity 50%; specificity 95%) and seminal vesicle invasion. In addition, MRI can detect or exclude perineural invasion of tumor with high accuracy, which is an important criterion when nerve-sparing surgery is considered. Overall sensitivity and specificity for the local staging of prostate cancer range between 51% and 89% and 67% and 87% respectively, with accuracy between 54% and 88% [34, 63-65]. The addition of spectroscopy to MRI may aid in the detection of prostatic cancer tissue and may increase the specificity of MRI. The addition of MRS to MRI improved staging accuracy significantly when used by an inexperienced reader compared with that by an experienced reader, and the latter was able to further improve the accuracy [63]. However, MRS is sensitive to technical artifacts and currently still considered investigational. It also needs to be emphasized that these data from MRI and MRS were generated by a few experienced institutions; it remains to be seen whether the reported high sensitivity and specificity can be reproduced once the technique enjoys widespread use.

[18F]FDG PET was found to have a low accuracy in primary staging of prostate cancer. FDG is slightly better in detecting metastatic prostate cancer than in primary lymph nodes [66]. Heicappell et al. [67] used [18F]FDG PET for preoperative imaging of pelvic lymph nodes in 17 newly diagnosed prostate cancer patients and then compared the findings with postoperative histopathology. [18F]FDG PET was able to diagnose metastatic lymph nodes accurately in 4 of 6 affected patients. The 2 false-negative results were attributed to the small size of the lesions (less than 5mm). There was no false-positive result in this study. Other small studies reported sensitivities ranging from 0 to 50% and specificities ranging from 72 to 90% for detection of nodal metastases [23, 68]. [18F]FDG uptake was noted in pelvic lymph nodes in patients with PET negative primary. [18F]FDG PET is variable in the detection of bone marrow metastases; however, the general belief is that it is more sensitive for the detection of bone metastases than local disease [69]. Yeh et al. [70] found that only 18% of bone lesions on the bone scan showed [18F]FDG uptake. Kao et al. [71] reported a high specificity of [18F]FDG PET in detection of bone marrow metastases. Nunez et al. [72] found better detection of cervical spine metastases by [18F]FDG PET than by bone scan. In another study Morris et al. [73] evaluated 154 bone lesions in 17 patients. Both FDG and the bone scan were positive in 71% lesions, 23% were seen only on bone scan, and 6% were seen only on PET scan. Schöder and Larson [42] suggested that FDG may selectively detect more aggressive tumors, which depend on higher glucose metabolism.

At summary, because [18F]FDG PET and PET/CT are not indicated for staging of primary prostate cancer, other radiotracers such as [11C]Choline, were investigated.

T-staging of prostate cancer involves the assessment of capsular penetration, seminal vesicles infiltration and/or infiltration of neurovascular bundles. More than 30% of men with clinically organ-confined disease have evidence of extracapsular extension on pathological analysis [74]. Thus the exact extent of the disease (extra-capsular extension, seminal vescicle invasion, pT4) should be clarified before surgery as precisely as possible, to plan rational therapy related to the clinical state and the prognosis of the prostate cancer.

Martorana et al. [51] investigated the role of [11C]Choline PET/CT in the assessment of extra-prostatic extension (EPE) comparing with the sensitivity/specificity of PET/CT and MRI. EPE was suspected in the presence of focal uptake outside the rim of the prostate and/or inside the seminal vesicles. Considering MRI results EPE was determined for each prostatic lobe, and considered positive in the presence of at least 1 area such as irregular capsular bulge, periprostatic fat infiltration, asymmetry, or direct involvement of the neurovascular bundles, obliteration of the rectoprostatic angle and seminal invasion (judged by low signal intensity within seminal vesicles). Histopathologically proven EPE was present in 19 (44%) patients, and in 32 of 86 (37%) prostate lobes; EPE/vesicular involvement was unilateral in 6 patients and bilateral in 13. Sensitivity of [11C]Choline PET/CT for this purpose was low in comparison with MRI (22% vs 63%, $p < 0.001$), as was the correct recognition (67% vs 83%, $P=0.004$). Unsurprisingly these data indicated that [11C]Choline PET/CT had poor sensitivity for the detection of extra-prostatic extension. This result can be partially attributed to the limitations of CT for detailed visualization of prostatic anatomy, in addition to the poor spatial resolution (5 mm) of PET.

Rinnab et al. [75] evaluated the staging capability of [11C]Choline PET in clinically localized prostate cancer. They compared images obtained by TRUS and PET with the histopathological findings after radical prostatectomy. The primary prostate cancer could be visualized by PET as a focal accumulation within the prostate in 52 of 53 patients (SUV= 4.19 ± 1.49). There was a slight correlation between SUV and the PSA level and there was no correlation between SUV with grading or Gleason score. In one patient both PET and TRUS failed to detect the tumour. Two patients had histologically confirmed lymph node metastasis; [11C]Choline PET identified one and in the second there were micrometastases in a pelvic lymph node. The overall mean (95% CI) staging accuracy by TRUS was 26% (15–40%), and for PET was 70 (56–82)%. [11C]Choline PET in the present study had a sensitivity of 66% and a PPV of 66% for predicting seminal vescicle invasion, and 27% and 57% for detecting extracapsular envolvement. For detecting a pT4 tumour the sensitivity was 25% and the PPV 100%. They concluded that PET was more sensitive than TRUS but in all cases both TRUS and PET tended to understage the prostate cancer. PET significantly more often determined the correct tumour stage than TRUS ($P<0.001$). In conclusion, [11C]Choline PET and TRUS might only assist in deciding the appropriate management for each patient and they may not be a base for treatment decisions alone because macroscopic disease beyond the prostatic capsule and into the periprostatic fat or in seminal vesicle is often accurately detected, but both imaging methods understage the microscopic extension of prostate cancer.

On the basis of literature results, [11C]Choline PET/CT is not appropriate for initial T staging due to limited spatial resolution of PET that does not enable the assessment of capsular infiltration and microscopic spread of local disease.

The presence of metastatic spread has therapeutic implications; frequently it renders surgical therapy inappropriate, anti-hormonal or radiation therapy regimens instead being favoured. Lymph node metastases are found in up to 25% of the patients with prostate cancer depending on the tumor stage and grade [76]. Lymph node involvement is correlated with progressive disease in most of the patients, and the 5-ys disease-free survival rate subsequently decreases from 85% (nonmetastatic) to approximately 50% for pN1 disease [76]. Determination of tumor involvement of regional lymph nodes is of key importance for the proper planning of treatment. Pelvic lymphadenectomy is the gold standard for lymph node staging in prostate cancer, but this invasive procedure is associated with morbidity [77]. Conventional imaging techniques such as CT and MRI have been shown to have a low sensitivity in determination of tumor involvement of pelvic lymph nodes, because nodal involvement is not always correlated with enlargement [78, 79]. For patient with high risk of lymph-node metastases, a pelvic lymphadenectomy is an independent procedure. For this group of patients, an accurate noninvasive imaging technique for lymph-nodal staging would be welcome because this could not only reduce morbidity and lead to faster planning of treatment but also be cost-effective.

Kotzerke et al. [23] explored the role of [11C]Choline PET in staging primary prostate cancer in 12 patients with proven prostate cancer; all patients presented focally increased [11C]Choline uptake suspicious for lymph-node metastases. Histophatological examination showed 9 patients true negative, 1 true positive, 1 false positive (probably due to bowel activity), 1 false negative (size smaller than 1 cm). This preliminary result suggested that [11C]Choline PET could be a promising new tool for the primary staging of prostate cancer, and that therapeutic management could be influenced by these results in that the technique may permit avoidance of surgical lymph node exploration.

Further investigation was made by de Jong et al. [24] with the aim to evaluate the strength and accuracy of [11C]Choline PET in preoperative noninvasive staging of pelvic lymph nodes in prostate cancer. They examined 67 consecutive patients with histologically proven prostate cancer with [11C]Choline PET and the results of PET were compared with the results of histology of the pelvic lymph nodes and with the follow-up data. [11C]Choline PET detected nodal metastases in 12 patients and false-negative [11C]Choline PET scans were found in 3 patients. In 1 patient a micrometastasis of 3 mm was not visualized. The failure to detect a micrometastasis could be due to the limitations of the resolution of the PET cameras, which is around 4-5 mm. However, next to the intrinsic resolution of the system, the signal-to-noise ratio (contrast) is also of importance in PET imaging. A low contrast will decrease the visualization: in the second patient, the uptake of [11C]Choline in an obturator nodal metastasis of 2 cm was not above the background activity of the pelvic region and in the third patient the activity in the small bowel interfered with evaluation of the pelvic area and masked a 2-cm lymph node metastasis with extra nodal extension located on the right external iliac vein. In 52 patients without lymph node metastases, 2 false-positive

[11C]Choline PET scans were seen: one case of inflammatory lymph node and one focal bowel. In conclusion [11C]Choline PET showed a sensitivity of 80%, a specificity of 96%, and an accuracy of 93%. This study was not designed to compare [11C]Choline PET with conventional imaging, but in this series the conventional axial imaging showed a sensitivity of 47% and a specificity of 98%.

Schiavina et al. [80] evaluated the accuracy of [11C]Choline PET/CT in lymph-node staging in intermediate and high-risk prostate cancer, compared to nomograms. Twenty seven patient at intermediate risk (cT1–2 and PSA between 10 and 20 ng/ml and/or Gleason score (Gs) and 30 at high risk (PSA≥20 and/or Gs=8–10 and/or cT3) underwent preoperative [11C]Choline PET/CT and radical prostatectomy with extended pelvic lymph-node dissection. Fifteen patients (26%) had lymph-node metastasis, and a total of 41 metastatsis were identified. On a patient analysis, sensitivity, specificity, positive predictive value, negative predictive value, and number of correctly recognized cases at PET/CT were 60.0%, 97.6%, 90.0%, 87.2%, and 87.7% while, on node analysis, these values were 41.4%, 99.8%, 94.4%, 97.2%, and 97.1%. The mean diameter (in mm) of the metastatic deposit of true-positive lymph-node metastasis was significantly higher than that of false-negative lymph-nodes (9.2 vs 4.2; p=0.001). Certainly the limited spatial resolution of PET/CT (~ 4–5 mm) may hamper the recognition of micrometastasis. However, 10 of 25 metastatic deposits >5 mm were not visualized by PET/CT, and thus the metabolic state of the tumor may play a major role in some of these cases. For these reasons, a negative [11C]Choline PET/CT actually is not sufficient to rule out a lymphadenectomy. Another factor that may influence the sensitivity of [11C]Choline PET/CT is the number of lymph-node metastasis in the single patient independently of the Gleason score, the PSA, or the tumor stage. [11C]Choline PET/CT performed better than clinical nomograms for lymph-node metastases detection, with equal sensitivity and better specificity [80]. The sensitivity of [11C]Choline-PET/CT was quite low at patient analysis (60%) but the diagnostic accuracy of this technique seems to be better than that of the conventional imaging techniques as reported in the literature [81].

With regard to metastatic spread, in the study by Scher et al. [52] PET showed a per-patient sensitivity of 81.8% (9/11) and produced no false positive findings. In 26 of the 37 patients with proven prostate cancer, metastatic spread could be ruled out on the basis of histopathological examination of resection specimens obtained at surgical procedures or a follow-up period of at least 6 months. These data suggested that [11C]Choline PET/CT, as a whole-body procedure, may prove to be a valuable diagnostic tool in the search for metastases.

Eschmann et al. [82] compared the diagnostic accuracy of [11C]Choline PET/CT and whole body magnetic resonance imaging (WB-MRI) for diagnostic work-up of prostate cancer. They evaluated retrospectively 42 patients with untreated prostate cancer (n=17), or increasing levels of prostate-specific antigen (PSA) after curative therapy (n=25) who had been investigated by both [11C]Choline PET/CT and WB-MRI. 88/103 detected lesions were considered as malignant by histology, follow-up, or consensus reading: 44 bone metastases, 22 local tumor, 15 lymph node metastases, 3 lung, and 3 brain metastases. Overall sensitivity, specificity and accuracy for PET/CT were 96.6%, 76.5%, and 93.3%, respectively, and for WB-MRI 78.4%, 94.1%, and 81.0%, respectively. For lymph-node metastasis sensitivity, specificity for PET/CT were 100% and 66.7% and for MRI were 80% and 88.9%; for bone metastasis were respectively 93.2%, 100% and 88.6% and 100%. Both proved high accuracy in the detection of bone and lymph node metastases of prostate cancer. While for the detection of bone metastases both methods are of similar accuracy, differences can be observed for the assessment of lymph node metastases. On one hand, [11C]Choline PET/CT has a high potential for the detection of lymph node metastases, combined with a lower specificity due to choline uptake in inflammatory lymphadenopathy. On the other hand, WB-MRI has a lower detection rate of lymph node metastases, paired with a higher specificity. The authors concluded that an optimal diagnostic accuracy for lymph node assessment might provided by combination of both imaging techniques.

In a technical study Luboldt et al. [83] evaluated the feasibility of fast diffusion-weighted imaging as a replacement for conventional magnetic resonance imaging sequences (STIR and T1 weighted SE) and the role of [11C]Choline PET/CT in the detection of skeletal metastases from prostate cancer. They examined 11 patients with prostate carcinoma and known bone metastasis. Imaging revealed 16 metastases. This pilot study suggested that diffusion weighted imaging is equally as sensitive as, if not more sensitive than, the conventional sequences in the detection of bone metastasis from prostate cancer. In addition the quality of diffusion weighted imaging in the depiction of bone metastasis even suggested that it was as effective as [11C]Choline PET/CT.

At present, the use of [11C]Choline PET/CT imaging for preoperative staging should cannot be recommended for clinical routine use. Large prospective studies including a cost-effectiveness analysis are needed to assess the real impact of [11C]Choline PET/CT in prostate cancer staging.

[11C]Choline PET and PET/CT studies on the staging of primary prostate cancer are summarized in Table 2.

[11C]Choline PET: Restaging of Prostate Cancer

In most cases, recurrent disease initially presents as biochemical recurrence (BCR), with an increase in serum levels of PSA. Approximately one-third of patients undergoing radical prostatectomy and a similar number of patients with radiation therapy will develop BCR. In the majority of cases, this early rise in PSA occurs without any symptom or other objective findings. The clinical behaviour of patients is extremely heterogeneous and it is common for patients to survive for 5 to 10 years with an elevated PSA as the only evidence of recurrent disease. Technical advances in surgery and radiation therapy have improved therapeutic efficacy and decreased treatment-associated morbidity and toxicity, respectively. However, despite these improvements, there is still a significant risk of cancer recurrence after therapy and up to 27-53% of all patients undergoing radiation therapy or radical prostatectomy will develop local or distant recurrences within 10 years after initial therapy and 16-35% of patients will receive second-line treatment within 5 years of initial therapy [84-88]. In asymptomatic patients, a disease-specific history and a serum PSA measurement supplemented by digital rectal exam are the recommended tests in routine follow-up. These tests should be performed at 3, 6, and 12 months after treatment, then every 6 months until 3 yr, and then annually. Nomograms have been established for predicting probability for PSA relapse and to determinate the likelihood for local recurrence vs metastatic disease. After radical prostatectomy, a serum PSA level > 0.2 ng/ml can be associated with residual or recurrent disease (grade B recommendation). After radiation therapy, a rising PSA level 2.0 ng/ml above the nadir value, rather than a specific threshold value, is the most reliable sign of persistent or recurrent disease [32].

Table 2. Principal [11C]Choline PET and PET/CT studies on staging of primary prostate cancer. (modified from Bouchelouche and Oehr [124])

Authors	N. Pts	% Sn	% Sp	% PPV	% NPV	Tomograph	Comments
Martorana et al. [51]	43	22	-	-	-	PET/CT	T-staging. comparison with MRI for extraprostatic extension. Correct recognition: 67% for PET/CT and 83% for MRI
Rinnab et al. [75]	55	36 (17-59)	-	73 (39-89)	-	PET	T-staging comparison with TRUS. PET accuracy: 70% (56-82%) and for TRUS 26% (15-40%).PET and TRUS tend to understage.
Kotzerke et al. [23]	12	-	-	-	-	PET	N-staging. Histology. 9 TN, 1 TP, 1 FP (bowel activity), 1 FN (< 1cm).
de Jong et al. [24]	67	80	96	-	-	PET	N-staging. Histology and follow-up. PET accuracy: 93%. Conventional imaging: Sn 47%, Sp 98%
Schiavina et al. [80]	57	60 41	98 100	90 94	87 97	PET/CT	N-staging. patient/lesion-based analyses. PET/CT better than clinical nomograms.
Scher et al. [52]	58	82	-	-	-	PET/CT	M-staging. Per ptanalysis.
Eschmann et al. [82]	42	97 100 93	76 67 100	- - -	- - -	PET/CT	comparison with MRI. PET/CT accuracy 93%. N and M MRI staging: Sn 78%, Sp 94%, accuracy 81%. N staging, MRI: Sn 80%, Sp 89%.Bone, MRI: Sn 89%, Sp 100%.
Luboldt et al. [83]	11	-	-	-	-	PET/CT,	Comparison with MRI sequences in detection of bone metastases.

Although PSA is the most sensitive tool for detecting recurrences, it cannot distinguish between local, regional or distant recurrence. Unfortunately, physical examination and conventional imaging techniques are not sensitive in the detection of recurrent disease and it is frequent to have patients with biochemical recurrence and negative conventional imaging studies. Local recurrence appears as a hypoechoic mass in the surgical bed with TRUS. However, up to 30% of recurrent tumors may be isoechoic and difficult to detect by TRUS [89]. The fibrosis after radiation distorts normal tissue planes and also makes it difficult to detect recurrence in this altered tissue. The overall sensitivity of TRUS alone following radiation or surgery is less than 50% [89, 90]. In these conditions sampling error is the major cause of false-negative results. TRUS is not useful for detection of distant metastases. CT has been used to assess metastastic disease in the prostate bed, nodes, bone, and visceral organs. The role of CT in the detection of local recurrence is limited particularly if the size of the recurrent tumor is smaller than 2 cm [37]. CT detection rate of local recurrence was only 36% even in patients with the tumors larger than 2 cm in a study by Kramer et al. [91]. CT currently is the most frequently used imaging modality in the evaluation of nodal metastases. CT relies on the nodal size of > 1 cm to discriminate metastases, therefore, the specificity is low; however, due to its easy accessibility and fast scanning, it is used most frequently. CT scans may be normal when the bone scan is positive for metastatic disease, therefore, the role of CT in the evaluation of bone metastases is limited and its routine use is not recommended. However, CT can be used to characterize a focal uptake in a bone scan as benign or malignant. Endorectal MRI/MRS have been proven capable of detecting local recurrence in many patients with rising PSA particularly in patients with no palpable tumor in the prostatic fossa. Silverman and Krebs [92] reported both sensitivity and specificity of 100%. A high efficacy (95% sensitivity and 100% specificity) also has been reported by Sella et al. [93]. Although the most common site of local recurrence is the perianastomotic site, 30% can occur elsewhere in the pelvis. All of these sites of recurrence also can be seen and detected by MRI. Endorectal MRI is still not in routine clinical use; however, it seems very promising. MRI has the potential to direct a transrectal biopsy to these sites and, when used in conjunction with pelvic phase array coil, it can evaluate pelvic lymph nodes and osseous structures, thus detecting all sites of pelvic relapse in a single examination. MRI is very good at evaluating bone marrow metastases; however, it is expensive, time consuming, and susceptible to motion artifacts, therefore, it is rarely used for evaluation of distant bone metastases.

In summary local recurrences are detected by transrectal ultrasound, CT or MRI with a sensitivity of approximately 76% unless endorectal coil-MRI is done (100% sensitivity) [92, 94]. Metastases to bone are evaluated by [99mTc]methylenediphosphonate (MDP) bone scintigraphy with high sensitivity but low specificity (approximately 50%) [95]. Lymph node involvement is evaluated by CT or MRI with approximately 65% sensitivity [96]. Detecting recurrence via diagnostic imaging is difficult, particularly for PSA increases smaller than 20 ng/ml yearly since a high incidence of false-negative findings is reported [94].

Improvement may be reached by imploying an imaging technique based on metabolism rather than an anatomic imaging technique. In this respect, PET may play a role. Overall the results of imaging recurrent prostate cancer were not successful with FDG. Seltzer et al. [97] found a detection rate of 50% with [18F]FDG PET and CT when the PSA was > 4 ng/ml. Herrmann et al. [98] found a detection rate of 35% in recurrent prostate cancer. They concluded, due to the low yield of true-positive findings, that [18F]FDG PET is not useful in patients with PSA less than 2.4 ng/ml after prostatectomy. Other PET tracers, in particular [11C]Acetate and [11C]Choline appear more promising in the detection of recurrent prostate cancer. [11C]Choline PET has been studied by several investigators, mainly in Europe, who have reported high success rate in identifying recurrent prostate cancer.

De Jong et al. [99] investigated the feasibility of [11C]Choline PET in the evaluation after radical prostatectomy or external beam radiotherapy in localized prostate cancer. A total of 36 patients participated in this study and all patients were without evidence of lymph node or distant metastases at the time of treatment. Patients on adjuvant hormonal therapy within six months prior to the study were excluded. From both treatment groups were selected patients with no biochemical evidence of disease. Patients with biochemical evidence of disease after radical prostatectomy were identified by a postoperative serum PSA > 0.2 ng/ml while patients with a biochemical evidence of disease after radiotherapy were identified by three consecutive rises of serum PSA after nadir according to the American Society for Therapeutic Radiology and Oncology criteria. The results of this study showed that in a total of 36 patients, in 14 patients with no biochemical evidence of disease [11C]Choline PET was true negative. 22 patients had a biochemical recurrence of prostate cancer (13 patiens of the group of radical prostatectomy and 9 of the group of radiotherapy). Of the 13 patients with an increased PSA after radical prostatectomy, a local recurrence was proven by biopsy in 3 patients and lymph node metastases were proven after lymphadenectomy in 4 patients. [11C]Choline PET identified all lesions, as did conventional imaging with TRUS and CT scan in these patients. On a patients basis [11C]Choline PET was true positive in 5/13 (38%). In 7 patients with an increased PSA after radical prostatectomy between 0.9 ng/ml and 4.9 ng/ml, [11C]Choline PET was false negative. Although no proof of recurrence was found by DRE, conventional imaging, biopsy and clinical follow up in these patients, a benign cause for a PSA increase after radical prostatectomy is extremely uncommon. This result is in corroboration with the experience reported by Kozerke et al. [23] using [11C]Choline PET in two patients with increasing PSA < 2 ng/ml after radical prostatectomy. Both [11C]Choline PET and conventional imaging failed to detect a biopsy-proven local recurrence in these patients. These false negative [11C]Choline PET scans could be due to the small volume of recurrent disease in this early stage. Failure to detective small volume disease is a known limitation of PET as well as of conventional imaging. Although the intrinsic resolution of the PET camera is around 5 mm, the signal-to-noise ratio (contrast) is also of importance in PET imaging. Seven/9 (78%) patients with increased PSA after radiotherapy, were correctly identified by [11C]Choline PET. A negative [11C]Choline PET scan in one patient with an increased PSA of 2.3 ng/ml is likely true negative as PSA further decreased on follow up. The second patient with false negative [11C]Choline PET scan showed PSA progression on follow-up. In conclusion this study shows that [11C]Choline PET is a feasible technique for the evaluation of treatment for localized prostate cancer. The site of recurrence was correctly detected in 78% of the patients after external beam radiotherapy compared to 38% of patients after radical prostatectomy. Moreover, no positive PET scans were observed in patients with a serum PSA < 5 ng/ml.

Picchio et al. [100] compared [11C]Choline PET with [18F]FDG PET for restaging prostate cancer in a group of 100 patients. Patients, referred to whole body [18F]FDG PET for clinical restaging after prostatectomy (n=77) or radiotherapy (n=23), were retrospectively evaluated. In the same day all patients were also referred to [11C]Choline PET. The indication for [18F]FDG PET was the suspicion of local recurrences or metastases due to increasing PSA on 3 measurements. Mean PSA was 6.57 ng/ml and in 93% of cases it was less than 20 ng/ml. Patients treated with radiotherapy in 6 months before PET or with prostatectomy in the previous 3 months were excluded from the study. The study showed that 47 of 100 patients had positive [11C]Choline PET findings versus 27 patients with positive [18F]FDG PET findings. None of the [18F]FDG PET positive cases was negative on [11C]Choline PET. A higher number of cases with local and/or distant lesions was detected by [11C]Choline PET compared with [18F]FDG PET in all patients. Of the 100 cases, 49 were positive on conventional imaging and agreement between [11C]Choline PET and conventional imaging was noted in 86 patients. Of the remaining 14 patients [11C]Choline PET-positive, conventional imaging-negative and [11C]Choline PET-negative, conventional imaging-positive findings were noted in 6 and 8, respectively. Of the 6 [11C]Choline PET-positive, conventional imaging-negative cases increased PSA followed by a decrease after hormonal therapy was observed in 2 suspicious for local recurrence. The 3 patients with suspicious lymph node involvement were suggested to have inflammatory lymph node disease on conventional imaging and serum PSA was lower than 1.2 ng/ml for the whole study period. The remaining patient presented with a hyper-metabolic vertebral lesion. However, [99mTc] methylenediphosphonate (MDP) bone scintigraphy and MRI were negative and serum PSA remained constant during follow-up. All 8 [11C]Choline PET-negative, conventional imaging-positive cases showed local recurrence on transrectal ultrasound and biopsy. In conclusion, [11C]Choline PET seems useful for restaging prostatectomy cases with increasing serum PSA. It is superior to [18F]FDG PET/CT and complementary to conventional imaging techniques but with the advantage of staging disease at a single step. By combining morphological and functional information, [11C]Choline PET/CT could be the first line examination for restaging prostatectomy cases with evidence of increasing PSA. When [11C]Choline PET/CT findings are positive, the patient could start the most appropriate therapy. In cases of [11C]Choline PET/CT negative scans, pelvic morphological imaging (MRI or transrectal ultrasound) should complete the diagnostic phase to exclude local recurrence.

In the last few years, a new technique has been introduced in clinical practice: PET/CT, combining morphological imaging technique with a metabolic diagnostic technique. The union of a procedure that identifies metabolic activity with an imaging technique increases the sensitivity in the diagnostic test, and can help us to find the exact location of the [11C]Choline uptake sites.

Rinnab et al. [101] used [11C]Choline PET/CT for detecting clinical recurrence after primary treatment for prostate cancer. All 50 patients with prostate cancer who had initial therapy (radical prostatectomy in 40, external beam radiotherapy (EBRT) in 3 and interstitial brachytherapy (IB) in 7) had [11C]Choline PET/CT in the presence of an increased PSA level. The mean (range) time to biochemical progression was 22 months (2-136 months). PSA level were determined in all patients at the time of examination. Patients with biochemical evidence of disease after RP were identified by a serum PSA level of > 0.2 ng/ml, and those after EBRT or IB were identify by three consecutive increase in serum PSA level after nadir. The criteria for diagnosing recurrent prostate cancer was an increased uptake of [11C]Choline retro-vesically and pre-rectally that had to be a least twice that of

[11C]Choline uptake in the periprostatic soft tissue or pararectal fat tissue, as determined by the standard uptake value. The results were correlated with the histopathology reports after targeted biopsy or surgery, and with the clinical follow-up. The mean PSA level in [11C]Choline PET/CT positive patients was 3.62 ng/ml (0.5-13.1 ng/ml) and in those [11C]Choline PET/CT negative was 0.9 ng/ml (0.41-1.40 ng/ml). Comparing the results of [11C]Choline PET/CT and histology for patients with different ranges of PSA level, the overall sensitivity and the positive predictive value (with 95% CI) of [11C]Choline PET/CT for detecting recurrence was 95% (83-99%) and 86% (73-95%), the overall specificity was 40% (12-74%) with a negative predictive value of 67% (22-96%) and an accuracy of 84% (71-93%). The sensitivity at a PSA level of < 2.5 ng/ml (including 30 patients) was 91% (71-99%) and the specificity 50% (16-84%), respectively. In conclusion, [11C]Choline PET/CT seems to be useful for restaging men with prostate cancer and increasing serum PSA levels, and should be recommended for PSA levels of 2.5 ng/ml.

Scattoni et al. [102] evaluated prospectively the accuracy of integrated [11C]Choline PET/CT in the diagnosis of lymph node recurrence in prostate cancer patients with biochemical failure after surgery, with the use of histologic results as the standard of reference, in patients undergoing retroperitoneal and/or pelvic lymph node dissection because of a rising PSA level and isolated evidence of nodal recurrence. All patients underwent a digital rectal examination, [11C]Choline PET/CT, bone scan, morphologic imaging (either CT or MR), and TRUS guided prostatic fossa biopsy to restage the disease. On the basis of clinical and radiologic data and evidence of lymph-node metastatic involvement on either [11C]Choline PET/CT and/or morphologic imaging (CT or MR), 25 patients were enrolled in the study and referred to surgical lymph node dissection. Mean PSA value at [11C]Choline PET/CT analysis was 4.01± 5.4 ng/ml. [11C]Choline PET/CT results were positive in 21 patients (13 at pelvic sites and 8 at retroperitoneum site), and conventional CT or MR imaging results were positive in 12 cases (9 at pelvic sites and 3 at retroperitoneum site). In 8 cases (67%) positive results were obtained by both PET/CT and conventional imaging. In 4 patients (37%) conventional imaging results were positive, whereas PET/CT results were negative. In 19 of 21 (90%) patients with positive [11C]Choline PET/CT, histologic examination confirmed the presence of nodal metastases of adenocarcinoma of the prostate. [11C]Choline PET/CT provided additional true positive information, in comparison to conventional CT/MR, in 11 patients (58%). The 4 patients with PSA recurrence with negative [11C]Choline PET/CT findings and positive CT/MR imaging results had no histologically proven nodal involvement of disease. In conclusion this study shows that integrated [11C]Choline PET/CT is a valuable diagnostic tool in patients with a PSA failure after radical retropubic prostatectomy, mainly because this modality is able to depict recurrent disease. Low NPV seems to depend on the limited capability of [11C]Choline PET/CT to detect microscopic or small-volume lesions. The high PPV of [11C]Choline PET/CT, even in patients with low PSA values, provides a basis for further treatment decisions. Patients with a negative PET scan with increasing PSA values should be further examined for the presence of microscopic disease or at least be strictly followed up, whereas patients with a positive PET scan should undergo therapeutic procedures.

Reske et al. [103] examined the accuracy and clinical impact of [11C]Choline PET/CT for detecting and localizing low volume local relapse of prostate cancer after radical prostatectomy. Thirty-six asymptomatic patients with biochemical evidence of relapse, histological evaluation of local recurrence of prostate cancer and a [11C]Choline PET/CT examination were included in the study (group A). All patients had histopathology from biopsy samples of the suspected local relapse. As a control group, 13 consecutive patients with a serum PSA < 0.2 ng/ml at the time of [11C]Choline PET/CT examination and a 1 year follow-up without significant increase in serum PSA and no clinical evidence of relapse were recruited (group B, controls). Inclusion criteria for all patients were adenocarcinoma of the prostate treated with radical prostatectomy, biochemical suspicion of relapse, and no evidence of metastatic disease as shown by conventional clinical work-up including DRE, TRUS, and bone scan. A PSA > 0.2 ng/ml was regarded as biochemical evidence of relapse. Androgen deprivation therapy at [11C]Choline PET/CT was allowed. In "group A" [11C]Choline PET/CT showed true positive focal nodular lesions with increased [11C]Choline uptake in the prostatic fossa in 23/33 (70%) patients with histological verification of local recurrence. In 3 patients, histology showed no evidence of recurrence. One of these patients was positive at [11C]Choline PET/CT, the other 2 were negative. SUV of the lesion and PSA were correlated. The SUV of patients with local recurrence was 3.0 (0.6-7.4). 10 patients with a positive histology had an unremarkable PET/CT with a SUV of 1.0 (0.6-1.5), which was significantly lower than the SUV of [11C]Choline PET/CT positive patients at 3.9 (1.8-7.4) ($p<0.0001$). PSA of patients with negative [11C]Choline PET/CT findings for local recurrence was significantly lower compared to that of patients with positive delineation of local recurrence by PET/CT. In "group B" [11C]Choline PET/CT was concordantly judged negative for local recurrence by both observers in 12/13 controls. In one of the 13 men in the control group, PET/CT showed a nodular, moderate focally increased radioactivity uptake in the prostatic fossa with a SUV of 1.4, which was visually judged as positive. In conclusion, [11C]Choline PET/CT can identify a subgroup of 70% of patients with biopsy proven low volume local recurrence of prostate cancer after radical prostatectomy. Thirty percent of patients failed to exhibit increased [11C]Choline uptake in local recurrence[103].

Krause et al. [104] assessed the potential of [11C]Choline PET/CT for the detection and localization of recurrent disease in patients with biochemical recurrence after primary therapy for prostate cancer and the dependence of the detection rate of [11C]Choline PET/CT on the PSA level. In this analysis were included 63 patients suffering from biochemical recurrence of prostate cancer and all patients had biopsy-proven prostate cancer. In patients after prostatectomy, a biochemical recurrence was defined as an increase of PSA value after primary treatment ≥ 0.2 ng/ml [105, 106]. Three consecutive increases in PSA defined a biochemical recurrence in patients after radiation therapy according to American Society for Therapeutic Radiology and Oncology Consensus Panel [107]. The mean interval between the primary therapy and the time of referral to [11C]Choline PET/CT was 47 ± 38 months. The mean serum PSA level of the patients was 5.9 ± 9.7 ng/ml. Results showed that 17 of the 63 patients, had anti-androgen therapy at the time of [11C]Choline PET/CT. Of the 63 patients, 35 (56%) had one or more localized areas of pathological [11C]Choline uptake indicating recurrent prostate cancer. For PSA values < 1.0 ng/ml, the detection rate of [11C]Choline PET/CT was 36% (8/22). Of the 35 patients, 18 (51%) had pathological [11C]Choline uptake in the prostate region, 9 (26%) showed abnormal uptake in abdominopelvic lymph nodes and

2 (6%) in periprostatic soft tissue. The detection rate of [11C]Choline PET/CT for recurrent prostate cancer (including local and abdominopelvic lymph nodes) showed a positive relationship with the serum PSA value. The detection rate was 36% (8/22) for a PSA value < 1 ng/ml, 43% (3/7) for PSA value < 2 ng/ml, 62% (5/8) for PSA < 3 ng/ml and 73% (19/26) for PSA levels ≥ 3 ng/ml. For the group of patients with a PSA level < 3 ng/ml, the detection rate was 47%. In conclusion, this study demonstrated a positive relationship between [11C]Choline PET/CT detection rate and serum PSA level in patients with recurrent prostate cancer. For serum PSA levels >3 ng/ml, the detection rate of [11C]Choline PET/CT is approximately 75%. [11C]Choline PET/CT could be useful in patients with prostate cancer in the case of biochemical recurrence even if the PSA value is <1 ng/ml. This implies potentially important clinical consequences because cure after radical prostatectomy with salvage radiotherapy is expected to be more successful in patients with low PSA level, local therapy is not appropriate once systemic disease has been shown and therapy could be individualised in case of detection of abdominopelvic lymph node involvement [104].

Schilling et al. [108] evaluated the potential of [11C]Choline PET/CT for planning surgery in patients with prostate cancer and PSA relapse after treatment with curative intent. To further delineate the role of [11C]Choline PET/CT in prostate cancer restaging, this study retrospectively reviewed the charts of 10 patients who had a laparoscopic lymphadenectomy for [11C]Choline PET/CT positive lymph nodes and with a PSA relapse after curative treatment for prostate cancer. Patients were referred for histological confirmation of the PET findings to determine further treatment. Recurrent prostate cancer after radical prostatectomy was defined as an increase in PSA level to >0.1 ng/ml. The results showed that in all, 22 positive lymph nodes were found by [11C]Choline PET/CT. On conventional imaging by CT, 14 enlarged lymph nodes had been identified. Comparison between PET and conventional imaging techniques showed concordance in 13 lymph nodes. 9 [11C]Choline PET/CT positive nodes had no correlating abnormal morphology on conventional CT, whereas one suspicious lesion on conventional imaging was not detected in the [11C]Choline PET/CT study. In conclusion [11C]Choline PET/CT has a limited PPV for locating pelvic lymph node metastases in recurrent prostate cancer. An important differential diagnosis for PET positive lesions is inflammatory changes in the true pelvis. Currently, the method represents the most promising tool in restaging patients with prostate cancer, as it combines conventional and functional imaging. However, [11C]Choline PET requires careful interpretation and consideration of clinical data, e.g. PSA levels, in clinical decision-making and therapeutic procedures. To date, there are no controlled studies with imaging before and after extensive salvage lymphadenectomy.

Tuncel et al. [109] recently assessed the potential of [11C]Choline PET/CT for staging and restaging patients suffering from advanced prostate cancer (n=45), with an emphasis on skeletal manifestations and to compare the diagnostic performance of PET, CT and PET/CT. PET/CT with a diagnostic CT revealed the highest detection rate: 295 lesions were detected with PET/CT (diagnostic CT). A total of 221 lesions were detected with diagnostic CT, 272 lesions with PET/CT (low-dose CT) and 178 lesions with PET alone, respectively. Two thirds of the lesions were located in the bone; one third in the prostate, lymph nodes, periprostatic tissue and soft tissue (lung, liver). Comparison of the diagnostic performance revealed that PET/CT with a diagnostic CT showed highest detection rate in staging and restaging of patients with advanced prostate cancer. An increase in detection efficiency of PET/CT vs. PET was in agreement with findings from other groups showing that the number of detected lesions increases with PET/CT, and more malignant lesions are detected[12]. In this study, the difference in detection rate of [11C]Choline PET/CT compared with PET was most pronounced in bone lesions. The added value of CT in [11C]Choline PET/CT for the characterization of skeletal manifestations is mostly related to localization of lesions (i.e., bone marrow, early cortical involvement). In fact [11C]Choline PET/CT results in a higher detection rate and improved lesion localization and characterization allowing precise whole-body restaging in a single imaging procedure. The superiority of [11C]Choline PET/CT in lesion localization and characterization can be explained by 1) the exclusion of interfering physiological uptake for example in the intestine and pancreas, 2) better identification of liver lesions because of the high tracer uptake in the liver interfering in the interpretation of PET alone, 3) better delineation of lymph nodes from other structures like vessels and bowel activity, 4) detection of bone lesions regardless of the metabolic activity of the lesion (i.e., after therapy) and 5) better characterization of lymph nodes and soft tissue lesions by using CT criteria (uptake of contrast agent, absence of fatty hilus or size). This improvement was achieved with a low-dose CT. This study revealed an added value of [11C]Choline PET/CT for skeletal manifestations with respect to localization and PET-negative bone lesions and may be helpful to differentiate active from inactive disease after therapy. In this patient group, [11C]Choline PET/CT changed therapeutic management in 11 (24%) of 45 patients. Subsequent to [11C]Choline PET/CT, three patients received radiation therapy, six received antihormonal therapy, one received chemotherapy, and one received chemotherapy and radiation therapy. During the clinical follow-up, in 10 of 11 patients PSA level decreased after initiation of the therapy.

In conlusion there is good agreement among several centres on the utility of PET/CT with [11C]choline for the detection of recurrence of disease after radical treatment in patients with biochemical failure. [11C]Choline PET/CT plays an important, emerging role in oncology because it is a noninvasive whole-body examination that allows evaluation of local and distant disease in one single step (Fig. 2 and Fig. 3).

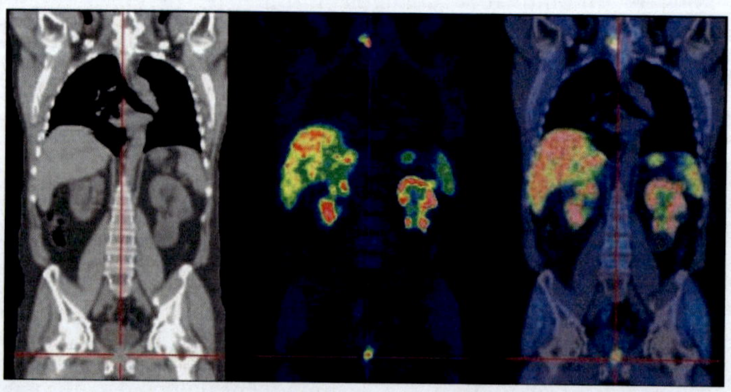

Figure 2. [11C]Choline PET/CT images of a patient with biochemical failure of prostate cancer after radical prostatectomy. CT, PET and PET/CT fused images are presented in coronal displays and demonstrate a neoplastic involvement in prostatic bed (red cross) and at a cervical vertebra.

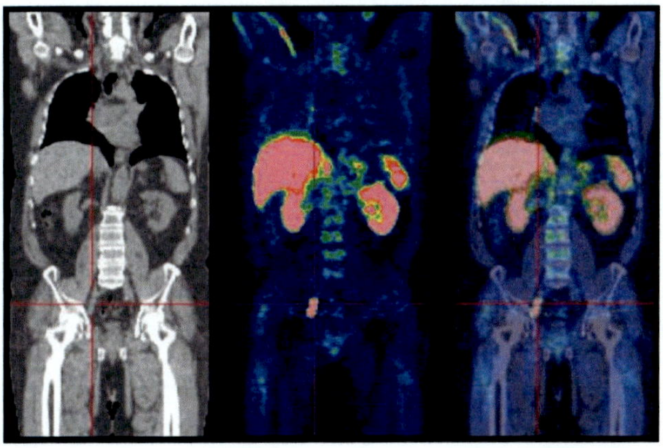

Figure 3. [11C]Choline PET/CT images of a patient with persistently increased PSA plasma levels, after radical prostatectomy. CT, PET and PET/CT fused images are presented in coronal displays., showing a lymph node metastasis in the right iliac region

[11C]Choline PET and PET/CT studies on the restaging of prostate cancer are summarized in Table 3.

Other Radiopharmaceuticals in Prostate Cancer

While [11C]Choline PET was proved a useful tool in different steps of management of prostate cancer, the short half-life (20 minutes) causes logistic problems when a cyclotron is not available on site. Due to the longer half-life of [18F] (110 minutes) [18F]labeled choline ([18F]FCH) was been introduced for PET imaging. The major advantage of [18F]FCH vs [11C]Choline is the substantially longer half-life, which allows the distribution of this tracer to PET centers without a cyclotron on site, much like the common distribution of [18F]FDG. Because urinary tracer excretion is low, these markers are advantageous in urological malignancies.

Recently, studies using [18F]FCH for prostate cancer imaging were performed. In the study by Schmid et al. [110] 10 patients with newly diagnosed prostate cancer and 9 suspected of having recurrence were evaluated with [18F]FCH PET. The study showed that differentiating malignant areas from benign prostatic hyperplasia was not possible with [18F]FCH PET.

Contrary to this finding, Kwee et al. [111] observed that the distribution of tumors in the prostate gland could be visualized with [18F]FCH PET. Prostate sextant positive for malignancy on biopsy demonstrated significantly higher SUV than biopsy negative sextants. Another study by Kwee et al. [112] showed that delayed or dual phase [18F]FCH PET may improve imaging of malignant areas of the prostate.

Dual phase [18F]FCH PET was also used in a study by Cimitan et al. [113] reporting that [18F]FCH PET was useful for detecting bone metastases. Schmid et al. [110] reported that [18F]FCH PET also may be promising for detecting local recurrence and lymph node metastases. Similar results were reported by Cimitan et al. [113] that showed that [18F]FCH PET was useful for detecting lymph node involvement, although a significant impact on therapeutic management of prostate cancer was shown only in patients with PSA greater than 4 ng/ml; the technique was also more useful in patients with poorly differentiated primary tumors (Gleason score greater than 7). In contrast, Hacker et al. [114] reported that [18F]FCH PET was not useful for preoperatively detecting regional lymph node metastases. Heinisch et al. [115] found that, when restaging cases of prostate cancer, [18F]FCH PET yielded positive findings even at PSA less than 5 ng/ml.

Table 3. Principal [11C]Choline PET and PET/CT studies on restaging of prostate cancer after primary treatment. (modified from Bouchelouche and Oehr [124])

Authors	N. Pts	% Sn	% Sp	% PPV	% NPV	Tomograph	Comments
De Jong et al. [99]	36	38-78	-	-	-	PET	No positive PET scans with a serum PSA < 5 ng/ml.
Picchio et al. [100]	100	-	-	-	-	PET,	Choline is superior to FDG PET and complementary to conventional imaging with the advantage of staging disease in a single session.
Rinnab et al [101]	50	95 (83-99)	40 (12-74)	86 (73-95)	67 (22-96)	PET/CT	Accuracy 84% (71-93%). Sn and Sp at a PSA level < 2.5ng/ml: 91% (71-99%) and 50% (16-84%). PET/CT recommended for PSA level ≥2.5ng/ml.
Scattoni et al. [102]	25	64	90	86	72	PET/CT	Accuracy 77% in diagnosis of LN recurrence. Histological results as the standard of reference.
Reske et al. [103]	49	73	88	-	-	PET/CT,	Recurrence detection in 71% of pts with a favourable biochemical response to local radiotherapy.
Krause et al. [104]	63	-	-	-	-	PET/CT	Detection rate: 36% for a PSA-value <1 ng/ml, 43% for a PSA-value 1–<2 ng/ml, 62% for a PSA-value 2–<3 ng/ml and 73% for a PSA-value ≥3 ng/ml. Anti-androgen therapy did not show a significant effect on the detection rate.
Schilling et al. [108]	10	-	-	-	-	PET/CT	Valuable tool for detecting recurrent cancer. Limited PPV.
Tuncel et al. [109]	45	-	-	-	-	PET/CT	Changes in disease management in 11/45 (24%).

Legends: Pt = patient, Sn = sensitivity, Sp = specificity, PPV = positive predictive value, NPV = negative predicitve value, P ca = prostate cancer, BPH = benign prostatic hyperplasia, FN = false negative, FP = false positive, TN = true negative, TP = true positive, LN = lymph-node, GS = Gleason score, pT = pathological T stage, MRI = magnetic resonance imaging, MRS = magnetic resonance spectroscopy, HGPIN = high-grade prostate intraepithelial neoplasm, TRUS = transrectal ultrasound, EBRT = external beam radiotherapy.

Another radiotracer for imaging prostate cancer is [11C]Acetate: cellular uptake of [11C]Acetate in tumor cells is proportional to lipid synthesis and it is incorporated into the cellular lipid pool [33]. An increase in fatty acid synthesis and over expression of the key enzyme fatty acid synthase have been demonstrated in prostate cancer. There seems to be no urinary excretion of [11C]Acetate.

Oyama et al. [116] reported marked [11C]Acetate uptake in prostate cancer lesions in 22 patients and found that [11C]Acetate PET was more sensitive for detecting prostate cancer than [18F]FDG PET.

Kato et al. [117] investigated the accumulation of [11C]Acetate in 30 subjects without prostate cancer, including 21 subjects with normal prostate and 9 with benign prostatic hyperplasia, and in 6 patients with prostate cancer. All patients with prostate cancer had positive findings on [11C]Acetate PET. However, the difference in SUV between patients with a normal prostate or benign prostatic hyperplasia and patients with prostate cancer was not statistically significant. In a pilot study Kotzerke et al. [23] investigated whether significant differences in prostate cancer imaging would be observed using [11C]Acetate and [11C]Choline PET. Twelve men in primary staging (n=2), with increasing concentration of tumour marker PSA (n=6) or following complete prostatectomy (n=4) were enrolled in this study; they underwent both [11C]Acetate PET and [11C]Choline PET. Normal [11C]Acetate distribution shortly after administration was characterized by increased uptake in the spleen, pancreas, lungs, liver, renal parenchyma, small bowel to varying extent, bone marrow, and minor uptake in heart and muscle. No tracer elimination via urine was observed during the period of acquisition resulting in a photopenic presentation of the urinary bladder. However, [11C]Choline uptake was much higher in liver and kidney parenchyma, similar in lungs, small bowel, bone marrow and muscle and reduced in spleen and pancreas compared to [11C]Acetate. Urinary bladder activity was noted in two patients. The potential of both radiotracers to detect known bone lesions, lymph node metastases, and imaging of the primary tumour was identical. However, both failed to detect a small local recurrence in two patients as well as to demonstrate lymph node involvement in one patient, which was confirmed by surgery. Failure to detect small metastatic sites was probably due to limited PET resolution no matter which radiotracer is applied. This preliminary study showed that uptake of both radiotracers in prostate cancer or in cancer metastases was nearly identical and none of them should be favoured.

Fricke et al. [118] compared [11C]Acetate PET with [18F]FDG PET for detecting lymph node involvement and found that [11C]Acetate PET was more useful for detecting regional lymph node metastases than [18F]FDG PET. In the skeletal district, lesion sensitivity of [11C]Acetate PET and [18F]FDG was 83% and 75%, respectively. [11C]Acetate PET has also shown promising results for the early detection of prostate cancer recurrence in patients with increasing PSA after initial radiotherapy or radical surgery.

In the study by Sandblom et al. [119] pathological uptake was seen in patients with PSA as low as 0.5 ng/ml after radical prostatectomy. However, in this study false-positive uptake was seen in 3 patients. Oyama et al. [120] investigated the effectiveness of PET with [11C]Acetate for evaluating patients with increasing PSA after radical prostatectomy or radiation therapy. The study demonstrated marked uptake in prostate cancer. [11C]Acetate PET had higher sensitivity than [18F]FDG PET for detecting recurrent prostate cancer. In this study [11C]Acetate PET was not able to detect recurrent tumors below PSA 3.0 ng/ml. In the study by Fricke et al. [11C]Acetate was more useful than [18F]FDG for detecting local recurrence and regional lymph node metastases. However, in the same study [18F]FDG appeared to be more accurate for visualizing distant metastasis [120]. Wachter et al. [121] investigated 50 patients with increased PSA after radical therapy. Image fusion changed the characterization of equivocal lesions as normal at 5 of 51 sites (10%) and abnormal at 9 (18%). It precisely defined the anatomical location of abnormal uptake at 37 of 51 sites (73%). [11C]Acetate PET findings influenced treatment in 14 of 50 patients (28%). The investigators concluded that retrospective fusion of [11C]Acetate and CT/MRI is feasible.

Albrecht et al. [122] investigated the diagnostic potential of [11C]Acetate PET in the early detection of prostate cancer recurrence. Patients were studied at biochemical relapse after initial radiotherapy (n=17, median PSA=6 ng/ml) or radical surgery (n=15, median PSA=0.4 ng/ml). In all postradiotherapy patients (n=6) in whom biopsies were obtained, histology confirmed the local recurrence. In the postsurgical group, [11C]Acetate PET was positive or equivocal for local recurrence in nine of 15 patients. In the subgroup of 11 patients with PSA values higher than 0.8 ng/ml, local recurrence was observed in a similar percentage. Albrecht et al. compared early (2 min after injection) and late (47 min after injection) PET scans in order to investigate the retention time in tumours and whether individual tumours remained positive [122]. Tumour recurrence remained positive or equivocal in a large majority of patients at the late scanning time, two physical half-lives after the injection of [11C]Acetate. Recently, Morris and Scher [123] have compared the publications on [11C]Acetate PET in prostate cancer. Bouchelouche and Oehr [124] also suggested that new prospective PET imaging trials are necessary, evaluating PET/CT in uro-oncology with different PET tracers. Other biochemical pathways in cancer cells used for tumor imaging in urology refer to amino acid transport and to some extent protein synthesis, e.g. [11C]Methionine and [18F]1-amino-3-fluorocyclobutane-1-carboxylic acid ([18F]FACBC), a tumor avid amino acid that is a synthetic l-leucine analogue. To detect accelerated cell division 2-[18F]Fluoroethyl-L-tyrosine ([18F]FLT) is currently under investigation.

Nunez et al. [72] found that [11C]Methionine is more effective than [18F]FDG PET for detecting bone metastasis in patients with prostate cancer. They combined [18F]FDG and [11C]Methionine PET in 12 patients with newly progressive metastatic cancer and compared the scans with conventional imaging. The sensitivity of [18F]FDG PET and [11C]Methionine PET was 48% (167 of 348 lesions) and 72% (251 of 348), respectively, with conventional imaging used as the gold standard. In this study a significant portion of lesions (26%) had no detectable metabolism of [18F]FDG or [11C]Methionine.

Other tumor biology related radiotracer approaches are androgen receptor binding by 16beta-18F-fluoro-5alpha-dihydrotestosterone ([18F]FDHT), hypoxic imaging agents such as [18F]fluoromisonidazole ([18F]FMISO) and scanning with [18F]Fluoride, which localize in regions with malignant bone lesions and increased bone turnover. [18F]FDHT, may be used for monitoring the treatment response [124]. Two groups have reported that [18F]FDHT PET can detect metastatic and recurrent prostate cancer. In 7 patients with a median PSA of 69 ng/ml Larson et al. [125] found that [18F]FDHT detected 78% of the lesions that were identified by conventional imaging, while [18F]FDG detected 97. Dehdashti et al. [126] investigated the feasibility of androgen receptor imaging with [18F]FDHT PET in patients with advanced prostate cancer and a mean PSA of 87 ng/ml. This study showed 63% sensitivity on a patient by patient basis (12 out of 19) and 86% sensitivity on a lesion by lesion basis (24 out of 28). However, in these 2 studies the androgen receptor content of prostate cancer tumors was not investigated. Further studies are needed to evaluate whether [18F]FDHT PET is predictive of the response to hormonal therapy in patients with prostate cancer. In a prospective study Even-Sapir et al. [127] performed bone scintigraphy, SPECT and [18F]fluoride PET/CT on the same day in 44 patients with high risk prostate cancer. The sensitivity, specificity, PPV and NPV of planar bone scintigraphy were 70%, 57%, 64% and 55%, respectively. For multiple field of view SPECT the values were 92%, 82%, 86% and 90%, respectively. For [18F]Fluoride PET the values were 100%, 62% 74% and 100%, respectively. For [18F]Fluoride PET/CT the values were 100% for all parameters. Despite the fact that [18F]Fluoride produces high quality bone scans localizing in regions with malignant bone lesions and increased bone turnover, the cost outweights its utility in comparison to conventional bone scan [13].

Conclusion

Currently, [11C]Choline PET/CT is not an established tool for detection and staging of primary prostate cancer, mainly because its low spatial resolution limiting evaluation of local extension and small lymph nodes and skeletal involvement. Some authors proposed the use of this technique as a second-line imaging method in patients at high risk for cancer and reiterative negative biopsies: [11C]Choline PET/CT-guided the additional biopsies should be directed at the region of abnormal uptake.

The reduction of prostate [11C]Choline uptake following anti-androgenic hormonal therapy indicates that PET/CT with [11C]Choline could be used to monitor the response to treatment. At the same time, this findings implies that, when this technique is being considered for pre-operative staging of disease, the scan should be performed after withdrawal of neo-adjuvant androgen deprivation therapy.

PET/CT with [11C]Choline is a valid tool for restaging patients with an increasing PSA level after radical therapy. The results from literature suggest that, in these patients, PET may accurately detect the presence of recurrences in lymoh nodes, in the skeleton, and, although with less accuracy, in prostate bed. The positive detection rate increases with increasing PSA values, but it can be used in selected patients at grater risk of disease even for PSA values lower than 1 ng/ml. [11C]Choline PET/CT is complementary to conventional imaging modalities, but with the advantage of restaging the disease in a single step. The greatest advantage of the technique lies in the capability of detecting disease in lymph nodes, while false negative may occur in prostatic bed. Integrated PET/CT imaging may accurately relate abnormal metabolic changes derived by PET to precise anatomic structures visualized by CT imaging. Clinical guidelines to proper indicate a PET/CT with [11C]Choline study in prostatectomized patients will be defined. New clinical applications of PET/CT in prostate cancer, such as target volume definition for radiotherapy treatment, are still under investigation.

References

[1] Phelps, M. E. (2000). Inaugural article: positron emission tomography provides molecular imaging of biological processes. *Proc Natl Acad Sci U S A*,97:9226-33.

[2] Phelps, M. E., Hoffman, E. J., Mullani, N. A. & Ter-Pogossian, M. M. (1975). Application of annihilation coincidence detection to transaxial reconstruction tomography. *J Nucl Med.*, 16:210-24.

[3] Blokland, J. A, Trindev, P., Stokkel, M. P. & Pauwels, E. K. (2002). Positron emission tomography: a technical introduction for clinicians. *Eur J Radiol.*, 44:70-5.

[4] Lindsay, M. J., Siegel, B. A., Tunis, S. R., Hillner, B. E., Shields, A. F., Carey, B. P. & Coleman, R. E. (2007). The National Oncologic PET Registry: expanded medicare coverage for PET under coverage with evidence development. *AJR Am J Roentgenol.*, 188:1109-13.

[5] Reske, S. N., Grillenberger, K. G., Glatting, G., Port, M., Hildebrandt, M., Gansauge, F. & Beger, H. G. (1997). Overexpression of glucose transporter 1 and increased FDG uptake in pancreatic carcinoma. *J Nucl Med.*, 38:1344-8.

[6] Rempel, A., Mathupala, S. P., Griffin, C. A., Hawkins, A. L. & Pedersen, P. L. (1996). Glucose catabolism in cancer cells: amplification of the gene encoding type II hexokinase. *Cancer Res.*, 56:2468-71.

[7] Brown, R. S. & Wahl, R. L. (1993). Overexpression of Glut-1 glucose transporter in human breast cancer. An immunohistochemical study. *Cancer.*, 72:2979-85.

[8] Kapoor, V., McCook, B. M. & Torok, F. S. (2004). An introduction to PET-CT imaging. *Radiographics.*, 24:523-43.

[9] Hamberg, L. M., Hunter, G. J., Alpert, N. M., Choi, N. C., Babich, J. W. & Fischman, A. J. (1994). The dose uptake ratio as an index of glucose metabolism: useful parameter or oversimplification? *J Nucl Med.*, 35:1308-12.

[10] Messa, C., Di Muzio, N., Picchio, M., Gilardi, M. C., Bettinardi, V. & Fazio, F. (2006). PET/CT and radiotherapy. Q J Nucl Med Mol Imaging., 50:4-14.

[11] Czernin, F. J. (2002). Dual modality PET/CT: an imaging technology that changes the care of cancer patients.http://www.iame.com/learning/PETimagrc/petimage_cintro.html

[12] Bar-Shalom, R., Yefremov, N., Guralnik, L., Gaitini, D., Frenkel, A., Kuten, A., Altman, H., Keidar, Z. & Israel, O. (2003). Clinical performance of PET/CT in evaluation of cancer: additional value for diagnostic imaging and patient management. *J Nucl Med.*, 44:1200-9.

[13] Fanti, S., Nanni, C., Ambrosini, V., Gross, M. D. & Rubello, D. (2007). Farsad M. PET in genitourinary tract cancers. *Q J Nucl Med Mol Imaging.*, 51:260-71.

[14] Zeisel, S. H. & Blusztan, J. K. (1994). Choline and human nutrition. *Annual review of nutrition.*, 14:269-96.

[15] Podo, F. (1999). Tumour phospholipid metabolism. *NMR in biomedicine.*, 12:413-39.

[16] Ramírez de Molina A, R. G. A, Gutiérrez, R., Martínez-Piñeiro, L., Sánchez, J., Bonilla, F., Rosell, R. & Lacal, J. (2000). Overexpression of choline kinase is a frequent feature in human tumor-derived cell lines and in lung, prostate, and colorectal human cancers. *Biochem Biophys Res Commun.*, 296:580-3.

[17] Ratnam, S. & Kent, C. (1995). Early increase in choline kinase activity upon induction of the H-ras oncogene in mouse fibroblast cell lines. *Arch Biochem Biophys.*, 323: 313-22.

[18] Katz-Brull, R., Koudinov, A. R. & Degani, H. (2005). Direct detection of brain acetylcholine synthesis by magnetic resonance spectroscopy. *Brain Res.*, 1048:202-10.

[19] Hara, T., Kosaka, N., Shinoura, N. & Kondo, T. (1997). PET imaging of brain tumor with [methyl-11C]choline. *J Nucl Med.*, 38:842-7.
[20] Kobori, O., Kirihara, Y., Kosaka, N. & Hara, T. (1999). Positron emission tomography of esophageal carcinoma using (11)C-choline and (18)F-fluorodeoxyglucose: a novel method of preoperative lymph node staging. *Cancer.*, 86:1638-48.
[21] Hara, T., Kosaka, N. & Kishi, H. (1998). PET imaging of prostate cancer using carbon-11-choline. *J Nucl Med.*, 39:990-5.
[22] Roivainen, A., Forsback, S., Gronroos, T., Lehikoinen, P., Kahkonen, M., Sutinen, E. & Minn, H. (2000). Blood metabolism of [methyl-11C]choline; implications for in vivo imaging with positron emission tomography. *Eur J Nucl Med.*, 27:25-32.
[23] Kotzerke, J., Prang, J., Neumaier, B., Volkmer, B., Guhlmann, A., Kleinschmidt, K., Hautmann, R. & Reske, S. N. (2000). Experience with carbon-11 choline positron emission tomography in prostate carcinoma. *Eur J Nucl Med.*, 27:1415-9.
[24] de Jong, I. J., Pruim, J., Elsinga, P. H., Vaalburg, W. & Mensink, H. J. (2003). Preoperative staging of pelvic lymph nodes in prostate cancer by 11C-choline PET. *J Nucl Med.*, 44:331-5.
[25] de Jong, I. J., Pruim, J., Elsinga, P. H., Vaalburg, W. & Mensink, H. J. (2002). Visualization of prostate cancer with 11C-choline positron emission tomography. *Eur Urol.*, 42:18-23.
[26] Breeuwsma, A. J., Pruim, J., Jongen, M. M., Suurmeijer, A. J., Vaalburg, W., Nijman, R. J. & de Jong, I. J. (2005). In vivo uptake of [11C]choline does not correlate with cell proliferation in human prostate cancer. *Eur J Nucl Med Mol Imaging.*, 32:668-73.
[27] Ohtani, T., Kurihara, H., Ishiuchi, S., Saito, N., Oriuchi, N., Inoue, T. & Sasaki, T. (2001). Brain tumour imaging with carbon-11 choline: comparison with FDG PET and gadolinium-enhanced MR imaging. *Eur J Nucl Med.*, 28:1664-70.
[28] Picchio, M., Treiber, U., Beer, A. J., Metz, S., Bossner, P., van Randenborgh, H., Paul, R., Weirich, G., Souvatzoglou, M., Hartung, R., Schwaiger, M. & Piert, M. (2006). Value of 11C-choline PET and contrast-enhanced CT for staging of bladder cancer: correlation with histopathologic findings. *J Nucl Med.*, 47:938-44.
[29] Khan, N., Oriuchi, N., Zhang, H., Higuchi, T., Tian, M., Inoue, T., Sato, N. & Endo, K. (2003). A comparative study of 11C-choline PET and [18F]fluorodeoxyglucose PET in the evaluation of lung cancer. *Nucl Med Commun.*, 24:359-66.
[30] Nanni, C., Zamagni, E., Cavo, M., Rubello, D., Tacchetti, P., Pettinato, C., Farsad, M., Castellucci, P., Ambrosini, V., Montini, G. C., Al-Nahhas, A., Franchi, R. & Fanti, S. (2007). 11C-choline vs. 18F-FDG PET/CT in assessing bone involvement in patients with multiple myeloma. *World J Surg Oncol.*, 5:68.
[31] Tian, M., Zhang, H., Oriuchi, N., Higuchi, T. & Endo, K. (2004). Comparison of 11C-choline PET and FDG PET for the differential diagnosis of malignant tumors. *Eur J Nucl Med Mol Imaging.*, 31:1064-72.
[32] Heidenreich, A., Aus, G., Bolla, M., Joniau, S., Matveev, V. B., Schmid, H. P. & Zattoni, F. (2008). EAU guidelines on prostate cancer. *Eur Urol.*, 53:68-80.
[33] Jana, S. & Blaufox, M. D. (2006). Nuclear medicine studies of the prostate, testes, and bladder. *Semin Nucl Med.*, 36:51-72.
[34] Yu, K. K. & Hricak, H. (2000). Imaging prostate cancer. *Radiol Clin North Am.*, 38:59-85, viii.
[35] Shinohara, K., Wheeler, T. M. & Scardino, P. T. (1989). The appearance of prostate cancer on transrectal ultrasonography: correlation of imaging and pathological examinations. *J Urol.*, 142:76-82.

[36] Purohit, R. S., Shinohara, K., Meng, M. V. & Carroll, P. R. (2003). Imaging clinically localized prostate cancer. *Urol Clin North Am.*, 30:279-93.

[37] Hricak, H., Schoder, H., Pucar, D., Lis, E., Eberhardt, S. C., Onyebuchi, C. N. & Scher, H. I. (2003). Advances in imaging in the postoperative patient with a rising prostate-specific antigen level. *Semin Oncol.*, 30:616-34.

[38] Kurhanewicz, J., Vigneron, D. B., Males, R. G., Swanson, M. G., Yu, K. K. & Hricak, H. (2000). The prostate: MR imaging and spectroscopy. Present and future. *Radiol Clin North Am.*, 38:115-38, viii-ix.

[39] Costello, L. C. & Franklin, R. B. (1997). Citrate metabolism of normal and malignant prostate epithelial cells. *Urology.*, 50:3-12.

[40] Narayan, P. K. J. (1992). Magnetic resonance spectroscopy in prostate disease: diagnostic possibilities and future developments. *Prostate Suppl.*, 4:43-50.

[41] Teegarden, D., Taparowsky, E. J. & Kent, C. (1990). Altered phosphatidylcholine metabolism in C3H10T1/2 cells transfected with the Harvey-ras oncogene. *J Biol Chem.*, 265:6042-7.

[42] Schoder, H. & Larson, S. M. (2004). Positron emission tomography for prostate, bladder, and renal cancer. *Semin Nucl Med.*, 34:274-92.

[43] Effert, P. J., Bares, R., Handt, S., Wolff, J. M., Bull, U. & Jakse, G. (1996). Metabolic imaging of untreated prostate cancer by positron emission tomography with 18fluorine-labeled deoxyglucose. *J Urol.*, 155:994-8.

[44] Hofer, C., Laubenbacher, C., Block, T., Breul, J., Hartung, R. & Schwaiger, M. (1999). Fluorine-18-fluorodeoxyglucose positron emission tomography is useless for the detection of local recurrence after radical prostatectomy. *Eur Urol.*, 36:31-5.

[45] Liu, I. J., Zafar, M. B., Lai, Y. H., Segall, G. M. & Terris, M. K. (2001). Fluorodeoxyglucose positron emission tomography studies in diagnosis and staging of clinically organ-confined prostate cancer. *Urology.*, 57:108-11.

[46] Oyama, N., Akino, H., Suzuki, Y., Kanamaru, H., Sadato, N., Yonekura, Y. & Okada, K. (1999). The increased accumulation of [18F]fluorodeoxyglucose in untreated prostate cancer. *Jpn J Clin Oncol.*, 29:623-9.

[47] Sutinen, E., Nurmi, M., Roivainen, A., Varpula, M., Tolvanen, T., Lehikoinen, P. & Minn, H. (2004). Kinetics of [(11)C]choline uptake in prostate cancer: a PET study. *Eur J Nucl Med Mol Imaging.*, 31:317-24.

[48] Yamaguchi, T., Lee, J., Uemura, H., Sasaki, T., Takahashi, N., Oka, T., Shizukuishi, K., Endou, H., Kubota, Y. & Inoue, T. (2005). Prostate cancer: a comparative study of 11C-choline PET and MR imaging combined with proton MR spectroscopy. *Eur J Nucl Med Mol Imaging.*, 32:742-8.

[49] Farsad, M., Schiavina, R., Castellucci, P., Nanni, C., Corti, B., Martorana, G., Canini, R., Grigioni, W., Boschi, S., Marengo, M., Pettinato, C., Salizzoni, E., Monetti, N., Franchi, R. & Fanti, S. (2005). Detection and localization of prostate cancer: correlation of (11)C-choline PET/CT with histopathologic step-section analysis. *J Nucl Med.*, 46:1642-9.

[50] Reske, S. N., Blumstein, N. M., Neumaier, B., Gottfried, H. W., Finsterbusch, F., Kocot, D., Moller, P., Glatting, G. & Perner, S. (2006). Imaging prostate cancer with 11C-choline PET/CT. *J Nucl Med.*, 47:1249-54.

[51] Martorana, G., Schiavina, R., Corti, B., Farsad, M., Salizzoni, E., Brunocilla, E., Bertaccini, A., Manferrari, F., Castellucci, P., Fanti, S., Canini, R., Grigioni, W. F. & D'Errico Grigioni, A. (2006). 11C-choline positron emission tomography/computerized

tomography for tumor localization of primary prostate cancer in comparison with 12-core biopsy. *J Urol.*, 176:954-60; discussion 60.

[52] Scher, B., Seitz, M., Albinger, W., Tiling, R., Scherr, M., Becker, H. C., Souvatzogluou, M., Gildehaus, F. J., Wester, H. J. & Dresel, S. (2007). Value of 11C-choline PET and PET/CT in patients with suspected prostate cancer. *Eur J Nucl Med Mol Imaging.*, 34:45-53.

[53] Testa, C., Schiavina, R., Lodi, R., Salizzoni, E., Corti, B., Farsad, M., Kurhanewicz, J., Manferrari, F., Brunocilla, E., Tonon, C., Monetti, N., Castellucci, P., Fanti, S., Coe, M., Grigioni, W. F., Martorana, G., Canini, R. & Barbiroli, B. (2007). Prostate cancer: sextant localization with MR imaging, MR spectroscopy, and 11C-choline PET/CT. *Radiology.*, 244:797-806.

[54] Giovacchini, G., Picchio, M., Coradeschi, E., Scattoni, V., Bettinardi, V., Cozzarini, C., Freschi, M., Fazio, F. & Messa, C. (2008). [(11)C]choline uptake with PET/CT for the initial diagnosis of prostate cancer: relation to PSA levels, tumour stage and anti-androgenic therapy. *Eur J Nucl Med Mol Imaging.*, 35:1065-73.

[55] Hellstrom, M., Haggman, M., Brandstedt, S., de la Torre, M., Pedersen, K., Jarlsfeldt, I., Wijkstrom, H. & Busch, C. (1993). Histopathological changes in androgen-deprived localized prostatic cancer. A study in total prostatectomy specimens. *Eur Urol.*, 24: 461-5.

[56] Mueller-Lisse, U. G., Swanson, M. G., Vigneron, D. B., Hricak, H., Bessette, A., Males, R. G., Wood, P. J., Noworolski, S., Nelson, S. J., Barken, I., Carroll, P. R. & Kurhanewicz, J. (2001). Time-dependent effects of hormone-deprivation therapy on prostate metabolism as detected by combined magnetic resonance imaging and 3D magnetic resonance spectroscopic imaging. *Magn Reson Med.*, 46:49-57.

[57] Swinnen, J. V. & Verhoeven, G. (1998). Androgens and the control of lipid metabolism in human prostate cancer cells. *J Steroid Biochem Mol Biol.*, 65:191-8.

[58] Yoshimoto, M., Waki, A., Obata, A., Furukawa, T., Yonekura, Y. & Fujibayashi, Y. (2004). Radiolabeled choline as a proliferation marker: comparison with radiolabeled acetate. *Nucl Med Biol.*, 31:859-65.

[59] Li, X., Liu, Q., Wang, M., Jin, X., Yao, S., Liu, S. & Li, J. (2008). C-11 choline PET/CT imaging for differentiating malignant from benign prostate lesions. *Clin Nucl Med.*, 33:671-6.

[60] Rifkin, M. D., Zerhouni, E. A., Gatsonis, C. A., Quint, L. E., Paushter, D. M., Epstein, J. I., Hamper, U., Walsh, P. C. & McNeil, B. J. (1990). Comparison of magnetic resonance imaging and ultrasonography in staging early prostate cancer. Results of a multi-institutional cooperative trial. *N Engl J Med.*, 323:621-6.

[61] Smith, J. A., Jr. Scardino, P. T., Resnick, M. I., Hernandez, A. D., Rose, S. C. & Egger, M. J. (1997). Transrectal ultrasound versus digital rectal examination for the staging of carcinoma of the prostate: results of a prospective, multi-institutional trial. *J Urol.*, 157:902-6.

[62] Hricak, H., Dooms, G. C., Jeffrey, R. B., Avallone, A., Jacobs, D., Benton, W. K., Narayan, P. & Tanagho, E. A. (1987). Prostatic carcinoma: staging by clinical assessment, CT, and MR imaging. *Radiology.*, 162:331-6.

[63] Yu, K. K., Hricak, H., Alagappan, R., Chernoff, D. M., Bacchetti, P. & Zaloudek, C. J. (1997). Detection of extracapsular extension of prostate carcinoma with endorectal and phased-array coil MR imaging: multivariate feature analysis. *Radiology.*, 202:697-702.

[64] Bartolozzi, C., Menchi, I., Lencioni, R., Serni, S., Lapini, A., Barbanti, G., Bozza, A., Amorosi, A., Manganelli, A. & Carini, M. (1996). Local staging of prostate carcinoma

with endorectal coil MRI: correlation with whole-mount radical prostatectomy specimens. *Eur Radiol.*, 6:339-45.

[65] Tempany, C. M., Zhou, X., Zerhouni, E. A., Rifkin, M. D., Quint, L. E., Piccoli, C. W., Ellis, J. H. & McNeil, B. J. (1994). Staging of prostate cancer: results of Radiology Diagnostic Oncology Group project comparison of three MR imaging techniques. *Radiology.*, 192:47-54.

[66] Shvarts, O., Han, K. R., Seltzer, M., Pantuck, A. J. & Belldegrun, A. S. (2002). Positron emission tomography in urologic oncology. *Cancer Control.*, 9:335-42.

[67] Heicappell, R., Muller-Mattheis, V., Reinhardt, M., Vosberg, H., Gerharz, C. D., Muller-Gartner, H. & Ackermann, R. (1999). Staging of pelvic lymph nodes in neoplasms of the bladder and prostate by positron emission tomography with 2-[(18)F]-2-deoxy-D-glucose. *Eur Urol.*, 36:582-7.

[68] Sanz, G., Robles, J. E., Gimenez, M., Arocena, J., Sanchez, D., Rodriguez-Rubio, F., Rosell, D., Richter, J. A. & Berian, J. M. (1999). Positron emission tomography with 18fluorine-labelled deoxyglucose: utility in localized and advanced prostate cancer. *BJU Int.*, 84:1028-31.

[69] Kumar, R., Zhuang, H. & Alavi, A. (2004). PET in the management of urologic malignancies. *Radiol Clin North Am.*, 42:1141-53, ix.

[70] Yeh, S. D., Imbriaco, M., Larson, S. M., Garza, D., Zhang, J. J., Kalaigian, H., Finn, R. D., Reddy, D., Horowitz, S. M., Goldsmith, S. J. & Scher, H. I. (1996). Detection of bony metastases of androgen-independent prostate cancer by PET-FDG. *Nucl Med Biol.*, 23:693-7.

[71] Kao, C. H., Hsieh, J. F., Tsai, S. C., Ho, Y. J. & Yen, R. F. (2000). Comparison and discrepancy of 18F-2-deoxyglucose positron emission tomography and Tc-99m MDP bone scan to detect bone metastases. *Anticancer Res.*, 20:2189-92.

[72] Nunez, R., Macapinlac, H. A., Yeung, H. W., Akhurst, T., Cai, S., Osman, I., Gonen, M., Riedel, E., Scher, H. I. & Larson, S. M. (2002). Combined 18F-FDG and 11C-methionine PET scans in patients with newly progressive metastatic prostate cancer. *J Nucl Med.*, 43:46-55.

[73] Morris, M. J., Akhurst, T., Osman, I., Nunez, R., Macapinlac, H., Siedlecki, K., Verbel, D., Schwartz, L., Larson, S. M. & Scher, H. I. (2002). Fluorinated deoxyglucose positron emission tomography imaging in progressive metastatic prostate cancer. *Urology.*, 59:913-8.

[74] Barocas, D. A., Han, M., Epstein, J. I., Chan, D. Y., Trock, B. J., Walsh, P. C. & Partin, A. W. (2001). Does capsular incision at radical retropubic prostatectomy affect disease-free survival in otherwise organ-confined prostate cancer? *Urology.*, 58:746-51.

[75] Rinnab, L., Blumstein, N. M., Mottaghy, F. M., Hautmann, R. E., Kufer, R., Hohl, K. & Reske, S. N. (2007). 11C-choline positron-emission tomography/computed tomography and transrectal ultrasonography for staging localized prostate cancer. *BJU Int.*, 99:1421-6.

[76] Danella, J. F., deKernion, J. B., Smith, R. B. & Steckel, J. (1993). The contemporary incidence of lymph node metastases in prostate cancer: implications for laparoscopic lymph node dissection. *J Urol.*, 149:1488-91.

[77] Paul, D. B., Loening, S. A., Narayana, A. S. & Culp, D. A. (1983). Morbidity from pelvic lymphadenectomy in staging carcinoma of the prostate. *J Urol.*, 129:1141-4.

[78] Platt, J. F., Bree, R. L. & Schwab, R. E. (1987). The accuracy of CT in the staging of carcinoma of the prostate. *AJR Am J Roentgenol.*, 149:315-8.

[79] Jager, G. J., Barentsz, J. O., Oosterhof, G. O., Witjes, J. A. & Ruijs, S. J. (1996). Pelvic adenopathy in prostatic and urinary bladder carcinoma: MR imaging with a three-dimensional TI-weighted magnetization-prepared-rapid gradient-echo sequence. *AJR Am J Roentgenol.*, 167:1503-7.

[80] Schiavina, R., Scattoni, V., Castellucci, P., Picchio, M., Corti, B., Briganti, A., Franceschelli, A., Sanguedolce, F., Bertaccini, A., Farsad, M., Giovacchini, G., Fanti, S., Grigioni, W. F., Fazio, F., Montorsi, F., Rigatti, P. & Martorana, G. (2008). (11)C-choline positron emission tomography/computerized tomography for preoperative lymph-node staging in intermediate-risk and high-risk prostate cancer: comparison with clinical staging nomograms. *Eur Urol.*, 54:392-401.

[81] Prando, A. (2004). Baseline staging of newly diagnosed prostate cancer: a summary of the literature. *Int Braz J Urol.*, 30:341-2.

[82] Eschmann, S. M., Pfannenberg, A. C., Rieger, A., Aschoff, P., Muller, M., Paulsen, F., Anastasiadis, A., Claussen, C. D., Bares, R. & Schlemmer, H. P. (2007). Comparison of 11C-choline-PET/CT and whole body-MRI for staging of prostate cancer. *Nuklearmedizin.*, 46:161-8; quiz N47-8.

[83] Luboldt, W., Kufer, R., Blumstein, N., Toussaint, T. L., Kluge, A., Seemann, M. D. & Luboldt, H. J. (2008). Prostate Carcinoma: Diffusion-weighted Imaging as Potential Alternative to Conventional MR and 11C-Choline PET/CT for Detection of Bone Metastases. *Radiology*.

[84] Grossfeld, G. D., Stier, D. M., Flanders, S. C., Henning, J. M., Schonfeld, W., Warolin, K. & Carroll, P. R. (1998). Use of second treatment following definitive local therapy for prostate cancer: data from the caPSURE database. *J Urol.*, 160:1398-404.

[85] Lu-Yao, G. L., Potosky, A. L., Albertsen, P. C., Wasson, J. H., Barry, M. J. & Wennberg, J. E. (1996). Follow-up prostate cancer treatments after radical prostatectomy: a population-based study. *J Natl Cancer Inst.*, 88:166-73.

[86] Fowler, F. J., Jr. Barry, M. J., Lu-Yao, G., Roman, A., Wasson, J. & Wennberg, J. E. (1993). Patient-reported complications and follow-up treatment after radical prostatectomy. The National Medicare Experience: 1988-1990 (updated June 1993). *Urology.*, 42:622-9.

[87] Partin, A. W., Pearson, J. D., Landis, P. K., Carter, H. B., Pound, C. R., Clemens, J. Q., Epstein, J. I. & Walsh, P. C. (1994). Evaluation of serum prostate-specific antigen velocity after radical prostatectomy to distinguish local recurrence from distant metastases. *Urology.*, 43:649-59.

[88] Bott, S. R. (2004). Management of recurrent disease after radical prostatectomy. *Prostate Cancer Prostatic Dis.*, 7:211-6.

[89] Parra, R. O., Wolf, R. M. & Huben, R. P. (1990). The use of transrectal ultrasound in the detection and evaluation of local pelvic recurrences after a radical urological pelvic operation. *J Urol.*, 144:707-9.

[90] Laufer, M., Pound, C. R., Carducci, M. A. & Eisenberger, M. A. (2000). Management of patients with rising prostate-specific antigen after radical prostatectomy. *Urology.*, 55:309-15.

[91] Kramer, S., Gorich, J., Gottfried, H. W., Riska, P., Aschoff, A. J., Rilinger, N., Brambs, H. J. & Sokiranski, R. (1997). Sensitivity of computed tomography in detecting local recurrence of prostatic carcinoma following radical prostatectomy. *Br J Radiol.*, 70:995-9.

[92] Silverman, J. M. & Krebs, T. L. (1997). MR imaging evaluation with a transrectal surface coil of local recurrence of prostatic cancer in men who have undergone radical prostatectomy. *AJR Am J Roentgenol.*, 168:379-85.

[93] Sella, T., Schwartz, L. H., Swindle, P. W., Onyebuchi, C. N., Scardino, P. T., Scher, H. I., Hricak, H. (2004). Suspected local recurrence after radical prostatectomy: endorectal coil MR imaging. *Radiology.*, 231:379-85.

[94] Moul, J. W. (2000). Prostate specific antigen only progression of prostate cancer. *J Urol.*, 163:1632-42.

[95] Rosenthal, D. I. (1997). Radiologic diagnosis of bone metastases. *Cancer.*, 80:1595-607.

[96] Hoh, C. K., Schiepers, C., Seltzer, M. A., Gambhir, S. S., Silverman, D. H., Czernin, J., Maddahi, J. & Phelps, M. E. (1997). PET in oncology: will it replace the other modalities? *Semin Nucl Med.*, 27:94-106.

[97] Seltzer, M. A., Barbaric, Z., Belldegrun, A., Naitoh, J., Dorey, F., Phelps, M. E., Gambhir, S. S. & Hoh, C. K. (1999). Comparison of helical computerized tomography, positron emission tomography and monoclonal antibody scans for evaluation of lymph node metastases in patients with prostate specific antigen relapse after treatment for localized prostate cancer. *J Urol.*, 162:1322-8.

[98] Schoder, H., Herrmann, K., Gonen, M., Hricak, H., Eberhard, S., Scardino, P., Scher, H. I. & Larson, S. M. (2005). 2-[18F]fluoro-2-deoxyglucose positron emission tomography for the detection of disease in patients with prostate-specific antigen relapse after radical prostatectomy. *Clin Cancer Res.*, 11:4761-9.

[99] de Jong, I. J., Pruim, J., Elsinga, P. H., Vaalburg, W. & Mensink, H. J. (2003). 11C-choline positron emission tomography for the evaluation after treatment of localized prostate cancer. *Eur Urol.*, 44:32-8; discussion 8-9.

[100] Picchio, M., Messa, C., Landoni, C., Gianolli, L., Sironi, S., Brioschi, M., Matarrese, M., Matei, D. V., De Cobelli, F., Del Maschio, A., Rocco, F., Rigatti, P. & Fazio, F. (2003). Value of [11C]choline-positron emission tomography for re-staging prostate cancer: a comparison with [18F]fluorodeoxyglucose-positron emission tomography. *J Urol.*, 169:1337-40.

[101] Rinnab, L., Mottaghy, F. M., Blumstein, N. M., Reske, S. N., Hautmann, R. E., Hohl, K., Moller, P., Wiegel, T., Kuefer, R. & Gschwend, J. E. (2007). Evaluation of [11C]-choline positron-emission/computed tomography in patients with increasing prostate-specific antigen levels after primary treatment for prostate cancer. *BJU Int.*, 100: 786-93.

[102] Scattoni, V., Picchio, M., Suardi, N., Messa, C., Freschi, M., Roscigno, M., Da Pozzo, L., Bocciardi, A., Rigatti, P. & Fazio, F. (2007). Detection of lymph-node metastases with integrated [11C]choline PET/CT in patients with PSA failure after radical retropubic prostatectomy: results confirmed by open pelvic-retroperitoneal lymphadenectomy. *Eur Urol.*, 52:423-9.

[103] Reske, S. N., Blumstein, N. M. & Glatting, G. (2008). [(11)C]choline PET/CT imaging in occult local relapse of prostate cancer after radical prostatectomy. *Eur J Nucl Med Mol Imaging.*, 35:9-17.

[104] Krause, B. J., Souvatzoglou, M., Tuncel, M., Herrmann, K., Buck, A. K., Praus, C., Schuster, T., Geinitz, H., Treiber, U. & Schwaiger, M. (2008). The detection rate of [(11)C]Choline-PET/CT depends on the serum PSA-value in patients with biochemical recurrence of prostate cancer. *Eur J Nucl Med Mol Imaging.*, 35:18-23.

[105] Boccon-Gibod, L., Djavan, W. B., Hammerer, P., Hoeltl, W., Kattan, M. W., Prayer-Galetti, T., Teillac, P. & Tunn, U. W. (2004). Management of prostate-specific antigen relapse in prostate cancer: a European Consensus. *Int J Clin Pract.*, 58:382-90.

[106] Pound, C. R., Partin, A. W., Eisenberger, M. A., Chan, D. W., Pearson, J. D. & Walsh, P. C. (1999). Natural history of progression after PSA elevation following radical prostatectomy. *JAMA.*, 281:1591-7.

[107] Consensus statement: guidelines for PSA following radiation therapy. (1997). American Society for Therapeutic Radiology and Oncology Consensus Panel. *Int J Radiat Oncol Biol Phys.*,37:1035-41.

[108] Schilling, D., Schlemmer, H. P., Wagner, P. H., Bottcher, P., Merseburger, A. S., Aschoff, P., Bares, R., Pfannenberg, C., Ganswindt, U., Corvin, S. & Stenzl, A. (2008). Histological verification of 11C-choline-positron emission/computed tomography-positive lymph nodes in patients with biochemical failure after treatment for localized prostate cancer. *BJU Int.*,102:446-51.

[109] Tuncel, M., Souvatzoglou, M., Herrmann, K., Stollfuss, J., Schuster, T., Weirich, G., Wester, H. J., Schwaiger, M. & Krause, B. J. (2008). [(11)C]Choline positron emission tomography/computed tomography for staging and restaging of patients with advanced prostate cancer. *Nucl Med Biol.*, 35:689-95.

[110] Schmid, D. T., John, H., Zweifel, R., Cservenyak, T., Westera, G., Goerres, G. W., von Schulthess, G. K. & Hany, T. F. (2005). Fluorocholine PET/CT in patients with prostate cancer: initial experience. *Radiology.*, 235:623-8.

[111] Kwee, S. A., Coel, M. N., Lim, J. & Ko, J. P. (2005). Prostate cancer localization with 18fluorine fluorocholine positron emission tomography. *J Urol.*, 173:252-5.

[112] Kwee, S. A., Wei, H., Sesterhenn, I., Yun, D. & Coel, M. N. (2006). Localization of primary prostate cancer with dual-phase 18F-fluorocholine PET. *J Nucl Med.*, 47:262-9.

[113] Cimitan, M., Bortolus, R., Morassut, S., Canzonieri, V., Garbeglio, A., Baresic, T., Borsatti, E., Drigo, A. & Trovo, M. G. (2006). [(18)F]fluorocholine PET/CT imaging for the detection of recurrent prostate cancer at PSA relapse: experience in 100 consecutive patients. *Eur J Nucl Med Mol Imaging.*, 33:1387-98.

[114] Hacker, A., Jeschke, S., Leeb, K., Prammer, K., Ziegerhofer, J., Sega, W., Langsteger, W. & Janetschek, G. (2006). Detection of pelvic lymph node metastases in patients with clinically localized prostate cancer: comparison of [18F]fluorocholine positron emission tomography-computerized tomography and laparoscopic radioisotope guided sentinel lymph node dissection. *J Urol.*, 176:2014-8; discussion 8-9.

[115] Heinisch, M., Dirisamer, A., Loidl, W., Stoiber, F., Gruy, B., Haim, S. & Langsteger, W. (2006). Positron emission tomography/computed tomography with F-18-fluorocholine for restaging of prostate cancer patients: meaningful at PSA < 5 ng/ml? *Mol Imaging Biol.*, 8:43-8.

[116] Oyama, N., Akino, H., Kanamaru, H., Suzuki, Y., Muramoto, S., Yonekura, Y., Sadato, N., Yamamoto, K. & Okada, K. (2002). 11C-acetate PET imaging of prostate cancer. *J Nucl Med.*, 43:181-6.

[117] Kato, T., Tsukamoto, E., Kuge, Y., Takei, T., Shiga, T., Shinohara, N., Katoh, C., Nakada, K. & Tamaki, N. (2002). Accumulation of [11C]acetate in normal prostate and benign prostatic hyperplasia: comparison with prostate cancer. *Eur J Nucl Med Mol Imaging.*, 29:1492-5.

[118] Fricke, E., Machtens, S., Hofmann, M., van den Hoff, J., Bergh, S., Brunkhorst, T., Meyer, G. J., Karstens, J. H., Knapp, W. H. & Boerner, A. R. (2003). Positron emission

tomography with 11C-acetate and 18F-FDG in prostate cancer patients. *Eur J Nucl Med Mol Imaging*., 30:607-11.

[119] Sandblom, G., Sorensen, J., Lundin, N., Haggman, M. & Malmstrom, P. U. (2006). Positron emission tomography with C11-acetate for tumor detection and localization in patients with prostate-specific antigen relapse after radical prostatectomy. *Urology*., 67:996-1000.

[120] Oyama, N., Miller, T. R., Dehdashti, F., Siegel, B. A., Fischer, K. C., Michalski, J. M., Kibel, A. S., Andriole, G. L., Picus, J. & Welch, M. J. (2003). 11C-acetate PET imaging of prostate cancer: detection of recurrent disease at PSA relapse. *J Nucl Med*., 44:549-55.

[121] Wachter, S., Tomek, S., Kurtaran, A., Wachter-Gerstner, N., Djavan, B., Becherer, A., Mitterhauser, M., Dobrozemsky, G., Li, S., Potter, R., Dudczak, R. & Kletter, K. (2006). 11C-acetate positron emission tomography imaging and image fusion with computed tomography and magnetic resonance imaging in patients with recurrent prostate cancer. *J Clin Oncol*., 24:2513-9.

[122] Albrecht, S., Buchegger, F., Soloviev, D., Zaidi, H., Vees, H., Khan, H. G., Keller, A., Bischof Delaloye, A., Ratib, O. & Miralbell, R. (2007). (11)C-acetate PET in the early evaluation of prostate cancer recurrence. *Eur J Nucl Med Mol Imaging*., 34:185-96.

[123] Morris, M. J. & Scher, H. I. (2007). (11)C-acetate PET imaging in prostate cancer. *Eur J Nucl Med Mol Imaging*., 34:181-4.

[124] Bouchelouche, K. & Oehr, P. (2008). Positron emission tomography and positron emission tomography/computerized tomography of urological malignancies: an update review. *J Urol*., 179:34-45.

[125] Larson, S. M., Morris, M., Gunther, I., Beattie, B., Humm, J. L., Akhurst, T. A., Finn, R. D., Erdi, Y., Pentlow, K., Dyke, J., Squire, O., Bornmann, W., McCarthy, T., Welch, M. & Scher, H. (2004). Tumor localization of 16beta-18F-fluoro-5alpha dihydrotestosterone versus 18F-FDG in patients with progressive, metastatic prostate cancer. *J Nucl Med*., 45:366-73.

[126] Dehdashti, F., Picus, J., Michalski, J. M., Dence, C. S., Siegel, B. A., Katzenellenbogen, J. A. & Welch, M. J. (2005). Positron tomographic assessment of androgen receptors in prostatic carcinoma. *Eur J Nucl Med Mol Imaging*., 32:344-50.

[127] Even-Sapir, E., Metser, U., Mishani, E., Lievshitz, G., Lerman, H. & Leibovitch, I. (2006). The detection of bone metastases in patients with high-risk prostate cancer: 99mTc-MDP Planar bone scintigraphy, single- and multi-field-of-view SPECT, 18F-fluoride PET, and 18F-fluoride PET/CT. *J Nucl Med*., 47:287-97.

In: Handbook of Prostate Cancer Cell Research...
Editor: Alan T. Meridith
ISBN: 978-1-60741-954-9

Chapter 3

The Stem Cell Paradigm and Its Application to Prostate Cancer— An Old and Yet still Young Idea

Krassimira Todorova and Soren Hayrabedyan
Institute of Biology and Immunology of Reproduction,
Bulgarian Academy of Sciences, Sofia, Bulgaria

Abstact

Prostate cancer (PCa) is initially regarded as an androgen-dependent disease, as therapeutic androgen ablation induces massive apoptosis of malignant cells and clinically leads to tumor regression. However, after a remission period of up to several years, mutations occur and tumors usually progress despite low circulating androgen levels. Currently, no therapeutic options are available for the treatment of androgen-independent PCa.

One relatively new paradigm in explaining the biology of the Prostate cancer is the possible existence of cancer stem cells (CSC). The mechanisms of progression of prostate cancer have been extensively studied, yet are poorly understood. To explain these mechanisms and the heterogeneity in prostate cancer, one of the concepts that has evolved is that cancer arises from the neoplastic transformation of normal prostatic epithelial stem cells or transit amplifying cells. Based on this stem cell model, understanding the stages of cell differentiation in normal prostate epithelium is essential for the identification of cells that are involved in prostate carcinogenesis and ablation-resistant prostate cancer. Normal tissues arise from stem cells that grow and differentiate to create progenitor and mature cell populations. For normal stem cells, the most important standard is that the cell can engage a multilineage repopulation of a tissue or organ. Both somatic stem cells and CSCs have the capacity for organogenesis, but somatic stem cells generate normal tissues and tumor stem cells generate tumor tissues. The most crucial standard for a CSC is its ability to re-initiate a serially transplantable tumor that recapitulates the original heterogeneity and tumor histology. The CSCs theory postulates that CSCs are rather a functional definition, inferring that the expression of a putative stem cell marker from specific cell population does not mean at all that these cells are stem cells. Using this gold standard, many so-called prostate CSCs reported so far are based on correlative studies rather than being true CSCs.

Recent evidence has shown that cancer stem cells (CSCs) are a subset of the tumor cells that are responsible for initiating and maintaining the disease. Understanding normal stem cells and CSCs may provide insight into the origin of and new therapeutics for prostate cancer. Normal stem cells and CSCs have been identified in prostate tissue by the use of several markers or techniques. Although research on stem cells has been limited by the lack of suitable in vitro systems, recent studies show that not only primary cells but also several established cell lines may exhibit stem cell properties.

Various in vitro culture systems to propagate normal prostate stem cells and prostate CSCs together with molecular markers are reviewed together with transgenic models of prostate cancer in conjunction with meta-analyses based on genomics and proteomics data supporting a novel hypothesis regarding the pathways and genes involved in the transformation of normal prostate cells into neoplastic and their subsequent loss of androgen-receptor (AR) sensitivity.

Based on CSC theory, CSCs are biologically distinct from non-CSCs in the overall population and are thought to be resistant to conventional chemotherapy or radiation therapy. Most prostate cancer cells express AR and PSA, and this phenotype mimicking luminal cells has been reported. AR signaling in luminal cells not only regulates PSA expression, but it is also required for survival, and this requirement is maintained in early cancers.

Therefore, AR has a crucial role in the progression and treatment of prostate cancer. Although prostate cancer cells usually express AR, AR expression is heterogeneous in most primary prostate cancers and their metastases.

The signal transduction pathways in androgen-dependent and -independent prostate cancer cell proliferation were also being reviewed. Some of them include Akt, extracellular-signal-regulated kinase (ERK), and mitogen-activated protein kinase members' family (MAPK) and are involved in poor clinical outcome of prostate cancers. Akt activation, but not ERK activation, is correlated with proliferation in human prostate tumors. In connection with these pathways, a role has been established of the mTOR, whose specific inhibition has a growth-inhibitory effect on androgen-independent prostate carcinoma cell lines. The latter is involved in the metabolic stress pathways in tumor cells, thus giving rise to the hypothesis regarding the "shift" to metabolic and regulatory pathways supporting the overall survival in prostate cancer cells.

Introduction

Prostate carcinoma is a molecularly, phenotypically and clinically heterogeneous disease. Considerable effort is currently being devoted to discovering the molecular events that lead to its development. In addition, it is becoming increasingly clear that the identification of the cell(s) of origin of prostate cancer represents an essential step for a full understanding of this disease [1].

The early understanding of solid tumors was that they represent a homogenous mass of neoplastically transformed cells, stochastically equal in potential, but even the early studies on cancer started to build another paradigm for the development and maintenance of the tumor growth [2]:

It has long been known starting from the light microscopic studies that both normal tissues and the tumors that develop within them comprise a heterogeneous collection of cell types, frequently including immune cells, a stroma consisting of various mesenchymal and endothelial cells, and a variety of normal or malignant cells specific to the tissue. [3] Cells within the tumor often seem to correspond to different stages of development. Epithelial cancers, for example, typically contain cells exhibiting divergent nuclear morphologies and differentiation features. Prevailing explanations for the observed tumor cell heterogeneity include influences of the microenvironment and genomic instability that generate the genetic and epigenetic changes, which prevent faithful and accurate replication and transmission of stable genotypes and phenotypes. Such instability could also explain why tumors typically contain a subset of cells that are refractory to most treatments.

However, an alternative (or complementary) emerging concept is that malignant cell populations may reflect the continuing operation of perturbed differentiation processes. Inherent to such a model is the formation of malignant populations consisting of a developmentally defined hierarchy of heterogeneous phenotypes derived from a small subset of "cancer stem cells".

How These Ideas Have Developed

Stem cells sit at the top of the lineage hierarchy. They are self-renewing and multipotential, proliferating and capable of differentiating into the cell types that make up an organ. The idea of involvement of stem cells in cancer is not entirely new. In 1855, Virchow postulated in his "embryonal rest hypothesis," based on the histological similarities between terato-carcinomas and embryonic tissue that the former originated from the latter [4, 5].

Later, in 1937 Furth and Kahn found evidence of the existence of tumor stem cells by showing that leukemia could be transferred by a single cell in a mouse model [6]. It became apparent that only a very small fraction (0.0001% to 1%) of all tumor cells possesses the capability to induce a tumor, as was shown by Park et al.[7] and by Hamburger and Salmon in 1977 [8]. Years later Bonnet and Dick showed in 1997, that single leukemia cell clones could induce leukemia phenotypically identical to the parent tumor after transplanting them into NOD-SCID mice [9]. Those tumor-inducing cells showed the surface marker expression pattern CD34+/CD38–, characteristic of hematopoietic stem cells.

From these observations and the hierarchical cell profile of leukemia, reminiscent of hematopoiesis, Bonnet and Dick concluded that these leukemic tumor stem cells developed out of hematopoietic stem cells through a leukemogenic effect [9].

The possibility of derivation of tumor stem cells not only from hematopoietic stem cells but also from further differentiated cell states after accumulation of additional mutations was suggested by Miyamoto et al. [10].

Since AML1/ETO-positive hematopoietic stem cells of patients in long-term remission differentiated into functioning, mature cell states *in vitro*, it was proposed that further mutations are necessary for inducing leukemia. Then a suggestion came for an ancestry of more differentiated cell states, based on the loss of the stem cell-specific surface marker Thy-

1 on the leukemic cells. Of course, loss of Thy-1 expression could also occur on the stem cell level by the malignancy-causing mutation itself [10].

Discovery of the Philadelphia chromosome also facilitated several key studies that have led to a paradigm shift in our understanding of cancer as a disease of stem cells. During chronic phase CML, the Philadelphia chromosome is found in many different blood cell lineages [11, 12]. This implies that it must arise either in a stem or progenitor cell capable of giving rise to each lineage or independently in cells of each lineage [13]. Philip Fialkow and colleagues addressed this issue in a clever series of experiments examining Philadelphia chromosome–positive (Ph+) leukemic cells for the expression of an X chromosome–linked gene, glucose-6-phosphate dehydrogenase [11, 12]. Each cell in the body only expresses one allele for this gene as a result of the inactivation of one X chromosome during development. Interestingly, all Ph+ cells in female patients heterozygous for the gene expressed the same allele [11, 12]. Although it is possible that X chromosome inactivation status may change in cancer cells during tumor progression, this finding strongly suggests human CML is clonal in origin and therefore must originate in a stem cell. Subsequent studies using surface antigens to purify HSCs have confirmed that the Philadelphia chromosome is present in HSCs [14, 15]. More recent studies have demonstrated that stem cells also sustain disease in later phases of malignancy.

Observations that only a minor subpopulation of cells in tumor samples have the capacity to initiate clonal growth in *in vitro* cultures or in *in vivo* transplant models had perplexed researchers for over 50 years [16, 17]. Two theories were proposed to explain this paradox. The stochastic theory suggested that all cancer cells are equally malignant but only clones that randomly possess favorable biological properties will grow upon transplantation. An alternative theory predicted that tumors are hierarchical like normal tissues and only the rare subpopulation of cells at the pinnacle of that hierarchy have the unique biological properties necessary for tumor initiation [16, 17]. Studies by John Dick and colleagues provided evidence for the hierarchy model. This group demonstrated that only the small subpopulation (0.1%–1.0%) of Lin–CD34+CD38– cells within human acute myelogenous leukemia samples were capable of initiating disease when transplanted into immune-deficient mice [9]. These cells possessed the same antigenic profile as normal human HSCs, which are at the pinnacle of the normal hematopoietic hierarchy. This population also had the unique capacity to self-renew to propagate the disease as well as differentiate to produce the many leukemic cell types represented in the original leukemia. Since these cancer cells possess properties unique to normal tissue stem cells, they have been termed "*cancer stem cells*" (CSCs). These studies make it clear that stem cells play critical roles in both the initiation and maintenance of human leukemia.

But does this mean stem cells are exclusively competent to perform these functions? Huntly et al. [18] showed that BCR-ABL can only initiate CML in mice when expressed in HSCs and not in committed progenitor cells. Inactivation of the AP-1 transcription factor (TF) JunB, a transcriptional regulator of myelopoiesis and potential tumor suppressor, is also only sufficient to induce disease in HSCs [19]. However, leukemia-associated fusion oncogenes such as MOZ-TIF2, MLL-ENL, and MLL-AF9 will induce disease from both HSCs as well as myeloid progenitors [18, 20, 21]. To determine why MOZ-TIF2 but not BCR-ABL can transform committed progenitors, Huntly et al. examined the in vitro self-renewal activity of progenitors expressing each oncogene and found that MOZ-TIF2 is sufficient to restore self-renewal in committed progenitors while BCR-ABL is not [18]. MLL-AF9 similarly can confer self-renewal activity in committed progenitors. Microarray studies comparing the expression profile of leukemic granulocyte-macrophage progenitors (GMPs) expressing MLL-AF9 with normal HSCs and progenitor cells showed that leukemic GMPs maintain an expression profile similar to that of normal GMPs but reacquire a self-renewal–associated transcription program characteristic of normal HSCs [21]. These studies demonstrate it is the self-renewal activity of stem cells that makes them more conducive to oncogenesis. Acquisition of this property by other cells, however, appears to enable them to support tumor initiation.

The Role of Stem Cells is Now Being Addressed in Many Solid Tissue Cancers

Recently, tumor stem cells were also found in solid tumors. Al-Hajj et al. showed that transplantation of only a few human breast cancer cells was needed to induce a tumor growth phenotypically identical to the original tumor. These tumor-inducing cells (CD44+/CD24-/low) differed from the non-inducing cells (CD44–/CD24+) by the expression pattern of the stem cell surface marker CD44. In contrast to the tumor cell mass, they had the capacity for self-renewal and massive proliferation [22]. Other groups reported the existence of neuronal tumor stem cells. Transplantation of just a few CD133+ human neuronal tumor cells into the NOD-SCID mouse model generated a phenocopy of the original tumor, while 100,000 CD133- tumor cells could not induce a tumor [4].

Although prostate cancer is now the most commonly diagnosed cancer in men, the etiology of this disease remains unclear.

Interestingly, several aspects of this disease are remarkably similar to CML. CML is most frequently diagnosed in adults and the elderly and first presents as a chronic yelogenous hyperplasia that eventually progresses to acute immature blast crisis following the accumulation of genetic and epigenetic alterations in addition to BCR-ABL. Prostate cancer is similarly a disease of aging that is thought to begin with the chronic accumulation of prostatic intraepithelial neoplasia (PIN) lesions that may eventually develop into adenocarcinoma, although no experimental evidence for this has yet been provided. Many adenocarcinomas also progress to a poorly differentiated, hormone-refractory prostate cancer (HRPC) consisting mostly of immature blast-like cells [13].

Another Cancer Paradigm is Shared between Haematogenic and Solid Tumors

Discovery and study of the Philadelphia chromosome have provided the foundation for many of the fundamental concepts about cancer today [23]. The Philadelphia chromosome is a reciprocal translocation between the long arms of chromosomes 9 and 22 [t(9;22)(q34;q11)] that is present in most human chronic myelogenous leukemia (CML) patients [24]. This translocation causes fusion of the breakpoint cluster region (BCR) with portions of the v-Abl Abelson murine leukemia viral oncogene homolog (ABL) gene [25]. The resulting chimeric oncoprotein, BCR-ABL, possesses aberrant tyrosine kinase activity [25]. These findings have demonstrated the role of genetic change, genomic instability, and tyrosine kinase signaling in tumorigenesis and have highlighted BCR-ABL as an important therapeutic target [13].

Arul Chinnaiyan and colleagues have recently identified a class of translocations that, like BCR-ABL, are present in the majority of prostate cancers [26]. These translocations result in fusion of the androgen-regulated promoter for TMPRSS2 with the coding regions of ETS family TFs that are associated with Ewing sarcoma and several types of human leukemia [27, 28].

Existing Evidence for the Cell of Origin of Prostate Cancer

Anatomy of the prostate gland

Prostatic epithelium is composed of two morphologically defined components, the basal and luminal layers, and three types of cells, basal, secretory and neuroendocrine cells. The different prostatic epithelial cells can be distinguished based on their localization, morphology and their phenotypic characteristics. Luminal cells form a layer just above the basal cells and secrete prostatic proteins into the lumenal space. They express androgen receptor (AR), prostate-specific antigen (PSA), prostatic acid phosphatase (PAP), CD57 [29] and a low molecular weight cytokeratin, such as Keratin (K) 8 and K18. Keratins belong to the group of intermediate filament proteins interlacing the cytoskeleton by heterodimeric interaction and polymerization [30]

On the other hand, basal cells are non-secretory and display CD44, [29] p63, [31] a high molecular weight cytokeratin—such as K5 and K14, but not AR—PSA and prostatic acid phosphatase.

Some cells were found to exhibit a keratin pattern intermediate between that of basal and luminal cells with expression of K5, K8 and K18 but not K14. The explanation of their possible origin has laid in the finding that the secretory cells are dependent on androgen for survival, while basal cells are independent.

The classical androgen cycling experiments by John Isaacs et al. suggested that the prostate epithelium must contain a stem cell population. When rodents are deprived of androgen by surgical or medicinal castration, the gland atrophies due to the apoptosis of terminally differentiated cells dependent on androgen for viability [32]. When androgen is replaced, the gland regenerates and resumes normal secretory function. Isaacs's group showed that this involution and regeneration could be repeated for many sequential cycles [33]. This showed that a stem cell population with profound capacity for self-renewal and differentiation must exist within the gland.

Based on this Isaacs and Coffey [34] proposed that androgen-independent stem cells give rise to two types of cells: *stem cells* and *androgen-independent transit amplifying cells*, which can divide rapidly with limited proliferative capacity and can differentiate into luminal cells through an intermediate phenotype.

In 1988, an additional cell population co-localizing K5 and K18 within the luminal epithelium was identified in rats [15]. As this cell population had an immunophenotype intermediate between basal and luminal cells, it potentially represented the amplifying cell type, originally described by Isaacs and Coffey, which was further supported by its appearance in androgen-induced regeneration of the prostate epithelium [3].

By serial-section analysis and immunofluorescent double/triple-staining, eventually two intermediate cell types were distinguished within the human prostate epithelium. The *first intermediate cell-type* is characterized by expression of K5 in the absence of K14 (K5 ++ / 18 +) in the basal compartment, while a *second intermediate subpopulation* co-localizes K5 and K18 (K5 + / 18 ++) within the luminal epithelium [35-38].

Neuroendocrine cells are distributed throughout the basal and luminal layers and express chromogranin A and a variety of peptide hormones, but lack AR and PSA [39, 40]. Neuroendocrine cells are also thought to arise from dendritic cells rather than from epithelial stem cells and have a complex expression pattern of basal and luminal markers [39]. They appear to play a role in growth and differentiation, but their function is largely unknown [40].

The prostate also contains several types of stromal cells including fibroblasts, myofibroblasts, and smooth muscle cells that guide the growth and differentiation of the epithelium. Blood cell elements as well as vascular and stromal endothelial cells are also present in the gland.

To date, several putative stem cell populations have been identified as prostate stem cells by means of clonal assay, identification of several cell surface markers or side population (SP) analysis [32, 34].

The Concept of Stem Cell Niche

Stem cell function depends on the interaction between intrinsic genetic programs and extrinsic regulatory cues derived from a stem cell's microenvironment or 'niche'. The concept of the stem cell niche was initially proposed by Schofield in the context of the mammalian blood system [41]. Thirty years on, the niche concept has been generalized to include many different stem cell systems and is considered to represent a defined anatomical location that preserves stem cells and affords their expansion protecting it from differentiation, loss of self-renewal capabilities, and apoptosis. This is regulated by direct cell contacts, matrix contacts, extracellular matrix components, cytokines, and growth factors. Tumor stem cells seem to reside in similar protective microenvironments. The limited availability of stem cells within niches also contributes to how the size of a stem cell compartment is determined [42].

The Embryonic Prostate Stem Cell Niche

Little is known of the embryonic prostate stem cell niche. In the 9-week old embryonic human prostate, condensation of the mesenchyme surrounding the urogenital sinus at the level of the mesonephric duct openings is the first sign of prostate development. Epithelial bud formation occurs in week 10. Although mesenchymal cells appear to have close contact with the basal lamina, the basement membrane remains continuous and separates the epithelium from the mesenchyme. Later in development, gaps in the basement membrane occur, allowing contact between the outgrowing epithelium and the underslying mesenchyme. In parallel, the testicular Leydig cells mature, exhibit characteristics of steroid-secreting cells, implying that androgens regulate prostate development. Thus, the human prostate embryonic niche requires signals from the mesenchyme and testicular androgens to initiate prostate development and facilitate epithelial cell differentiation [43].

Nuclear androgen receptor (AR) labeling was initially detected in mesenchymal cells between embryonic (E) day E14.5 and E16.5, whereas epithelial AR expression appeared at 10 days postnatally with the onset of differentiation. By 10 weeks of age, epithelial AR labeling became more prominent while mesenchymal labeling declined [43].

Using the defective androgen receptor blocking by androgen signaling in the developing male testicular feminization mouse model (Tfm/Y), it was suggested that organogenesis and outgrowth of the prostatic buds is mediated by signals from AR+ mesenchymal cells. Postnatally, epithelial AR regulates the final stages of differentiation and secretory function of the prostatic epithelium [43].

Human Prostate Development

Epithelial bud formation occurs during week 10 of development and the beginning of cannulization at the proximal ends is observed by the end of week 11. Initially, the epithelium resembles that of the stratified urethral epithelium; however they appear to contain more organelles. Cells with neuroendocrine characteristics also appear in the basal regions of the buds [44]. The number of epithelial outgrowths increases between weeks 11 and 14, developing into tubulo-acinar glands. At this stage, they consist of three to five layers of cells, most of which are apolar and round in shape. The prostatic epithelium expresses primarily non-secretory acid phosphatase in lysosomes at week 10. However, some cells contain acid phosphatase-positive apical granules, suggesting the beginning acid phosphatase secretion [43].

Aumüller et al. did not observe secretory acid phosphatase or prostate specific antigen (PSA) expression in the prenatal prostate [44]. The secretions also stain positive for PAS and Alcian blue, indicating that epithelial cells synthesize neutral and acidic mucopolysaccharides [44]. Limited polarization of apical cells and the appearance of dense secretory granules occur at week 13, and during weeks 15 to 16, epithelial outgrowth increase. The basal lamina becomes discontinuous in some places and presumably under the influence of the mesenchyme, epithelial cell differentiation occurs. Triangular-shaped cells similar to postnatally-mature basal cells appear in the basal region of the gland [44]. Connective tissue surrounding the prostatic epithelium increases and the inner zone forms the lamina propria of the acinar epithelium. Immature prostatic glands are retained until the onset of puberty [44]. Stratified epithelium in the fetal and prepubertal human prostate appears to stain positive for all cytokeratins until after puberty when only secretory (or luminal) and basal cells are present [43, 44].

Intermediate forms between basal and luminal cells were not observed in the developing human prostate [44]. By 14 years of age, secretory PAP and PSA immunoreactivity are detected in prostatic epithelium and PAS-positive cells gradually become less and are absent in the normal adult prostate [43].

Localization of Stem Cells in the Prostate

The preferential survival of basal cells following androgen ablation has led to the traditionally held hypothesis that prostate stem cells (PSCs) reside within the basal cell layer of the gland [32]. This is supported by several findings:

After androgen-deprived prostate gland involution, the subsequent androgen replacement results in total reconstitution of the organ. These regression/regeneration studies have shown that even after more than 30 rounds, androgen treatment could still restore secretory glandular structure. Basal cells preferentially survive androgen ablation whereas 90% of luminal epithelial cells are lost through programmed cell death. The surviving epithelial cells appear androgen independent; however they remain responsive to exogenous androgen treatment. Basal cells exhibit higher proliferation rates than luminal cells in normal and hyperplastic acini [42, 43].

Based on these observations, Isaacs and Coffey developed a prostate stem-cell model [50]. According to this model, androgen-independent stem cells give rise to progenitor cells that are androgen-sensitive, but not androgen-dependent, which then, under the influence of androgen, differentiate into androgen-dependent cells of the prostate epithelium. Because the basal cell layer remains intact during androgen ablation-caused prostate involution, the prostate stem cells were thought to reside in this compartment [33, 34]. Accordingly, immunohistochemical studies on expression of the stem-cell antigen Sca-1 in murine prostate showed a distinct accumulation of Sca-1 in cells with high regenerative potential in the proximal duct. The few cells with high regenerative potential in the distal duct were also Sca-1-positive.

Collins et al. could show that human basal cells from prostate, when transplanted into immunodeficient mice, could induce the formation of new tissue that was phenotypically and morphologically identical to the original tissue. They could also demonstrate a characteristic expression pattern of the surface markers integrin and CD133, with prostate stem cells showing an integrin $\alpha 2\beta 1^{hi}$/CD133+ pattern [45, 46].

Recent findings revealed a high proteasomal turnover rate of functional AR in a CD133-subset of $\alpha 2\beta 1^{hi}$-enriched basal cell population, resulting in androgen-responsiveness at very low to undetectable cellular levels of AR protein [45].

Additional evidence implicates the basal layer as an important source of prostate stem cells. Mice with a targeted deletion of the basal cell marker p63 show agenesis of the prostate and mammary gland, and complementation of $p63^{-/-}$ blastocysts with $p63^{+/+}$ β-galactosidase-positive ESCs generates chimeric animals with prostates composed of β-galactosidase-positive basal and luminal epithelial cells [47, 48].

Null mice for the basal cell marker p63 are born without the prostate or mammary gland [31, 47-49]. Chimeric animals created from p63–/– blastocysts complemented with p63+/+

β-galactosidase–positive ES cells also produce only β-galactosidase–positive luminal cells [50]. Human basal cells also express BCL-2, an antiapoptotic protein that is commonly expressed by tissue stem cells [35]. BrdU pulse chase experiments performed by Tsujimura et al. [51], however, show that long-term label-retaining cells localize to a specific region of the gland that is proximal to the urethra. This region is characterized by a thick band of morphologically distinct smooth muscle cells that secrete high levels of TGF-β, making it a compelling location for the PSC niche since this factor is known to promote stem cell quiescence [52].

Interestingly, both basal and luminal cells in the proximal region retain the label after many cycles of androgen ablation and add-back, suggesting stem cells may not be restricted to the basal layer. Cells from this region possess enhanced growth activity both in vitro and in vivo, and serial transplant studies have shown that these cells can self-renew to regenerate prostate tissue de novo for at least 4 passages [51, 53].

Although the developing human prostate has been extensively analyzed, the prostate stem cells which initiate this process are currently investigated to be identified. It is still not certain whether the prostate stem cells originate in the basal cell compartment or whether each epithelial lineage arises from a distinct stem/progenitor cell.

Models of Prostate Epithelial Differentiation

The traditional model for prostate epithelial differentiation proposes that PSCs residing in the basal cell layer give rise to intermediate, transit-amplifying cells that produce large numbers of terminally differentiated secretory luminal cells [54]. This model implies a linear differentiation scheme in which basal and luminal cells comprise one lineage and basal cells are essentially luminal cell progenitors (Figure 1). This hypothesis is supported by the existence of cells of intermediate phenotype that express both basal- and luminal cell–specific cytokeratins in both fetal and adult stages of prostate development [36, 37, 55]. Intermediate cells can also be identified in *in vitro* cultures of primary prostate epithelium [36, 56-58]. Several studies have also suggested basal cells can differentiate into luminal cells in vitro [29, 59, 60].

It is generally thought that PSCs do not express AR or p63. It was found that the loss of Pten results in the expansion of a Sca-1+/Bcl-2+ prostate stem/progenitor-like cell population. Approximately 70% of the cells expressing proliferation-associated antigens such as Ki-67 also were phenotypically basal cells. In archival human tissues representing primary prostate carcinoma and high grade prostatic epithelial neoplasia (PIN), De Marzo et al. identified an intermediate layer of $p27^{Kip1}$-negative cells sandwiched between $p27^{Kip1}$-positive basal and luminal cells and hypothesized that these cells were competent for entry into the cell cycle. Other intermediate cell markers (summarized in Figure 1) have been identified in the adult prostate as well as in cell culture [42].

Alternative theories for prostate epithelial differentiation propose basal and luminal cells may represent separate epithelial lineages (Figure 2). This is similar to prevailing models for epithelial differentiation in the mammary gland, a tissue that is anatomically and functionally analogous to the prostate [43].

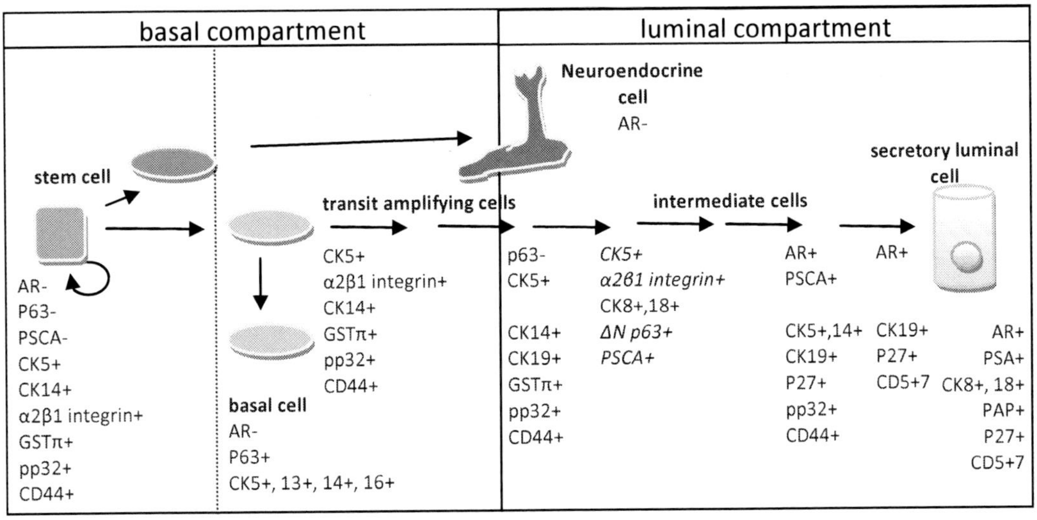

Figure 1. Models for ontogeny of the epithelial lineages in the prostate. The epithelial lineages are derived from PSCs which reside in the basal cell compartment [43].

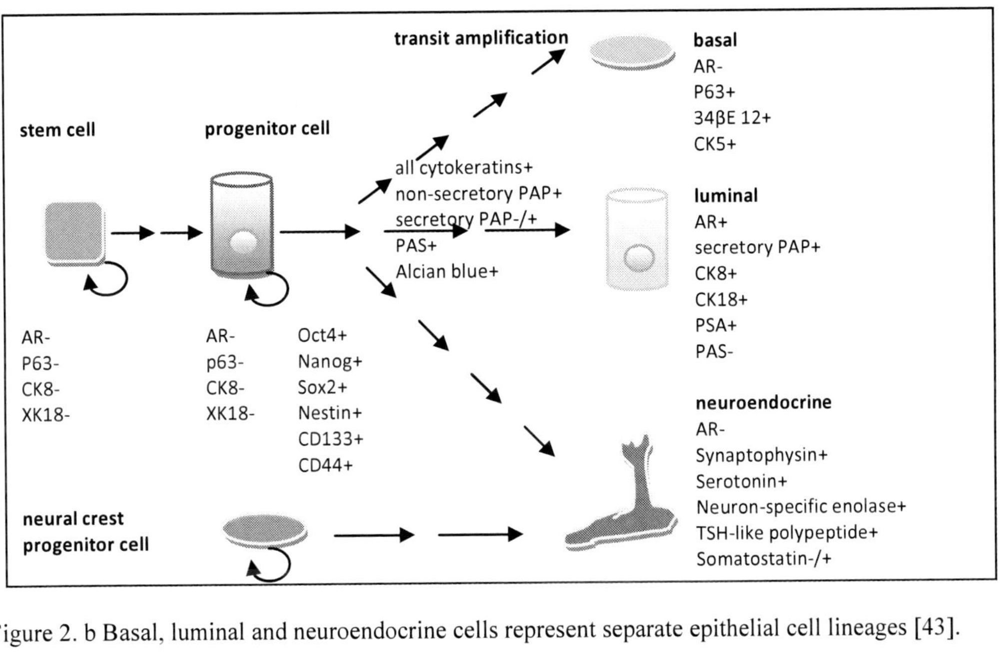

Figure 2. b Basal, luminal and neuroendocrine cells represent separate epithelial cell lineages [43].

In the prenatal human prostate, primitive epithelial cells are present from bud formation onward and neuroendocrine-like cells appear at the basal regions of the developing buds around the same time. Fetal and prepubertal human stratified epithelium appeared positive for all cytokeratin antibodies and intermediate phenotypes were not observed in the developing human prostate [42].

Recent studies by Gerald Cunha and colleagues showed that fetal urogenital sinus tissue from p63-null mice can regenerate prostate ductal tissue following implantation in immunodeficient mice [61]. Regenerated prostate tissue lacked phenotypically identifiable basal cells but did contain cells that expressed typical luminal cell markers and had secretory function. This presented the possibility that luminal cells arise in the absence of basal cells. However, there are other explanations for these observations, one example being that loss of p63 expression causes differentiation defects that make basal cells unidentifiable in regenerated tissue. BrdU-labeling experiments by Tsujimura et al. [51] have demonstrated that the dramatic growth of the murine gland that occurs during puberty is due to the rapid proliferation of a subset of luminal cells, although the majority of basal cells do not divide. This suggests that basal and luminal cell progenitors may also exist that can respond independently to growth cues. This would also explain why subsets of luminal cells survive in the absence of androgen after castration [32] and would correspond with recent demonstration by Asselin-Labat et al. [62] of a luminal-restricted progenitor population in the mammary gland. One limitation of each of these models is that they do not address the ontogeny of neuroendocrine cells in the prostate. Neuroendocrine cells may represent a subset of the basal cell lineage, a unique epithelial lineage, or even a lineage produced from a separate stem cell population [63].

Candidate prostate stem/progenitor cell populations have been identified by their preferential expression of the surface markers CD44, CD133, $\alpha 2\beta 1^{hi}$, CK5/14, CK8/18, SCA-1, CD49f, and ABCG2. CD44 is present in most basal cells, whereas CD133, $\alpha 2\beta 1^{hi}$, CK5/14, CK8/18 and ABCG2 are expressed in ~1% of the basal cell populations, and there is overlap between the $\alpha 2\beta 1^{hi}$, $CD44^{+}$, and $ABCG2^{+}$ cell populations [42].

Leong et al. published this year (2008) in Nature that they have identified CD117 (c-kit, stem cell factor receptor) as a new marker of a rare adult mouse PSC population, and demonstrate that a single stem cell defined by the phenotype Lin-Sca-1+CD133+CD44+CD117+ can generate a prostate after transplantation *in vivo*. CD117 expression is predominantly localized to the region of the mouse prostate proximal to the urethra and is upregulated after castration-induced prostate involution—two characteristics consistent with that of a PSC marker. CD117+ PSCs can generate functional, secretion-producing prostates when transplanted *in vivo*. Moreover, CD117+ PSCs have long-term self-renewal capacity, as evidenced by serial isolation and transplantation *in vivo*. They established that single cells in the adult mouse prostate with multipotent, self-renewal capacity are defined by a $Lin^{-}/Sca\text{-}1^{+}/CD133^{+}/CD44^{+}/CD117^{+}$ phenotype [64]. The authors were able also to find the same marker in human prostate stem cells.

Cancer stem cells paradigm and its implication in the prostate cancer pathology...

Both somatic stem cells and cancer stem cells (CSCs) have the capacity for organogenesis, but somatic stem cells generate normal tissues and tumor stem cells generate tumor tissues. Normal tissues arise from stem cells that grow and differentiate to create progenitor and mature cell populations.

CSCs theory postulates that CSCs are a functional definition, that is, that a cell or a population of cells expresses a putative stem cell marker does not mean at all that this cell is a stem cell. For normal stem cells, the most important standard is that the cell can engage a multilineage repopulation of a tissue or organ.

The most crucial standard for a CSC is its ability to re-initiate a serially transplantable tumor that recapitulates the original heterogeneity and tumor histology [3, 65].

Using this gold standard, many so-called prostate CSCs reported so far are based on correlative studies rather than being true CSCs. [45, 66]The current postulate of CSCs theory is that tumors contain a small population of stem-like cells, which are often called CSCs or tumor-initiating cells. In fact, recent evidence suggests that many tumors may actually arise from the transformation of progenitor cells rather than stem cells. The first evidence for CSCs has been shown in the context of acute myeloid leukemia [9, 67]. This principle has also been implicated in solid tumors, such as brain tumor, breast cancer and prostate cancer [45].

Most prostate cancer cells express AR and PSA, and this phenotype mimicking luminal cells has been reported. Initially, cancer was thought to be a homogeneous mass, which consisted of highly proliferative cells as a stochastic model, and most cancer treatment was designed to eliminate these cells. Based on CSC theory, CSCs are biologically distinct from non-CSCs in the overall population and are thought to be resistant to conventional chemotherapy or radiation therapy.

AR signaling in luminal cells not only regulates PSA expression, but it is also required for survival, and this requirement is maintained in early cancers. Therefore, AR has a crucial role in the progression and treatment of prostate cancer. AR activity has been investigated in a number of cell lines or animal models as a main target of therapeutics in prostate cancer. It has been postulated that many mechanisms, including AR gene amplification or AR gene mutations, may promote the growth of ablation-resistant prostate cancer cells [68]. However, the mechanism of this progression still remains elusive. In fact, the majority of men will eventually have a relapse to an ablation-resistant state, yet androgen ablation therapy remains the primary therapy for patients with metastatic prostate cancer. Although prostate cancer cells usually express AR, AR expression is heterogenous in most primary prostate cancers and their metastases [69]. Understanding this heterogeneity is a key to elucidate prostate cancer progression and to develop the future treatment paradigms, it may support, in part, the CSCs theory in the prostate. Several recent studies have postulated the existence of CSCs in the prostate using primary prostate cancer cells, [45] prostate cancer cell lines [66, 70, 71] or animal models [72, 73].

However, this hypothetical model for the hierarchical organization of prostate cancer cells remains unproven because of the lack of appropriate in vitro and in vivo models.

The Cell of Origin of Prostate Cancer

Most speculation over the cellular origin of human prostate cancer comes from correlative studies examining the cellular composition of prostate tumors. In human prostate adenocarcinomas, the majority of cancer cells express luminal cell–specific markers such as cytokeratin 8 (CK8), CK18, and prostate-specific antigen [74]. Cells that solely express basal cell markers such as CK5, CK14, and p63 are rarely observed. This has led some to suggest that prostate cancers are derived from a luminal cell progenitor or mature luminal cell that has acquired self-renewal activity through mutation. Some reports, however, have identified intermediate cells that coexpress both basal and luminal cell markers within prostate cancers [35]. PSCA, a putative marker of normal late-intermediate prostate cells, is also often upregulated in prostate cancers [60, 75]. These data indicate the disease may originate in an intermediate or transit-amplifying epithelial cell that precedes luminal cell differentiation.

Several properties of human prostate cancers suggest the disease may instead arise from a PSC. The progression to androgen-independent HRPC during androgen ablation therapy has led to speculation that prostate tumors may contain a small population of androgen-independent cells that survive and can expand in the absence of androgen [54]. Since normal PSCs are androgen independent, it is reasonable to suspect they may be the source of these cells. Recent efforts to identify and characterize prostate CSCs also support this hypothesis. Collins et al. [45] demonstrated that the primary human prostate cancer cell subpopulation with the highest in vitro proliferative potential is negative for androgen receptor (AR) expression, as is suspected for normal PSCs. This population also possesses a $CD44^{+}\alpha2\beta1^{hi}CD133^{+}$ antigenic profile characteristic of normal human PSCs and predominantly expresses the basal cell cytokeratins [45, 76]. Using several human prostate xenograft tumors and cell lines, Patrawala et al. [70] have also demonstrated that the CD44+ population within these models displays enhanced proliferative activity in vitro and increased tumor-initiating and metastatic activity in vivo. These CD44+ cells are likewise AR– and express higher mRNA levels of several "stemness" genes, including OCT3/4, BMI1, β-CATENIN, and SMOOTHENED.

The lineage status of CSCs has also been addressed by several other groups using human *telomerase* reverse transcriptase–immortalized primary human prostate cancer cell lines [66, 77]. Gu et al. [77] showed that, as with CSCs, these clonally derived cell lines could regenerate prostate tumors in mice that resembled the original patient tumor with respect to histopathology and Gleason score. Regenerated tumors also contained basal, luminal, and neuroendocrine-like cancer cells, suggesting the clone of origin of the lines had multilineage differentiation capacity. Analysis of these lines showed they are predominantly AR– and p63– and express the pluripotency markers Oct4, Nanog, and Sox2, as well as the progenitor markers CD44, CD133, and CD117 (also known as c-Kit) [13].

An interesting question generated from these findings is whether the TMPRSS2-ETS fusion gene is expressed in CSCs if they are in fact AR–. Fusion of the androgen-regulated TMPRSS2 promoter with the coding region of various ETS TFs is thought to result in TF overexpression (16). ETS-regulated target genes have numerous potentially oncogenic functions, including the regulation of proliferation, differentiation, apoptosis, angiogenesis, and transformation. If PSCs and CSCs are AR–, it is likely they do not express the fusion transcript.

Does this mean that PSCs are not the origin of human prostate cancer, or that the fusion gene is not an initiating event but instead a secondary event generated through genomic instability?

Recent identification of the TMPRSS2-ETS fusion gene in approximately 20% of PIN lesions suggests that its formation may be an early event in tumorigenesis [78]. Perhaps it serves as a "first hit" that takes place in a primitive AR– cell but does not take effect until it is expressed in differentiated progeny that are AR+. Several studies have shown that some HSCs may not express BCR-ABL even though they contain the fusion gene. Expression of the fusion protein in the progeny of such HSCs is, however, cancer promoting. Research to answer these critical questions will provide invaluable insight into the etiology of prostate cancer.

Several laboratories have begun to investigate prostate cancer initiation using *mouse models of the disease*. As in human prostate cancer, disease in the Pten prostate-specific knockout mouse model progresses from PIN to invasive adenocarcinoma and metastasis [79]. Although Cre mediates Pten deletion in both basal and luminal cells, disease progression is associated with a disproportionate expansion of basal cells, indicating disease in this model may initiate and be sustained by primitive cells within the basal cell population [80]. Zhou et al. [81] also suggest that disease in their Rb–/–p53–/– prostate-specific knockout model initiates from primitive cells. The earliest invasive neoplasms that develop in these mice are observed in the proximal region of the gland and express stem cell antigen-1 (Sca-1), which is characteristic of murine prostate stem and progenitor cells. In contrast to the PTEN knock- out model, the cells that comprise these lesions express luminal and neuroendocrine markers but are negative for basal cell markers like CK5. The authors of this study further found that both proximal and distal cells will regenerate tumors when engrafted under the kidney capsule of immunodeficient mice. This suggests that microenvironmental factors unique to the proximal region of the gland may promote cancer formation in this model [13].

Functional studies to directly compare the tumorigenic potential of different prostate cell types are necessary to resolve this debate over the cell of origin of prostate cancer. Several technical advancements are necessary for these studies: (a) development of reliable in vitro and in vivo assays for prostate epithelial cell growth; (b) identification of PSCs and epithelial lineage hierarchy; and (c) development of methods to model prostate cancer from dissociated populations of prostate cells.

Identification and Isolation of Tumor-Initiating Cells

The fraction of tumor-initiating cells in a tumor is probably less than 0.1%. Thus, for a successful isolation, the capability to exactly identify those cells is essential [4].

Identification can be accomplished by

(1) characteristic cell surface markers,
(2) the cells' ability to expel dyes, and
(3) by DNA labeling.

Investigations on expression and down-regulation of cell-surface proteins specific to stem cells or tumors yielded distinct characteristic expression patterns for several tumor types. It is unclear, however, whether those expression patterns can be used for isolation of pure cultures or rather just to enrich tumor stem cells. Many of the surface proteins used are not specific in that they are also expressed on physiological stem cells and on tumor-initiating cells of other tumor types. Additionally, identification using antibodies can lead to false-positive results due to non-specific cross-reactivities.

Another way of isolating tumor-initiating cells is based on detection of a "side population" (SP) of cells characterized by being able to actively transport lipophilic dyes (Hoechst 33342) out of the cells by drug-transporter proteins (mainly ABCG2) [65]. The specificity of this method for tumor stem cell isolation is questionable because some more differentiated progenitor cells also have this ability.

The characterization of tumor stem cells by DNA labeling depends on the label-retaining characteristics of the seldom dividing stem cells [65]The fact that tumor stem cells and physiological stem cells can both be labeled with this technique shows that it also lacks absolute specificity.

Development of Prostate Epithelial Stem Cell Assays

The in vivo dissociated prostate cell regeneration system

Most in vivo assays for the growth of normal adult rodent epithelial cells are based on the tissue recombination methods developed by Cunha and Lung [82]. In this system, tissue fragments of fetal urogenital sinus mesenchyme are used to support the growth of normal prostate epithelial tissue fragments when implanted in collagen under the renal capsule of immunodeficient mice. In order to evaluate the growth activities of different prostate cell subpopulations, we have modified this system using mechanical and enzymatic digestion to dissociate both the urogenital sinus mesenchyme and adult murine prostate tissue into single cell suspensions [72]. Dissociated prostate epithelia regenerate ductal structures that histologically resemble normal murine prostate. Prostate tissue can also be regenerated when dissociated cells are implanted s.c. in Matrigel or in collagen onto the flanks of immunodeficient mice [46, 83]. Experiments utilizing mixed populations of differentially marked epithelial cells have demonstrated that the tubules regenerated using these protocols are clonal, as chimeric tubules expressing both markers are rarely observed [13, 73, 83]. Immunohistochemical analyses have shown that each duct contains both basal and luminal cell lineages [13, 72, 73, 83]. Since regenerated ducts are clonal, this suggests that tubule-regenerating cells possess the capacity for multilineage differentiation.

To investigate PSC self-renewal, Lynette Wilson and colleagues have utilized a serial transplantation approach analogous to the serial reconstitution method used to measure HSC self-renewal [53]. This group demonstrated that regenerated prostate tissue can be dissociated and transplanted to regenerate prostate tissue at least three times, indicating regenerated ducts contain cells capable of self-renewing to support several rounds of prostate regeneration. Interestingly, the regenerative activity of transplanted cells drops almost 100-fold from generation one to three. This may be because many cells within regenerated tissue represent short-term progenitors with limited regenerative activity that eventually expires during serial transplantation. Alternatively, the diminution in regenerative activity may be due to loss of regenerating cells during tissue digestion at each passage.

In vitro prostate colony assays

Epithelial cells from many organs such as the skin and mammary gland can be grown as colonies in two dimensional culture conditions [56, 84]. Human prostate epithelial cells can also be cultivated in this manner when cocultured with irradiated fibroblast feeder layers in low-calcium, serum-free medium containing growth factors such as FGF and EGF, which are known to promote stem cell maintenance [54]. Under these conditions, murine prostate epithelial cells form colonies of cells that express epithelial cytokeratins when cultured with irradiated 3T3 feeder cells [13]. These colonies appear to be clonogenic, as differential marking experiments have shown that chimeric colonies rarely appear. Castration was also shown to enrich for colony-forming cells, suggesting that colonies arise predominantly from primitive prostate epithelia. These findings demonstrate that this assay can be used to quantitatively compare stem cell enrichment strategies.

In vitro prostate sphere assays

Although colonies appear to be derived from primitive cells, they do not passage efficiently since this culture condition appears to promote differentiation. Stem cells from the central nervous system and mammary gland have been maintained by culturing them in nonadherent conditions as three-dimensional spheres [85, 86]. Human prostate cancer cells lines and immortalized lines of normal cells can form spheroid structures when embedded in Matrigel – a solubilized basement membrane preparation [87-89]. A murine prostate sphere–forming assay was developed using similar conditions (64). The majority of spheres generated in these conditions are composed of several epithelial cell layers that surround a hollow center. We speculate that this is analogous to the canalization process that occurs during prostate duct development.

Like the colony assay, differential marking and castration studies have shown that prostate spheres are of clonal origin and derive from primitive cells [13]. Spheres can be dissociated and passaged for over 10 generations, suggesting they contain a population of cells with extensive self-renewal capacity [13]. The self-renewing population, however, appears to be a minority population, as only 1%–2% of sphere cells form new spheres upon replating. Implantation of sphere cells in the renal capsule regeneration assay has shown that both primary and passaged sphere cells can regenerate prostate ductal structures containing both basal and luminal cells [13]. These data suggest that murine PSCs are capable of extensive self-renewal and that this activity can be maintained using this prostate sphere assay.

Identification and Characterization of Orostate Epithelial Stem Cells

Traditional approaches for isolation of tissue stem cells have utilized cell surface markers and cell-sorting technology. Using this approach, Collins and colleagues [46] demonstrated that prostate basal cells that rapidly adhere to type I collagen in vitro express high levels of α2β1 integrin, possess greater colony-forming efficiency in vitro, and can produce prostate glandular structures when grafted onto the flanks of athymic mice. Subdivision of the α2β1 hi basal cell compartment using CD133, a neural and hematopoietic stem cell marker, results in an 11-fold increase in colony formation [76]. These CD133+α2β1 hi cells can also give rise to glandular tissue in vivo.

Independent studies by Witte et al. (here) and Burger et al. [90] have shown that Sca-1, a marker used to enrich for hematopoietic, mammary, and pulmonary stem cells, can be used to enrich for murine prostate cells with enhanced tubule-forming activity in vivo [73, 90-93]. Sca-1+ cells possess several properties of stem cells, including multilineage differentiation capacity and replication quiescence [73]. Sca-1 also appears to be present on the majority of primitive cells that remain following castration-induced apoptosis of mature prostate cells. Sca-1 was also found to be constitutively expressed by all epithelial cells of the fetal prostate, indicating that the marker is present on PSCs from very early stages in their development [94].

Wilson and colleagues demonstrated that Sca-1+ prostate cells express other stem cell markers such as BCL-2 and CD49f (integrin α6) [90]. CD49f is a particularly prevalent stem cell marker that is expressed by hematopoietic, neural, embryonic, skin, and mammary epithelial stem cells (76–80). Magnetic bead sorting for CD49f+ cells enriches for cells with enhanced regenerative activity in vivo (30). It was shown that combining these markers to sort for cells with a Lin (CD45/CD31/Ter119)–Sca-1+CD49f+ (LSC) antigenic profile results in a 60-fold enrichment for colony- and sphere-forming cells in vitro [13]. These cells can self-renew to form spheres for many generations and can differentiate to produce ductal structures containing both basal and luminal cells in vivo. These studies estimate that PSCs can be isolated to a purity of up to 1 in 35 by using this antigenic profile.

Immunohistochemical staining of sorted LSC cells indicate that they possess a basal cell cytokeratin expression profile. LSC cells also localize to the basal cell layer within the proximal region of the gland, integrating both of the proposed models for PSC localization. The ability of LSC cells to produce CK8+ luminal cells in vivo also provides the first functional in vivo evidence for the traditional linear model for prostate epithelial cell differentiation, which proposes that basal cells give rise to luminal cells.

Murine Prostate Cancer Regeneration Models

Human cancers initiate from a wide range of genetic and epigenetic alterations. Studies in leukemia have shown that oncogenes induce tumorigenesis in some cell types more efficiently than in others, depending on their ability to confer self-renewal properties in the target cell [16]. Since the cellular source of prostate cancer may likewise be oncogene dependent, a long-term goal is to compare the tumorigenic activity of different prostate cell subpopulations following exposure to different oncogenic stimuli. Toward this end, the dissociated prostate regeneration assay is used to investigate the tumorigenic effects of several prostate cancer–associated genetic alterations in PSCs as well as in more differentiated prostate cell populations.

The PTEN/AKT Pathway

Loss of function of PTEN and activation of AKT are highly correlated with human prostate cancer [95, 96]. Transgenic animals that express a constitutively active form of AKT under regulation of the prostate-specific probasin promoter develop PIN lesions [97]. PTEN prostate-specific null animals also develop high-grade prostate adenocarcinomas and metastases [79, 98]. It was found that lentivirus-mediated expression of constitutively active AKT in dissociated prostate cells results in the regeneration of prostate tubules containing PIN lesions that progress to frank carcinoma in some experiments [73, 99]. PIN lesions also develop when prostate cells are infected with lentivirus carrying siRNA for targeted knockdown of PTEN [73].

The AR

Alterations in AR signaling causing receptor hypersensitivity, promiscuity, or androgen-independent receptor transactivation are also common features of prostate cancer [100]. Interestingly, it was found that overexpression of AR does not cause pathological growth but instead seems to reduce the tubule-forming activity of dissociated prostate cells in this assay [99]. This is consistent with in vitro studies reporting that overexpression of AR reduces proliferation and promotes differentiation [57, 101]. Lentivirus-mediated cell-autonomous overexpression of both AR and AKT, however, is sufficient to induce initiation and progression to frank carcinoma in this system. These regenerated tumors also progress to androgen independence, as continued growth is observed after several weeks of castration-induced androgen deprivation.

Investigation of tumorigenic activity of prostate cell subpopulations

In preliminary efforts to compare the tumorigenic activity of primitive and mature prostate cell populations, Sca-1+ and Sca-1– cells were isolated and their response compared to the overexpression of AKT1. The Sca-1+ cells were found to regenerate ductal structures that contain PIN, while Sca-1– cells give rise predominantly to normal ducts [99]. AKT-regenerated tumors are marked by a 10-fold expansion of the Sca-1+ population compared with normal regenerated prostate tissue. This expansion is also observed in prostate tumors that develop in PTEN-null animals [80]. A future goal is to compare the oncogenic potential of LSCs and more differentiated cells following perturbation of the PTEN/AKT pathway and AR signaling. Efforts to develop other models of prostate cancer are also underway using other prostate cancer–associated genetic alterations.

Regulatory Mechanisms of Stem Cells and Tumor Stem Cells

Epithelial cell maintenance - role of telomerase activity and further use in mouse model studies

Tissue homeostasis requires an equilibrium of cellular proliferation and apoptosis. Prostate epithelial basal and luminal cells reveal supplementary proliferative and apoptotic activity with proliferation occurring predominantly in the basal cell compartment and apoptosis in the luminal cell layer [102-104].

The resistance of basal cells to apoptosis results at least partially from their high telomeraseactivity and bcl-2 levels. *Cell senescence* is associated with progressive shortening of (chromosomal-) telomeres and upon a critical telomere length activation of the apoptotic cascade will result in cell death. Reduction in telomere length is prohibited by telomerase, a riboprotein complex that delays the occurrence of apoptosis by maintenance of telomeres; accordingly an important role in malignant transformation has been proposed and indeed in a number of human cancers telomerase activity is found [105].

Within the prostate epithelium telomerase activity is high in basal cells and low in the luminal epithelium indicating basal cells are less accessible to apoptosis [106]. While expression of the onco-protein bcl-2 in basal cells provides protection from apoptotic stimuli, its absence in the luminal epithelium predisposes exocrine- and neuro-endocrine cells to undergo apoptosis [107-109].

Progression through the cell cycle requires activated complexes of cyclins and cyclin-dependent kinases (cdk), which can be blocked by cdk inhibitors such as $p21^{cip1/waf}$ and $p27^{kip1}$. The cdk inhibitor $p27^{kip1}$ preferentially inactivates the cyclin E/cdk2 complex preventing cell cycle progression from G1 into S phase.

Within the non-malignant prostate epithelium $p27^{kip1}$ is expressed at high levels in luminal cells but is absent in subgroups of basal cells and cells with a morphologically intermediate position in between the basal and luminal cell compartment [43]. This observation corresponds with the high proliferative activity in basal and intermediate cells, while most luminal cells have reached a state of terminal differentiation [39, 107, 110, 111].

Testosteron is the principal male androgenic hormone secreted by the testis and is converted into dihydrotestosteron (DHT) through the action of the enzyme 5α-reductase [112]. DHT directly affects luminal prostate epithelium via interaction with AR, while activation of mesenchymal cells results in indirect growth factor release targeting respective receptors within the basal cell compartment [113].

The dynamics of prostate epithelial maintenance have been investigated in castration and androgen-restoration studies in rodents and in patients who had undergone adjuvant androgen-deprivation for prostate carcinoma. In both models androgen-withdrawal results in apoptosis of luminal epithelium, while basal cells largely remain intact [32, 114, 115].

A subpopulation of luminal cells, however, also remains viable after androgen-deprivation indicating that both basal cells and a subgroup of luminal cells are independent of androgens for their survival. During androgen-depletion the luminal pros- tate epithelium consists of K5-positive as well as - negative cells [115]. This implies that phenotypically differentiated luminal cells ($K8^{++}/18^{++}$) can be sub-divided into an androgen-dependent and - independent group. Van Leenders and Schalken consider the K5-negative luminal cells that are viable after androgen-deprivation as late intermediate cells While enhanced proliferative activity of luminal epithelium during androgen-deprivation initially suggested distinct origins of both cell layers, according these authors it rather reflects an enrichment of intermediate cells with proliferative activity after disappearance of post-mitotic, terminally differentiated cells [115].

How signaling pathways contribute to cancer

At least three layers of complexity are intrinsic to cancer development. First, there are direct analogies to the logic of normal development, which is incompletely understood but includes the idea that outcomes depend on multiple progressive cellular communications and the way in which internal pathways are reset by these communications. Second, the genome of the prospective cancer cell changes during development. Hence, the normal developmental plan is modified several times during cancer ontogeny in a way that is not preordained and therefore potentially highly variable from one cancer to the next. Third, there are extensive selection processes at work so the observer only see a tiny, highly selected subset of the products of the developmental processes and genome changes that occurred during cancer development.

Thus, cancer development is a much more complicated process than normal development and, worst of all, it is neither predictably reproducible nor readily observed.

In stem cells, regulatory pathways and proteins ensure a balance between self-renewal and apoptosis.

Especially important are the pathways involving the androgen receptor and the proteins Sonic hedgehog (Shh), Wnt, Notch, Pten, Bmi-1, and Bcl-2. The knowledge gleaned from stem cells led to investigations of the same pathways in tumor stem cells. It is interesting to note that disruption of these pathways or the loss or overexpression of those proteins plays a role in many tumors [4].

Wnt Pathway

The Wnt pathway directs embryonic growth, governing processes such as cell fate specification, proliferation, polarity and migration. It is also implicated in maintenance of stem cell populations [116]. It is conserved in organisms from worms to mammals and its mechanism was first elucidated through analysis of Wingless signalling, its Drosophila homolog [117].

Wnt proteins—a large family of cysteine-rich secreted ligands—bind to a class of seven-pass transmembrane receptors encoded by the frizzled genes transducing a signal to the cytoplasmic protein Dishevelled (Dvl), which is recruited to the membrane, forms a complex with Axin and induces its dephosphorylation [118]. Axin (product of the gene AXIN1) acts as a scaffold protein maintaining the configuration of a complex involving APC (encoded by the Adenomatous Polyposis Coli gene) and β-catenin (encoded by CTNNB1) and facilitating phosphorylation of both APC and β-catenin by glycogen synthase kinase 3b (GSK3b) .

The central molecule of the pathway is β-catenin, which exists in three cellular pools — at the membrane [119] (associated with E-cadherin, a-catenin and other molecules, involved in cell adhesion), cytoplasmic and nuclear (Figure 3). In the absence of a Wnt signal, a multiprotein complex including APC, GSK3b and the scaffolding axin phosphorylates free cytoplasmic β-catenin, marking it for degradation by ubiquitin [120, 121].

The presence of a Wnt ligand at the transmembrane receptor Frizzled activates the cytoplasmic protein Dishevelled, which dephosphorylates axin. This decreases the capacity of axin to form complexes with APC and β-catenin. Conductin, a structural homolog of axin, encoded by AXIN2, can function in a similar way to axin in this protein complex [122]. Phosphorylation of β-catenin by GSK3b, and hence degradation of β-catenin by ubiquitin, is decreased. Cytoplasmic β-catenin thus accumulates and is translocated to the nucleus where it associates with members of the T-cell factor (TCF) and lymphoid enhancer factor (LEF) family of transcriptional factors [123]. The β-catenin–TCF/LEF complex activates transcription of target genes including c-MYC, c-jun and fra-1 (components of the AP-1 transcription complex) and uPAR (the urokinase-type plasminogen activator receptor), the metalloproteinases matrilysin and MMP-26 and cyclin D1. [124].

In summary, activation of the pathway by the presence of a Wnt molecule on the cell surface receptor decreases phosphorylation of β-catenin by the APC/axin/GSK3b complex. β-catenin is thus not degraded, and accumulates in the nucleus, where it stimulates transcription of a variety of cancer-associated genes [4].

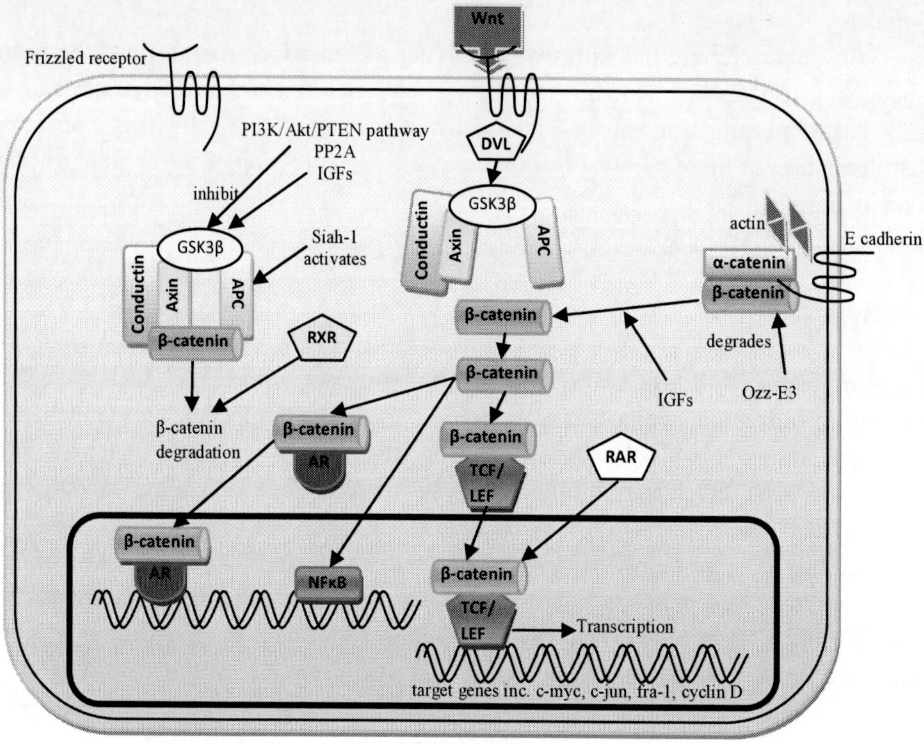

Figure 3. Wnt/beta-catenin signaling: Wnt family members are secreted glycoproteins that bind to cell surface receptors such as Frizzled **Error! Reference source not found.**.

The wnt pathway and prostate cancer

Dysregulation of the Wnt pathway-related E-cadherin can also lead to tumorigenesis and metastasis. The expression of E-cadherin, β-catenin and GSK3b is diminished in prostate cancer cell lines of greater invasive potential. Expression of β-catenin in bone metastases appears downregulated, compared with that seen in corresponding primary tumours in patients with untreated prostate cancer. β-catenin mutations were also found in prostate cancer specimens [125]. The affected CTNNB1 exon corresponds with the site of phosphorylation of β-catenin by GSK3b so such mutations would make β-catenin invulnerable to degradation, allowing it to accumulate and effect gene transcription. These authors demonstrated that β-catenin mutations occur *focally*, suggesting for possible underestimation of the mutation frequency and suggesting for higher impact [116].

Additionally to β-catenin exon mutations, *APC mutations* were found in the xenografts. All mutations were mutually exclusive, consistent with their equivalent effects on β-catenin stability. It was suggested that alterations in the function of Wnt members other than β-catenin may play a role in prostatic carcinogenesis, since the hallmark of Wnt activation – nuclear immunohistochemical β-catenin staining pattern was found in a heterogeneous fashion in 24% of metastatic specimens, but the CTNNB1 exon 3 mutations were in only 5% of the samples [126].

In support to the *hypothesis that alterations in the Wnt pathway may contribute to progression of prostate cancer to androgen independence* was the observation of abnormal β-catenin immunohistochemical pattern hormone-refractory prostate cancer samples. Expression of a mutant form of β-catenin, which is less susceptible to degradation in a mouse model, produced lesions with histological appearances similar to prostatic intraepithelial neoplasia (PIN), the putative precursor of prostate cancer, but these lesions did not progress to invasion or metastasis in the animals. Sublines of apoptosis-resistant LNCaP cells were developed by repeated brief exposure to apoptotic stimuli. These showed increased nuclear β-catenin expression and increased TCF/LEF transcription compared to the parent lines [116].

β-catenin and the androgen receptor (AR)

Studies using prostate cancer cells have demonstrated another function of nuclear β-catenin as an activator of the androgen receptor (AR). Along with mutation and amplification of the AR gene, changes in the influences of AR coactivators are considered important in the conversion to lethal androgen-independent disease [127].

In cell lines expressing the androgen receptor, β-catenin significantly enhances androgen-stimulated transcriptional activity by the AR and diminishes its antagonism by bicalutamide (AR antagonist). A further study has confirmed this observation, noting enhanced androgen-mediated transcription on overexpression of β-catenin in prostate cancer cells. This suggests that the role of β-catenin in prostate oncogenesis may not be limited to transcriptional activation of TCF/LEF and further links β-catenin with development of androgen insensitivity [116].

A specific protein–protein interaction has been demonstrated between β-catenin and AR in both prostatic and neuronal cells. A proportion of the β-catenin translocated from the cytoplasm to the nucleus may be shuttled bound to AR, stimulated by androgen, independent of Wnt molecules [128]. β-catenin/TCF-related transcription in prostate cancer cell lines is inhibited by androgen treatment. This inhibition is AR-dependent, as it only occurs in cells expressing AR endogenously or transiently, and is abrogated by AR antagonists. Androgen-induced suppression of TCF/LEF transcription may be due to competition between AR and TCF/LEF for β-catenin. Such competition has been demonstrated in prostate cells and colorectal cells transiently transfected with an AR expression construct. However in a study of a panel of AR regulators in prostate cancer cell lines, xenografts and clinical specimens using quantitative real-time RT-PCR, overexpression of β-catenin was not found [116].

Crosstalk between the Wnt pathway, the AR and the phosphatidylinositol 3-kinase/Akt pathway in prostate cancer cell lines

The tumour suppressor PTEN, which is frequently mutated in prostate cancer, inhibits the PI3K/Akt pathway. PTEN has been shown to modulate androgen-induced prostate cancer cell growth and AR-mediated transcription [129]. Work using a synthetic PI3K inhibitor, and re-expression of PTEN in a PTEN-null prostate cancer cell line has shown that the mechanism of this involves the Wnt pathway: GSK3b, a downstream effector of PI3K/Akt, also participates in the Wnt pathway; PTEN phosphorylates and inactivates GSK3b via PI3K/Akt; GSK3b- dependent inactivation of cytoplasmic β-catenin is attenuated; β-catenin shuttles to the nucleus and augments ligand-stimulated transcription by AR (Figure 4 and 5) [130].

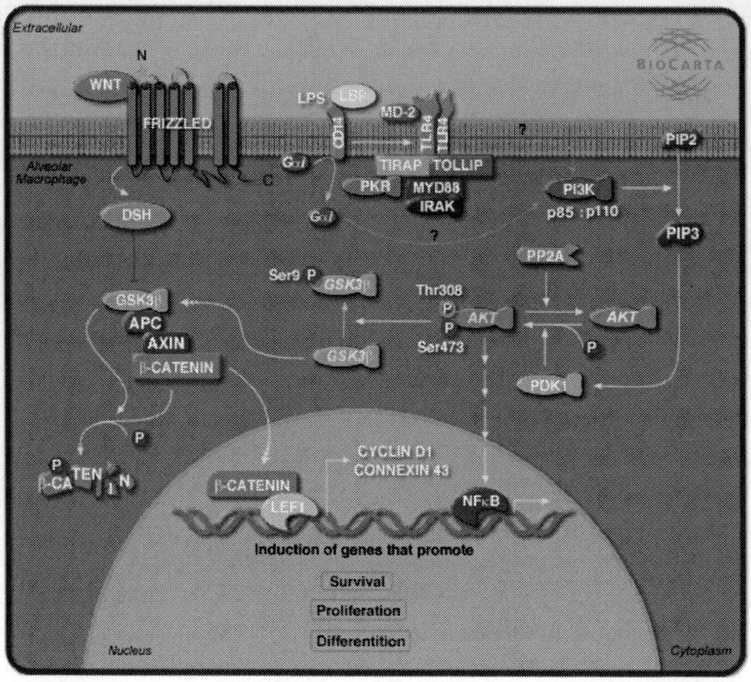

Figure 4. Biocarta pathway (www.biocarta.com) of Wnt signaling: Wnt family members are secreted glycoproteins who bind to cell surface receptors such as Frizzled. Wnt members can play a role in the expression of many genes by interacting with multiple disparate signaling pathways. Shown is the Wnt/beta-catenin pathway.

It was shown that increased activation of Akt, a downstream effector of phosphoinositide 3-kinase (PI3K) together with decreased activation of extracellular-signal-regulated kinase (ERK), a member of the mitogen-activated protein kinase (MAPK) family, predicted poor clinical outcome in prostate cancer [131]. Paramita et al. (Kreisberg group) later showed that Akt activation, but not ERK activation, is correlated with proliferation in human prostate tumors as estimated by the expression of the cell proliferation antigen Ki67 [132]. They in vitro validated these results, using the androgen-dependent prostate cancer cell line LNCaP and its androgen-independent clone C4-2 as models of prostate cancer of good and poor clinical outcome, respectively. C4-2 cells expressed higher Akt activation, lower ERK activation and increased proliferation compared with LNCaP cells, similar to cases of poor clinical outcome. The PI3K inhibitor LY294002, but not the MAPK/ERK kinase inhibitor PD98059, induced growth arrest in both cell lines. Transient transfection with constitutively active Akt increased proliferation while dominant negative Akt decreased it, thus showing that Akt plays an important role in prostate cancer proliferation. Akt regulates the expression and activation of the androgen receptor. Androgen receptor inhibition with Casodex induced growth arrest in LNCaP cells, but not in C4-2 cells. Another PI3K downstream effector, p70 S6 kinase, requires prior phosphorylation by mammalian target of rapamycin (mTOR) for complete activation. Activation of p70 S6 kinase was higher in C4-2 compared with LNCaP cells. Rapamycin, an mTOR inhibitor, had a growth-inhibitory effect in C4-2 cells, but not in LNCaP cells. They suggested that a shift from a Casodex-sensitive proliferation pathway in LNCaP cells to a rapamycin- sensitive pathway in C4-2 cells exist [132].

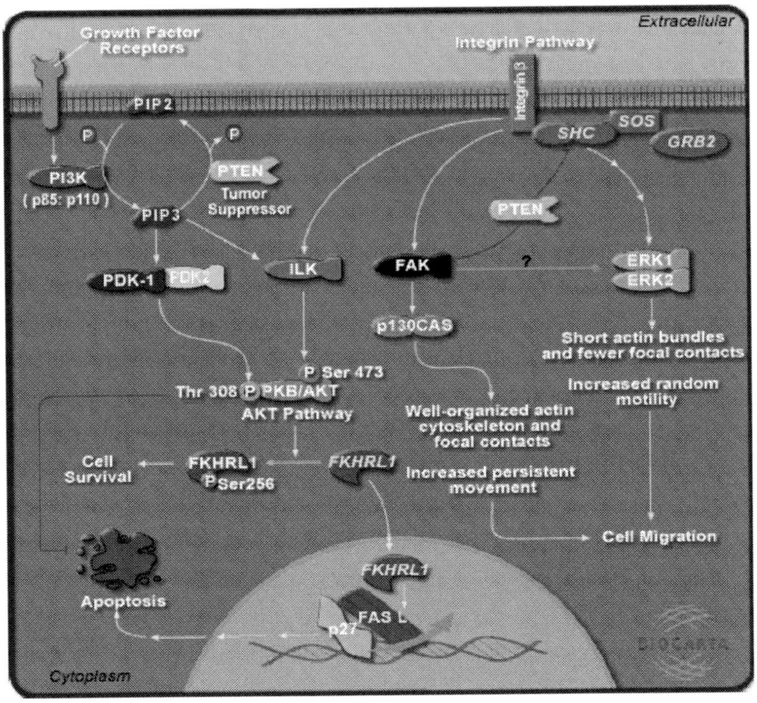

Figure 5. Biocarta (www.biocarta.com) pathway: PTEN dependent cell cycle arrest and apoptosis. PTEN is a tumor suppressor gene. Recombinant PTEN is capable of dephosphorylating phosphatidylinositol 3,4,5-triphosphate[PI(3,4,5)P3], the product of phosphatidylinositol 3 -kinase. Many of the cancer-related mutations have been mapped to the phosphatase catalytic domain, it has been suggested that the phosphatase activity of PTEN is required for its tumor suppressor function. The activation of PKB/AKT is regulated in a complex manner via phosphorylation of AKT on Thr308 and Ser473 by PDK1 and ILK(integrin-linked kinase) respectively. Inactivation of PTEN will constitutively activate PKB/AKT pathway. In addition to its role in regulating the PI 3-K/AKT cell survival pathway, PTEN also inhibits growth factor-induced Shc phosphorylation and suppresses the mitogen-activated protein (MAP) kinase signaling pathway. PTEN also interact with FAK, a key molecule implicated in integrin signaling pathways, and it directly dephosphorylates tyrosine-phosphorylated FAK. PTEN down-regulation of p130CAS through FAK results in inhibition of cell migration and spreading.

Sonic Hedgehog

Of the three intracellular signaling proteins of the hedgehog family found in mammals, Sonic hedgehog (Shh) is the best known variant. It is involved in development of CNS, lungs, and intestine during embryogenesis and controls differentiation and self-renewal in adult stem cells. Shh cancels inhibition of the transmembrane protein Smoothened in multi-receptor complexes on the target cell. This leads to modification of the Gli1 protein complex inside the cell, resulting in translocation to the nucleus, where Gli1 functions as a transcriptional activator. Without Shh, the Patched protein inhibits Smoothened, resulting in Gli1 acting as a transcriptional inhibitor. Disruption of Shh signaling leads to aberrant differentiation and proliferation. In Shh-deficient mice, defects in separation of the 2 brain hemispheres in the frontal cortex lead to cyclopia. In several tumors, including urothelial carcinoma, glioma, and esophageal adenocarcinoma, either an up- or a down-regulation of the Hedgehog pathway has been found [4].

Hedgehog involvement in forming basal cell hyperplasia and its progressing towards cancer, presumably by transforming the normal basal stem cells into the cancer stem cells where persistent Hedgehog activation might be mandatory for tumorigenesis. Dysregulation of Hedgehog signaling pathway has been implicated in tumorigenesis of human prostate cancer and its animal models [133]. Hedgehog signaling activation was found in human prostate cancer cell lines and in advanced human prostate cancer specimens. Hedgehog signaling activation was also demonstrated to promote prostate xenograft tumor growth. Inhibition of Hedgehog signaling activity was shown to block prostate cancer proliferation and metastasis. In animal models, selective Hedgehog protein was found increased in the LADY model and it was shown that Hedgehog overexpression could induce prostate tumorigenesis starting from the normal status [133].

Hedgehog signaling dysregulation was shown to mediate skin basal cell hyperplasia formation, leading to basal cell carcinoma in the mice. Until recently, however, there has been no direct evidence indicating Hedgehog involvement in prostate basal cell hyperplasia and basal cell carcinoma.

The Hedgehog signaling was compared among cells of normal, hyperplastic, transit-amplifying, and cancer status. Hedgehog signaling members including PTCH1, GLI1, GLI2, and GLI3 were found co-localized with p63 expression in most hyperplastic basal cells, but rarely in normal basal cells, suggesting Hedgehog promotion of hyperplasia and transit-amplifying cell accumulation. Both CK-14 and CK-8 markers were found co-localized in the majority of hyperplastic basal cells, but relatively few in the normal basal cells, indicating a Hedgehog-promoted transitory differentiation. Furthermore, CK-14 and PTCH1 were found co-localized with CD44 in the hyerplastic basal cells, in a way similar to the CD44 co-localization with PTCH1 and GLI1 in the cancer cells, implying persistent Hedgehog involvement in hyperplastic basal cells progression towards cancer stem cells. Authors hypothesize that their data supported the presence of prostate cancer stem cells originating from the normal basal stem cells, as a result of persistent Hedgehog signaling activation [134].

Both Hedgehog and p63 had been implicated in regulation of epithelial stem cells [80] and p63 was recently shown to induce Sonic hedgehog expression in the epidermis. The same scenario might be applicable in the prostate basal stem cells, as p63 and Hedgehog were co-localized. Interestingly, p63 expression did not persist in all cells under either basal cell hyperplasia or cancer conditions, resulting in co-existence of three populations (positive or negative or intermediate for p63 expression). Such co-existence might be reflective of the heterogeneous nature of prostate cancer. The loss of p63 expression in the more advanced CaP seemed not easy to reason, since p63 had been implicated in maintaining basal cell proliferation of keratinocytes. On the contrary, Hedgehog expression persisted all the way during transforming basal cells from the normal to the hyperplastic and further towards the tumorigenic conditions. Hedgehog signaling was known to regulate cell proliferation by induc- ing cyclin D and cyclin E and to upregulate Bcl-2, which might compensate p63 loss in the more advanced CaP. The transitory differentiation status of the hyperplastic basal cells appeared most likely to be promoted by Hedgehog signaling, since most of its members were found active in the BCH cells [134].

A Pten deletion condition was shown to promote p63-positive prostate basal cell proliferation without blocking differentiation, a condition similar to the basal cell hyperplasia described here in this study. Moreover, with Pten deletion, a Bcl-2-positive and Sca-1-positive subpopulation was promoted with tumor initiation [80], in a way similar to the enrichment of CD44-positive cells that were regarded as prostate cancer stem cells. Since Bcl-2 can be upregulated through Hedgehog signaling, persistent Hedgehog activation could promote establishment of prostate cancer stem cells *in vivo* as they were transformed from hyperplastic basal cells [134].

Notch Pathway

The Notch pathway is a juxtacrine signaling cascade involved in embryogenesis and in regulation of differentiation, proliferation, and self-renewal in adult stem cells.

Membrane-bound ligands (δ-like-1, -3, -4, and Jagged-1 and -2) induce cleavage of the intracellular domain of the Notch receptors (Notch1– 4) by the γ-secretase complex. After association of the intracellular part of Notch (icN) with the transcription factor CSL and additional proteins (Mastermind-like-1, -2, -3) a multiprotein complex is formed in the nucleus leading to transcription of several target genes. Depending on the tissue and context, activation of the Notch pathway can lead either to an accumulation of immature progenitor cells or to terminal differentiation. The role of dysregulation of the pathway in tumorigenesis must therefore always be viewed in the context of the specific tissue [4].

In the mouse, it regulates differentiation of epidermal cells; Notch ablation leads to hyperplasia and basal cell carcinoma. In another mouse model, integration of the retroviral mouse mammary tumor virus (MMTV) in the vicinity of Notch4 led to Notch4 over-expression and neoplastic transformation of mammary and salivary glands. A high-level expression of Notch1 was correlated in human breast cancer with a significantly lower 5-year survival rate. For several other tumors, including kidney cancer and prostate cancer, over-expression of proteins of the Notch signaling cascade has been shown.

In prostate cancer, a down-regulation of Jagged-1 led to cell-cycle arrest in S phase. A deregulation of the Notch signaling cascade by over-expression of Jagged-1 has been found in advanced prostate cancers. In clear cell renal cell carcinomas, over-expression of Notch1 and Jagged-1 could be shown in vitro and in vivo. By blocking the Notch signaling cascade, a significant inhibition of tumor growth could be achieved in a mouse model [4].

PTEN Pathway

The PTEN tumor suppressor is involved in cell-cycle regulation in almost all tissues. By inhibiting uncontrolled growth and cell division, it can inhibit tumorigenesis. It is also involved in cell migration, cell adhesion, and angiogenesis. Growth factors activate phosphatidylinositol 3-ki-nase (PI 3-kinase), leading to synthesis of the second messenger phosphatidylinositol (3,4,5)-triphosphate (PIP3). The phosphatase PTEN catalyzes the opposite reaction, dephosphorylating PIP3 to yield PIP2. PIP3 can activate Akt via the protein kinase PDK1, the Akt activation in turn leading to inactivation of proapoptotic proteins and to increased cell proliferation. Thus, the loss of PTEN function results in an increased concentration of PIP3 and in Akt hyperactivation, leading to protection from various apoptotic stimuli [4].

PTEN is also connected with various other tumor-relevant signaling pathways. PTEN-inactivating mutations have been found in various tumors, including endometrial carcinoma, glioblastoma, astrocytoma, melanoma, and in bladder and kidney cancer. In bladder cancer, a correlation between the decrease of PTEN protein expression and tumor stage and grade was shown. In advanced renal cell carcinoma, PTEN expression is frequently reduced [4].

PTEN seems to be the most frequently mutated gene in prostate cancer and is predominantly associated with aggressive and advanced tumors. Additionally, an association between androgen-independent tumor growth and a PTEN mutation has been shown. While PTEN mutations lead to a predisposition for prostate cancer in mouse models, such a connection could not yet be shown for humans. PTEN loss-of-function mutations are also often associated with resistance to conventional chemotherapy by activating MDM2-mediated anti-apoptotic mechanisms [4].

Polycomb Family

Bmi-1 is a member of the Polycomb family of transcriptional repressors involved in accurate cell differentiation, proliferation, and self-renewal. Overexpression of Bmi-1 leads to repression of INK4a (p16), resulting in an inactivation of Rb. This prompts the cell to enter the cell cycle and to proliferate. At the same time, Arf is repressed, leading to a reduction of p53-mediated apoptosis. Dysregulation of the Polycomb protein complex by mutation of Bmi-1 leads to increased tumor cell proliferation. Leukemic stem cells from Bmi-$1^{+/+}$ mice could, in contrast to leukemic stem cells of Bmi-1 knockout animals, induce leukemia in host mice [4].

Androgen Receptor Signaling

Androgens are an important growth factor for the normal prostate, signaling through the androgen receptor (AR) localized in the cytoplasm of stromal and secretory epithelial cells [135]. AR is a member of the steroid hormone receptor transcription factor family. Upon ligand binding, AR enters the nucleus and activates panoply of target genes involved in diverse biological processes such as proliferation, differentiation, apoptosis and secretion. Androgens therefore balance proliferation and survival versus death within the prostate and control prostate homeostasis (see Figure 6).

AR in PCa

The role of AR in prostate cancer has been extensively studied. AR is only rarely mutated in the early stages of the disease, however, a high percentage of tumors obtained from patients who had failed androgen ablation overexpress AR. Approximately 30% of this overexpression is associated with genetic amplification and 10– 40% with AR mutations.

Androgen receptor pathway can be hyperactivated through several mechanisms, which can be divided in three main categories: (1) Hypersensitivity in the androgen receptor through genomic amplification, mutations or increased systemic androgen levels, (2) Promiscuous activation of the receptor by mutation or co-regulators alterations, and (3) Receptor activation by alternative pathways such as growth-factors, receptor tyrosine-kinases or Akt [137] (see Figure 6).

Taken together, the acquisition of oncogenic mutations in the AR gene appears to be responsible in many patients. Most of these mutations occur after androgen ablation therapy, indicating the strong selective pressure to maintain AR activity for the survival of prostate cells. Also the ligand-binding domain appears to be the more commonly mutated causing AR to be responsive to other molecules, such as other hormone-like ligands, growth factors or antiandrogens. To date, many mutations have been identified in the AR gene [for a detailed database of mutations consult http://www.mcgill.ca/androgendb/ [138]]; however, only a few have been studied in any detail. Recently, two common AR mutations have been directly related to prostate cancer progression, E231G and T877A [139].

Moreover, in the cases where AR is not hyperactivated, several investigators have shown that alternative survival pathways are activated (see Figure 7). Bcl2 overexpression or combined loss of important tumor suppressors such as p53/Rb or PTEN/Nkx3.1 have been strongly implicated in androgen independence [140]. In addition, there is increasing evidence to suggest that androgen-independent progenitors/stem cells may contribute in important ways to the initiation of prostate cancers and in the development of androgen independence [65]. However, although there have been several reports of pluripotency in precursor-like populations, markers that identify the true prostate cancer stem cell have not yet been described.

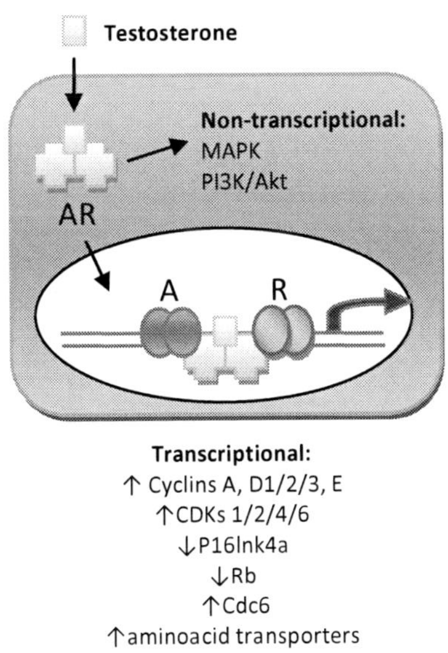

Figure 6. Androgen receptor signaling [136].

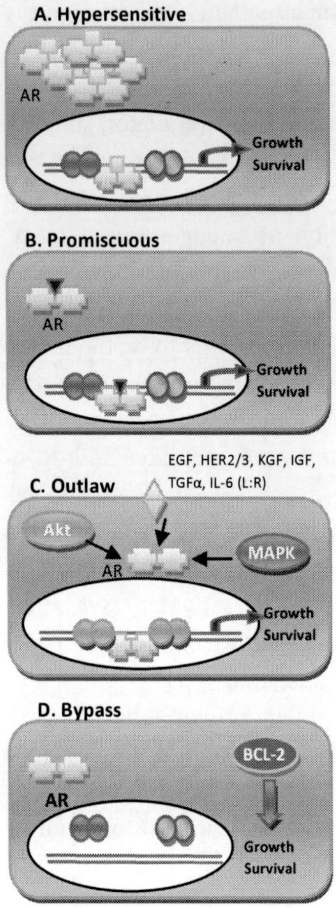

Figure 7. State of activity and interaction of the androgen receptor complex **Error! Reference source not found.**.

Evolution of the androgen receptor pathway during progression of prostate cancer

Whereas the androgen receptor is a key regulator for prostatic function and survival in the normal prostate epithelium, the androgen receptor has been converted into an inducer of uncontrolled cell growth in prostate carcinoma [141]. The recent identification of frequent chromosomal rearrangement in prostate cancer that results in fusion of TMPRSS2 with ETS transcription factor genes explains this conversion [26]. The ETS family members ERG and ETV1 are normally not or lowly expressed in prostate tissues. Fusion of these genes to the strong androgen-regulated TMPRSS2 promoter results in androgen-induced expression of these proto-oncogenes, which likely cause prostate tumorigenesis [142].

That enhanced androgen receptor activity of itself is unlikely to be sufficient for causing aggressive growth that could be deduced from current knowledge on androgen receptor function in the normal prostate. In vitro experiments indicated that androgens induce differentiation and thereby inhibit growth of prostate epithelial cells. In a transgenic mouse model,overexpression of the wild-type (WT) androgen receptor in prostatic epithelial cells resulted in intraepithelial neoplasia but not in carcinomas. The growth-stimulating effect of androgens on normal epithelial cells is at least partly mediated by the prostatic stromal cells, which produce and secrete growth factors, the andromedins, in response to androgen exposure. During prostatic carcinogenesis, the balance between the differentiation- and proliferation-inducing functions of the androgen receptor might be disturbed. Such a transition would likely be reflected by changes within the androgen receptor pathway genes [143].

To identify androgen receptor pathway genes expression Hendriksen et al. performed a microarray analyses on the prostate carcinoma cell line LNCaP and on a panel of 13 prostate carcinoma xenografts. In total, 200 androgen receptor pathway genes were identified. To test whether the androgen receptor pathway is changed during prostate carcinoma progression, the expression of these 200 genes was assessed in both xenografts and clinical prostate carcinoma specimens. The results indicate that prostate carcinomas down-regulate part of the androgen receptor pathway before acquiring the ability to metastasize. This finding was further supported by quantitative reverse transcription-PCR (RT-PCR) on a selection of genes on an independent set of prostate carcinoma samples [143].

Approximately half of the androgen receptor pathway genes were up-regulated in well-differentiated prostate cancer compared with normal prostate. Functionally distinct parts of the androgen receptor pathway were specifically down-regulated in high-grade cancers. Unexpectedly, metastases have down-regulated the vast majority of androgen receptor pathway genes. The significance of this progressive down-regulation of androgen receptor pathway genes was shown for a few androgen receptor–regulated genes. Lower mRNA expression of HERPUD1, STK39, DHCR24, and SOCS2 in primary prostate tumors was correlated with a higher incidence of metastases after radical prostatectomy. HERPUD1 mRNA expression predicted the occurrence of metastases almost perfectly. In vitro experiments showed that overexpression of the stress response gene HERPUD1 rapidly induces apoptosis [143].

Based on the functions of the genes within the distinct subsets, authors proposed the following *model*: Enhanced androgen receptor activity is involved in the early stages of prostate cancer. In well-differentiated prostate cancer, the androgen receptor activates growth-promoting as well as growth-inhibiting and cell differentiation genes resulting in a low growth rate. The progression from low-grade to high-grade prostate carcinoma and metastases is mediated by a selective down-regulation of the androgen receptor target genes that inhibit proliferation, induce differentiation, or mediate apoptosis.

AR in prostate stem cells!

The response of prostate cancer to androgen deprivation is mediated by the androgen receptor (AR). Castrate-resistant disease is marked by a gain-of-function in AR and AR reactivation. The stem cell hypothesis of cancer would therefore predict that AR should be expressed in the prostate cancer stem cell, since genetic selection for gain-of-function changes in AR, such as AR gene amplification, should occur at the level of the stem cell population. Initial reports characterizing prostate cancer stem cells suggest that AR is not expressed in this population, which is an apparent conundrum. Sharifi et al. examined the CD44+/24− LNCaP putative stem cell population by *in-cell Western* and has shown that AR is expressed at the protein level. Their findings suggest that at least a subset of prostate cancers express AR in the putative stem cell population. In support to this finding is also the observation of high proteasomal turnover rate of functional AR in a CD133- subset of $\alpha2\beta1^{hi}$-enriched basal cell population, resulting in androgen-responsiveness at very low to undetectable cellular levels of AR protein.

Genome-Wide Expression Profiles Combined with Pathway Analysis

In the short history of genetic profiling with DNA micro-arrays, high throughput examination of multiple cancer types has accumulated data with the ability to guide clinicians and predict outcome. Despite the breadth and depth of this data, the variety of experimental platforms and output profiles makes direct comparison difficult.

In general, DNA microarrays are based on the arrangement of thousands or tens of thousands of "probes" on a solid support in a distinct pattern such that each feature may be identified by its location. Total RNA is isolated from samples, converted to anti-sense RNA (aRNA) or complementary DNA (cDNA), labeled with a fluorescent tag, and hybridized to the microarray of "probes." Based on this general principle, multiple platforms have been developed, each with unique characteristics based on control group choice, method of array formation, and data output [144].

Large-scale meta-analysis of cancer microarray data identifies common transcriptional profiles of neoplastic transformation and progression

Many studies have used DNA microarrays to identify the gene expression signatures of human cancer, yet the critical features of these often unmanageably large signatures remain elusive. To address this, the group of Arul Chinnaiya's group developed a statistical method, comparative metaprofiling, which identifies and assesses the intersection of multiple gene expression signatures from a diverse collection of microarray data sets [145]. They collected and analyzed 40 published cancer microarray data sets, comprising 38 million gene expression measurements from >3,700 cancer samples. From this, they characterized a common transcriptional profile that is universally activated in most cancer types relative to the normal tissues from which they arose, likely reflecting essential transcriptional features of neoplastic transformation. In addition, they also characterized a transcriptional profile that is commonly activated in various types of undifferentiated cancer, suggesting common molecular mechanisms by which cancer cells progress and avoid differentiation. Finally, these transcriptional profiles were validated on independent data sets.

The purpose of meta-profiling is to address the hypothesis that a selected set of differential expression signatures shares a significant intersection of genes (a meta-signature), thus inferring a biological relatedness. Although many sophisticated analytical and statistical approaches have been applied to microarray normalization and differential expression analysis, they used a single approach that was simple in application yet robust to the heterogeneous data formats used (Affymetrix microarrays, or dual-channel ratio data, usually corresponding to spotted cDNA microarrays), experimental platforms, and experimental designs. After a global normalization procedure to all data sets, they defined two class differential expression analyses potentially relevant to the processes of neoplastic transformation and progression – these included cancer vs. respective normal tissue, undifferentiated vs. differentiated cancer, poor outcome vs. good outcome cancer, metastatic vs. primary cancer, and cancer subtype 1 (e.g., androgen receptor positive) vs. subtype 2 (e.g., androgen receptor negative) and were identified in 34 data sets.

Since a general agreement exists that microarray data from distinct experimental platforms, often using distinct reference samples, are not directly comparable, they developed a method that instead compares statistical measures (Q values) generated independently from each data set (25). For comparing statistical measures across data sets, the method required that analogous hypotheses have been tested in each data set (e.g., genes differentially expressed between normal tissue and cancer tissue). To identify and assess the intersection of multiple differential expression signatures, so-called *meta-signatures*, they applied an automated method, called *comparative meta-profilin.*

Meta-Signature of Neoplastic Transformation: A meta-profiling of 36 neoplastic transformation signatures from 21 data sets (overexpressed in cancer relative to respective normal tissue, $Q < 0.10$), which spaned 12 tissue types including breast, prostate, colon, lung, liver, brain, ovary, pancreas, uterus, salivary gland, bladder, and B lymphocytes. Many of these genes have had previously been associated with cancer; however, often associations have only been made with one specific type of cancer or in cell lines, and not with cancer in general. As is defined by the Gene Ontology Consortium (www.geneontology.org), the meta-signature contains genes involved in the *cell cyle* (CDKN3, CKS2, E2F5, PTMA, PLK, CCT4), *invasion* (MMP9), *transcriptional regulation* (E2F5, SOX4, HDAC1, CBX3, SMARCA4), *protein folding* (HSPD1, HSPE1, CCT4), and the *proteasome* (PSMA1, PSMC4, PSME2). The genes in this signature can be further explored with their web tool ONCOMINE (www.oncomine.org/meta).

The existence of a general cancer meta-signature may not be entirely surprising, because all cancer types share the common features of unregulated cell proliferation and invasion, and it would follow that the genes that are essential to these processes would be highly expressed in multiple cancer types. On the other hand, however, it is interesting that a small number of genes are almost universally activated, given the vast array of transforming mechanisms that are known to initiate cancer and the variety of tissue types represented in this analysis. Activation of these genes may represent convergence on the essential transcriptional features of neoplastic transformation.

Undifferentiated cancers of different tissue types all fail to recapitulate their normal tissue architecture, instead maintaining a disordered state of increased cellular proliferation and invasion. Furthermore, undifferentiated cancers are associated with aggressive behavior and poor patient outcomes. It was hypothesized that if an undifferentiated meta-signature existed, it might suggest common transcriptional mechanisms by which cancer cells avoid differentiation, or dedifferentiate (here). Meta-profiling was performed on seven ''undifferentiated vs. well differentiated'' signatures spanning six cancer types (overexpressed in undifferentiated cancers relative to well differentiated cancers, $Q < 0.10$). Interestingly, a fraction of genes in this meta-signature overlap with the meta-signature of neoplastic transformation. These genes are predominantly associated with *proliferation* (TOP2A, MCM3, CDC2, RFC4, etc.), the overlap likely owing to the parallel increase in proliferation in cancer relative to normal tissues and in undifferentiated cancer relative to differentiated cancer [145].

Of note, three genes unique to the undifferentiated meta-signature have a demonstrated role in *chromatin remodeling* and broad spectrum transcriptional regulation, including the *polycomb group* protein EZH2, which is involved in *transcriptional memory* (33), and the histone variant proteins, H2AFX and H2AFZ, which are known to *control the euchromatin–heterochromatin transition* (34).

The ability of these genes to modulate the expression of tens or hundreds of genes suggests that they may play a role in maintaining the undifferentiated cellular state of high-grade cancer.

It was also found that EZH2 is involved in the metastatic progression of prostate cancer and in the invasive breast cancer phenotype, and another study demonstrated amplification of the EZH2 gene locus in several primary tumor types.

Other genes present in the meta-signature whose function suggests a role in the undifferentiated phenotype include MELK, a kinase with a demonstrated role in early mammalian embryogenesis, and BIRC5 (survivin), an inhibitor of apoptosis (IAP family), which may allow undifferentiated cancer cells to overcome apoptotic checkpoints favoring aberrant progression through mitosis [145].

TMPRSS2 Fusions with Oncogenic ETS Factors in Prostate Cancer Involve Unbalanced Genomic Rearrangements and Are Associated with HDAC1 and Epigenetic Reprogramming

Fusion transcripts arising from translocations of a prostate-specific TMPRSS2 gene with oncogenic ETS factors, ERG, ETV1, or ETV4, were recently identified in up to half of all human prostate cancers. Although both translocation partners have been previously studied in prostate cancer, the identification of a translocation between the two genes provides new insights into the mechanisms of androgen-dependent prostate tumorigenesis. TMPRSS2 encodes for an androgen-dependent transmembrane serine protease which is strongly expressed in both normal and prostate cancer tissues. ETS transcription factors are characterized by a conserved DNA binding domain and they have been previously implicated in oncogenic translocations in Ewing's sarcomas, leukemias, lymphomas, fibrosarcomas, and secretory breast carcinomas. ETS factors regulate the expression of genes with important cancer-relevant biological processes, such as cell growth, differentiation, and transformation. However, their specific roles in prostate cancer development and progression have still remained unclear.

Kristiina Iljin et al. studied advanced prostate cancers for ETS factor alterations, using RT-PCR and DNA and RNA array technologies, and identified putative ERG downstream gene targets from the microarray data of 410 prostate samples. They found that out of the 27 ETS factors, ERG was most frequently over-expressed. Seven cases showed TMPRSS2:ERG gene fusions, whereas the TMPRSS2:ETV4 fusion was seen in one case. In five out of six tumors with high ERG expression, array-CGH analysis revealed interstitial 2.8 Mb deletions between the TMPRSS2 and ERG loci, or smaller, unbalanced rearrangements.

The molecular consequences of the TMPRSS2:ERG fusion gene formation are unknown. Translocations fusing the strong androgen-responsive gene, TMPRSS2, with ERG or other oncogenic ETS factors may facilitate prostate cancer development.

The ERG gene is the most consistently over-expressed oncogene in malignant epithelial cells of the prostate. To identify potential ERG target genes and deregulated biological processes in vivo, the authors analyzed *in silico* gene coexpression data from three different prostate data sets, altogether consisting of 410 human prostate tissue samples.

In silico analysis of the ERG gene co-expression patterns revealed an association with high expression of the histone deacetylase 1 gene (HDAC1), and low expression of its target genes. Histone deacetylase 1 (HDAC1) was the only gene among the top ERG coexpressed genes in all three data sets and therefore was the most consistent feature of ERG-overexpressing prostate cancers. By RT-PCR validation of the HDAC1 expression levels, all ERG-positive prostate cancers were strongly HDAC1-positive, whereas the ERG-negative tumors showed more variable, but significantly lower ($P = 0.0001$) expression levels.

HDAC1 catalyzes the deacetylation of lysine residues of the core histones and other proteins, leading to epigenetic silencing of target genes. HDAC1 has been previously shown to be strongly expressed in hormone-refractory prostate cancers. Currently, it remains unclear if HDAC1 is a direct transcriptional target of ERG or whether HDAC1 up-regulation results from other changes occurring in ERG-overexpressing tumors. ERG has been shown to interact indirectly with HDAC1 via SETDB1 methyl transferase. However, its functional role in prostate cancer development and progression has not yet been clearly determined.

Results from *gene ontology analyses* indicated that the ''organogenesis'' as well as ''cell growth and maintenance'' were the most significantly ($P = 0.001$ and 0.02) overrepresented gene ontology terms in ERG-positive tumors.

Gene set enrichment analysis (GSEA) results indicated that the WNT and PITX2 pathways were among the most highly enriched pathways in ERG-overexpressing tumors. The WNT pathway controls organogenesis by inducing, e.g., PITX2 transcription factor, which serves as an important modulator of growth control genes. HDAC1 itself is linked to these two pathways.

The down-regulated gene sets in ERG-overexpressing tumors included CCR5, cell death, and tumor necrosis factor/FAS. Authors were excited that the HDAC pathway with known HDAC-target genes and regulators was also highlighted by this analysis, suggesting that the up-regulation of HDAC1 in ERG-positive tumors led, as one might expect to the down-regulation of HDAC-target genes.

In summary, the TMPRSS2:ERG translocation is common in advanced prostate cancer and occurs by virtue of unbalanced genomic rearrangements. Activation of ERG by fusion with TMPRSS2 may lead to epigenetic reprogramming, WNT signaling, and down-regulation of cell death pathways, implicating ERG in several hallmarks of cancer.

Integrative biomedical knowledge approaches for modeling pathways

The identification of key transition points in the progression of prostate cancer relies on profiling precursor lesions and "pure" cell populations. Utilizing laser-capture microdissection (LCM) to isolate 101 cell populations, a more specific genetic profile of progression from benign epithelium to metastatic disease was obtained. This laser-capture profile was analyzed in the context of the molecular concepts map (MCM), a compendium of over 20,000 molecular concepts including other expression profiles of prostate cancer, to obtain an integrative molecular model of progression [144].

A "molecular concepts map" was developed to calculate the degree of overlap among "molecular concepts," defined as any biological concept such as disease state, drug treatment, pathway, etc. that can be represented by a molecular signature of genes or proteins [142]. The MCM is based on a context of approximately 20,000 "molecular concepts" derived from twelve data-bases and 500 high-throughput data sets and populated by three general types of concepts: (1) gene and protein annotations from external databases (chromosomal locations, protein domains, molecular functions, cellular locations, biological processes, signaling pathways, and protein-protein interaction networks), (2) microarray-based gene expression profiles (in-vitro profiles collected from the literature and normal and cancer profiles from the Oncomine 64 database collection of over 270 profiling studies and 20,000 microarray experiments) (3) computationally - derived regulatory networks(transcription factor promoters and conserved promoter and untranslated region elements). By comparing the actual and the expected degree of overlap between concepts, the molecular concepts map would suggest a significant overlap and similarity between the "concepts" [144].

Molecular concepts map (MCM) analysis of PCA progression

LCM was utilized to isolate 101 specific cell populations of histological transition points from 44 patients and built signatures reflecting PCA progression by analysis on 20,000 element cDNA microarrays. To identify molecular correlates of PCA progression, the expression signatures were generated with *Oncomine* for subsequent analysis with the MCM.

LCM allowed them to overcome the difficulties of specific cell isolation and stromal contamination and provided further analysis of the impact of stroma on PCA progression. After confirming minimal stromal contamination in epithelial samples, direct comparison between transition points identified robust progression signatures comprised of genes whose expression increased or decreased during progression from benign epithelium to PIN to localized PCA to metastatic PCA. These lists were corrected for multiple hypothesis testing (Q value) and limited to a false discovery value of 5%.76 These signatures were then queried within the MCM to determine concepts associated with the progression of PCA. The *"overexpressed in progression"* signature enriched with concepts related to cell proliferation such as the GeneOntology (GO) process 'cell cycle' (p = 1.6 x 10-6) and the literature concept 'differentially expressed genes in HeLa cells during the cell cycle (p = 1.7 x 10-10). The *"underexpressed in progression"* signature demonstrated enrichment with androgen-signaling concepts [144].

As an example of MCM analysis, the enrichment of genes regulated by androgen in the "overexpressed in PIN vs. benign epithelia" and "underexpressed in progression" signatures is consistent with an increase in androgen signaling during the transition from benign epithelia to PIN lesions and subsequent decreased androgen signaling during the progression to hormone-resistant metastatic disease.

Another concept associated with the "underexpressed in progression" signature included the KEGG pathway 'Glutathione metabolism' (p = 5.1E-6) containing the gene GSTP1, which has been shown to be under-expressed in up to 90% of prostate cancer through promoter hypermethylation [146].

Specific transitions from one histologic type to another were also examined by paired comparison within the MCM. Far fewer genes were differentially expressed between PIN and PCA (1.2% of measured features) than between benign epithelium and PIN (13.4% of measured features) or between localized and metastatic PCA (15.7% of measured features), suggesting that the key transitions may occur with the development of PIN precursor lesions. They identified robust expression signatures based on 1,213 and 1,335 genes over- and under-expressed in PIN compared with benign epithelium. MCM analysis identified enrichment of concepts related to protein biosynthesis (GO process 'protein biosynthesis,' p = 1.3 x 10-7; KEGG pathway 'ribosome,' p = 4.9 x 10-5) and concepts related to the Transfac promoter binding site for E26 transformation-specific (ETS) transcription factor ELK1 (p = 2.1 x 10-9).

The potential involvement of ETS transcription factors in PCA development has recently been documented with gene-fusions of ETS family members in the majority of prostate cancers.78,79 However, the timing and causality of these ETS gene fusions in the transition from PIN to PCA are currently unknown. Previous studies have documented a subset of prostate cancers with mutually exclusive overexpression of ETS-members ERG, ETV1, and ETV4 by fusion to the androgen regulated gene TMPRSS2.

The localized tumors from this study and two additional studies were divided [147] into those with overexpression of ETS members and those without overexpression in order to determine if the ETS dysregulation reflects a distinct prostate cancer subtype.

MCM analysis of genes overexpressed in ETS tumors compared with non-ETS tumors demonstrated enrichment between studies, suggesting overlap of signatures, and with the chromosome sub-arm 6q21. Taken together, this suggests either a cooperative amplification at 6q21 for ETS tumors or a loss of 6q21 in non-ETS tumors, a conclusion suggested by the documented loss of 6q21 in approximately half of localized PCA [144].

Analyzing the transition from localized prostate cancer to metastatic prostate cancer has been hindered by the paucity of studies profiling early metastatic disease. To overcome the limitation of sample size, MCM analysis of hormone-naïve patients from multiple studies detected two interaction networks centered on protein biosynthesis and proliferation. Progression from benign to localized PCA to hormone-naïve (HN) metastatic cancer was more strongly associated with protein biosynthesis while signatures comparing localized and HN PCA were more strongly associated with proliferation concepts. Protein biosynthesis was previously shown to associate with PIN and localized PCA compared with benign epithelium, suggesting increased proliferation may distinguish HN metastatic disease from localized PCA. To determine signatures related to the transition to a hormone-refractory (HR) state, they compared the same HN concepts with hormone-refractory concepts from the LCM tissues and other studies. While there was enrichment of proliferation networks with the 'progression to HR PCA' and the 'HR vs. localized PCA' signatures, the strongest enrichment noted was between the signature 'underexpressed in HR vs. localized PCA' and the concepts associated with androgen signaling. These enrichment networks did not include signatures for HN metastatic disease, suggesting that HN and HR share increased proliferation but HR metastatic cancers have a marked decrease in androgen signaling and protein biosynthesis. The decrease in androgen signaling concepts in the progression to the HR state is consistent with reports of a 'castrated' state selecting cancerous cells with the ability to maintain androgen signaling through receptor amplification and mutations [144].

Thus, MCM analysis suggests a model in which genetic changes drive the transition to HN metastatic cancer through increased proliferation and protein biosynthesis, whereas androgen ablation selects for lesions with the ability to restore aspects of androgen signaling in the castrated state.

Comparison of CD133 Gene Expression Profiles

Genome-wide of expression profiles of CD133+ cells from human tumors were analyzed, presenting about 100 candidate-genes, which were differentially expressed in the CD133+-fraction [148, 149]. An overlap consisting of just 10 differentially regulated between the studies genes was show. Despite the relatively small datasets and the different ways of analyses in the original studies, the public available *Search Tool for the Retrieval of Interacting Genes/Proteins* (*STRING*) provided evidence that 9 out of the 10 candidate genes have been shown to form an interactive network directly interact with each other. Upon expansion of the network it becomes obvious that these target genes are positioned at the interface between *proliferative pathways* (JAK/STAT) and *differentiating pathways* (HOX, PBX, MEIS, GATA2). Of importance in this cascade is the gene KIT. This gene encodes the receptor for stem cell factor (SCF) which is the human homolog of the protooncogene C-KIT. Mutations in KIT are associated with the outcome and progression in gliomas and acute myelogenous leukemia [40–42]. It was very recently published the ability of stem-like cells expressing c-kit (CD117) to produce entire prostate gland when transplanted in nude mice [64].

When DAVID [150] annotation was performed together with KEGG [151] pathway analysis on the respective datasets significantly enriched scores of numerous pathways were revealed. Amongst these are *Glycolisis/Gluconeogenesis* (HSA00010), *Cell Communication* (HSA01430), *Proteasome* (HSA03050), *Cell-Cycle* (HSA04110), *WNT-Signaling* (HSA04310), *TGF-β-Signaling* (HSA04350) and *Apoptosis* (HSA04210) [152].

Ajay P. Singh et al have previously developed a unique *in vitro* LNCaP cell model, which includes androgen-dependent (LNCaP- C33), androgen-independent (LNCaP-C81) and an intermediate phenotype (LNCaP-C51) cell lines resembling the stages of prostate cancer progression to hormone independence. This claim that this model has advantages in overcoming the heterogeneity associated with the prostate cancer up to a certain extent [153].

They characterized and compared the gene expression profiles in the androgen-dependent(LNCaP-C33) vs. androgen-independent (LNCaP-C81) cells using Affymetrix GeneChip array analyses. Multiple genes were identified exhibiting differential expression during androgen-independent progression. Among the important genes upregulated in androgen-independent cells were PCDH7, TPTE, TSPY, EPHA3, HGF, MET, EGF, TEM8, etc., whereas many candidate tumor suppressor genes (HTATIP2, CDKN2A, CDKN2B, CDKN1C, TP53, TP73, ICAM1, SOCS1/2, SPRY2, PPP2CA, PPP3CA, etc.) were decreased.

Among the selected over-expressed genes in androgen-independent (C81) cells, two (TPTE and TSPY) have previously been shown to exhibit a testis-specific restricted pattern of expression in normal conditions. TPTE (transmembrane phosphatase with tensin homology) encodes a PTEN-related tyrosine phosphatase (also referred as PTEN2) with four potential transmembrane domains [154]. In contrast to this finding, another study has reported recurrent loss of a gene cluster on Y-chromosome involving TSPY (tumor-specific protein-Y-linked) in prostate cancer [128]. This further supports the heterogeneous nature of AI prostate cancer as previously described [3] and suggests the existence of different signaling mechanisms among subtypes of metastatic prostate cancer. Nonetheless, a recent study has shown the oncogenic potential of TSPY in HeLa and NIH3T3 cells [155].

Other important over-expressed genes were HGF (hepatocyte growth factor) and MET/HGFR (hepatocyte growth factor receptor). HGF and its receptor, c-Met, have been demonstrated to elicit a number of key functions in numerous tissues that are important in the progression, invasion and metastasis of cancer. Involvement of HGF-Met signalling in metastatic prostate cancer development has also been reported.

In this study, Angiopoietin 2 (ANGP2), whose expression has been shown to correlate with increasing grade of prostate cancer, was found to be down-regulated in androgen-independent prostate cancer cells. Thrombospondin 1 (THBS1) is another important gene that was highly repressed in androgen-independent prostate cancer cells. In previous studies, it has been shown to possess antiangiogenic activity. The intratumoral administration of a THBS1 expression vector into androgen-independent DU-145 prostate cancer xenografts was shown to inhibit tumor growth in vivo through inhibition of tumor angiogenesis. Androgens have been shown to repress the expression of thrombospondin 1 in normal and neoplastic prostate tissues. HTATIP2, another tumor and metastasis suppressor gene, was also repressed in androgen-refractory PC cells. Ectopic expression of HTATIP2 in v-SCLC cells induced expression of a number of genes including the apoptosis-related genes Bad and Siva, as well as metastasis suppressor NM23-H2 [156].

SPRY2, an inhibitor of receptor tyrosine kinase (RTK) signaling, was found to be under-expressed and it has been shown to be epigenetically inactivated in prostate cancer [31]. Progressive loss of ALCAM, which is downregulated in androgen-independent prostate cancer cells, has been reported in high-grade prostate carcinomas [157]. Among other genes of importance exhibiting downregulated expression in androgen-independent prostate cancer cells were CDKN2B/TP53, CDKN1C/TP57, TP73. These genes are known tumor suppressors and are frequently silenced in cancers [158].

In silico pathway prediction analysis has identified important gene networks associated with growth-promoting and apoptotic signaling that were perturbed during *androgen-independent progression*. Further investigation of one of the genes, PPP2CA, which encodes the catalytic subunit of a serine phosphatase PP2A, a potent tumor suppressor, revealed that its expression was decreased in prostate cancer compared to adjacent normal/benign tissue. Furthermore, the downregulated expression of PPP2CA was significantly correlated with tumor stage and Gleason grade. Ingenuity pathway analysis was used to analyze the functional relations in androgen-independent progression. The signaling networks identified, suggested that perturbation of these molecular pathways is associated with the gain of androgen-refractory phenotype. In other studies, these pathways have been shown to be linked with various aspects of tumor progression [36–40]. Most of the identified factors in the above study were repressed in androgen-refractory prostate cancer cells. Importantly, many (TP53, TP73, p16 and TRAIL) among these downregulated pathways are associated with growth inhibitory cell signaling [153].

PPP2CA, which encodes for the catalytic subunit (alpha isoform) of the protein phosphatase 2A (PP2AC), was one of several genes of interest that exhibited a downregulated expression in this androgen-independent prostate cancer cells analysis. PP2A is a potent tumor suppressor and is involved in broad cellular functions. In light of recent observations, it seems to be of potential interest in prostate cancer. *Data mining analysis* using Oncomine cancer profiling database (www.oncomine.org) demonstrate that PPP2CA is significantly downregulated in prostate cancer compared to normal/benign prostate (four studies) and in metastatic prostate cancer compared to primary prostate carcinoma (three studies). Interestingly, in a recent report the expression of the beta isoform of catalytic subunit of PP2A (PPP2CB) is also shown to be downregulated in prostate cancer. Further, the downregulated expression of PP2A in prostate cancer was validated immunohistochemically in normal/benign and matched prostate cancer tissues. However, this study was not able to demonstrate the expression of PP2A during progression to AI phenotype on patient-matched, androgen-dependent and androgen-independent specimens.

Conclusion

Overview of Cancer Stem Cell Paradigm

CSC definition

The tumor stem cell is the original cell of the tumor, responsible for tumorigenesis, tumor differentiation, tumor maintenance, and also for tumor spread and tumor relapse. This is analogous to a stem cell being the original cell of an organ, responsible for organogenesis and organ maintenance.

Characteristics/properties of the CSC

Cancer stem cells have, similar to stem cells, a variety of specific properties.

1. (Unlimited) self-renewal is the central function of (tumor) stem cells allowing the maintenance of the (tumor) stem-cell pool by symmetric cell division. The regulation of this function demands tightly coordinated pathways. For some tumor types, pathological changes in those pathways have already been shown.
2. Differentiation is the capability of stem cells to produce specific tissue types and to maintain tissue homeostasis by asymmetric cell division. Analogously, the aberrant differentiation of tumor stem cells leads to formation of the heterogeneous bulk tumor.
3. Self-preservation of (tumor) stem cells is achieved by strategies including activation of antiapoptotic pathways, increased activity of membrane transporters, and enhanced DNA-repair activity.
4. (Tumor) stem cells have the ability to induce a phenotypic copy of the original tumor after transplantation.

Origin of the cancer stem cell

The cancer stem cell's origin is still not established. The term "cancer stem cell" owes its existence to the stem cell-like properties described above. This does not imply that the cancer stem cell develops directly from a stem cell. At which level of the cell lineage a cell is turned into a cancer stem cell by one or more mutations is still unknown. Initially, three hypotheses were formulated. The enormous plasticity of stem cells, however, makes the three hypotheses seem too static and simplified. New insights into cell fusion and cell transdifferentiation require a more dynamic way of understanding the origin of these cells, both in a single tumor and in different tumor types.

1. The cancer stem cell could directly arise from an aberrant stem cell, because only stem cells have a life span long enough to accumulate enough mutations to cause malignant transformation. Support for this theory comes from expression of stem cell-specific surface markers and similar central regulatory mechanisms.
2. The idea that the tumor stem cell arises, after an initial, not tumor-inducing mutation of a stem cell, from accumulation of further mutations on the progenitor cell level, was put forward by Miyamoto et al. after investigations on the chromosomal t(8;21) translocation in acute myeloid leukemia (AML).
3. The concept of the tumor stem cell being a differentiated progenitor cell that has regained stem-cell properties was suggested by experimental evidence from different laboratories. Huntley et al. could induce AML by transfecting progenitor cells with the MOZ-TIF2 oncogene. Cozzio et al. could induce leukemia identical to a leukemia arisen from a stem cell by transfecting progenitor cells with the fusion oncogene MLL-ENL.
4. A tumor stem cell could also arise from fusion of a bone marrow-derived stem cell with a differentiated, tissue-specific mutated cell, as described by Houghton et al.
5. Bone marrow-derived stem cells can transdifferentiate into function-specific cell types of the host tissue after transplantation, even across germ layers. After induction of a malignant transformation in bone marrow-derived stem cells, those cells formed diverse tumors by trans- differentiation, including epithelial, muscular, and neuronal tumors. The complexity of stem cell biology explains why the term "tumor stem cell" is being increasingly replaced by the term "tumor-initiating cell".

Cell division in tumor stem cells

The present theories have been developed based on the facts known about cell division of the physiological stem cell. While symmetric division of a stem cell gives rise to 2 new stem cells, asymmetric division gives rise to a new stem cell and a daughter cell that enters the differentiation process. Stem cells can alternate between symmetric and asymmetric division. Asymmetric cell division is regulated by intrinsic factors (e.g., specific arrangements of cell polarity factors and/or cell fate factors like Numb or PAR-aPKC), and by extrinsic (e.g., stem cell niche) mechanisms. It is believed that symmetric cell division is more prone to induce new tumors, while asymmetric cell division seems to play a role in tumor suppression. When regulation of asymmetric stem cell division by cell polarity determinants like PAR-aPKC or Partner of Inscrutable (PINS) is disrupted in Drosophila, symmetric cell division and tumor formation result. Expression of a constitutively active form of atypical protein kinase C (aPKC) in neuroblasts also leads to a large increase in symmetrical cell division.

Following the initial identification of hematopoietic tumor stem cells, such cells were also found in several solid tumor types. In urology, cancer stem cells have only been found in prostate tumors so far. The concept and detection of tumor stem cells rely heavily on findings derived from stem cell research. Therefore, in addition to identifying and characterizing urologic tumor stem cells, research in uro-oncology should also aim at better understanding the stem-cell biology of urologic organs. Insights in similarities and differences gleaned from these studies could be used to develop strategies for targeted destruction of tumor stem cells while sparing the physiological stem cells.

References

[1] Signoretti, S. & Loda, M. (2007). Prostate stem cells: from development to cancer. *Semin Cancer Biol.*, 17(3): 219-24.

[2] Miki, J. & Rhim, J. S. (2007). Prostate cell cultures as in vitro models for the study of normal stem cells and cancer stem cells. *Prostate Cancer Prostatic Dis.*, 11(1): 32-39.

[3] Clarke, M. F. et al., (2006). Cancer stem cells--perspectives on current status and future directions: AACR Workshop on cancer stem cells. *Cancer Res.*, 66(19): 9339-44.

[4] Moltzahn, F. R. et al., (2008). Cancer stem cells-Lessons from Hercules to fight the Hydra. *Urol Oncol.,* 26(6): 581-9.

[5] Virchow, R. (1855). Cellular pathology. *Arch Pathol Anat Physiol KlinMed.*, 8: 3-39.

[6] Furth, J. & Kahn, M. (1937). The transmission of leukemia in mice with a single cell. *Am J Cancer.*, 31: 276-82.

[7] Park, C. H., Bergsagel, D. E. & McCulloch, E. A. (1971). Mouse myeloma tumor stem cells: A primary cell culture assay. *J Natl Cancer Inst.*, 46: 411-22.

[8] Hamburger, A. W. & Salmon, S. E. (1977). Primary bioassay of human tumor stem cells. *Science.*, 197: 461-3.

[9] Bonnet, D. & Dick, J. E. (1997). Human acute myeloid leukemia is organized as a hierarchy that originates from a primitive hematopoietic cell. *Nat Med.,* 3(7): 730-7.

[10] Miyamoto, T., Weissman, I. L. & Akashi, K. (2000). AML1/ETO-expressing nonleukemic stem cells in acute myelogenous leukemia with 8;21 chromosomal translocation. *Proc Natl Acad Sci U S A*, 97(13): 7521-6.

[11] Fialkow, P. J. (1974). Human genetic markers as tracers of tumour histogenesis. *J Clin Pathol Suppl.* (R Coll Pathol), 7: 11-5.

[12] Fialkow, P. J., Jacobson, R. J. & Papayannopoulou, T. (1977). Chronic myelocytic leukemia: clonal origin in a stem cell common to the granulocyte, erythrocyte, platelet and monocyte/macrophage. *Am J Med.,* 63(1): 125-30.

[13] Lawson, D. A. & Witte, O. N. (2007). Stem cells in prostate cancer initiation and progression. *J Clin Invest.*, 117(8): 2044-50.

[14] Kabarowski, J. H. & Witte, O. N. (2000). Consequences of BCR-ABL expression within the hematopoietic stem cell in chronic myeloid leukemia. *Stem Cells.*, 18(6): 399-408.

[15] Takahashi, N. et al., (1998). Lineage Involvement of Stem Cells Bearing the Philadelphia Chromosome in Chronic Myeloid Leukemia in the Chronic Phase as Shown by a Combination of Fluorescence-Activated Cell Sorting and Fluorescence In Situ Hybridization. *Blood*. 4758-4763.

[16] Huntly, B. J. & Gilliland, D. G. (2005). Leukaemia stem cells and the evolution of cancer-stem-cell research. *Nat Rev Cancer.*, 5(4): 311-21.

[17] Reya, T. et al., (2001). Stem cells, cancer, and cancer stem cells. *Nature*, 414(6859): 105-11.

[18] Huntly, B. J. et al., (2004). *MOZ-TIF2,* but not *BCR-ABL,* confers properties of leukemic stem cells to committed murine hematopoietic progenitors. *Cancer Cell.*, 6(6): 587-96.

[19] Passegue, E., Wagner, E. F. & Weissman, I. L. (2004). JunB deficiency leads to a myeloproliferative disorder arising from hematopoietic stem cells. *Cell.,* 119(3): 431-43.

[20] Cozzio, A. et al., (2003). Similar MLL-associated leukemias arising from self-renewing stem cells and short-lived myeloid progenitors. *Genes Dev.*, 17(24): 3029-35.

[21] Krivtsov, A. V. et al., (2006). Transformation from committed progenitor to leukaemia stem cell initiated by MLL-AF9. *Nature.*, 442(7104): 818-22.

[22] Al-Hajj, M. et al., (2003). Prospective identification of tumorigenic breast cancer cells. *Proc Natl Acad Sci U S A*, 100(7): 3983-8.

[23] Nowell, P. C. & Hungerford, D. A. (1960). Chromosome studies on normal and leukemic human leukocytes. *J Natl Cancer Inst.*, 25: 85-109.

[24] Rowley, J. D. (1973). Letter: A new consistent chromosomal abnormality in chronic myelogenous leukaemia identified by quinacrine fluorescence and Giemsa staining. *Nature.*, 243(5405): 290-3.

[25] Wong, S. & Witte, O. N. (2004). The BCR-ABL story: bench to bedside and back. *Annu Rev Immunol.*, 22: 247-306.

[26] Tomlins, S. A. et al., (2005). Recurrent fusion of TMPRSS2 and ETS transcription factor genes in prostate cancer. *Science.*, 310(5748): 644-8.

[27] Bohlander, S. K. (2005). *ETV6:* a versatile player in leukemogenesis. *Semin Cancer Biol.*, 15(3): 162-74.

[28] Kovar, H. (2005). Context matters: the hen or egg problem in Ewing's sarcoma. *Semin Cancer Biol.*, 15(3): 189-96.

[29] Liu, A. Y. et al., (1997). Cell-cell interaction in prostate gene regulation and cytodifferentiation. *Proc Natl Acad Sci U S A*, 94(20): 10705-10.

[30] Moll, R. et al., (1982). The catalog of human cytokeratins: patterns of expression in normal epithelia, tumors and cultured cells. *Cell.*, 31(1): 11-24.

[31] Signoretti, S. et al., (2000). p63 is a prostate basal cell marker and is required for prostate development. *Am J Pathol.*, 157(6): 1769-75.

[32] English, H. F., Santen, R. J. & Isaacs, J. T. (1987). Response of glandular versus basal rat ventral prostatic epithelial cells to androgen withdrawal and replacement. *Prostate.,* 11(3): 229-42.

[33] Isaacs, J. T. (1985). Control of cell proliferation and cell death in the normal and neoplastic prostate: a stem cell model. *In: Benign prostatic hyperplasia.* Isaacs, J. T. 1985. Control of cell proliferation and cell death in the normal and neoplastic prostate:

a stem cell model. In: *Benign prostatic hyperplasia*. C.H. Rodgers et al., editors. Department of Health and Human Services. Washington, DC, USA. NIH. Bethesda, Maryland, USA. 85-94., ed. R. C.H. Department of Health and Human Services. Washington, DC, USA. NIH. Bethesda, Maryland, USA. 85-94.

[34] Isaacs, J. T. & Coffey, D. S. (1989). Etiology and disease process of benign prostatic hyperplasia. *Prostate Suppl.*, 2: 33-50.

[35] Verhagen, A. et al., (1992). Colocalization of basal and luminal cell-type cytokeratins in human prostate cancer. *Cancer Res.*, 52(22): 6182-7.

[36] van Leenders, G. et al., (2000). Demonstration of intermediate cells during human prostate epithelial differentiation in situ and in vitro using triple-staining confocal scanning microscopy. *Lab Invest.,* 80(8): 1251-8.

[37] Xue, Y. et al., (1998). Identification of intermediate cell types by keratin expression in the developing human prostate. *Prostate.*, 34(4): 292-301.

[38] Hudson, D.L. et al., (2001). Epithelial cell differentiation pathways in the human prostate: identification of intermediate phenotypes by keratin expression. *J Histochem Cytochem.*, 49(2): 271-8.

[39] Bonkhoff, H., Stein, U. & Remberger, K. (1995). Endocrine-paracrine cell types in the prostate and prostatic adenocarcinoma are postmitotic cells. *Hum Pathol.*, 26: 167-170.

[40] di Sant'Agnese, P. A. (1998). Neuroendocrine cells of the prostate and neuroendocrine differentiation in prostatic carcinoma: a review of morphologic aspects. *Urology.*, 51: 121-124.

[41] Schofield, R. (1978). The relationship between the spleen colony-forming cell and the haemopoietic stem cell. *Blood Cells.*, 4(1-2): 7-25.

[42] Dzierzak, E. & Enver, T. (2008). Stem cell researchers find their niche. *Development.*, 135(9): 1569-73.

[43] Kasper, S. (2008). Exploring the origins of the normal prostate and prostate cancer stem cell. *Stem Cell Rev.*, 4(3): 193-201.

[44] Aumuller, G., Seitz, J. & Bischof, W. (1983). Immunohistochemical study on the initiation of acid phosphatase secretion in the human prostate. Cytochemistry and biochemistry of acid phosphatases IV. *J Androl.*, 4(3): 183-91.

[45] Collins, A.T. et al., (2005). Prospective identification of tumorigenic prostate cancer stem cells. *Cancer Res.*, 65(23): 10946-51.

[46] Collins, A.T. et al., (2001). Identification and isolation of human prostate epithelial stem cells based on alpha(2)beta(1)-integrin expression. *J Cell Sci.*, 114(Pt 21): 3865-72.

[47] Mills, A. A., Qi, Y. & Bradley, A. (2002). Conditional inactivation of p63 by Cre-mediated excision. *Genesis.*, 32(2): 138-41.

[48] Mills, A.A. et al., (1999). *p63* is a *p53* homologue required for limb and epidermal morphogenesis. *Nature.*, 398(6729): 708-13.

[49] Yang, A. et al., (1999). *p63* is essential for regenerative proliferation in limb, craniofacial and epithelial development. *Nature.*, 398(6729): 714-8.

[50] Signoretti, S. et al., (2005). *p63* regulates commitment to the prostate cell lineage. *Proc Natl Acad Sci U S A*, 102(32): 11355-60.

[51] Tsujimura, A. et al., (2002). Proximal location of mouse prostate epithelial stem cells: a model of prostatic homeostasis. *J Cell Biol.*, 157(7): 1257-65.

[52] Salm, S. N. et al., (2005). *TGF-{beta}* maintains dormancy of prostatic stem cells in the proximal region of ducts. *J Cell Biol.*, 170(1): 81-90.

[53] Goto, K. et al., (2006). Proximal prostatic stem cells are programmed to regenerate a proximal-distal ductal axis. *Stem Cells.*, 24(8): 1859-68.

[54] Litvinov, I. V. et al., (2006). Low-calcium serum-free defined medium selects for growth of normal prostatic epithelial stem cells. *Cancer Res.*, 66(17): 8598-607.

[55] Wang, Y. et al., (2001). Cell differentiation lineage in the prostate. *Differentiation.*, 68(4-5): 270-9.

[56] Uzgare, A. R., Xu, Y. & Isaacs, J. T. (2004). In vitro culturing and characteristics of transit amplifying epithelial cells from human prostate tissue. *J Cell Biochem.*, 91(1): 196-205.

[57] Garraway, L. A. et al., (2003). Intermediate basal cells of the prostate: in vitro and in vivo characterization. *Prostate.*, 55(3): 206-18.

[58] Tokar, E. J. et al., (2005). Stem/progenitor and intermediate cell types and the origin of human prostate cancer. *Differentiation.*, 73(9-10): 463-73.

[59] Robinson, E. J., Neal, D. E. & Collins, A. T. (1998). Basal cells are progenitors of luminal cells in primary cultures of differentiating human prostatic epithelium. *Prostate.*, 37(3): 149-60.

[60] Tran, C. et al., (2002). Prostate stem cell antigen is a marker of late intermediate prostate epithelial cells. Mol *Cancer Res.*, 1(2): 113-21.

[61] Kurita, T. et al., (2004). Role *of p63* and basal cells in the prostate. *Development.*, 131(20): 4955-64.

[62] Asselin-Labat, M. L. et al., (2007). *Gata-3* is an essential regulator of mammary-gland morphogenesis and luminal-cell differentiation. *Nat Cell Biol.*, 9(2): 201-9.

[63] Abrahamsson, P. A. (1999). Neuroendocrine differentiation in prostatic carcinoma. *Prostate.*, 39(2): 135-48.

[64] Leong, K. G. et al., (2008). Generation of a prostate from a single adult stem cell. *Nature*.

[65] Tang, D. G. et al., (2007). Prostate cancer stem//progenitor cells: identification, characterization, and implications. *Mol Carcinog.*, 46: 1-14.

[66] Miki, J. et al., (2007). Identification of putative stem cell markers, CD133 and CXCR4, in hTERT-immortalized primary nonmalignant and malignant tumor-derived human prostate epithelial cell lines and in prostate cancer specimens. *Cancer Res.*, 67(7): 3153-61.

[67] Lapidot, T. et al., (1994). A cell initiating human acute myeloid leukaemia after transplantation into SCID mice. *Nature.*, 367: 645-648.

[68] Nelson, W. G., De Marzo, A. M. & Isaacs, W. B. (2003). Prostate cancer. *N Engl J Med.*, 349: 366-381.

[69] Shah, R. B. et al., (2004). Androgen-independent prostate cancer is a heterogeneous group of diseases: lessons from a rapid autopsy program. *Cancer Res.*, 64: 9209-9216.

[70] Patrawala, L. et al., (2006). Highly purified CD44+ prostate cancer cells from xenograft human tumors are enriched in tumorigenic and metastatic progenitor cells. *Oncogene.*, 25(12): 1696-708.

[71] Patrawala, L. et al., (2007). Hierarchical organization of prostate cancer cells in xenograft tumors: the CD44+alpha2beta1+ cell population is enriched in tumor-initiating cells. *Cancer Res.*, 67(14): 6796-805.

[72] Xin, L. et al., (2003). In vivo regeneration of murine prostate from dissociated cell populations of postnatal epithelia and urogenital sinus mesenchyme. *Proc Natl Acad Sci U S A*, 100 Suppl 1: 11896-903.

[73] Xin, L., Lawson, D. A. & Witte, O. N. (2005). The Sca-1 cell surface marker enriches for a prostate-regenerating cell subpopulation that can initiate prostate tumorigenesis. *Proc Natl Acad Sci U S A*, 102(19): 6942-7.

[74] Okada, H. et al., (1992). Keratin profiles in normal/hyperplastic prostates and prostate carcinoma. *Virchows Arch A Pathol Anat Histopathol.*, 421(2): 157-61.

[75] Reiter, R. E. et al., (1998). Prostate stem cell antigen: a cell surface marker overexpressed in prostate cancer. *Proc Natl Acad Sci U S A*, 95(4): 1735-40.

[76] Richardson, G. D. et al., (2004). *CD133,* a novel marker for human prostatic epithelial stem cells. *J Cell Sci.*, 117(Pt 16): 3539-45.

[77] Gu, G. et al., (2007). Prostate cancer cells with stem cell characteristics reconstitute the original human tumor in vivo. *Cancer Res.*, 67(10): 4807-15.

[78] Cerveira, N. et al., (2006). *TMPRSS2-ERG* gene fusion causing *ERG* overexpression precedes chromosome copy number changes in prostate carcinomas and paired HGPIN lesions. *Neoplasia.*, 8(10): 826-32.

[79] Wang, S. et al., (2003). Prostate-specific deletion of the murine Pten tumor suppressor gene leads to metastatic prostate cancer. *Cancer Cell.*, 4(3): 209-21.

[80] Wang, S. et al., (2006). Pten deletion leads to the expansion of a prostatic stem/progenitor cell subpopulation and tumor initiation. *Proc Natl Acad Sci U S A*, 103(5): 1480-5.

[81] Zhou, Z., Flesken-Nikitin, A. & Nikitin, A. Y. (2007). Prostate cancer associated with p53 and Rb deficiency arises from the stem/progenitor cell-enriched proximal region of prostatic ducts. *Cancer Res.*, 67(12): 5683-90.

[82] Cunha, G. R. and Lung, B. (1978). The possible influence of temporal factors in androgenic responsiveness of urogenital tissue recombinants from wild-type and androgen-insensitive (Tfm) mice. *J Exp Zool.*, 205(2): 181-93.

[83] Azuma, M. et al., (2005). A quantitative matrigel assay for assessing repopulating capacity of prostate stem cells. *Biochem Biophys Res Commun.*, 338(2): 1164-70.

[84] Hudson, D. L. et al., (2000). Proliferative heterogeneity in the human prostate: evidence for epithelial stem cells. *Lab Invest.*, 80(8): 1243-50.

[85] Reynolds, B. A. & Weiss, S. (1996). Clonal and population analyses demonstrate that an EGF-responsive mammalian embryonic CNS precursor is a stem cell. *Dev Biol.*, 175(1): 1-13.

[86] Dontu, G. et al., (2003). In vitro propagation and transcriptional profiling of human mammary stem/progenitor cells. *Genes Dev.*, 17(10): 1253-70.

[87] Bello-DeOcampo, D. et al., (2001). Laminin-1 and alpha6beta1 integrin regulate acinar morphogenesis of normal and malignant human prostate epithelial cells. *Prostate.*, 46(2): 142-53.

[88] Lang, S. H. et al., (2001). Prostate epithelial cell lines form spheroids with evidence of glandular differentiation in three-dimensional Matrigel cultures. *Br J Cancer.*, 85(4): 590-9.

[89] Webber, M. M. et al., (1997). Acinar differentiation by non-malignant immortalized human prostatic epithelial cells and its loss by malignant cells. *Carcinogenesis.*, 18(6): 1225-31.

[90] Burger, P. E. et al., (2005). Sca-1 expression identifies stem cells in the proximal region of prostatic ducts with high capacity to reconstitute prostatic tissue. *Proc Natl Acad Sci U S A*, 102(20): 7180-5.

[91] Welm, B. E. et al., (2002). Sca-1(pos) cells in the mouse mammary gland represent an enriched progenitor cell population. *Dev Biol.*, 245(1): 42-56.

[92] Spangrude, G. J., Heimfeld, S. & Weissman, I. L. (1988). Purification and characterization of mouse hematopoietic stem cells. *Science.*, 241(4861): 58-62.

[93] Kim, C. F. et al., (2005). Identification of bronchioalveolar stem cells in normal lung and lung cancer. *Cell.*, 121(6): 823-35.

[94] Lawson, D. A. et al., (2005). Prostate stem cells and prostate cancer. *Cold Spring Harb Symp Quant Biol.*, 70: 187-96.

[95] Deocampo, N. D., Huang, H. & Tindall, D. J. (2003). The role of PTEN in the progression and survival of prostate cancer. *Minerva Endocrinol.*, 28(2): 145-53.

[96] Sun, M. et al., (2001). AKT1/PKBalpha kinase is frequently elevated in human cancers and its constitutive activation is required for oncogenic transformation in NIH3T3 cells. *Am J Pathol.*, 159(2): 431-7.

[97] Majumder, P. K. et al., (2003). Prostate intraepithelial neoplasia induced by prostate restricted Akt activation: the MPAKT model. *Proc Natl Acad Sci U S A*, 100(13): 7841-6.

[98] Ma, X. et al., (2005). Targeted biallelic inactivation of Pten in the mouse prostate leads to prostate cancer accompanied by increased epithelial cell proliferation but not by reduced apoptosis. *Cancer Res.*, 65(13): 5730-9.

[99] Xin, L. et al., (2006). Progression of prostate cancer by synergy of AKT with genotropic and nongenotropic actions of the androgen receptor. *Proc Natl Acad Sci U S A*, 103(20): 7789-94.

[100] Feldman, B. J. & Feldman, D. (2001). The development of androgen-independent prostate cancer. *Nat Rev Cancer.*, 1(1): 34-45.

[101] Berger, R. et al., (2004). Androgen-induced differentiation and tumorigenicity of human prostate epithelial cells. *Cancer Res.*, 64(24): 8867-75.

[102] van Leenders, G. J. & Schalken, J. A. (2003). Epithelial cell differentiation in the human prostate epithelium: implications for the pathogenesis and therapy of prostate cancer. *Crit Rev Oncol Hematol.*, 46 Suppl: S3-10.

[103] Helmut Bonkhoff, N. W. G. D. K. R. (1991). *Relation of endocrine-paracrine cells to cell proliferation in normal, hyperplastic, and neoplastic human prostate*. 91-98.

[104] Krajewska, M. et al., (1997). Immunohistochemical analysis of in vivo patterns of expression of CPP32 (Caspase-3), a cell death protease. *Cancer Res.*, 57(8): 1605-13.

[105] Kim, N. W. et al., (1994). *Specific association of human telomerase activity with immortal cells and cancer. Science.*, 266(5193): 2011-5.

[106] Sommerfeld, H. J. et al., (1996). Telomerase activity: a prevalent marker of malignant human prostate tissue. *Cancer Res.*, 56(1): 218-22.

[107] Xue, Y. et al., (1997). Prostatic neuroendocrine cells have a unique keratin expression pattern and do not express Bcl-2: cell kinetic features of neuroendocrine cells in the human prostate. *Am J Pathol.*, 151(6): 1759-65.

[108] McDonnell, T. J. et al., (1992). Expression of the protooncogene bcl-2 in the prostate and its association with emergence of androgen-independent prostate cancer. *Cancer Res.*, 52(24): 6940-4.

[109] Colombel, M. et al., (1993). Detection of the apoptosis-suppressing oncoprotein bc1-2 in hormone-refractory human prostate cancers. *Am J Pathol.*, 143(2): 390-400.

[110] Bonkhoff, H., Stein, U. & Remberger, K. (1994). The proliferative function of basal cells in the normal and hyperplastic human prostate. *Prostate.*, 24(3): 114-8.

[111] De Marzo, A. M. et al., (1998). Prostate stem cell compartments: expression of the cell cycle inhibitor p27Kipl in normal, hyperplastic, and neoplastic cells. *Am J Pathol.*, 153(3): 911-9.

[112] Coffey, D. S. (1993). The molecular biology of the prostate, *In: Prostate diseases*, H. Lepor and R.K. Lawson, Editors., Saunders: Philadelphia. 28-56.

[113] Byrne, R. L., Leung, H. & Neal, D. E. (1996). Peptide growth factors in the prostate as mediators of stromal epithelial interaction. *Br J Urol.*, 77(5): 627-33.

[114] Evans, G. S. & Chandler, J. A. (1987). Cell proliferation studies in rat prostate. I. The proliferative role of basal and secretory epithelial cells during normal growth. *Prostate.*, 10(2): 163-78.

[115] van der Kwast, T. H. et al., (1998). Cycling activity of benign prostatic epithelial cells during long-term androgen blockade: evidence for self-renewal of luminal cells. *J Pathol.*, 186(4): 406-9.

[116] Yardy, G. W. & Brewster, S. F. (2005). Wnt signalling and prostate cancer.*Prostate Cancer and Prostatic Diseases.,* 8:119-126.

[117] Wodarz, A. & Nusse, R. (1998). Mechanisms of Wnt signaling in development. *Annu Rev Cell Dev Biol.,* 14: 59-88.

[118] Willert, K., Shibamoto, S. & Nusse, R. (1999). Wnt-induced dephosphorylation of axin releases beta-catenin from the axin complex. *Genes Dev.*, 13(14): 1768-73.

[119] Chesire, D. R. & Isaacs, W. B. (2003). Beta-catenin signaling in prostate cancer: an early perspective. *Endocr Relat Cancer.,* 10(4): 537-60.

[120] Fearnhead, N. S., Britton, M. P. & Bodmer, W. F. (2001). The ABC of APC. *Hum Mol Genet.*, 10(7): 721-33.

[121] Dajani, R. et al., (2003). Structural basis for recruitment of glycogen synthase kinase 3beta to the axin-APC scaffold complex. *Embo J.*, 22(3): 494-501.

[122] Behrens, J. et al., (1998). Functional interaction of an axin homolog, conductin, with beta-catenin, APC, and GSK3beta. *Science.*, 280(5363): 596-9.

[123] Roose, J. & Clevers, H. (1999). TCF transcription factors: molecular switches in carcinogenesis. *Biochim Biophys Act*a., 1424(2-3): M23-37.

[124] Crawford, H.C. et al., (1999). The metalloproteinase matrilysin is a target of beta-catenin transactivation in intestinal tumors. *Oncogene.*, 18(18): 2883-91.

[125] Voeller, H. J., Truica, C. I. & Gelmann, E. P. (1998). Beta-catenin mutations in human prostate cancer. *Cancer Res.*, 58(12): 2520-3.

[126] Chesire, D. R. et al., (2002). In vitro evidence for complex modes of nuclear beta-catenin signaling during prostate growth and tumorigenesis. *Oncogene.*, 21(17): 2679-94.

[127] Culig, Z. et al., (1998). Expression, structure, and function of androgen receptor in advanced prostatic carcinoma. *Prostate.*, 35(1): 63-70.

[128] Mulholland, D. J. et al., (2002). The androgen receptor can promote beta-catenin nuclear translocation independently of adenomatous polyposis coli. *J Biol Chem.*, 277(20): 17933-43.

[129] Li, P., Nicosia, S. V. & Bai, W. (2001). Antagonism between PTEN/MMAC1/TEP-1 and androgen receptor in growth and apoptosis of prostatic cancer cells. *J Biol Chem.*, 276(23): 20444-50.

[130] Sharma, M., Chuang, W. W. & Sun, Z. (2002). Phosphatidylinositol 3-kinase/Akt stimulates androgen pathway through GSK3beta inhibition and nuclear beta-catenin accumulation. *J Biol Chem.*, 277(34): 30935-41.

[131] Kreisberg, J. I. et al., (2004). Phosphorylation of Akt (Ser473) is an Excellent Predictor of Poor Clinical Outcome in Prostate Cancer. *Cancer Res.*, 5232-5236.

[132] Ghosh, P. M. et al., (2005). Signal transduction pathways in androgen-dependent and -independent prostate cancer cell proliferation. 119-134.

[133] Chen, B. Y. et al., (2006). A mouse prostate cancer model induced by Hedgehog overexpression. *Journal of Biomedical Science.*, 13(3): 373-384.

[134] Chen, B. Y. et al., (2007). Hedgehog is involved in prostate basal cell hyperplasia formation and its progressing towards tumorigenesis. *Biochemical and Biophysical Research Communications.*, 357(4): 1084-1089.

[135] Chatterjee, B. (2003). The role of the androgen receptor in the development of prostatic hyperplasia and prostate cancer. *Molecular and Cellular Biochemistry.*, 253(1): 89-101.

[136] Nieto, M. et al., (2007). Prostate cancer: Re-focusing on androgen receptor signaling. *The International Journal of Biochemistry & Cell Biology.*, 39(9): 1562-1568.

[137] Feldman, B. J. & Feldman, D. *The development of androgen-independent prostate cancer.*

[138] Bruce Gottlieb, L. K. B. J. H. W. M. T. (2004). *The androgen receptor gene mutations database (ARDB): 2004 update.*, 527-533.

[139] Han, G. et al., (2005). *Mutation of the androgen receptor causes oncogenic transformation of the prostate.*, 1151-1156.

[140] Zhou, Z., et al., (2006). *Synergy of p53 and Rb Deficiency in a Conditional Mouse Model for Metastatic Prostate Cancer.*, 7889-7898.

[141] Heinlein, C. A. & Chang, C. (2004). *Androgen Receptor in Prostate Cancer.*, 276-308.

[142] Tomlins, S. A. et al., (2007). Integrative molecular concept modeling of prostate cancer progression. *Nat Genet.*, 39(1): 41-51.

[143] Hendriksen, P. J. M. et al., (2006). *Evolution of the Androgen Receptor Pathway during Progression of Prostate Cancer.*, 5012-5020.

[144] Morris, D. S. et al., (2007). *Integrating biomedical knowledge to model pathways of prostate cancer progression.*, 1177-87.

[145] Rhodes, D. R. et al., (2004). *Large-scale meta-analysis of cancer microarray data identifies common transcriptional profiles of neoplastic transformation and progression.*, 9309-9314.

[146] Lee, W. H. et al., (1994). Cytidine methylation of regulatory sequences near the pi-class glutathione S-transferase gene accompanies human prostatic carcinogenesis. *Proc Natl Acad Sci U S A*, 91(24): 11733-7.

[147] Glinsky, G. V. et al., (2004). Gene expression profiling predicts clinical outcome of prostate cancer. *J Clin Invest.*, 113(6): 913-23.

[148] He, X. et al., (2005). Differential gene expression profiling of CD34+ CD133+ umbilical cord blood hematopoietic stem progenitor cells. *Stem Cells Dev.*, 14(2): 188-98.

[149] Huang, T. S., et al., (2008). Functional network reconstruction reveals somatic stemness genetic maps and dedifferentiation-like transcriptome reprogramming induced by *GATA2. Stem Cells.*, 26(5): 1186-201.

[150] Dennis, G., Jr. et al., (2003). DAVID: Database for Annotation, Visualization, and Integrated Discovery. *Genome Biol.,* 4(5): P3.

[151] Wixon, J. & Kell, D. (2000). The Kyoto encyclopedia of genes and genomes--KEGG. *Yeast.*, 17(1): 48-55.

[152] Regenbrecht, C. R., Lehrach, H. & Adjaye, J. (2008). Stemming cancer: functional genomics of cancer stem cells in solid tumors. *Stem Cell Rev.*, 4(4): 319-28.

[153] Singh, A. P. et al., (2008). Genome-wide expression profiling reveals transcriptomic variation and perturbed gene networks in androgen-dependent and androgen-independent prostate cancer cells. *Cancer Lett.*, 259(1): 28-38.

[154] Wu, Y. et al., (2001). *PTEN 2, a Golgi-associated Testis-specific Homologue of the PTEN Tumor Suppressor Lipid Phosphatase.*, 21745-21753.

[155] Oram, S. et al., (2006). *TSPY potentiates cell proliferation and tumorigenesis by promoting cell cycle progression in HeLa and NIH3T3 cells.*, 154.

[156] Xiao, H. et al., (2000). *TIP30* has an intrinsic kinase activity required for up-regulation of a subset of apoptotic genes. *Embo J.*, 19(5): 956-63.

[157] Glen Kristiansen, C. P. C. W. C. S. L. W. V. L. S. L. M. D. A. R. (2003). *ALCAM/CD166 is up-regulated in low-grade prostate cancer and progressively lost in high-grade lesions.*, 34-43.

[158] Lodygin, D. et al., (2005). Functional Epigenomics Identifies Genes Frequently Silenced in Prostate Cancer., 4218-4227.

In: Handbook of Prostate Cancer Cell Research...
Editor: Alan T. Meridith
ISBN: 978-1-60741-954-9

Chapter 4

Interactions between the Androgen Receptor and Other Signaling Pathways Implicated in Prostate Cancer Growth and Development

Oluwakemi Obajimi
Department of Medicine, University of Medicine and Dentistry of New Jersey, Robert Wood Johnson Medical School, 401 Haddon Avenue, Camden, NJ 08103, USA; Cooper Cancer Institute, Cooper University Hospital, Camden, NJ 08103, USA

Abstract

Prostate cancer is the second most prevalent type of cancer diagnosed in men in the western world and the leading cause of cancer mortality in older men. Since early-stage development of prostate cancer depends on androgens for growth and survival, androgen deprivation by surgical or chemical castration has been used as the main therapeutic intervention in the treatment of the disease. Despite initial responsiveness to androgen ablation therapy, tumor cells invariably progress to an androgen-independent and anti-androgen resistance that is untreatable, ultimately leading to mortality. It is well established that androgen receptor mediated signaling is critical not only in normal prostate functioning but also in the development and progression of prostate cancer. Given its pivotal role in prostate cancer progression, it is important to delineate its interaction with other signaling pathways that may contribute to prostate cancer growth and development. This is particularly necessary since the signaling mechanisms involved in the transition from normal to androgen-dependent and then to androgen-independent prostate cancer are poorly understood. It has become increasingly evident that crosstalk exists between the androgen receptor and other signaling pathways. This suggests the existence of a coordinated mechanism where there is a converging point for generated signals, thereby driving prostate cancer growth and aggressiveness, downstream. This review therefore focuses on the role that the androgen receptor plays in the major signaling pathways implicated in prostate cancer development and progression.

Introduction

Androgens play important roles in the development and progression of prostate cancer. Early-stage prostate cancer and normal prostate tissues depend on androgens for growth and survival. Androgen functions by binding to its receptor, the androgen receptor (AR), thus inducing transcriptional activation of its target genes [1,2]. AR belongs to the group of nuclear receptor superfamily of ligand-controlled transcription factors [3]. In the absence of the ligand, AR exists within the target cell in a transcriptionally inactive form. Upon binding to the ligand, it undergoes a conformational change and associates with specific DNA sequences. The transcriptional activity of AR is influenced by a diverse array of coactivators, corepressors and interactions with other transcription factors [4-6]. Growth, differentiation and maintenance of the prostate depend on AR and androgens, in particular testosterone and dihydrotestosterone (DHT). In the prostate gland, testosterone is converted into DHT by 5-α reductase 1. DHT is the more active form of testosterone in this gland and has a binding affinity for AR by approximately 10-fold more than testosterone [7].

Because AR-mediated signaling plays a pivotal role in the progression of prostate cancer [8], androgen ablation therapy either surgically or biochemically has been used in the treatment of the disease at the initial stages [9]. However, androgen independent or hormone refractory prostate cancer eventually develops. Most [10-13], though not all [10] androgen-independent prostate cancers are AR-expressing, and despite the low androgen ligand environment resulting from androgen ablation therapy, AR signaling remains functional [14,15]. Several hypotheses have been postulated to explain the phenomenon by which AR-mediated signaling affects growth and survival of normal and cancerous (both androgen-dependent and androgen-independent forms) [16-18] prostate cells, but the exact mechanism is yet to be defined. During normal development of the prostate, AR signaling is associated with growth arrest and differentiation of prostatic epithelial cells [19]. On the other hand, AR-expressing prostate cancer cells somehow continue to possess the ability to proliferate. The underlying reason for this disparity in cellular response to AR-mediated signaling is poorly understood but begs this question: "what is the exact mechanism that is responsible for the role of AR in the development and/or progression of prostate cancer?" The mechanisms of androgen dependence are important in the understanding of the progression of prostate cancer to androgen independence. However, there is increasing evidence demonstrating that other signaling pathways significantly contribute to prostate cancer development and progression besides AR signaling [20-23]. Therefore, this chapter focuses on current evidence of the existence of crosstalk between AR and other signaling pathways as a coordinated mechanism driving prostate cancer development, growth and progression.

AR Signaling

Androgens are steroid hormones necessary for the development and maturation of the human prostate gland while AR is the mediator of its action. AR is a ligand-controlled transcription factor and a member of the nuclear/steroid receptor gene family. It is a fundamental requirement for androgen signaling within the prostate gland. The AR gene contains eight exons encoding four functional domains viz: (1) the amino (NH_2)-terminal regulatory domain (NTD); (2) the DNA-binding domain (DBD); (3) a hinge domain and (4) the carboxy (COOH) terminal ligand binding domain (LBD). In its inactive form, AR is mainly localized in the cytoplasm, unbound to the ligand and associated with heat shock proteins (HSPs). Upon binding of the androgen to the LBD of AR, a cascade of events is initiated, causing conformational changes in AR: phosphorylation, homodimerization and dissociation of AR from HSPs. Ligand-bound AR is then translocated into the nucleus [24-26].

Unlike untreated prostate cancers, several studies have shown that the AR gene is amplified in about 30% of hormone refractory prostate carcinomas [27-29]. Even without AR gene amplification, AR protein appears to be overexpressed in hormone refractory tumors [27,30]. AR is a DNA-binding transcription factor that becomes activated in the ligand-bound form. Mutations that can alter the ligand specificity of AR and also allow the formation of alternative steroid ligands that do not bind to the wild-type AR have however been detected in both tumors and cancer cell lines of the prostate [31,32]. Once in the nucleus, AR binds to a palindromic DNA sequence called an androgen-responsive element (ARE) [33]. The resulting AR-androgen complex modifies gene expressions by further interacting either directly or indirectly with other transcriptional cofactors. The function of AR is regulated by co-activators [34] or co-repressors [20]. While the array of genes regulated by AR is not entirely known, some AR target genes such as the prostate specific antigen (PSA) [35] have been well characterized [36]. These target genes have the potential for exploitation as biological targets for therapy and as diagnostic or prognostic markers in the treatment of human prostate cancer (Table 1). Currently, there are also drugs utilized against prostate cancer growth but targeting major signaling mechanisms that interact with the AR signaling pathway. For example, the drug PI-103 targets the PI3K/Akt pathway and CP-751, 871 targets the growth factor, IGF1 receptor signaling pathway (Table 2).

There are numerous studies suggesting that AR signaling is important for the normal proliferation and functional activities of prostate cells [37,38]. In most primary and metastatic prostate cancers, cells continue to express functional AR independent of the androgen [39]. For instance, in androgen-dependent LNCaP cells, the cell cycle regulators, CDK2 and CDK4 are androgen responsive [40]. Also, the association of the retinoblastoma protein (RB), involved in tumor suppression, with AR may cause a cascade of events which in turn accelerates cell proliferation by promoting the transition from the G1 phase of the cell cycle to the S phase [41]. Not only does AR play a central role in the developmental stage of prostate cancers, it is a determinant in their response to therapy. The relationship existing between androgens and the prostate gland (both in the normal and cancer states) indicates that the AR signaling pathway is an important regulator of prostate growth and tumorigenesis and continues to be an inviting therapeutic target in prostate cancer treatment. It has become increasingly evident that crosstalk exists between AR signaling and other pathways which have the ability to influence and modulate generated AR signals downstream. These points of crosstalk may shed more light on the behavior of the prostatic cells in the cancer state.

Table 1. A list of targetable AR-regulated genes and the mode of action of a few anti-cancer drugs

AR Target gene	Gene implications in prostate cancer	Drug/mode of action	References
Caspase-2	In vivo, AR interacts with an ARE in intron 8 of the caspase-2 gene. High levels of DHT results in reduced caspase-2 levels in prostate cancer cells.	Docetaxel. It exerts anti-tumor activity via multiple mechanisms. It induces apoptosis by a mechanism involving the activation of caspase-2.	[163,164]
Kallikrein 2 (KLK2)	KLK-2 is a serine protease, androgen-regulated and it is positively correlated with prostate cancer grade and progression.	Silymarin. It is an anti-cancer agent that suppresses the secretion of KLK2 and inhibits androgen-dependent cell proliferation in prostate cancer cells. It also inhibits AR function by reducing its translocation into the nucleus.	[165,166]
p21	It a cyclin dependent kinase (CDK) inhibitor. The upregulation of p21 gene expression by androgen suggests that it may have an anti-apoptotic function in prostatic epithelial cells.	Genistein. It has been demonstrated to induce the expression of the p21 tumor suppressor gene and promote cell cycle arrest in prostate cancer cells.	[167-169]
Prostate specific antigen (PSA)	It is an AR responsive gene used as a biomarker of prostate cancer and to monitor response to treatment. It is negatively correlated with prostate cancer grade and progression.	Genistein. This drug transcriptionally downregulates AR and inhibits PSA expression in prostate cancer cells.	[170]
Prostatic acid phosphatase (PAcP)	It dephosphorylates and deactivates HER-2/neu and in turn decreases growth and tumorigenicity in prostate cancer.	1□25-dihydroxyvitamin D3 (1,25D). It has been shown to inhibit growth of human prostate cancer cells by upregulating PAcP.	[171]

Table 2. A list of drugs targeting major signaling pathways that crosstalk with AR signaling as a measure against prostate cancer

Drug	Drug target	Signaling pathway	Mode of action/Status
PI-103 (Pyridinylfuranopyrimidine inhibitor)	PI3K	PI3K/Akt	PI-103 is a potent, cell-permeable, ATP-competitive inhibitor of the PI3K family members with selectivity toward DNA-PK, PI3K and mTOR [172]. This drug is still undergoing clinical investigation.
ZSTK474 [2-(2-difluoromethylbenzimidazol-1-yl)-4,6-dimorpholino-1,3,5-triazine]	PI3K	PI3K/Akt	The *s*-triazine derivative, ZSTK474, strongly inhibits the growth of tumor cells by targeting PI3K [173].
CP-751,871	Epidermal growth factor receptor	IGFI-R	CP-751, 871 is a human antibody against type I insulin-like growth factor receptor. It has been shown to suppress the secretion of PSA in prostate cancer [174]. This drug is currently under investigation.
Gefitinib (*N*-(3-chloro-4-fluorophenyl)-7-methoxy-6-[3-4-morpholin) propoxy])	Epidermal growth factor receptor	ErbB	Gefitinib is an anilinoquinazoline, a selective epidermal growth factor receptor tyrosine kinase inhibitor. However, the drug had limited therapeutic efficacy in hormone-refractory prostate cancer [175].
Pertuzumab is also known as 2C4	Epidermal growth factor receptor	ErbB	Pertuzumab is a second-generation anti–human EGFR monoclonal antibody. Upon binding, it inhibits the dimerisation of EGFR which in turn exerts anti-proliferative properties in tumor cells [176]. However, it failed in phase II clinical trials [176].
U0126 (1,4-diamino-2,3-dicyano-1,4-bis(2-aminophynyltio)butadiene)	MEK	MAPK	U0126 is a MEK inhibitor that causes target gene specific decreases in AR activity in prostate cancer cells [111].
TPEN (N,N,N',N',-tetrakis(2-pyridylmethyl) ethylenediamine)	XIAP (X-linked Inhibitor of Apoptosis Protein)	Zinc	TPEN is a membrane-permeable zinc chelator shown to promote the activation of caspases-3 and -9. It also sensitizes prostate cancer cells to TRAIL-mediated apoptosis [177].

Some of the major pathways that interact with the AR signaling pathway and have been implicated as contributors to prostate cancer development and progression are the mitogen-activated protein kinase (MAPK)- [42], the Wnt-[43] and the growth factor/cytokine- [44] signaling pathways. Therefore, it is the intent of this chapter to focus on the increasing body of evidence of molecular interactions between these signaling pathways and AR signaling in order to highlight the implications of such crosstalk for prostate cancer biology and treatment.

Pathways to Androgen Independence in Prostate Tumorigenesis

The molecular mechanisms responsible for and those associated with the transition of prostate cancer from androgen dependence to independence are poorly understood. Since the majority of androgen-independent tumors remain AR- and PSA- expressing [12], a plausible explanation for this phenomenon is that the AR signaling pathway is intact but with a dysregulated mechanism in prostate cancer. Whether androgen dependency in prostate cancer results from the capability to proliferate at low levels of the androgen or if it is the diminished androgen concentration that yields the selection of cells for continuous growth at such low androgen levels, is unclear. Nonetheless, different mechanisms have been proposed to explain the activation of AR at depleted androgen levels. These mechanisms include: AR gene amplification or protein upregulation [8]; mutations of the AR gene [45]; AR activation by non-steroid ligands such as growth factors [46] or non-androgenic steroids [47]; alterations in tumor suppressor genes such as p53 and protein RB [48,49]; and changes in the expressions of AR co-activators [50] or co-repressors [51]. Whether one or more combinations of these mechanisms occur sequentially or simultaneously in the progression of prostate cancer cells to androgen independency remains unknown. Importantly, if multiple mechanisms contribute to the development of androgen independency, targeting a single pathway as a means of controlling prostate cancer advancement may be insufficient to inhibit prostate tumor growth and progression.

Crosstalk between AR and Other Cell Signaling Pathways

Prostate cancer cells utilize a diverse array of pathways to grow, thrive and survive especially under condition of low androgen levels following initial anti-androgen therapeutic strategies. These pathways include modulations of the p53 and RB tumor suppressor genes [41,52]. While extensive studies have shown that AR signaling can be regulated by different cofactors in benign prostate epithelial cells [53], very little is known of their role in prostate cancer. Hence, this chapter will review the evidence that other signaling pathways can influence the AR signaling pathway, acting as contributors to the development and progression of prostate cancer. Among the numerous pathways that have been reported to interact with AR signaling in the progression of prostate cancer, this chapter will focus on growth factor/cytokine signaling, P13K/Akt signaling, mitogen-activated protein kinase (MAPK) signaling, Wnt signaling and a few co-activators and co-repressors of AR.

Growth Factor/Cytokine Signaling Alters AR Activity

There is considerable evidence in support of the activation of AR by its interaction with polypeptide growth factors and cytokines. Although this phenomenon is poorly understood, it has been implicated as a mechanism critical for cell proliferation by prostate cancer cells at castrate androgen levels. In this chapter, this process will be referred to as signaling crosstalk between AR and polypeptide growth hormones. Some of the growth factors that have been found to activate AR include the epidermal growth factor (EGF), insulin-like growth factor-1 (IGF-1) and fibroblast growth factor (FGF) [46]. Since signaling crosstalk between AR and polypeptide growth hormones has been discussed in detail elsewhere [2], this chapter will focus on the growth factors that have been reported to interact with AR.

Initial studies in the androgen-sensitive prostate cancer cell line, LNCaP, showed that in the absence of androgens, each of the growth hormones, IGF-1, the keratinocyte growth factor (KGF), EGF and the cytokine interleukin-6 (IL-6) is capable of stimulating the transactivation function of AR [39,54]. Also, studies have shown that the level of IL-6 is higher in the serum of prostate cancer patients and tissue extracts from specimen samples of prostate tumors who failed androgen ablation therapy [55]. In agreement with these observations, studies in androgen-independent DU145 prostate cancer cells transfected with AR cDNA and different reporters, further demonstrated the activation of AR by the growth hormones IGF-1, EGF and KGF independent of androgens [46]. KGF is a member of the FGF family of growth factors. Previous studies have shown that stable transfections of PC-3, AR-negative [56], androgen- insensitive prostate cancer cells with AR cDNA enhances the proliferative effects of EGF [57]. This supports the existence of crosstalk between polypeptide growth hormones and the AR signaling pathway by which androgens stimulate prostate cancer cell growth. The EGF receptor molecule HER-2/neu, also known as erbB2, was shown to activate AR in the absence of the ligand (androgens) and even synergized with low androgen levels causing an upregulation in the AR signaling pathway [58] in LNCaP cells stably transfected with the HER-2/neu expression plasmid. The protein tyrosine kinase HER-2/neu is a member of the EGF receptor family. Normally, this protein is expressed at low levels in human prostate epithelial cells [58,59]. Although the protein has been reported in a number of studies to be overexpressed and the gene amplified in a subset of prostate cancer patients [60-62], whether the cellular levels of HER-2/neu is increased in all forms of prostate cancer is yet to be defined. However, the expression of the epidermal growth factor receptor correlates with the Gleason score and prostate cancer progression from hormone-sensitive to hormone-insensitive prostate cancer [63]. In spite of this evidence, phase II clinical trials of compounds targeting the ErbB signaling pathway have failed in the treatment of hormone refractory prostate cancers (Table 2).

Due to controversial reports, it is not clear whether IL-6 acts as an activator or repressor of cell proliferation in prostate cancer [64-66]. Hobisch and colleagues [54], demonstrated that IL-6 interacts with the AR signaling pathway causing an increase in the activation of AR and inducing PSA gene expression in both AR-expressing LNCaP cells and DU145 (AR-negative cells) that were transfected with AR cDNA. Since the treatment of LNCaP prostate cancer cells with IL-6 has been shown to induce phosphorylation of the receptor tyrosine kinase HER-2/neu through the MAPK signaling pathway [67] and because HER-2/neu induces PSA expression through AR activation [68], Yeh and colleagues [69] proposed that the mechanism of AR activation by HER-2/neu is mediated by MAPK signaling [69]. In addition, unlike normal prostate tissues which showed little or no MAPK activation, phosphorylation of MAPK is increased in aggressive prostate cancer cell lines and also in tissues of patients with hormone refractory prostate cancer [70]. The mechanism of AR activation via HER-2/neu has been addressed in many reports [21,71]. The receptor tyrosine kinase HER-2/neu has also been reported to activate Akt (protein kinase B) in order to promote prostate cancer cell proliferation in the absence of androgen. This is based on studies utilizing an inhibitor of Akt or a dominant-negative Akt [21]. In this study, blockage of the Akt pathway inhibited HER-2/neu-induced AR signaling and in turn suppressed cell proliferation both in the absence and presence of testicular androgen [21].

IGF-1 is another growth hormone that has been implicated in the regulation of AR in prostate cancer. While there are studies suggesting that high IGF-1 levels in the human serum is related to the incidence of prostate cancer [72,73], others [74-76] report the lack of an association between levels of circulating IGF-1 and the risk of the disease. In summary, the nature of its role in the development and/or progression of prostate cancer is still controversial. Nevertheless, strategies targeting IGF-1 are being developed (see Table 2) probably because the downstream targets of IGF-1 tyrosine kinase receptor include Akt and MAPK among many others [77,78]. For instance, Lin and colleagues demonstrated that IGF-1 enhances AR degradation by a mechanism involving the activation of the Akt pathway [79]. This report also found that the cytokine IL-6 can reduce AR protein expression levels through the Akt but not the MAPK signaling pathway. This demonstrates that AR signaling utilizes multiple pathways in the differential control of AR activity probably at varying stages of prostate cancer in order to support growth, survival and progression of the devastating illness.

PI3K/Akt Signaling Regulates AR Activation

The majority of the growth factor effects on AR signaling described above have been reported to involve the PI3K/Akt signaling pathway [21,77,79]. The AR is a phosphoprotein with potential phosphorylation sites for a variety of kinases. Importantly, these kinases include Akt and MAPK. Since several investigators have found that protein Akt is overexpressed in prostate cancer both in cellular models and specimen samples [80-82], a plausible explanation for the involvement of Akt in regulating the transactivation functions of AR is that Akt is capable of differentially modulating the activity of AR and the effects triggered downstream. This in turn promotes growth and survival of prostate cancer cells. The depletion of androgen levels causes apoptotic cell death in normal and malignant androgen-sensitive prostate epithelial cells. In contrast, when the more aggressive androgen-independent prostate cancer develops, androgen withdrawal fails to result in this well characterized process of programmed cell death [83]. Since the activation of the serine-threonine kinase, Akt, triggers anti-apoptotic signals promoting cell survival and since this protein has a potential phosphorylation site on AR, the Akt pathway may play a pivotal role in the transition to androgen-independent prostate cancer.

Akt, also known as protein kinase B (PKB), is a downstream target of phosphatidylinositol-3-kinase (PI3K). In its inactive form, Akt is mainly localized in the cytoplasm. While PI3K phosphorylates phosphatidyl inositol (PI) to generate PI-3,4,5-triphosphate (PIP_3), PTEN, is a phosphatase, that dephosphorylates PIP_3 back to PI-4,5-bisphosphate [84]. PIP_3 is a second messenger that is required for the translocation of Akt from the cytoplasm to the plasma membrane. Here, Akt becomes fully activated by phosphorylation at two sites: Threonine-308 and Serine-473. When Akt is completely activated, it regulates cellular processes such as metabolism and proliferation as well as apoptosis [85]. PTEN is a tumor suppressor protein that acts to deactivate the Akt signaling pathway causing apoptotic cell death and cell growth arrest [86]. The loss or inactivation of PTEN permits a continued generation of PIP_3 and in turn induces the Akt pathway (Figure 1). Several studies have shown that PTEN function is lost in advanced-stage prostate cancers [87,88]. Based on this molecular abnormality in prostate cancer one can say that PTEN antagonizes AR function through the deactivation of protein Akt, since the latter is for the most part, anti-apoptotic and induces prostate cell growth. That PTEN and AR play opposing roles have been previously demonstrated [83,89]. This was further characterized by Nan et al. [90], who demonstrated that independent of the cell line studied, PTEN inhibits AR function by preventing the activation of Akt. In contrast however, Lin et al. [91] showed that AR function is inhibited by Akt signaling. Although Akt appears to have differential regulatory effects on AR signaling, these findings suggest the likelihood that the activation of AR in prostate cancer is regulated by the stimulation of a downstream target of the PI3K/Akt signaling pathway coupled to the inactivation of PTEN.

In the normal prostate, the rate of cell proliferation is at equilibrium with the rate of programmed cell death also known as apoptosis [83]. Therefore, it is reasonable to state that increased cell growth rates such as that observed in prostate cancer results owing to an imbalance between the rate of growth and the rate of apoptosis. Although a vast body of evidence exists to support the statement that Akt exerts anti-apoptotic effects [91] as well as promotes cell proliferation and survival, the interacting mechanisms contributing to its effect on AR signaling have only recently begun to become less obscure. To date, no altered or mutated form of the Akt gene has been reported in mammalian cells. Importantly, PI3K can be activated by the cytokine, IL-6, and factors capable of activating the EGF receptor molecule, HER-2/neu [92]. These factors in turn activate Akt. Several mechanisms have been proposed to contribute to the communication link between the PI3K/Akt pathway and the AR signaling pathway (see Figure 1). The identification of downstream targets of the PI3K/Akt pathway especially those involved in negatively regulating cellular apoptotic responses such as the forkhead family of transcription factors (for a detailed review see [93]), has shed more light on the possible mechanisms involved leading to cell survival upon Akt activation. Therefore, when PTEN function is lost in prostate cancer cells possessing a functional AR signaling pathway, the modulation of AR activity by the activation of the Akt pathway provides a conducive environment that is suitable even at such low androgen levels for the promotion of cell survival and the progression from androgen-dependent to androgen-independent prostate cancer states. However, the functional relevance of this in terms of prostate tumors in vivo and the treatment of the disease is yet to be clarified.

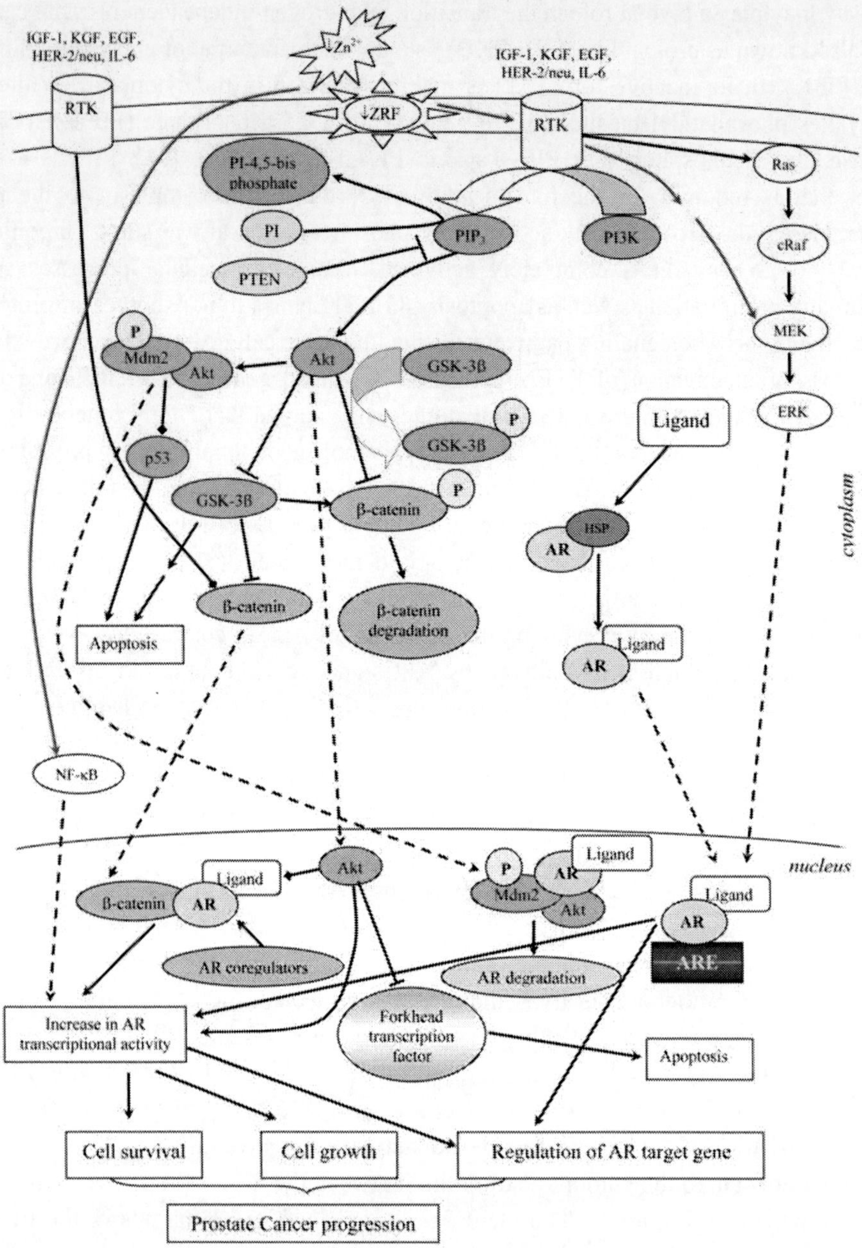

Figure 1. Crosstalk between the AR signaling pathway and major signaling pathways in prostate cancer. Beyond the myriad of factors and cascades that have been implicated as contributors to the development and progression of prostate cancer, there are a few that channel the cell proliferation signal. Several signaling pathways converge on AR and drive cell growth and survival in prostate cancer. Among these pathways are: (i) Activation of the receptor tyrosine kinases (RTK) stimulates the PI3K/Akt signaling pathway. Activation of Akt by PI3K, phosphorylates and inactivates GSK-3β by causing its degradation. This in turn inhibits GSK-3β-induced apoptotic cell death and concomitantly stabilizing β-

catenin, a coactivator of AR and an important component of the canonical Wnt signaling pathway. This enables the translocation of β-catenin to the nucleus where it binds to AR, increases the transactivation function of AR and promotes prostate cancer cell growth. In addition, the binding of β-catenin to AR allows the recruitment of other AR coregulators resulting in the promotion of prostate cancer growth and progression. Furthermore, PTEN is a negative regulator of AR because it suppresses the activation of the Akt signaling pathway causing cell death and cell cycle arrest. (ii) Once activated, Akt is translocated into the nucleus. There, Akt inhibits apoptosis by inhibiting the forkhead family of transcription factors. Moreover, Akt interacts with Mdm2 and phosphorylates it. The phosphorylation of Mdm2 in turn inhibits Mdm2-induced p53 degradation. This stabilizes p53, making this pathway functional but only in prostate cancer cells possessing the uncompromised form of protein p53. (iii) Growth factor/cytokine binding to their respective receptors at the surface of the cell activates the MAPK pathway, stabilizes β-catenin and regulates AR signaling. Activation of the MAPK pathway enhances the binding of the liganded AR complex to the ARE in the nucleus and regulates the activity of AR target genes. **(iv)** Diminished extracellular zinc levels are sensed and regulated by the zinc regulating factors (ZRF) such as the zinc sensing receptors or the hZIP proteins on the surface of the cell. This triggers the MAPK signaling pathway. Once this pathway is activated, it may enhance the translocation of the liganded AR into the nucleus, its binding to the ARE and the regulation of AR target genes. The lowering zinc/ZRF response may also regulate the NF-κB pathway and in turn promote prostate cancer. Please refer to the main text for additional information.

As shown in several studies, not only does the PI3K/Akt pathway regulate the transcriptional activity of AR it also controls its stability and its expression [21,43,91]. The PI3K/Akt pathway inhibitor, LY294002, inhibits interleukin-4 (IL-4)-induced PSA expression and AR signaling in prostate cancer LNCaP cells [94]. This demonstrates that IL-4 promotes the expression of the AR-target gene, PSA, via the activation of both PI3K/Akt and the AR signaling pathways [94]. Also, in AR-negative PC-3, androgen-independent prostate cancer cells transfected with the AR expression plasmid, the activation of the PI3K/Akt pathway suppressed the expression of the cyclin-dependent protein kinase inhibitor, p21. p21 is another AR-target gene (Table 1) and a pro-apoptotic factor involved in the regulation of cell cycle arrest [91]. Lin and colleagues [91] demonstrated similar correlations in LNCaP prostate cancer cells which express functional AR. However, stable transfections of LNCaP cells with a kinase deficient mutant form of Akt resulted in a 50% decrease in the activation of Akt which in turn caused a significant increase in the expression of p21 in response to DHT [91].

Co-immunoprecipitation studies demonstrate that Akt interacts with AR in vivo [91]. In contrast to the reports demonstrating the positive regulation of AR by Akt, studies have also shown that the PI3K/Akt pathway has the ability to negatively regulate the transactivation function of wild-type but not the mutant form of AR in COS-1 cells [91]. Although AR mutations were observed in only approximately 10% of patients with prostate cancer who failed hormone therapy, there was no correlation between prostate cancer progression and AR mutations [45]. Nevertheless, its occurrence in prostate cancer may confer a growth advantage at castrate androgen levels. The phosphorylation of AR is important for its degradation. It has been shown that the growth factor, IGF-1, can phosphorylate AR via the PI3K/Akt pathway in COS-1cells [79,91]. This further implicates Akt as a contributory factor in the regulation of AR degradation.

Not only does Akt interact with Mdm2, a protein involved in the degradation of the p53 tumor suppressor protein, it also phosphorylates it [95]. The mitogen-induced activation of the PI3K/Akt pathway results in the phosphorylation of Mdm2 which in turn stimulates the translocation of Mdm2 from the cytoplasm to the nucleus [95]. By co-immunoprecipitation studies utilizing COS-1 cells, it was further demonstrated that Akt regulates the degradation of AR by the formation of a complex with Mdm2 and AR [79]. Therefore, Akt differentially regulates the transcriptional function of AR either by inducing cell growth arrest or promoting cell proliferation and survival possibly depending on the cells microenvironment [96].

On the other hand, it has been shown in mouse vas deferens epithelial cells (VDEC) and in LNCaP human prostate cancer cells, that the PI3K/Akt pathway is essential for basal and DHT-induced expression of the AR protein [97]. Furthermore, Akt stabilizes β-catenin, a coactivator of AR, known to increase the transcriptional function of AR, in turn increasing PSA secretion and promoting cell growth and survival at low androgen levels in the progression of prostate cancer [43]. Taken together, crosstalk between Akt and AR appears to involve other signaling pathways and to primarily drive the survival of cells in prostate cancer in the presence of little or no androgen. Therefore the identification of other pathways crosslinked to the AR signaling mechanism and involved in the promotion of prostate cancer cell survival and progression from androgen-dependence to independence will be highly beneficial particularly in the identification of potential biomarkers and the development of novel drug targets for the treatment of prostate cancer. The PI3K/Akt pathway therefore presents a promising kinase target for drug development against prostate cancer. On this note, new and improved inhibitors of the PI3K/Akt signaling pathway are currently undergoing clinical trials and so far have demonstrated positive pharmacologic activities against prostate cancer cell growth (see Table 2).

MAPK Signaling and Its Interaction with the AR Signaling Pathway

In mammalian cells, the MAPK signaling pathway regulates a wide variety of cellular processes including gene expression, cell survival, growth, death, cell cycle and cell motility (see review [98]). In these cells, the groups of MAPKs that have been extensively characterized include: (1) the extracellular signal-related kinases (ERK1/2); (2) the jun-amino-terminal kinases (JNK1/2/3); (3) the extracellular signal-related kinase 5 (ERK 5) and (4) the p38 MAPKs. All these MAPKs are activated by phosphorylation and are inactivated by MAPK phosphatases [98]. Generally, ERK1/2 are activated by mitogens and involved in cell proliferative responses whereas the JNKs and the p38 MAPKs are activated upon cellular stress and not by mitogens. Once activated by specific MAPK kinases e.g. MEK1/2 (for ERK1/2), MKK3/6 (for p38 MAPKs), JNKK1/2 (for the JNKs) and MEK 5 (for ERK 5), these proteins are translocated into the nucleus where they regulate several cellular processes ranging from gene expressions to cell death [98]. The MAPK activation pathway is a complicated system whereby each of the MAPKKs is activated by upstream MAPKK kinases (MAPKKKs). Even though MAPK is hyperphosphorylated in advanced-stage prostate cancer cell lines and in tissue samples from patients with hormone refractory prostate cancers [70], the role of the MAPK signaling pathway in AR phosphorylation is incompletely understood. It has been reported that AR activity in prostate cancer cells can be mediated by factors such as HER-2/neu and IL-6 via the MAPK signaling pathway [69,99]. In addition, the expression of the MAPK phosphatase 1 (MKP-1) has been reported to be upregulated in the early-stage form of prostate tumors yet to be treated by androgen ablation [100]. However, its expression is reduced with disease advancement [100]. This phenomenon suggests that MAPKs play critical roles in the developmental stages of prostate cancer. Studies by MagiGalluzzi and colleagues [100] concluded that MKP-1 functions to prevent apoptotic cell death in prostate cancer based on the observation that the overexpression of MKP-1 was concomitant with a decrease in JNK-1 but not ERK-1 enzyme activity.

Although the human prostatic acid phosphatase (PAcP) is not a designated MAPK phosphatase, it regulates the activity of EGFR and HER-2/neu by dephosphorylating these tyrosine kinases. This in turn decreases the activation of the extracellular signal-regulated kinase (ERK)/MAPK at the same time decreasing cell proliferation [101,102]. PAcP may play a critical role in the regulation of growth and androgen dependency in prostate cancer cells [101]. In addition, its activity is elevated in the serum of prostate cancer patients and this level decreases with disease progression [103]. These findings that are suggestive of the existence of crosstalk between MAPK signaling and AR via PAcP regulation may be important for the progression of prostate cancer cells to androgen-independence. On the one hand, androgen-insensitive prostate cancer cell lines have been shown to possess little or no cellular form of PAcP, independent of the expression level of AR [104,105]. In contrast however, androgen-dependent prostate cancer cells express AR, have higher PAcP levels and exhibit slower growth rates [104,105]. Importantly, similar results were obtained from clinical studies utilizing prostate cancer specimen samples [106,107].

Several investigators have suggested that the phosphorylation of AR upon the activation of the ERK/MAPK signaling pathway increases the activity of AR in an androgen-independent manner [69,102,104]. This in turn increases the rate of cell proliferation in androgen-insensitive prostate cancers [69,102,104]. These studies provide a mechanistic basis for the clinical observation that the more aggressive prostate tumors continue to express functional AR while they are insensitive to the androgen. The androgen-independent prostate cancer cells exhibit the following characteristics: they express little or no cellular level of PAcP [104,105,108]; they show no amplification of the HER-2/neu gene while the protein is overexpressed [109,110]. At the same time, the phosphorylation level of the ERK/MAPK protein is elevated [70]. Relevant to these findings are studies showing that the MAPK signaling pathway [Ras activation of a MAPKKK (c-Raf) which in turn activates MEK (a MAPKK) and then finally activates ERK] (Figure 1) is a key player in the crosstalk between the AR signaling pathway and cytokine/peptide growth factor signaling in prostate cancer cells [39,54]. Also, a recent study by Agoulnik et al. [111] demonstrated that the exposure of prostate cancer cells to U0126, a MEK inhibitor, suppressed the activation of MAPK and at the same time decreased the transcriptional activity of AR (Table 2). Treatment with U0126 whether in the presence or absence of the synthetic androgen, R1881, reduced the expression of AR mRNA and AR protein levels in prostate cancer cells cultured in media supplemented with charcoal-stripped serum [111].

Taking all these findings into consideration, the MAPK signaling pathway plays an important role in AR signaling via the activation of non-steroidal compounds such as the EGF receptor family of the protein tyrosine kinase, HER-2/neu or the cytokine, IL-6. Since the MAPK cascade acts in response to many different stimuli leading to a wide variety of cellular responses, it raises the question of whether it presents a converging point for the AR signaling pathway particularly, at low cellular androgen levels. Perhaps this point is determined by the nature of the stimulus which in turn drives prostate cancer cell growth and aggressiveness, downstream. Although such a point may be quite complex, it presents an attractive target for exploitation in prostate cancer therapy.

The WNT Pathway Interacts with AR Signaling

The Wnt signaling pathway is a complex network of proteins that is associated with many cellular processes including embryogenesis and tumor development of various origins including the prostate (see review [43]). The Wnt proteins are members of a highly conserved family of secreted growth factors. Intracellular Wnt signaling has been classified into two groups: canonical (i.e. β-catenin-dependent) and non-canonical (i.e. β-catenin-independent) Wnt signaling pathways. The canonical Wnt signaling pathway will be the focus in this chapter because of its involvement with AR signaling in prostate cancer.

The β-catenin-dependent Wnt signaling pathway regulates gene transcription especially for those cellular reactions that rely on the T cell-specific transcription factor/lymphoid enhancer factor-1 (TCF/Lef-1) for gene expression. In relevance to this chapter, AR signaling is known to be associated with the β-catenin-dependent Wnt signaling pathway because it is a direct target of TCF/Lef-1 as demonstrated in experiments utilizing the androgen-dependent human prostate cancer cell line, LNCaP [112]. Unlike AR mRNA expression levels which were significantly increased by canonical Wnt signaling, the expression of AR protein levels were markedly inhibited in LNCaP prostate cancer cells [112].

One of the mechanisms proposed to be responsible for an increase in AR signaling at castrate androgen levels is an increase in the transcriptional activity of AR owing to its interaction with other coactivators [113]. Apart from being a component of the Wnt signaling pathway, β-catenin is a well-known coactivator of AR [43]. Besides its recruitment of other coactivators such as the protein p160, β-catenin is also capable of altering the ligand specificity of AR. For instance, in the presence of mutant β-catenin, the addition of rather weak AR ligands such as estradiol, significantly increased the transcriptional activity of AR in the TSU-Prl human prostate cancer cell line transfected with wild-type AR (ARwt) even above that induced by R1881 hormone (a synthetic androgen) or the potent testicular androgen, testosterone. Importantly, the same concentration of estradiol had no effect on ARwt without β-catenin [114].

Co-immunoprecipitation of AR with β-catenin in androgen-dependent LNCaP cells demonstrated that the two molecules co-exist in the same complex [114]. This suggests that β-catenin influences AR signaling and may therefore be a crucial factor in prostate cancer progression to androgen-independence. The report showing β-catenin mutations (encoded in a region of the gene controlling its stability) in 5% of primary prostate tumors [115] strengthens the possibility of a link between β-catenin activity and prostate cancer progression. Moreover, high levels of β-catenin and Wnt-1 (a member of the Wnt family of proteins) have been correlated with advanced-stage metastatic androgen-independent prostate cancer. In fact, these molecules can serve as biomarkers of the progression of prostate cancer [116].

Studies by Verras and Sun [117] showed that β-catenin is involved in the peptide growth factor (IGF-1)-mediated regulation of the androgen receptor in LNCaP cells. They demonstrated that IGF-1 augments the stability of β-catenin and that a decrease in cellular β-catenin levels reduces IGF-1R mediated activation of AR [117]. More recently, it was demonstrated in PC-3 and LNCaP cells that the interaction between AR and β-catenin promotes prostate cancer cell proliferation at low androgen levels [118]. These findings further suggest that crosstalk between AR signaling and the canonical Wnt signaling pathway contributes to the progression of prostate cancer from androgen-sensitivity to androgen-independence [119]. Furthermore, differential gene expression studies by cDNA microarray analysis revealed that the Wnt inhibitory factor 1 (WIF1) gene was down-regulated in 64% of prostate tumor samples in comparison to normal epithelial cells [120]. However, there was no correlation between the down-regulation of the WIF1 gene and prostate cancer progression even though WIF1 protein expression was reduced in 23% of the prostate tumors that were analyzed [120].

In summary, β-catenin-dependent Wnt signaling together with the AR signaling pathway clearly play central roles in prostate cancer development and/or progression from androgen-dependent to -independent disease states. This is based on the following observations: (i) the activation of canonical Wnt pathway enhances the transactivation function of AR [117,121]; (ii) there is an upregulation of Wnt expression in prostate cancer versus normal prostate cells [116]; (iii) AR (wild-type or mutant) recruits β-catenin and this association in turn promotes prostate cancer cell growth [122]; and (iv) WIF1 protein expression is significantly down-regulated in prostate cancer tumors in comparison to normal prostate cells [120].

AR Coactivators and Corepressors Regulate AR Signaling

That the transcriptional activity of AR can be influenced by several coactivators and corepressors (collectively called coregulators) via various mechanisms has been demonstrated by numerous studies. The expression levels of many of these coregulators are altered in prostate cancer in comparison to the normal prostate [50,123]. While coactivators are factors that can interact with AR and enhance its transcriptional activity, corepressors diminish its activity on interaction with the receptor molecule. Both coregulatory types are important for efficient control and optimal functioning of the AR signaling pathway. However, when these coregulators become mutated or their expression levels are altered, changes in the transactivation capabilities of AR may occur and therefore contribute to prostate disease states [8]. Such changes have been implicated in prostate cancer and its progression from androgen-sensitivity to non-dependence on the androgen [115,120,124]. As yet, several coregulators of AR such as the type II transmembrane serine protease (TMPRSS2), β-catenin [43], PTEN [125] and glycogen synthase kinase-3β [126] have been identified and new discoveries are currently being reported [127,128] in the literature from time-to-time. Therefore, reference will be given to excellent reviews in the literature because an exhaustive discussion of all known AR-interacting coregulators is beyond the scope of this chapter [113,123,126]. Here-in, a current knowledge of selected coregulators that have been implicated as contributors to the development and/or progression of prostate cancer in the AR signaling pathway will be summarized.

- ***The p160 family of coactivators:*** There are three distinct members of the family of 160 kDa protein steroid nuclear/receptor coactivators that have been identified. The family includes the steroid receptor coactivator 1 (SRC-1), the transcriptional intermediary factor 2 (TIF2)/its mouse homolog glucocorticoid receptor interacting protein 1 (GRIP1) and the amplified in breast cancer protein (AIB1/SRC3) [123]. Upon binding to the NTD of AR, the p160 coactivators regulate the transactivation function of AR and also enable the recruitment of secondary coactivators, such as the creb binding protein and its homologue, p300 (CBP/p300), some of which are involved in chromatin remodeling [123]. In fact, the protein level of CBP/p300 has been associated with prostate cancer progression [123]. It is not clear if alterations in p160 or CBP/p300 expression levels are dependent or independent of the androgen. However, levels of CBP and p300 were reported to be expressed at similar levels in early-stage prostate cancers whether in the benign prostatic hyperplasia, treated or untreated hormone refractory prostate cancer [129].
- ***β-catenin, a coactivator of AR:*** Studies have shown that levels of β-catenin increases with increasing prostate cancer grades [119]. In addition, mutations in β-catenin increase the sensitivity of AR even at low androgen levels [130]. At the same time, it broadens the responsiveness of AR to rather weak ligands such as estradiol or androstenedione as agonists of AR signaling [43]. The binding of β-catenin, be it wild-type or mutant form, to the LBD of AR is androgen-dependent and promotes the translocation of β-catenin to the nucleus. Since the binding of β-catenin is critical also for the TCF family of transcription factors, there is the possibility of some competition between AR and TCF for the available β-catenin. For more information, the implications of this competition for prostate cancer have previously been discussed [131,132].
- ***ARA55 is a coactivator of AR:*** The ligand-dependent AR-associated protein, ARA55, binds to the LBD of AR in an androgen-dependent manner, and by so doing, increases the activity of AR [133]. Studies utilizing whole tissues have shown that the expression of ARA55 is reduced in prostate tumors in comparison to normal prostate tissues [134]. Also, protein ARA55, regulates androgen-dependent KGF expression in the stromal but not epithelial compartments of prostate cells [135].
- ***Protein RB positively regulates AR:*** RB is a tumor suppressor protein. It interacts directly with the androgen receptor by binding to the NTD of AR in a ligand-independent manner [41]. In prostate cancer, RB plays a critical role in the proliferative response of cells to androgens because its inactivation is required for cell progression through the cell cycle [136]. It enhances the transactivation function of AR in prostate cancer cells by acting as a coactivator of AR [41]. Its upregulation increases the transcriptional activity of AR [137]. The contribution of RB to prostate cancer development and/or progression is evident because its function has been reported to be lost in some human prostate carcinomas [138]. Loss of functional

activity of protein RB inhibits AR [137]. Also, alterations in the mRNA expression levels of RB1 have been associated with tumor grades in a subset of human prostate cancer [139]. The lower the expression levels of protein RB1, the higher the prostate tumor grade [139]. Complementing this report is another study demonstrating that RB1 mRNA expression levels are abnormally low in 12 of 33 prostate cancer patients (i.e. in 36%) who failed combined androgen blockade therapy [140]. On the contrary, only 6 of 48 prostate cancer patients (i.e. 12.5%) who are treatment-naïve, showed such abnormally low mRNA levels of protein RB1 [140]. In addition, the depletion of RB in prostate cancer cells caused an increase in cell proliferation upon androgen ablation [141]. Furthermore, by fluorescent in situ hybridization (FISH) analysis, the loss of RB, occurs in prostate cancer by almost four times more frequently, after androgen ablation therapy than prior to this treatment [142]. Taken together, these findings demonstrate that the functional loss of protein RB is involved in the progression to hormone-refractory prostate cancers.

- ***Cyclin D1 is a negative regulator of AR:*** The levels of cyclin D1 is similarly expressed across different prostate cancer grades. It is not associated with disease progression whether in benign prostatic hyperplasia, treated or untreated hormone-insensitive prostate cancers [129]. Cyclin D1 binds to the hinge domain of AR and inhibits the formation of AR mutations and polymorphisms thus repressing androgen-dependent transactivation functions of AR in prostate cancer cells [143]. Interactions of cyclin D1 with histone deacetylase 3 (HDAC3) is critical for its negative regulation of AR [144].
- ***Akt acts a corepressor of AR:*** Akt is an oncoprotein and a serine/threonine protein kinase that is important in the PI3K mediated signaling pathway. It phosphorylates and deactivates AR upon binding and suppresses the recruitment of AR coactivators such as ARA70 [91].
- ***FoXo1 represses AR activation:*** Foxo1 is a member of the Foxo subfamily of the forkhead transcription factor family. It is a downstream target of the PI3K/Akt pathway. Upon its binding to the LBD of AR, in vitro studies showed that it suppresses androgen-mediated transactivation functions of AR and in turn suppresses the proliferation of prostate cancer cells [145].
- ***Glycogen synthase kinase-3β (GSK-3β):*** GSK-3β is a serine/threonine protein kinase [146]. Studies have shown that GSK-3β can act both as a corepressor and a coactivator of AR. On the one hand, it is a corepressor of AR in prostate cancer cells by phosphorylating AR at the N-terminus, repressing the transactivation function of AR and in turn inhibiting cell growth [147]. GSK-3β suppresses interactions between the NTD and LBD of AR [147]. In prostate cancer cells GSK-3β negatively regulates both the PI3K/Akt and the canonical Wnt signaling pathways [43,148]. This protein kinase (GSK-3β) regulates the phosphorylation and degradation of β-catenin. However, this process is inhibited by the phosphorylation and inactivation of GSK-3β by Akt [43]. Inactivation of GSK-3β stabilizes β-catenin. This facilitates the translocation of β-catenin into the nucleus where it has the ability to increase the transcriptional activity of AR [149]. On the other hand, GSK-3β can also act as a coactivator on AR [148]. This is based on experiments in prostate cancer cells

demonstrating that the inhibition of GSK-3β, either by utilizing GSK-3β inhibitors or small interference RNA (SiRNA) gene silencing techniques, did not block androgen-dependent translocation of AR into the nucleus. In addition, it decreased the expression of PSA, an AR target gene, in an androgen-dependent manner [148]. These findings demonstrate that functional interactions exists between GSK-3β and AR and provides yet another evidence that other pathways such as the Wnt and the PI3K/Akt pathways, are cross-linked to the AR signaling pathway (Figure 1).

The diverse array of AR coregulators (some of which are capable of acting as both coactivators and corepressors) makes the regulation of AR rather complex. One of the intriguing features of prostate cancer is the ability of AR coregulators to differentially affect the transactivation function of AR in order to modulate the activity of specific AR target genes involved in cell survival, programmed cell death, cell cycle progression, cell proliferation, invasion and tumor metastasis. Justifiably so, these factors are increasingly becoming potential targets for exploitation as prognostic/diagnostic biomarkers and as pharmacotherapies in the treatment of prostate cancer. Most importantly, increasing our understanding of the mechanisms intertwined in AR regulation versus AR dysregulation, in response to changes in the microenvironment of the cell, which drive prostate cancer development and/or progression will be highly beneficial in the management of the disease.

Is There a Relationship between Depleting Zinc Levels and AR Signaling in Prostate Cancer?

Intracellular zinc concentrations in the normal (non-cancerous) prostate are among the highest in the human body. In prostate cancer however, cellular zinc levels are significantly reduced [150]. Aside from being unique, this observation is the most consistent phenomenon in prostate malignancy. Intracellular zinc levels continue to decrease during the stages of prostate cancer from its development to androgen-independent tumor growth. Moreover, growth rates in cancer cells that continue to express high intracellular zinc levels are significantly slower [151]. Several studies have shown that the accumulation of high levels of zinc in the mitochondria of normal prostate cells inhibits the mitochondrial aconitase enzyme activity. This inhibition is at the expense of isocitrate oxidation via the Krebs cycle and this in turn decreases the total cellular energy production in these cells. In the prostate, cells that have undergone this metabolic transformation from zinc/citrate accumulating normal prostate to zinc-diminished/citrate-oxidizing prostate cancer cells are able to meet cellular energy requirements to support malignancy [152]. Furthermore, the accumulation of intracellular zinc in prostate cells triggers the release of cytochrome C from the mitochondria followed by a caspase-dependent apoptotic response and growth inhibition in these cells [153].

During tumorigenesis, significant changes in extracellular zinc levels are observed in the human prostate [150]. The uptake and release of cellular zinc include mechanisms that import zinc into the cell, release or sequester zinc internally and those involving the extracellular zinc sensing receptors [154-157]. Interestingly, downregulation of the zinc uptake transporters (hZIP1 and hZIP2) has been associated with prostate cancer in African American men [158]. In addition, recent studies have shown that prostate metastatic cells are impermeable to zinc but extracellular zinc has the ability to trigger a zinc sensing receptor. This in turn regulates prostate cancer cell growth **[156]**. In spite of the consistent evidence supporting the notion that high intracellular zinc levels in prostate cells may be protective against prostate cancer, a signaling role for zinc in the development and progression of prostate cancer is yet to be established.

Despite the fact that AR signaling is a crucial pathway contributing to the development and progression of prostate cancer, it is not known if AR signaling crosstalks with zinc-related signals in the prostate gland. Neither is it known whether such interactions are contributory factors generating downstream signals in the development and/or progression of prostate cancer. More recently, Golovine and colleagues reported that the depletion of intracellular zinc levels increases the expression of cytokines such as IL-6, IL-8 and the vascular endothelial growth factor (VEGF) in prostate cancer cells by a mechanism involving NF-κB [159]. Importantly, all these factors associated with depleting zinc levels [159] play critical roles in prostate cancer. Therefore, by virtue of the following observations, some of which have been discussed earlier-on in this chapter, it is reasonable to suggest that systems involved in diminishing zinc levels in prostate cancer crosstalk with the AR signaling pathway. (1) Intracellular zinc levels continue to diminish with the advancement of prostate cancer. (2) IL-6 levels are higher in the serum of patients with androgen-independent prostate cancer [55]. (3) IL-6 coregulates AR and induces PSA gene expression [54]. (4) IL-6 is capable of reducing the protein expression of AR [79]. (5) Increased VEGF levels enhances prostate tumor growth [160]. (6) IL-8 signaling increases AR protein expression and promotes AR activation independent of the ligand [161]. (7) Increased activation of NF-κB is associated with prostate cancer progression to androgen-independence [96,162].

Furthermore, Dubi and colleagues [156] demonstrated that both the MAPK and the PI3K/Akt pathways are activated by extracellular zinc in PC-3, zinc-dependent but androgen-insensitive prostate cancer cells (Figure 1). These investigators also showed that crosstalk between the MAPK and the PI3K/Akt pathways play major roles in mediating zinc-dependent responses affecting proliferation in prostate cancer cells [156]. While the causal relationship between these events remains unknown, it appears likely that the pathway controlling intracellular zinc levels and the AR signaling pathway are linked in prostate tumorigenesis. However given the consistency in the literature regarding both phenomena, it is clear that more research is required to identify the mechanisms underlying the involvement of zinc depletion in the AR signaling pathway and the contributory effects of such interactions to the development and/or progression of prostate cancer.

Conclusion

The treatment of prostate cancer with chemicals or surgical procedures that block AR function and/or deplete the amount of circulating androgens is only a temporary measure against prostate cancer. This is because there is a high probability of disease reoccurrence in a more aggressive form due to cancer resistance to therapy. In general, it is clear that tumor cells employ multiple pathways in order to avoid cell death. Although not discussed above, alterations in tumor suppressor genes such as p53 have been associated with AR signaling and thus, have implications in the progression of prostate cancer [68]. The significant amount of evidence available to support the interactions existing between androgen receptor signaling and many other signaling pathways substantiates the notion that multilinked signals are critical for the maintenance of androgen sensitivity in prostate cancer. Hence, other pathways besides those previously discussed above may crosstalk with the AR signaling pathway causing the activation of AR-mediated proliferation of prostate cancer cells, independent of the androgen levels in the system.

The existence of AR coactivators and corepressors that can influence the transcriptional activity of AR in response to agonist and antagonists in the cells microenvironment makes the AR signaling pathway even more complex. This further highlights the significant role played by the AR signaling pathway in growth, survival and progression from androgen dependent to androgen independent prostate cancers. Overall, these various forms of interactions underscore how crosstalk between AR and other signaling pathways as well as AR coregulation by a diverse array of cofactors may lead to the dysregulation in cell growth as that observed in prostate cancer. The cause of the dysregulation in prostate cancer remains yet to be defined. It is therefore important to identify whether the signals generated from these interconnected pathways converge, downstream or upstream of the androgen receptor and in turn drive prostate cancer development and/or progression. Although many questions remain open, a better understanding of these contributory mechanisms will lead to the development of more targeted and effective therapies and possibly preventive measures against human prostate cancer. The pharmacological challenge will be to put into consideration the prosurvival contributions of these various cross-linked signaling pathways during the early-, middle- and advanced/late-stage of prostate cancer that enable the escape of some cells from androgen dependence to androgen-insensitive and aggressive prostate tumor cell growth. Therefore differential genomic and proteomic studies that compare prostate samples from the different cancer stages will provide further insight into individual, synergistic or even parallel signaling pathways at play before and after hormone ablation therapy. This may unveil novel means by which combination therapies can be differentially utilized for the treatment of different prostate cancer patients at varying stages of the disease.

Aknowledgments

This work was supported by the Cooper Cancer Institute of New Jersey, Cooper University Hospital, Camden, NJ 08103. Dr. Raymond Ochs of the Dept. of Pharmaceutical Sciences, St. John's University, NY and Dr. Jonathan Brody of the Dept. of Surgery at Thomas Jefferson University, PA are thanked for reading this review and for their helpful comments and suggestions.

References

[1] Heinlein, C. A. & Chang, C. (2002). The Roles of Androgen Receptors and Androgen-Binding Proteins in Nongenomic Androgen Actions. *Mol Endocrinol*, *16*, 2181-2187.

[2] Zhu, M. L. & Kyprianou, N. (2008). Androgen receptor and growth factor signaling partnerships in prostate cancer cells. *Endocr Relat Cancer*, ERC-08-0084.

[3] Gronemeyer, H., Gustafsson, J. A. & Laudet, V. (2004). Principles for modulation of the nuclear receptor superfamily. *Nature Reviews Drug Discovery*, 3, 950-964.

[4] Chang, C. Y. & McDonnell, D. P. (2005) Androgen receptor-cofactor interactions as targets for new drug discovery. *Trends in Pharmacological Sciences*, 26, 225-228.

[5] Singh, P., Hallur, G., Anchoori, R. K., Bakare, O., Kageyama, Y., Khan, S. R. & Isaacs, J. T. (2008). Rational design of novel antiandrogens for neutralizing androgen receptor function in hormone refractory prostate cancer. *The Prostate*, *68*, 1570-1581.

[6] Devlin, H. L. & Mudryj, M. (2008). Progression of prostate cancer: Multiple pathways to androgen independence. *Cancer Letters*, doi:10.1016/j.canlet.2008.06.007.

[7] Suzuki, H., Ueda, T., Ichikawa, T. & Ito, H. (2003). Androgen receptor involvement in the progression of prostate cancer. *Endocrine-Related Cancer*, *10*, 209-216.

[8] Chen, C. D., Welsbie, D. S., Tran, C., Baek, S. H., Chen, R., Vessella, R., Rosenfeld, M. G. & Sawyers, C. L. (2004). Molecular determinants of resistance to antiandrogen therapy. *Nature Medicine*, *10*, 33-39.

[9] Debes, J. D. & Tindall, D. J. (2004). Mechanisms of Androgen-Refractory Prostate Cancer. *N Engl J Med*, *351*, 1488-1490.

[10] van Bokhoven, A., Varella-Garcia, M., Korch, C., Johannes, W. U., Smith, E. E., Miller, H. L., Nordeen, S. K., Miller, G. J. & Lucia, M. S. (2003). Molecular characterization of human prostate carcinoma cell lines. *The Prostate*, *57*, 205-225.

[11] Sadi, M. V., Walsh, P. C. & Barrack, E. R. (1991). Immunohistochemical study of androgen receptors in metastatic prostate-cancer - comparison of receptor content and response to hormonal-therapy. *Cancer*, *67*, 3057-3064.

[12] Ruizeveld de Winter, J. A., Trapman, J., Vermey, M., Mulder, E., Zegers, N. D. & van der Kwast, T. H. (1991). Androgen receptor expression in human tissues: an immunohistochemical study. *J Histochem Cytochem*, *39*, 927-36.

[13] Chodak, G. W., Kranc, D. M., Puy, L. A., Takeda, H., Johnson, K. & Chang, C. (1992). Nuclear-localization of androgen receptor in heterogeneous samples of normal, hyperplastic and neoplastic human prostate. *Journal of Urology*, *147*, 798-803.

[14] Tararova, N. D., Narizhneva, N., Krivokrisenko, V., Gudkov, A. V. & Gurova, K. V. (2007). Prostate cancer cells tolerate a narrow range of androgen receptor expression and activity. *The Prostate*, *67*, 1801-1815.

[15] Litvinov, I. V., De Marzo, A. M. & Isaacs, J. T. (2003). Is the Achilles' heel for prostate cancer therapy a gain of function in androgen receptor signaling? *Journal of Clinical Endocrinology and Metabolism*, *88*, 2972-2982.

[16] Bruckheimer, E. M., Spurgers, K., Weigel, N. L., Logothetis, C. & McDonnell, T. J. (2003). Regulation of Bcl-2 Expression by Dihydrotestosterone in Hormone Sensitive LNCaP-FGC Prostate Cancer Cells. *The Journal of Urology*, *169*, 1553-1557.

[17] Kirschenbaum, A., Liu, X. H., Yao, S., Narla, G., Friedman, S. L., Martignetti, J. A. & Levine, A. C. (2006). Sex steroids have differential effects on growth and gene expression in primary human prostatic epithelial cell cultures derived from the peripheral versus transition zones. *Carcinogenesis*, *27*, 216-224.

[18] Wang, G., Reed, E. & Li, Q. Q. (2004). Apoptosis in prostate cancer: Progressive and therapeutic implications (Review). *International Journal of Molecular Medicine*, *14*, 23-34.

[19] Bauman, D. R., Steckelbroeck, S., Peehl, D. M. & Penning, T. M. (2006). Transcript profiling of the androgen signal in normal prostate, benign prostatic hyperplasia, and prostate cancer. *Endocrinology*, *147*, 5806-5816.

[20] Takayama, K., Horie-Inoue, K., Ikeda, K., Urano, T., Murakami, K., Hayashizaki, Y., Ouchi, Y. & Inoue, S. (2008). FOXP1 is an androgen-responsive transcription factor that negatively regulates androgen receptor signaling in prostate cancer cells. *Biochemical and Biophysical Research Communications*, *374*, 388-393.

[21] Wen, Y., Hu, M. C. T., Makino, K., Spohn, B., Bartholomeusz, G., Yan, D. H. & Hung, M. C. (2000). HER-2/neu Promotes Androgen-independent Survival and Growth of Prostate Cancer Cells through the Akt Pathway. *Cancer Res*, *60*, 6841-6845.

[22] Zha, S., Gage, W. R., Sauvageot, J., Saria, E. A., Putzi, M. J., Ewing, C. M., Faith, D. A., Nelson, W. G., De Marzo, A. M. & Isaacs, W. B. (2001). Cyclooxygenase-2 Is Up-Regulated in Proliferative Inflammatory Atrophy of the Prostate, but not in Prostate Carcinoma. *Cancer Res*, *61*, 8617-8623.

[23] Shah, R. B., Mehra, R., Chinnaiyan, A. M., Shen, R., Ghosh, D., Zhou, M., MacVicar, G. R., Varambally, S., Harwood, J., Bismar, T. A., Kim, R., Rubin, M. A. & Pienta, K. J. (2004). Androgen-Independent Prostate Cancer Is a Heterogeneous Group of Diseases: Lessons from a Rapid Autopsy Program. *Cancer Res*, *64*, 9209-9216.

[24] Shi, X. B., Xue, L., Yang, J., Ma, A. H., Zhao, J., Xu, M., Tepper, C. G., Evans, C. P., Kung, H. J. & White, R. W. D. (2007). An androgen-regulated miRNA suppresses Bak1 expression and induces androgen-independent growth of prostate cancer cells. *Proceedings of the National Academy of Sciences of the United States of America*, *104*, 19983-19988.

[25] Shang, Y. F., Myers, M. & Brown, M. (2002). Formation of the androgen receptor transcription complex. *Molecular Cell*, *9*, 601-610.

[26] Louie, M. C., Yang, H. Q., Ma, A. H., Xu, W., Zou, J. X., Kung, H. J. & Chen, H. W. (2003). Androgen-induced recruitment of RNA polymerase II to a nuclear receptor-p160 coactivator complex. *Proceedings of the National Academy of Sciences of the United States of America*, *100*, 2226-2230.

[27] Koivisto, P., Kononen, J., Palmberg, C., Tammela, T., Hyytinen, E., Isola, J., Trapman, J., Cleutjens, K., Noordzij, A., Visakorpi, T. & Kallioniemi, O. P. (1997). Androgen receptor gene amplification: A possible molecular mechanism for androgen deprivation therapy failure in prostate cancer. *Cancer Research*, *57*, 314-319.

[28] Bubendorf, L., Kononen, J., Koivisto, P., Schraml, P., Moch, H., Gasser, T. C., Willi, N., Mihatsch, M. J., Sauter, G. & Kallioniemi, O. P. (1999). Survey of gene amplifications during prostate cancer progression by high throughput fluorescence in situ hybridization on tissue microarrays. *Cancer Research*, *59*, 803-806.

[29] Visakorpi, T., Hyytinen, E., Koivisto, P., Tanner, M., Keinanen, R., Palmberg, C., Palotie, A., Tammela, T., Isola, J. & Kallioniemi, O. P. (1995). In-vivo amplification of the androgen receptor gene and progression of human prostate-cancer. *Nature Genetics*, *9*, 401-406.

[30] Kinoshita, H., Shi, Y., Sandefur, C., Meisner, L. F., Chang, C. S., Choon, A., Reznikoff, C. R., Bova, G. S., Friedl, A. & Jarrard, D. F. (2000). Methylation of the androgen receptor minimal promoter silences transcription in human prostate cancer. *Cancer Research*, *60*, 3623-3630.

[31] Dehm, S. M. & Tindall, D. J. (2005). Regulation of androgen receptor signaling in prostate cancer. *Expert Review of Anticancer Therapy*, **5**, 63-74.

[32] Javidan, J., Deitch, A. D., Shi, X. B. & White, R. W. D. (2005). The androgen receptor and mechanisms for androgen independence in prostate cancer. *Cancer Investigation*, *23*, 520-528.

[33] Nelson, C. C., Hendy, S. C., Shukin, R. J., Cheng, H., Bruchovsky, N., Koop, B. F. & Rennie, P. S. (1999). Determinants of DNA sequence specificity of the androgen, progesterone, and glucocorticoid receptors: Evidence for differential steroid receptor response elements. *Molecular Endocrinology*, *13*, 2090-2107.

[34] Janne, O. A., Palvimo, J. J., Kallio, P. & Mehto, M. (1993). Androgen receptor and mechanism of androgen action. *Annals of Medicine*, *25*, 83-89.

[35] Jenster, G. (1999). The role of the androgen receptor in the development and progression of prostate cancer. *Seminars in Oncology*, *26*, 407-421.

[36] Mari Kaarbø, T. I. K. F. S. (2007). Androgen signaling and its interactions with other signaling pathways in prostate cancer. *BioEssays*, *29*, 1227-1238.

[37] Brinkmann, A. O., Blok, L. J., de Ruiter, P. E., Doesburg, P., Steketee, K., Berrevoets, C. A. & Trapman, J. (1999). Mechanisms of androgen receptor activation and function. *Journal of Steroid Biochemistry and Molecular Biology*, *69*, 307-313.

[38] Brinkmann, A. O. (2001). Lessons to be learned from the androgen receptor. *European Journal of Dermatology*, *11*, 301-303.

[39] Culig, Z., Klocker, H., Bartsch, G. & Hobisch, A. (2002). Androgen receptors in prostate cancer. *Endocrine-Related Cancer*, *9*, 155-170.

[40] Lu, S., Tsai, S. Y. & Tsai, M. J. (1997). Regulation of Androgen-dependent Prostatic Cancer Cell Growth: Androgen Regulation of CDK2, CDK4, and CKI p16 Genes. *Cancer Res*, *57*, 4511-4516.

[41] Yeh, S., Miyamoto, H., Nishimura, K., Kang, H., Ludlow, J., Hsiao, P., Wang, C., Su, C. & Chang, C. (1998). Retinoblastoma, a Tumor Suppressor, Is a Coactivator for the Androgen Receptor in Human Prostate Cancer DU145 Cells. *Biochemical and Biophysical Research Communications*, *248*, 361-367.

[42] Abreu-Martin, M. T., Chari, A., Palladino, A. A., Craft, N. A. & Sawyers, C. L. (1999). Mitogen-activated protein kinase kinase kinase 1 activates androgen receptor-dependent transcription and apoptosis in prostate cancer. *Molecular and Cellular Biology*, *19*, 5143-5154.

[43] Terry, S., Yang, X. Z., Chen, M. W., Vacherot, F. & Buttyan, R. (2006). Multifaceted interaction between the androgen and Wnt signaling pathways and the implication for prostate cancer. *Journal of Cellular Biochemistry*, *99*, 402-410.

[44] Zhu, M. L., Partin, J. V., Bruckheimer, E. M., Strup, S. E. & Kyprianou, N. (2008). TGF-beta signaling and androgen receptor status determine apoptotic cross-talk in human prostate cancer. *Prostate*, *68*, 287-295.

[45] Taplin, M. E., Rajeshkumar, B., Halabi, S., Werner, C. P., Woda, B. A., Picus, J., Stadler, W., Hayes, D. F., Kantoff, P. W., Vogelzang, N. J. & Small, E. J. (2003). Androgen receptor mutations in androgen-in dependent prostate cancer: Cancer and Leukemia Group B Study 9663. *Journal of Clinical Oncology*, *21*, 2673-2678.

[46] Culig, Z., Hobisch, A., Cronauer, M. V., Radmayr, C., Trapman, J., Hittmair, A., Bartsch, G. & Klocker, H. (1994). Androgen Receptor Activation in Prostatic Tumor-Cell Lines by Insulin-like Growth-factor-1, Keratinocyte Growth-factor, and Epidermal Growth-factor. *Cancer Research*, *54*, 5474-5478.

[47] Veldscholte, J., Risstalpers, C., Kuiper, G., Jenster, G., Berrevoets, C., Claassen, E., Vanrooij, H. C. J., Trapman, J., Brinkmann, A. O. & Mulder, E. (1990). A mutation in the ligand-binding domain of the androgen receptor of human LNCaP cells affects steroid binding characteristics and response to anti-androgens. *Biochemical and Biophysical Research Communications*, *173*, 534-540.

[48] Zhou, Z., Flesken-Nikitin, A., Corney, D. C., Wang, W., Goodrich, D. W., Roy-Burman, P. & Nikitin, A. Y. (2006). Synergy of p53 and Rb Deficiency in a Conditional Mouse Model for Metastatic Prostate Cancer. *Cancer Res*, *66*, 7889-7898.

[49] Nesslinger, N. J., Shi, X. B. & deVere White, R. W. (2003). Androgen-independent Growth of LNCaP Prostate Cancer Cells Is Mediated by Gain-of-Function Mutant p53. *Cancer Res*, *63*, 2228-2233.

[50] Gregory, C. W., He, B., Johnson, R. T., Ford, O. H., Mohler, J. L., French, F. S. & Wilson, E. M. (2001). A Mechanism for Androgen Receptor-mediated Prostate Cancer Recurrence after Androgen Deprivation Therapy. *Cancer Res*, *61*, 4315-4319.

[51] Belandia, B., Powell, S. M., Garcia-Pedrero, J. M., Walker, M. M., Bevan, C. L. & Parker, M. G. (2005). Hey1, a Mediator of Notch Signaling, Is an Androgen Receptor Corepressor. *Mol. Cell. Biol.*, *25*, 1425-1436.

[52] Eastham, J. A., Stapleton, A. M., Gousse, A. E., Timme, T. L., Yang, G., Slawin, K. M., Wheeler, T. M., Scardino, P. T. & Thompson, T. C. (1995). Association of p53 mutations with metastatic prostate cancer. *Clin Cancer Res*, *1*, 1111-1118.

[53] Muller, J. M., Isele, U., Metzger, E., Rempel, A., Moser, M., Pscherer, A., Breyer, T., Holubarsch, C., Buettner, R. & Schule, R. (2000). FHL2, a novel tissue-specific coactivator of the androgen receptor. *Embo Journal*, *19*, 359-369.

[54] Hobisch, A., Eder, I. E., Putz, T., Horninger, W., Bartsch, G., Klocker, H. & Culig, Z. (1998). Interleukin-6 Regulates Prostate-specific Protein Expression in Prostate Carcinoma Cells by Activation of the Androgen Receptor. *Cancer Res*, *58*, 4640-4645.

[55] Twillie, D. A., Eisenberger, M. A., Carducci, M. A., Hseih, W. S., Kim, W. Y. & Simons, J. W. (1995). Interleukin-6: A candidate mediator of human prostate cancer morbidity. *Urology*, *45*, 542-549.

[56] Kaighn ME, N. K., Ohnuki, Y., Lechner, J. F., Jones, L. W. (1979). Establishment and characterization of a human prostatic carcinoma cell line (PC-3). *Investigative urology*, *17*, 16-23.

[57] Brass, A. L., Barnard, J., Patai, B. L., Salvi, D. & Rukstalis, D. B. (1995). Androgen Up-Regulates Epidermal Growth Factor Receptor Expression and Binding Affinity in PC3 Cell Lines Expressing the Human Androgen Receptor. *Cancer Res*, *55*, 3197-3203.

[58] Craft, N., Shostak, Y., Carey, M. & Sawyers, C. L. (1999). A mechanism for hormone-independent prostate cancer through modulation of androgen receptor signaling by the HER-2/neu tyrosine kinase. *Nature Medicine*, *5*, 280-285.

[59] Robinson, D., He, F., Pretlow, T. & Kung, H. J. (1996). A tyrosine kinase profile of prostate carcinoma. *Proceedings of the National Academy of Sciences of the United States of America*, *93*, 5958-5962.

[60] Mellon, K., Thompson, S., Charlton, R. G., Marsh, C., Robinson, M., Lane, D. P., Harris, A. L., Horne, C. H. W. & Neal, D. E. (1992). p53, c-ErbB-2 and the epidermal growth-factor receptor in the benign and malignant prostate. *Journal of Urology*, *147*, 496-499.

[61] Ross, J. S., Sheehan, C., HaynerBuchan, A. M., Ambros, R. A., Kallakury, B. V. S., Kaufman, R., Fisher, H. A. G. & Muraca, P. J. (1997). HER-2/neu gene amplification status in prostate cancer by fluorescence in situ hybridization. *Human Pathology*, *28*, 827-833.

[62] Myers, R. B., Srivastava, S. S., Oelschlager, D. K. & Grizzle, W. E. (1994). Expression of p160(ErbB-3) and p185(ErbB -2) in prostatic intraepithelial neoplasia and prostatic adenocarcinoma. *Journal of the National Cancer Institute*, *86*, 1140-1145.

[63] Di Lorenzo, G., Tortora, G., D'Armiento, F. P., De Rosa, G., Staibano, S., Autorino, R., D'Armiento, M., De Laurentiis, M., De Placido, S., Catalano, G., Bianco, A. R. & Ciardiello, F. (2002). Expression of Epidermal Growth Factor Receptor Correlates with Disease Relapse and Progression to Androgen-independence in Human Prostate Cancer. *Clin Cancer Res*, *8*, 3438-3444.

[64] Degeorges, A., Tatoud, R., FauvelLafeve, F., Podgorniak, M. P., Millot, G., deCremoux, P. & Calvo, F. (1996). Stromal cells from human benign prostate hyperplasia produce a growth-inhibitory factor for LNCaP prostate cancer cells, identified as interleukin-6. *International Journal of Cancer*, *68*, 207-214.

[65] Serve, H., Steinhauser, G., Oberberg, D., Flegel, W. A., Northoff, H. & Berdel, W. E. (1991). Studies on the interaction between interleukin-6 and human-malignant nonhematopoietic cell-lines. *Cancer Research*, *51*, 3862-3866.

[66] Okamoto, M., Lee, C. & Oyasu, R. (1997). Interleukin-6 as a paracrine and autocrine growth factor in human prostatic carcinoma cells in vitro. *Cancer Research*, *57*, 141-146.

[67] Qiu, Y., Ravi, L. & Kung, H. J. (1998). Requirement of ErbB2 for signalling by interleukin-6 in prostate carcinoma cells. *Nature*, *393*, 83-85.

[68] Grossmann, M. E., Huang, H. J. & Tindall, D. J. (2001). Androgen receptor signaling in androgen-refractory prostate cancer. *Journal of the National Cancer Institute*, *93*, 1687-1697.

[69] Yeh, S. Y., Lin, H. K., Kang, H. Y., Thin, T. H., Lin, M. F. & Chang, C. S. (1999). From HER2/Neu signal cascade to androgen receptor and its coactivators: A novel pathway by induction of androgen target genes through MAP kinase in prostate cancer

cells. *Proceedings of the National Academy of Sciences of the United States of America*, *96*, 5458-5463.

[70] Gioeli, D., Mandell, J. W., Petroni, G. R., Frierson, H. F. & Weber, M. J. (1999). Activation of mitogen-activated protein kinase associated with prostate cancer progression. *Cancer Research*, *59*, 279-284.

[71] Gregory, C. W., Whang, Y. E., McCall, W., Fei, X., Liu, Y., Ponguta, L. A., French, F. S., Wilson, E. M. & Earp, H. S., III (2005). Heregulin-Induced Activation of HER2 and HER3 Increases Androgen Receptor Transactivation and CWR-R1 Human Recurrent Prostate Cancer Cell Growth. *Clin Cancer Res*, *11*, 1704-1712.

[72] Chan, J. M., Stampfer, M. J., Ma, J., Gann, P., Gaziano, J. M., Pollak, M. & Giovannucci, E. (2002). Insulin-like growth factor-I (IGF-I) and IGF binding protein-3 as predictors of advanced-stage prostate cancer. *Journal of the National Cancer Institute*, *94*, 1099-1106.

[73] Stattin, P., Rinaldi, S., Biessy, C., Stenman, U. H., Hallmans, G. & Kaaks, R. (2004). High levels of circulating insulin-like growth factor-I increase prostate cancer risk: A prospective study in a population-based nonscreened cohort. *Journal of Clinical Oncology*, *22*, 3104-3112.

[74] Severi, G., Morris, H. A., MacInnis, R. J., English, D. R., Tilley, W. D., Hopper, J. L., Boyle, P. & Giles, G. G. (2006). Circulating insulin-like growth factor-I and binding protein-3 and risk of prostate cancer. *Cancer Epidemiology Biomarkers & Prevention*, *15*, 1137-1141.

[75] Morris, J. K., George, L. M., Wu, T. & Wald, N. J. (2006). Insulin-like growth factors and cancer: no role in screening. Evidence from the BUPA study and meta-analysis of prospective epidemiological studies. *British Journal of Cancer*, *95*, 112-117.

[76] Chen, C., Lewis, S. K., Voigt, L., Fitzpatrick, A., Plymate, S. R. & Weiss, N. S. (2005). Prostate carcinoma incidence in relation to prediagnostic circulating levels of insulin-like growth factor I, insulin-like growth factor binding protein 3, and insulin. *Cancer*, *103*, 76-84.

[77] Myers, M. G., Grammer, T. C., Wang, L. M., Sun, X. J., Pierce, J. H., Blenis, J. & White, M. F. (1994). Insulin-receptor substrate-1 mediates phosphatidylinositol 3'-kinase and p70(S6K) signaling during insulin, insulin-like growth-factor-I, and interleukin-4 stimulation. *Journal of Biological Chemistry*, *269*, 28783-28789.

[78] Leroith, D., Werner, H., Beitnerjohnson, D. & Roberts, C. T. (1995). Molecular and cellular aspects of the insulin-like growth-factor-i receptor. *Endocrine Reviews*, *16*, 143-163.

[79] Lin, H. K., Wang, L., Hu, Y. C., Altuwaijri, S. & Chang, C. S. (2002). Phosphorylation-dependent ubiquitylation and degradation of androgen receptor by Akt require Mdm2 E3 ligase. *Embo Journal*, *21*, 4037-4048.

[80] Srivastava, R., Chen, X. & Thakkar, H. (2001). Constitutively active Akt is an important regulator of TRAIL sensitivity in prostate cancer. *Clinical Cancer Research*, *7*, 3718S-3719S.

[81] Liao, Y. D., Grobholz, R., Abel, U., Trojan, L., Michel, M. S., Angel, P. & Mayer, D. (2003). Increase of AKT/PKB expression correlates with Gleason pattern in human prostate cancer. *International Journal of Cancer*, *107*, 676-680.

[82] Ghosh, P. M., Malik, S., Bedolla, R. & Kreisberg, J. I. (2003). Akt in prostate cancer: Possible role in androgen-independence. *Current Drug Metabolism*, *4*, 487-496.

[83] Denmeade, S. R., Lin, X. S. & Isaacs, J. T. (1996). Role of programmed (apoptotic) cell death during the progression and therapy for prostate cancer (vol 28, pg 251, 1996). *Prostate*, *28*, U6-U6.

[84] Vivanco, I. & Sawyers, C. L. (2002). The phosphatidylinositol 3-kinase-AKT pathway in human cancer. *Nature Reviews Cancer*, *2*, 489-501.

[85] Vanhaesebroeck, B. & Alessi, D. R. (2000). The P13K-PDK1 connection: more than just a road to PKB. *Biochemical Journal*, *346*, 561-576.

[86] Cantley, L. C. (2002). The phosphoinositide 3-kinase pathway. *Science*, *296*, 1655-1657.

[87] Dong, J. T., Li, C. L., Sipe, T. W. & Frierson, H. F. (2001). Mutations of PTEN/MMAC1 in primary prostate cancers from Chinese patients. *Clinical Cancer Research*, *7*, 304-308.

[88] Suzuki, H., Freije, D., Nusskern, D. R., Okami, K., Cairns, P., Sidransky, D., Isaacs, W. B. & Bova, G. S. (1998). Interfocal heterogeneity of PTEN/MMAC1 gene alterations in multiple metastatic prostate cancer tissues. *Cancer Research*, *58*, 204-209.

[89] Li, P. F., Nicosia, S. V. & Bai, W. L. (2001). Antagonism between PTEN/MMAC1/TEP-1 and androgen receptor in growth and apoptosis of prostatic cancer cells. *Journal of Biological Chemistry*, *276*, 20444-20450.

[90] Nan, B., Snabboon, T., Unni, E., Yuan, X. J., Whang, Y. E. & Marcelli, M. (2003). The PTEN tumor suppressor is a negative modulator of androgen receptor transcriptional activity. *Journal of Molecular Endocrinology*, *31*, 169-183.

[91] Lin, H. K., Yeh, S. Y., Kang, H. Y. & Chang, C. S. (2001). Akt suppresses androgen-induced apoptosis by phosphorylating and inhibiting androgen receptor. *Proceedings of the National Academy of Sciences of the United States of America*, *98*, 7200-7205.

[92] Culig, Z. (2004). Androgen receptor cross-talk with cell signalling pathways. *Growth Factors*, *22*, 179-184.

[93] Nicholson, K. M. & Anderson, N. G. (2002). The protein kinase B/Akt signalling pathway in human malignancy. *Cellular Signalling*, *14*, 381-395.

[94] Lee, S. O., Lou, W., Hou, M., Onate, S. A. & Gao, A. C. (2003). Interleukin-4 enhances prostate-specific antigen expression by activation of the androgen receptor and Akt pathway. *Oncogene*, *22*, 7981-7988.

[95] Mayo, L. D. & Donner, D. B. (2001). A phosphatidylinositol 3-kinase/Akt pathway promotes translocation of Mdm2 from the cytoplasm to the nucleus. *Proceedings of the National Academy of Sciences of the United States of America*, *98*, 11598-11603.

[96] Wang, Y., Kreisberg, J. I. & Ghosh, P. M. (2007). Cross-talk between the androgen receptor and the phosphatidylinositol 3-kinase/akt pathway in prostate cancer. *Current Cancer Drug Targets*, *7*, 591-604.

[97] Manin, M., Baron, S., Goossens, K., Beaudoin, C., Jean, C., Veyssiere, G., Verhoeven, G. & Morel, L. (2002). Androgen receptor expression is regulated by the phosphoinositide 3-kinase/Akt pathway in normal and tumoral epithelial cells. *Biochemical Journal*, *366*, 729-736.

[98] Chang, L. & Karin, M. (2001). Mammalian MAP kinase signalling cascades. *Nature*, *410*, 37-40.

[99] Lin, D. L., Whitney, M. C., Yao, Z. & Keller, E. T. (2001). Interleukin-6 Induces Androgen Responsiveness in Prostate Cancer Cells through Up-Regulation of Androgen Receptor Expression. *Clin Cancer Res*, *7*, 1773-1781.

[100] MagiGalluzzi, C., Mishra, R., Fiorentino, M., Montironi, R., Yao, H., Capodieci, P., Wishnow, K., Kaplan, I., Stork, P. J. S. & Loda, M. (1997). Mitogen-activated protein kinase phosphatase 1 is overexpressed in prostate cancers and is inversely related to apoptosis. *Laboratory Investigation*, *76*, 37-51.

[101] Zhang, X. Q., Lee, M. S., Zelivianski, S. & Lin, M. F. (2001). Characterization of a Prostate-specific Tyrosine Phosphatase by Mutagenesis and Expression in Human Prostate Cancer Cells. *J. Biol. Chem.*, *276*, 2544-2550.

[102] Lee, M. S., Igawa, T. & Lin, M. F. (2004). Tyrosine-317 of p52Shc mediates androgen-stimulated proliferation signals in human prostate cancer cells. *Oncogene*, *23*, 3048-3058.

[103] Huggins, C. & Hodges, C. V. (1972). Studies on Prostatic Cancer: I. The Effect of Castration, Of Estrogen and of Androgen Injection on Serum Phosphatases in Metastatic Carcinoma of the Prostate. *CA Cancer J Clin*, *22*, 232-240.

[104] Meng, T. C., Lee, M. S. & Lin, M. F. (2000). Interaction between protein tyrosine phosphatase and protein tyrosine kinase is involved in androgen-promoted growth of human prostate cancer cells. *Oncogene*, *19*, 2664-2677.

[105] Lin, M. F. & Meng, T. C. (1996). Tyrosine phosphorylation of a 185 kDa phosphoprotein (pp185) inversely correlates with the cellular activity of human prostatic acid phosphatase. *Biochemical and Biophysical Research Communications*, *226*, 206-213.

[106] Sakai, H., Shiraishi, K., Minami, Y., Yushita, Y., Kanetake, H. & Saito, Y. (1991). Immunohistochemical prostatic acid-phosphatase level as a prognostic factor of prostatic-carcinoma. *Prostate*, *19*, 265-272.

[107] Abrahamsson, P. A., Lilja, H., Falkmer, S. & Wadstrom, L. B. (1988). Immunohistochemical distribution of the 3 predominant secretory proteins in the parenchyma of hyperplastic and neoplastic prostate-glands. *Prostate*, *12*, 39-46.

[108] Lin, M. F., Lee, M. S., Zhou, X. W., Andressen, J. C., Meng, T. C., Johansson, S. L., West, W. W., Taylor, R. J., Anderson, J. R. & Lin, F. F. (2001). Decreased expression of cellular prostatic acid phosphatase increases tumorigenicity of human prostate cancer cells. *Journal of Urology*, *166*, 1943-1950.

[109] Osman, I., Scher, H. I., Drobnjak, M., Verbel, D., Morris, M., Agus, D., Ross, J. S. & Cordon-Cardo, C. (2001). HER-2/neu (p185neu) protein expression in the natural or treated history of prostate cancer. *Clinical Cancer Research*, *7*, 2643-2647.

[110] Signoretti, S., Montironi, R., Manola, J., Altimari, A., Tam, C., Bubley, G., Balk, S., Thomas, G., Kaplan, I., Hlatky, L., Hahnfeldt, P., Kantoff, P. & Loda, M. (2000). Her-2-neu expression and progression toward androgen independence in human prostate cancer. *Journal of the National Cancer Institute*, *92*, 1918-1925.

[111] Agoulnik, I. U., Bingman, W. E., III, Nakka, M., Li, W., Wang, Q., Liu, X., Brown, M. & Weigel, N. L. (2008). Target Gene Specific Regulation of Androgen Receptor Activity by p42/p44 MAPK. *Mol Endocrinol*, doi:10.1210/me.2007-0481.

[112] Yang, X., Chen, M. W., Terry, S., Vacherot, F., Bemis, D. L., Capodice, J., Kitajewski, J., de la Taille, A., Benson, M. C., Guo, Y. & Buttyan, R. (2006). Complex regulation of human androgen receptor expression by Wnt signaling in prostate cancer cells. *Oncogene*, *25*, 3436-3444.

[113] Rahman, M., Miyamoto, H. & Chang, C. (2004). Androgen Receptor Coregulators in Prostate Cancer: Mechanisms and Clinical Implications. *Clin Cancer Res*, *10*, 2208-2219.

[114] Truica, C. I., Byers, S. & Gelmann, E. P. (2000). {beta}-Catenin Affects Androgen Receptor Transcriptional Activity and Ligand Specificity. *Cancer Res*, *60*, 4709-4713.

[115] Voeller, H. J., Truica, C. I. & Gelmann, E. P. (1998). {beta}-Catenin Mutations in Human Prostate Cancer. *Cancer Res*, *58*, 2520-2523.

[116] Chen, G., Shukeir, N., Potti, A., Sircar, K., Aprikian, A., Goltzman, D. & Rabbani, S. A. (2004). Up-regulation of Wnt-1 and beta-catenin production in patients with advanced metastatic prostate carcinoma. *Cancer*, *101*, 1345-1356.

[117] Verras, M. & Sun, Z. (2005). {beta}-Catenin Is Involved in Insulin-Like Growth Factor 1-Mediated Transactivation of the Androgen Receptor. *Mol Endocrinol*, *19*, 391-398.

[118] Schweizer, L., Rizzo, C. A., Spires, T. E., Platero, J. S., Wu, Q., Lin, T. A., Gottardis, M. M. & Attar, R. M. (2008). The androgen receptor can signal through Wnt/beta-Catenin in prostate cancer cells as an adaptation mechanism to castration levels of androgens. *Bmc Cell Biology*, *9*, 15.

[119] Chesire, D. R., Ewing, C. M., Gage, W. R. & Isaacs, W. B. (2002). In vitro evidence for complex modes of nuclear beta-catenin signaling during prostate growth and tumorigenesis. *Oncogene*, *21*, 2679-2694.

[120] Wissmann, C., Wild, P. J., Kaiser, S., Roepcke, S., Stoehr, R., Woenckhaus, M., Kristiansen, G., Hsieh, J. C., Hofstaecter, F., Hartmann, A., Knuechel, R., Rosenthal, A. & Pilarsky, C. (2003). WIF1, a component of the Wnt pathway, is down-regulated in prostate, breast, lung, and bladder cancer. *Journal of Pathology*, *201*, 204-212.

[121] Verras, M., Brown, J., Li, X., Nusse, R. & Sun, Z. (2004). Wnt3a Growth Factor Induces Androgen Receptor-Mediated Transcription and Enhances Cell Growth in Human Prostate Cancer Cells. *Cancer Res*, *64*, 8860-8866.

[122] Masiello, D., Chen, S. Y., Xu, Y., Verhoeven, M. C., Choi, E., Hollenberg, A. N. & Balk, S. P. (2004). Recruitment of {beta}-Catenin by Wild-Type or Mutant Androgen Receptors Correlates with Ligand-Stimulated Growth of Prostate Cancer Cells. *Mol Endocrinol*, *18*, 2388-2401.

[123] Chmelar, R., Buchanan, G., Need, E. F., Tilley, W. & Greenberg, N. M. (2007). Androgen receptor coregulators and their involvement in the development and progression of prostate cancer. *International Journal of Cancer*, *120*, 719-733.

[124] Niu, Y., Yeh, S., Miyamoto, H., Li, G., Altuwaijri, S., Yuan, J., Han, R., Ma, T., Kuo, H. C. & Chang, C. (2008). Tissue Prostate-Specific Antigen Facilitates Refractory Prostate Tumor Progression via Enhancing ARA70-Regulated Androgen Receptor Transactivation. *Cancer Res*, *68*, 7110-7119.

[125] Lin, H. K., Hu, Y. C., Lee, D. K. & Chang, C. (2004). Regulation of Androgen Receptor Signaling by PTEN (Phosphatase and Tensin Homolog Deleted on Chromosome 10) Tumor Suppressor through Distinct Mechanisms in Prostate Cancer Cells. *Mol Endocrinol*, *18*, 2409-2423.

[126] Wang, L., Hsu, C. L. & Chang, C. (2005). Androgen receptor corepressors: An overview. *The Prostate*, *63*, 117-130.

[127] Kaulfuss, S., Grzmil, M., Hemmerlein, B., Thelen, P., Schweyer, S., Neesen, J., Bubendorf, L., Glass, A. G., Jarry, H., Auber, B. & Burfeind, P. (2008). Leupaxin, a Novel Coactivator of the Androgen Receptor, Is Expressed in Prostate Cancer and Plays a Role in Adhesion and Invasion of Prostate Carcinoma Cells. *Mol Endocrinol*, *22*, 1606-1621.

[128] Peng, Y., Chen, F., Melamed, J., Chiriboga, L., Wei, J., Kong, X., McLeod, M., Li, Y., Li, C. X., Feng, A., Garabedian, M. J., Wang, Z., Roeder, R. G. & Lee, P. (2008). Distinct nuclear and cytoplasmic functions of androgen receptor cofactor p44 and association with androgen-independent prostate cancer. *Proceedings of the National Academy of Sciences*, *105*, 5236-5241.

[129] Linja, M. J., Porkka, K. P., Kang, Z. K., Savinainen, K. J., Janne, O. A., Tammela, T. L. J., Vessella, R. L., Palvimo, J. J. & Visakorpi, T. (2004). Expression of androgen receptor coregulators in prostate cancer. *Clinical Cancer Research*, *10*, 1032-1040.

[130] Cronauer, M. V., Schulz, W. A., Ackermann, R. & Burchardt, M. (2005). Effects of WNT/beta-catenin pathway activation on signaling through T-cell factor and androgen receptor in prostate cancer cell lines. *International Journal of Oncology*, *26*, 1033-1040.

[131] Chesire, D. R. & Isaacs, W. B. (2002). Ligand-dependent inhibition of beta-catenin/TCF signaling by androgen receptor. *Oncogene*, *21*, 8453-8469.

[132] Mulholland, D. J., Read, J. T., Rennie, P. S., Cox, M. E. & Nelson, C. C. Functional localization and competition between the androgen receptor and T-cell factor for nuclear [beta]-catenin: a means for inhibition of the Tcf signaling axis. *Oncogene*, *22*, 5602-5613.

[133] Sampson, E. R., Yeh, S. Y., Chang, H. C., Tsai, M. Y., Wang, X., Ting, H. J. & Chang, C. (2001). Identification and characterization of androgen receptor associated coregulators in prostate cancer cells. *Journal of Biological Regulators and Homeostatic Agents*, *15*, 123-129.

[134] Mestayer, C., Blanchere, M., Jaubert, F., Dufour, B. & Mowszowicz, I. (2003). Expression of androgen receptor coactivators in normal and cancer prostate tissues and cultured cell lines. *Prostate*, *56*, 192-200.

[135] Heitzer, M. D. & DeFranco, D. B. (2007). Hic-5/ARA55 a prostate stroma-specific AR coactivator. *Steroids*, *72*, 218-220.

[136] Knudsen, K. E., Arden, K. C. & Cavenee, W. K. (1998). Multiple G1 Regulatory Elements Control the Androgen-dependent Proliferation of Prostatic Carcinoma Cells. *J. Biol. Chem.*, *273*, 20213-20222.

[137] Lu, J. & Danielsen, M. (1998). Differential Regulation of Androgen and Glucocorticoid Receptors by Retinoblastoma Protein. *J. Biol. Chem.*, *273*, 31528-31533.

[138] Bookstein, R., Rio, P., Madreperla, S. A., Hong, F., Allred, C., Grizzle, W. E. & Lee, W. H. (1990). Promoter deletion and loss of retinoblastoma gene-expression in human prostate carcinoma. *Proceedings of the National Academy of Sciences of the United States of America*, *87*, 7762-7766.

[139] Tricoli, J. V., Gumerlock, P. H., Yao, J. L., Chi, S. G., Dsouza, S. A., Nestok, B. R., White, R. W. D., French, F. S., Gelmann, E. P., Pretlow, T. G. & Stearns, M. E. (1996). Alterations of the retinoblastoma gene in human prostate adenocarcinoma. *Genes Chromosomes & Cancer*, *15*, 108-114.

[140] Mack, P. C., Chi, S. G., Meyers, F. J., Stewart, S. L., deVere-White, R. W. & Gumerlock, P. H. (1998). Increased RB1 abnormalities in human primary prostate cancer following combined androgen blockade. *The Prostate*, *34*, 145-151.

[141] Sharma, A., Comstock, C. E. S., Knudsen, E. S., Cao, K. H., Hess-Wilson, J. K., Morey, L. M., Barrera, J. & Knudsen, K. E. (2007). Retinoblastoma Tumor Suppressor Status Is a Critical Determinant of Therapeutic Response in Prostate Cancer Cells. *Cancer Res*, *67*, 6192-6203.

[142] Kaltz-Wittmer, C., Klenk, U., Glaessgen, A., Aust, D. E., Diebold, J., Lohrs, U. & Baretton, G. B. (2000). FISH analysis of gene aberrations (MYC, CCND1, ERBB2, RB, and AR) in advanced prostatic carcinomas before and after androgen deprivation therapy. *Laboratory Investigation*, *80*, 1455-1464.

[143] Reutens, A. T., Fu, M. F., Wang, C. G., Albanese, C., McPhaul, M. J., Sun, Z. J., Balk, S. P., Janne, O. A., Palvimo, J. J. & Pestell, R. G. (2001). Cyclin D1 binds the androgen receptor and regulates hormone-dependent signaling in a p300/CBP-associated factor (P/CAF)-dependent manner. *Molecular Endocrinology*, *15*, 797-811.

[144] Lin, H. M., Zhao, L. & Cheng, S. Y. (2002). Cyclin D1 is a ligand-independent co-repressor for thyroid hormone receptors. *Journal of Biological Chemistry*, *277*, 28733-28741.

[145] Fan, W., Yanase, T., Morinaga, H., Okabe, T., Nomura, M., Daitoku, H., Fukamizu, A., Kato, S., Takayanagi, R. & Nawata, H. (2007). Insulin-like Growth Factor 1/Insulin Signaling Activates Androgen Signaling through Direct Interactions of Foxo1 with Androgen Receptor. *J. Biol. Chem.*, *282*, 7329-7338.

[146] Cohen, P., Nimmo, H. G. & Proud, C. G. (1978). How does insulin stimulate glycogen synthesis? *Biochem Soc Symp*, 69-95.

[147] Wang, L., Lin, H. K., Hu, Y. C., Xie, S., Yang, L. & Chang, C. (2004). Suppression of Androgen Receptor-mediated Transactivation and Cell Growth by the Glycogen Synthase Kinase 3{beta} in Prostate Cells. *J. Biol. Chem.*, *279*, 32444-32452.

[148] Liao, X., Thrasher, J. B., Holzbeierlein, J., Stanley, S. & Li, B. (2004). Glycogen Synthase Kinase-3{beta} Activity Is Required for Androgen-Stimulated Gene Expression in Prostate Cancer. *Endocrinology*, *145*, 2941-2949.

[149] Mulholland, D. J., Dedhar, S., Wu, H. & Nelson, C. C. (2006). PTEN and GSK3[beta]: key regulators of progression to androgen-independent prostate cancer. *Oncogene*, *25*, 329-337.

[150] Costello, L. & Franklin, R. (2006). The clinical relevance of the metabolism of prostate cancer; zinc and tumor suppression: connecting the dots. *Molecular Cancer*, *5*, 17.

[151] Franklin, R. B., Ma, J., Zou, J., Guan, Z., Kukoyi, B. I., Feng, P. & Costello, L. C. (2003). Human ZIP1 is a major zinc uptake transporter for the accumulation of zinc in prostate cells. *Journal of Inorganic Biochemistry*, *96*, 435-442.

[152] Costello, L. C., Franklin, R. B. & Feng, P. (2005). Mitochondrial function, zinc, and intermediary metabolism relationships in normal prostate and prostate cancer. *Mitochondrion*, *5*, 143-153.

[153] Pei Feng, Li, T. L., Guan, Z. X., Franklin, R. B. & Costello, L. C. (2002). Direct effect of zinc on mitochondrial apoptogenesis in prostate cells. *The Prostate*, *52*, 311-318.

[154] Golovine, K., Makhov, P., Uzzo, R. G., Shaw, T., Kunkle, D. & Kolenko, V. M. (2008). Overexpression of the Zinc Uptake Transporter hZIP1 Inhibits Nuclear Factor-{kappa}B and Reduces the Malignant Potential of Prostate Cancer Cells In vitro and In vivo. *Clin Cancer Res*, *14*, 5376-5384.

[155] Costello, L. C., Liu, Y., Zou, J. & Franklin, R. B. (1999). Evidence for a Zinc Uptake Transporter in Human Prostate Cancer Cells Which Is Regulated by Prolactin and Testosterone. *J. Biol. Chem.*, *274*, 17499-17504.

[156] Dubi, N., Gheber, L., Fishman, D., Sekler, I. & Hershfinkel, M. (2008). Extracellular zinc and zinc-citrate, acting through a putative zinc-sensing receptor, regulate growth and survival of prostate cancer cells. *Carcinogenesis*, *29*, 1692-1700.

[157] Hershfinkel, M., Silverman, W. F. & Sekler, I. (2007). The zinc sensing receptor, a link between zinc and cell signaling. *Molecular Medicine*, *13*, 331-336.

[158] Rishi, I., Baidouri, H., Abbasi, J. A., Bullard-Dillard, R., Kajdacsy-Balla, A., Pestaner, J.P., Skacel, M., Tubbs, R. & Bagasra, O. (2003). Prostate cancer in African American men is associated with downregulation of zinc transporters. *Applied Immunohistochemistry & Molecular Morphology*, *11*, 253-260.

[159] Golovine, K., Uzzo, R. G., Makhov, P., Crispen, P. L., Kunkle, D. & Kolenko, V. M. (2008). Depletion of intracellular zinc increases expression of tumorigenic cytokines VEGF, IL-6 and IL-8 in prostate cancer cells via NF-kappaB-dependent pathway. *The Prostate*, *68*, 1443-1449.

[160] Heinrich, E., Schally, A. V., Buchholz, S., Rick, F. G., Halmos, G., Mile, M., Groot, K., Hohla, F., Zarandi, M. & Varga, J. L. (2008). Dose-dependent growth inhibition in vivo of PC-3 prostate cancer with a reduction in tumoral growth factors after therapy with GHRH antagonist MZ-J-7-138. *The Prostate*, In press.

[161] Seaton, A., Scullin, P., Maxwell, P. J., Wilson, C., Pettigrew, J., Gallagher, R., O'Sullivan, J. M., Johnston, P. G. & Waugh, D. J. J. (2008). Interleukin-8 signaling promotes androgen-independent proliferation of prostate cancer cells via induction of androgen receptor expression and activation. *Carcinogenesis*, *29*, 1148-1156.

[162] Sweeney, C., Li, L., Shanmugam, R., Bhat-Nakshatri, P., Jayaprakasan, V., Baldridge, L. A., Gardner, T., Smith, M., Nakshatri, H. & Cheng, L. (2004). Nuclear Factor-{kappa}B Is Constitutively Activated in Prostate Cancer In vitro and Is Overexpressed in Prostatic Intraepithelial Neoplasia and Adenocarcinoma of the Prostate. *Clin Cancer Res*, *10*, 5501-5507.

[163] Rokhlin, O. W., Taghiyev, A. F., Guseva, N. V., Glover, R. A., Chumakov, P. M., Kravchenko, J. E. & Cohen, M. B. (2005). Androgen regulates apoptosis induced by

TNFR family ligands via multiple signaling pathways in LNCaP. *Oncogene*, *24*, 6773-6784.

[164] Fabbri, F., Amadori, D., Carloni, S., Brigliadori, G., Tesei, A., Ulivi, P., Rosetti, M., Vannini, I., Arienti, C., Zoli, W. & Silvestrini, R. (2008). Mitotic catastrophe and apoptosis induced by docetaxel in hormone-refractory prostate cancer cells. *Journal of Cellular Physiology*, *217*, 494-501.

[165] Zhu, W., Zhang, J. S. & Young, C. Y. F. (2001). Silymarin inhibits function of the androgen receptor by reducing nuclear localization of the receptor in the human prostate cancer cell line LNCaP. *Carcinogenesis*, *22*, 1399-1403.

[166] Xie, X. M., Zhao, X. Q., Liu, Y. F., Young, C. Y. F., Tindall, D. J., Slawin, K. M. & Spencer, D. M. (2001). Robust prostate-specific expression for targeted gene therapy based on the human kallikrein 2 promoter. *Human Gene Therapy*, *12*, 549-561.

[167] Lu, S., Liu, M., Epner, D. E., Tsai, S. Y. & Tsai, M. J. (1999). Androgen Regulation of the Cyclin-Dependent Kinase Inhibitor p21 Gene through an Androgen Response Element in the Proximal Promoter. *Mol Endocrinol*, *13*, 376-384.

[168] Majid, S., Kikuno, N., Nelles, J., Noonan, E., Tanaka, Y., Kawamoto, K., Hirata, H., Li, L. C., Zhao, H., Okino, S. T., Place, R. F., Pookot, D. & Dahiya, R. (2008). Genistein Induces the p21WAF1/CIP1 and p16INK4a Tumor Suppressor Genes in Prostate Cancer Cells by Epigenetic Mechanisms Involving Active Chromatin Modification. *Cancer Res*, *68*, 2736-2744.

[169] Banerjee, S., Li, Y., Wang, Z. & Sarkar, F. H. (2008). Multi-targeted therapy of cancer by genistein. *Cancer Letters*, *269*, 226-242.

[170] Davis, J. N., Kucuk, O. & Sarkar, F. H. (2002). Expression of prostate-specific antigen is transcriptionally regulated by genistein in prostate cancer cells. *Molecular Carcinogenesis*, *34*, 91-101.

[171] Stewart, L. V., Lyles, B., Lin, M. F. & Weigel, N. L. (2005). Vitamin D receptor agonists induce prostatic acid phosphatase to reduce cell growth and HER-2 signaling in LNCaP-derived human prostate cancer cells. *The Journal of Steroid Biochemistry and Molecular Biology*, *97*, 37-46.

[172] Prevo, R., Deutsch, E., Sampson, O., Diplexcito, J., Cengel, K., Harper, J., O'Neill, P., McKenna, W.G., Patel, S. & Bernhard, E.J. (2008). Class I PI3 Kinase Inhibition by the Pyridinylfuranopyrimidine Inhibitor PI-103 Enhances Tumor Radiosensitivity. *Cancer Res*, *68*, 5915-5923.

[173] Yaguchi, S. i., Fukui, Y., Koshimizu, I., Yoshimi, H., Matsuno, T., Gouda, H., Hirono, S., Yamazaki, K. & Yamori, T. (2006). Antitumor Activity of ZSTK474, a New Phosphatidylinositol 3-Kinase Inhibitor. *J. Natl. Cancer Inst.*, *98*, 545-556.

[174] de Bono, J. S., Attard, G., Adjei, A., Pollak, M. N., Fong, P. C., Haluska, P., Roberts, L., Melvin, C., Repollet, M., Chianese, D., Connely, M., Terstappen, L. W. M. M. & Gualberto, A. (2007). Potential Applications for Circulating Tumor Cells Expressing the Insulin-Like Growth Factor-I Receptor. *Clin Cancer Res*, *13*, 3611-3616.

[175] Boccardo, F., Rubagotti, A., Conti, G., Battaglia, M., Cruciani, G., Manganelli, A., Ricci, S. & Lapini, A. (2008). Prednisone plus gefitinib versus prednisone plus placebo in the treatment of hormone-refractory prostate cancer: A randomized phase II trial. *Oncology*, *74*, 223-228.

[176] de Bono, J. S., Bellmunt, J., Attard, G., Droz, J. P., Miller, K., Flechon, A., Sternberg, C., Parker, C., Zugmaier, G., Hersberger-Gimenez, V., Cockey, L., Mason, M. & Graham, J. (2007). Open-Label Phase II Study Evaluating the Efficacy and Safety of Two Doses of Pertuzumab in Castrate Chemotherapy-Naive Patients With Hormone-Refractory Prostate Cancer. *J Clin Oncol*, *25*, 257-262.

[177] Makhov, P., Golovine, K., Uzzo, R. G., Rothman, J., Crispen, P. L., Shaw, T., Scoll, B. J. & Kolenko, V. M. (2008). Zinc chelation induces rapid depletion of the X-linked inhibitor of apoptosis and sensitizes prostate cancer cells to TRAIL-mediated apoptosis. *Cell Death Differ*, *15*, 1754-1751.

In: Handbook of Prostate Cancer Cell Research...
Editor: Alan T. Meridith
ISBN: 978-1-60741-954-9

Chapter 5

Prostate Stem Cells and Cell Differentiation

Guadalupe Aparicio Gallego[1], Silvia Díaz Prado[1,2], Moisés Blanco Calvo[1], Javier Cassinello Espinosa[3], Enrique Grande Pulido[4], Luís Antón Aparicio[2,5,*]

[1]Biomedical Research Institute of A Coruña (INIBIC), A Coruña, Spain
[2]Medicine Department, Universtiy of A Coruña, A Coruña, Spain
[3]Medical Oncology Service, Guadalajara University Hospital, Guadalajara, Spain
[4]Medical Oncology Service, Ramón y Cajal University Hospital, Madrid, Spain
[5]Medical Oncology Service, A Coruña University Hospital, A Coruña, Spain

Abstract

Normal prostate gland development requires many coordinated cellular processes through prostate stem cells, including epithelial proliferation, mesenchymal-epithelial interaction, ductal branching morphogenesis, and ductal canalization.

Stem cells are defined functionally as cells that have the capacity to self-renew as well as the ability to generate differentiated cells. Stem cells generate daughter cells identical to their mother (self-renewal) as well as produce progeny with more restricted potential (differentiation). Self-renewal is achieved by symmetrical cell division while maintaining pluripotency; and this can be modulated by extrinsic factors, transcriptional regulator, and effectors.

Even though current research is focused on the identification, characterization, and isolation of stem cells from the adult, to date, most adult stem cells have restricted potential, and achieving indefinite proliferation and expansion of the stem cells in culture is still not routine.

One of the critical questions in the adult stem cell field concerns the mechanisms that regulate the decision between self-renewal and differentiation. During

* Corresponding author: Medical Oncoloy Service, A Coruña University Hospital. As Xubias, 84. ZIP: 15006. A Coruña. Spain, @Luis.M.Anton.Aparicio@sergas.com

embryogenesis, cells are initially proliferative and pluripotent; they only gradually become restricted to a different cell fate. Adult stem cells have two fundamental properties: a long-term capacity to divide and the ability to produce daughter cells that either retain stem cell identity or initiate differentiation along the appropriate lineage(s). The balance between self-renewal and initiation of differentiation is crucial.

In regard to prostate cells, although signals from the luminal layer specify self-renewal or stem cells identity, other signals from the microenvironment also play roles in regulating the fate of daughters of stem cell decisions. Interactions between progenitor cells (basal layer) and mature cells (luminal layer) appear to restrict expression of stem cell characteristics and ensure that stem cell daughters displaced away from the basal cells for initiating differentiation. Thus, the luminal cells play an important guardian role, preventing accumulation of basal cells with ambiguous identity.

The signal that controls which daughter cell of an adult stem cell remains as a stem cell and which begins the process of differentiation may be mediated through a number of signaling pathways including, Wnt, Hedgehog (Hh), Notch, Oct4, Bone Morphogenetic Proteins (BMPs), JAK/Stat, or other signaling pathways.

Introduction

The regenerative capacity of prostate gland has been attributed to stem/progenitor cells within adult prostatic epithelium. It was hypothesized that the adult prostate contains stem, transit-amplifying, and postmitotic cells and that the stem cells are androgen-independent for survival.

In the normal prostate gland epithelium exist a spectrum of cells at many stages expressing a number of differentiation markers, with distinct biological properties, all of them beginning from stem/progenitor cells through multiple intermediate cell types along different lineages to terminally differentiated cells.

The traditional simplistic division of prostate epithelium into *basal cells* and *luminal cells* is no longer appropriate because all basal cells and all luminal cells are apparently not alike [1].

One of the critical questions in the adult stem cell field concerns the mechanisms that regulate the decision between self-renewal and differentiation. Adult stem cells have two fundamental properties: a long term capacity to divide and the ability to produce daughter cells that either retain stem cell identity or initiate differentiation along the appropriate lineage(s). The balance between self-renewal and initiation of differentiation is crucial. Self-renewal is achieved by symmetrical cell divisions through which is maintained pluripotency; this may be modulated by extrinsic factors, transcriptional regulators, and effectors (Figure 1).

Asymmetric cell division, another characteristic of stem cells, allows generation of daughter cells that enter the differentiation pathway. Although signals from the mature luminal epithelial prostate cells specify self-renewal of stem cell identity, other signals from the microenvironment also play roles in regulating the fate of daughters from stem cell decisions. Interactions between stem cells (basal layer) and mature cells (luminal layer) appear to restrict expression of stem cell characteristics and ensure that stem cell daughters displaced away from the basal cells to initiate differentiation. Thus, the luminal cells play an important guardian role, preventing accumulation of basal cells with ambiguous identity.

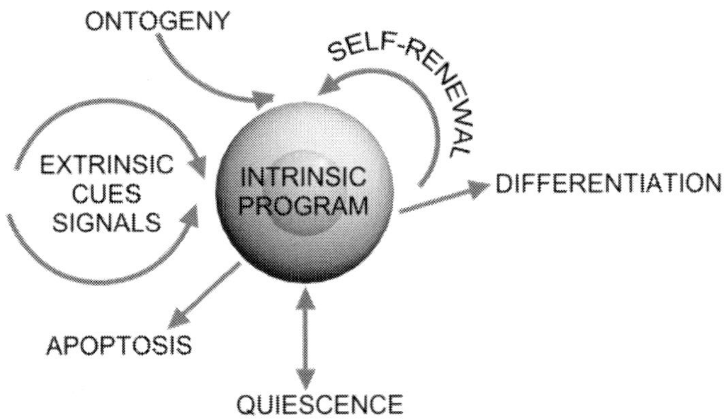

Figure 1. Stem cell fate. Stem cell behavior is regulated by both extrinsic signal and intrinsic programs.

Regulation of Stem Cell Fate

In the embryo, stem cells function as units of organogenesis and tissue formation, whereas in adulthood stem cells are perhaps more appropriately considered as units of tissue maintenance or regeneration.

Systems with a tremendous daily turnover, as the blood system, give an example where stem cells provide a critical maintenance function. Conversely, stem cells in tissues with low turnover, such as muscles, may function in repair/regeneration. Other tissues, including the epidermis, epithelial lining of the gut, liver, pancreas, or prostatic gland that exhibit secretion in the normal biologic situation must be also renewed.

Adult somatic stem cells have historically been considered as stringently tissue-restricted in their differentiation potential. However, publication of several reports led to an alternative view in which adult stem cells may possess some kind of developmental plasticity. Adult stem cells must presumably themselves derive from differentiation of other stem cells during ontogeny. Information regarding the precise developmental origins of adult stem cells is, by and large, rather limited. In regard to prostate stem cells, however, little has been learned over the past years. It is now generally accepted that the first prostate stem cells arise intra-embryonically in the urogenital sinus (UGS), and the adult prostate stem cells arise in the basal layer.

Once formed, prostate adult stem cells must satisfy a lifetime´s demand for the mature effector cells of the prostate secretory gland. Meeting such demands requires the fine balancing of stem cells self-renewal and differentiation. In adult stem cells-based systems, which are composed of many different cell types, lineage specification is an additional feature of the differentiation process. However, the extent in which these processes are coupled or are independent at a mechanistic level remains unclear. Similarly, it has been hotly debated the relative extent in which extrinsic factors, transcriptional regulators and effectors (Figure 2) are involved in the decision between self-renewal and differentiation. In the case of prostate epithelium, several extrinsic factors that have well characterized activities in early developmental processes have been shown to affect self-renewal of the prostate adult stem cell compartment. These include members of the Wnt, Notch, BMP, and Hh families. Much less attention has been focused on the role of transcription regulators in the ontogeny, self-renewal, and differentiation/lineage commitment of stem cells. The various transcriptional factors identified in these and other pathways are likely to represent only a subset of transcriptional factors involved in regulation of the stem cell compartment. In summary, a small amount of substantial information has been accumulated about the role of transcription factors within the prostate gland epithelium generation.

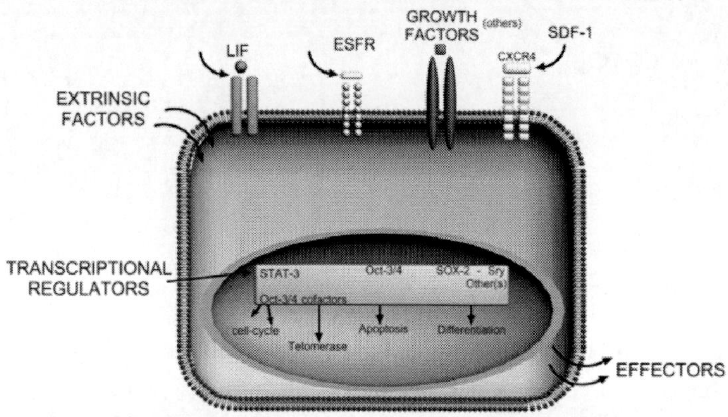

Figure 2. Molecular mechanisms governing stem cells self-renewal. The mechanism can be divided into three categories: extrinsic factors, transcriptional regulation, and effectors. SDF-1: Stromal cell-Derived Factor-1. LIF: cytokine inducing differentiation and preventing self-renewal. ESRF: Embryonic Stem cell Renewal Factor. STAT-3: activated by LIF, as a transcriptional regulator in nuclei; they might activate the expression of a partner of Oct-3/4. Oct-3/4: Governing self-renewal and differentiation of totipotent and pluripotent stem cells. SOX-2: member of the Sry related transcription factor family that works as cofactor of Oct-3/4; their expression is regulated by Oct-3/4. Sry: Master switch that determines the differentiation factor family, that work as cofactor of Oct-3/4; their expression is regulated by Oct-3/4. Sry: Master switch that determines the differentiation of the bipotential genital ridge; expression of Sry sets in motion a cascade of made-specific gene expression such as Sox9, Sox8, and Fgl9.

Ontogeny of Adult Stem Cells

The origin of most adult stem cells is poorly understood, because they classically are defined as cells with capacity to self-renew and to generate daughter cells that can differentiate down several cell lineages to form all of the cell types that are found in the mature tissue. The basic characteristic of adult stem cells is that a single cell (clonal) is able to self-renew itself and generate differentiated cells [2].

Stemness refers to the common molecular processes underlying the core stem cell properties of self-renewal and generation of differentiated progeny. Although stem cells in different cellular microenvironments or niches necessarily have different physiological demands and therefore distinct molecular programs, there are likely certain genetic characteristics specifically shared by all stem cells. Through transcriptional profiling, many of the genes with some role in stem cells have been identified [3-5].

The transcriptional profiling of stem cells has suggested that they share distinct molecular characteristics. Stem cells appear to have the capacity to use a broad range of growth factors and signaling molecules and to express many of the downstream signaling components involved in the transduction pathways present and perhaps active in stem cells, including Wnt, Hh, Nocth, TGF-β, and JAK/STAT family members. Stem cells also express many components involved in establishing their specialized cell cycles, either related with maintaining cell cycle arrest in G_1, for most quiescent adult stem cells, or connected with progression through cell cycle checkpoints promoting rapid cycling, as in the case for mobilized adult stem cells [6, 7].

Studies designated to elucidate the ontogeny of adult stem cells could help to reveal their lineage specific relationship and could shed light on their plasticity and potential. Information on the origins of adult stem cells would also help to define the molecular programs involved in lineage determination, which may in turn provide insights into methods for manipulating their differentiation.

Carefully designed experiments promise to answer questions important not only for stem cell biology but also for development. These include questions about features of the developmental program that are intrinsic of individual cells, such as what differentiation and patterning signals act exclusively to instruct specific cell fates, and how developmental changes in cell-intrinsic programs restrict the response of progenitor to cell-extrinsic signals.

Two concepts about the fate or potential of stem cells have moved to the forefront of adult stem cell research. The first is plasticity, the idea that restriction in cell fate is not permanent but is flexible and reversible. A related concept is trans-differentiation, i.e. the generation of functional cells of a tissue, organ, or lineage that is distinct from that of the founding stem cell [8].

Niche

The concept of a stem cell niche was first proposed 30 years ago [9], as a specific micro-environment in which adult stem cells reside in their tissue of origin. Adult organs need lifelong reservoir of somatic stem cells to ensure tissue turnover. These stem cells are localized in specific areas, known as "niches", that provide the right environment for the viability and function of the stem cells [10].

Within the niche, stem cells are able to maintain their ability for self-renewal, as well as their multi-potentiality, and consequently, detachment from the niche compartment induces stem cell differentiation and loss of self-renewal.

Various lines of evidence suggest that once a stem cell compartment is formed in a tissue/organ, stem cells take up long-term residence there. The ability of stem cells to reside within niches is an evolutionarily conserved phenomenon, and hence it is not surprising to see that the basic molecular features of stem cell retention are broadly utilized across the eukaryotic kingdom.

Stem cells can become transiently amplified by producing daughter cells, also called progenitors, which have high proliferative activity and short-term self-renewal capacity. In organs with a rapid turnover, stem cells inside the niches are normally quiescent and divide slowly. Stem cells produced in excess are expulsed from the niche to a new environment and then, differentiated. In other organs with a low turnover, the quiescent stem cells can be mobilized and activated by specific signals.

Adult stem cells are often localized to specific niches, where they utilize many, but not necessarily all, external and intrinsic cues used by the embryonic counterparts in selecting a specific fate. Locating and analyzing stem cell niches and elucidating the molecules that orchestrate specific developmental programs are important steps in determining the key components of the environment that legislate differentiation commitments and stem cell regulation.

The niche is critical in maintaining the intrinsic self-renewal; undifferentiated character of the resident stem cells in the niche's microenvironments is both proliferation- and differentiation-inhibitory. The normal microenvironment, established by signals from the various other cells (stroma) that normally surround the niche seen to be important in maintaining the slow-cycling properties of labeled-retaining cells (LRCs) and keeping in reserve. When stem cells cannot be identified or isolated in a particular organ, their existence may be inferred from kinetics studies of 5′-bromo-2′-deoxyuridine (BrdU) incorporation. Because stem cells are believed to be slowly dividing, the presence of LRCs can identify the anatomical location of a stem cell niche.

The regulation of stem cell niche can therefore directly dictate the characteristic of an organ, and it is common that the regulation of the stem cell niche have a major influence on the function and morphology of an organ. This flexible regulation of the stem cell niche could have been a relatively easy way to acquire radically different cells types during evolution. Stem cells, residing within a niche, are surrounded by stromal cells and their extracellular matrix, which provide an optimal microenvironment for the stem cells function. In the niche, various factors such as cytokines, extracellular matrix interactions, and intercellular adhesion molecules, are thought to work cooperatively to maintain the undifferentiated stem cell phenotype [11].

There is now strong evidence that the stem cells of many tissues reside in physically delineated as well as physiologically specialized structures termed niches. Inside the niche, stem cells are often quiescent; outside the niche, stem cells must either possess sufficient intrinsic factors to overcome differentiation or succumb to the fate. The signal that controls which daughter cell of an adult stem cell remains as stem cell and which begins the process of differentiation may be mediated through a number of signaling pathways including Wnt, Hh, Nocth, BMP, Oct-4, SDF-1, SCF-1, VCAM and others (Figure 3).

The Wnt pathway directs embryonic growth, governing processes such as cell fate, differentiation, proliferation, cell polarity and migration. It is also implicated in maintenance of stem cell populations. The Hh pathway plays a critical role in embryonic development. The Hh signaling has been widely recognized as one of the major mechanisms for cell-cell communication and tissue polarity. Many works have revealed that active niche supports self-renewal of stem cells via activation of the Janus-kinase (JAK)-signal transducer and activator of transcription (STAT) pathway within the adjacent stem cells [12, 13]. JAKs are non-receptor tyrosine kinases that mediate signaling downstream of many mammalian cytokyne and growth factors receptors, in part by phosphorylation and activation of STAT [14]. In summary, the signal for stem cells self-renewal is transduced from the activated JAK via STAT pathways.

Exactly when and how most somatic stem cell niche develops is still a mystery, and in the world of stem cell niches, there are considerable variations in niche design. The identification of a niche within any tissue involves knowledge of the location of the stem cells. According to literature reported, to prove that a niche is present, the stem cell must be removed and subsequently repopulated while the niche persists, providing support to the remaining exogenous cells [15].

Early evidence for prostatic stem cell niche came from a set of experiments involving androgen withdrawal from both human and rat prostate [16-18]. It has been reported that two populations of cells capable of regenerating prostatic tissue in an *in vivo* prostate reconstitution assay are present in different regions of prostatic ducts. The first population (with considerable growth potential) resides in the proximal region of ducts and in the urethra, and the survival of these cells does not require the presence of androgens. The second population (with more limited growth potential) is found in the remaining ductal regions and requires androgens for survival (figure 4) [17-20].

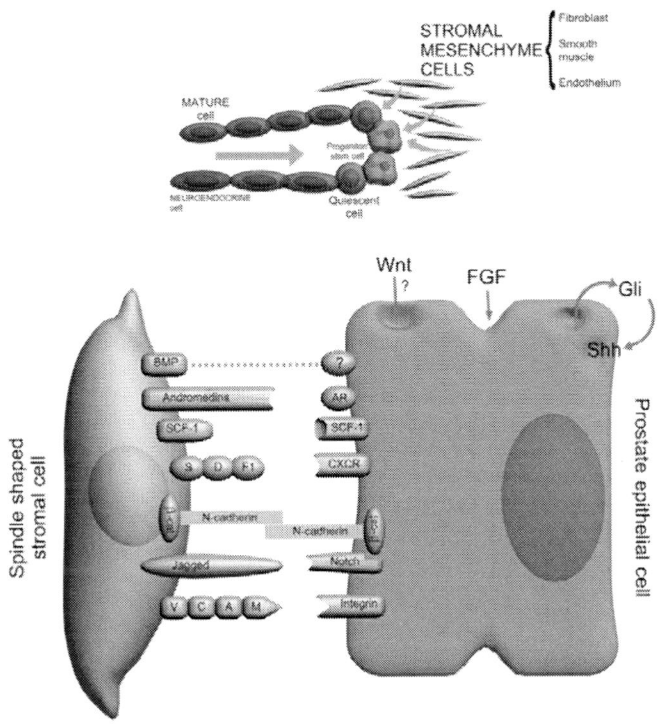

Figure 3. Multiple growth factor and cytokins are involved in stem-niche interactions. These include SCF/Kit, SDF-1/CXCR4, Jagged/Notch, Androgen/Receptor, etc. BMP4 is expressed in stromal cells, but the type of receptor expressed in prostate stem cells is unknown. The Wnt signal is important for stem cell self-renewal, but the Wnts present in the niche are unknown. The source is true for FGF and Hh.

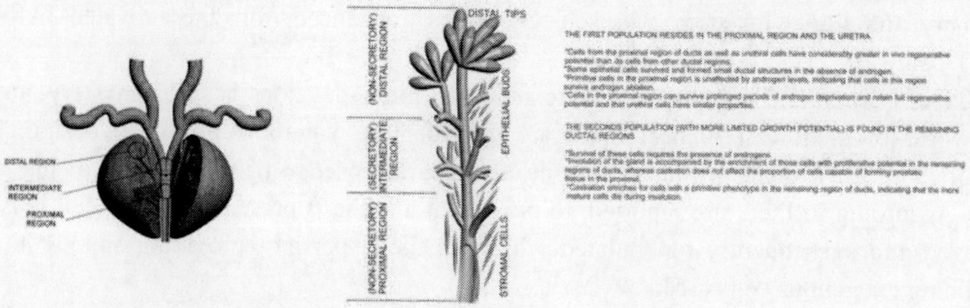

Figure 4. Location of stem cells: niche.

The proximal region of mouse prostatic ducts is enriched in cells exhibiting stem cell-like properties, namely quiescence, high proliferative potential, and ability to generate progeny of more than one lineage [21]. These cells express Sca-1 [22, 23], and are maintained in a quiescent state by high levels of active transforming growth factor β (TGF-β) [24]. The different regions of prostatic ducts are heterogeneous in terms of morphology, telomerase expression and levels of TGF-β signaling [19, 24, 25].

Cell lineage

Self-renewal and multilineage differentiation are two key features of stem cells. The self-renewal property of stem cells varies in different tissue types *in vivo*. Whereas the adult organs or cells systems display extensive and continuous self-renewal capacity (i.e. skin, gastrointestinal tract, hematopoietic] and certain organs show massive self-renewal during early development but become more quiescent in adulthood (i.e. nervous system], in others such as the prostate gland the self-renewal activity of stem cells is hormonally regulated [26]. The self-renewal capacity of prostate stem cells has been shown by castration and androgen cycling experiments performed in rodents [26] and demonstrated at clonal levels [27].

Adult prostate epithelium consist mainly of secretory luminal epithelial cells and basal cells aligned along the basement membrane, and a rare neuroendocrine cells [28], which are presumed to differentiate from common progenitor/stem cells.

Progenitor/stem cells are highly concentrated in the embryonic urogenital sinus epithelium from which prostatic epithelial buds develop, and also are present within the basal compartment of adult prostatic epithelium; the spectrum of differentiation markers of both embryonic and adult progenitor/stem cells will be similar (Table 1) [29].

At all stages (embryonic to adult) prostatic epithelial progenitor/stem cells maintain a profile of differentiation markers similar to the original embryonic progenitor of the prostate. Adult prostate basal and luminal cells are proposed to be derived from a common pluripotent progenitor/stem cells located in the basal compartment that maintains its embryonic profile of differentiation markers from embryonic to adult stages. These progenitor/stem cells differentiate into mature basal cells by maintaining CK5, CK14, CK19, p63 and GSTpi; they can differentiate into luminal cells by maintaining CK8, CK18 and losing all other markers [29, 30] (Figure 5).

Table 1. Progenitor/stem cells in embryonic urogenital sinus epithelium and adult prostate epithelium with their respective cell lineage markers.

Stem cell type	Cell type	Markers
UGS epithelium (embryonic)	Luminal cell	CK8, CK18
	Basal cell	CK5, CK14, p63
	Transitional/ Intermediate	CK19, GSTpi
Prostate epithelium (adult)	Luminal cell/predominant	CK8, CK18, PSA, AR
	Neuroendocrine cell	Chromogranin-A, Synaptophysin, Neuron-specific enolase, Serotonin
	Basal cell/predominant subpopulation	CK5, CK14, CK19, p63, GSTpi
	Basal cell /small fraction	CK5, CK8, CK14, CK18, CK19, P63, GSTpi
	Basal cell / subset	CK15, CK17, CK19
CK: cytokeratin; GSTPi: Glutathione-S-transferase-pi.		

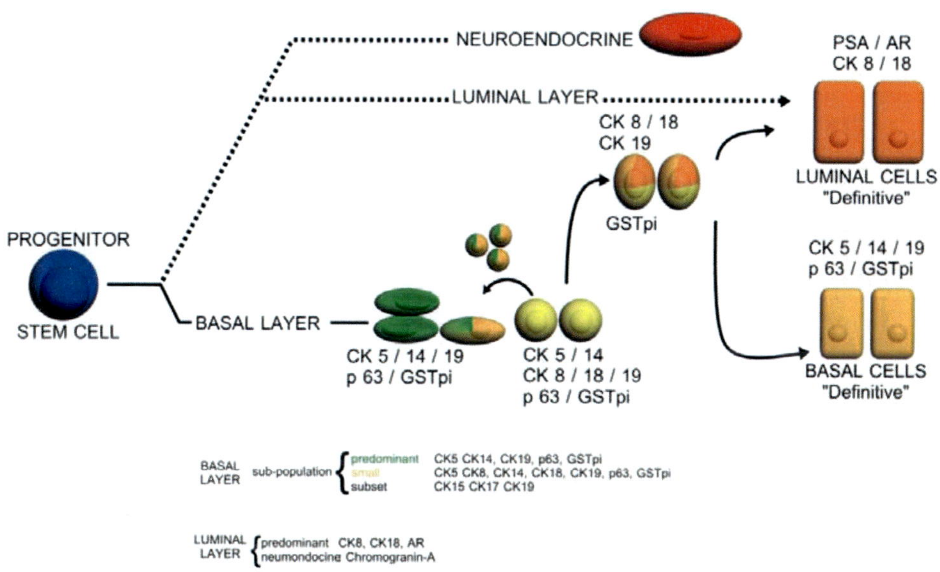

Figure 5. Cell lineage relationships in adult prostate: basal cells (CK5/14/p63/CK19/GSTpi+/CK8/CK18-); definitive luminal cells (CK8/18+/CK5/14/p63-); embryonic-like cells co-expressing both luminal and basal cell markers (CK8/18/14/5/p63/CK19/GSTpi+).

Neuroendocrine cells (NE) of the prostate, commonly found in the basal cell layer, are of different origin from other prostatic epithelial cells, representing a cell lineage in itself, being of neurogenic origin and therefore distinct from the urogenital sinus-derived prostate secretory and basal cells [31-34]. There are two types: the open cells with extensions at their apex that connect with the lumen, and closed cells with dendritic processes that extend between adjacent cells, resting on the basal lamina and in close topographical relationship with nerves (Figure 6).

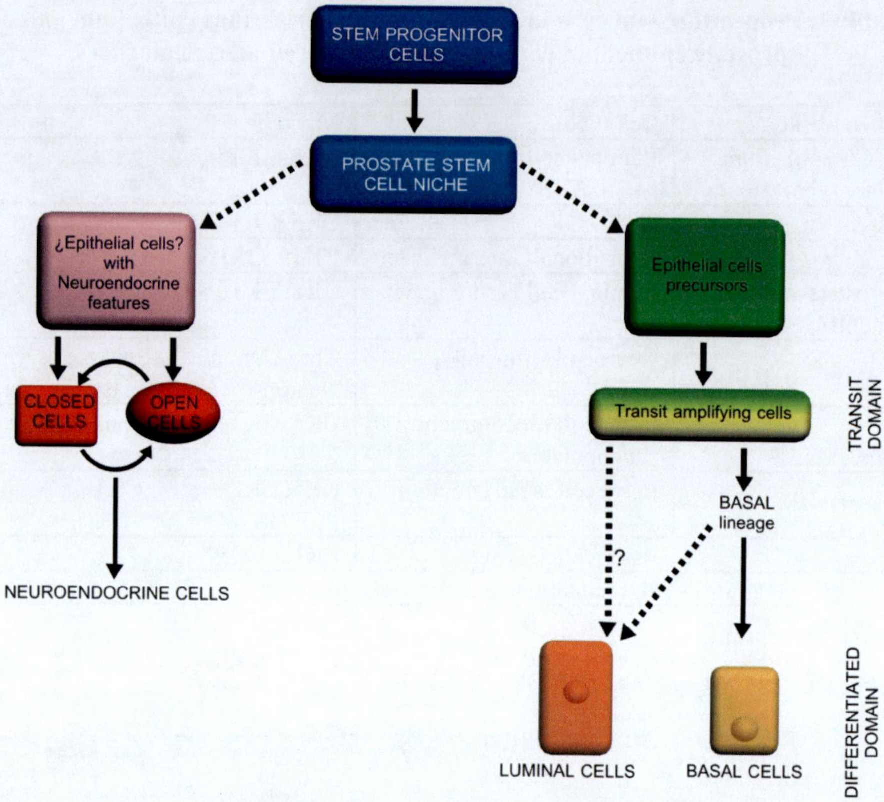

Figure 6. Developmental proposed model that implicates the different prostate gland epithelial cells genesis from stem/progenitor cells.

Study from murine prostate stem cells demonstrated self-renewal and multi-lineage differentiation from single adult prostate stem/progenitors cells in a specific *in vitro* microenvironment and in tissue fragment recombination assay [35].

Multi-lineage differentiation from a single prostatic cell or a clonally derived cell population has been recently demonstrated [36]. Adult mouse prostatic progenitor/stem cells exhibit self-renewal and multi-lineage differentiation. *In vivo* clonal populations of adult prostatic cells demonstrated tissue-specific multi-lineage differentiation exhibiting their capacity to generate organized prostate ductal structures with luminal and basal cell layers, when grafted into mice kidney in the presence of fetal urogenital mesenchyme. *In vitro*, these undifferentiated prostatic cells expressed markers associated with prostate stem cells (Table 2).

The differentiation program of the prostate stem cells remains unclear. Human prostate basal cells have been shown to differentiate into luminal cells *in vitro* culture systems [46, 47]. These data suggest that prostate stem cells may differentiate in a linear fashion in which stem cells generate basal cells, which in turn differentiate into luminal cells [48]. Alternatively, stem cells for each lineage could separately undergo further maturation. This theory is supported by the observation that human and rodent embryonic prostate epithelial progenitors cells express lineage markers of both basal and luminal cells [49, 50].

Table 2. Cells present in prostate with characteristics and associated.cell markers.

Cell type	Characteristics	Cell markers	Refs.
Luminal epithelial cells	Androgen-dependent, more differentiated, low proliferative capacity, high apoptotic index.	AR^+, PSA, CK8, CK18	36, 37, 38
Basal epithelial cells	Androgen-independent, undifferentiated profile, high proliferative capacity, anti-apoptotic protein.	AR^-, CK5, CK14, CK15, p63, Bcl-2	36, 39, 40, 41
Murine prostate	Stem cells	Sca-1, CK5, p63, CD4af	22, 23, 41, 42
Human prostate	Stem cells	CK5, CK14, CD133, CD44, CD4af	43, 44, 45
Prostate progenitor/ /stem cells			
Stem cell markers		$Sca\text{-}1^+$, $CD4af^+$, $CD34^-$	36
Basal cell markers		CK5, CK14, p63	36
Luminal cell markers		CK8, CK18, AR	36

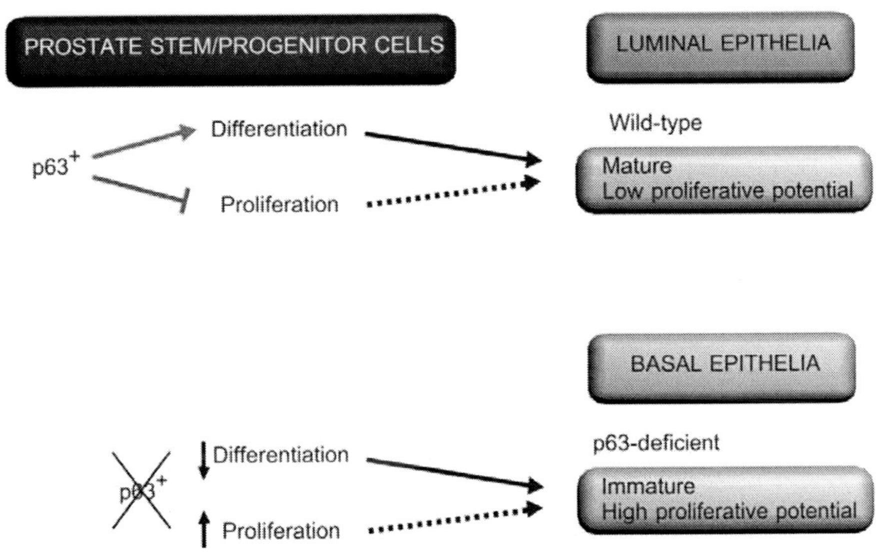

Figure 7. Regulation of prostate stem/progenitor cell function by p63. p63 coordinates signals controlling the correct differentiation of prostate stem and progenitor cells into mature prostate cells as well as the inhibition of stem and progenitor cell proliferation.

The ability of progenitor/stem cells to either proliferate or differentiate must be tightly controlled for proper functionality, and some signaling pathways are known to control specific sets of mammalian adult stem cells [51, 52]. The p53 homolog, p63, has a controversial role in prostatic progenitor stem cells (Figure 7). p63 is expressed in basal cells of the prostate [36-38] and has a pivotal role in prostate development.

Stem/progenitor cells

Early evidence or stem cells in the prostate came from a set of experiments involving androgen withdrawal from both human and rat prostate [17, 53, 54]. The existence of prostate stem cells is probably best illustrated by animal studies investigating the effect of androgen on the prostate. Castration leads to rapid involution of the gland, but once androgen levels are restored, the gland completely regenerates. As this cycle of involution-regeneration can be repeated many times, a population of long-lived, prostatic epithelial stem cells must exist [55].

Evidence for the stem cell hierarchy (Figure 8) in prostate epithelium came from the study of the expression patterns of various markers of differentiation [43, 56-58]. Normal prostate have a basal cell phenotype with expression of CD133, high levels of β1-integrin, and keratins K5 and K14. These cells give rise to transit-amplifying populations that still have high integrin expression but no longer express CD133. Basal cells express keratins K5 and K14 together and the anti-apoptotic protein, Bcl-2, whereas luminal cells express K8 with K18, CD57, AR (androgen receptor) and the secretory proteins, prostatic specific antigen (PSA), and prostatic acid phosphatase. Neuroendocrine cells arise from differentiation of normal basal cells and have a complex marker expression pattern of basal and luminal keratins, together with secretion of chromogranin A and serotonin. CD133 expression was restricted to a small population of cells within the basal layer, but was randomly located throughout acini [45]. High surface expression of integrin $\alpha_2\beta_1$ has been shown to be associated with stem cells in the prostate. In human prostate, stem cells can be distinguished from transit amplifying cells (daughter cells that have a more limited proliferative capacity), by two to three fold higher surface levels of integrin $\alpha_2\beta_1$ [43]. A subpopulation of $\alpha_2\beta_1$ basal cells expresses the antigen CD133 and this expression correlates with a high proliferative potential and ability to regenerate a fully differentiated prostatic epithelium. These cells (which make up 3% of the total basal population) are also randomly located throughout acini [43]. In contrast, others found that stem cells (label-retaining cells) in the rodent prostate are concentrated in the proximal region of the prostatic ducts [21, 59, 60].

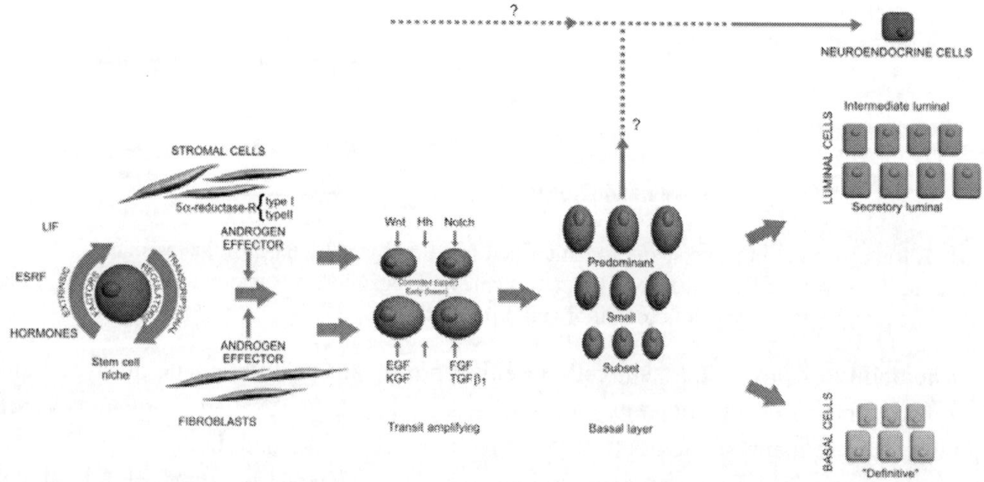

Figure 8. Prostate stem cell hierarchy.

In contrast, for other researchers stem/progenitor cells express a number of distinct markers. Thus, for stem cells the makers proposed are $CD133^+$, $CD44^+$, $K14^+$, $K5^+$, c-met^-, AR^-, and $K18^-$. For transit amplifying/progenitor cells is carried out a subdivision between early and late developing cells. Thus, the early progenitor cells display a marker profile with $CD133^-$, $CD44^+$, $K14^-$, $K5^+$, c-met^+, AR^- and $K18^-$, whereas the late progenitor cells display a marker profile with $CD133^-$, $CD44^-$, $K14^-$, $K5^-$, c-met^+, $AR^{+/-}$, and $K18^{+/-}$. In the case of differentiated prostatic cells, they have been classified in two groups attending to its origin as neuroendocrine or exocrine cells which display distinct marker profile. In this way, neuroendocrine cells express $CD133^-$, $K14^-$, $K5^+$, c-met^-, AR^-, $K18^-$ and NE^+ whereas exocrine cells show a marker profile with $CD133^-$, $K14^-$, $K5^+$, c-met^-, AR^+ and $K18^+$ [61].

Regulation of Prostate Epithelial Stem Cells

What coaxes stem cells to become prostate epithelium rather than bladder epithelium is one of the most fascinating questions in the urogenital stem cell field. Many of the answers still have been not obtained, but they have begun to emerge in the past years.

It has long been known that in embryonic prostate development as in postnatal time, this decision to coax cells toward the prostate fate involves critical cross-talk between epithelial and mesenchymal cells.

In embryonic development of prostate, Wnt, Hh, Notch and BMP signaling pathways have arose as two major players that are decisive for normal morphogenesis. In this chapter we propose carry out a review about the main aspects of knowledge so far gained in the study of these pathways in relation with normal prostate development.

The development of the prostate is induced by androgens that the fetal testicles begin to produce in humans around 9 weeks after of birth. Prostatic cells are sensitive to both testosterone and dihydrotestosterone (DHT), a derived compound of testosterone synthesized locally by the prostatic cells itself and that it is up to 10 times more potent than testosterone. The action of androgens on the prostate may regulate the expression of a number of genes involved in its development, among which are genes included within Wnt, Hh, Notch and BMP pathways. These signaling pathways are essential in many tissues for the maintenance of the stem cell populations as well as in their differentiation to several mature cell types. Thus, they also act in both messenchyme and epithelium of developing prostate exchanging signals for promoting and directing prostatic developing and achieve the morphological structure of mature prostate.

So far there are few evidences that connect the signaling pathways with the development of healthy prostate. However, in contrast there are a lot of studies focused on the alterations of expression that undergo multiple components of these signaling pathways in prostate cancer as well as their interactions. Thus, through these observations, it is easy understand that signaling pathways have important roles both in prostate development and in the maintenance of its normal physiology. In this chapter, we combine both approaches to try depicting the molecular mechanism involved in development of healthy prostate (for overview see table 3).

Table 3. Overview of expression data and interactions for distinct components of signaling pathways involved in prostate development.

Pathway	Component	Function	Sample	Expression	Refs.
Wnt	WNT11	Ligand	Prostate tumor cell lines	Androgen--dependent: high with androgens; inhibited without androgens	71
	SFRP2	Ligand inhibitor	Mouse prostate, early/advanced stages of development	Activation/ down-regulation in early/advanced stages respectively	63
	SFRP4	Ligand inhibitor	Rat prostate ablation and DHT-dependent re-growth	Repressed	64
BMP	BMP2	Ligand	Normal/tumor prostate tissues	Expressed in both tissues	73
	BMP2	Ligand	Benign/tumor tissues	Decreased in tumor tissue	78
	BMP3	Ligand	Normal/tumor prostate tissues	Expressed in both samples	73
	BMP4	Ligand	Normal prostate development	Expressed	75
	BMP4	Ligand	Normal/tumor prostate tissues	Expressed in both but predominantly expressed in normal prostate	73
	BMP4	Ligand	Prostate tumor cell lines	Induction under WNT3A, WNT5 administration or DKK1 deletion	72
	BMP5	Ligand	Normal/benign prostate tissues	Upregulated in benign tissue	74
	BMP6	Ligand	Prostate tumor cell lines	Induction under WNT3A, WNT5 administration or DKK1 deletion	72
	BMP6	Ligand	Normal/tumor prostate tissues	Expressed in both tissues	73
	BMP7	Ligand	Normal prostate development	Expressed	82
	BMP9	Ligand	Prostate tumor	Decreased or absent	79
	SMAD4	Intracellular effector	Benign/tumor tissues	Decreased in tumor tissue	78
	SMAD8	Intracellular effector	Normal/benign/ tumor tissues	Expressed in normal/benign; absent in tumor	78
Notch	NOTCH1	Receptor	Normal prostate development	Overexpressed in response to BMP7	82
	NOTCH1	Receptor	Rat prostate ablation and DHT-dependent re-growth	Repressed in rat ventral prostate	83

Table 3. (Continued)

Pathway	Component	Function	Sample	Expression	Refs.
	NOTCH1	Receptor	Human normal prostate cell lines: PrEC, BPH-1, PNT2, RWPE-1, PZHPV-1	Expressed	83
	NOTCH1	Receptor	Mouse prostate ablation and androgen-dependent re-growth	Induction after castration. Return to normal levels during re-growth	85
	NOTCH2	Receptor	Human normal prostate cell line: PNT2	Expressed	83
	NOTCH2	Receptor	Rat prostate development	Expressed in specific areas of stromal mesenchyme	88
	NOTCH3	Receptor	Human normal prostate cell line: PNT2	Not expressed	83
	NOTCH4	Receptor	Human normal prostate cell line: PNT2	Not expressed	83
	JAG1	Ligand	Human normal prostate cell lines: PrEC, PNT2	Expressed	83
	JAG1	Ligand	Rat prostate ablation and DHT-dependent re-growth	Repressed in rat ventral prostate	64
	JAG2	Ligand	Human normal prostate cell lines: PrEC, PNT2	Expressed	83
	DLL1	Ligand	Human normal prostate cell lines: PrEC, PNT2	Expressed	83
	DLL1	Ligand	Rat prostate development	Expressed in specific areas of stromal mesenchyme	88
	DLL4	Ligand	Human normal prostate cell line: PrEC	Not expressed	83
	SEL1L	Inhibitor	Rat prostate development in response to DHT	Induced in rat ventral prostate	64
	DTX (Deltex)	Intermediary in signaling	Human normal prostate cell line: PNT2	Not expressed	83
	HES1	Intermediary target	Human normal prostate cell line: PNT2	Expressed	83
	HES1	Intermediary target	Normal prostate development	Inhibited in presence of BMP7	82

Table 3. (Continued)

Pathway	Component	Function	Sample	Expression	Refs.
	HEY1	Intermediary target	Human normal prostate cell lines: PrEC, RWPE-1, PZHPV-1	Expressed	83
Hedgehog	SHH	Ligand	Mouse UGS	Expressed in epithelial cells	80
	SHH	Ligand	Developing mouse prostate	Expressed in epithelial cells in tips of buds	91
	SHH	Ligand	Human fetal prostates	Expressed in epithelial ducts	97
	IHH	Ligand	Mouse prostate development	Up-regulated in SHH nulls	95
	DHH	Ligand	Human fetal prostates	Expressed in epithelial ducts	97
	PTCH1	Receptor	Mouse developing prostate	Expressed in mesen-chyme around tip buds	91
	PTCH1	Receptor	Rat developing prostate	Expressed in mesenchyme stromal cells adjacent to epithelium of tip buds	92
	PTCH1	Receptor	Human fetal prostates	Expressed in epithelial ducts	97
	PTCH1	Receptor	Adult prostate	Not expressed in epithelial basal cells	98
	PTCH2	Receptor	Rat developing prostate	Expressed in mesenchyme stromal cells adjacent to epithelium of tip buds	92
	PTCH2	Receptor	Human fetal prostates	Expressed in epithelium and stroma	97
	SMO	Trans- -membrane effector	Human fetal prostates	Expressed in epithelial ducts	97
	GLI1	Intracellular effector	Mouse developing prostate	Mesenchyme around tip buds	91
	GLI1	Intracellular effector	Human fetal prostates	Expressed in epithelial ducts	97
	GLI1	Intracellular effector	Adult prostate	Not expressed in epithelial basal cells	98
	GLI2	Intracellular effector	Mouse developing prostate	Expressed in mesen-chyme around tip buds	91
	GLI2	Intracellular effector	Adult prostate	Not expressed in epithelial basal cells	98
	GLI3	Intracellular effector	Mouse developing prostate	Expressed throughout mesenchyme	91
	GLI3	Intracellular effector	Adult prostate	Not expressed in epithelial basal cells	98

Wnt signaling pathway

Regarding Wnt pathway, it has been demonstrated that this pathway regulates the morphogenesis of epithelial branching and controls the differentiation process of the luminal secretory cells in developing rat prostate cultures, through regulation of proliferative activity in prostate epithelial progenitor cells. The activity of Wnt pathway is measured through expression of *AXIN2* gene, a well established target of Wnt pathway. Thus, it was observed that Wnt pathway is active in developing prostate and is downregulated in adult individuals. Also, it was proved that upon castration by androgen deprivation in adult prostate the *AXIN2* levels are increased, and therefore the Wnt pathway activity is upregulated, whereas the androgen restoration results in a downregulation of *AXIN2* gene and in recovery of normal Wnt activity levels [62]. In addition, it was depicted how *SFRP2* (secreted frizzled related protein 2), a Wnt antagonist protein, was highly expressed in early stages and downregulated in advanced stages of prostate development in a murine model [63]. Additionally, it was depicted that *SFRP4* (secreted frizzled related protein 4), another Wnt antagonist, was repressed in rat prostates after castration and subsequent re-growth under androgen administration [64]. These results, in principle, may appear conflictive, but however simply may be the result of the complex interplay between the multiple pathways at multiple levels of regulation. However, SFRP4 must be a decisive player in prostate development modulating Wnt-mediated signaling because it is able to inhibit proliferation and reduce the metastatic potential of prostate cancer cells [65]. On the other hand, it was depicted that *AXIN2*, and therefore Wnt pathway, is upregulated in some human prostate cancer cell lines. Consequent with this observation, it was also observed that another Wnt inhibitor, Dickkopf-1 (DKK1), is able to inhibit the growth of prostate cancer cells in culture. Together, these data are indicative that Wnt pathway plays an important role in prostate development.

Indeed, in other paper it was recently depicted how in a murine model, Wnt signaling pathway is active in a fraction of epithelial progenitor cells what demonstrates that activation of Wnt pathway may lead to enhancing the growth of prostate confirming its role in maintenance and proliferation of the prostate progenitor cells. In this study, it was also depicted how *AXIN2* expression is upregulated by androgens what is indicative of the existence of a link between androgen stimulation and Wnt pathway activation in prostate [66]. This is not a novel finding because already in 2002, it was demonstrated the existence of a connection between β-catenin, the intracellular effector of Wnt pathway, and androgen pathway in prostate cancer cells. In this way, it was depicted that there is a strong and specific interaction between β-catenin and the intracellular AR and thus, through this interaction, β-catenin is able to increase the ligand-dependent activity of AR on target genes in prostate cancer cells [67]. Even though these observation were made in a tumor context, they provide relevant information for understand the role of androgens in normal prostatic development. In line with the observations previously discussed, it was also depicted that human *AR* gene is a target of Lef/Tcf [lymphoid enhancer-binding factor/transcription factor (T-cell specific, HMG-box)] cofactors, that together with β-catenin, act activating the transcription of target genes of Wnt pathway. Conversely, it has been demonstrated that, under certain conditions, glycogen synthase kinase-3 β (GSK3B), an intermediary enzyme in Wnt pathway that modulates β-catenin levels through its phosphorylation and subsequent tagging for degradation, is also able to phosphorylate AR and inhibit its ability to activate the transcription on target genes (reviewed in [68]). Taking together, these data suggest that there is a complex interaction between Wnt and androgen pathway and that this interaction can regulate the proliferative activity of prostate progenitor cells during prostate development under androgen stimulation.

In this way, in a recent paper was demonstrated that WNT3A (a Wnt family member) is able to recruit the AR to the promoters of *MYC* and cyclin D1 genes which are targets for Wnt pathway. Indeed, it was demonstrated that AR and β-catenin are recruited to the promoter of *PSA* gene, a known AR target [69]. Therefore, it is possible to venture that at least some components of Wnt pathway and AR can work as cofactors in the regulation of both AR and Wnt target genes and, thus, promote the growth of prostate cells both during normal development and tumorigenesis. Other evidence of the relevance of roles that Wnt pathway plays on the cellular proliferation was provided through the observations that conditional inactivation of *APC*, one of the proteins intermediaries into the Wnt signaling pathway, is able to induce the formation of prostate tumors in mouse. These tumors have levels abnormally highs of β-catenin (because without APC, the cells are unable to tag the β-catenin for its degradation) and this drives the excessive proliferation which is a hallmark for cancer [70]. Additionally, it was also provided information about the *WNT11* gene behavior in relation to androgens and androgen receptor in prostate cancer cells so that it is possible to draw a model in which *WNT11* expression is inhibited by androgens and is restored when the prostate tumor cells are treated with an inhibitor of androgen function. Additionally, it was demonstrated that the expression of *WNT11* is able to inhibit the transcriptional activities of androgens and therefore, the androgen-dependent cell growth. Furthermore, this action of WNT11 is accompanied by an inhibition of canonical Wnt pathway [71]. These data suggest a hypothetical role for WNT11 in normal development of prostate in relation with androgens. Besides in prostate cancer cell lines, they were recently obtained data about the connection between Wnt pathway and BMP pathway through which the administration of WNT3A and WNT5A, or the deletion of *DKK1* (an inhibitor of Wnt signal) induces the expression of *BMP4* and *BMP6* [72]. This is comprehensible as in front of proliferative activity of Wnt pathway, the cells search to oppose the anti-proliferative and pro-differentiation effects of BMP pathway.

BMP signaling pathway

In this way, BMP signal has a fundamental role on developing prostate. The first evidence for expression of *BMP*s in prostate was provided in 1994 when it was analyzed the expression of several BMP cytokines in distinct normal and tumor prostatic tissues from humans and rats and in prostate cancer cell lines. In this study, it was demonstrated that in normal human prostate are expressed *BMP2*, *BMP3*, *BMP4* and *BMP6*, but the *BMP4* mRNA is the expressed predominantly [73]. In addition, it was shown that *BMP5* is upregulated in benign prostate hyperplasia versus normal tissue which may be indicative of some role for this molecule in control of prostate development [74]. Later, it was possible demonstrate a functional connection between *BMP4* expression and normal prostate development as it was shown that *BMP4* expression in mouse prostate changes from a wide distribution throughout the mesenchyme in prostate pre-bud stage to an expression circumscribed to the area of mesenchyme immediately surrounding the locations for nascent buds and branches. Thus, addition of exogenous *BMP4* to urogenital sinus cultured *in vitro* triggers inhibition of both proliferation of epithelial cells in extending tips and budding formation. These data provide information about the role for *BMP4* expression in budding and branching morphogenesis and determination [75]. Additionally, several papers provide data about distinct alterations for *BMP6* expression in prostate cancer [76, 77] and, therefore, it is possible that *BMP6* have also a role in normal prostate development.

On the other hand, the expression of *BMP2* and *SMAD4* (which encode intermediaries in BMP signaling) is decreased in prostate cancer versus benign prostate tissue. Also, *SMAD8* (which encodes other intracellular intermediary for BMP pathway) is expressed in normal and benign prostatic tissue but is absent in prostate tumor tissue [78]. Therefore, these data suggest that *BMP*2, *SMAD8* and *SMAD4* expression is associated with normal proliferative behavior in prostate. In line with these observations, it has been demonstrated that the expression of *BMP9* is decreased or absent in prostate cancer and that its expression is able to trigger apoptosis and prevent growth, migration and invasion in prostate cancer cells [79]. In addition, it was shown how Noggin (a co-secreted inhibitor of BMP signaling) has a critical role during mouse prostate development through its opposition to BMP4-mediated anti-proliferative effect [80].

But the function of BMP pathway must be understood in a complex context of multiple interactions with other pathways. Thus, it has been proved that BMP pathway interacts with MAPK (mitogen-activated protein kinase) signaling to trigger distinct effects on normal and tumor prostate development. In this way, SMAD1, an intracellular effector in BMP pathway that is activated through BMP-dependent C-terminal phosphorylation, can be further phosphorylated in its linker region by MAPK pathway components. This phosphorylation leads to an inhibitory action for SMAD1 on androgen-activated AR. Thus, BMP pathway through SMAD1 interaction with MAPK and androgen pathways may arrest the proliferation androgen-dependent in prostate cells [81].

In another interesting paper it was demonstrated that, in addition to *BMP4*, *BMP7* expression is also able to inhibit branching morphogenesis in the epithelium of mouse developing prostate. Also, the authors demonstrated how the *BMP7* levels change dynamically throughout the distinct phases of prostate development, changes probably controlled by a number of signals such as Wnt, Hh or androgens what provides more evidences for a complex interplay between distinct pathways. Additionally, these changes reflect the role of BMP7 for determination of sites in prebud epithelium which later undergo ductal and branching morphogenesis. Furthermore, the expression of *HES1* (a target for NOTCH1 receptor and, therefore, a reporter for NOTCH1 activity) was analyzed in relation with *BMP7* expression in prostate development. In this way, *BMP7* overexpression is able to inhibit the expression of *HES1* gene and NOTCH1 signal transduction. Conversely, *BMP7* deletion in mice promotes branching morphogenesis in prostate associated with an increase of areas with NOTCH1 activity in urogenital epithelium. In consonance with the previous data, it was also demonstrated that during embryogenesis and postnatal development of prostate it is possible detect high NOTCH1 activity in prebud epithelium as well as in the tips of initial buds and extending branches [82].

Taken together, these findings support a model thereby NOTCH1 activity is associated with epithelial cell proliferation in tips of forming prostate buds and branches. Likewise,

BMP7 signaling from mesenchyme inhibits the formation of new buds and branches through control of number of areas with high NOTCH1 activity.

Notch signaling pathway

So, it is clear that Notch pathway plays a major role in prostate development. In this way, it has been obtained data for expression of distinct components of Notch pathway in several normal prostate cell lines and in normal/benign prostatic tissues. Thus, in cell lines, it was reported expression for genes *JAG1*, *JAG2*, and *DLL*, which encode Notch ligands; for the genes *NOTCH1* and *NOTCH2*, which encode receptors; and for the Notch targets, *HES1* and *HEY1*. But, it was not possible detect expression for the genes of the ligand *DLL4*, of the receptors *NOTCH3* and *NOTCH4*, and of *DTX*, an ubiquitin ligase involved in non-canonical Notch signaling pathway. Additionally, *NOTCH1* expression was also detected by imnunohistochemistry only in basal cells from human adult benign prostate tissue. However, the Notch target gene Hey1 was detected only in secretory luminal cells. On other hand, *NOTCH3* expression was undetectable in normal human prostate tissue (for an extensive review, see [83]).

Other studies have been focused on assigning a concrete functional role for Notch pathway components in prostate development. In this way, they have been carried out a number of assays in which Notch pathway is inhibited by addition of γ-secretase (protease necessary for triggering Notch signaling) inhibitors to cultures of developing prostates. The blocking of Notch signaling enhances the proliferation in epithelial cells and causes its maintenance in a state undifferentiated in which there isn't separation between luminal and basal cells, besides reducing branching morphogenesis. Additionally, it was also studied the effect of *NOTCH1* conditional deletion on prostate development in mouse, obtaining similar results that in cultured model with an increase in proliferation of undifferentiated epithelial cells, triggering branch morphogenesis, typical of a hyperplastic phenotype [84]. In previous studies, it was already demonstrated, causing selective apoptosis of *NOTCH1*-expressing cells, that *NOTCH1* and the cells where is expressed are essential for correct development of mouse prostate and for its re-growth after resection and hormonal treatment with androgens. Thus, it was shown that *NOTCH1*-expressing cells correspond to progenitor cells and that these cells are located among basal epithelial cells [85, 86].

In other paper, it was depicted that *HEY1* and *HEY2* proteins, whose genes are targets for activation by Notch signaling, are able to inhibit the transcription of androgen target genes and that, therefore, are inhibitors for AR-mediated transcriptional activity [87]. Curiously, in other work is reported how *NOTCH1* and its ligand *JAG1* are repressed in ventral area of rat prostates after DHT administration and previous orchiectomy. Additionally, *SEL1L*, an inhibitor of Notch signaling, was induced [64]. These data are conflictive with others where microarray experiments in mice prostates showed that castration leads to an increased level of *NOTCH1* expression, and androgen replacement reversed this change to normal levels [85]. Recently, they were provided further data about Notch signaling during rat prostate development through analysis of expression for *NOTCH2* and for the ligand *DLL1* in normal prostate stromal layer. Thus, it was shown that the expression of *NOTCH2* and *DLL1* is restricted to specific areas of stroma during prostate development. Additionally, it also was demonstrated that Notch signaling inhibition gives rise to a loss of stromal tissue in cultures of both rat prostate and rat female ventral mesenchymal pad and that this effect only occurs in presence of testosterone [88].

Hedgehog signaling pathway

Hedgehog signaling is another of pathways decisive in prostate development (for review see [89]). Of the ligands participants in this pathway, Sonic hedgehog (SHH) is the most widely expressed during development of several tissues and in the prostate morphogenesis it was the most extensively studied. SHH was first depicted as expressed in the mouse UGS epithelium during the formation of initial prostatic ducts and its antibody-mediated blockade interferes with normal ductal morphogenesis [90] what represented the starting point for further and most deep investigations.

Later, it was determined that ductal formation in mouse prostate depends of a molecular conversation between urogenital epithelium and urogenital mesenchyme which exchange signals from Hh pathway and its components. In this way, it was shown that the expression of SHH is focused on the epithelial cells located in the tips of buds and growing branches and that this expression occurs in response to testosterone stimulation. Conversely, the genes encoding for the SHH receptor *PTCH1* (Patched1) and the hedgehog intracellular effectors, *GLI1* (glioma-associated oncogene homolog 1) and *GLI2* (GLI-Kruppel family member GLI2), are expressed in the mesenchyme located immediately around the prostate buds while the remaining effector *GLI3* (GLI-Kruppel family member GLI3) is expressed diffusely throughout all mesenchyme. The expression of *GLI1* and *PTCH1* in mesenchyme, but not the expression of *GLI2* and *GLI3*, depends on SHH stimulation what is indicative for the existence of a Hedgehog paracrine signaling loop between epithelium and mesenchyme. Additionally, the importance of Hedgehog pathway in prostate morphogenesis was highlighted by the fact that chemical inhibition of SHH signaling leads to arrest of duct formation [91].

In line with the model for Hedgehog paracrine signaling, it was depicted that the Hh receptors *PTCH1* and *PTCH2* (Patched2) are expressed in the stromal cells located adjacent to the epithelium where is produced SHH. However, in controversial with previous findings, it was also shown that SHH is able to inhibit the cell proliferation while promotes the differentiation of luminal cells in rat prostate cultures. Additionally, it was demonstrated that SHH inhibits prostatic branching through its action on stromal mesenchymal cells which, in response to SHH signaling, produce anti-proliferative and pro-differentiation cytokines of TGF-β (Transforming Growth Factor-beta) superfamily [92]. Besides in controversial with previous data, it was published that Hh signaling inhibition (through construction of mouse *SHH* null mutant or using the chemical inhibitor cyclopamine with rat explants of urogenital sinus) do not affects to prostatic budding induction in response to testosterone [93] but instead, in more advanced stages of development, produces increase in ductal tip number, reduced proliferation and reduced size of prostates. In line with this, in other work it was also shown that Hedgehog signaling inhibition by cyclopamine is unable to block mouse prostate formation in culture although causes morphological abnormalities what means that Hedgehog pathway has some role in ductal generation [94]. These findings mean that SHH signaling is not essential for initial prostatic induction although is important in posterior events during prostate development (prostate growth, proliferation and branching morphogenesis).

But all this lot of contradictory and varied data obtained from distinct sources and procedures have been called into doubt after the findings by which was proved the existence of functional redundancy between Hh ligands (concretely when *SHH* gene is lost, *IHH*, Indian hedgehog, the gene for other ligand in Hh pathway, is upregulated) and intracellular effectors so that each one is able to substitute functionally other(s) that may be altered either naturally or artificially during the studies of pathway. On the other hand, it was demonstrated that the Hh intracellular effector GLI2 is fundamental for proper prostate development while GLI1 or GLI3 effectors does not are necessary [95]. This finding was further supported by the fact that ectopic expression of *GLI2* in normal prostatic epithelial cells leads to acceleration of cell cycle and increase of proliferation what is a hallmark for develop of cancer [96].

In other study, expression of several components of Hh pathway was analyzed in specimens of human fetal prostates and it was obtained that *SHH*, *DHH* (desert hedgehog homolog (Drosophila): other ligand in Hh pathway), *SMO* (Smoothened homolog (Drosophila): transmembrane effector in Hh pathway), *PTCH1* (receptor in Hh pathway), and *GLI1* (intracellular effector of Hh pathway) were expressed in developing prostatic epithelial ducts while *PTCH2* (receptor in Hh pathway) was located both in epithelium and stroma [97]. Additionally, expression of *PTCH1*, *GLI1*, *GLI2* and *GLI3* do not was located in adult normal epithelial basal cells but it was instead detected in hyperplasia of basal cells co-localized with the p63 marker for stemness and proliferation [98]. This is indicative of an important and specific role for Hh signaling in early prostate development, role that is restricted in adult prostate and whose alteration may lead to cancer.

Conclusion

Prostate cancer is the neoplastic condition which has the highest number of new cases diagnosed in men in developed countries, and the second in the whole world [99] and, despite the fact that its associated mortality has been reduced in developed countries, it is still a major health problem.

In recent years, they are beginning to develop and implement a number of new biological therapies in the treatment of cancer. Some of these therapies are based on the use of monoclonal antibodies for the blocking of certain signaling pathways which are altered in the tumor cell so that they are used to circumvent proliferative controls to which they must be subjected. Nowadays, gene therapy is being used to try to repair the genetic alterations which are present in tumors and that give rise to cancer. In this regard, there are clinical trials underway using RNAi.

However, the next step is to try to achieve a cure for cancer through the eradication of cancer stem cells. In recent years, an idea gaining interest is that tumors are not homogeneous masses of cells but, as in normal tissue, are composed of a number of different cell types. Most of the tumor cells are cells that can be considered differentiated, with a limited or without any capacity to divide. However, there are other types of cells with characteristics similar to normal stem cells, called cancer stem cells or tumor-initiating cells, that have virtually an unlimited capacity for division and proliferation, and that are thought to be responsible for generation of remaining tumor cells and, therefore, they are also responsible for tumor growth, aggressiveness and ability to migrate and cause distant metastases. These cells, like normal stem cells, have a cell cycle with slow progression which makes them invulnerable to conventional chemotherapy that targets the cells with intense mitotic activity. Thus, after chemotherapy only more differentiated tumor cells which are the majority are eliminated, so that there is a partial or total response of tumor, but the risk of re-growth is not avoided, as the cancer stem cells were not removed. Thus, in order to cure cancer, hypothetically it would be necessary to eliminate the cancer stem cells. However, our knowledge about these cells is still far from the minimum necessary to design an effective therapy against them. Because of the similarity between the cancer stem cells and normal stem cells, an avenue to reach a better understanding of the former is to first know the latter. Due to this fact, in recent years a lot of articles have been published about stem cells, the special features that differentiate them from other cells, as well as the molecular basis of these peculiarities. In this way, the study of signaling pathways has undergone significant progress in relation to stem cells. In the prostate, the stem cells appear to be located among the epithelial cells of the basal layer. From these cells are generated other basal cells and luminal epithelial cells which are responsible for the secretory function of the prostate. However, isolating these cells is still far from being a reality as the various researchers do not agree on the markers that define these stem cells and which would serve for sorting them. This is complicated because in the prostate, as in other organs, the classification between differentiated cells and stem cells is not so clear, but there are a number of cells in intermediate stages which without being stem cells, are neither fully differentiated.

It is increasingly clear that in maintaining the characteristics of stem cells a number of intercellular signals that are the mechanism through which cells communicate are involved in a decisive way. It is thought that stem cells are organized in a so-called niche, a set of cells where the stem cells are surrounded by accessory cells which send the signals necessary to maintain the stem cells in an undifferentiated state.

Into the niche, stem cells are in a state virtually quiescent, in which they are not divided. Through transitory stimuli, the stem cells are forced to be divided again to quickly return to the state of quiescence. When a stem cell divides, it may lead to two types of cells: a stem cell similar to the original, which will remain undifferentiated and contribute to the maintenance of the number of stem cells necessary to ensure the survival of tissue and organ; or a progenitor cell which is expulsed out of the niche and that, although originally resembling a stem cell, has a limited number of divisions that can be carried out. When this cell progresses in the number of divisions will gradually acquire features of a mature cell until finally replacing the damaged or dead cells by fully functional cells. It appears that both in maintaining the undifferentiated state of stem cells and in the decision of division, as well as in the progression of progenitor cells toward differentiation, there are involved a large number of cellular pathways including the Wnt, Notch, Hh and BMP signaling pathways. In general, BMP signal is anti-proliferative so that it is specialized in the maintenance of the stem cells in their niche in a quiescent state. This same signal is pro-differentiation so that is what forces progenitor cells once expulsed from the niche to generate mature cells for tissue maintenance. On the other hand, the Wnt and Hh pathways provide proliferative signals which may act in conjunction with others to temporarily force the division of stem cells within the niche to lead to either new stem cells or new tissue mature after cell differentiation. The Notch signaling is specialized in defining the fate of each progenitor cell to give rise to distinct cell types that form the mature tissue through cell-cell interactions. In the prostate, it appears that adult stem cells are located among the cells of basal layer of no secretory epithelial cells. During development, the combination of multiple signals, which act on the urogenital sinus both from surrounding mesenchyme and from within the epithelium, lead to the generation of the primary branches and buds and, then, to the secondary and tertiary branches. Also, these signals are responsible for making inside the epithelial cordons, initially solid, the formation of ducts through which the secretions will be evacuated, and that are covered with secretory epithelial cells fully differentiated and functional. It appears that during the developing prostate, the predominant signal from the mesenchyme is an anti-proliferative signal mediated by, at least, the BMP cytokines, BMP4 and BMP7. These BMP act on the epithelium for assuring that it not be expanded randomly but form branched structures which constitute the normal morphology of the prostate. Apparently, Hh signaling (both autocrine, from the same cell, and paracrine, from the surrounding mesenchymal cells) on epithelial cells located in the tips of prostate branches is associated with the production of BMP cytokines from the mesenchyme. This BMP signal assures that, at a given moment, the branch be split into two new branches. Similarly, Notch signaling is located in the areas of epithelium which later will undergo budding and will lead to branches. The number of branches appears to be together controlled by Notch and BMP signaling, as the *BMP7* deletion led to an increase in branching morphogenesis which previously displayed high NOTCH1 activity. Furthermore, in combination with the signaling pathways discussed previously, work other pathways with important roles, such as the FGF or MAPK pathways. In concrete, MAPK interferes with the BMP-mediated signal through phosphorylation of the intermediary SMAD1 and inhibition of androgen receptor-mediated activity. Thus, in view of a proliferative signal as those from MAPK pathway, the cell is opposed with an anti-proliferative signal.

Also not insignificant is the role of the androgen pathways, which during the embryonic development of males are produced by the testes at a given moment forcing the urogenital sinus to grow and differentiate into the prostate. Also, during adulthood androgens have an important role in maintaining the structure and function of the prostate. In fact, the prostate is able to regenerate itself after a surgical procedure of resection through the actions of androgens. However, in mouse models subjected to orchiectomy, and therefore without capacity for generating androgens, if the prostate is resectioned, re-growth is not possible. Currently, research is focused on the molecular anomalies that occur in the adult prostate during development and progression of neoplastic disease. However, knowledge about the molecular mechanisms that act on the prostate is still very poor.

Undoubtedly, in the future we should focus our efforts on the study of molecular, genetic and signaling processes that act during prostate development, since such knowledge is the door to unraveling the key to the aberrant behavior of stem cells in prostate tumor development and to learn how to repair it, so it be possible to achieve a hypothetical cure for prostate cancer.

References

[1] Tokar, E. J., Ancrile, B. B., Cuhna, G. R. & Webber, M. M. (2005). Stem/progenitor and intermediate cell types and the origin of human prostate cancer. *Differentiation*, Dec 73, 463- 473.

[2] Blau, H. M., Brazelton, T. R. & Weiman, J. M. (2001). The evolving concept of a stem cell: entity or function? *Cell*, Jun 105, 829-841.

[3] Ivanova, N. B., Dimos, J. T., Schaniel, C., Hackney, J. A., Moore, K. A. & Lemischka, I. R. (2002). A stem cell molecular signature. *Science*, Oct 298, 601-604.

[4] Ramalho-Santos, M., Yoon, S., Matsuzaki, Y., Mulligan, R. C. & Melton, D. A. (2002). "Stemness": Transcriptional profiling of embryonic and adult stem cells. *Science*, Oct 298, 597-600.

[5] Tanaka, T. S., Kunath, T., Kimber, W. L., Jaradat, S. A., Stagg, C. A., Usuda, M., Yolota, T., Niwa, H., Rossant, J. & Ko, M. S. (2002). Gene expression profiling of embryo-derived stem cells reveals candidate genes associated with pluripotency and lineage specificity. *Genome Res*, Dec 12, 1921-1928.

[6] Burdon, T., Chambers, I., Stracey, C., Niwa, H. & Smith, A. (1999). Signalling mechanisms regulating self-renewal and differentiation of pluripotent embryonic stem cells. *Cells Tiss Organs*, 165, 131-143.

[7] Savatier, P., Lapillonne, H., Jirmanova, L., Vitelli, L. & Samarut, J. (2002). Analysis of the cell cycle in mouse embryonic stem cells. *Methods Mol Biol*, 185, 27-33.

[8] Liu, Y. & Rao, M. S. (2003). Transdifferentiation: Fact or artefact. *J Cell Biochem*, Jan 88, 29-40.

[9] Schofield, R. (1978). The relationship between the spleen colony-forming cell and the hematopoietic stem cell. *Blood Cells*, 4, 7-25.

[10] Fuchs, E., Tumbar, T. & Guasch, G. (2004). Socializing with the neighbors: stem cells and their niche. *Cell*, Mar 116, 769-778.

[11] Conti, L., Pollard, S. M., Gorba, T., Reitano, E., Toselli, M., Biella, G., Sun, Y., Sanzone, S., Ying, Q. L., Cattaneo, E. & Smith, A. (2005). Niche-independent symmetrical self-renewal of a mammalian tissue stem cells. *Plos Biol*, Sep 3, e283.

[12] Tulina, N. & Matrevis, E. (2001). Control of stem cell self-renewal in Drosophila spermatogenesis by JAK-STAT signalling. *Science*, Dec 294, 2546-2549.

[13] Kiger, A. A., White-Cooper, H. & Fuller, M. T. (2000). Somatic support cells restrict germ line stem cell self-renewal and promote differentiation. *Nature*, Oct 407, 750-754.

[14] Ihle, J. N., Stravapodis, D., Parganas, E., Thierfelder, W., Feng, J., Wang, D. & Teglund, S. (1998). The roles of JAKs and STATs in cytokine signalling. *Cancer J Sci Am*, May 4(suppl), 84-91.

[15] Sprandling, A., Drummond-Barbosa, D., Kai, T. Stem cells find their niche. *Nature*, 2001 Nov 14, 98-104.

[16] English, H. F., Santen, R. J. & Isaacs, J. T. (1987). Response of glandular versus basal rat ventral prostatic epithelial cell to androgen withdrawal and replacement. *Prostate*, 11, 229-242.

[17] Bonkhoff, H. & Remberger, K. (1996). Differentiation pathways and histogenetic aspects of normal and abnormal prostatic growth: a stem cell model. *Prostate*, Feb 28, 98-106.

[18] Sugimura, Y., Cuhna, G. R. & Donjacour, A. A. (1986). Morphological study of castration-induced degeneration and strogen-induced regeneration in the mouse prostate. *Biol Reprod*, Jun 34, 973-983.

[19] Isaacs, J. T., Schulze, H. & Coffey, D. S. (1987). Development of androgen resistance in prostatic cancer. *Prog Clin Biol Res*, 243, 21-23.

[20] Goto, K., Salm, S. N., Coetzee, S., Xiong, X., Burger, P. E., Shapiro, E., Lepor, H., Moscatelli, D. & Wilson, E. L. (2006). Proximal prostatic stem cells are programmed to regenerate a proximal-distal ductal axis. *Stem Cells*, Aug 24, 1859-1868.

[21] Tsujimura, A., Koikawa, Y., Salm, S., Takao, T., Coetzee, S., Moscatelli, D., Shapiro, E., Lepor, H., Sun, T. T. & Wilson, E. L. (2002). Proximal location of mouse prostate epithelial stem cells: a model of prostatic homeostasis. *J Cell Biol*, Jun 157, 1257-1265.

[22] Burger, P. E., Xiong, X., Coetzee, S., Salm, S. N., Moscatelli, D., Goto, K. & Wilson, E. L. (2005). Sca-1 expression identifies stem cells in the proximal region of prostatic ducts with high capacity to reconstitute prostatic tissue. *Proc Natl Acad Sci USA*, May 102, 7180-7185.

[23] Xin, L., Lawson, D. A. & Witte, O. N. (2005). The Sca-1 cell surface marker enriches for a prostate-regenerating cell subpopulation than can initiate prostate tumorigenesis. *Proc Natl Acad Sci USA*, May 102, 6942-6947.

[24] Salm, S. N., Burger, P. E., Coetzee, S., Goto, K., Moscatelli, D. & Wilson, E. L. (2005). TGF-β maintains dormancy of prostatic stem cells in the proximal regions of ducts. *J Cell Biol*, Jul 170, 81-90.

[25] Rouleau, M., Leger, J. & Tenniswood, M. (1990). Ductal heterogeneity of cytokeratins, gene expression, and cell death in the rat ventral prostate. *Mol Endocriol*, Dec 4, 2003-2013.

[26] Isaacs, J. T. (1987). Control of cell proliferation and death in the normal and neoplastic prostate. A stem cell model. In: C. H. Rodgers, D. S. Coffey, G. Cuhna, et al. eds. *Benign Prostatic Hyperplasia*. Bethesda, MD. Report nº NIH 87-2881. Department of Health and Human Services. National Institutes of Health., 85-94.

[27] Xin, L., Lukacs, R. U., Lawson, D. A., Cheng, D. & Witte, O. N. (2007). Self-renewal and multilineage differentiation in vitro from murine prostate stem cells. *Stem Cells*, Nov 25, 2760- 2769.

[28] Abate-Shen, C. & Shen, M. M. (2000). Molecular genetics of prostate cancer. *Genes Dev*, Oct 14, 2410-2434.

[29] Wang, Y., Hayward, S. W., Cao, M., Thayer, K. A. & Cuhna, G. R. (2001). Cell differentiation lineage in the prostate. *Differentiation*, Oct 68, 270-279.

[30] Signoretti, S. & Loda, M. (2006). Defining cell lineages in the prostate epithelium. *Cell Cycle*, Jan 5, 138-141.

[31] di Sant'Agnese, P. A., de Mesy Jensen, K. L., Churukian, C. J. & Agarwal, M. M. (1985). Human prostatic endocrine-paracrine (APUD) cells. Distributional analysis with a comparison of serotonin and neuron-specific enolase immunoreactivity and silver stains. *Arch Pathol Lab Med*, Jul 109, 607-612.

[32] Abrahamsson, P. A. (1999). Neuroendocrine cells in tumour growth of the prostate. *Endocr Relat Cancer*, Dec 6, 503-519.

[33] Vashchenko, N. & Abrahamsson, P. A. (2005). Neuroendocrine differentiation in prostate cancer: implications for new treatment modalities. *Eur Urol*, Feb 47, 147-155.

[34] Hudson, D. L. (2004). Epithelial stem cells in human prostate growth and disease. *Prostate Cancer Prostatic Dis*, 7, 188-194.

[35] Cuhna, G. R. & Lung, B. (1978). The possible influence of temporal factors in androgenic responsiveness of urogenital tissue recombinant from wild-type and androgen-insensitive (Tfm) mice. *J Exp Sool*, Aug 205, 181-193.

[36] Barclay, W. W., Axanova, L. S., Chen, W., Romero, L., Maund, S. L., Soker, S., Lees, C. J. & Cramer, S. D. (2008). Characterization of adult prostatic progenitor/stem cells exhibiting self-renewal and multilineage differentiation. *Stem Cells*, Mar 26, 600-610.

[37] Aumüller, G. (1983). Morphologic and endocrine aspects of prostatic function. *Prostate*, 4, 195-214.

[38] Liu, A. Y., True, L. D., LaTray, L., Nelson, P. S., Ellis, W. J., Vessella, R. L., Lange, P. H., Hood, L. & van den Engh, G. (1997). Cell-cell interaction in prostate gene regulation and cytodifferentiation. *Proc Natl Acad Sci U S A*, Sep 94,10705-10710.

[39] Magi-Galluzzi, C. & Loda, M. (1996). Molecular events in the early phases of prostate carcinogenesis. *Eur Urol*, 30, 167-176.

[40] Signoretti, S., Pires, M. M., Lindauer, M., Horner, J. W., Grisanzio, C., Dhar, S., Majumder, P., McKeon, F., Kantoff, P. W., Sellers, W. R. & Loda, M. (2005). p63 regulates commitment to the prostate cell lineage. *Proc Natl Acad Sci U S A*, Aug 102, 11355-11360.

[41] Wang, S., Garcia, A. J., Wu, M., Lawson, D. A., Witte, O. N. & Wu, H. (2006). PTEN deletion leads to the expansion of a prostatic stem/progenitor cell subpopulation and tumor initiation. *Proc Natl Acad Sci U S A*, Jan 103, 1480-1485.

[42] Lawson, D. A., Xin, L., Lukacs, R. U., Cheng, D. & Witte, O. N. (2007). Isolation and functional characterization of murine prostate stem cells. *Proc Natl Acad Sci U S A*, Jan 104, 181-186.

[43] Collins, A. T., Habib, F. K., Maitland, N. J. & Neal, D. E. (2001). Identification and isolation of human prostate epithelial stem cells based on alpha(2)beta(1)-integrin expression. *J Cell Sci*, Nov 114, 3865-3872.

[44] Miki, J., Furusato, B., Li, H., Gu, Y., Takahashi, H., Egawa, S., Sesterhenn, I. A., McLeod, D. G., Srivastava, S. & Rhim, J. S. (2007). Identification of putative stem cell markers, CD133 and CXCR4, in hTERT-immortalized primary nonmalignant and malignant tumor-derived human prostate epithelial cell lines and in prostate cancer specimens. *Cancer Res*, Apr 67, 3153-3161.

[45] Richardson, G. D., Robson, C. N., Lang, S. H., Neal, D. E., Maitland, N. J. & Collins, A. T. (2004). CD133, a novel marker for human prostatic epithelial stem cells. *J Cell Sci*, Jul 117, 3539-3545.

[46] Tran, C. P., Lin, C., Yamashiro, J. & Reiter, R. E. (2002). Prostate stem cell antigen is a marker of late intermediate prostate epithelial cells. *Mol Cancer Res*, Dec 1, 113-121.

[47] Isaacs, J. T. & Isaacs, W. B. (2004). Androgen receptor outwits prostate cancer drugs. *Nat Med*, Jan 10, 26-27.

[48] Litvinov, I. V., Vander Griend, D. J., Xu, Y., Anthony, L., Dalrymple, S. L. & Isaacs, J. T. (2006). Low-calcium serum-free defined medium selects for growth of normal prostatic epithelial stem cells. *Cancer Res*, Sep 66, 8598-8607.

[49] Wang, Y., Hayward, S., Cao, M., Thayer, K. & Cunha, G. (2001). Cell differentiation lineage in the prostate. Differentiation, Oct 68, 270-279.

[50] Xue, Y., Smedts, F., Debruyne, F. M., de la Rosette, J. J. & Schalken, J. A. (1998). Identification of intermediate cells types by keratin expression in the developing human prostate. *Prostate*, Mar 34, 292-301.

[51] Pirrotta, V. & van Lohuizen, M. (2006). Differentiation and gene regulation. Genomic programs and differentiation: editorial overview. *Curr Opin Genet Dev*, Oct 16, 443-446.

[52] Nystul, T. G. & Spradling, A. C. (2006). Breaking out of the mold: diversity within adult stem cells and their niches. *Curr Opin Genet Dev*, Oct 16, 463-468.

[53] English, H. F., Santen, R. J. & Isaacs, J. T. (1987). Response of glandular versus basal rat ventral prostatic epithelial cell to androgen withdrawal and replacement. *Prostate*, 11, 229-242.

[54] Sugimura, Y., Cuhna, G. R. & Donjancour, A. A. (1986). Morphological and histological study of castration-induced degeneration and androgen-induced generation in the mouse prostate. *Biol Reprod*, Jun 34, 973-983.

[55] Isaacs, J. T., Sxchulze, H. & Coffey, D. S. (1987). Development of androgen resistance in prostatic cancer. *Prog Clin Biol Res*, 243, 21-23.

[56] Alam, T. N., O'Hare, M. J., Laczkó, I., Freeman, A., Al-Beidh, F., Masters, J. R. & Hudson, D. L. (2004). Differential expression of CD44 during human prostate epithelial cell differentiation. *J Histochem Cytochem*, Aug 52, 1083-1090.

[57] Verhagen, A. P., Aalders, T. W., Ramaekers, F. C., Debruyne, F. M. & Schalken, J. A. (1988). Differential expression of keratins in the basal and luminal compartments of rat prostatic epithelium during degeneration and regeneration. *Prostate*, 13, 25-38.

[58] Hudson, D. L., Guy, A. T., Fry, P., O'Hare, M. J., Watt, F. M. & Masters, J. R. (2001). Epithelial cell differentiation pathways in the human prostate: identification of intermediate phenotypes by keratin expression. *J Histochem Cytochem*, Feb 49, 271-278.

[59] Bhatt, R. I., Brown, M. D., Hart, C. A., Gilmore, P., Ramani, V. A., George, N. J. & Clarke, N. W. (2003). Novel method for the isolation and characterization of the putative prostatic stem cell. *Cytometry A*, Aug 54, 89-99.

[60] Hudson, D. L., O'Hare, M., Watt, F. M. & Masters, J. R. (2000). Proliferative heterogeneity in the human prostate: evidence for epithelial stem cells. *Lab Invest*, Aug 80, 1243-1250.

[61] Algaba, F., Bostwick, D., Isaacs, J., Shiraishi, T., Umbas, R. & Watson, B. (2006). *New developments in the pathobiology of prostate disease*. In: J. McConnell, L. Denis, H. Akaza, S. Khoury, J. Schalken (Eds). Prostate cancer, from 6th International Consultation on New Developments in Prostate Cancer and Prostate Disease.

[62] Wang, B. E., Wang, X. D., Ernst, J. A., Polakis, P. & Gao, W. Q. (2008). Regulation of epithelial branching morphogenesis and cancer cell growth of the prostate by Wnt signaling. *PLoS ONE*, May 3, e2186.

[63] Zhang, T. J., Hoffman, B. G., Ruiz de Algara, T. & Helgason, C. D. (2006). SAGE reveals expression of Wnt signalling pathway members during mouse prostate development. *Gene Expr Patterns*, Mar 6, 310-324.

[64] Nantermet, P. V., Xu, J., Yu, Y., Hodor, P., Holder, D., Adamski, S., Gentile, M. A., Kimmel, D. B., Harada, S., Gerhold, D., Freedman, L. P. & Ray, W. J. (2004). Identification of genetic pathways activated by the androgen receptor during the induction of proliferation in the ventral prostate gland. *J Biol Chem*, Jan 279, 1310-1322.

[65] Horvath, L. G., Lelliott, J. E., Kench, J. G., Lee, C. S., Williams, E. D., Saunders, D. N., Grygiel, J. J., Sutherland, R. L. & Henshall, S. M. (2007). Secreted frizzled-related protein 4 inhibits proliferation and metastatic potential in prostate cancer. *Prostate*, Jul 67, 1081-1090.

[66] Ontiveros, C. S., Salm, S. N. & Wilson, E. L. (2008). Axin2 expression identifies progenitor cells in the murine prostate. *Prostate*, Sep 68, 1263-1272.

[67] Yang, F., Li, X., Sharma, M., Sasaki, C. Y., Longo, D. L., Lim, B. & Sun, Z. (2002). Linking beta-catenin to androgen-signaling pathway. *J Biol Chem*, Mar 277, 11336-11344.

[68] Terry, S., Yang, X., Chen, M. W., Vacherot, F. & Buttyan, R. (2006). Multifaceted interaction between the androgen and Wnt signaling pathways and the implication for prostate cancer. *J Cell Biochem*, Oct 99, 402-410.

[69] Schweizer, L., Rizzo, C. A., Spires, T. E., Platero, J. S., Wu, Q., Lin, T. A., Gottardis, M. M. & Attar, R. M. (2008). The androgen receptor can signal through Wnt/beta-Catenin in prostate cancer cells as an adaptation mechanism to castration levels of androgens. *BMC Cell Biol*, Jan 9, 4.

[70] Bruxvoort, K. J., Charbonneau, H. M., Giambernardi, T. A., Goolsby, J. C., Qian, C. N., Zylstra, C. R., Robinson, D. R., Roy-Burman, P., Shaw, A. K., Buckner-Berghuis, B. D., Sigler, R. E., Resau, J. H., Sullivan, R., Bushman, W. & Williams, B. O. (2007). Inactivation of Apc in the mouse prostate causes prostate carcinoma. *Cancer Res*, Mar 67, 2490-2496.

[71] Zhu, H., Mazor, M., Kawano, Y., Walker, M. M., Leung, H. Y., Armstrong, K., Waxman, J. & Kypta, R. M. (2004). Analysis of Wnt gene expression in prostate cancer: mutual inhibition by WNT11 and the androgen receptor. *Cancer Res*, Nov 64, 7918-7926.

[72] Dai, J., Hall, C. L., Escara-Wilke, J., Mizokami, A., Keller, J. M. & Keller, E. T. (2008). Prostate cancer induces bone metastasis through Wnt-induced bone morphogenetic protein-dependent and independent mechanisms. *Cancer Res*, Jul 68, 5785-5794.

[73] Harris, S. E., Harris, M. A., Mahy, P., Wozney, J., Feng, J. Q. & Mundy, G. R. (1994). Expression of bone morphogenetic protein messenger RNAs by normal rat and human prostate and prostate cancer cells. *Prostate*, Apr 24, 204-211.

[74] Luo, J., Dunn, T., Ewing, C., Sauvageot, J., Chen, Y., Trent, J. & Isaacs, W. (2002). Gene expression signature of benign prostatic hyperplasia revealed by cDNA microarray analysis. *Prostate*, May 51, 189-200.

[75] Lamm, M. L., Podlasek, C. A., Barnett, D. H., Lee, J., Clemens, J. Q., Hebner, C. M. & Bushman, W. (2001). Mesenchymal factor bone morphogenetic protein 4 restricts

ductal budding and branching morphogenesis in the developing prostate. *Dev Biol*, Apr 232, 301-314.

[76] Haudenschild, D. R., Palmer, S. M., Moseley, T. A., You, Z. & Reddi, A. H. (2004). Bone morphogenetic protein (BMP)-6 signaling and BMP antagonist noggin in prostate cancer. *Cancer Res*, Nov 64, 8276-8284.

[77] Yuen, H. F., Chan, Y. P., Cheung, W. L., Wong, Y. C., Wang, X. & Chan, K. W. (2008). The prognostic significance of BMP-6 signaling in prostate cancer. *Mod Pathol*, advance online publication 17 October, doi: 10.1038/modpathol.2008.94.

[78] Horvath, L. G., Henshall, S. M., Kench, J. G., Turner, J. J., Golovsky, D., Brenner, P. C., O'Neill, G. F., Kooner, R., Stricker, P. D., Grygiel, J. J. & Sutherland, R. L. (2004). Loss of BMP2, Smad8, and Smad4 expression in prostate cancer progression. *Prostate*, May 59, 234-242.

[79] Ye, L., Kynaston, H. & Jiang, W. G. (2008). Bone morphogenetic protein-9 induces apoptosis in prostate cancer cells, the role of prostate apoptosis response-4. *Mol Cancer Res*, Oct 6, 1594-1606.

[80] Cook, C., Vezina, C. M., Allgeier, S. H., Shaw, A., Yu, M., Peterson, R. E. & Bushman, W. (2007). Noggin is required for normal lobe patterning and ductal budding in the mouse prostate. *Dev Biol*, Dec 312, 217-230.

[81] Qiu, T., Grizzle, W. E., Oelschlager, D. K., Shen, X. & Cao, X. (2007). Control of prostate cell growth: BMP antagonizes androgen mitogenic activity with incorporation of MAPK signals in Smad1. *EMBO J*, Jan 26, 346-57.

[82] Grishina, I. B., Kim, S. Y., Ferrara, C., Makarenkova, H. P. & Walden, P. D. (2005). BMP7 inhibits branching morphogenesis in the prostate gland and interferes with Notch signaling. *Dev Biol*, Dec 288, 334-347.

[83] Leong, K. G. & Gao, W. Q. (2008). The Notch pathway in prostate development and cancer. *Differentiation*, Jul 76, 699-716.

[84] Wang, X. D., Leow, C. C., Zha, J., Tang, Z., Modrusan, Z., Radtke, F., Aguet, M., de Sauvage, F. J. & Gao, W. Q. (2006). Notch signaling is required for normal prostatic epithelial cell proliferation and differentiation. *Dev Biol*, Feb 290, 66-80.

[85] Wang, X. D., Shou, J., Wong, P., French, D. M. & Gao, W. Q. (2004). Notch1-expressing cells are indispensable for prostatic branching morphogenesis during development and re-growth following castration and androgen replacement. *J Biol Chem*, Jun 279, 24733-24744.

[86] Shou, J., Ross, S., Koeppen, H., de Sauvage, F. J. & Gao, W. Q. (2001). Dynamics of notch expression during murine prostate development and tumorigenesis. *Cancer Res*, Oct 61, 7291-7297.

[87] Belandia, B., Powell, S. M., García-Pedrero, J. M., Walker, M. M., Bevan, C. L. & Parker, M. G. (2005). Hey1, a mediator of notch signaling, is an androgen receptor corepressor. *Mol Cell Biol*, Feb 25, 1425-1436.

[88] Orr, B., Grace, O. C., Vanpoucke, G., Ashley, G. R. & Thomson, A. A. (2008). A role for Notch signalling in stromal survival and differentiation during prostate development. *Endocrinology*, first published ahead of print September 18, as doi:10.1210/en.2008-0383.

[89] Shaw, A. & Bushman, W. (2007). Hedgehog signaling in the prostate. *J Urol*, Mar 177, 832-838.

[90] Podlasek, C. A., Barnett, D. H., Clemens, J. Q., Bak, P. M. & Bushman, W. (1999). Prostate development requires Sonic hedgehog expressed by the urogenital sinus epithelium. *Dev Biol*, May 209, 28-39.

[91] Lamm, M. L., Catbagan, W. S., Laciak, R. J., Barnett, D. H., Hebner, C. M., Gaffield, W., Walterhouse, D., Iannaccone, P. & Bushman, W. (2002). Sonic hedgehog activates mesenchymal Gli1 expression during prostate ductal bud formation. *Dev Biol*, Sep 249, 349-66.

[92] Wang, B. E., Shou, J., Ross, S., Koeppen, H., De Sauvage, F. J. & Gao, W. Q. (2003). Inhibition of epithelial ductal branching in the prostate by sonic hedgehog is indirectly mediated by stromal cells. *J Biol Chem*, May 278, 18506-18513.

[93] Freestone, S. H., Marker, P., Grace, O. C., Tomlinson, D. C., Cunha, G. R., Harnden, P. & Thomson, A. A. (2003). Sonic hedgehog regulates prostatic growth and epithelial differentiation. *Dev Biol*, Dec 264, 352-362.

[94] Berman, D. M., Desai, N., Wang, X., Karhadkar, S. S., Reynon, M., Abate-Shen, C., Beachy, P. A. & Shen, M. M. (2004). Roles for Hedgehog signaling in androgen production and prostate ductal morphogenesis. *Dev Biol*, Mar 267, 387-398.

[95] Doles, J., Cook, C., Shi, X., Valosky, J., Lipinski, R. & Bushman, W. (2006). Functional compensation in Hedgehog signaling during mouse prostate development. *Dev Biol*, Jul 295, 13-25.

[96] Thiyagarajan, S., Bhatia, N., Reagan-Shaw, S., Cozma, D., Thomas-Tikhonenko, A., Ahmad, N. & Spiegelman, V. S. (2007). Role of GLI2 transcription factor in growth and tumorigenicity of prostate cells. *Cancer Res*, Nov 67, 10642-10646.

[97] Zhu, G., Zhau, H. E., He, H., Zhang, L., Shehata, B., Wang, X., Cerwinka, W. H., Elmore, J. & He, D. (2007). Sonic and desert hedgehog signaling in human fetal prostate development. *Prostate*, May 67, 674-684.

[98] Chen, B. Y., Liu, J. Y., Chang, H. H., Chang, C. P., Lo, W. Y., Kuo, W. H., Yang, C. R. & Lin, D. P. (2007). Hedgehog is involved in prostate basal cell hyperplasia formation and its progressing towards tumorigenesis. *Biochem Biophys Res Commun*, Jun 357, 1084-1089.

[99] From American Cancer Society. Cancer Statistics for 2007., Global Cancer Facts and Figures 2007: www.cancer.org/docroot/STT/stt_0_2007.asp?sitearea=STT&level=1

In: Handbook of Prostate Cancer Cell Research...
Editor: Alan T. Meridith
ISBN: 978-1-60741-954-9

Chapter 6

Organoselenium Compounds: A Promising Prospect in Prostate Cancer Therapy

Carmen Sanmartín and Juan Antonio Palop
Department of Organic and Pharmaceutical Chemistry, University of Navarra, Irunlarrea 1, E-31008, Pamplona, Spain

Abstract

Prostate cancer is the most frequently diagnosed cancer and a leading cause of cancer deaths. Epidemiological and clinical intervention studies support the protective role of selenium against the development of prostate cancer. Earlier studies have shown that inorganic forms of selenium significantly decrease the expression of the androgen receptor (AR) and prostate-specific antigen (PSA), which are widely used clinically for prostate cancer diagnostics and as indicators of therapeutic efficacy and recurrence. More recently, organoselenium compounds were found to exhibit differential effects on growth, PSA secretion, oxidative stress and selective gene responses in human prostate cancer. Furthermore, literature reports have consistently shown that the dose and chemical forms of selenium are important factors in anticancer properties. However, the mechanisms responsible for the inhibitory activity by this micronutrient remain elusive and could involve separate pathways. One hypothesis concerns selenium metabolites, especially methylselenol, which are implicated as active metabolites for the anticancer effect. Other possible mechanisms involve changes in enzyme activities and/or mRNA levels of glutathione transferase (GST), apoptosis induction, caspase-3 activation, protein kinase regulation and cell cycle modulation. The information outlined above supports the idea that organoselenium compounds could be important new agents in prostate cancer treatment. In recent years our team has designed and synthesised several organoselenium derivatives that have shown interesting antiproliferative activity on the PC-3 cell line. We present here a review of significant organic structures that contain selenium and have marked prostate cancer antitumoural activity.

1. Introduction

Selenium (Se) is an essential trace element for humans, animals and some bacteria and it is important for many cellular processes, cardiovascular disease, central nervous system pathologics and may prevent cancer [1-3]. The evidence that Se is a cancer preventive agent includes that from geographic, animal, prospective and intervention studies. Prostate cancer (PC) represents a major clinical public health challenge, is the most commonly diagnosed malignancy in males and is the second leading cause of cancer death. Yet, the aetiology of prostate cancer remains largely unknown. Dietary factors, dietary supplements, and physical activity might be important in the prevention of the disease. Treatment options for advanced PC –including hormone ablation therapy, radiation and surgery– do not offer cure but delay the inevitable recurrence of the lethal hormone-refractory disease. Chemotherapy using available anticancer drugs, with the exception of the taxane drug docetaxel, for late stage PC does not offer any survival benefit. All of these treatments are costly and have significant side effects including impotence and incontinence, which negatively affect the quality of life of the patients. Epidemiological and clinical intervention studies support the protective role of selenium against the development of prostate cancer [4,5]. Furthermore, literature reports have consistently shown that the dose and form of selenium [6,7] are important factors in cancer chemoprevention. These include the forms of selenium present in the diet and in the body, their functions and mechanisms of action, and methods employed in assessing an individual's selenium nutritional status –both in general and in epidemiological studies into the risk of cancer in relation to diet, as well as in connection with long-term trials for investigating the disease-preventive potential of selenium supplementation [8]. The rapid advance in the knowledge of different selenoproteins and their biological functions has opened up new possibilities to increase our understanding of the biological effects of selenium supplementation [9]. Selenoprotein deficiency leads to the accelerated development of lesions associated with prostate cancer progression, implicating selenoproteins in cancer risk and development and raising the possibility that selenium prevents cancer by modulating the levels of these selenoproteins [10]. Moreover, additional trials of this type are needed to define the benefits and risks of different types and doses of selenium supplements that in the future may be implemented for public health reasons. Another necessary focus for future research is a better understanding of the mechanisms by which selenium interferes with the carcinogenesis process. The direction of future studies lies in clarifying the effects of these products and exploring the biological mechanisms responsible for the prevention of prostate cancer.

2. Antiandrogen Activity

The development of prostate cancer and its progression to a hormone-refractory state is highly dependent on androgen receptor (AR) expression. AR regulates the expression of genes involved in the proliferation and differentiation of prostate cancer cells. Due to differential interactions with coactivators and corepressors, AR activation results in the stimulation of a mitogenic response or in the expression of secretory proteins. AR is functional in advanced carcinoma of the prostate, as evidenced by studies of mutant receptors and ligand-independent activation. AR point mutations appear in advanced prostate cancer more frequently than in organ-confined disease. In addition, androgen signalling pathway activity increases during long-term androgen ablation. AR coactivator complexes might be a target for novel therapies for prostate cancer. Recent studies have shown that the selenium-based compound methylseleninic acid (MSeA, Figure 1) can disrupt AR signalling in prostate cancer cells [11] by reaction with reduced glutathione within the prostate cancer cell. On the other hand, it was observed that a combination of MSeA with α-tocopheryl succinate produced a greater than additive effect in suppressing AR signalling compared with the single agent [12]. Sodium selenite (Figure 1) is another compound that has been studied in relation to prostate cancer and it may modulate the androgen receptor through the repression of interleukin-6 (IL-6) [13].

Methylselenocysteine (MeSeCys) (Figure 2) significantly inhibited LNCaP tumour growth and also reduced AR expression in tumour tissues [14] and serum prostate specific antigen (PSA) levels in mice treated with this compound.

Pinto [15] later carried out a study aimed at comparing the effects of naturally occurring methylselenocysteine, selenomethionine (SeMet) and synthetic 1,4-phenylenebis (methylene)selenocyanate (*p*-XSC) and *p*-xylylbis(methylselenide) (*p*-XMS) (Figure 3) organoselenium compounds in androgen responsive LNCaP and its androgen-independent clone LNCaP C4-2. Depending on the structure, these molecules exhibit differential effects on growth and PSA secretion. For example, SeMet and MeSeCys led to a significant decrease in growth at high doses whereas *p*-XSC and *p*-XMS gave an effect at low doses. The largest decreases in PSA levels were obtained with *p*-XMS, whereas MeSeCys had no effect.

Sinha [8] later put forward the hypothesis that the inhibition of cell growth for SeMet and *p*-XSC is due, in part, to selenium interaction with redox-sensitive proteins.

Methylseleninic acid Sodium selenite

Figure 1.

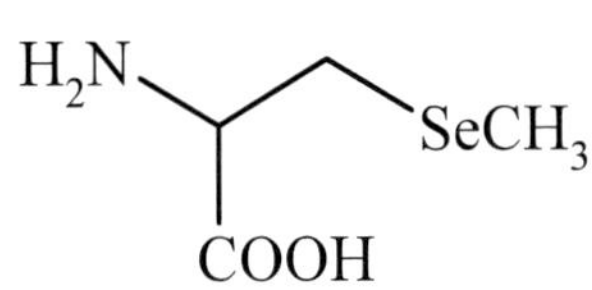

Methylselenocysteine

Figure 2.

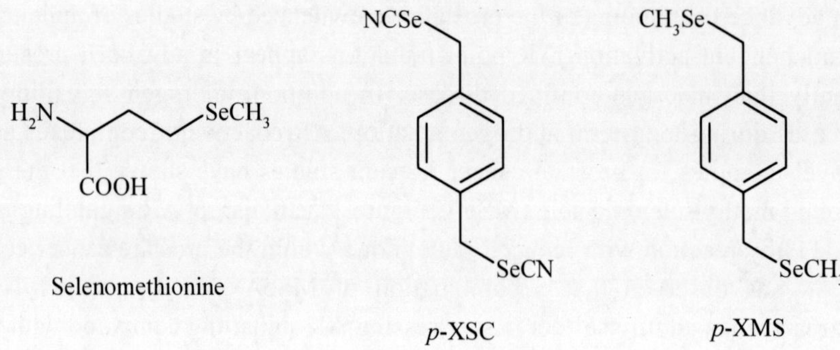

Figure 3.

3. Antioxidant Function

Selenium compounds appear to achieve cancer chemoprevention by multifaceted mechanisms, depending on the dosage and chemical form of selenium and the nature of the carcinogenic insult. Evidence from epidemiological, experimental (animal), and molecular biology studies on the antitumour effects of selenium indicates that there are two modes of action by which selenium affects cancer risk: firstly, selenium can function as an essential nutrient that provides the catalytic centres for a number of selenoenzymes, including some with antioxidant and redox functions; secondly, such compounds can serve as a source of selenium metabolytes that affect carcinogenesis in other ways. Based on clinical findings, it is likely that the antioxidant function of one or more selenoproteins [16] is responsible for the chemopreventive effect, although low molecular weight seleno-compounds have also been proposed to induce selective apoptosis in transformed cells. Furthermore, selenium supplementation has effects on antioxidant selenoprotein expression in prostate cells [17]. Some protective enzymes, such as glutathione peroxidase members (GPX_1 and GPX_4), are highly induced by supplemental selenium. These GPXs essentially contain selenium atoms in the active domain and reduce oxidative stress in cells. Thus, sodium selenite (Figure 1) has been shown to be most effective both *in vitro* and *in vivo*. Recent studies have shown that selenite is not only an antioxidant, but that it also possesses oxidising properties in the presence of specific substrates. The hypothesis in this case is that the selenium compounds with the greatest anticarcinogenic potency are likely to be sodium selenite, with Se in the +4 oxidation state, and methylseleninic acid. These compounds exert their cancer chemopreventive effects by directly oxidising critical thiol-containing cellular substrates [18] and they are more effective than the more frequently used supplements selenomethionine and Se-methylselenocysteine, which lack oxidation capability. Selenate (+6 Se) is the immediate precursor of selenite (+4 Se) and it can be metabolically reduced. Selenate also appears to be a more effective anticarcinogen than organic forms of dietary selenium, even though it is less potent than the (+4 Se) compounds cited above. On the other hand, there are numerous studies that concern the induction of a protective hepatic enzyme related to gluthathione-S-tranferase and gluthathione peroxidase for 2-substituted selenazolidine-4(*R*)-carboxylic acids [19,20] (Figure 4), selenocystine (Figure 4), selenomethionine, sodium selenite and methylselenocysteine and it seemed of interest to evaluate these compounds in prostate cancer cells.

Terada et al. [21] presented a hypothetical mechanism in which selenite, hydrogen selenide (H_2Se), and selenodiglutathione induced superoxide production in the presence of reduced glutathione (GSH) and oxygen in a cell-free system. A series of selenium compounds have been reported to generate superoxide *in vitro* through the oxidation of GSH and other thiols and this was investigated using a lucigenin-dependent chemiluminescent assay as an indirect measure of superoxide generation. These selenium compounds readily form the selenide anion (RSe^-) by GSH reduction and are capable of generating $O_2^{\bullet -}$. Furthermore, the forms and concentrations of Se compounds are important factors for the activity. In several studies it has been found that the ratio between GSH and Se may play an important role in Se catalysis and toxicity for the production of reactive oxygen species (ROS) [22]. ROS can react with biomacromolecules, including DNA and proteins, leading to the degeneration of these molecules. Various selenium derivatives may be candidates as ROS scavengers. Among the compounds reported in this respect are selenites, selenocysteine and some diselenides (Figure 5). Although the generation of superoxide by selenocystamine and diselenodipropionic acid in the presence of GSH is less rapid than that from selenite, the catalytic reaction is both continuous and sustainable. The catalytic nature of these and other reduced diselenides, as well as selenocyanates, is an important feature in the pharmaceutical application of these organic Se compounds and this characteristic is not shared with the inorganic Se compound selenite.

Other organoselenium compounds that have also effectively scavenged anionic superoxide *in vitro* are 2-amino-1,3-selenazoles and bis-(2-amino-5-selenazolyl)ketone derivatives [23] (Figure 6), where R, R' and R" are alkyl, aryl, cycloalkyl and heterocycloalkyl.

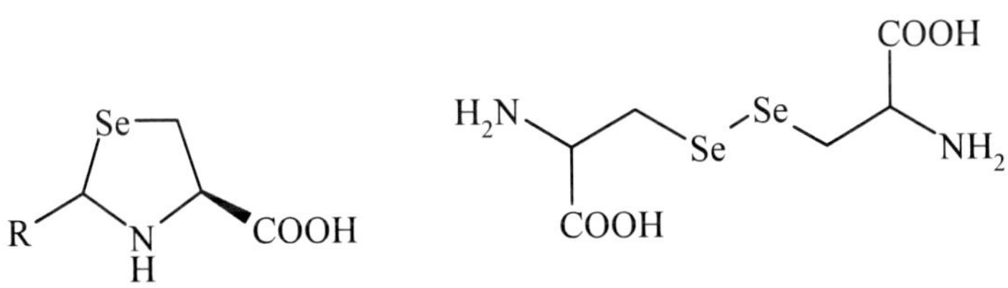

Selenazolidines Selenocystine

Figure 4.

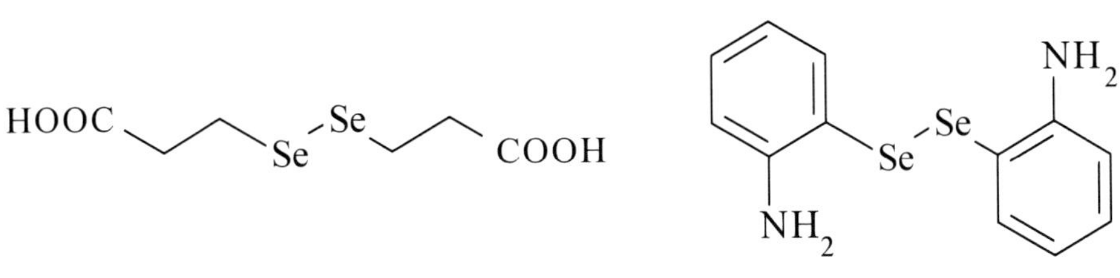

Diselenodipropionic acid 2-aminophenyldiselenide

Figure 5.

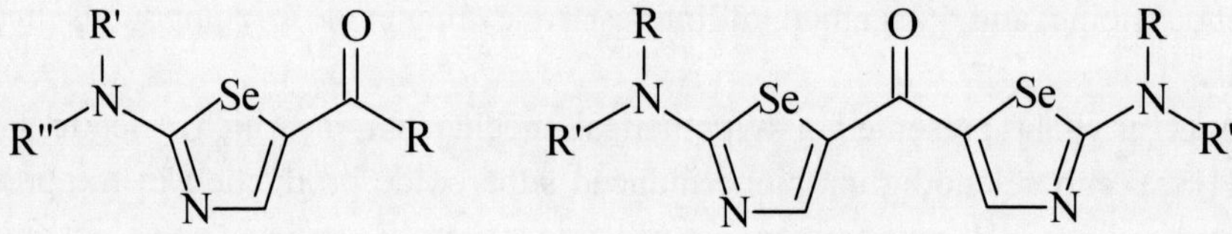

Figure 6.

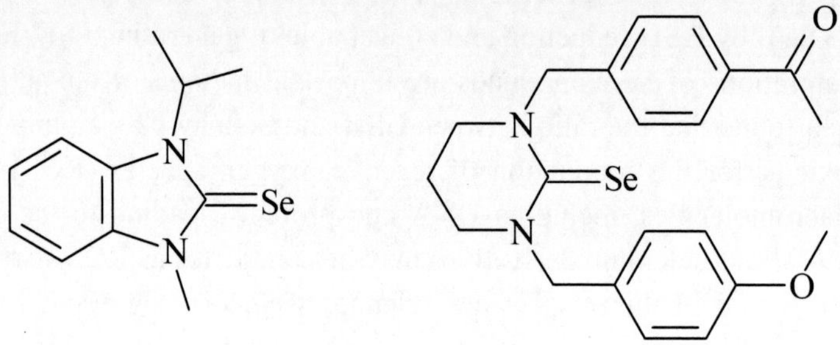

Figure 7.

Finally, in some recent hypotheses [24] certain selenone derivatives of benzimidazole and pyrimidine (Figure 7) have been proposed as antioxidative defence systems against oxidative damage.

4. DNA Damage and Repair

The DNA damage response is a key factor in the maintenance of genome stability. The vast majority of DNA damage affects the primary structure of the double helix; that is, the bases themselves are chemically modified. These modifications can in turn disrupt the regular helical structure of the molecules by introducing non-native chemical bonds or bulky adducts that do not fit within the standard double helix. Unlike proteins and RNA, DNA usually lacks tertiary structure and therefore damage or disturbance does not occur at that level. The integrity of genomic DNA is challenged by genotoxic stress originating during normal cellular metabolism or by external insults. Cellular responses to DNA damage involve elegant checkpoint cascades that enforce cell cycle arrest, damage repair, apoptosis or cellular senescence. The loss or alteration of genes involved in the damage response pathways have been reported in many cancer susceptibility syndromes and in sporadic tumours. Furthermore, this surveillance pathway is activated during early tumourigenesis, presumably due to uncontrolled replicative cycles, and has been recognised as one of the main barriers against the development of cancer. Numerous models of prostate carcinogenesis suggest that increasing chromosomal instability with the acquisition of mutations and chromosomal aberrations drives progression from preneoplasia to neoplasia. Furthermore, increased levels of chromosomal aberrations can be associated with decreased telomere length in high-grade prostatic intraepithelial neoplasia, and acquired centrosome dysfunction is associated with prostate cancer progression and dissemination. Human cells have therefore evolved complex signalling responses to both endogenous and exogenous DNA damage in order to preserve genomic integrity. Tumour progression in a number of epithelial malignancies has been associated with the sequential loss of gene functions that normally protect against DNA damage. For example, in response to DNA double-strand breaks, the ATM (ataxia telangiectasia mutated) protein is activated and stabilises the p53 tumour suppressor protein. This leads to the up-regulation of p53-dependent genes (e.g.: *p21WAF, Bax*, and *Gadd45*) and post-translational modifications of the CHK2, BRCA1, and NBS1 proteins. These proteins act together to induce G_1, S, and G_2 cell cycle arrest, DNA repair, and/or activation of cell death pathways (e.g.: apoptosis, mitotic catastrophe, or terminal growth arrest) depending on the cellular context. Defective DNA repair as a determinant of prostate cancer progression has not been extensively studied. Several groups have observed defective mismatch-repair in prostate cancer cell lines. Other data support DNA polymorphism in the *Xrcc1*, *Ogg1*, and DNA polymerase-β genes (involved in base excision repair or DNA single-strand break repair) as risk factors for prostate cancer. However, there is a lack of data concerning the homologous recombination and nonhomologous recombination (i.e., end-joining) pathways, which are involved in the repair of DNA double-strand breaks [25,26].

The compounds that have been evaluated in greatest detail are the presumed methylselenol precursors such as methylseleninic acid (MSeA) and methylselenocysteine (MeSeCys) in comparison with selenomethionine (SeMet) and selenite [27]. This difference is because methylselenol has been implicated as an active anticancer selenium (Se) metabolite. Investigation of the DNA integrity to detect genotoxic effects of Se treatments showed that MSeA and MeSeCys exerted a dose-dependent inhibition of DU-145 xenograft growth and both compounds were more potent than SeMet and selenite. Selenite treatment was found to increase DNA single-strand breaks in peripheral lymphocytes, whereas the other Se forms did not. Li [28] established that superoxide is a primary mediator for selenite in the induction of DNA-SSBs and p53 activation.

Other compounds proposed for the prevention of DNA damage are inorganic oxo compounds. This is the case for some selenates (Figure 8) [29,30] that may interact with alkylating carcinogens to protect DNA from alkylation. The explanation of the mechanism concerns the propensity of these compounds to form oxo species in solution. Presumably the active anionic oxo species behave as potent nucleophiles towards electrophilic attack and these authors suggest that the design of new compounds for cancer prevention should incorporate reactive oxo groups with high anionic charge density in order to accept alkylating toxins [30].

Na_2SeO_4 $[(C_6H_5)_4P]_3(O_3SeOCH_2OSeO_3)(HSeO_4)$

Figure 8.

D-501036 [31], 2,5-bis(5-hydroxymethyl-2-selenienyl)-3-hydroxymethyl-*N*-methyl pyrrole (Figure 9), has been identified as a novel antineoplastic agent with a broad spectrum of antitumoural activity against several human cancer cells and has an IC_{50} value in the nanomolar range. This compound induces cell death associated with the DNA damage-mediated induction of ataxia telangiectasia-mutated activation without interfering without topoisomerase-I and topoisomerase-II function.

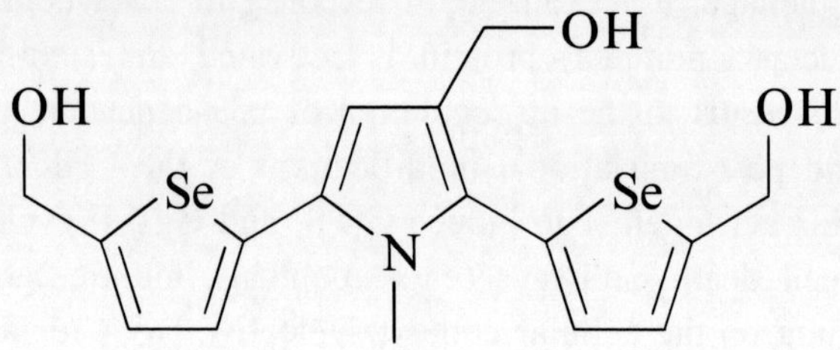

Figure 9.

Another mechanism that has been proposed for the activity of D-501036 is angiogenesis inhibition. Although anti-angiogenesis strategies have generated a great deal of enthusiasm for therapeutic applications, it is still unknown whether these systems would be feasible for prevention. The possibility of interfering very early in tumour progression by modulating the cancer angiogenic switch is appealing, though there is increasing evidence for close correlation between inflammation, the micro-environment and tumour-associated neo-angiogenesis causing the adverse outcomes of prostate cancer [32]. In particular, methylseleninic acid [33] in human prostate cancer DU-145 inhibits tumour growth by inhibition of the mitogen-stimulated proliferation of endothelial cells that may therefore contribute to the inhibition of tumour angiogenesis. On the other hand, sodium selenate [34] in PC-3 prostate tumour cells retards the growth of primary prostatic tumours and the development of retroperitoneal lymph node metastases, a process that was associated with a decrease in angiogenesis.

5. Selenium Compounds and Protein Kinases

Protein kinases are of particular interest in the treatment of human diseases because of their enzymatic activity and susceptibility to successful therapeutic targeting. The molecular targets that can be modulated by selenium include different kinases such as the three major MAPK pathways, JNK, extracellular signalling-regulated kinase (ERK), and p38 kinase. For example, sodium selenite has recently been shown to induce the phosphorylation of p53 and p38 to activate caspase-independent cell death and could also significantly activate anti-apoptotic PI3K/AKT and ERK pathways to protect prostate cancer cells [35]. These authors thus suggest that selenite could differentially modulate protein phosphorylation to regulate various cellular functions. Whereas the anticancer mechanisms have not been clearly defined, one hypothesis relates to selenium metabolites, especially the monomethyl selenium pool, generated under supranutritional selenium supplementation. In an effort to highlight potential molecular targets to mediate the chemopreventive activity, Jiang et al. [36] have studied the effects of methylseleninic acid (MSeA), a novel precursor of methylselenol, versus sodium selenite, a representative of the hydrogen selenide metabolite pool, on apoptosis execution, cell cycle distribution, and selected protein kinases in DU-145 human prostate cancer cells. Exposure of DU-145 cells to MSeA led to a profound G_1 arrest at 24 h, and exposure to greater concentrations led not only to G_1 arrest but also to DNA fragmentation and caspase-mediated cleavage of poly(ADP-ribose) polymerase (PARP), both of which are biochemical hallmarks of apoptosis. In addition, G_1 arrest was associated with increased expression of p27kip1 and p21cip1. A similar process occurs with selenoproteins that can act as antioxidants and inhibit signalling enzymes, such as protein kinase C (PKC), that play a crucial role in tumour promotion [37]. It was observed that selenocompounds can inactivate PKC by oxidising the vicinal thiols present within the catalytic domain. Methylselenol, the postulated cancer-preventive metabolite, has no direct effect on PKC activity. However, methylseleninic acid, locally generated by the reaction of membrane methylselenol with PKC-bound tumour-promoting fatty acid hydroperoxides, selectively inactivates PKC. The redox-mediated inactivation of PKC may, at least in part, be responsible for the antioxidant-induced inhibition of tumour promotion and cell growth, as well as for the induction of cell death. Selenoproteins also regulate vascular endothelial growth factor (VEGF), a finding that correlates with a diminished invasive capacity of tumour cells [38]. Other kinases studied include the κ kinases that have been investigated as targeting multiple signalling pathways as a strategy for managing prostate cancer [39]. Sodium selenite and methylseleninic acid [40] inhibited NF-κB DNA binding and IκB kinase activation, IκBα phosphorylation and degradation induced by TNF-α. Park [41] later described how this same acid blocks tumour invasion by inhibiting pro-matrix metalloproteinase-2 (MMP-2) activation mediated by suppression of membrane type-1 (MT1-MMP) expression, which is regulated by the nuclear factor-κB (NF-κB) signal. Previously, Jiang [42] reported that MeSeA reduced the phosphorylation level of mitogen-activated protein kinases (MAPKs) and suggested that a pool of methylselenol, an active metabolite from methylseleninic acid, might inhibit the expression of MMP-2 by vascular endothelial cells and of vascular endothelial growth factor (VEGF) by cancer epithelial cells. These two proteins are critical for angiogenesis and its regulation. Finally, although there has not been a common consensus on the association of COX-2 with disease stage, it is generally agreed that COX-2 over expression is associated with the development of various cancers, including prostate and colon cancers. Hwang [43] found that AMP-activated protein kinase (AMPK), which functions as a cellular energy sensor, mediates anticancer effects of selenium through a cyclooxygenase-2 (COX-2)/prostaglandin E signalling pathway. More recently, Siddiqui [44] and Vunta [45] showed that COX-2, mRNA and protein levels were over-expressed in prostate cancer tissue compared with paired benign tissue.

6. Other Potentially Anticarcinogenic Selenium Compounds

Interest in selenium compounds has increased markedly in recent years. As a result, these compounds have become of interest to research groups as building blocks in the synthesis of selenium-containing heterocycles such as tetrazoles [46], triazoles [47], and 1,3-

selenazinanones [48] (Figure 10). Indeed, the anticarcinogenic activity of many examples of these compounds is under investigation.

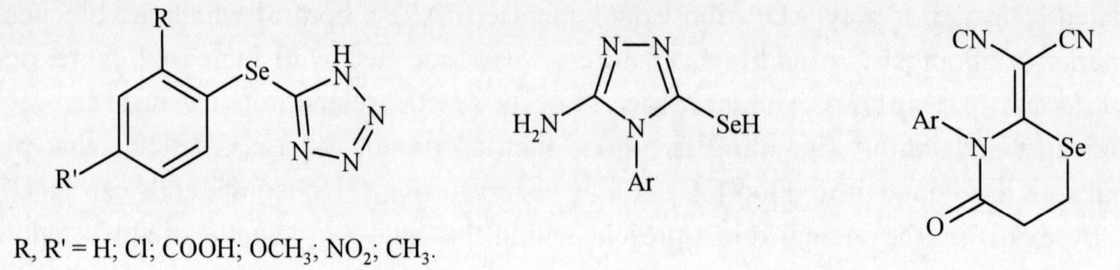

Figure 10.

For example, the mitochondrial apoptotic pathway responses a wide range of apoptotic originating from other subcellular compartments (e.g.: the nucleus, lysosomes, the endoplasmic reticulum, or the cytosol) that converge on mitochondria where they favour mitochondrial membrane permeabilisation (MMP) [53]. This process represents a crucial check-point in the cascade of events that lead to cell death and cause the release of a number of apoptotic factors present in the mitochondrial intermembrane space. Upon permeabilisation of the mitochondrial outer membrane (MOM), intermembrane space (IMS) proteins, which include caspase activators such as cytochrome c (Cyt c), Omi/HtrA2 (Omi stress-regulated endoprotease/High temperature requirement protein A 2) and Smac/DIABLO (second mitochondria-derived activator of caspase/direct IAP binding protein with a low pI), as well as caspase-independent death effectors like apoptosis-inducing factor (AIF) and endonuclease G (EndoG), are released into the cytosol. Cyt c promotes the activation of the initiator caspase-9 in a direct fashion through the assembly of the apoptosome (together with the apoptosis protease activating factor-1, i.e. APAF-1, and ATP/dATP), while Omi/HtrA2 and Smac/Diablo favour the caspase cascade indirectly by antagonizing the activity of endogenous inhibitors of caspases, i.e. the inhibitor of apoptosis proteins (IAPs). AIF and EndoG translocate to the nucleus where they mediate chromatin condensation and large-scale DNA fragmentation, independently of caspases. The integrity of mitochondrial membranes is largely under the control of members of the Bcl-2 family. The Bcl-2 family [54] includes both the pro- and anti-apoptotic members that play a central regulatory role in apoptosis. Currently, at least 20 members of the Bcl-2 family have been identified and all of these share at least one Bcl-2 homology (BH) domain. The family may be divided into three groups, one of which contains five anti-apoptotic proteins, Bcl-x_L, Bcl-w, A1, Mcl1, and Bcl-2 itself. Two further subfamilies are pro-apoptotic proteins; the Bax family has BH1-3 domains similar to those in Bcl-2, whereas the other pro-apoptotic proteins have only the BH3 domain. Commitment to apoptosis is typically governed by opposing factions of the Bcl-2 family of cytosolic proteins. Initiation of the proteolytic cascades requires assembly of certain caspase precursors on a scaffold protein, and the Bcl-2 family determines whether or not this complex can form. The pro-survival members of this family can act by sequestering the scaffold protein and/or preventing the release of apoptogenic molecules like cytochrome c from mitochondria. In contrast, the pro-apoptotic members act as sentinels for cellular damage as cytotoxic signals induce their translocation to the organelles, where they bind to their pro-survival relatives, promote organellar damage and trigger apoptosis. Cytochrome c is released either through channels created by integration of Bax and Bak in the mitochondrial membrane or following opening of the mitochondrial permeability transition (MPT) [55]. This channel is composed of proteins of both inner and outer mitochondrial membranes together with proteins of the intermembrane space, and opening of the channel results in an influx of ions such as calcium, which in turn causes swelling of the mitochondria. This swelling produces breaks in the outer membrane, although the inner membrane remains intact, and cytochrome c escapes through these outer membrane breaks. Another particular modality of cell death associated with mitochondrial membrane permeabilisation (MMP) is the permeability transition, which manifests itself as a loss of the inner transmembrane potential ($\Delta\Psi_m$), matrix swelling and the long-lasting opening of the permeability transition pore complex (PTPC) at the contact sites. The PTPC is composed of a minimum unit of three proteins, namely the voltage-dependent anion channel (VDAC) [56] in the mitochondrial outer membrane (MOM), the adenine nucleotide translocase (ANT) in the inner membrane (IM) and cyclophilin D in the matrix (CypD). These three proteins can interact within the contact site between the inner and the outer mitochondrial membranes (MIM and MOM, respectively). The minimum unit is regulated in its opening probability by the metabolic state of the cells at the cytosol/intermembrane/matrix interphases, as well as by a panoply of additional proteins whose expression can be cell-type specific. At the level of the MOM, homo- or hetero-oligomers composed of cytosolic proteins (e.g.: Bax, Bid) and/or constitutive mitochondrial proteins (e.g.: Bak, VDAC) or local ruptures have been proposed to mediate the release of intermembrane space proteins. Depending on the model, caspases and Bax/Bcl-2 family members can initiate, participate and/or regulate the mitochondrial phase of apoptosis. In addition, it appears that a large number of pro-apoptotic effectors, either proteins or non-proteaceous in nature, can translocate to mitochondria and exert MMP-modulatory functions, thus influencing cell fate at the level of this organelle. The two pathways, intrinsic and extrinsic, converge in a common executor mechanism that involves activated proteases (caspases) and DNA endonucleases, which cleave regulatory and structural molecules and lead to cellular death.

7. Selenium Compounds, Apoptosis and Prostate Cancer

7.1. Introduction

Apoptosis is a specific process that leads to programmed cell death through the activation of evolutionarily conserved intracellular pathways. Besides enabling malignant transformation and tumour progression, defects in apoptosis can result in resistance to cytotoxic cancer therapies. Apoptotic cell death is characterised by a series of unique biochemical and morphological events [49-51], such as cleavage of multiple proteins, phosphatidyl serine exposure, cell shrinkage, and condensation and fragmentation of chromatin. There are classically two main mammalian pathways of apoptosis; the “extrinsic” pathway involves signals transduced through cell-surface death receptors such as

CD95/APO1/Fas, tumour necrosis factor receptor 1 (TNF-R1), or TNF-related apoptosis-inducing ligand (TRAIL) death receptors DR4 and DR5 located on a cell's surface; and the "intrinsic" pathway channels through mitochondrial mechanisms before activation of the common caspases. The regulation of these pathways is complex [52] and involves a number of anti-apoptotic and pro-apoptotic molecules, which are depicted and referenced in Figure 11.

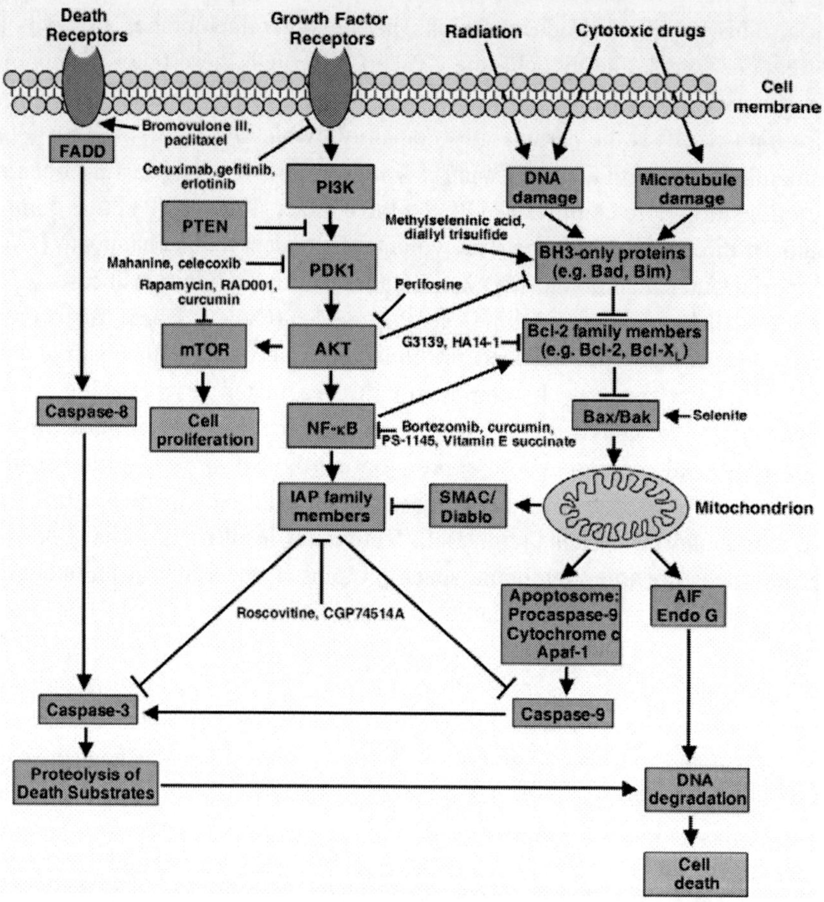

Figure 11. The two, extrinsic and intrinsic, main pathways of apoptosis.

The response of traditional prostate cell lines varies according to different apoptotic stimuli, although both androgen-dependent and -independent lines contain intact cell death programs [57]. Expression of these apoptosis-related genes has not been evaluated rigorously in prostate cancer. However, a few trends are apparent and there is some controversy concerning the expression of Bcl-2, Bax, NF-κB, etc.

The diverse results discussed above indicate that apoptosis-related signalling in prostate cancer is heterogeneous and ultimately may depend upon the differing activation levels of the various interacting pathways, many of which combine on apoptotic end effectors.

Epidemiological studies, preclinical investigations and clinical intervention trials support the role of selenium compounds as potent cancer chemopreventive agents; the dose and the form of selenium are critical factors in cancer prevention. Induction of apoptosis and inhibition of cell proliferation are considered to be important cellular events that could account for the cancer preventive effects of selenium. Prior to apoptosis induction, selenium compounds alter the expression and/or activities of a number of cell cycle regulatory proteins, signalling molecules, proteases, mitochondrial associated factors, transcriptional factors, tumour suppressor genes and polyamine and glutathione levels. Depending on the form, selenium compounds can target separate pathways, although further research is needed to learn about the disruption of different pathways that converge to apoptosis. Numerous selenium compounds are known to inhibit carcinogenesis in several animal models but not all of these have been examined for their efficacy in inducing apoptosis or vice versa in the corresponding target organ. Studies aimed at investigating the effects of selenium compounds on apoptosis in the target organ both *in vivo* and *in vitro* are limited. On the basis of information provided in this chapter, several novel synthetic organoselenium compounds need to be examined both *in vitro* and *in vivo* for their potential to induce apoptosis; such an investigation may provide improved and mechanism-based cancer chemoprevention as well as chemotherapeutic agents.

In the following sections we describe selenium-containing structures that induce apoptosis and discuss their mechanisms of action and the potential for further development into novel therapeutics agents.

7.2. Selenium and Modulation of Caspases

Caspases are a conserved family of cysteine proteases that coordinate apoptosis, a process in which the cell is dismantled by targeting a panoply of proteins. The mammalian caspase family contains 14 members, with a subset participating in cellular demise and the remaining ones involved in the processing of pro-inflammatory cytokines. Defects in the expression or activity of these caspases are related to certain pathological conditions and these include neurodegenerative disorders, autoimmune diseases and cancer. Some selenium compounds induce apoptosis through this mechanism. Organoselenocyanates represent an important class of chemopreventive agents [15] that possess antioxidative, antimutagenic and cancer chemopreventive properties. For example, diphenylmethylselenocyanate (Figure 12) has been evaluated [58] for its ability to induce apoptosis against 7,12-dimethylbenz[*a*]anthracene-croton oil and it was observed that caspase-3, which contributes in part to the process of cellular apoptosis to prevent further cellular differentiation, was elevated significantly. However, this compound did not show any modulatory effect on hepatocellular apoptosis [58] caused by acute doses of carbon tetrachloride.

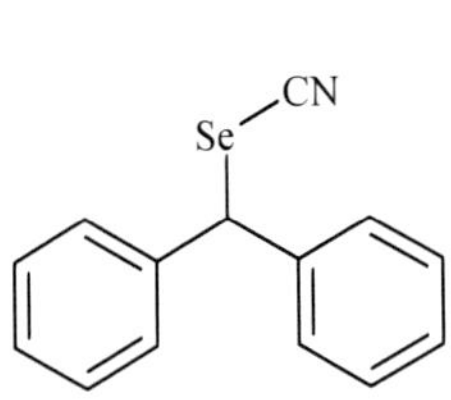

Diphenylmethylselenocyanate

Figure 12.

2-(4-Methylphenyl)-1,3-selenazol-4-one (Figure 13) inhibits the proliferation of and induces apoptosis [59] in a human ovarian cancer cell line (SKOV3) and an acute myelocytic leukemia cell line (HL-60), with caspase-3 activation also apparent in the latter. These results suggest that the mechanism of apoptotic induction may differ between the two cell lines.

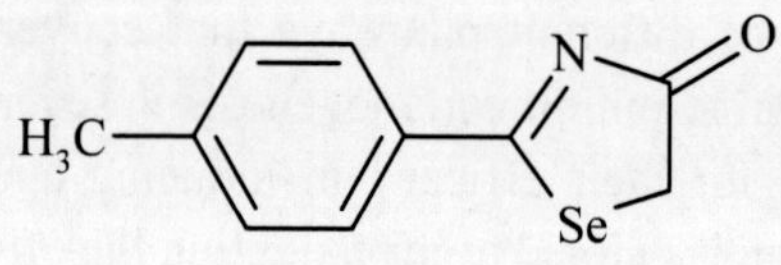

Figure 13

Methylseleninic acid (Figure 1) was found to enhance apoptosis induced by cancer therapeutic drugs in DU-145 cells through MSeA and some kinases that amplify the caspase-8 [60] activation cascade and, together with α-tocopherol ether-linked acetic acid analogue [2,5,7,8-tetramethyl-2R-(4R,8R-12-trimethyltridecyl)chroman-6-yloxyacetic acid (α-TEA)] or trans-resveratrol (t-RES) [61], enhance levels of apoptosis in human prostate (LnCaP, PC-3, and DU-145) cancer cell lines by activation of caspases 8 and 9 and downregulation of pro-survival proteins FLIP and survivin. In addition, MSeA sensitises hormone refractory prostate cancer (HRPCa) cells to apoptosis induced by paclitaxel (Taxol) by enhancing multiple caspases [62].

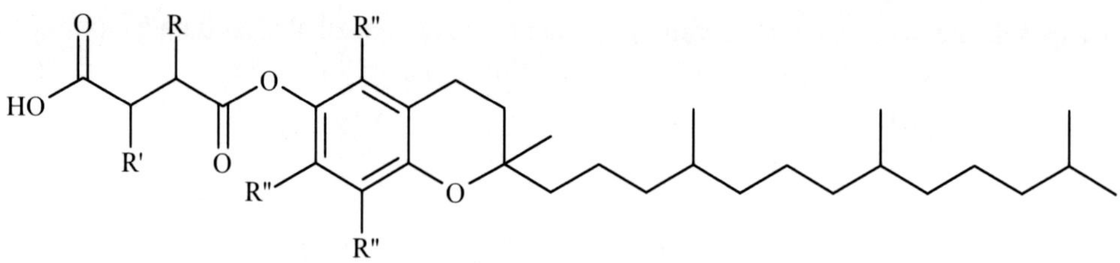

R, R' = H, SePh; R" = H, CH_3 Tocopherol derivative

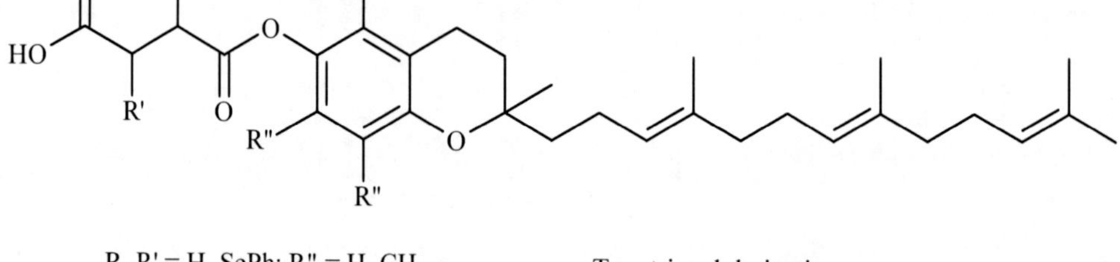

R, R' = H, SePh; R" = H, CH_3 Tocotrienol derivative

Figure 14.

On the other hand, selenomethionine (Figure 3) –when combined with methioninase [63], which generates methylselenol from selenomethionine– induces caspase activation and loss of cell adherence in melanoma cells by altering integrin expression. Several years ago, vitamin E and selenium were reported [64] to be effective in the prevention of prostate cancer. Structurally, vitamin E is related to phenolic compounds called tocopherols and tocotrienols. The anticancer properties of tocotrienols were also evaluated in a variety of animal models and there is evidence for the possible role of γ-tocotrienol as an anticancer agent in prostate tumours [65]. In order to gain more of an insight into the synergism between selenium and vitamin E, as well as to enhance the apoptotic properties of the latter compound, Vraka et al. [66] developed the synthesis of 2-phenylselenyl succinates from tocopherols and tocotrienols (Figure 14).

It has already been shown in several cell lines that α-tocopheryl succinate induces apoptosis through the activation of the caspase cascade. It was of interest to assess the replacement of succinate with a phenylselenyl succinate moiety within tocopheryl/tocotrienyl monoesters. This change elicited specific apoptotic cascades through the activation of caspases and controlled the lipophilicity, conformational flexibility, size, and acidity of molecules within the functional domain in a study carried out in prostate cell lines PC-3 and DU-145.

7.3. Apoptosis and Redox Process

The reactive oxygen species (ROS) can damage nucleic acids. The oxidative modification of DNA constitutes the fundamental molecular event in carcinogenesis and for this reason there is a great deal of interest in studying the involvement of ROS in that process. On the other hand, oxidative DNA damage-induced mutagenesis is widely hypothesised to be a frequent event in the normal human cell. There is an enormous amount of evidence to suggest that ROS play an important role in the expansion and progression of tumour clones, which makes them a relevant class of carcinogens. ROS usually include [67] the superoxide anion ($O_2 \cdot^-$), hydrogen peroxide (H_2O_2), and the highly reactive by-product of H_2O_2, hydroxyl radicals (OH·), all of which are capable of reacting with and damaging DNA, proteins, and lipids. Aside from the damaging activity of ROS, low levels of intracellular ROS have also been identified as second messengers involved in a variety of signalling pathways and serve as transcription regulators. These diverse activities of intracellular ROS in initiating and/or amplifying death signals or in the regulation of apoptosis have been well studied and extensively reviewed. Recent data seem to suggest that $O_2 \cdot^-$ may affect pathways involved in cell death and proliferation in a way that is distinct from H_2O_2.

On the other hand, cells in multicellular organisms are exposed to both endogenous oxidative stresses generated metabolically and to oxidative stresses that originate from neighbouring cells and from other tissues. To protect themselves from oxidative stress [68], cells are equipped with reducing buffer systems (glutathione/GSH and thioredoxin/ thioredoxin reductase) and have developed several enzymatic mechanisms against oxidants, including catalase, superoxide dismutase and glutathione peroxidase. Other major extrinsic defences (from the diet) include ascorbic acid, β-carotene and other carotenoids, and selenium. Recent evidence indicates that in addition to their antioxidant function, several of these redox species and systems are involved in the regulation of biological processes, including cellular signalling, transcription factor activity, and apoptosis in normal and cancer cells. The survival and overall well-being of the cell is dependent upon the balance between the activity and the intracellular levels of these antioxidants as well as their interaction with various regulatory factors, including Ref-1, nuclear factor-κB, and activating protein-1. In addition, it is well known that reduced glutathione (GSH) is closely involved in the metabolism and bioactivity of Se.

The monomethylated selenium compound methylseleninic acid (MSeA) [69] has been shown to cause global thiol redox modification of proteins with an accumulation of aberrantly folded proteins in the endoplasmic reticulum (ER). This change triggers a defined set of transducers to correct the defects in PC-3 human prostate cancer cells or commit the cells to apoptosis if the rescue effort is exhausted. The apoptotic markers increased proportionally with the dose of MSeA, whereas the rescue markers failed to keep pace with the increasing challenge from MSeA.

On the other hand, sodium selenite is able to alter the intracellular redox state, as measured by cellular GSH and GSSG levels, and after acute treatment [70] in LNCaP cells, selenite causes mitochondrial damage and apoptosis induction. Li et al. [28] later reported that selenite exhibits distinct effects in two prostate cancer cells lines. LNCaP cells are more sensitive than PC-3 cells to selenite-induced apoptosis and there is a rapid and high level of superoxide production in the former but only low levels in PC-3. Among the organic compounds, Ebselen [2-phenyl-1,2-benzoisoselenazol-3(2*H*)-one] (Figure **15**) was revealed to possess strong antioxidant properties [71]. Sung et al. [72] reported that this compound blocked the expression of the disintegrin and metalloprotease ADAM9 in LNCaP or C4-2 prostate cancer cells through inhibition of ROS production.

However, this compound does have some drawbacks, such as low glutathione peroxidase (GPX) activity and water insolubility. For this reason, Lv et al. [73] synthesised 6A,6A'-dicyclohexylamine-6B,6B'-diselenide-bis-β-cyclodextrin (6-CySeCD) in order to imitate the antioxidant enzyme glutathione peroxidase (GPX). This compound exhibits better GPX activity than 6,6'-diselenide-bis-cyclodextrin (6-SeCD), 2-selenium-bridged β-cyclodextrin (2-SeCD) and 2-phenyl-1,2-benzoisoselenazol-3(2*H*)-one (Ebselen) in the reduction of H_2O_2, tert-butyl hydroperoxide and cumenyl hydroperoxide by glutathione. However, Sun et al. [74] synthesised a novel GPX mimic (2-selenium-bridged β-cyclodextrin, 2-SeCD) (Figure 16) that has an antioxidant efficiency superior to Ebselen and is non-toxic to NIH3T3 cells.

Recently, twenty seven selenium compounds [75] were tested for quinone reductase and glutathione-S-transferase activity in murine hepatoma cells. The most potent compounds were methylseleninic acid (Figure 1), Ebselen (Figure 15), some diselenides (dimethyl, diphenyl, dibenzyl), benzeneselenol, benzeneseleninic acid and triphenylselenonium chloride (Figure 17). The cell growth inhibition levels correlated with selenol metabolites.

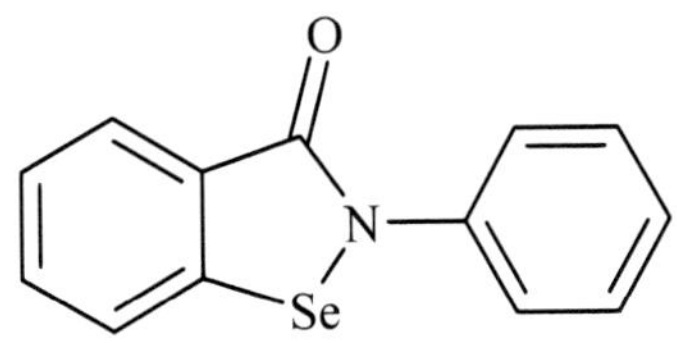

Ebselen

Figure 15.

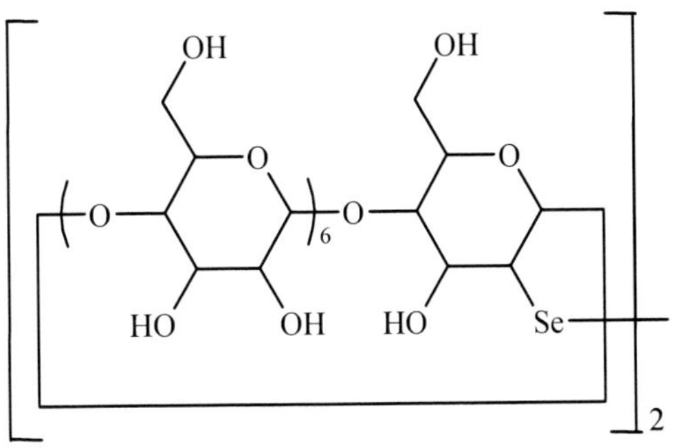

2-SeCD

Figure 16.

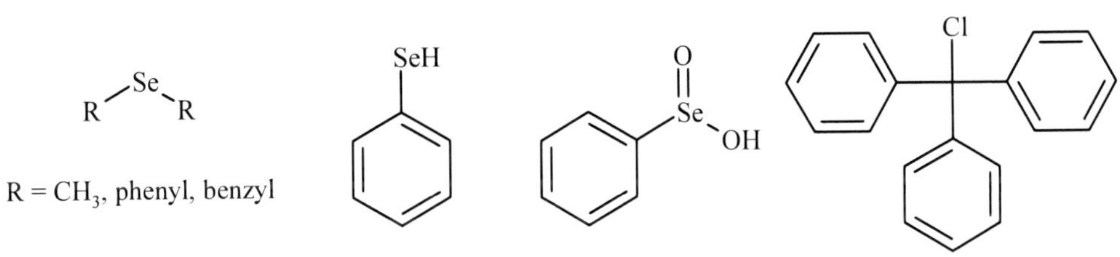

Figure 17.

7.4. Apoptosis and Cell Cycle

It is well known that there is a relationship between the cell cycle and apoptosis. Among the selenium compounds that are relevant in this area is selenomethionine (Figure **3**), mainly because this is the major organic selenium compound found in the diet. Flow cytometry analysis [76] showed that selenomethionine induced G_2/M arrest in LNCaP by modulating transcript levels of genes involved in the cell cycle. Methylseleninic acid [33] induced altered expression in the cell cycle and this was mainly arrest in the G_1 phase through relevant active anticancer selenium metabolites in DU-145 and in PC-3. Finally, the novel organoselenium compound 1,2-bis-[1,2-benzisoselenazolone-3(2*H*)-ketone]ethane (BBSKE) (Figure 18), which has shown an inhibitory effect on the growth of a variety of human cancer cells, provokes S phase arrest [77,78] accompanied by increases in the protein levels of cyclin A, E and p21 and decreases in levels of cyclin B1, D1 and Cdk4.

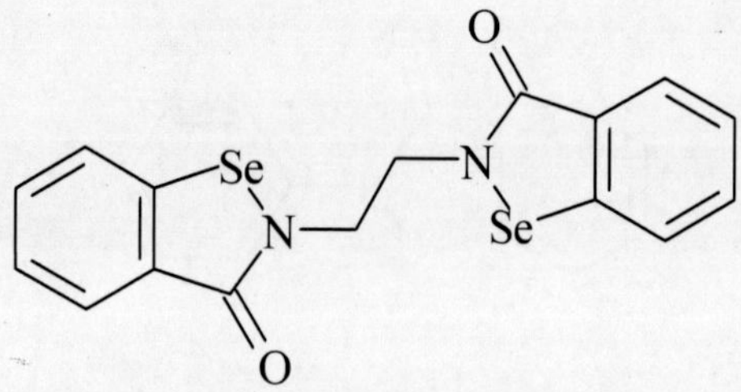

BBSKE

Figure 18

7.5. Apoptosis and Mitochondria

Li et al. [28] found that sodium selenite induced apoptosis signalling in prostate cancer cells through the following route: Selenite metabolism → ROS (superoxide, hydroperoxide) → DNA strand breaks approximately p53 phosphorylation → p53 and Bax translocation to mitochondria → mitochondrial leakage → caspase-activation →nucleosomal fragmentation. The role of superoxide and mitochondria in selenite-induced apoptosis in human prostate cancer cells (LNCaP) was subsequently investigated [79]. The results suggest that selenite induced cancer cell death and apoptosis by producing superoxide radicals, and also caused a decrease in mitochondrial membrane potential, the release of cytochrome c into the cytosol, and activation of caspases 9 and 3. More recently, Fickova et al. [80] described the cytotoxic and antiproliferative activities of new derivatives of kojic acid (Figure 19). These compounds were studied in human skin carcinoma (A431) and in human breast carcinoma and the results indicated that injury of intracellular mitochondria and lysosomes precedes the loss of membrane integrity.

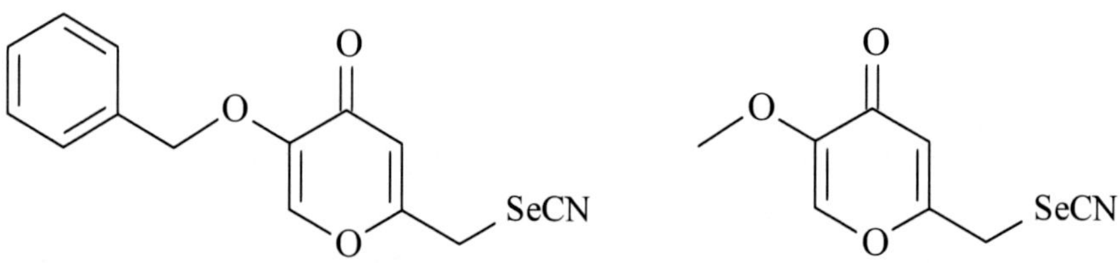

Figure 19.

7.6. Apoptosis and Proteins

The protein kinase (PK) family of enzymes is a family of kinases that are involved in the transduction of signals for cell proliferation, differentiation, apoptosis and angiogenesis. Not surprisingly, disruption of PK regulation is implicated in tumourigenesis and drug resistance. Akt, also known as PKB, is a cellular homologue of the oncogene product v-Akt encoded by AKT8 retrovirus and belongs to a family of serine/threonine kinases. This compound acts downstream of phosphoinositide 3-kinase (PI3-K) and plays a critical role in cell survival and oncogenesis. Over the last few years, the Akt pathway has become a major target due to its role as a signalling pathway in which modulation of substrates prevents apoptosis. The involvement of Akt in the cell survival pathway is a complex process that requires extensive machinery for intracellular events. It has also been found that Akt plays an important role [81] in regulating apoptosis sensitivity of LNCaP cells and DU-145 prostate cancer cells against methylseleninic acid. It was observed that MSeA decreases phospo-Akt activity [82] in a time- and concentration-dependent manner. Akt also inhibits P13-K activity and phosphoinositide-dependent kinase 1 (PDK1). The inhibition of PI3-K could prevent tumour invasion and metastasis and the mechanism proposed for the inhibition involves interference as a potent redox modulator [83] in the global protein thiol/disulfide interchange. However, these kinases did not significantly regulate apoptosis induced by sodium selenite in LNCaP. Other evidence concerning MSeA [35] indicates cell death mediated by protein kinase ERK. These findings support the differential involvement of these protein kinase pathways in regulating apoptosis induction by different forms of selenium. Other kinases that coordinate various extracellular signals to regulate cell proliferation, differentiation and cell survival are mitogen-activated protein kinases (MAPK), a family of structurally-related serine/threonine protein kinases [84]. These proteins have been altered by methylseleninic acid [85] in murine prostate cells. Another strategy that has been proposed for pro-apoptotic signal modulation is the coadministration of high doses of salicylates or other nonsteroidal anti-inflammatory drugs with organoselenium compounds [86]. The aim of this approach is to decrease NF-κB activity, which is elevated in a high proportion of cancers –particularly advanced cancers that have been treated previously. This factor promotes transcription of a large number of proteins that inhibit both the intrinsic and extrinsic apoptosis pathways. In the case of BBSKE (Figure **18**) a complementary mechanism for apoptosis has also been postulated and is based on the inhibition of the activity of thioredoxin reductase, leading to the accumulation of oxidised thioredoxin [87]; the change in the redox state of thioredoxin results in a decrease in the NF-κB DNA-binding activity, which in turn down-regulates the expression of some anti-apoptosis genes. Finally, other kinases implicated in some cancers are Src-kinases, although these have not been studied in relation to prostate cancer and selenium compounds. The activity of these systems can be regulated in several ways, with autophosphorylation site Y419 and negative regulatory phosphorylation site Y530 being the most representative examples [88]. Several protein tyrosine phosphatases (PTP) are capable of activating Src by dephosphorylating Y530 and, among these, PTP1B contributes to the progression in human colon cancer cells [88] and breast cancer [89,90]. Recent studies [90] have demonstrated that ablation of PTP1B does not have any effect on breast tumourigenesis but promising results have been obtained on lung tumour.

7.7. Apoptosis and p53 Regulation

Mutation of p53, a classical tumour suppressor, is frequently associated with oncogenesis. Cellular functions modulated by the p53 protein include DNA synthesis, DNA repair, cell cycle arrest, gene transcription, senescence and apoptosis. The most conserved function of the p53 protein is tumour suppression through the induction of apoptosis. Elevated expression of pro-apoptotic genes with promoters containing the p53 responsive element represents one mechanism through which p53-dependent apoptosis is induced. Intracellular concentrations of p53 are increased markedly by stresses such as DNA damage, ionizing radiation, UV radiation, hypoxia, heat shock, growth factor withdrawal, oncogene activation and exposure to cytotoxic agents. Additionally, the translocation of a specific polymorphic form of p53 from the cytosol to the mitochondria is associated with the induction of apoptosis. In response to the genotoxic insult, both transcription-dependent and transcription-independent cell death may be modulated concurrently by p53 with the resultant amplification of the apoptotic signal. Among the selenium compounds, sodium selenite [91] induced a time- and concentration-dependent phosphorylation of p53 with concomitant activation and nuclear localisation of cell cycle inhibitor p21. On the other hand, selenomethionine (Figure **3**) modulates p53 activity by causing redox regulation of key p53 cysteine residues [92]. In addition, treatment with methylseleninic acid gave rise to phosphorylation of one or more p53 threonine residues, but did not affect any known serine phosphorylation sites [92]. On the other hand, selenium dioxide [93] in leukemia cell lines and sodium selenite [94] in cultured cortical neurones showed up-regulation for p53 and superoxide production mediated by p53 [95] in LNCaP. Recently, a selenophene derivative has emerged [96] that induces apoptosis through activation of p53 in a time-dependent manner. Furthermore, p53 was found to be translocated from the cytosol to the mitochondria in response to drug treatment for 12 h.

8. A Case Study: Novel Selenocompounds as Cytotoxic Agents and Apoptosis Inducers in Prostate Cancer Cells

8.1. Introduction

In the last four years, several articles [97-104] have been published by our research group related to the design, synthesis and biological evaluation of novel compounds as cytotoxic agents and apoptosis inducers. In addition, mounting evidence suggests that selenium (Se) works by inhibiting important early steps in carcinogenesis in a variety of experimental models and the anticancer activity is dependent on the chemical form of selenium. Se occurs in both organic and inorganic forms. Based on these findings we envisaged a new investigation that involves the synthesis of new compounds that incorporate the selenium-containing moiety.

8.2. Structures and Biological Results

Initially, the rationale behind the design of these compounds was to maintain molecular symmetry, a structural property that is frequently present in cytotoxic and pro-apoptotic drugs [101]. The structures synthesised correspond to molecules with a central nucleus made up of an alkyl imidothiocarbamate (alkyl isothiourea) or alkyl imidoselenocarbamate (alkyl isoselenourea) connected by a carbonyl group on each side to two identical lateral aromatic or heteroaromatic rings (Figure 20). The sulfur and selenium substituents were varied (methyl, ethyl and isopropyl) to determine the effect of the alkyl chain length and the ramifications that this has on the activity [105].

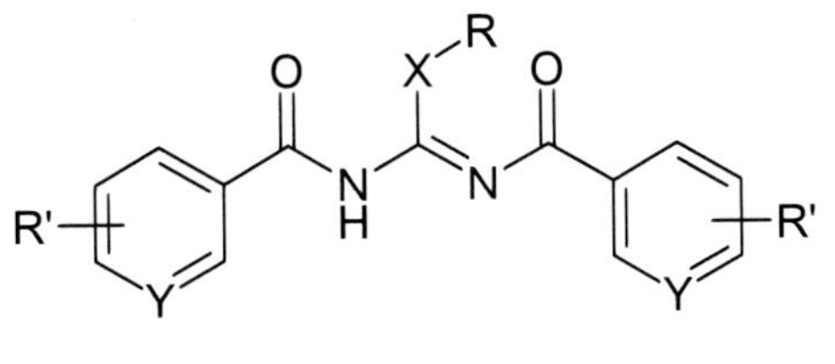

X = S, Se; Y = C, N; R = alkyl; R' = H, alkyl, methyloxy, chloro, cyano, nitro, trifluoromethyl

Figure 20.

The best results in PC-3 were obtained for the compound with X = Se, Y = C, R = CH_3 and R' = 4-CH_3. This compound was the most potent (IC_{50} = 1.85 μM) and was 4.5 times more active than standard methylselenic acid (IC_{50} = 8.38 μM) and 3.2 times more active than etoposide (IC_{50} = 13.6 μM), an agent used in the treatment of prostate cancer. In addition, the novel compound was less toxic than the reference and apoptotic inducer in MCF-7 and CCRF-CEM [106].

During the course of this work a great number of different structural classes of selenocompounds were reported [107,108]. For this reason, and in order to improve the potency of our compounds, we decided to introduce some structural modifications. For example, there is current interest in heterocyclic compounds that contain a selenium atom in the ring [19,23,31,48,59,71,72,77,78]. Bearing this fact in mind, and as a continuation of our previous work, we proposed the synthesis 1,2,5-selenadiazolo[3,4-*d*]pyridines and 1,2,5-benzoselenadiazolo derivatives (Figure 21).

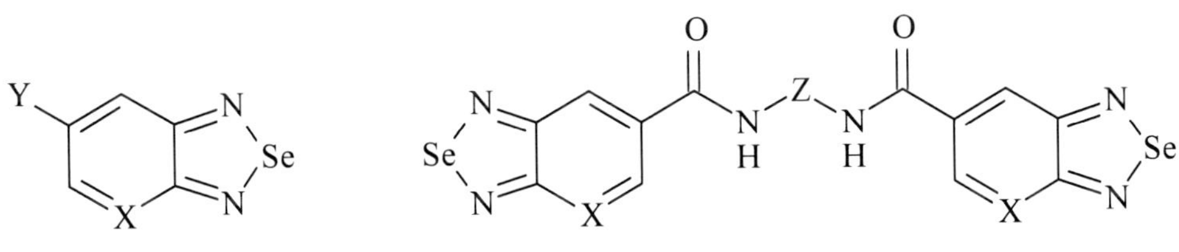

X = C, N; Y = X, alkyl, aryl, COOH, CN etc.; Z = alkyl

Figure 21.

The most promising molecule was a pyridine derivative (results will be published shortly). The preliminary results from the biological screening of these novel compounds are very encouraging and these systems could offer an excellent framework in this field and may ultimately lead to discovery of potent antitumour agent. Thus, the search for new drugs with selenium continues to be a great challenge in medical science.

Conclusion

It is clear from the studies discussed above that selenium compounds do have effects on growth, cell cycle and apoptosis and that such compounds offer great promise as anticancer and apoptotic agents in many tumoural processes, mainly for prostate cancer. As a result of these studies, selenium derivatives are rapidly emerging as valid chemotherapeutic agents. However, various organic and inorganic selenium compounds used in some studies have produced variable results when they are tested in animal models and human subjects and more investigations are urgently needed in order to ascertain the safety in their use. This chapter mainly covers the general chemical structures that incorporate selenium published in the last few years, with the aim of reviewing the most recent articles in this field. We have included some structures that have not yet been evaluated in prostate cancer cells because we believe that the study of these compounds would be of interest. Although several possible mechanisms have been proposed to explain the anticancer and apoptotic properties of selenium compounds [5,107], the results described here suggest the following preliminary considerations:

1. The chemical form is a determinant factor for the activity. The methylated forms, for example, methylselenocysteine, methylseleninic acid or selenomethionine, require metabolism to methylselenol for anticarcinogenic activity.
2. The effect of some selenium compounds mainly depends on the dose and the oxidation state of selenium. For inorganic selenium compounds the +4 oxidation state gives the highest anticarcinogenic properties and for organic selenium compounds the activity is mainly observed for the +2 oxidation state.
3. The existence of diverse responses for the same chemical structure suggests several mechanisms of action. For example, sodium selenite induces apoptosis by redox processes, p53 regulation and protein kinase-mediated increase in the DNA single-strand breaking mechanism amongst others. Methylseleninic acid provokes disruption of AR signalling, inhibits mitogen-stimulated proliferation of endothelial cells, and reduces the phosphorylation level of mitogen-activated protein kinases (MAPK) amongst other mechanisms. The expectation of a broad therapeutic benefit from agents that target only one member of either pathway may be overly simplistic due to the complex interrelated network governing apoptosis.
4. Experimental evidence shows that molecular symmetry, as a broad concept, could be a positive factor for cancer prevention and apoptosis (sodium selenite, methylseleninic acid, *p*-XSC, *p*-XMS, selenocystine, diselenodipropionic acid, 2-aminophenyl diselenide, bis-(2-amino-5-selenazolyl) ketones, diphenylmethyl-selenocyanate, 2-SeCD, diaryl and dialkylselenides, BBSKE). The importance of molecular symmetry in cytotoxic and pro-apoptotic activities was reported by us [101] in 2006. Recently, we described [105] a new series of symmetrical organoselenium compounds that are potent as cytotoxic agents in prostate cancer cells.

This class of compound offers a great deal of promise to broaden significantly the horizons of modern apoptosis and anticancer drug discovery for the potential treatment of prostate cancer. Animal data, epidemiological data, and intervention trials have shown a clear role for selenium derivatives in both the prevention of specific cancers and antitumourigenic effects in postinitiation phases of cancer through apoptosis induction. Accordingly, in recent years there has been substantial interest directed toward the synthesis of selenium-containing derivatives that could be used as cytotoxic, cancer chemopreventive and apoptotic agents. However, a great deal of further research is needed to unravel the precise manner in which selenium compounds act.

Acknowledgments

This work was supported by the University of Navarra Research Plan (Plan de Investigación de la Universidad de Navarra, PIUNA) and for CAJA NAVARRA Foundation.

References

[1] Bleys, J., Navas-Acien, A. & Guallar, E. (2008). Serum selenium levels all-cause, cancer and cardiovascular mortality among US adults. *Arch. Intern. Med.*, *168*, 404-410.

[2] Letavayova, L., Vlasakova, D., Spallholz, J. E., Bromanova, J. & Chovanec, M. (2008). Toxicity and mutagenicity of selenium compounds in Saccharomyces cerevisiae. *Mutat. Res.*, *638*, 1-10.

[3] Patrick, L. (2004). Selenium biochemistry and cancer: A Review of the Literature. *Altern. Med. Rev.*, *9*, 239-258.

[4] Bemis, D. L., Katz, A. E., Buttyan, R. (2006). Clinical trials of natural products as chemopreventive agents for prostate cancer. *Expert Opin. Investig. Drugs*, *15*, 1191-1200.

[5] Rikiishi, H. (2007). Apoptotic cellular events for selenium compounds involved in cancer prevention. *J Bioenerg Biomembr.*, *39*, 91-98.

[6] Li, F., Wang, F. F., Yu, F., Fang, Y., Xin, Z. H., Yang, F. M., Xu, J., Zhao, L. Y. & Hu, Q. H. (2008). *In vitro* antioxidant and anticancer activities of ethanolic extract of selenium-enriched green tea. *Food Chem.*, *111*, 165-170.

[7] Ueno, H., Hasegawa, G., Ido, R., Okuno, T. & Nakamuro, K. (2008). Effects of selenium status and supplementary seleno-chemical sources on mouse T-cell mitogenesis. *J. Trace Elem. Med. Biol.*, *22*, 9-16.

[8] Sinha, R., Pinto, J. T., Facompre, N., Kilheffer, J., Baatz, J. E. & El-Bayoumy, K. (2008). Effects of naturally occurring and synthetic organoselenium compounds on protein profiling in androgen responsive and androgen independent human prostate cancer cells. *Nutr. Cancer*, *60*, 267-275.

[9] Rebsch, C. M., Penna, F. J. & Copeland, P. R. (2006). Selenoprotein expresión is regulated at multiple levels in prostate cells. *Cell Res.*, *16*, 940-948.

[10] Diwadkar-Navsariwala, V., Prins, G. S., Swanson, S. M., Birch, L. A., Ray, V. H., Hedayat, S., Lantvit, D. L. & Diamond, A. M. (2006). Selenoprotein deficiency accelerates prostate carcinogenesis in a transgenic model. *Proc. Natl. Acad. Sci. USA*, *103*, 8179-8184.

[11] Husbeck, B., Bhattacharyya, R. S., Feldman, D. & Knox, S. J. (2006). Inhibition of androgen receptor signalling by selenite and methylseleninic acid in prostate cancer cells: two distinct mechanisms of action. *Mol. Cancer Ther.*, *5*, 2078-2085.

[12] Zhang, H., Wu, Y., Malewicz, B., Lu, J., Li, S., Marshall, J., Ip, C. & Dong, Y. (2006). Augmented suppression of androgen receptor signalling by a combination of alfha-tocopheryl succinate and methylseleninic acid. *Cancer*, *107*, 2942-2948.

[13] Gazi, M. H., Gonz, A., Donkena, K. V. & Young, C. Y. (2007). Sodium selenite inhibits interleukin-6-mediated androgen receptor activation in prostate cancer cells via upregulation of c-Jun. *Clin. Chim. Acta*, *380*, 145-150.

[14] Lee, S. O., Yeon Chun, J., Nadiminty, N., Trump, D. L., Ip, C., Dong, Y. & Gao, A. C. (2006). Monomethylated selenium inhibits growth of LNCaP human prostate cancer xenograpft accompanied by a decrease in the expression of androgen receptor and prostate-specific antigen (PSA). *Prostate*, *66*, 1070-1075.

[15] Pinto, J. T., Sinha, R., Papp, K., Facompre, N. D., Desai, D. & El-Bayoumy, K. (2007). Differential effects of naturally occurring and synthetic organoselenium compounds on biomarkers in androgen independent human prostate carcinoma cells. *Int. J. Cancer*, *120*, 1410-1417.

[16] Thirunavukkarasu, C., Premkumar, K., Sheriff, A. K. & Sakthisekaran, D. (2008). Sodium selenite enhances glutathione peroxidase activity and DNA strand breaks in hepatoma induced by *N*-nitrosodiethylamine and promoted by phenobarbital. *Mol. Cell. Biochem.*, *310*, 129-139.

[17] Foster, C. B., Aswath, K., Chanock, S. J., Mckay, H. F. & Peters, U. (2006). Polymorphism analysis of six selenoprotein genes: support for a selective sweep at the glutathione peroxidase I locus (3p2I) in Asian populations. *Bmc Gent.*, *7*, 1-20.

[18] Drake, E. N. (2006). Cancer chemoprevention: selenium as a prooxidant, not an antioxidant. *Med. Hypotheses*, *67*, 318-322.

[19] Franklin, M. R., Moos, P. J., El-Sayed, W. M., Aboul-Fadl, T. & Roberts, J. C. (2007). Pre- and post-initiation chemoprevention activity of 2-alkyl/arylselenazolidine-4(*R*)-carboxylic acids against tobacco-derived nitrosamine(NNK)-induced lung tumours in the A/J mouse. *Chem. Biol. Interact.*, *168*, 211-220.

[20] El-Sayed, W. M., Aboul-Fadl, T., Lamb, J. G., Roberts, J. C. & Franklin, M. R. Effect of selenium-containing compounds on hepatic chemoprotective enzymes in mice. *Toxicology*, (2006). *220*, 179-188.

[21] Terada, A., Yoshida, M., Seko, Y., Kobayashi, T., Yoshida, K., Nakada, M., Nakada, K., Echizen, H., Ogata, H. & Rikihisa, T. (1999). Active oxygen species generation and celular damage by additives of parenteral preparations: Selenium and sulfhydryl compounds. *Nutrition*, *15*, 651-655.

[22] Chen, J. J., Boylan, L. M., Wu, C. K. & Spallholz, J. E. (2007). Oxidation of glutathione and superoxide generation by inorganic and organic selenium compounds. *Biofactors*, *31*, 55-66.

[23] Sekiguchi, A., Nishina, A., Kimura, H., Fukumoto, R. H., Kogami, M., Ishihara, H. & Koketsu, M. (2006). Bis-(2-amino-5-selenazoyl)ketone as a superoxide anion-scavenger. *Biol. Pharm. Bull.*, *29*, 1404-1407.

[24] Talas, Z. S., Ozdemir, I., Yilmaz, I. & Gok, Y. (2008). Antioxidative effects of novel synthetic organoselenium compound in rat lung and kidney. *Ecotoxicol. Environ. Saf.*, (in press).

[25] Lee, J. T., Lehmann, B. D., Terrian, D. M., Chappell, W. H., Stivala, F., Libra, M., Martelli, A. M., Steelman, L. S. & McCubrey, J. A. (2008). Targeting prostate cancer based on signal transduction and cell cycle pathways. *Cell Cycle*, *7*, 1745-1762.

[26] Hallstrom, T. M. K. A. & Laiho, M. (2008). Genetic changes and DNA damage responses in the prostate. *Prostate*, *68*, 902-918.

[27] Li, G. X., Lee, H. J., Wang, Z., Hu, H., Liao, J. D., Watts, J. C., Combs, G. F. & Lu, J. (2008). Superior *in vivo* inhibitory efficacy of methylseleninic acid against human prostate cancer over selenomethionine or selenite. *Carcinogenesis*, *29*, 1005-1012.

[28] Li, G. X., Hu, H. B., Jiang, C., Schuster, T. & Lu, J. X. (2007). Differential involvement of reactive oxygen species in apoptosis induced by two classes of selenium compounds in human prostate cancer cells. *Int. J. Cancer*, *120*, 2034-2043.

[29] Hamilton, E. E. & Wilker, J. J. (2004). Inhibition of DNA alkylation damage with inorganic salts. *J. Biol. Inorg. Chem.*, *9*, 894-902.

[30] Hamilton, E. E., Fanwick, P. E. & Wilker, J. J. (2006). Alkylation of inorganic oxo compounds and insights on preventing DNA damage. *J. Am. Chem. Soc.*, *128*, 3388-3395.

[31] Juang, S. H., Lung, C. C., Hsu, P. C., Hsu, K. S., Li, Y. C., Hong, P. C., Shiah, H. S., Kuo, C. C., Huang, C. W., Wang, Y. C., Huang, L., Chen, T. S., Chen, S. F., Fu, K. C., Hsu, C. L., Lin, M. J., Chang, C. J., Ashendel, C. L., Chan, T. C. K., Chou, K. M. & Chang, J.Y. (2007). D-501036, a novel selenophene-based triheterocycle derivative, exhibits, potent *in vitro* and *in vivo* antitumoral activity which involves DNA damage and ataxia telangiectasia-mutated nuclear protein kinase activation. *Mol. Cancer. Ther.*, *6*, 193-202.

[32] Araldi, E. M. V., Dell'Aica, I., Sogno, I., Lorusso, G., Garbisa, S. & Albini, A. (2008). Natural and synthetic agents targeting inflammation and angiogenesis for chemoprevention of prostate cancer. *Curr. Cancer Drug Targets*, *8*, 146-155.

[33] Wang, Z., Hu, H., Li, G., Lee, H. J., Jiang, C., Kim, S. H. & Lü, J. (2008). Methylseleninic acid inhibits microvascular endothelial G_1 cell cycle progression and decreases tumor microvessel density. *Int. J. Cancer*, *122*, 15-24.

[34] Corcoran, N. M., Najdovska, M. & Costello, A. J. (2004). Inorganic selenium retards progression of experimental hormone refractory prostate cancer. *J. Urol.*, *171*, 907-910.

[35] Hu, H. B., Jiang, C., Li, G. X. & Lu, J. X. (2005). PKB/AKT and ERK regulation of caspase-mediated apoptosis by methylseleninic acid in LNCaP prostate cancer cells. *Carcinogenesis*, *26*, 1374-1381.

[36] Jiang, C., Wang, Z., Ganther, H. & Lü, J. (2002). Distinct effects of methylseleninic acid versus selenite on apoptosis, cell cycle, and protein kinase pathways in DU145 human prostate cancer cells. *Mol. Cancer Ther.*, *1*, 1059-1066.

[37] Gopalakrishna, R. & Gundimeda, U. (2002). Antioxidant regulation of protein kinase C in cancer prevention. *J. Nutr.*, *132*, 3819S-3823S.

[38] Cat, B., Stuhlmann, D., Steinbrenner, H., Alili, L., Holtkötter, O., Sies, H. & Brenneisen, P. (2006). Enhancement of tumor invasion depends on transdifferentiation of skin fibroblasts mediated by oxygen species. *J. Cell Sci.*, *119*, 2727-2738.

[39] McCarty, M. F. (2004). Targeting multiple signalling pathways as a strategy for managing prostate cancer: multifocal signal modulation therapy. *Integr. Cancer Ther.*, *3*, 349-380.

[40] Gasparian, A. V., Yao, Y. J., Lü, J., Yemelyanov, A. Y., Lyakh, L. A., Slaga, T. J. & Budunova, I. V. (2002). Selenium compounds inhibit IκB kinase (IKK) and nuclear factor-κB (NF-κB) in prostate cancer cells. *Mol. Cancer Ther.*, *1*, 1079-1087.

[41] Park, J. M., Kim, A., Oh, J. H. & Chung, A. S. (2007). Methylseleninic acid inhibits PMA-stimulated pro-MMP-2 activation mediated by MTI-MMP expression and further tumor invasion through suppression of NF-κB activation. *Carcinogenesis*, *28*, 837-847.

[42] Jiang, C., Ganther, H. & Lu, J. X. (2000). Monomethyl selenium-specific inhibition of MMP-2 and VEGF expression: Implications for angiogenic switch regulation. *Mol. Carcinog.*, *29*, 236-250.

[43] Hwang, J. T., Kim, Y. M., Surh, Y. J., Baik, H. W., Lee, S. K., Ha, J. & Park, O. J. (2006). Selenium regulates cyclooxygenase-2 and extracellular signal-regulated kinase signalling pathways by activating AMP-activated protein kinase in colon cancer cells. *Cancer Res.*, *66*, 10057-10063.

[44] Siddiqui, I. A., Afaq, F., Adhami, V. M. & Mukhtar, H. (2008). Prevention of prostate cancer through custom tailoring of chemopreventive regimen. *Chem. Biol. Interact.*, *171*, 122-132.

[45] Vunta, H., Davis, F., Palempalli, U. D., Bhat, D., Arner, R. J., Thompson, J. T., Peterson, D. G., Reddy, C. C. & Prabhu, K. S. (2007). The anti-inflammatory effects of selenium are mediated through 15-deoxy-Δ(12,14)-prostaglandin J(2) in macrophages. *J. Biol. Chem.*, *282*, 17964-17973.

[46] Özkan, H., Yavuz, S., Diçli, A., Yildirir, Y. & Türker, L. (2007). Synthesis of novel 5-aryl-1*H*-tetrazoles. *Heteroatom. Chem.*, *18*, 255-258.

[47] Sommen, G. L., Linden, A. & Heimgartner, H. (2007). Selenium-containing heterocycles from isoselenocyanates: synthesis of 5-amino-2,4-dihydro-3*H*-1,2,4-triazole-3-selones. *Helv. Chim. Acta*, *90*, 641-651.

[48] Sommen, G. L., Linden, A. & Heimgartner, H. (2007). Selenium-containing heterocycles from isoselenocyanates: synthesis of 1,3-selenazinane and 1,3-selenazinanone derivatives. *Helv. Chim. Acta*, *90*, 472-487.

[49] Meiler, J. & Schuler, M. (2006). Therapeutic targeting of apoptotic pathways in cancer. *Current Drug Targets*, *7*, 1361-1369.

[50] Kim, R., Emi, M. & Tanabe, K. (2005). Caspase-dependent and –independent cell death pathways alter DNA-damage. *Oncol. Rep.*, *14*, 595–599.

[51] Ghobrial, I. M., Witzig, T. E. & Adjei, A. A. (2005). Targeting apoptosis pathways in cancer therapy. *CA Cancer J. Clin.*, *55*, 178–194.

[52] Uzzo, R. G., Haas, N. B., Crispen, P. L. & Kolenko, V. M. (2008). Mechanisms of apoptosis resistance and treatment strategies to overcome them in hormone-refractory prostate cancer. *Cancer*, *112*, 1660-1671.

[53] Galluzzi, L., Zamzami, N., Rouge, T. D., Lemaire, C., Brenner, C. & Kroemer, G. (2007). Methods for the assessment of mitochondrial membrane permeabilization in apoptosis. *Apoptosis*, *12*, 803-813.

[54] Kumar, K. A. & Babu, P. P. (2002). Mitochondrial anomalies are associated with the induction of intrinsic cell death proteins Bcl2, Bax, cytochrome c and p53 in mice brain during experimental fatal murine cerebral malaria. *Neurosci. Lett.*, *329*, 319-323.

[55] Scarabelli, T. M., Knight, R., Stephanou, A., Townsend, P., Chen-Scarabelli, C., Lawrence, K., Gottlieb, R., Latchman, D. & Narula, J. (2006). Clinical implications of apoptosis in ischemic myocardium. *Curr. Probl. Cardiol.*, *31*, 181-264.

[56] Tarze, A., Deniaud, A., Le Bras, M., Maillier, E., Molle, D., Larochette, N., Zamzami, N., Jan, G., Kroemer, G. & Brenner, C. (2007). GAPDH, a novel regulator of the pro-apoptotic mitochondrial membrane permeabilization. *Oncogene*, *26*, 2606-2620.

[57] Joshua, A. M., Evans, A., Van der Kwast, T., Zielenska, M., Meeker, A. K., Chinnaiyan, A. & Squire, J. A. (2008). Prostatic preneoplasia and beyond. *Biochim. Biophys. Acta*, *1785*, 156-181.

[58] Das, R. K., Hossain, S. U. & Bhattacharya, S. (2007). Protective effect of diphenylmethyl selenocyanate against carbon tetrachloride-induced hepatic injury. *J. Appl. Toxicol.*, *27*, 527-537.

[59] Ahn, H. J., Koketsu, M., Yang, E. M., Kim, Y. M., Ishihara, H. & Yang, H. O. (2006). 2-(4-methylphenyl)-1,3-selenazol-4-one induces apoptosis by different mechanisms in SKOV3 and HL 60 cells. *J. Cell. Biochem.*, *99*, 807-815.

[60] Hu, H., Jiang, C., Ip, C., Rustum, Y. M. & Lü, J. (2005). Methylseleninic acid potentiates apoptosis induced by chemotherapeutic drugs in androgen-independent prostate cancer cells. *Clin. Cancer. Res.*, *11*, 2379-2388.

[61] Snyder, R. M., Yu, W. P., Jia, L., Sanders, B. G. & Kline, K. (2008). Vitamin E analog α-TEA, methylseleninic acid, and trans-resveratrol in combination synergistically inhibit human breast cancer cell growth. *Nutr. Cancer*, *60*, 401-411.

[62] Hu, H., Li, G. X., Wang, L., Watts, J., Combs, G. F. & Lu, J. X. (2008). Methylseleninic acid enhances taxane drug efficacy against human prostate cancer and down-regulates antiapoptotic proteins Bcl-XL and survivin. *Clin. Cancer Res.*, *14*, 1150-1158.

[63] Kim, A., Oh, J. H., Park, J. M. & Chung, A. S. (2007). Methylselenol generated from selenomethionine by methioninase downregulates integrin expression and induces caspase-mediated apoptosis of B16F10 melanoma cells. *J. Cell. Physiol.*, *212*, 386-400.

[64] Brawley, O. W. & Parnes, H. (2000). Prostate cancer prevention trials in the USA. *Eur. J. Cancer*, *36*, 1312-1315.

[65] Constantinou, C., Papas, A. & Constantinou, A. I. (2008). Vitamin E and cancer: An insight into the anticancer activities of vitamin E isomers and analogs. *Int. J. Cancer*, *123*, 739-752.

[66] Vraka, P. S., Drouza, C., Rikkou, M. P., Odysseos, A. D. & Keramidas, A. D. (2006). Synthesis and study of the cancer cell growth inhibitory properties of α-, γ-tocopheryl and γ-tocotrienyl 2-phenylselenyl succinates. *Bioorg. Med. Chem.*, *14*, 2684-2696.

[67] Shen, H. M. & Pervaiz, S. (2006). TNF receptor superfamily-induced cell death: redox dependent execution. *Faseb J.*, *20*, 1589-1598.

[68] McEligot, A. J., Yang, S. & Meyskens, F. L. (2005). Redox regulation by intrinsic species and extrinsic nutrients in normal and cancer cells. *Annu. Rev. Nutr.*, *25*, 261-295.

[69] Wu, Y., Zhang, H. T., Dong, Y., Park, Y. M. & Ip, C. (2005). Endoplasmic reticulum stress signal mediators are targets of selenium action. *Cancer Res.*, *65*, 9073-9079.

[70] Zhong, W. X. & Oberley, T. D. (2001). Redox-mediated effects of selenium on apoptosis and cell cycle in the LNCaP human prostate cancer cell line. *Cancer Res.*, *61*, 7071-7078.

[71] Nakamura, Y., Fena, Q., Kumagai, T., Torikai, K., Oigais, H., Osawa, T., Noguchi, N., Niké, E. & Uchida, K. (2002). Ebselen, a glutathione peroxidase mimetic seleno-organic compound, as a multifunctional antioxidant. Implication for inflammation-associated carcinogenesis. *J. Biol. Chem.*, *277*, 2687-2694.

[72] Sung, S. Y., Kubo, H., Shigemura, K., Arnold, R. S., Logani, S., Wang, R. X., Konaka, H., Nakagawa, M., Mousses, S., Amin, M., Anderson, C., Johnstone, P., Petros, J. A., Marshall, F. F., Zhau, H. Y. E. & Chung, L. W. K. (2006). Oxidative stress induces ADAM9 protein expression in human prostate cancer cells. *Cancer Res.*, *66*, 9519-9526.

[73] Lv, S. W., Wang, X. G., Mu, Y., Zang, T. Z., Ji, Y. T., Liu, J. Q., Shen, J. C. & Luo, G. M. (2007). A novel dicyclodextrinyl diselenide compound with glutathione peroxidase activity. *Febs J.*, *274*, 3846-3854.

[74] Sun, Y., Mu, Y., Ma, S., Gong, P., Yan, G., Liu, J., Shen, J. & Luo, G. (2005). The molecular mechanism of protecting cells against oxidative stress by 2-selenium-bridged β-cyclodextrin with glutathione peroxidase activity. *Biochim. Biophys. Acta*, *1743*, 199-204.

[75] Xiao, H. & Parkin, K. L. (2006). Induction of phase II enzyme activity by various selenium compounds. *Nutr. Cancer*, *55*, 210-223.

[76] Zhao, H. & Brooks, J. D. (2007). Selenomethionine induced transcriptional programs in human prostate cancer cells. *J. Urol.*, *177*, 743-750.

[77] Shi, C., Yu, L., Yang, F., Yan, J. & Zeng, H. (2003). A novel organoselenium compound induces cell cycle arrest and apoptosis in prostate cancer cell lines. *Biochem. Biophys. Res. Commun.*, *309*, 578-583.

[78] Shi, C. J., Zeng, H. H., Li, H. W., Yang, F. G., Wu, X. Q. & Yu, L. Z. (2003). Induction of apoptosis in prostate cancer cell line PC-3 by BBSKE, a novel organoselenium compound, and its effect *in vivo*. *Zhonghua Yi Xue Za Zhi.*, *83*, 1984-1988.

[79] Xiang, N., Zhao, R. & Zhong, W. (2008). Sodium selenite induces apoptosis by generation of superoxide via the mitochondrial-dependent pathway in human prostate cancer cells. *Cancer Chemother. Pharmacol.*, (in press).

[80] Fickova, M., Pravdova, E., Rondhal, L., Uher, M. & Brtko, J. (2008). *In vitro* antiproliferative and cytotoxic activities of novel kojic acid derivatives: 5-benzyloxy-2-selenocyanatomethyl- and 5-methoxy-2-selenocyanatomethyl-4-pyranone. *J. Appl. Toxicol.*, *28*, 554-559.

[81] Wang, Z., Jiang, C., Ganther, H. & Lü, J. (2001). Antimitogenic and proapoptotic activities of methylseleninic acid in vascular endothelial cells and associated effects on PI3K-AKT, ERK, JNK and p38 MAPK signalling. *Cancer Res.*, *61*, 7171-7178.

[82] Wu, Y., Zu, K., Warren, M. A., Wallace, P. K. & Ip, C. (2006). Delineating the mechanism by which selenium deactivates Akt in prostate cancer cells. *Mol. Cancer Ther.*, *5*, 246-252.

[83] McClung, J. P., Roneker, C. A., Mu, W., Lisk, D. J., Langlais, P., Liu, F. & Lei, X. G. (2004). Development of insulin resistance and obesity in mice overexpressing cellular glutathione peroxidase. *Proc. Natl. Acad. Sci. USA*, *101*, 8852-8857.

[84] Basu, A. & Sivaprasad, U. (2007). Protein kinase Cε makes the life and death decision. *Cell. Sign.*, *19*, 1633-1642.

[85] González-Moreno, O., Segura, V., Serrano, D., Nguewa, P., Rivas, J. D. L. & Calvo, A. (2007). Methylseleninic acid enhances the effect of etoposide to inhibit prostate cancer growth *in vivo*. *Int. J. Cancer*, *121*, 1197-1204.

[86] McCarty, M. F. & Block, K. I. (2006). Preadministration of high-dose salicylates, suppressors of NF-κB activation, may increase the chemosensitivity of many cancers: an example of proapoptotic modulation therapy. *Integrative Cancer Therapies*, *5*, 252-268.

[87] Lan, L., Zhao, F., Wang, Y. & Zeng, H. (2007). The mechanism of apoptosis induced by a novel thioredoxin reductase inhibitor in A549 cells: possible involvement of nuclear factor-κB-dependent pathway. *Eur. J. Pharmacol.*, *555*, 83-92.

[88] Zhu, S., Bjorge, J. D. & Fujita, D. J. (2007). PTP1B contributes to the oncogenic properties of colon cancer cells through Src activation. *Cancer Res.*, *67*, 10129-10137.

[89] Tonks, N. K. & Muthuswamy, S. K. (2007). A brake becomes an accelerator: PTP1B-a new therapeutic target for breast cancer. *Cancer Cell.*, *11*, 214-216.

[90] Julien S. G., Dubé, N., Read, M., Penney, J., Paquet, M., Han, Y., Kennedy, B. P., Muller, W. J. & Tremblay, M. L. (2007). Protein tyrosine phosphatase 1B deficiency or inhibition delays ErbB2-induced mammary tumorigenesis and protects from lung metastasis. *Nat. Genet.*, *39*, 338-346.

[91] Rudolf, E., Rudolf, K. & Cervinka, M. (2008). Selenium activates p53 and p38 pathways and induces caspase-independent cell death in cervical cancer cells. *Cell Biol. Toxicol.*, *24*, 123-141.

[92] Smith, M. L., Lancia, J. K., Mercer, T. I. & Ip, C. (2004). Selenium compounds regulate p53 by common and distinctive mechanisms. *Anticancer Res.*, *24*, 1401-1408.

[93] Wang, X. H., Wei, Y. M., Bai, H., Ou, J. F., Lu, J. H. & Zheng, R. L. (2004). Apoptosis and regulation of expressions of apoptosis-related gene Bcl-2 and p53 induced by selenium dioxide in three leukemia cell lines. *Di Yi Jun Yi Da Zue Xue Bao*, *24*, 1160-1163.

[94] Xiao, R., Qiao, J. T., Zhao, H. F., Liang, J., Yu, H. L., Liu, J., Guo, A. M. & Wang, W. (2006). Sodium selenite induces apoptosis in cultured cortical neurons with special concomitant changes in expression of the apoptosis-related genes. *Neurotoxicology*, *27*, 478-484.

[95] Zhao, R., Xiang, N., Domann, F. E. & Zhong, W. (2006). Expression of p53 enhances selenite-induced superoxide production and apoptosis in human prostate cancer cells. *Cancer Res.*, *66*, 2296-2304.

[96] Shiah, H. S., Lee, W. S., Juang, S. H., Hong, P. C., Lung, C. C., Chang, C. J., Chou, K. M. & Chang, J. Y. (2007). Mitochondria-mediated and p53-associated apoptosis

induced in human cancer cells by a novel selenophene derivative, D-501036. *Biochem. Pharmacol.*, *73*, 610-619.

[97] Sanmartín, C., Echeverría, M., Mendívil, B., Cordeu, L., Cubedo, E., García-Foncillas, J., Font, M. & Palop, J. A. (2005). Synthesis and biological evaluation of new symmetrical derivatives as cytotoxic agents and apoptosis inducers. *Bioorg. Med. Chem.*, *13*, 2031-2044.

[98] Sanmartín, C., Ardaiz, E., Cordeu, L., Cubedo, E., García-Foncillas, J., Font, M. & Palop, J. A. (2005). *Lett. Drug Des. Discov.*, *2*, 341-354.

[99] Font, M., Ardaiz, E., Cordeu, L., Cubedo, E., García-Foncillas, J., Sanmartín, C. & Palop, J. A. (2006). Structural characteristics of novel symmetrical diaryl derivatives with nitrogenated functions. Requirements for cytotoxic activity. *Bioorg. Med. Chem.*, *14*, 1942-1948.

[100] Echeverría, M., Mendívil, B., Cordeu, L., Cubedo, E., García-Foncillas, J., Font, M., Sanmartín, C. & Palop, J. A. (2006). Synthesis and biological evaluation of heteroaryldiamides and heteroaryldiamines as cytotoxic agents, apoptosis inducers and caspase-3 activators. *Arch. Pharm.*, *339*, 182-192.

[101] Sanmartín, C., Font, M. & Palop J. A. (2006). Molecular symmetry: A structural property frequently present in new cytotoxic and proapoptotic drugs. *Mini Rev. Med. Chem.*, *6*, 639-650.

[102] Cubedo, E., Cordeu, L., Bandrés, E., Rebollo, A., Malumbres, R., Sanmartín, C., Font, M., Palop, J. A. & García-Foncillas, J. (2006). New symmetrical quinazoline derivatives selectively induce apoptosis in human cancer cells. *Cancer Biol. & Ther.*, *5*, 850-859.

[103] Cordeu, L., Cubedo, E., Bandrés, E., Rebollo, A., Sáenz, X., Chozas, H., Domínguez, M. V., Echeverría, M., Mendívil, B., Sanmartín, C., Palop, J. A., Font, M. & García-Foncillas, J. (2007). Biological profile of new apoptotic agents based on 2,4-pyrido [2,3-*d*]pyrimidine derivatives. *Bioorg. Med. Chem.*, *15*, 1659-1669.

[104] Sanmartín, C., Domínguez, M. V., Cordeu, L., Cubedo, E., García-Foncillas, J., Font, M. & Palop, J. A. (2008). Synthesis and biological evaluation of 2,4,6-functionalized derivatives of pyrido[2,3-*d*]pyrimidines as cytotoxic agents and apoptosis inducers. *Arch. Pharm.*, *341*, 28-41.

[105] Plano, D., Sanmartín, C., Moreno, E., Prior, C., Calvo, A. & Palop, J. A. (2007). Novel potent organoselenium compounds as cytotoxic agents in prostate cancer cells. *Bioorg. Med. Chem. Lett.*, *17*, 6853-6859.

[106] Plano, D., Moreno, E., Font, M., Encío, I., Palop, J. A. & Sanmartín, C. Development of potent organoselenium compounds as antiproliferative and cytotoxic agents with apoptotic properties. *Bioorg. Med. Chem. Lett.*, (submitted for publication).

[107] Sanmartín, C., Plano, D. & Palop, J. A. (2008). Selenium compounds and apoptotic modulation: a new perspective in cancer therapy. *Mini Rev. Med. Chem.*, *8*, 1020-1031.

[108] Naithani, R. (2008). Organoselenium compounds in cancer chemoprevention. *Mini Rev. Med. Chem.*, *8*, 657-668.

(a) E.J. Wolvetang, K.L. Johnson, K. Krauser, S.J. Ralph and A.W. Linnane, FEBS Lett. 339 (1994), pp. 40–44.

In: Handbook of Prostate Cancer Cell Research...
Editor: Alan T. Meridith
ISBN: 978-1-60741-954-9

Chapter 7

Origin of Oscillations in Molecular Mechanism of Radioresistance of Prostate Carcinoma Cells: Implications for Successful Therapy

Petar M. Mitrasinovic*
Belgrade Institute of Science and Technology, Serbia

Abstract

Prostate cancer is the second leading cause of cancer-related mortality in the human male population. Conventional external beam doses below 68 Gy have a limited curative potential for the locally advanced stages of prostate cancer (LAPC) because two-thirds are hypothesized to experience disease relapse every 5 years after a treatment. Since it is well-established that ionizing radiation (IR) induces cellular toxicity through mechanisms involving generation of intracellular reactive oxygen species (ROS), the fate of irradiated cells is believed to be controlled by networks of signaling elements leading to different modes of cell death or cell survival. In this light, molecular mechanisms for the LAPC radiation resistance are not quite clear. In this chapter, a combined approach, based on experimental cell biology studies and systems biology strategies, has been shown to be of value for identifying signaling molecules with therapeutic relevancy, thus making more localized and efficient treatments of various ROS-mediated pathologies substantially facilitated. Since morphological, physiological and molecular studies on death of cells in tissues are usually carried out on isolated cell populations due to known difficulties manifested by interactions with surrounding cells, systems biology is herein considered and established to be a means of overcoming these difficulties. The practical usefulness of a systems biology approach in rationalizing the ability of cells to interpret the same TNF-induced signal action in distinct fashions, survival vs. death signal transduction, has been confirmed. Taking into account the fact that the low-LET-IR effects are

* petar.mitrasinovic@gmail.com

primarily reflected by ROS generation, in the context of the radioresistance of PC-3 cells, the IR-induced expression both of antioxidant enzymes (MnSOD, CuZnSOD, catalase) and of the transcriptional nuclear factor-kappa B (NF-κB) has been discussed. Besides, a systems biology model, aimed to shed more light on how MnSOD affects the biological state of PC-3 cells depending on the production of H_2O_2, has been described. Due to an excellent agreement between the experimental results and the systems biology insights, the radioresistance of PC-3 cells has been suggested to be a consequence of the positive, feed-forward vicious circle established between the H_2O_2-mediated elevation of MnSOD expression. The inclusion of some other important pathways, such as the hydroxyl radical (·OH)-mediated activation of NF-κB, has been demonstrated to bring oscillations in the model, thus indicating a broad range of possible instabilities that might have undesirable consequences on the development of successful therapies. Although the NF-κB signaling pathway in the model consists of 26 molecular components linked by 64 unidirectional reactions, a phase plane interpretation of the model has indicated that a much lower dimensional representation of the system of 64 kinetic equations could narrow down the range of possible instabilities that the oscillatory signaling may exhibit.

Keywords: prostate cancer, PC-3 cells, antioxidant enzymes, radioresistance, systems biology

Introduction

In the human male population, prostate cancer as the most commonly diagnosed malignancy is the second leading cause of cancer-related mortality [1]. Since much progress has been made in reducing mortality rate, the risk of prostate cancer death is only 3%, while the current lifetime risk of a prostate cancer diagnosis is 17% [2]. Based on statistical reports including cancer incidence, mortality, and survival by site, sex, race/ethnicity, education, geographic area, and calendar year, as well as the proportionate contribution of selected sites to the overall trends, prostate cancer accounts for more deaths than heart disease in men under age 85 years [3]. Further progress can be accomplished by supporting and applying novel discoveries across all segments of the male population.

Existing therapeutic methods for treating cancers can be generally divided into two groups being traditional and more recent respectively. Traditional are chemotherapy, radiotherapy, and surgery, while more recent are cancer stem cells-based therapy, oxidant-based therapy, and therapy based on light-activated molecules killing cancer cells. Cancer is most commonly treated by chemotherapy. Such treatment is relatively non-specific. Besides targeting tumor cells, the therapeutic drugs acting intravenously may cause general systemic distribution with deleterious side effects on surrounding healthy cells. Therefore, in the search for a novel drug-targeted delivery method, it is necessary to reduce both the extent of systemic distribution of the cytotoxic drug and unwanted side effects, and to reduce the dosage influenced by more localized targeting of the drug. Regardless of the progress accomplished in chemotherapy, cancer is still a leading life-threatening pathology. Consequently, there is a need for new therapeutic strategies based on our current knowledge of cancer biology.

Normal cells perfectly fit their environment by responding to external signals via precisely regulated pathways that either trigger or repress growth. Cancer arises when a cell, for a variety of reasons, begins to divide in an uncontrolled fashion. Tumorgenesis is conceivable as a multistep mechanism that reflects the genetic alterations driving progressively a normal tissue to malignancy. The key standpoint of developing successful therapy is to focus on cellular processes under oxidative stress conditions. In the context of prostate cancer, a combined approach based on cell biology studies and systems biology strategies is herein shown to be of vital importance for the elucidation of some critical phenomena, such as the radioresistance of PC-3 cells.

A key factor influencing outcome for clinically localized prostate cancer is tumor stage. Tumors extending palpably beyond the prostatic capsule were recognized as ones carrying a prognostic importance long before the advent of prostate-specific antigen (PSA). Radiation therapy was recognized as the mainstay of treatment for these locally advanced stages. The long-term results attained by radiation therapy in the pre-PSA era were relatively poor. Reports on outcome after radiation therapy properly emphasized the significance of the pretreatment PSA level when PSA was used as an endpoint [4]. Conventional external beam doses below 68 Gy were established to have limited curative potential for the locally advanced stages of prostate cancer because about two thirds were hypothesized to experience disease relapse every 5 years after a treatment is undertaken [5]. Molecular mechanisms of the radiation resistance of the locally advanced stages of prostate cancer, which are associated with reactive oxygen species production [6], are not currently well understood.

Cellular Processes under Oxidative Stress Conditions: From Experiment to Systems Biology

To understand the role of oxidative damage in disease, it is important to focus on principal mechanisms that are responsible for cellular functions under oxidative stress conditions. Due to an explosive progress in systems biology in recent years, biology has arguably established itself as one of the most exciting scientific fields [7]. The principal challenge facing systems biology is complexity in terms of investigating the interrelationships of all of the components in a functioning system in order to elucidate how the system works. With more than 30000 genes in the human genome the study of all relationships simultaneously becomes an incredibly difficult task [8]. Bearing in mind that biological phenomena cannot be predicted with a level of numerical accuracy as in classical physics, computational model building and signaling pathway analysis are anticipated to play a major role in the search for organizing principles [9]. The concept of cancer as a system failure and the potential of using systems biology approaches to understand the disease are generating significant debate in the literature [10]. System level understanding, the approach advocated in systems biology, requires a change in our notion of "what to look for" in biology [11]. While understanding role of individual genes and proteins continues to be important, the focus is now shifted towards understanding a system structure, function and dynamics. Systems-level kinetic models are thus expected to explain dynamic behavior and go far beyond the static pictures of the topologies of signaling pathways [12]. Since morphological, physiological and molecular studies on death of cells in tissues are usually carried out on isolated cell populations due to known difficulties manifested by interference and interaction with surrounding cells, systems biology is currently being considered as a means of overcoming the difficulties [13,14].

Various pathologies, including cancer, Alzheimer's disease, Parkinson's disease, sepsis, osteoporosis, diabetes, multiple sclerosis, rheumatoid arthritis, and inflammatory bowel diseases are closely related to genetically programmed cell death (PCD), a phenomenon that plays an important role in normal development, morphogenesis, tissue remodeling, and immune regulation [15]. A remarkable work [16], dating back to early seventies, has identified two pathways of PCD, apoptosis and necrosis, as well as a less likely variant of necrotic death. While apoptosis was characterized by condensation of nucleus and cytoplasm [17], necrosis was rationalized by abundant autophagic vacuoles with minimal or effectively no nuclear changes [18]. The unique features of an apoptotically dying cell have remained unchanged for different cell types and genotypes, and none of them has been observed in cells dying by necrosis. Noteworthy is that apoptosis and necrosis can be observed in the same cell. Therefore, a subtle balance between the two pathways, based on the relative expression level of several genes, may be suggested [17]. Molecular details of the overall machinery directing apoptotic pathways have been established in various types of cells [15,17]. The key issue that remains to be elucidated is fundamental background of the ability of cells to interpret the same signal action in distinct fashions, e.g., survival vs. death signal transduction [19].

Over the past decade, there has been a considerable progress towards identifying constituent components of molecular mechanisms that trigger programmed cell death. An intrinsic component or a modulator of a particular pathway is capable of interacting with other signaling pathways within a cell. For instance, transcription factors acting on gene promoters are able to trigger several possible mechanisms, each of which could lead to different outcomes in specific biological contexts. It is believed that general principles governing cellular functions, such as signal transduction, may be discovered by functional transition from molecular to modular cell biology [20,21]. This transition is of vital importance for understanding a system function in terms of defining functional modules, which are conceivable as groups of individual molecular components acting as a whole within a complex signaling scenario. Signaling pathways underlying the modules can be viewed as predictive kinetic connectivity maps, which are context and disease specific, including cell type, organ and tissue. The main feature of the kinetic connectivity maps, being both qualitative in terms of identifying critical molecules responsible for an outcome and quantitative in terms of values of concentrations and binding constants, is their ability to be validated using traditional laboratory techniques. In this context, applications of the state-of-the-art systems biology methods, derived form engineering and computer science, and stronger interactions between experiment and theory in cell biology may be quite relevant for understanding bioprocesses of cancer cell therapeutics.

Reactive Oxygen Species-Mediated Pathways with Relevant Therapeutic Targets: TNFα-Induced Cross-talk between NF-κB and JNK

To make more localized treatment of a ROS-mediated pathology facilitated, identifying key molecular components and understanding their role in signaling pathways underlying a specific outcome (cell survival or cell death) are of vital importance for knowing new therapeutic targets. Since 1984, when human tumor necrosis factor (TNF) was purified and its encoding cDNA was cloned and expressed [15], TNF-mediated apoptosis has attracted a tremendous attention of many multidisciplinary research groups worldwide. The principal reason lies in a permanent interest of the researchers to gain more insights into the anticancer activity, now known as TNF signaling. Systems biology is shown below to be quite an effective tool for the elucidation of the ability of cells to interpret the same TNF-induced signal action in distinct fashions, e.g., survival vs. death signal transduction.

TNF signaling is a complex scenario of intracellular events generated with help from the interaction of cytokine TNF, TNFα, with two receptors, TNF-R1 and TNF-R2. Especially effective is the stimulation of TNF-R1 being a driving force for several signal transmissions via protein interactions. In this way, the TNFα-induced formation of a multiprotein-signaling complex at the cell membrane is observed from a biochemical standpoint. Among other possible outcomes, the binding of TNFα to TNF-R1 results in the activation of two transcription factors, nuclear factor κB (NF-κB) and c-Jun. These transcription factors are crucial for the expression of genes implicated in a variety of biological processes, such as cell growth and death, development, oncogenesis, and immune, inflammatory, and stress responses. As far as additional key outcomes are concerned, the TNFα-to-TNF-R1 signal transduction results in apoptosis as well as in the activation of c-Jun NH_2-terminal kinase (JNK) [15]. NF-κB transcription factors are well established to act against the TNFα-induced apoptosis under inflammatory conditions [22]. The TNF signaling network is uniquely determined by intricacy of extensive cross-talk between the NF-κB and JNK pathways [15]. Even though most of the molecular actors participating in a series of the intracellular events have been characterized by means of both biochemical and genetic techniques, numerous intriguing questions in conjunction with the TNFα-induced NF-κB/JNK cross-talk seek answers.

The actions of ROS, such as singlet oxygen ($^1O_2$), hydrogen peroxide (H_2O_2), superoxide anion (O_2•-), hydroxyl radical (•OH), and hydroperoxyl radical (HOO•), disturb the oxidant-antioxidant balance in favor of the former, thus influencing many pathological conditions. Up to 10^3 oxidative damaging events upon the DNA of each cell in the human body have been estimated to occur every day [6]. In eukaryotes, most oxygen radicals stem from mitochondria [22]. Cells contain efficient strategies to antagonize ROS, e.g., by synthesizing superoxide dismutases (SODs), catalase, and glutathione peroxidases (Gpxs) from the family of enzymes capable of converting ROS into less-reactive species, as well as by limiting the cellular availability of transition metals, primarily iron [23]. Both suppressing the accumulation of ROS and controlling the activation of the JNK cascade can account for the antiapoptotic activity of NF-κB. Although some fundamental progress on the issue has been made using genetic means [22], the mechanisms responsible for inhibition of ROS proliferation by NF-κB remain puzzled. There has also been a controversy on the role of JNK activation in apoptosis. The JNK pathway has been hypothesized to function in a cell-type and stimulus-dependent manner [24]. In this context, an indirect way of halting the JNK cascade via an additional link between the NF-κB and JNK pathways appears to be unclear [22]. Whether JNK is an intrinsic component or a modulator of the death pathway, influenced by other signaling pathways such as NF-κB pathway, is the question of vital importance [24].

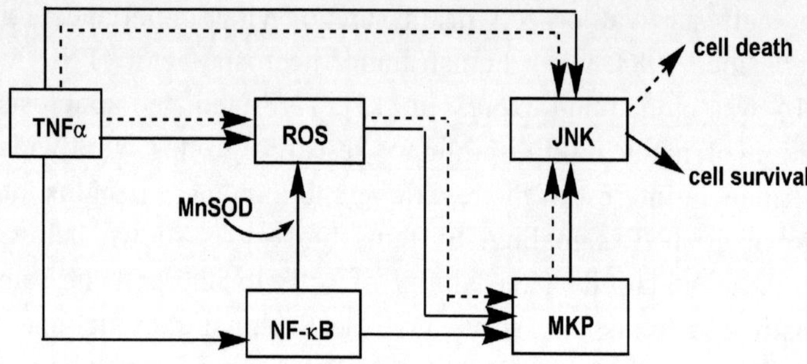

Figure 1. TNFα-induced NF-κB/JNK cross-talk. Transient JNK activation in NF-κB-rich cells (solid line) and sustained JNK activation in NF-κB-deficient cells (dashed line).

Attractive signaling pathways for ROS-promoted cell death (Figure 1) have been experimentally proposed in the literature [25]. It has been shown that TNFα-induced ROS generation can either be suppressed by NF-κB targets or lead to programmed cell death by inactivating JNK phosphatase from the mitogen-activated protein kinase (MAPK) phosphatase family (MKP). NF-κB therefore acts as a modulator of the TNFα-induced death response by controlling both ROS accumulation and MKP activity. This means that transient JNK activation is observed in cells with a negative regulator of PCD such as NF-κB (solid line in Figure 1), while sustained JNK activation is observed in NF-κB-deficient cells (dashed line in Figure 1). From a more biochemical perspective, ROS inhibit the JNK-inactivating phosphatase by both oxidizing and converting its catalytic cysteine to sulfenic acid. The induction of mitochondrial superoxide dismutase (MnSOD) has some impact on the prevention of ROS accumulation by NF-κB. A possible mechanism for the TNFα-induced ROS generation is unknown [25]. To decipher the key puzzles of the ROS-promoted NF-κB/JNK cross-talk, a simultaneous interplay of the TNFα-induced NF-κB and JNK pathways in terms of the connectivity of their constituent components has been observed.

Systems Biology View at TNFα-Induced Cross-Talk between NF-κB and JNK: Cell Death and Cell Survival

We have extended the outline shown in Figure 1 by including other molecular details of the signaling pathways that can be found in the literature. The TNFα-induced NF-κB pathway has been based upon the model of Ihekwaba *et al.* [26], while the TNFα-induced JNK pathway has been considered as the mitogen-activated protein kinase cascade of Kholodenko [27]. Besides, the two signaling routes have been linked by a series of three additional connections, being between NF-κB and ROS, NF-κB and MKP, and ROS and MKP, respectively. Since possible mechanisms of the TNFα-induced ROS generation are note quite clear [25], the TNFα → ROS segment has not been considered. Hence, the model, consisting of 36 molecules (IκBα, NF-κB, IκBα-NF-κB, IκBβ, IκBβ-NF-κB, IκBε, IκBε-NF-κB, IKKIκBα, IKKIκBα-NF-κB, IKK, IKKIκBβ, IKKIκBβ-NF-κB, IKKIκBε, IKKIκBε-NF-κB, NF-κB$_n$, IκBα$_n$, IκBα$_n$- NF-κB$_n$, IκBβ$_n$, IκBβ$_n$-NF-κB$_n$, IκBε$_n$, IκBε$_n$-NF-κB$_n$, *Source*, IκBα-t, *Sink*, IκBβ-t, IκBε-t, ROS, MKP, JNK, JNK-PP, MKK, MKK-P, MKK-PP, MKKK, MKKK-P, TRAF2) participating in a series of 77 unidirectional reactions, is shown schematically in Figure 2a. Even though the stimulation of IKK by TNFα has been imagined to proceed via the TNFα→TNF-R1→TRADD→RIP→IKK signaling, raising the concentration of IKK to 0.1 μM, after an equilibration period of 2000 minutes has simulated the stimulation [26]. The dashed line in Figure 2a shows the feedback effect of JNK on the rate of MKKK phosphorylation. An extracellular signal for switching off the TRAF2-to-68 signaling has been assumed to contribute to transient JNK activation. The TRAF2-to-68 signaling without the feedback (dashed line in Figure 2a) has been an activator of the protein kinase cascade in the context of sustained JNK activation. It is clear that a predominant role of the protein kinase cascade over the NF-κB signaling pathway (Figure 2a), that is, a predominant role of the Object module over the Controller module (Figure 2b) is associated with the sustained activation JNK leading to cell death. This specific outcome is closely related to cells with an insufficient amount of NF-κB, and therefore suggests a functional classification of cells, such as NF-κB-rich and NF-κB-deficient cells (Figure1) corresponding to cell survival and cell death.

Understanding the role of NF-κB in the regulation of ROS accumulation is related to a cell type. A biological context both of NF-κB-rich cells and of NF-κB-deficient cells is therefore a determinative factor of programmed cell death. Essentially, a simple IκB-NF-κB signaling model of Hoffmann *et al.* [28] has been employed to regulate ROS production. According to this model, NF-κB is maintained inactive in the cytoplasm of a non-stimulated cell by three IκB isoforms, IκBα, IκBβ, and IκBε. By stimulating a cell and activating IKK complex, phosphorylation and ubiquitination of the IκB proteins are observed. Consequently, activation of genes including IκBα occurs by moving unbound NF-κB into the nucleus. Synthesizing IκBβ and IκBε at a steady rate provides complex temporal control of NF-κB with a local negative feedback [28]. The particular context of the Controller module in terms of the system function is nicely visible in Figure 2b illustrating the overall modular representation of the signaling pathways given in Figure 2a. To evaluate control performances of the biological scenario, it is necessary to observe the actual time-dependent responses associated with JNK activation. The ultimate goal is thus to determine the JNK concentration vs. time plots underlying different outcomes, such as cell death and cell survival, and to address the question of how the plots correlate with the concentrations of NF-κB. As the NF-κB pathway of Ihekwaba *et al.* [26] had been based on the IκB-NF-κB signaling pathway of Hoffmann *et al.* [28] modeling the experimental data reasonably well, the oscillatory concentrations of NF-κB in the nucleus (NF-κB$_n$) is the most realistic choice to be explored. However, the correlation between JNK activation type (transient and sustained) and temporal control of concentrations of some important molecules (other than NF-κB) needs to be highlighted as well.

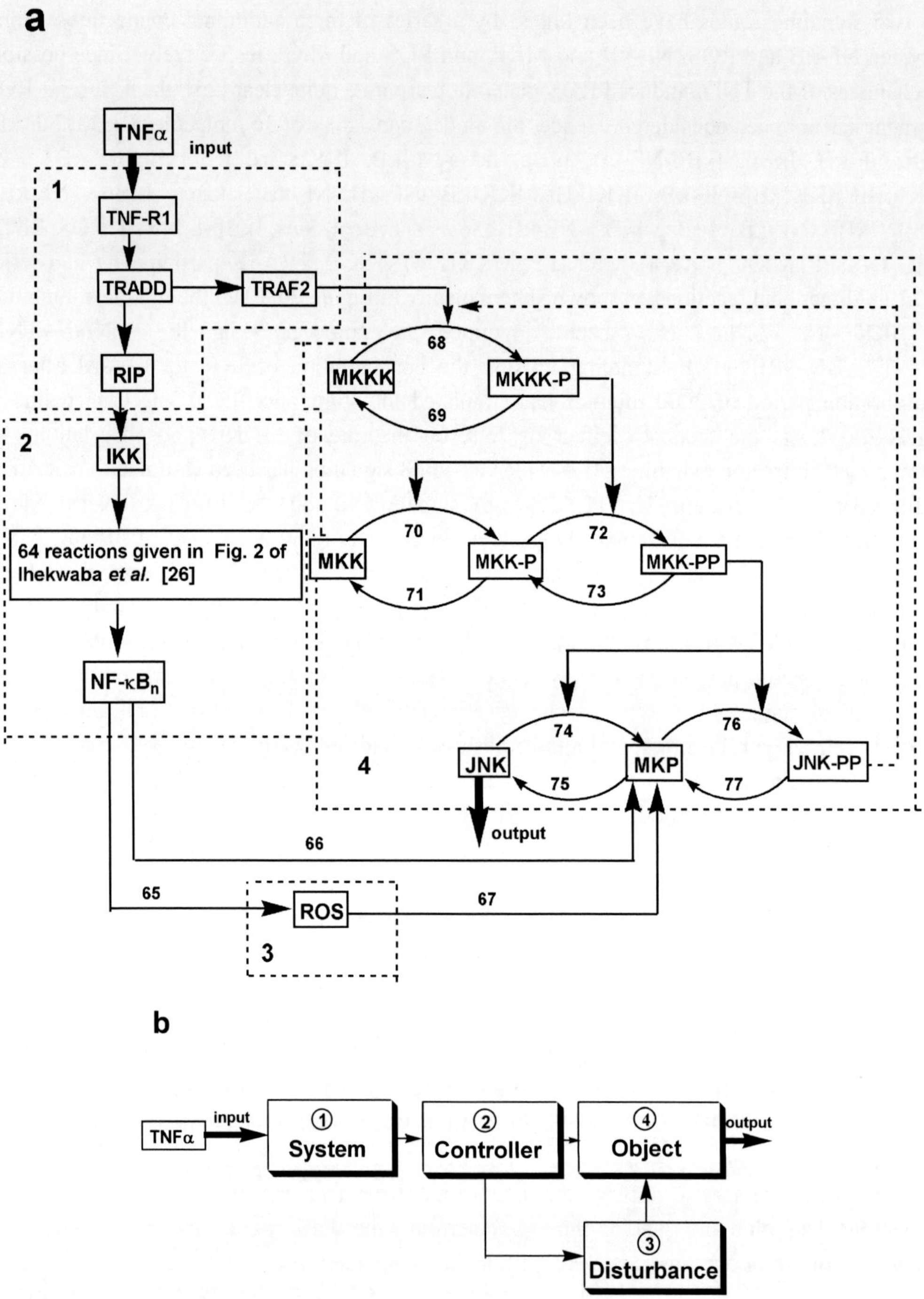

Figure 2. (a) Kinetic scheme describing ROS-mediated NF-κB/JNK cross-talk. Numbering corresponds to individual reaction steps. The dashed line gives feedback effect of JNK on the rate of MKKK phosphorylation. Note that an extracellular signal for switching off the TRAF2-to-68 signaling corresponds to transient JNK activation. The TRAF2-to-68 signaling without the feedback (dashed line) corresponds to sustained JNK activation. (b) Modular representation of the system function corresponding to the kinetic scheme.

All reactions including possible transport, consummation and complex formation rates must be knowable before attempting to describe changes in the concentration of a reaction component. Even though connection maps, such as that in Figure 2a, are qualitative in nature, quantitative information (initial concentrations, binding and kinetic rate constants) on the identity of the molecular components and their interactions is needed in order to approach predictive models [29]. The initial concentrations for all molecules other than ROS have been found in the literature [26,27]. The initial concentrations of ROS have been those used in [25], that is, 0.3, 1, 3, and 10 μM. Relevant kinetic data for the reactions 1-64 have been taken from table 2 of Ihekwaba *et al.* [26], while those for the reactions 68-77 have been taken from table 2 of Kholodenko [27]. The modeling environment GEPASI 3.30, running under the MS-Windows operating system, has been employed to carry out the simulations of biochemical pathways [30]. The GEPASI 3.30 suite of programs is capable of performing time-course and steady-state simulations. Its scan parameter capability is one feature of the software package that has been exploited extensively. It allowed varying specific parameters, selecting the range, and defining the extent of the variation. In this way, unknown values of the mass action (irreversible) constants for the reactions 65, 66 and 67 (Figure 2a) have been estimated to be 2e-5, 4e-5, and 1e-3 s^{-1}, respectively.

To examine regulation performances of the NF-κB pathway in controlling the TNFα-induced generation of ROS, the time-dependent concentration of nuclear NF-κB, in terms of the IκB-NF-κB signaling model of Hoffmann *et al.* [28], has been kept unchanged for any initial concentration of ROS. The time course of NF-κB$_n$ is shown schematically in Figure 3. This oscillatory plot illustrates the TNFα-induced NF-κB pathway in cells that do not contain two isoforms of IκB, IκBβ and IκBε, thus reducing the possibility of having the IκB-mediated inhibition of NF-κB.

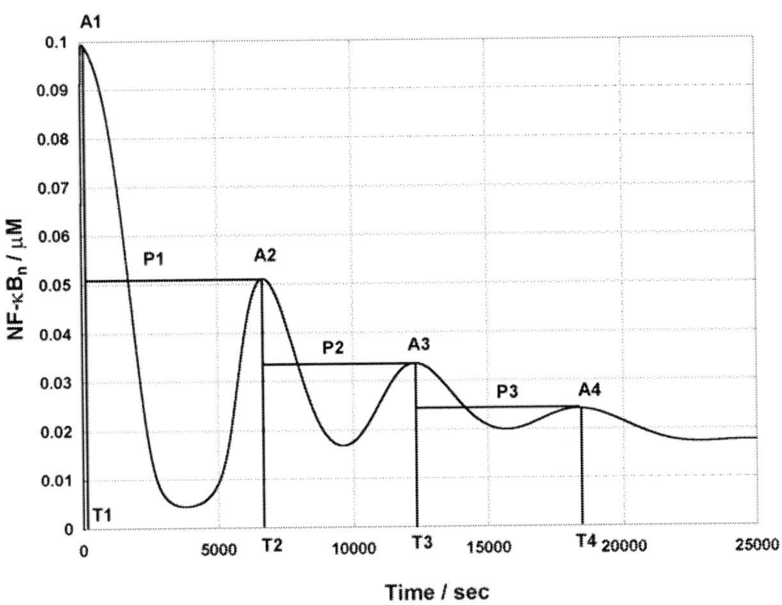

Figure 3. Nuclear NF-κB (NF-κBn) vs. time. The definitions of the amplitudes (A), times (T) and periods (P) characterize nuclear NF-κB as a function of time.

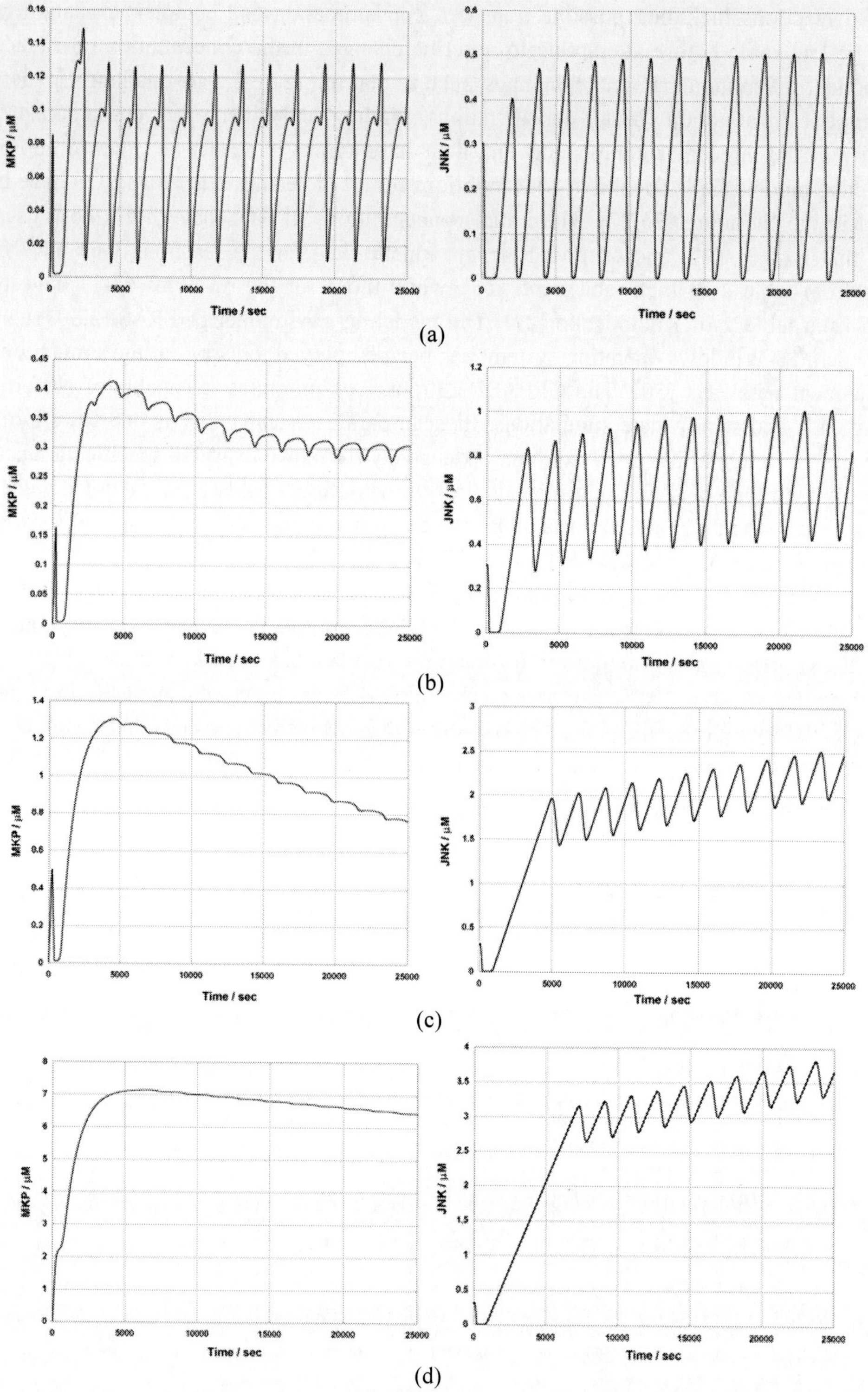

Figure 4. MKP vs. time (left) and JNK vs. time (right) for the following initial concentrations of ROS: (a) 0.3 μM, (b) 1 μM, (c) 3 μM, and (d) 10 μM. Note that the dynamic picture of JNK is associated with the transient activation of JNK.

In addition to controlling ROS accumulation, NF-κB controls the TNFα-induced death response by influencing MKP activity [25]. Hence, the time course of MKP associated with the time-dependent response of JNK for various initial concentrations of ROS is given in Figure 4 too. For an initial ROS concentration of 0.3 μM (Figure 4a), both MKP and JNK display a clear oscillatory behavior, that of JNK being quite similar to the time course of MAPK that has been reported by Kholodenko [27]. This means that the negative feedback (dashed line in Figure 2a) likely brings oscillations in the mitogen-activated protein kinase cascade, which has the prevailing impact on JNK as a function of time relative to the NF-κB signaling pathway. As far as the external activation of an initial kinase and the subsequent activation of a terminal cascade kinase are concerned, the substantial negative feedback can be usable in turning off the activation of a cascade, as given in the caption of Figure 2. The dynamic plot of JNK is in line with a transient response to a stimulus [31]. Following an increase in the initial concentration of ROS, from 0.3 to 10 via 1 and 3 μM, the presence of oscillations in the dynamic response of JNK is preserved, but the oscillations get more localized in terms of their amplitude range (Figure 4b-d). The preserved oscillatory behavior of JNK is associated with a clear disappearance of oscillations in the MKP activity (Figure 4b-d), thus accounting for a more pronounced interplay between the NF-κB and JNK signaling pathways.

A previous finding has pointed out MKP as a prospective therapeutic target that substantially contribute to the possibility of having quite a satisfactory localization of potential therapies [25]. However, the systems biology strategy described in this chapter has been based on a system of 64 differential (kinetic) equations defining the dynamic behavior of the NF-κB signaling pathway consisting of 26 species linked by 64 unidirectional reactions [13]. The mathematical model of the Controller module (Figure 2) is essentially associated with the extreme non-linearity of the entire system, thus indicating a broad range of possible instabilities that could have therapeutically undesirable consequences on reducing ROS generation by mitochondrially-targeted antioxidants. On the basis of an extensive sensitivity analysis of parameters controlling the oscillatory signaling, Ihekwaba *et al.* [26] have shown that the mathematical representation of the TNFα-induced NF-κB signaling pathway can be reduced to 9 most important reactions/parameters. Interestingly, these 9 parameters directly affect the concentrations of only 2 reactants other than NF-κB$_n$, IKK and IκBα. To learn more on the involvement of these molecules into the signaling scenario shown in Figure 2a, the phase plane plot of the time-dependent relationship between the concentrations of IKK, IκBα and JNK is shown schematically in Figure 5. The three stated variables, plotted against each other, reveal an intimate involvement of the mediators in the oscillatory behavior of JNK. This might indicate that a much lower dimensional representation of the NF-κB signaling pathway, in terms of the number of differential equations, is able to depict the oscillatory signaling in the overall NF-κB/JNK cross-talk. Even though the reduced representation does not fit the biological scenario formally, it does provide a simplistic picture limiting the range of possible instabilities associated with the oscillations. In light of several results reported in the literature [32,33], additional modeling investigations are needed to explore the stimulus-dependent functioning of the signaling pathways.

In contrast to the negative feedback (dashed line in Figure 2a), a cascade with a strong positive feedback can show a dynamics corresponding to one of the three different steady states: stable "off" and stable "on" states with low and high phosphorylation levels, respectively, separated by an unstable state determining a threshold level [34]. When the kinase activity is larger than the threshold, the cascade switches to its "on" state. The time course of the concentration of JNK, when the cascade activation (1e-005 μM) exists and the negative feedback is not present (see the caption of Figure 2), is shown schematically in Figure 6. This dynamic picture corresponds to a sustained activation, as opposed to a transient response [31]. The dynamic behavior of MKP associated with the sustained activation of JNK is given in Figure 6 as well. Since MKP has been previously hypothesized as a potential therapeutic target [25], it is important to note that a very characteristic, transient profile of MKP is associated with sustained activation of JNK.

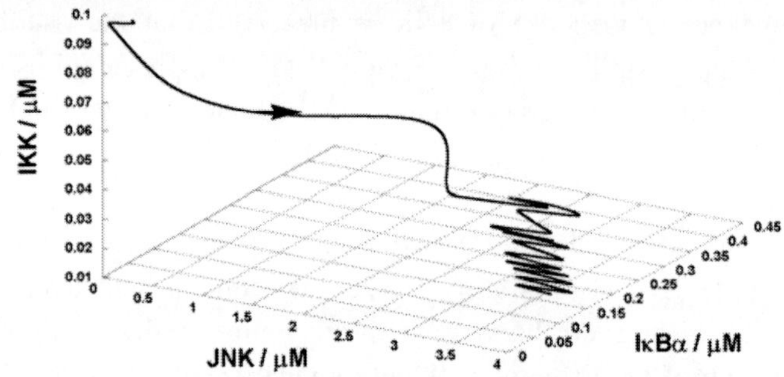

Figure 5. Phase plane plot between the time-dependent concentrations of IKK, IκBα and JNK in the context of transient activation of JNK. Initial concentration of ROS is 3 μM.

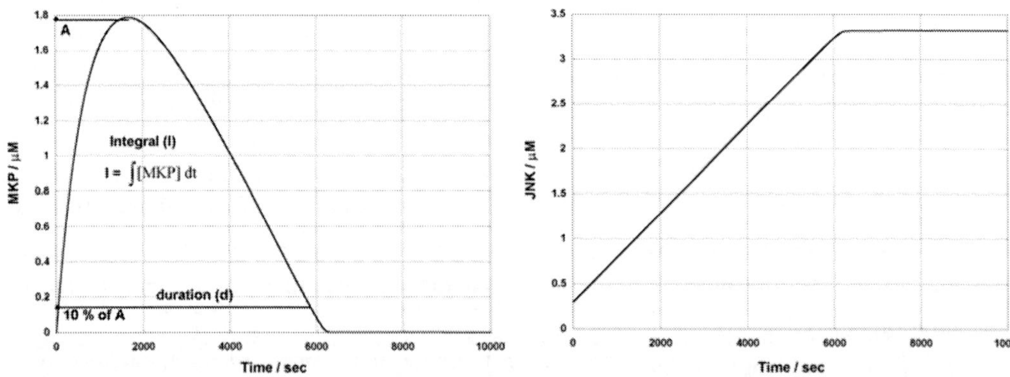

Figure 6. MKP vs. time (left). Amplitude (A), duration (d) and integrated response (I) depict the transient profile of MKP. Sustained activation of JNK (right). Initial concentration of ROS is 3 μM.

Based on the discussions presented by now, the practical usefulness of a systems biology approach in rationalizing the ability of cells to interpret the same TNF-induced signal action in distinct fashions, survival vs. death signal transduction, has been confirmed. The considered strategy has been capable of determining transient and sustained JNK activations underlying cell survival and cell death respectively. Moreover, a functional classification of cells, such as NF-κB-rich and NF-κB-deficient cells has been identified as a key factor influencing the two different outcomes. In other words, the computational strategy has been demonstrated to be of value for identifying signaling molecules with therapeutic relevancy, thus suggesting supplementing experimental laboratory techniques by molecular modeling in targeted therapeutic technologies.

Radioresistance of PC-3 Cells and Implications for Successful Therapy: From Experiment to Systems Biology

The DNA-binding of NF-κB plays an important role in the androgen-independent prostate tumor cells PC-3 by correlating with disease progression [35]. NF-κB, as a major signaling molecule, promotes malignant behavior by suppressing apoptosis and stimulating transcription of proteins involved in cell cycle progression [36]. NF-κB is believed to be influenced by overexpression of antioxidant enzymes, primarily superoxide dismutases (SODs) [37], since NF-κB-binding sequences have been found in the promoters of SODs [38-40]. Metastatic lesions of prostate cancer contain higher levels of antioxidant enzymes, such as SODs and catalase, but significantly lower levels of glutathione peroxidases with respect to primary tumors, suggesting that antioxidant enzymes in metastatic and primary tumors may be modulated in different fashions [41,42]. Although antioxidant enzymes are able to scavenge oxidants [43], ROS escaping from the enzymes are still capable of damaging cells to a certain extent [44]. Taking into account that various constitutive enzyme equipments of the metastatic PC-3 cell line offer the possibilities of investigating the antitumor effects of ionizing radiatin (IR), a therapeutic dose of 2 Gy as well as cumulative doses between 10 and 20 Gy have been used in our previous study [45]. In addition to the expressions of cellular antioxidant enzymes and NF-κB the IR-induced changes in cellular parameters, such as cell viability, proliferative index, apoptosis, and cell cycle phases have been observed [45]. In this section, the radioresistance of PC-3 cells is elaborated using our measurements [45] in combination with a systems biology model based on the previously discussed NF-κB signaling module [13,46].

A short account of the experimental conditions under which our cell culture studies [45] have been carried out is as follows. Androgen-independent (PC-3) human prostate carcinoma cell line has been provided by the ATCC - American Type Culture Collection (CRL 1435, Roekville, MD). Monolayer cell cultures have grown in RPMI 1640 and supplemented with 10% fetal bovine serum, 2 mM L-glutamine, 100 IU/ mL penicillin / 100μg/mL streptomycin at 37°C in a humidified atmosphere with 5% CO_2, and subcultured once a week. To study the radiation-induced effects, $1.2x10^6$ cells have been seeded in 75 cm^2 culture flasks. In the exponential phase of cell growth, cell radiation has been performed using 2, 10 and 20 Gy of gamma-rays from a ^{60}Co source at a fixed dose rate of 20 Gy/h. The cell viability has been determined by trypan blue exclusion (TBE) assay. The ratio between viable cell numbers in IR and control cells has been reported as the Viability index (Vi). For cell cycle analysis, cells have been fixed with 70% ethanol, treated with 50 μg/mL RNAse-A, stained with 50 μg/mL propidium iodide (PI) and analyzed with a FACS Calibur flow cytometer (Becton Dickinson, San Jose, CA, USA). The determination of cell cycle distribution has been established using ModFIT software (Verity Software House, Inc., Topsham, ME, USA). For the cell death analysis, 10^5 PC-3 cells have been stained with 5 μg/mL Annexin V-FITC and 5 μg/mL PI (Travigen Inc., Gaithersburg, MD, USA), and multiparameter cytometric measurements have been done using a minimum of 20.000 events for each sample. The data were processed with the LYSIS II software (Becton Dickinson). The trypsinized and washed cells have been lysed with a buffer containing 0.32M sucrose, 10mM TrisHCl pH 7.4, 5mM $MgCl_2$, and 1% Triton X-100. Aliquots of lysates have been denatured, separated by electophoresis on 10% acrylamide gel, and transferred to nitrocellulose membranes. The membranes have been blocked using 10 mM Tris buffer pH 7.4 supplemented with 150 mM NaCl, 1% BSA and 0.1% Tween-20, and separately incubated with rabbit anti-NF-κB or actin antibody (NF-κB p65, C-20; SC-372, or C-11; SC-1615, Santa Cruz Biotechnology, USA), rabbit anti-human MnSOD or CuZnSOD antibody (SOD-101 or SOD-100, Stressgen Biotechnologies, Victoria, BC, Canada), and rabbit anti-human catalase antibody (Calbiochem, USA). A secondary goat anti-rabbit IgG Horseradish Peroxidase Conjugate SAB-300 (Stressgen Biotechnologies, Victoria, BC, Canada) has been employed for detection. The quantification of specific antigen bands has been performed by computer image processing (Image J). The level of detected proteins is expressed in arbitrary units (AU/mg of protein). The reported values have been in the form of mean ± SD or mean ± SEM of two separate experiments with N representing the number of measurements in every single experiment. Statistical significance has been evaluated using a one-way analysis of variance (ANOVA) and the Tukey post-hoc test. The probability thresholds, $^{***}p<0.001$, $^{**}p<0.01$, and $^{*}p<0.05$ have been considered significant throughout the work [45].

In order to analyze expression of cellular antioxidant enzymes as a function of ionizing radiation, the levels of MnSOD, CuZnSOD and catalase have been measured in cell lysates 72 h after the treatment. The Western immunoblotting analysis has shown that the MnSOD level has been unchanged at 2 Gy, but it has experienced a 2-fold increased at 10 and 20 Gy. The expression of CuZnSOD has significantly increased at all radiation doses. The expression of catalase has slightly increased after IR, but these alterations have not been of statistical significance. Noteworthy is the expression of NF-κB that has not been altered in PC-3 cells after the IR treatment, although its transcriptional activity has indirectly been detected through the transactivation of both MnSOD and CuZnSOD. Reactive oxygen species are thus suggested to activate SODs, while NF-κB is suggested to mediate the antiproliferative and cytotoxic action of ionizing radiation in PC-3 cells, most likely by both scavenging a part of reactive oxygen species and attenuating the direct oxidation of cellular targets. The attenuating mechanism may be associated with high radioresistance of PC-3 cells reflected by high IC_{50} and sustained viability. In this context, the high-ionizing radiation resistance of human prostate cancer PC-3 cells has been detected by $IC_{50} \approx 8.5$ Gy and $EC_{50} >$ 20 Gy, and therefore may be ascribed to the high constitutive and IR-induced expression of mitochondrial MnSOD and cytosolic CuZnSOD. In the absence of catalase, the expression enhancement of MnSOD and CuZnSOD has led to the relatively successful adaptation of PC-3 cells to prooxidative conditions induced by ionizing radiation [45]. These measurements of viability, proliferation and cell cycle profile in irradiate (IR) PC-3 cells are fully in agreement with the previous results [47-49], indicating the high IR resistance of metastatic prostate cancer cell lines. Besides, PC-3 cells have responded to a high dose of IR with quite a low apoptosis and an increased necrosis. The increase in percentage of necrotic cells has also been low, even at rather high IR doses. A dominant response of PC-3 cells to IR has been a dose-dependent inhibition of cell proliferation, inhibited by 50 % (IC_{50}) at 8 Gy, while 50 % of cell cytotoxicity (EC_{50}) has not been attainable within the 2-20 Gy range [45]. This is in accordance with other studies [50,51]. Since the viability of the surviving cells is rather high [45], an IR treatment is inclined to induce an adaptive response by generating a tolerance to the IR-imposed oxidative stress in the surviving of PC-3 cells.

The DNA-binding of NF-κB has been reported to be constitutively activated in PC-3 cells by the high levels of SODs and catalase [35,52]. As a consequence of non-functional p53 protein [47], the adaptation of PC-3 cells to oxidative stress occurs during differentiation from the primary tumor [39-42,53]. In the absence of functional p53, the G2-arrest in PC-3 cells is dependent upon the p21WAF1/Cip1 protein, a member of the Cip/Kip family of cyclin-dependent kinase inhibitors [54,55] that is sensitive to H_2O_2 [56]. MnSOD overexpression has been suggested to alter the intracellular redox environment by means of changes in mitochondrial H_2O_2 [57], and thereby initiating downstream signaling cascades [58]. However, molecular mechanisms by which MnSOD affects IR-induced adaptive responses are currently unknown. On the basis of several contributions demonstrating that the pro- and anti-apoptotic responses can be controlled by the NF-κB-mediated expression of MnSOD [59,60], our further efforts have been dedicated to the modeling of the molecular mechanisms using a systems biology approach [13,46].

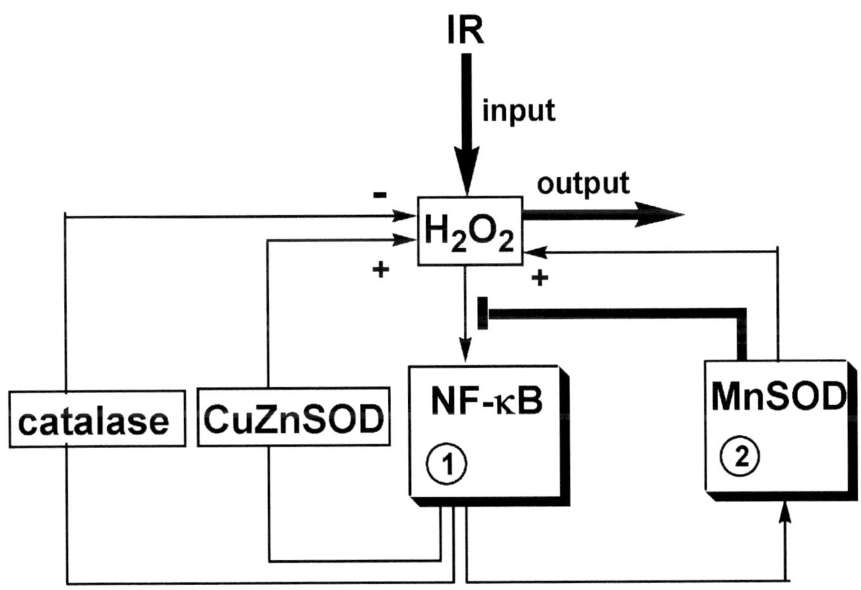

Figure 7. Modular representation of the systems biology model describing the IR-induced expression of MnSOD, which is associated with the radioresistance of PC-3 cells. The corresponding kinetic connectivity map of the signaling pathways is given in Figure 8. ("With kind permission from Springer Science+Business Media: <Annals of Biomedical Engineering, Experimental and Systems Biology Studies of the Molecular Basis for the Radioresistance of Prostate Carcinoma Cells, vol. 36, 2008, p. 834, Ana Niciforovic, Jelena Djordjevic, Miroslav Adzic, Vesna Vucic, Petar M. Mitrasinovic, and Marija B. Radojcic, Figure 2a, 0090-6964/08/0500-0831/0 © 2008 Biomedical Engineering Society>.")

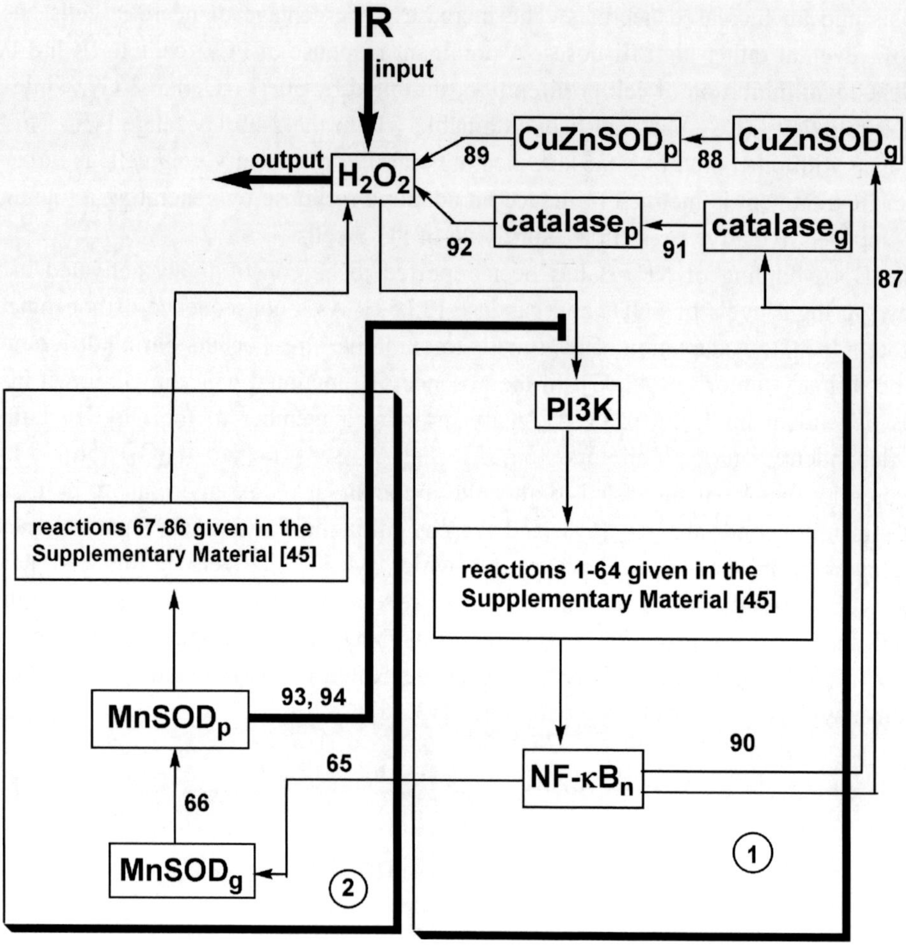

Figure 8. Kinetic connectivity map modeling the IR-induced expression of MnSOD, which is associated with the radioresistance of PC-3 cells. The "p" subscript denotes the "protein" term, the "g" subscript denotes the "gene" term, and $NF\text{-}\kappa B_n$ means nuclear NF-κB. The numbers denote specific reactions listed in the Supplementary Material of [45]. ("With kind permission from Springer Science+Business Media: <Annals of Biomedical Engineering, Experimental and Systems Biology Studies of the Molecular Basis for the Radioresistance of Prostate Carcinoma Cells, vol. 36, 2008, p. 834, Ana Niciforovic, Jelena Djordjevic, Miroslav Adzic, Vesna Vucic, Petar M. Mitrasinovic, and Marija B. Radojcic, Figure 2b, 0090-6964/08/0500-0831/0 © 2008 Biomedical Engineering Society>.")

The main functional modules of the systems biology model, which sheds more light on how MnSOD affects the biological state of cells depending upon the production of H_2O_2, are given in Figure 7. The kinetic connectivity map underlying the functional modules is schematically shown in Figure 8. Reaction steps, 1-64 and 67-86 within the NF-κB and MnSOD modules respectively, are given in the Supplementary Material of ref. 45.

The NF-κB module is essentially the IκB-NF-κB signaling pathway of Hoffmann *et al.* modeling the experimental data successfully [28]. After an equilibration period of 2000 min, raising the concentration of PI3K to 0.1 μM was chosen to simulate the stimulation of a cell by ionizing radiation, as previously reported for IKK by Ihekwaba *et al.* [26]. The first 64 reactions taking place within this module are the irreversible mass action reactions, as given in the Auxiliary Material of [45]. The MnSOD module is based on a set of 20 unidirectional reactions (67-86) taken form the literature [61]. Because varying levels of the mitochondrial form of SOD, $MnSOD^{III}$, were considered in the previous work [61], $MnSOD_p$ shown in Figure 8 corresponds to $MnSOD^{III}$ from [61]. The NF-κB and MnSOD modules are linked by the mass action reaction 65 running between nuclear NF-κB (NF-κB_n) and $MnSOD_g$, while $MnSOD_g$ and $MnSOD_p$ are connected with the reaction 66 (Figure 8). The inhibitory effect of MnSOD on IκB-α degradation and NF-κB activation is associated with the reactions 93 and 94, as previously suggested in the literature [62,63]. Besides the IR-induced expression of mitochondrial MnSOD, the IR-induced expression of cytosolic CuZnSOD and catalase was initially considered. Taking into account a set of the reactions 67-86 for MnSOD reported in the literature [61], two similar sets of kinetic reactions for CuZnSOD and catalase are not available. In this context, the problem was simplified to a certain extent. The CuZnSOD and catalase feedback mechanisms, from NF-κB_n to H_2O_2, were considered to proceed via the 87-88-89 and 90-91-92 reaction pathways respectively (Figure 8). Since the capacity of the system for removing H_2O_2 was maintained constant, the inclusion of a primarily peroxisomal enzyme, such as catalase, in the H_2O_2-removing system would make no difference in the simulation performances of the model [61]. The modeling environment GEPASI 3.30, running under the MS-Windows operating system, was employed to carry out the simulations of biochemical pathways [30]. Scan parameter capability of GEPASI 3.30 was employed to evaluate unknown values of the kinetic parameters for the reactions 65, 66, and 87-94. Estimated values of the kinetic parameters given in the Auxiliary Material of [45] were of the same order of magnitude as those recommended in the literature [64,65]. As we are particularly interested in the IR-induced expression of MnSOD, the initial concentrations of both $CuZnSOD_p$ and catalase were chosen to be fixed, 0 μM. The time-dependent concentrations of H_2O_2 for various levels of $MnSOD_p$, such as 0.7, 1, 4, 7, and 10 μM, were observed. The specific amounts of MnSOD and an initial concentration of 0 μM for H_2O_2 were taken from the literature [61].

The time course of [NF-κB_n] is shown schematically in Figure 9, while the compounded plot of the time-dependent concentrations of H_2O_2 for various initial levels of $MnSOD_p$ is given in Figure 10. On the basis of the model, the 0.1-1.3 μM concentration range of H_2O_2 generated by the input IR between 2- and 20 Gy corresponds to the 0.6-2.8 μM concentration range of H_2O_2 that is associated with the MnSOD induction of 0.7-10 μM (Figure 10). Figure 10 shows that the adaptation of prostate cancer cells to IR is assocciated with an increasing trend of the concentrations of H_2O_2 due to MnSOD induction. We thus suggest that the radioresistance of PC-3 cells may be closely related to the positive, feed-forward vicious circle established between the H_2O_2-mediated elevation of MnSOD expression (Figure 7). This proposal is believed to have some more profound implications on a conceptual framework upon which a future therapeutic strategy for treating prostate cancers with metastases should rely [45].

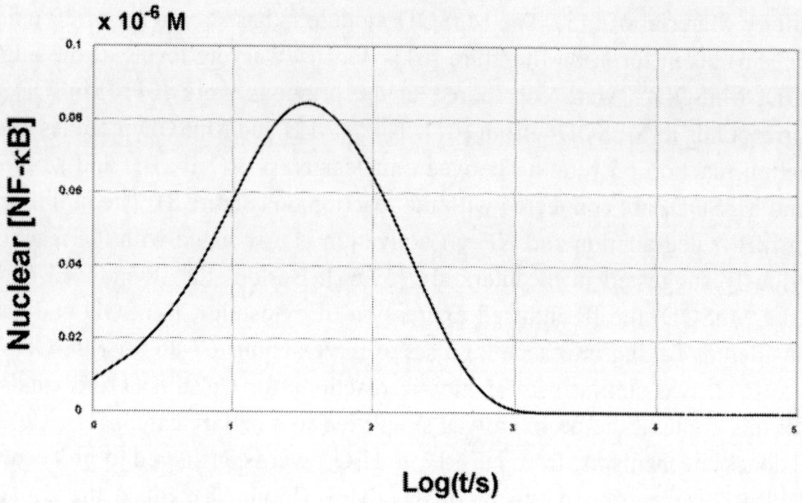

Figure 9. Time-dependent concentration of nuclear NF-κB for any initial concentration of MnSOD. ("With kind permission from Springer Science+Business Media: <Annals of Biomedical Engineering, Experimental and Systems Biology Studies of the Molecular Basis for the Radioresistance of Prostate Carcinoma Cells, vol. 36, 2008, p. 835, Ana Niciforovic, Jelena Djordjevic, Miroslav Adzic, Vesna Vucic, Petar M. Mitrasinovic, and Marija B. Radojcic, Figure 3, 0090-6964/08/0500-0831/0 © 2008 Biomedical Engineering Society>.")

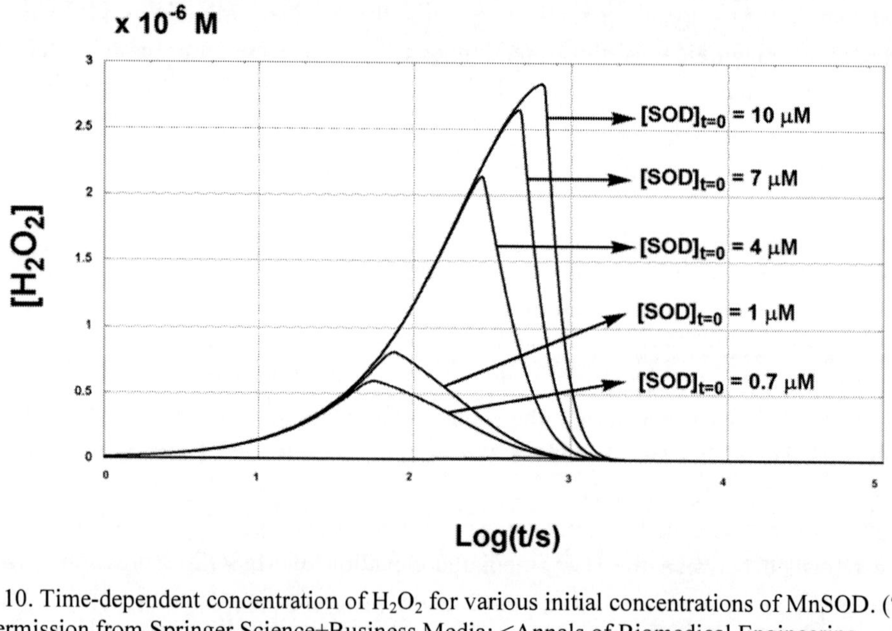

Figure 10. Time-dependent concentration of H_2O_2 for various initial concentrations of MnSOD. ("With kind permission from Springer Science+Business Media: <Annals of Biomedical Engineering, Experimental and Systems Biology Studies of the Molecular Basis for the Radioresistance of Prostate Carcinoma Cells, vol. 36, 2008, p. 835, Ana Niciforovic, Jelena Djordjevic, Miroslav Adzic, Vesna Vucic, Petar M. Mitrasinovic, and Marija B. Radojcic, Figure 4, 0090-6964/08/0500-0831/0 © 2008 Biomedical Engineering Society>.")

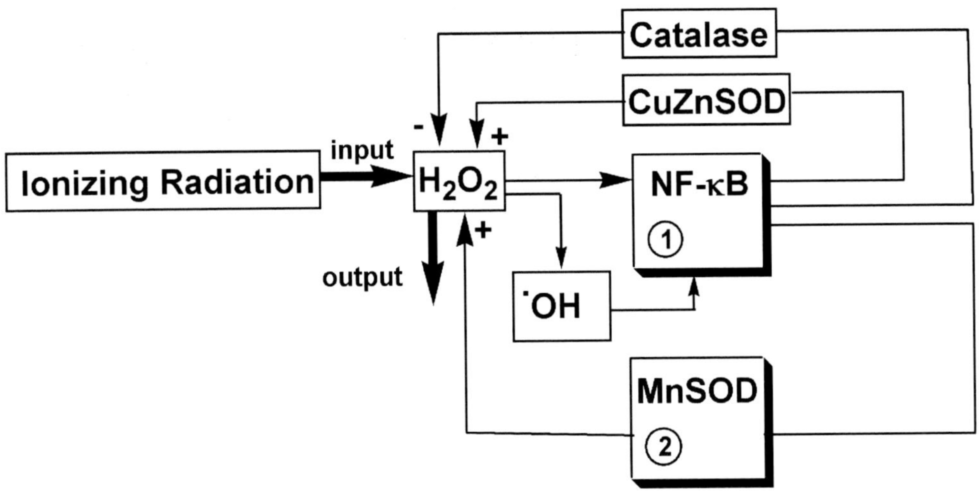

Figure 11. Modular representation of the systems biology model including the hydroxyl radical (·OH)-mediated NF-κB activation. The corresponding kinetic connectivity map of the signaling pathways associated with the radioresistance of PC-3 cells is given in Figure 12.

Origin of Oscillations in Molecular Mechanism of Radioresistance of PC-3 Cells

We have previously discussed a systems biology model showing that the adaptation of prostate cancer cells (PC-3) to ionizing radiation is based on the increasing trend of the effective concentrations of H_2O_2 due to MnSOD induction. As the model has been verified experimentally [45], the radioresistance of PC-3 cells has been suggested to be associated with the positive, feed-forward loop established between the H_2O_2-mediated activation of NF-κB and the elevated MnSOD activity. It is important to note that hydroxyl radical (·OH) has been shown to be primarily responsible for the activation of NF-κB, and that the hydroxyl radical-mediated NF-κB activation is not limited to particular cell types [66]. The question, how the ·OH-induced NF-κB activation influences the radioresistance of PC-3 cells, is addressed in this section.

The model is based on two main modules, NF-κB and MnSOD modules (Figure 11). The NF-κB module is a modified version of that previously reported [13,46], which is essentially the IκB-NF-κB signaling pathway of Hoffmann *et al.* [28]. The MnSOD module is based on a set of 20 unidirectional reactions (67-86) taken form the literature [61]. All kinetic details for the reactions 1-94 are available as the Auxiliary Material of [45]. Two reactions, 95 and 96, are now added to model the NF-κB activation by hydroxyl radical (Figure 12). These reactions are two irreversible, mass action reactions with kinetic constants of 4 and $2 \times 10^{-3}\ s^{-1}$, respectively. The GEPASI 3.3 suite of programs was employed to perform time-course simulations [30].

The interplay between nuclear NF-κB and H_2O_2 is nicely visible in Figure 13. Clearly, the inclusion of the hydroxyl radical-mediated NF-κB activation into the model of the radioresistance of PC-3 cells brings oscillations into the time-dependent concentrations of nuclear NF-κB and H_2O_2 for various inital levels of MnSOD (Figure 13) relative to those shown in figs. 9 and 10. Note that an MnSOD induction of 0.7 μM (Figure 13b) caused both a substantial increase in the amplitude of the first maximum of H_2O_2 and a decrease in the amplitude of the first maximum of nuclear NF-κB with respect to those wihout MnSOD (Figure 13a). Raising the initial level of MnSOD from 0.7 μM to 10 μM via 1, 4, and 7 μM is associated with a further increase in the amplitude of the first maximum of H_2O_2 (Figure 13c-f) , which is quite similar to that shown in Figure 10. The equilibrium time corresponding to that until the first maximum of nuclear NF-κB is between 2 and 3 sec for any initial level of MnSOD (Figure 13). By following an increase of MnSOD induction in the 0.7-10 μM range, noteworthy is a more frequent and conspicuous distribution of the oscillations in the time course of nuclear NF-κB after the equilibrium period of time.

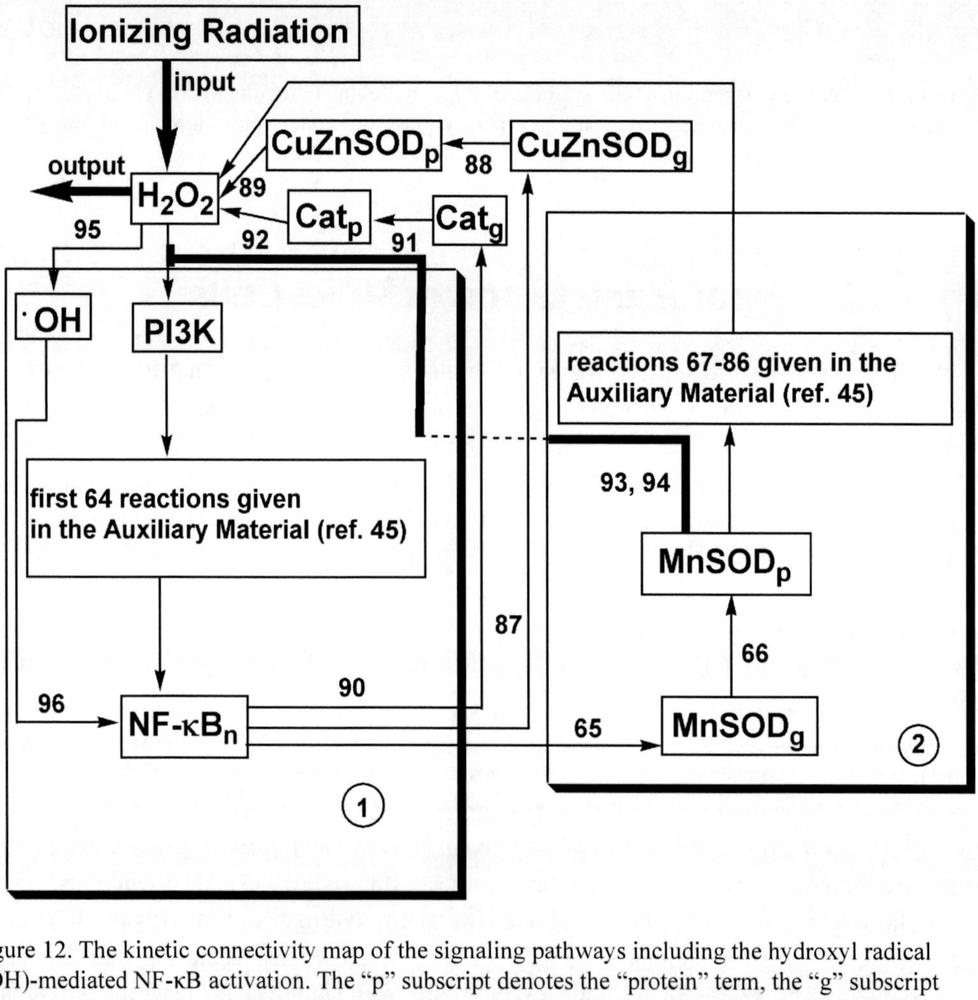

Figure 12. The kinetic connectivity map of the signaling pathways including the hydroxyl radical (·OH)-mediated NF-κB activation. The "p" subscript denotes the "protein" term, the "g" subscript denotes the "gene" term, the "Cat" term denotes catalase, and $NF\text{-}\kappa B_n$ means nuclear NF-κB.

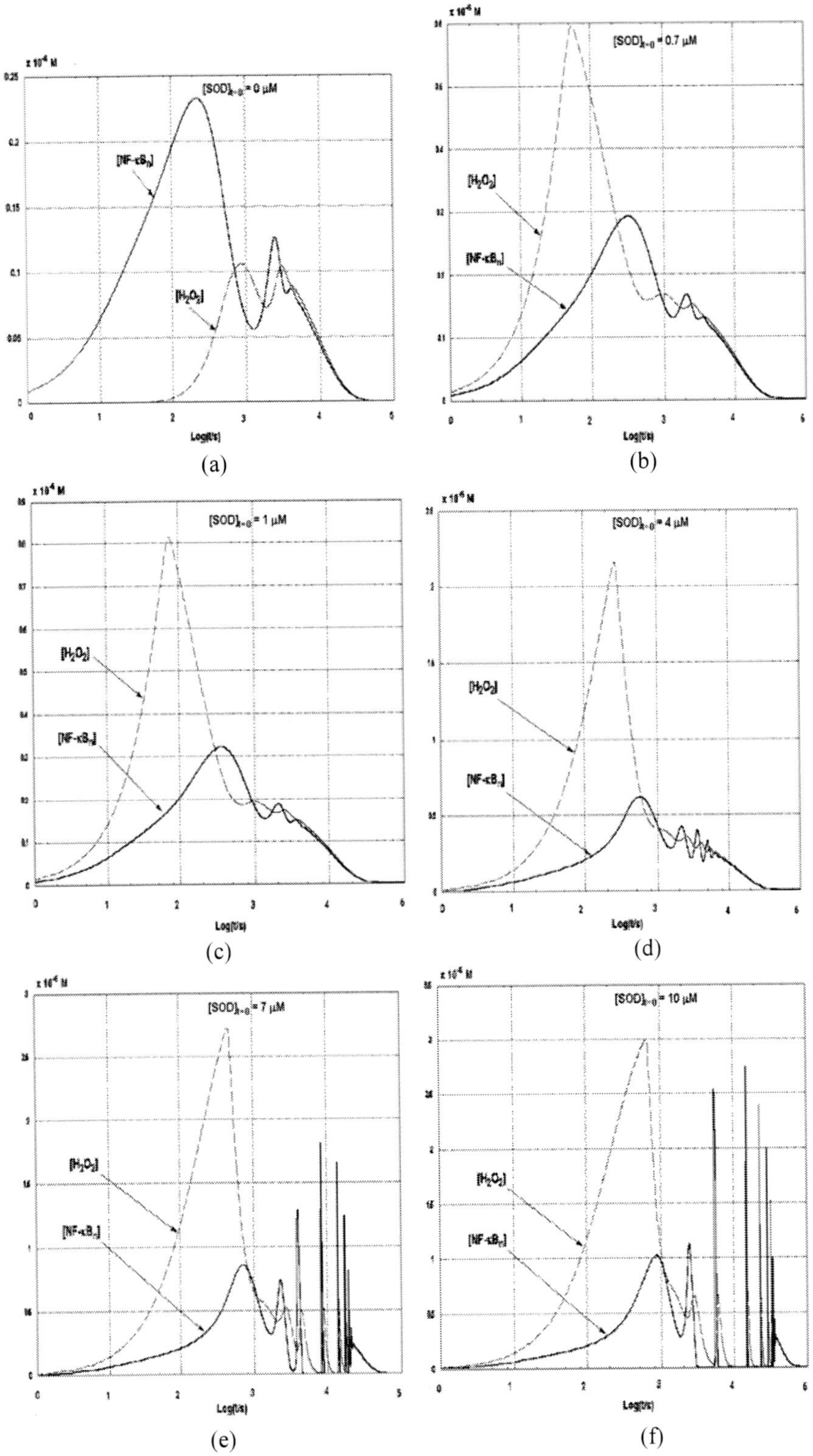

Figure 13. The time-dependent concentrations of NF-κB_n (solid line) and H_2O_2 (dashed line). Initial levels of MnSOD are: (a) 0, (b) 0.7, (c) 1, (d) 4, (e) 7, and (f) 10 μM.

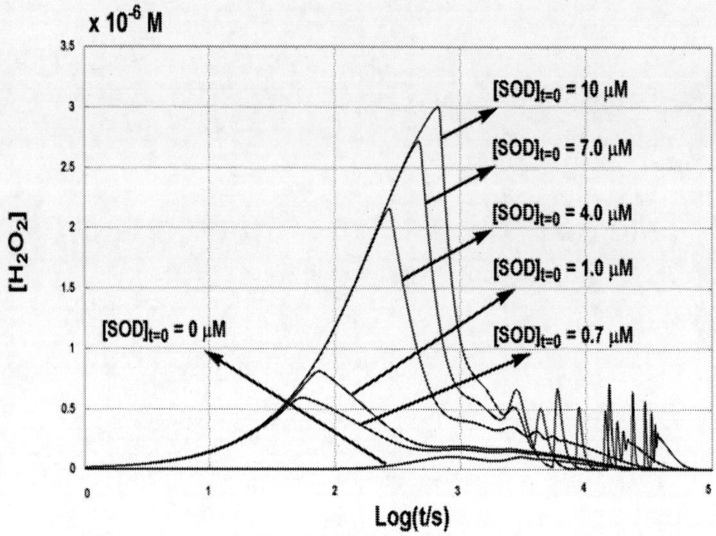

Figure 14. The time-dependent concentration of H_2O_2 for various initial levels of MnSOD.

To rationalize the relationship between the oscillatory behavior of nuclear NF-κB and that of H_2O_2, the compounded plot of the time-dependent concentrations of H_2O_2 for various initial levels of MnSOD is given in Figure 14. The H_2O_2 oscillations taking place between 3 and 5 sec along the X axis are nicely visible. It is important to note that the oscillations occur in the 0.2-0.7 μM concentration range of H_2O_2 corresponding the 0.7-10 μM range of MnSOD induction (Figure 14). Based on the present observations and on the literature data, we designed the systems biology model for the adaptation to oxidative stress/resistance by IR, which is associated with the MnSOD/NF-κB induction/activation by ROS (H_2O_2). The model indicates that the input IR between 2- and 20 Gy generates the 0.1-1.3 μM concentration range of H_2O_2 which gets decreased by the MnSOD induction of 0.7-10 μM to the 0.2-0.7 μM concentration range of H_2O_2. In other words, a possible mechanism for the adaptation of prostate cancer cells to IR is assocciated with a decreasing trend of the effective concentrations of H_2O_2 due to MnSOD induction. Consequences of the detoxifying/scavanging activity of MnSOD, which are manifested by both intensive survival of the cells and a low death rate at high-dose IR, are thus supported by our measurements [45]. Hence, the inclusion of some inhibitors with an adjuvant effect on the SOD and/or NF-κB activity might be implicated in the search for an effective IR therapy. However, there is a need of being aware of an apparent padox. Since inhibition of NF-κB by overexpression of IκB induces insignificant effects on the radiosensitivity of PC-3 prostate cancer cells [63], inhibition of NF-κB does not always lead to increased cell radiosensitivity. This indicates that IR-induced signaling elements interacting with each other form a network of signaling pathways providing a by-pass in order to keep cell survival. Also a variety of signaling functions regulated by transcription factors depend on the metabolic state of different cell lines at the time of treatment. A possible treatment with new drugs blocking MnSOD and CuZnSOD expression could down-regulate a group of IR-induced genes by increasing cell radiosensitivity [67]. In this context, fibroblasts derived from CuZnSOD knockout mice were shown to proliferate more slowly (75%) than control cells, thus emphasizing the importance of CuZnSOD in cell growth and survival [68].

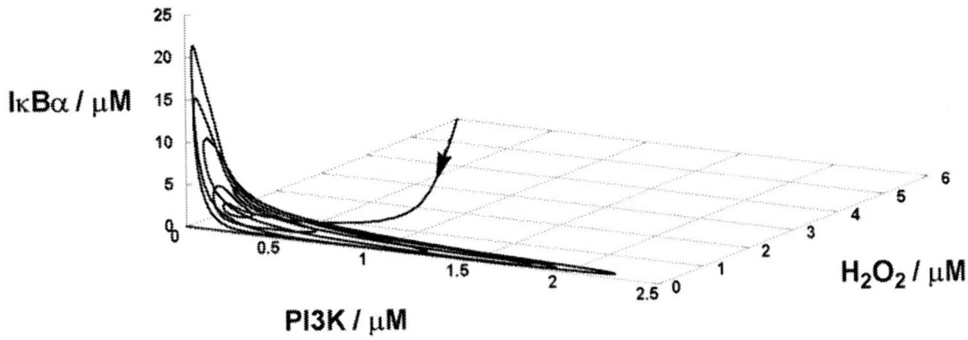

Figure 15. Phase plane plot between the time-dependent concentrations of [PI3K], [IκBα], and [H2O2].

An extensive sensitivity analysis [26] of parameters controlling oscillatory signaling has shown that the NF-κB signaling pathway consisting of 26 species linked by 64 unidirectional reactions can be reduced to 9 most important reactions/parameters. Interestingly, these 9 parameters directly affect the concentrations of only 2 reactants other than NF-κB_n, PI3K and IκBα. To learn more on the involvement of these molecules into the signaling scenario, the phase plane plot of the time-dependent relationship between the concentrations of PI3K, IκBα and H_2O_2 is shown schematically in Figure 15. The three stated variables, plotted against each other, reveal an intimate involvement of the mediators in the oscillatory behavior of H_2O_2. Even though the reduced representation does not fit the biological scenario formally, it demonstrates that a much lower-dimensional representation of the system of 64 kinetic equations could narrow down the range of possible instabilities that the oscillatory signaling may exhibit. This indication is of vital importance in the search for novel therapeutic means, since the broad range of possible instabilities may have undesirable consequences on the development of successful therapies.

Conclusion

Due to a tremendous progress accomplished in systems biology in recent years, biology has arguably established itself as one of the most exciting scientific fields. Even though the concept of cancer as a system failure is currently generating significant debate in the literature, this chapter demonstrates the potential of using the state-of-the-art systems biology approaches, derived from engineering and computer science, to understand the disease. System level understanding, the approach advocated in systems biology, requires a change in our notion of "what to look for" in biology. While understanding role of individual genes and proteins continues to be important, the focus is now shifted towards understanding a system structure, function and dynamics. In the context of prostate cancer, systems-level kinetic models have been shown to be able to explain dynamic behavior and go far beyond the static (and more traditional) pictures of the topologies of signaling pathways. Since morphological, physiological and molecular studies on death of cells in tissues are usually carried out on isolated cell populations due to known difficulties manifested by interference and interaction with surrounding cells, systems biology has been considered as a means of overcoming the difficulties. We therefore propose a conceptual framework, based on experimental, mechanistic cell biology studies and systems biology strategies, to be indispensable for understanding bioprocesses of PC-3 cell therapeutics leading to more localized and efficient treatment of the disease.

References

[1] Jemal, A., Thomas, A., Murray, T. & Thun, M. (2002). Cancer statistics. *CA Cancer J. Clin., 52,* 23-47.

[2] Lucia, M. S., Darke, A. K., Goodman, P. J., La Rosa, F. G., Parnes, H. L., Ford, L. G., Coltman, C. A., Jr. & Thompson, I. M. (2008). Pathologic characteristics of cancers detected in the prostate cancer prevention trial: implications for prostate cancer detection and chemoprevention. *Cancer Prevention Research, 1,* 167-173.

[3] Jemal, A., Siegel, R., Ward, E., Hao, Y., Xu, J., Murray, T. & Thun, M. J. (2008). Cancer statistics. *CA Cancer J. Clin., 58,* 71-96.

[4] Zagars, G. K., Pollack, A. & Smith, L. G. (1999). Conventional external-beam radiation therapy alone or with androgen ablation for clinical stage III (T3, NX/N0, M0) adenocarcinoma of the prostate. *Int. J. Radiation Oncology Biol. Phys., 44,* 809–819.

[5] Kaplan, I. D., Cox, R. S. & Bagshaw, M. A. (1993). Prostate specific antigen after external beam radiotherapy for prostatic cancer: follow-up. *J. Urol., 149,* 519-522.

[6] Mitrasinovic, P. M. (2005). Cross-linking between thymine and indolyl radical: possible mechanisms for cross-linking of DNA and tryptophan-containing peptides. *Bioconjugate Chemistry, 6,* 588-597.

[7] Sansom, C. (2005). Systems biology: will it work? *IEE Proceedings of Systems Biology, 2,* 1-4.

[8] Mesarovic, M. D., Sreenath, S. N. & Keene, J. D. (2004). Search for organizing principles: understanding in *systems biology. IEE Proceedings of Systems Biology, 1,* 19-27.

[9] Sontag, E. D. (2004). Some new directions in control theory inspired by *systems biology. IEE Proceedings of Systems Biology, 1,* 9-18.

[10] Wigle, D. & Jurisica, I. (2007). Cancer as a system failure. *Cancer Informatics, 2,* 10-18.

[11] Kitano, H. (2002). Systems biology: a brief overview. *Science, 295,* 1662-1664.

[12] Sakaki, Y., Kholodenko, B. N., Hatakeyama, M., Kitano, H., Kolch, W., De Meyts, P., Yarden, Y., Westerhoff, H. V. & Wiley, H. S. (2005). International consortium on systems biology of receptor tyrosine kinase regulatory networks. *IEE Proceedings of Systems Biology, 152*, 53-60.

[13] Mitrasinovic, P. M. & Mitrasinovic, O. (2006). On possible mechanisms for reactive oxygen species (ROS)-mediated cross-talk between NF-κB and JNK: a systems biology view. *GESTS Int'l Trans. Computer Science and Engr., 27,* 41-52.

[14] Mitrasinovic, P. M. & Mihajlovic, M. L. (2008). Recent advances in radiation therapy of cancer cells: a step towards an experimental and systems biology framework. *Current Radiopharmaceuticals, 1,* 22-29.

[15] Chen, G. & Goeddel, D. V. (2002). TNF-R1 signaling: a beautiful pathway. *Science, 296,* 1634-1635.

[16] Schweichel J. U. & Merker, H. J. (1973). The morphology of various types of cell death in prenatal tissues. *Teratology, 7,* 253-266.

[17] Fiers, W., Beyaert, R., Declercq, W. & Vandenabeele, P. (1999). More than one way to die: apoptosis, necrosis and reactive oxygen damage. *Oncogene, 18,* 7719-7730.

[18] Kitanaka, C. & Kuchino, Y (1999). Caspase-independent programmed cell death with necrotic morphology. *Cell Death. Differ., 6,* 508-515.

[19] Davis, R. J. (2000). Signal transduction by the JNK group of MAP kinases. *Cell, 103,* 239-252.

[20] Hartwell, L. H., Hopfield, J. J., Leibler, S. & Murray, A. W. (1999). From molecular to modular cell biology. *Nature (Supp), 402,* C47-C52.

[21] Freeman, B. C. & Yamamoto, K. R. (2001). Continuous recycling: a mechanism for modulatory signal transduction. *Trends in Biochemical Sciences*, 26, 285-290.

[22] Pham, C. G., Bubici, C., Zazzeroni, F., Papa, S., Jones, J., Alvarez, K., Jayawardena, S., De Smaele, E., Cong, R., Beaumont, C., Torti, F. M., Torti, S. V. & Franzoso, G. (2004). Ferritin heavy chain upregulation by NF-κB inhibits TNFα-induced apoptosis by suppressing reactive oxygen species. *Cell, 119,* 529-542.

[23] Curtin, J. F., Donovan, M. & Cotter, T. G. (2002). Regulation and measurement of oxidative stress in apoptosis. *J. Immunol. Methods, 265,* 49-72.

[24] Lin, A. (2002). Activation of the JNK signaling pathway: breaking the brake on apoptosis. *BioEssays, 25,* 17-24.

[25] Kamata, H., Honda, S., Maeda, S., Chang, L., Hirata, H. & Karin, M. (2005). Reactive oxygen species promote TNFα-induced death and sustained JNK activation by inhibiting MAP kinase phosphatases. *Cell, 120,* 649-661.

[26] Ihekwaba, A. E. C., Broomhead, D. S., Grimley, R. L., Benson, N. & Kell, D. B. (2004). Sensitivity analysis of parameters controlling oscillatory signaling in the NF-κB pathway: the roles of IKK and IκBα. *Systems Biology, 1,* 93-103.

[27] Kholodenko, B. N. (2000). Negative feedback and ultrasensitivity can bring about oscillations in the mitogen-activated protein kinase cascades. *Eur. J. Biochem., 267,* 1583-1588.

[28] Hoffmann, A., Levchenko, A., Scott, M. L. & Baltimore, D. (2002). The IκB-NF-κB signaling module: temporal control and selective gene activation. *Science, 298,* 1241-1245.

[29] Neves, S. R. & Iyengar, R (2002). Modeling of signaling networks. *BioEssays, 24,* 1110-1117.

[30] Mendes, P. & Kell, D. (1998). Non-linear optimization of biochemical pathways: applications to metabolic engineering and parameter estimation. *Bioinformatics, 14,* 869-883.

[31] Marshall, C. J. (1995). Specificity of receptor tyrosine kinase signaling: transient versus sustained extracellular single-regulated kinase activation. *Cell, 80,* 179-185.

[32] Covert, M. W., Leung, T. H., Gaston, J. E. & Baltimore, D. (2005). Achieving stability of lipopolysaccharide-induced NF-κB activation. *Science, 309,* 1854-1857.

[33] Werner, S. L., Barken, D. & Hoffmann, A. (2005). Stimulus specificity of gene expression programs determined by temporal control of IKK activity. *Science, 309,* 1857-1861.

[34] Ferrel, J. E., Jr & Machleder, E. M. (1998). The biochemical basis of an all-or-none cell fate switch in Xenopus Oocytes. *Science, 280,* 895-898.

[35] Shukla, S., MacLennan, G. T., Fu, P., Patel, J., Marengo, S. R., Resnick, M. I., & Gupta, S. (2004). Nuclear factor-kappaB/p65 (Rel A) is constitutively activated in human prostate adenocarcinoma and correlates with disease progression. *Neoplasia, 6,* 390-400.

[36] Lin, A. & Karin, M. (2003). NF-kappaB in cancer: a marked target. *Semin. Cancer. Biol., 13,* 107-114.

[37] Manna, S. K., Zhang, H. J., Yan, T., Oberley, L. W. & Aggarwal, B. B. (1998). Overexpression of manganese superoxide dismutase suppresses tumor necrosis factor-induced apoptosis and activation of nuclear transcription factor-B and activated protein-1. *J. Biol. Chem., 273,* 13245-13254.

[38] Kim, H. T., Kim, Y. H., Nam, J. W., Lee, H. J., Rho, H. M. & Jung, G. (1994). Study of 5'-flanking region of human Cu/Zn superoxide dismutase. *Biochem. Biophys. Res. Commun., 201,* 1526-1533.

[39] Xu, Y., Kiningham, K. K., Devalaraja, M. N., Yeh, C. C., Majima, H., Kasarskis, E. J. & St Clair, D. K. (1999). An intronic NF-kappaB element is essential for induction of the human manganese superoxide dismutase gene by tumor necrosis factor-alpha and interleukin-1beta. *DNA Cell Biol., 18,* 709-722.

[40] Zhou, L. Z., Johnson, A. P. & Rando, T. A. (2001). NF kappa B and AP-1 mediate transcriptional responses to oxidative stress in skeletal muscle cells. *Free Radic. Biol. Med., 31,* 1405-1416.

[41] Jung, K., Seidel, B., Rudolph, B., Lein, M., Cronauer, M. V., Henke, W., Hampel, G., Schnorr, D. & Loening, S. A. (1997). Antioxidant enzymes in malignant prostate cell lines and in primary cultured prostatic cells. *Free Radic. Biol. Med., 23,* 127-133.

[42] Oberley, T. D., Zhong, W., Szweda, L. I. & Oberley, L. W. (2000). Localization of antioxidant enzymes and oxidative damage products in normal and malignant prostate epithelium. *Prostate, 44,* 144-155.

[43] Gajewska, J., Szczypka, M., Izbicki, T., Klepacka, T. & Laskowska-Klita, T. (1996). Antioxidant and glutathione-associated enzymes in Wilms' tumor after chemotherapy. *J. Cancer Res. Clin. Oncol., 122,* 483-488.

[44] Chen, Q. M., Liu, J. & Merrett, J. B. (2000). Apoptosis or senescence-like growth arrest: influence of cell-cycle position, p53, p21 and bax in H_2O_2 response of normal human. *Biochem. J., 347,* 543-551.

[45] Niciforovic, A., Djordjevic, J., Adzic, M., Vucic, V., Mitrasinovic, P. M. & Radojcic, M. B. (2008). Experimental and systems biology studies of the molecular basis for the radioresistance of prostate carcinoma cells. *Annals of Biomedical Engineering, 36,* 831-838.

[46] Mitrasinovic, P. M. (2006). Systems biology studies of reactive oxygen species (ROS)-mediated apoptosis. *Proceedings of the 8th International Conference on Fundamental and Applied Aspects of Physical Chemistry, 1,* 279-281.

[47] Isaacs, W. B., Carter, B. S. & Ewing, C. M. (1991). Wild-type p53 suppresses growth of human prostate cancer cells containing mutant p53 alleles. *Cancer Res., 51,* 4716-4720.

[48] Raffoul, J. J., Wang, Y., Kucuk, O., Forman, J. D., Sarkar, F. H. & Hillman, G. G. (2006). Genistein inhibits radiation-induced activation of NF-kappaB in prostate cancer cells promoting apoptosis and G2/M cell cycle arrest. *BMC Cancer, 6,* 107.

[49] Vucic, V., Isenovic, E. R., Adzic, M., Ruzdijic, S. & Radojcic, M. B. (2006). Effects of gamma-radiation on cell growth, cycle arrest, death, and superoxide dismutase expression by DU 145 human prostate cancer cells. *Braz. J. Med. Biol. Res., 39,* 227-236.

[50] Joiner, M. C., Marples, B., Forman, J. D. & Sarkar, F. H. (2004). Genistein potentiates inhibition of tumor growth by radiation in a prostate cancer orthotopic model. *Mol Cancer Ther., 3,* 1271-1279.

[51] Pawlik, T. M. & Keyomarsi, K. (2004). Role of cell cycle in mediating sensitivity to radiotherapy. *Int J Radiat Oncol Biol Phys, 59,* 928-942.

[52] Sweeney, C., Li, L., Shanmugam, R., Bhat-Nakshatri, P., Jayaprakasan, V., Baldridge, L. A., Gardner, T., Smith, M., Nakshatri, H. & Cheng, L. (2004). Nuclear factor-kappaB is constitutively activated in prostate cancer in vitro and is overexpressed in prostatic intraepithelial neoplasia and adenocarcinoma of the prostate. *Clin. Cancer Res., 10,* 5501-5507.

[53] Bunz, F., Dutriaux, A., Lengauer, C., Waldman, T., Zhou, S., Brown, J. P., Sedivy, J. M., Kinzler, K. W. & Vogelstein, B. (1998). Requirement for p53 and p21 to sustain G2 arrest after DNA damage. *Science, 282,* 1497-150.

[54] Lim, S. D., Sun, C., Lambeth, J. D., Marshall, F., Amin, M., Chung, L., Petros, J. A. & Arnold, R. S. (2005). Increased Nox1 and hydrogen peroxide in prostate cancer. *Prostate, 62,* 200-207.

[55] Roninson, I. B. (2002). Tumor senescence as a determinant of drug response in vivo. *Drug Resist. Updat., 5,* 204-208.

[56] Barnouin, K., Dubuisson, M. L., Child, E. S., Fernandez de Mattos, S., Glassford, J, Medema, R. H., Mann, D. J. & Lam, E. W. (2002). H_2O_2 induces a transient multi-phase cell cycle arrest in mouse fibroblasts through modulating cyclin D and p21Cip1 expression. *J Biol Chem., 277,* 13761-13770.

[57] Li, S., Yan, T., Yang, J. Q., Oberley, T. D. & Oberley, L. W. (2000). The role of cellular glutathione peroxidase redox regulation in the suppression of tumor cell growth by manganese superoxide dismutase. *Cancer Res., 60,* 3927-3939.

[58] Rhee, S., Bae, Y., Lee, S. R. & Kwon, J. (2000). Hydrogen peroxide: a key messenger that modulates protein phosphorylation through cysteine oxidation. *Sci STKE., 53,* PE1.

[59] Bernard, D., Monte, D., Vandenbunder, B. & Abbadie, C. (2002). The c-Rel transcription factor can both induce and inhibit apoptosis in the same cells via the upregulation of MnSOD. *Oncogene, 21,* 4392-4402.

[60] Wang, C. Y., Mayo, M. W. & Baldwin, A. S., Jr (1996). TNF- and cancer therapy-induced apoptosis: potentiation by inhibition of NF-kappaB. *Science, 274,* 784-787.

[61] Buettner, G. R., Ng. F., Chin, M., Wang, V., Rodgers, G. J. & Schafer, F. Q. (2006). A new paradigm: manganese superoxide dismutase influences the production of H_2O_2 in cells and thereby their biological state. *Free Radic. Biol. Med., 41,* 1338-1350.

[62] Manna, S. K., Zhang, H. J., Yan, T., Oberley, L. W. & Aggarwal, B. B. (1998). Overexpression of manganese superoxide dismutase suppresses tumor necrosis factor-induced apoptosis and activation of nuclear transcription factor-B and activated protein-1. *J. Biol. Chem., 273,* 13245-13254.

[63] Pajonk, F., Pajonk, K. & McBride, W. H. (1999). Inhibition of NF-kappaB, clonogenicity, and radiosensitivity of human cancer cells. *J. Natl. Cancer Inst., 91,* 1956-1960.

[64] Fersht, A. (1985). *Enzyme structure and mechanism*. NewYork, NY: W. H. Freeman and Company, p. 475.

[65] Vilenchik, M. M. & Knudson, A. G., Jr. (2000). Inverse radiation dose-rate effects on somatic and germ-line mutations and DNA damage. *Proc. Natl Acad. Sci. USA, 97,* 5381-5386.

[66] Shi, X., Dong, Z., Huang, C., Ma, W., Liu, K., Ye, J., Chen, F., Leonard, S. S., Ding, M., Castranova, V. & Vallyathan, V. (1999). The role of hydroxyl radical as a messenger in the activation of nuclear transcription factor NF-kappaB. *Mol. Cell Biochem., 194,* 63-70.

[67] Guo, G., Yan-Sanders, Y., Lyn-Cook, B. D., Wang, T., Tamae, D., Ogi, J., Khaletskiy, A. Li. Z., Weydert, C., Longmate, J. A., Huang, T. T., Spitz, D. R., Oberley, L. W. & Li, J. J. (2003). Manganese superoxide dismutase-mediated gene expression in radiation-induced adaptive responses. *Mol. Cell Biol., 23,* 2362-2378.

[68] Huang, T. T., Yasunami, M., Carlson, E. J., Gillespie, A. M., Reaume, A. G., Hoffman, E. K., Chan, P. H., Scott, R. W. & Epstein, C. J. (1997). Superoxide-mediated cytotoxicity in superoxide dismutase-deficient fetal fibroblasts. *Arch. Biochem. Biophys., 344,* 424-432.

In: Handbook of Prostate Cancer Cell Research... ISBN: 978-1-60741-954-9
Editor: Alan T. Meridith

Chapter 8

Post-Translational Regulation of Androgen Receptor in Proliferation of Prostate Cancer Cells

Shigetsugu Hatakeyama*
Department of Biochemistry, Hokkaido University Graduate School of Medicine, N15, W7, Kita-ku, Sapporo, Hokkaido 060-8638, Japan

Abstract

Transcription is currently one of the most prominent fields in cell biology, endocrinology and cancer biology, with the ubiquitin-proteasome system in particular being an area of intensive study. The androgen receptor (AR) is a transcription factor belonging to the family of nuclear receptors that mediate the action of androgen. The AR plays an important role in normal development of the prostate as well as in the progression of prostate cancer. The AR is regulated by several post-translational modifications, including phosphorylation, acetylation and ubiquitination. Activation of the AR affects the cell growth of prostate cancer and the expression and secretion of prostate-specific antigen (PSA), one of the most reliable diagnostic markers for prostate cancer. Furthermore, the AR has been reported to regulate several cell cycle regulators such as cyclins, retinoblastoma tumor suppressor (Rb) and cyclin-dependent kinase inhibitor (CKI), suggesting that AR is directly involved in cell cycle regulation, especially in prostate cancer cells. Recently, several enzymes for modifications of the AR have been identified and the molecular mechanism for AR activation is gradually being elucidated. It is important to clarify the mechanism of AR-dependent cell proliferation for identifying a novel diagnostic tool and a potentially therapeutic target for prostate cancer.

* Phone: +81 11 706 5899; Fax: +81 11 706 5169; E-mail address: hatas@med.hokudai.ac.jp

Introduction

Androgens are important for male sexual differentiation and development as well as for maintenance of the prostate or testis in adults [1]. Androgens are also important for the development and progression of benign prostatic hyperplasia and prostate cancer in men [2]. Prostate cancer is the most frequently diagnosed male cancer and the second or third leading cause of cancer-related death in men in most western countries [3]. Therefore, prostate cancer will remain a major problem in social medicine in the future.

Prostate cancer is usually treated on the basis of its biological characteristic, in which prostate cancer cells are dependent on androgen for growth and development [4, 5]. Androgen ablation causes cell death or cell cycle arrest of prostate cancer cells. Therefore, many studies on prostate cancer have focused on androgens and androgen-mediated signaling to find ways to decrease androgens in blood and to inhibit the androgen receptor (AR) [6]. Intracellular AR signalings are implicated in proliferation and differentiation of prostate cancer cells. Thus, biochemical studies on signaling via the AR have a pivotal role even after androgen ablation [7].

The AR is a ligand-dependent transcription factor of the nuclear receptor superfamily [8]. Transcriptional activity of the AR is mainly controlled by ligand binding and post-translational modifications [9]. Phosphorylation at tyrosines and serine/threonine residues, and other modifications such as acetylation, ubiquitination and sumoylation on the AR have been identified in recent years [10]. These modifications are likely to affect receptor stability and activity and to provide mechanisms for tissue or gene-specific regulation. A better understanding of the effects of post-translational modifications on AR function should provide informations about the identification of drugs (chemicals) for improved clinical approaches.

In addition to AR signaling with transcriptional regulation, it has recently been shown that other signaling pathways, including nongenomic regulation by androgen, are involved in prostate cancer proliferation and progression [11]. The AR may also mediate important cellular functions in the cytoplasm, independent of its function as a transcription factor. The AR has been shown to affect rapid signaling cascades through direct interaction with the c-Src tyrosine kinase [12]. This nongenomic signaling, followed by activation of the MAP kinase pathway, may play a role in cell proliferation in prostate cancers [13].

We present an overview of androgen and AR signaling in prostate cancer cells, focusing on recent findings about post-translational modifications of the AR. We then review the involvement of cell cycle regulation in prostate cancer, focusing on cross-talk between the AR and cell cycle regulators.

1. Structural Characteristics of Androgen Receptor

It has been reported that the nuclear receptor superfamily has 48 members in humans [8]. The sub-family of steroid hormone receptors is composed of the androgen receptor (AR), two estrogen receptors (ERs) named ERα and ERβ, the progesterone receptor (PR), the glucocorticoid receptor (GR) and the mineralocorticoid receptor (MR). In the absence of hormones, steroid hormone receptors are mainly found in the cytoplasm complexed with chaperone proteins such as Hsp90. Following ligand binding, they undergo conformational changes and move to the nucleus, followed by activation or repression of target genes as homodimers or heterodimers. Most of the steroid hormone receptors are organized in four main domains [14]: an N-terminal region bearing important transactivation functions, a DNA-binding domain (DBD) composed of two zinc fingers, a hinge region with the nuclear localization signal and a ligand-binding domain (LBD) with additional transactivation function. The LBD undergoes extensive conformational changes following binding by the hormone. The carboxy-terminal α-helix folds over to close the hydrophobic pocket into which the ligand binds, and creates a space for recognition by cofactors, which are necessary for recruitment of the general transcription machinery and activation of target genes. DNA response elements recognized by steroid hormone receptors are usually organized as semi-palindromic hexamers [15, 16]. In some cases, repressive functions of steroid receptors on several target genes have also been reported [17].

The three-dimensional structures of peptides representing the AR have been determined by x-ray crystallography [18]. The N-terminal domain (NTD) of the AR, conversely, is unstructured in solution and thus has been recalcitrant to structural determination. Nevertheless, several critical functional domains within the AR NTD have been described and characterized. Furthermore, post-translational modifications of the AR, including phosphorylation, acetylation, ubiquitination and sumoylation, add additional layers of regulation and are likely to influence the structure and function of these domains [10].

2. Androgen Signaling Pathway in Transcriptional Regulation

Androgen exhibits biological effects through the AR, which acts as a ligand-dependent transcription factor [19]. Testosterone is the most abundant androgens in serum, but in prostatic epithelia it is converted to a more potent androgen, dihydrotestosterone (DHT), via the enzymatic action of 5α-reductase [20]. Without ligand binding, the AR is inactive and associated with heat shock proteins and is unacceptable to DNA binding. Ligand binding causes the release of the inhibitory heat shock proteins, and then the AR rapidly translocates to the nucleus. Afterwards, the AR as a homodimer binds to androgen-responsive elements (AREs) within the regulatory regions of target genes [9]. Then recruitment of coactivators that have histone acetylase activities facilitates chromatin remodeling and transcriptional activation of AR-dependent genes [21]. So far, five major classes of coactivators have been identified: the p160 subfamily including SRC-1/N-CoA1, SRC-2/GRIP-2/TIF-2 and SRC-3/ACTR/AIB1/p/CIP/RAC3/TRAM-1, which serve mainly as adapters recruiting other complexes [22], histone acetyltransferases (HATs) such as CBP/p300 and p/CAF [23], histone arginine methyltransferases (HMTs) such as CARM1 and PRMT1 [24], chromatin remodeling complexes such as SWI/SNF [25], and mediator coactivator complexes that tether between NRs and the basal transcriptional machinery [26].

It is likely that the AR regulates the expression of more than 100 genes, including *prostate specific antigen (PSA), kallikrein, keratinocyte growth factor (KGF), p21,* and *ornithine decarboxylase (ODC)* genes and the *AR* gene itself. For example, the *PSA* gene is regulated by two AREs in the proximal promoter and a cluster of four AREs in an enhancer region upstream of the transcription start site [27]. The AR can up-regulate or down-regulate gene expression, and the direction is dependent on the interaction with coactivators or corepressors [28].

Recently, the concept of a "transcriptional clock" has been proposed [29]. Nuclear steroid hormone receptors are first recruited to the responsive promoter followed by HATs and HMTs which are essential for the construction of complexes involved in chromatin remodeling. Then molecular events facilitate the arrival of RNA Polymerase II and general transcription factors. This mechanism raised the idea that cyclic interaction with transcription complexes is a common feature of promoters activated by nuclear receptors, including the AR. It has been proposed that this cyclic recruitment of nuclear receptors and their cofactors on a target promoter might be directed by cyclic changes in chromatin structure and in the phosphorylation/acetylation/methylation status of histones, nuclear receptors, and cofactors [30]. However, recent studies have indicated that some of this process is likely to be regulated by the ubiquitin-proteasome pathway.

Several groups have demonstrated that nuclear receptors, as well as coactivators, corepressors and components of the transcriptional machinery, are ubiquitinated and degraded by the 26S proteasome, dependent on ligand binding [31, 32]. It has been suggested that the functional role of ubiquitin-proteasome pathway is attenuation of the transcriptional signal via protein degradation. It has been proposed that degradation by the ubiquitin-proteasome pathway may function as an efficient mechanism for regulating the cyclic interaction of nuclear receptors with the promoter [33].

3. AR Siganaling and Cell Cycle

Activation of the AR induces a series of gene expressions that promote cell cycle progression. During genital development, it is obvious that the AR plays a major role in stimulating proliferation of normal epithelial cells and prostate cancer cells. Adequate combinations of cofactors recruited to AREs are also likely to be important for tissue-specific and ligand-specific gene expression. It is clear that the AR endorses prostate cancer proliferation and survival of prostate cancer cells by way of these complicated regulations.

Androgen is a critical regulator of G1-S transition in the cell cycle of prostate cancer cells. In prostate cancer cells, androgen depletion causes cell cycle arrest in early G1 phase and induces loss of cyclin D1 and cyclin D3 expression, followed by attenuation of CDK4 kinase activity and hypophosphorylation of Rb protein [34]. Androgen appears to induce cyclin D expression through mTOR-dependent translation [35].

In contrast to cyclin D, cyclin E levels remain relatively unchanged by androgen ablation, suggesting that change of cyclin E expression is not a crucial mechanism for androgen-mediated cell cycle regulation [34, 35]. However, cyclin A expression levels and CDK2 activity are reduced by androgen withdrawal, indicating that androgen depletion activates the inhibitory activity of RB, since cyclin A is one of the targets of RB-mediated transcriptional repression.

Furthermore, androgen withdrawal results in a high expression level of p27, which inhibits CDK2 activity [34]. Expression of p27 is associated with adverse pathological parameters, tumor cell proliferation, and increased risk of recurrence. Clinically, the combined loss of PTEN and p27 is likely to increase the risk of the development of prostatic carcinomas [36]. By stimulation with androgen, p27 is degraded probably by the ubiquitin-proteasome system [37]. Conversely, p21 expression is lost by androgen ablation in prostate cancer cells *in vitro* and in human tumor specimens [34, 38]. Therefore, p21 is related to androgen stimulation in prostate cancer. Actually, the *p21* gene is regulated as a direct AR target gene [39]. Clinical studies in prostate cancer have shown that p21 expression level is high in tumors which are associated with a higher proliferative index and higher Gleason grade [40, 41]. These findings suggest that androgen induces cyclin D1 expression through mTOR, promotes activation of cyclin D1/CDK4 through assembly by potentially p21, and facilitates the phosphorylation and degradation of p27 and that activation of cyclin D1/CDK4 causes RB phosphorylation, followed by progression into S-phase. Hence, androgen regulates the activity of cyclinD1/CDK4 in cancer cells and contributes to progression of cancer cells. Consequently, it is important in androgen ablation therapy that molecular crosstalk between the AR and cell cycle regulators, including RB, cyclins and CKI, is involved in the androgen-dependent proliferation of prostate cancer cells.

Many proteins have been reported to be functionally associated with the AR with regard to the cell cycle. The expression or activity of cell cycle regulators, including RB, cyclin D1 and CDK4, is related to progression to G1 phase and progression from G1 to S phase or G2 phase. Therefore, it is likely that the AR activity is controlled during the cell cycle via interactions with these proteins. Moreover, the AR may be modulated during the cell cycle by transcriptional, translational or post-translational modifications, including phosphorylation, acetylation and ubiquitination, which might affect transcriptional activity or stability.

It has been reported that direct interactions between the AR and RB were identified by a pull-down assay and use of a mammalian two-hybrid system [42, 43]. RB binds to the amino-terminal domain of the AR. Overexpression of RB enhanced AR transcriptional activity, while AR transcriptional activity was lost in RB-negative cells. Moreover, RB interacts with the GR, and overexpression of RB can enhance GR transcriptional activity as well [44]. GR activation by RB is dependent on Brm, and complex formation with RB and Brm is likely to be required for GR coactivation [44]. Therefore, in contrast to the AR, the GR shows its transcriptional activity in RB-deficient cells [42]. It is not clear whether Brm regulates RB-dependent coactivation of the AR.

Modulational changes in AR transcriptional activity during the cell cycle have been reported [45]. AR transcriptional activity in L929 cells, which express an endogenous AR, was examined using ARE-regulated reporter genes (MMTV-CAT). It was shown in that study that AR transcriptional activity was decreased at G1/S phase and recovered during S-phase. On the other hand, GR activity is not affected at G1/S phase. AR protein level was slightly reduced at G1/S phase and a potential mechanism for the decreased AR expression during G1/S phase may be regulated by increased E2F1, which has been shown to suppress the *AR* gene via direct binding to the AR promoter [46].

A histone deacetylase (HDAC) inhibitor such as TSA can partially restore AR transcriptional activity at G1/S transition without an increase in AR protein, suggesting that coactivators with HAT activity are recruited to the AR at G1/S transition and cause a decrease in the acetylation of histone or the AR. Although histone acetylation enhances transcription, the role of acetylation of the AR is not clear [47]. It has been reported that HDAC inhibitors reduce the expression level of the AR via inhibition of Hsp90, which must be deacetylated by HDAC6 for activity [48]. However, it is difficult to clarify their roles in regulating AR activity during the cell cycle.

Loss of AR transcriptional activity at G1/S transition may be due to cyclin D1, which is expressed during G1/S phase and likely serves as an AR corepressor. Furthermore, RB, which can function as an AR coactivator, undergoes hyperphosphorylation at G1/S, potentially resulting in a decrease in AR activity. Cyclin E may also serve as an AR coactivator, and its increased expression in S-phase may contribute to AR reactivation. However, further studies are required to confirm the relation between AR activity and the cell cycle.

Flow cytometric analysis has clarified the change of AR protein expression during the cell cycle using prostate cancer cell lines such as LNCaP, CWR22Rv1, and LAPC-4 [49]. The binding amount of Hoechst dye is smallest in cells that have just exited mitosis. Flow cytometric analysis has shown that prostate cancer cell lines with the lowest level of Hoechst dye staining had no detectable AR. Immunoblot analysis with anti-AR antibody was performed using cells sorted with the lowest level of Hoechst dye binding and showed that sorted cells has extremely expressed low level of AR. Immunohistochemical analysis using cell lines *in vitro* and *in vivo* prostate cancers showed loss of AR protein expression in mitotic phase. These findings suggest that the AR in mitotic phase is a licensing factor for DNA replication and that loss of AR protein is required for the next round of DNA replication. This decline is specific to prostate cancer cells, which are androgen-sensitive and express high levels of the AR. Further studies are required to establish a link between AR expression and DNA licencing. Furthermore, the loss of AR protein can be restored with a proteasome inhibitor such as MG132, demonstrating that AR protein level is regulated by proteasomal degradation. These findings suggest that the AR undergoes post-translational modifications, including ubiquitination, which induces its degradation especially during mitosis. So far, several E3 ubiquitin ligases for the AR have been identified: E6AP, CHIP, and MDM2. However, the relation between their activities toward the AR and cell cycle regulation has not been clarified. Interestingly, it has been reported that CDK1 stabilizes AR protein, but it is not clear whether this stabilization of the AR is dependent on direct phosphorylation on the AR [50]. Therefore, the weakening of CDK1 activity at the end of mitosis may contribute to AR degradation in G1/S phase.

Clinical therapies for prostate cancer focus on regulating the action of the AR, because AR activity is required for carcinogenesis and progression of prostate cancer. Unfortunately, androgen ablation therapy is likely to result in a bypass mechanism to activate the receptor even in the absence of a ligand and thus restore AR-dependent cellular proliferation. Consequently, it is important to elucidate the mechanisms by which the AR regulates cell cycle progression in order to develop new procedures for treating recurrent prostate cancer. It is clear that the liganded and activated AR promotes G1-S progression, indicating that crosstalk between the AR and cell cycle regulators has a critical role in regulation of androgen-mediated cell signaling. Abnormal regulation by potential mutations of the AR may facilitate androgen-independent responses including cancer cell proliferation, consequently followed by the development of recurrent tumors. Results of future studies on molecular mechanisms of AR-cell cycle crosstalk in prostate cancer cells should provide useful information for therapy against recurrent cancers.

4. AR Function in Prostate Cancer

Prostate cancer is the most common male cancer and second or third leading cause of male cancer deaths in western countries [3]. The role of the AR in prostate cancer development and progression has recently been reviewed [51]. A role of the AR in the pathogenesis of prostate cancers was established by Nobel Prize winners Charles Huggins and Clarence Hodges [4], who showed that bilateral orchiectomy or administration of estrogen was effective treatment. Since their works, androgen ablation has been used as successful treatment for locally advanced or metastatic prostate cancer or for relapsed cancer after prostatectomy. Recently, gonadotropin-releasing hormone (GnRH) agonists such as leuprolide and/or AR antagonists such as bicalutamide has been used for androgen ablation therapy [52]. Much attention has been paid to the importance of the AR in the pathogenesis of prostate cancer. The amount of PSA in serum has been used to screen prostate cancer [53, 54], because the gene expression of PSA is governed by AR-mediated transcription. Actually, serum PSA concentration has been used for monitoring the success after radical prostatectomy or radiation therapy, disease recurrence after treatment, response to androgen ablation, and change to phenotype as androgen-independent or androgen-refractory prostate cancer [55]. Androgen-independent prostate cancer is resistant to further hormonal administration as well as other known treatments. Therefore, researchers have focused on comprehending the molecular networks of this stage of prostate cancer. Several experimental and clinical data studies have shown roles of the AR in proliferation and survival of androgen-independent prostate cancer cells. Androgen-independent prostate cancers often have high levels of AR expression, and serum PSA is highly expressed at this stage of prostate cancer [56]. Experimental data about *PSA* promoter activity has shown that androgen-dependent prostate cancer xenografts exhibited a rapid loss of AR activity upon castration but that AR activity to the PSA promoter was completely restored after progression to an androgen-independent stage [57].

As already described, current prostate cancer therapies are based on androgen deprivation. This may be achieved either by castration or by using antiandrogens. Although they cause regression of the tumor, it usually relapses as a result of the selection of androgen-independent cells. Consequently, the prostate cancer comes back and is androgen-independent or androgen-refractory. Procedures using ribozymes, antisense oligonucleotides, and small-interference RNA can inhibit AR expression, and AR inhibition causes decreases in PSA expression and cell proliferation in androgen-independent/refractory prostate cancers [58, 59, 60]. These findings suggest that androgen-independent/refractory prostate cancers can be controlled by AR-dependent regulation but not by androgen. Several mechanisms by which androgen-independent cancer remains AR-dependent have been described or proposed [7, 61]. Briefly, the AR at this stage can be activated via ligand-independent mechanisms, high sensitization to androgens, or agonistic activity of adrenal androgens or activity even towards non-androgenic steroids such as progesterone and estradiol. Activation of the AR may occur due to gene mutation of the *AR*, overexpression or post-translational modification of the AR, activation of AR-associated coregulators, or inhibition of corepressors. Although serum androgen level is decreased as a result of androgen ablation therapy, a low level of androgen may be sufficient to bring about AR activation in recurrent prostate cancer [62]. Therefore, it may be important to measure concentrations of androgens in local regions surrounding androgen-independent prostate cancer. Several studies have suggested that abnormal metabolic cascades promote increase in local synthesis of testosterone from adrenal androgens [63] or decrease in synthesis of DHT [64]. Further therapeutic interventions will be established by taking advantage of the fact that refractory prostatic cancers are androgen-independent but remain AR-dependent. Recently, phytochemicals, including genistein and curcumin, has been used in anti-cancer therapy for prostate cancer. Moreover, one of the natural compounds of *Rheum palmatum,* emodin, seems to be a more effective phytochemical than curcumin and genistein for suppressing prostate cancer growth [65]. It is likely that emodin attenuates the stability of the AR by interfering with the association of the AR with Hsp90 and by enhancing the association of the AR with Mdm2 [65]. An attractive model using artificial molecules that induce degradation of AR protein by PROteolysis Targeting Chimeric molecules (PROTACs) has been reported [66]. PROTAC-5 consists of DHT, two aminocaproic acids as a linker, ALAPYIP as an amino acid sequence recognized by E3 ligase, and a poly-D-arginine tag to confer cell permeability [66]. PROTAC-5 can degrade AR molecules in HEK293 cells through proteasome-dependent proteolysis [66]. The Hsp90 chaperone inhibitor geldanamycin can induce degradation of the AR. Cytoplasmic AR is complexed with Hsp90 in its inactive state, and geldanamycin can inhibit the complex formation, resulting in proteasomal degradation of the AR [67]. A series of geldanamycin-testosterone-linked hybrids has been synthesized and evaluated for activity against prostate cancer cell lines. The hybrid compounds have potent and selective cytotoxicity against prostate cancer cells with the androgen receptor [68].

5. Post-Translational Modification of AR

The transcriptional activity of the AR is regulated not only by ligand binding but also by post-translational modifications. So far, the modification sites of AR and related modifying enzymes have been reported. These chemical or non-chemical covalent bindings may affect receptor stability, subcellular localization and interactions with other proteins including coactivators or corepressors. However, the relations between the modifications and each function at the molecular level are very complicated. Thus, it is important to clarify the roles of phosphorylation, acetylation, ubiquitination and sumoylation in AR functions.

6. AR Phosphorylation

A report about the phosphorylation sites of AR has shown that Serine (S) 94 is constitutively phosphorylated, whereas the phosphorylations of S16, S81, S256, S308, S424 and S650 are elevated by ligand binding [69]. Another group showed that S650 is constitutively phosphorylated and that S650 phosphorylation is dependent on the phosphorylation of S515 [70]. However, phosphorylations of S94, S515 and S650 may not have a significant role in AR activity, because mutations of them did not affect transcriptional activity of the AR [70]. MAP kinase and Akt are candidates of kinases for the AR [71]. Akt causes phosphorylation of S213, and immunochemical analysis using a phosphorylation-specific antibody indicated highly S213 phosphorylation in prostate epithelial cells [72]. These findings suggest that Akt-mediated phosphorylation can regulate AR function. However, the phosphorylation is dependent on passage number in LNCaP cells [73]. Another study has shown that the phosphorylation of AR at S210 and S790 by Akt kinase suppresses AR transactivation through the inhibition of AR-ARA70 interaction [74]. ARA70 as an AR coactivator is implicated in AR translocation to the nucleus. Akt promotes phosphorylation of the AR, and then Mdm2 mediates ubiquitination of the AR, followed by degradation via a proteasome-dependent pathway [75].

It has been reported that phosphorylation of the AR at S650 may be important for nuclear export [70]. Glycogen synthase kinase 3β (GSK3β) is a proline-directed serine/threonine kinase and is inhibited by the PI3K/Akt pathway. GSK3β can phosphorylate AR at S650, which is localized in the hinge region, and suppress transcriptional activity of the AR [76]. However, others have reported that AR phosphorylation of S650 by GSK3β attenuates AR transcription by increase of AR export to the cytoplasm [77].

PP2A can dephosphorylate the N-terminal domain of the AR, resulting in a loss of its transcriptional activity [78]. Furthermore, it has been shown that interaction of the E3 ubiquitin ligase CHIP with AR is dependent on phosphorylation of the hinge region of the AR, but these sites have not been reported as important phosphorylated residues of the AR [79].

7. AR Acetylation

The KXKK motif (630–633 amino acids residue) in the hinge region of the AR is likely to be acetylated [80]. Histone acetylaces including Tip60, P/CAF and p300 mediate acetylation of the AR in prostate cancer cells [81, 82, 83]. Mutation of lysine residues to alanine in this KXKK motif markedly attenuates the transcriptional activity of the AR, resulting in the recruitment of a corepressor even in the presence of androgen [84]. This study suggests that acetylation may be an important modification of the AR for recruitment of coactivators or corepressors in some cases. It has been shown that mutation of acetylated sites of the AR exhibited both delayed nuclear entry and ligand-dependent misfolding and aggregation [84]. Expression with constructs mimicking acetylation by neutral polar substitution (K630Q and K630T) enhanced the binding to p300 and attenuated the binding to corepressors including N-CoR/HDAC/Smad3, whereas substitution to a charged residue (K630R) caused reduced binding to p300 and increased binding to the corepressor.

Furthermore, the AR (K630Q and K630T) promoted survival and growth of prostate cancer cells and increased transcription of growth control target genes [80]. Moreover, it has been reported that AR acetylation mutants exhibit a decrease in phosphorylation of the AR and that a phosphorylation mutant (S94A) of the AR impairs the activation of p300 [85].

8. AR Ubiquitination

Ubiquitination is a three-step biochemical process involving a ubiquitin-activating enzyme (E1), a ubiquitin-conjugating enzyme (E2) and an E3 ubiquitin ligase [86]. Only one E1 protein for ubiquitination has been identified, and approximately twenty E2 and several hundred putative E3 family members exist in humans. Polyubiquitination generally functions as a marker for protein degradation via the proteasome machinery, whereas mono- or multi-ubiquitination likely affects subcellular localization or protein–protein interactions [87]. Furthermore, approximately one hundred deubiquitinating enzymes exist in humans [88]. Almost all steroid receptors undergo ubiquitination, and several enzymes involved in ubiquitination have so far been identified. However, identification of ubiquitinated sites of steroid hormone receptors is very difficult, probably due to the degeneracy of ubiquitinated sites or the instability of polyubiquitinated proteins. It has been suggested that rapid degradation of transcription factors is required for the regulation of transcriptional activity. As well as ubiquitin-mediated degradation, ubiquitination of steroid hormone receptors, which is not linked to proteolysis, is necessary for the activation of transcription [89]. It has recently been proposed as a "transcriptional clock" model for the control of transcriptional events.

Many nuclear receptors, including steroid hormone receptors, seem to undergo rapid turnover in the presence of a ligand. This model has been established for GR, TR, RAR, PR and ER. However, the AR is one of the exceptions in the group of nuclear receptors, since the AR is not generally degraded after ligand binding. In the contrast, AR is upregulated in the presence of androgen, suggesting that the relation between ligand binding and stability of AR protein is more complicated. It has recently been reported that the AR is ubiquitinated after androgen treatment [90]. Administration of the synthetic androgen R1881 to LNCaP cells enhanced monoubiquitination of the AR, causing robust stability for 2 hours, and then the stability of the AR decreased in accordance with the accumulation of polyubiquitinated forms of the AR [90].

UbcH7, one of the E2 enzymes, regulates the transcriptional activity of several nuclear receptors, including the AR [91], suggesting that rapid turnover of the AR is required for resetting and recovery of full transcriptional activity by the AR. However, Mdm2 E3 ligase promotes poly-ubiquitination of the AR by stimulation via Akt, followed by recruitment of the AR to the proteasome [75]. Moreover, it has been shown that a complex containing the AR and Mdm2 recruits HDAC1, which causes a decrease in transcriptional activity of the AR [90]. CHIP, another E3 ubiquitin ligase, also attenuates AR activity by binding to the N-terminal domain of the AR [92]. Another study has shown that CHIP interacts with the hinge region of the AR dependent on phosphorylation of the AR [79].

Transducin-β-like protein (TBL1) and TBL1-related protein (TBLR1) may recruit the ubiquitin-proteasome system into transcription complexes including nuclear receptors and may replace nuclear receptor corepressors with coactivators after ligand binding [93]. After androgen treatment, TBL1 is recruited to the promoter of *kallikrein 2* and *PSA* genes and is required for AR-dependent transcription. Furthermore, TBL1 and TBLR1 can recruit ubiquitin-conjugating enzyme UbcH5 and E6AP to the transcriptional complex. It has also been reported that E6AP and ubiquitin-conjugating enzyme UbcH7 increase transcriptional activity of the AR after ligand binding [94]. E6AP likely activates transcriptional activity of the AR by degrading NcoR via ubiquitin-mediated proteolysis [95]. E6AP is expressed at high level in the normal prostate, where E6AP is required for regulation of the AR, whereas the expression level of E6AP is reduced in prostate cancer.

Furthermore, it has been shown that the small nuclear RING finger protein (SNURF), which has ubiquitin E3 ligase activity, interacts with the AR in a ligand-dependent manner [96, 97]. SNURF binds to the hinge region of the AR and enhances AR-dependent transcription as a scaffold protein and by promoting nuclear import of the AR. The RING finger domain of SNURF likely interacts with general transcription factors, including TATA-binding protein (TBP).

After ligand binding, the AR translocates to the nucleus and is recruited to the ARE sequences of androgen-responsive genes, followed by association with coactivators. Ubiquitination of the AR may occur in the nucleus and cytoplasm. The interaction with tumor susceptibility gene 101 (TSG101) blocks AR degradation by keeping the AR in a monoubiquitinated state and promoting the initiation of transcription. Actually, androgen treatment causes monoubiquitination of the AR and monoubiquitination is vital for transcriptional activation of the AR. Then the monoubiquitinated AR is stabilized by TSG101 [98]. In the presence of apoptosis-antagonizing transcription factor (AATF), TSG101 interacts with the monoubiquitinated AR and prevents the change to polyubiquitination of the AR. Therefore, TSG101 facilitates the accumulation of monoubiquitinated ARs, leading to the stabilization and activation of AR-mediated transcription.

Activity of RNA Polymerase II and general transcription factors may be partially regulated by the ubiquitination machinery. It has been reported that the TAFII250 subunit of the general transcription factor TFIID has both E1 ubiquitin-activating enzyme activity and E2 ubiquitin-conjugating enzyme activity and causes the monoubiquitination of histone H1 [99]. Moreover, it has been reported that p300, a well-characterized coactivator with histone acetylase activity, has also ubiquitin ligase activity [100]. Conversely, USP10, a ubiquitin-specific protease that has deubiquitinating activity, facilitates transcriptional activity of the AR possibly by inhibiting degradation [101].

TRIM68 is a member of the tripartite motif-containing protein (TRIM) family defined by the presence of a common domain structure composed of a RING-finger, a B-box and a coiled-coil motif [102]. In addition to these motifs, TRIM68 possesses a carboxy-terminal PRY/SPRY domain at the carboxy-terminus. TRIM family proteins are involved in a broad range of biological processes and, consistently, their alterations result in diverse pathological conditions such as genetic diseases, viral infection and cancer development [103]. It has been reported that TRIM68 is one of the autoantigens associated with Sjögren's syndrome (SjS) and is a new diagnostic marker for SjS [104]. TRIM21, which is structurally similar to

TRIM68 and is also one of autoantigens associated with SjS, has been found to bind DNA and has been suggested to act as a transcription factor regulating gene expression [105]. TRIM21 has also been shown to be involved in cellular proliferation and cell death [106]. Furthermore, TRIM68 has been shown to be highly expressed in the prostate compared with its expression levels in other normal tissues. Recently, it has been reported that TRIM68 interacts with the AR and enhances transcriptional activity of the AR in the presence of DHT and that TRIM68 interacts with TIP60, one of the coactivators of the AR, and synergizes in transactivation of the AR [107]. Overexpression of TRIM68 in prostate cancer cells caused an increase in secretion of PSA, whereas knock-down of TRIM68 attenuated the secretion of PSA and inhibited cell growth. Moreover, immunohistochemical analysis showed that TRIM68 expression is significantly upregulated in human prostate cancers compared with the expression in adjacent normal tissues. These results indicate that TRIM68 also functions as a cofactor for AR-mediated transcription and is likely to be a novel diagnostic tool and a potentially therapeutic target for prostate cancer.

These findings suggest that the ubiquitin-proteasome system regulates nuclear receptors, including the AR, in order to reset the activity of the promoter and to reinitate transcription.

9. AR Regulation by Proteasome

It has recently been shown that proteasome activity is also critical for modulating transcriptional activation of the AR [108]. It has been suggested that 26S proteasome is involved in the transcription step and that the 19S subunits affect transcriptional initiation and elongation [109]. The 19S subunit that contains ATPase activity is recruited to the promoter for induction of transcription, without 20S proteasome components. Sug1 is one of the six ATPases of the 19S subunit, and forms a complex with nuclear receptors, but interaction of Sug1 with the AR has not yet been checked [110]. It has been shown that the S1 subunit of the 19S proteasome is recruited to the *PSA* promoter and then the AR is released from the *PSA* promoter [111]. Overexpression of proteasome alpha subunit 7 (PSMA7) of the 20S proteasome causes an increase in transcriptional activity of the AR following androgen-induced nuclear translocation of the AR [108]. Another study has demonstrated that the proteasome inhibitor PS-341 attenuates AR-mediated transcriptional activity but does not change the subcellular localization of the AR [112], indicating that 20S proteasome may also be a crucial factor for AR-dependent transcription. Furthermore, MG132, another proteasome inhibitor, can interfere with interaction of the AR with coactivators ARA70 and TIF2, which is one of the SRC coactivators [108]. ARA70 exhibits AR-mediated transcription by enhancing nuclear localization of the AR [113]. Three important regions or motifs of the AR including the nuclear localization signal (628-669), KLKK sequence (630-633) and PEST sequence (638-658) may be ubiquitinated and its modification likely functions as a signal for recruiting to the 20S proteasome and coregulators to increase nuclear translocation of the AR. Furthermore, deletion of amino acids 628-646 increases transcriptional activity of the AR and induces degradation of the AR via proteasome-mediated proteolysis, suggesting that the hinge region is involved in regulating transcriptional activity of the AR [114]. Moreover, mutation of lysine residues in the KLKK sequence of the AR caused a delay in androgen-dependent translocation, and the mutated AR colocalized with proteasome, suggesting that proteasome is involved in nuclear translocation.

10. AR Sumoylation

Sumoylation, one of ubiquitin-like protein modifications, is a three-step process involving an E1-activating enzyme (Aos/Uba2), an E2-conjugating enzyme (Ubc9) and an E3 ligase including PIAS families [115]. A SUMO protein is conjugated onto lysine residues in the consensus ψKxE motif by covalent binding, and this modification is also reversed by SUMO-specific proteases. Sumoylation likely affects several biological activities, including subcellular localization and transcriptional activity. Sumoylation sites have been identified in almost all steroid receptors. Slight differences in the effects of sumoylation on steroid receptors have been reported, and it has been suggested that sumoylation is pivotal for fine tuning of the activity of these factors.

It has been reported that the AR is also modified by sumoylation, and the modification sites has been identified: K386 and K520 [97]. AR sumoylation is a ligand-dependent reaction and is mediated by Ubc9 and PIAS family members, mainly PIAS1, PIASxα and hZimp10 [116, 117]. Sumoylation has generally repressive effects, but this depends on each situation. Mutation of SUMO acceptor sites caused stimulation of AR activity [118]. Sumoylation of acceptor sites (K386 and K520) seems to inhibit AR activity through inhibition of the interaction of SMRT with the AR. Additionally, Daxx, another corepressor, affects the AR transactivation, dependent on AR sumoylation [119]. However, overexpression of enzymes such as Ubc9 and PIASxα shows different phenotypes in various ARE-containing promoters [120]. These findings suggest that other factors or modifications control the sumoylation-dependent transcriptional activity of the AR.

Conclusion

It is clear that multiple post-translatioal modifications may affect many steroid receptors and alter protein-protein interactions, protein stability and subcellular localization for exquisite regulation of gene transcription. Furthermore, many cofactors such as coactivators and corepressors also undergo post-translational modification, which may affect their activity. Detailed investigations of these biochemical processes, including investigation of characteristics of enzymes involved in steroid receptors and the crosstalk between many different pathways, are important for establishment of novel, more selective drugs for hormone-dependent cancers such as mammary cancers and prostate cancers.

Regulation of AR transcriptional activity appears to be very complicated, and the regulation is related to interactions between the AR and cofactors. Post-translational modifications, including phosphorylation, acetylation, ubiquitination and sumoylation, are likely to regulate intricate associations and dissociations of the AR with cofactors. Further investigations for therapy of prostate cancer must be advanced to identify and characterize the coregulatory proteins that associate with the AR and to elucidate the roles of post-translational modifications of the AR.

References

[1] Wilson, J. D., George, F. W. & Griffin, J. E. (1981). *The hormonal control of sexual development. Science*, *211*, 1278-1284.

[2] Grossmann, M. E., Huang, H. & Tindall, D. J. (2001). Androgen receptor signaling in androgen-refractory prostate cancer. *J Natl Cancer Inst.*, *93*, 1687-1697.

[3] Jemal, A., Siegel, R., Ward, E., Murray, T., Xu, J. & Thun, M. J. (2007). Cancer statistics, 2007. *CA Cancer J Clin.*, *57*, 43-66.

[4] Huggins, C. & Hodges, C. V. (1972). Studies on prostatic cancer. I. The effect of castration, of estrogen and androgen injection on serum phosphatases in metastatic carcinoma of the prostate. *CA Cancer J Clin.*, *22*, 232-240.

[5] Jenster, G. (1999). The role of the androgen receptor in the development and progression of prostate cancer. *Semin Oncol.*, *26*, 407-421.

[6] Singh, P., Uzgare, A., Litvinov, I., Denmeade, S. R. & Isaacs, J. T. (2006). Combinatorial androgen receptor targeted therapy for prostate cancer. *Endocr Relat Cancer*, *13*, 653-666.

[7] Feldman, B. J. & Feldman, D. (2001). The development of androgen-independent prostate cancer. *Nat Rev Cancer*, *1*, 34-45.

[8] Lamb, D. J., Weigel, N. L. & Marcelli, M. (2001). Androgen receptors and their biology. *Vitam Horm*, *62*, 199-230.

[9] Gelmann, E. P. (2002). Molecular biology of the androgen receptor. *J Clin Oncol*, *20*, 3001-3015.

[10] Faus, H. & Haendler, B. (2006). Post-translational modifications of steroid receptors. *Biomed Pharmacother*, *60*, 520-528.

[11] Foradori, C. D., Weiser, M. J. & Handa, R. J. (2008). Non-genomic actions of androgens. *Front Neuroendocrinol*, *29*, 169-181.

[12] Migliaccio, A., Castoria, G., Di Domenico, M., de Falco, A., Bilancio, A., Lombardi, M., Barone, M. V., Ametrano, D., Zannini, M. S., Abbondanza, C. & Auricchio, F. (2000). Steroid-induced androgen receptor-oestradiol receptor beta-Src complex triggers prostate cancer cell proliferation. *EMBO J*, *19*, 5406-5417.

[13] Unni, E., Sun, S., Nan, B., McPhaul, M. J., Cheskis, B., Mancini, M. A. & Marcelli, M. (2004). Changes in androgen receptor nongenotropic signaling correlate with transition of LNCaP cells to androgen independence. *Cancer Res.*, *64*, 7156-7168.

[14] Kumar, R., Johnson, B. H. & Thompson, E. B. (2004). Overview of the structural basis for transcription regulation by nuclear hormone receptors. *Essays Biochem*, *40*, 27-39.

[15] Claessens, F. & Gewirth, D. T. (2004). DNA recognition by nuclear receptors. *Essays Biochem.*, *40*, 59-72.

[16] Geserick, C., Meyer, H. A. & Haendler, B. (2005). The role of DNA response elements as allosteric modulators of steroid receptor function. *Mol Cell Endocrinol.*, *236*, 1-7.

[17] Moehren, U., Eckey, M. & Baniahmad, A. (2004). Gene repression by nuclear hormone receptors. *Essays Biochem.*, *40*, 89-104.

[18] Matias, P. M., Donner, P., Coelho, R., Thomaz, M., Peixoto, C., Macedo, S., Otto, N., Joschko, S., Scholz, P., Wegg, A., Basler, S., Schafer, M., Egner, U. & Carrondo, M. A. (2000). Structural evidence for ligand specificity in the binding domain of the human androgen receptor. Implications for pathogenic gene mutations. *J Biol Chem.*, *275*, 26164-26171.

[19] Evans, R. M. (1988). The steroid and thyroid hormone receptor superfamily. *Science*, *240*, 889-895.

[20] Russell, D. W. & Wilson, J. D. (1994). Steroid 5 alpha-reductase: two genes/two enzymes. *Ann Rev Biochem.*, *63*, 25-61.

[21] Heinlein, C. A. & Chang, C. (2002). Androgen receptor (AR) coregulators: an overview. *Endocr Rev.*, *23*, 175-200.

[22] Xu, J. & Li, Q. (2003). Review of the in vivo functions of the p160 steroid receptor coactivator family. *Mol Endocrinol.*, *17*, 1681-1692.

[23] Lee, J. W., Lee, Y. C., Na, S. Y., Jung, D. J. & Lee, S. K. (2001). Transcriptional coregulators of the nuclear receptor superfamily: coactivators and corepressors. *Cell Mol Life Sci.*, *58*, 289-297.

[24] Lee, D. Y., Teyssier, C., Strahl, B. D. & Stallcup, M. R. (2005). Role of protein methylation in regulation of transcription. *Endocr Rev.*, *26*, 147-170.

[25] Belandia, B. & Parker, M. G. (2003). Nuclear receptors: a rendezvous for chromatin remodeling factors. *Cell*, *114*, 277-280.

[26] Taatjes, D. J., Marr, M. T. & Tjian, R. (2004). Regulatory diversity among metazoan co-activator complexes. *Nat Rev Mol Cell Biol.*, *5*, 403-410.

[27] Huang, W., Shostak, Y., Tarr, P., Sawyers, C. & Carey, M. (1999). Cooperative assembly of androgen receptor into a nucleoprotein complex that regulates the prostate-specific antigen enhancer. *J Biol Chem.*, *274*, 25756-25768.

[28] Asirvatham, A. J., Schmidt, M., Gao, B. & Chaudhary, J. (2006). Androgens regulate the immune/inflammatory response and cell survival pathways in rat ventral prostate epithelial cells. *Endocrinology*, *147*, 257-271.

[29] Metivier, R., Penot, G., Hubner, M. R., Reid, G., Brand, H., Kos, M. & Gannon, F. (2003). Estrogen receptor-alpha directs ordered, cyclical, and combinatorial recruitment of cofactors on a natural target promoter. *Cell*, *115*, 751-763.

[30] Demarest, S. J., Martinez-Yamout, M., Chung, J., Chen, H., Xu, W., Dyson, H. J., Evans, R. M. & Wright, P. E. (2002). Mutual synergistic folding in recruitment of CBP/p300 by p160 nuclear receptor coactivators. *Nature*, *415*, 549-553.

[31] Kopf, E., Plassat, J. L., Vivat, V., de The, H., Chambon, P. & Rochette-Egly, C. (2000). Dimerization with retinoid X receptors and phosphorylation modulate the retinoic acid-induced degradation of retinoic acid receptors alpha and gamma through the ubiquitin-proteasome pathway. *J Biol Chem.*, *275*, 33280-33288.

[32] Lonard, D. M. & O'Malley, B. W. (2005). Expanding functional diversity of the coactivators. *Trends Biochem Sci.*, *30*, 126-132.

[33] Reid, G., Hubner, M. R., Metivier, R., Brand, H., Denger, S., Manu, D., Beaudouin, J., Ellenberg, J. & Gannon, F. (2003). Cyclic, proteasome-mediated turnover of

unliganded and liganded ERα on responsive promoters is an integral feature of estrogen signaling. *Mol Cell, 11*, 695-707.

[34] Knudsen, K. E., Arden, K. C. & Cavenee, W. K. (1998). Multiple G1 regulatory elements control the androgen-dependent proliferation of prostatic carcinoma cells. *J Biol Chem., 273*, 20213-20222.

[35] Xu, Y., Chen, S. Y., Ross, K. N. & Balk, S. P. (2006). Androgens induce prostate cancer cell proliferation through mammalian target of rapamycin activation and post-transcriptional increases in cyclin D proteins. *Cancer Res., 66*, 7783-7792.

[36] Halvorsen, O. J., Haukaas, S. A. & Akslen, L. A. (2003). Combined loss of PTEN and p27 expression is associated with tumor cell proliferation by Ki-67 and increased risk of recurrent disease in localized prostate cancer. *Clin Cancer Res., 9*, 1474-1479.

[37] Ye, D., Mendelsohn, J. & Fan, Z. (1999). Androgen and epidermal growth factor down-regulate cyclin-dependent kinase inhibitor $p27^{Kip1}$ and costimulate proliferation of MDA PCa 2a and MDA PCa 2b prostate cancer cells. *Clin Cancer Res., 5*, 2171-2177.

[38] Kolar, Z., Murray, P. G., Scott, K., Harrison, A., Vojtesek, B. & Dusek, J. (2000). Relation of Bcl-2 expression to androgen receptor, $p21^{WAF1/CIP1}$, and cyclin D1 status in prostate cancer. *Mol Pathol., 53*, 15-18.

[39] Lu, S., Liu, M., Epner, D. E., Tsai, S. Y. & Tsai, M. J. (1999). Androgen regulation of the cyclin-dependent kinase inhibitor p21 gene through an androgen response element in the proximal promoter. *Mol Endocrinol, 13*, 376-384.

[40] Aaltomaa, S., Lipponen, P., Eskelinen, M., Ala-Opas, M. & Kosma, V. M. (1999). Prognostic value and expression of $p21^{waf1/cip1}$ protein in prostate cancer. *Prostate, 39*, 8-15.

[41] Baretton, G. B., Klenk, U., Diebold, J., Schmeller, N. & Lohrs, U. (1999). Proliferation- and apoptosis-associated factors in advanced prostatic carcinomas before and after androgen deprivation therapy: prognostic significance of p21/WAF1/CIP1 expression. *Br J Cancer, 80*, 546-555.

[42] Lu, J. & Danielsen, M. (1998). Differential regulation of androgen and glucocorticoid receptors by retinoblastoma protein. *J Biol Chem., 273*, 31528-31533.

[43] Yeh, S., Miyamoto, H., Nishimura, K., Kang, H., Ludlow, J., Hsiao, P., Wang, C., Su, C. & Chang, C. (1998). Retinoblastoma, a tumor suppressor, is a coactivator for the androgen receptor in human prostate cancer DU145 cells. *Biochem Biophys Res. Commun., 248*, 361-367.

[44] Singh, P., Coe, J. & Hong, W. (1995). A role for retinoblastoma protein in potentiating transcriptional activation by the glucocorticoid receptor. *Nature, 374*, 562-565.

[45] Martinez, E. D. & Danielsen, M. (2002). Loss of androgen receptor transcriptional activity at the G(1)/S transition. *J Biol Chem., 277*, 29719-29729.

[46] Davis, J. N., Wojno, K. J., Daignault, S., Hofer, M. D., Kuefer, R., Rubin, M. A. & Day, M. L. (2006). Elevated E2F1 inhibits transcription of the androgen receptor in metastatic hormone-resistant prostate cancer. *Cancer Res., 66*, 11897-11906.

[47] Popov, V. M., Wang, C., Shirley, L. A., Rosenberg, A., Li, S., Nevalainen, M., Fu, M. & Pestell, R. G. (2007). The functional significance of nuclear receptor acetylation. *Steroids, 72*, 221-230.

[48] Kovacs, J. J., Murphy, P. J., Gaillard, S., Zhao, X., Wu, J. T., Nicchitta, C. V., Yoshida, M., Toft, D. O., Pratt, W. B. & Yao, T. P. (2005). HDAC6 regulates Hsp90 acetylation and chaperone-dependent activation of glucocorticoid receptor. *Mol Cell*, *18*, 601-607.

[49] Litvinov, I. V., Vander Griend, D. J., Antony, L., Dalrymple, S., De Marzo, A. M., Drake, C. G. & Isaacs, J. T. (2006). Androgen receptor as a licensing factor for DNA replication in androgen-sensitive prostate cancer cells. *Proc Natl Acad Sci USA*, *103*, 15085-15090.

[50] Chen, S., Xu, Y., Yuan, X., Bubley, G. J. & Balk, S. P. (2006). Androgen receptor phosphorylation and stabilization in prostate cancer by cyclin-dependent kinase 1. *Proc Natl Acad Sci USA*, *103*, 15969-15974.

[51] Culig, Z. & Bartsch, G. (2006). Androgen axis in prostate cancer. *J Cell Biochem.*, *99*, 373-381.

[52] Scherr, D., Swindle, P. W. & Scardino, P. T. (2003). National comprehensive cancer network guidelines for the management of prostate cancer. *Urology*, *61*, 14-24.

[53] Ablin, R. J., Bronson, P., Soanes, W. A. & Witebsky, E. (1970). Tissue- and species-specific antigens of normal human prostatic tissue. *J Immunol*, *104*, 1329-1339.

[54] Ablin, R. J., Soanes, W. A., Bronson, P. & Witebsky, E. (1970). Precipitating antigens of the normal human prostate. *J Reprod Fertil.*, *22*, 573-574.

[55] Balk, S. P., Ko, Y. J. & Bubley, G. J. (2003). Biology of prostate-specific antigen. *J Clin Oncol.*, *21*, 383-391.

[56] Buchanan, G., Irvine, R. A., Coetzee, G. A. & Tilley, W. D. (2001). Contribution of the androgen receptor to prostate cancer predisposition and progression. *Cancer Metastasis Rev.*, *20*, 207-223.

[57] Zhang, L., Johnson, M., Le, K. H., Sato, M., Ilagan, R., Iyer, M., Gambhir, S. S., Wu, L. & Carey, M. (2003). Interrogating androgen receptor function in recurrent prostate cancer. *Cancer Res.*, *63*, 4552-4560.

[58] Zegarra-Moro, O. L., Schmidt, L. J., Huang, H. & Tindall, D. J. (2002). Disruption of androgen receptor function inhibits proliferation of androgen-refractory prostate cancer cells. *Cancer Res.*, *62*, 1008-1013.

[59] Eder, I. E., Culig, Z., Ramoner, R., Thurnher, M., Putz, T., Nessler-Menardi, C., Tiefenthaler, M., Bartsch, G. & Klocker, H. (2000). Inhibition of LncaP prostate cancer cells by means of androgen receptor antisense oligonucleotides. *Cancer Gene Ther.*, *7*, 997-1007.

[60] Liao, X., Tang, S., Thrasher, J. B., Griebling, T. L. & Li, B. (2005). Small-interfering RNA-induced androgen receptor silencing leads to apoptotic cell death in prostate cancer. *Mol Cancer Ther.*, *4*, 505-515.

[61] Pienta, K. J. & Bradley, D. (2006). Mechanisms underlying the development of androgen-independent prostate cancer. *Clin Cancer Res.*, *12*, 1665-1671.

[62] Titus, M. A., Schell, M. J., Lih, F. B., Tomer, K. B. & Mohler, J. L. (2005). Testosterone and dihydrotestosterone tissue levels in recurrent prostate cancer. *Clin Cancer Res.*, *11*, 4653-4657.

[63] Stanbrough, M., Bubley, G. J., Ross, K., Golub, T. R., Rubin, M. A., Penning, T. M., Febbo, P. G. & Balk, S. P. (2006). Increased expression of genes converting adrenal

androgens to testosterone in androgen-independent prostate cancer. *Cancer Res.*, *66*, 2815-2825.

[64] Ji, Q., Chang, L., Stanczyk, F. Z., Ookhtens, M., Sherrod, A. & Stolz, A. (2007). Impaired dihydrotestosterone catabolism in human prostate cancer: critical role of AKR1C2 as a pre-receptor regulator of androgen receptor signaling. *Cancer Res.*, *67*, 1361-1369.

[65] Cha, T. L., Qiu, L., Chen, C. T., Wen, Y. & Hung, M. C. (2005). Emodin down-regulates androgen receptor and inhibits prostate cancer cell growth. *Cancer Res.*, *65*, 2287-2295.

[66] Schneekloth, J. S., Jr., Fonseca, F. N., Koldobskiy, M., Mandal, A., Deshaies, R., Sakamoto, K. & Crews, C. M. (2004). Chemical genetic control of protein levels: selective in vivo targeted degradation. *J Am Chem Soc.*, *126*, 3748-3754.

[67] Vanaja, D. K., Mitchell, S. H., Toft, D. O. & Young, C. Y. (2002). Effect of geldanamycin on androgen receptor function and stability. *Cell Stress Chaperones*, *7*, 55-64.

[68] Kuduk, S. D., Harris, T. C., Zheng, F. F., Sepp-Lorenzino, L., Ouerfelli, Q., Rosen, N. & Danishefsky, S. J. (2000). Synthesis and evaluation of geldanamycin-testosterone hybrids. *Bioorg Med Chem Lett*, *10*, 1303-1306.

[69] Gioeli, D., Ficarro, S. B., Kwiek, J. J., Aaronson, D., Hancock, M., Catling, A. D., White, F. M., Christian, R. E., Settlage, R. E., Shabanowitz, J., Hunt, D. F. & Weber, M. J. (2002). Androgen receptor phosphorylation. Regulation and identification of the phosphorylation sites. *J Biol Chem.*, *277*, 29304-29314.

[70] Wong, H. Y., Burghoorn, J. A., Van Leeuwen, M., De Ruiter, P. E., Schippers, E., Blok, L. J., Li, K. W., Dekker, H. L., De Jong, L., Trapman, J., Grootegoed, J. A. & Brinkmann, A. O. (2004). Phosphorylation of androgen receptor isoforms. *Biochem J*, *383*, 267-276.

[71] Wen, Y., Hu, M. C., Makino, K., Spohn, B., Bartholomeusz, G., Yan, D. H. & Hung, M. C. (2000). HER-2/neu promotes androgen-independent survival and growth of prostate cancer cells through the Akt pathway. *Cancer Res.*, *60*, 6841-6845.

[72] Taneja, S. S., Ha, S., Swenson, N. K., Huang, H. Y., Lee, P., Melamed, J., Shapiro, E., Garabedian, M. J. & Logan, S. K. (2005). Cell-specific regulation of androgen receptor phosphorylation in vivo. *J Biol Chem.*, *280*, 40916-40924.

[73] Lin, H. K., Hu, Y. C., Yang, L., Altuwaijri, S., Chen, Y. T., Kang, H. Y. & Chang, C. (2003). Suppression versus induction of androgen receptor functions by the phosphatidylinositol 3-kinase/Akt pathway in prostate cancer LNCaP cells with different passage numbers. *J Biol Chem.*, *278*, 50902-50907.

[74] Lin, H. K., Yeh, S., Kang, H. Y. & Chang, C. (2001). Akt suppresses androgen-induced apoptosis by phosphorylating and inhibiting androgen receptor. *Proc Natl Acad Sci USA*, *98*, 7200-7205.

[75] Lin, H. K., Wang, L., Hu, Y. C., Altuwaijri, S. & Chang, C. (2002). Phosphorylation-dependent ubiquitylation and degradation of androgen receptor by Akt require Mdm2 E3 ligase. *EMBO J*, *21*, 4037-4048.

[76] Salas, T. R., Kim, J., Vakar-Lopez, F., Sabichi, A. L., Troncoso, P., Jenster, G., Kikuchi, A., Chen, S. Y., Shemshedini, L., Suraokar, M., Logothetis, C. J., DiGiovanni,

J., Lippman, S. M. & Menter, D. G. (2004). Glycogen synthase kinase-3 beta is involved in the phosphorylation and suppression of androgen receptor activity. *J Biol Chem.*, *279*, 19191-19200.

[77] Gioeli, D., Black, B. E., Gordon, V., Spencer, A., Kesler, C. T., Eblen, S. T., Paschal, B. M. & Weber, M. J. (2006). Stress kinase signaling regulates androgen receptor phosphorylation, transcription, and localization. *Mol Endocrinol.*, *20*, 503-515.

[78] Yang, C. S., Vitto, M. J., Busby, S. A., Garcia, B. A., Kesler, C. T., Gioeli, D., Shabanowitz, J., Hunt, D. F., Rundell, K., Brautigan, D. L. & Paschal, B. M. (2005). Simian virus 40 small t antigen mediates conformation-dependent transfer of protein phosphatase 2A onto the androgen receptor. *Mol Cell Biol.*, *25*, 1298-1308.

[79] Rees, I., Lee, S., Kim, H. & Tsai, F. T. (2006). The E3 ubiquitin ligase CHIP binds the androgen receptor in a phosphorylation-dependent manner. *Biochim Biophys Acta*, *1764*, 1073-1079.

[80] Fu, M., Wang, C., Reutens, A. T., Wang, J., Angeletti, R. H., Siconolfi-Baez, L., Ogryzko, V., Avantaggiati, M. L. & Pestell, R. G. (2000). p300 and p300/cAMP-response element-binding protein-associated factor acetylate the androgen receptor at sites governing hormone-dependent transactivation. *J Biol Chem.*, *275*, 20853-20860.

[81] Gong, J., Zhu, J., Goodman, O. B., Jr., Pestell, R. G., Schlegel, P. N., Nanus, D. M. & Shen, R. (2006). Activation of p300 histone acetyltransferase activity and acetylation of the androgen receptor by bombesin in prostate cancer cells. *Oncogene*, *25*, 2011-2021.

[82] Fu, M., Rao, M., Wang, C., Sakamaki, T., Wang, J., Di Vizio, D., Zhang, X., Albanese, C., Balk, S., Chang, C., Fan, S., Rosen, E., Palvimo, J. J., Janne, O. A., Muratoglu, S., Avantaggiati, M. L. & Pestell, R. G. (2003). Acetylation of androgen receptor enhances coactivator binding and promotes prostate cancer cell growth. *Mol Cell Biol.*, *23*, 8563-8575.

[83] Gaughan, L., Logan, I. R., Cook, S., Neal, D. E. & Robson, C. N. (2002). Tip60 and histone deacetylase 1 regulate androgen receptor activity through changes to the acetylation status of the receptor. *J Biol Chem.*, *277*, 25904-25913.

[84] Thomas, M., Dadgar, N., Aphale, A., Harrell, J. M., Kunkel, R., Pratt, W. B. & Lieberman, A. P. (2004). Androgen receptor acetylation site mutations cause trafficking defects, misfolding, and aggregation similar to expanded glutamine tracts. *J Biol Chem.*, *279*, 8389-8395.

[85] Halkidou, K., Gnanapragasam, V. J., Mehta, P. B., Logan, I. R., Brady, M. E., Cook, S., Leung, H. Y., Neal, D. E. & Robson, C. N. (2003). Expression of Tip60, an androgen receptor coactivator, and its role in prostate cancer development. *Oncogene*, *22*, 2466-2477.

[86] Hershko, A. & Ciechanover, A. (1998). The ubiquitin system. *Ann Rev Biochem.*, *67*, 425-479.

[87] Haglund, K. & Dikic, I. (2005). Ubiquitylation and cell signaling. *EMBO J*, *24*, 3353-3359.

[88] Nijman, S. M., Luna-Vargas, M. P., Velds, A., Brummelkamp, T. R., Dirac, A. M., Sixma, T. K. & Bernards, R. (2005). A genomic and functional inventory of deubiquitinating enzymes. *Cell*, *123*, 773-786.

[89] Muratani, M. & Tansey, W. P. (2003). How the ubiquitin-proteasome system controls transcription. *Nat Rev Mol Cell Biol.*, *4*, 192-201.

[90] Gaughan, L., Logan, I. R., Neal, D. E. & Robson, C. N. (2005). Regulation of androgen receptor and histone deacetylase 1 by Mdm2-mediated ubiquitylation. *Nucleic Acids Res.*, *33*, 13-26.

[91] Verma, S., Ismail, A., Gao, X., Fu, G., Li, X., O'Malley, B. W. & Nawaz, Z. (2004). The ubiquitin-conjugating enzyme UBCH7 acts as a coactivator for steroid hormone receptors. *Mol Cell Biol.*, *24*, 8716-8726.

[92] He, B., Bai, S., Hnat, A. T., Kalman, R. I., Minges, J. T., Patterson, C. & Wilson, E. M. (2004). An androgen receptor NH2-terminal conserved motif interacts with the COOH terminus of the Hsp70-interacting protein (CHIP). *J Biol Chem.*, *279*, 30643-30653.

[93] Perissi, V., Aggarwal, A., Glass, C. K., Rose, D. W. & Rosenfeld, M. G. (2004). A corepressor/coactivator exchange complex required for transcriptional activation by nuclear receptors and other regulated transcription factors. *Cell*, *116*, 511-526.

[94] Nawaz, Z., Lonard, D. M., Smith, C. L., Lev-Lehman, E., Tsai, S. Y., Tsai, M. J. & O'Malley, B. W. (1999). The Angelman syndrome-associated protein, E6-AP, is a coactivator for the nuclear hormone receptor superfamily. *Mol Cell Biol.*, *19*, 1182-1189.

[95] Zhang, J., Guenther, M. G., Carthew, R. W. & Lazar, M. A. (1998). Proteasomal regulation of nuclear receptor corepressor-mediated repression. *Genes Dev*, *12*, 1775-1780.

[96] Moilanen, A. M., Poukka, H., Karvonen, U., Hakli, M., Janne, O. A. & Palvimo, J. J. (1998). Identification of a novel RING finger protein as a coregulator in steroid receptor-mediated gene transcription. *Mol Cell Biol.*, *18*, 5128-5139.

[97] Poukka, H., Karvonen, U., Janne, O. A. & Palvimo, J. J. (2000). Covalent modification of the androgen receptor by small ubiquitin-like modifier 1 (SUMO-1). *Proc Natl Acad Sci USA*, *97*, 14145-14150.

[98] Burgdorf, S., Leister, P. & Scheidtmann, K. H. (2004). TSG101 interacts with apoptosis-antagonizing transcription factor and enhances androgen receptor-mediated transcription by promoting its monoubiquitination. *J Biol Chem.*, *279*, 17524-17534.

[99] Pham, A. D. & Sauer, F. (2000). Ubiquitin-activating/conjugating activity of TAFII250, a mediator of activation of gene expression in Drosophila. *Science*, *289*, 2357-2360.

[100] Grossman, S. R., Deato, M. E., Brignone, C., Chan, H. M., Kung, A. L., Tagami, H., Nakatani, Y. & Livingston, D. M. (2003). Polyubiquitination of p53 by a ubiquitin ligase activity of p300. *Science*, *300*, 342-344.

[101] Faus, H., Meyer, H. A., Huber, M., Bahr, I. & Haendler, B. (2005). The ubiquitin-specific protease USP10 modulates androgen receptor function. *Mol Cell Endocrinol*, *245*, 138-146.

[102] Reymond, A., Meroni, G., Fantozzi, A., Merla, G., Cairo, S., Luzi, L., Riganelli, D., Zanaria, E., Messali, S., Cainarca, S., Guffanti, A., Minucci, S., Pelicci, P. G. & Ballabio, A. (2001). The tripartite motif family identifies cell compartments. *EMBO J*, *20*, 2140-2151.

[103] Meroni, G. & Diez-Roux, G. (2005). TRIM/RBCC, a novel class of 'single protein RING finger' E3 ubiquitin ligases. *Bioessays*, *27*, 1147-1157.

[104] Billaut-Mulot, O., Cocude, C., Kolesnitchenko, V., Truong, M. J., Chan, E. K., Hachula, E., de la Tribonniere, X., Capron, A. & Bahr, G. M. (2001). SS-56, a novel cellular target of autoantibody responses in Sjogren syndrome and systemic lupus erythematosus. *J Clin Invest.*, *108*, 861-869.

[105] Frank, M. B. (1999). Characterization of DNA binding properties and sequence specificity of the human 52 kDa Ro/SS-A (Ro52). zinc finger protein. Biochem *Biophys Res Commun.*, *259*, 665-670.

[106] Espinosa, A., Zhou, W., Ek, M., Hedlund, M., Brauner, S., Popovic, K., Horvath, L., Wallerskog, T., Oukka, M., Nyberg, F., Kuchroo, V. K. & Wahren-Herlenius, M. (2006). The Sjogren's syndrome-associated autoantigen Ro52 is an E3 ligase that regulates proliferation and cell death. *J Immunol.*, *176*, 6277-6285.

[107] Miyajima, N., Maruyama, S., Bohgaki, M., Kano, S., Shigemura, M., Shinohara, N., Nonomura, K. & Hatakeyama, S. (2008). TRIM68 regulates ligand-dependent transcription of androgen receptor in prostate cancer cells. *Cancer Res.*, *68*, 3486-3494.

[108] Lin, H. K., Altuwaijri, S., Lin, W. J., Kan, P. Y., Collins, L. L. & Chang, C. (2002). Proteasome activity is required for androgen receptor transcriptional activity via regulation of androgen receptor nuclear translocation and interaction with coregulators in prostate cancer cells. *J Biol Chem.*, *277*, 36570-36576.

[109] Gonzalez, F., Delahodde, A., Kodadek, T. & Johnston, S. A. (2002). Recruitment of a 19S proteasome subcomplex to an activated promoter. *Science*, *296*, 548-550.

[110] vom Baur, E., Zechel, C., Heery, D., Heine, M. J., Garnier, J. M., Vivat, V., Le Douarin, B., Gronemeyer, H., Chambon, P. & Losson, R. (1996). Differential ligand-dependent interactions between the AF-2 activating domain of nuclear receptors and the putative transcriptional intermediary factors mSUG1 and TIF1. *EMBO J*, *15*, 110-124.

[111] Kang, Z., Pirskanen, A., Janne, O. A. & Palvimo, J. J. (2002). Involvement of proteasome in the dynamic assembly of the androgen receptor transcription complex. *J Biol Chem.*, *277*, 48366-48371.

[112] Ikezoe, T., Yang, Y., Saito, T., Koeffler, H. P. & Taguchi, H. (2004). Proteasome inhibitor PS-341 down-regulates prostate-specific antigen (PSA) and induces growth arrest and apoptosis of androgen-dependent human prostate cancer LNCaP cells. *Cancer Sci.*, *95*, 271-275.

[113] Hu, Y. C., Yeh, S., Yeh, S. D., Sampson, E. R., Huang, J., Li, P., Hsu, C. L., Ting, H. J., Lin, H. K., Wang, L., Kim, E., Ni, J. & Chang, C. (2004). Functional domain and motif analyses of androgen receptor coregulator ARA70 and its differential expression in prostate cancer. *J Biol Chem.*, *279*, 33438-33446.

[114] Tanner, T., Claessens, F. & Haelens, A. (2004). The hinge region of the androgen receptor plays a role in proteasome-mediated transcriptional activation. *Ann N Y Acad Sci.*, *1030*, 587-592.

[115] Seeler, J. S. & Dejean, A. (2003). Nuclear and unclear functions of SUMO. *Nat Rev Mol Cell Biol.*, *4*, 690-699.

[116] Nishida, T. & Yasuda, H. (2002). PIAS1 and PIASxalpha function as SUMO-E3 ligases toward androgen receptor and repress androgen receptor-dependent transcription. *J Biol Chem.*, *277*, 41311-41317.

[117] Sharma, M., Li, X., Wang, Y., Zarnegar, M., Huang, C. Y., Palvimo, J. J., Lim, B. & Sun, Z. (2003). hZimp10 is an androgen receptor co-activator and forms a complex with SUMO-1 at replication foci. *EMBO J*, *22*, 6101-6114.

[118] Callewaert, L., Verrijdt, G., Haelens, A. & Claessens, F. (2004). Differential effect of small ubiquitin-like modifier (SUMO)-ylation of the androgen receptor in the control of cooperativity on selective versus canonical response elements. *Mol Endocrinol.*, *18*, 1438-1449.

[119] Lin, D. Y., Fang, H. I., Ma, A. H., Huang, Y. S., Pu, Y. S., Jenster, G., Kung, H. J. & Shih, H. M. (2004). Negative modulation of androgen receptor transcriptional activity by Daxx. *Mol Cell Biol.*, *24*, 10529-10541.

[120] Geserick, C., Meyer, H. A., Barbulescu, K. & Haendler, B. (2003). Differential modulation of androgen receptor action by deoxyribonucleic acid response elements. *Mol Endocrinol.*, *17*, 1738-1750.

In: Handbook of Prostate Cancer Cell Research...
Editor: Alan T. Meridith
ISBN: 978-1-60741-954-9

Chapter 9

Cytotoxic Effects of Rhenium (I) Carbonyl Complexes on Prostate Cancer Cell Lines

Dejene K. Orsa,[a] Columbus R. Nettles,[a] Saroj K. Pramanik,[b] Maurice O. Iwunze,[a] George E. Greco,[c] Jeanette A. Krause,[d] and Santosh K. Mandal[a]

[a]Morgan State University, Department of Chemistry, Baltimore, MD 21251, USA
[b]Morgan State University, Department of Biology, Baltimore, MD 21251, USA
[c]Goucher College, Baltimore, MD 21204, USA
[d]University of Cincinnati, Cincinnati, Ohio 45221, USA

Abstract

The drug, cisplatin (*cis*-diamminedichloroplatinum(II)) is routinely used worldwide to cure testicular, ovarian and many other cancers. Cisplatin and its analogs DWA-2114R {2-aminomethylpyrrolidine(1,1-cyclobutanedicarboxylato) platinum(II) and JM216 (cis-(ammine)(cyclohexylamine)dichloro-*trans*-(diacetato) platinum(IV)) are also found to be active against prostate cancer. Unfortunately, these and other chemotherapeutic drugs are toxic. Recently, some organorhenium(I) complexes have been reported to be cytotoxic against various cancer cell lines including prostate cancer cell lines. Surprisingly, these complexes are not toxic to prostate and other cancer cell lines according to these researchers.

This prompted us to synthesize a wide variety of mono-, di-, and tetrarhenium (diphosphine) and (diimine) carbonyl complexes. They were characterized by spectroscopic techniques and X-ray crystal structure determinations. As expected, the rhenium(I) (α-diimine) tricarbonyl complexes exhibit fluorescence. Both rhenium(I) (diphosphine) and (diimine) carbonyl complexes are cytotoxic against prostate cancer cell lines. However, the rhenium(I) (α-diimine) tricarbonyl complexes are preferable to diphosphines due to their fluorescence properties.

Introduction

There are approximately twelve major cancers that affect most Americans (Rennie, 1996). Prostate cancer is the second leading cause of death among American males. A survey has revealed that the incidence is rapidly growing and the frequency is higher among black Americans compared to white Americans. The incidence is surprisingly low within the Asian population but relatively higher among those settled in the US. It is estimated that about 19 of every 100 men born today will be diagnosed with prostate cancer in their lifetime. Every 2.5 minutes, a new case of prostate cancer is diagnosed. Although environmental stimulus is suspected to be one of the causes, genetic predisposition, dietary habits, and venereal infections are believed to be the potential reason behind the development of prostatic neoplasms. Prostate cancer typically manifests between the ages of 60 – 80. Digital rectal examination, transrectal ultrasonography, and measurement of serum Prostate Specific Antigen (PSA) are commonly used to diagnose the onset and the progression of the disease (Balk et al., 2003).

Prostate cancer spreads less rapidly in men with low-grade tumors as compared to higher grade tumors. Three approaches exist in the treatment of prostate cancer: surgery, radiation therapy, and chemotherapeutic treatment. Localized prostate cancer is curable using surgery and radiation while advanced stage prostate cancer is presently not curable with treatment focused on palliation. Several strategies are currently used for controlling prostate cancer which includes: (i) androgen ablation by castration or using hormonal antagonists, (ii) chemotherapeutic drugs of natural or synthetic origin, and (iii) modulation of the intracellular signaling pathway to stop malignant cell proliferation or inducing apoptosis. Chemotherapeutic agents used in the treatment of advanced prostate cancer are primarily hormonal, such as, luteinizing hormone-releasing hormone agonists, androgen antagonists, and estrogens (Callery and Gannett, 2002). Non-hormonal chemotherapeutic agents may be utilized as salvage therapy in hormone-refractory prostate cancer (Roth, 1996). However, none has fully realized efficacy to hasten or cure metastasized prostatic tumors.

Numerous inorganic anticancer agents have been synthesized and only a few of these have been found to be effective against prostate cancer. The antitumor drug, cisplatin, *cis*-$[Pt(NH_3)_2Cl_2]$, **13**, is currently being used in hospitals all over the world in the treatment of testicular cancer (Pil and Lippard, 1997). It is also highly effective against ovarian cancers and contributes to the treatment of bladder, prostate, lung, head and neck, certain cancers in children, and in other genitor-urinary cancers. The cisplatin analog, *DWA-2114R* {2-aminomethylpyrrolidine(1,1-cyclobutanedicarboxylato)platinum(II)} has been demonstrated to be cytotoxic to breast, ovarian, prostate, and other cancer cells (Weiss and Christian, 1993). The antitumor drug, JM216 {{cis-(ammine)(cyclohexylamine)dichloro-*trans*-(diacetato)platinum(IV)} is an oral platinum drug that is showing promise in clinical trials against prostate cancer (Hall, et. al., 2004). However, these drugs are very toxic. Recently, several mono, di-, and trirhenium carbonyl complexes have been reported to exhibit cytotoxicity against numerous cancer cell lines including prostate cancer (Yan, et. al., 2000; Wang, et. al., 2002; Zhang, et. al., 2002).

During the last several years we have been studying the chemistry of manganese and rhenium tricarbonyl complexes. Recently, we developed several easy and convenient synthetic methods for manganese and rhenium (tricarbonyl) diphosphine and diimine complexes: examples include perrhenato (Brown, et. al., 2002; Kimari, et. al., 2006), chloro (Orsa et. al., 2008), alkoxo and related complexes (Johnson, et. al., 2001). Many of these complexes are cytotoxic against a variety of cancer cell lines. In addition, rhenium (tricarbonyl) diimine complexes exhibit fluorescence (Stufkens and Vlček, 1998). Hence our interest in exploring the cytotoxic properties of manganese and rhenium (tricarbonyl) diphosphine and diimine complexes. Recently, we published our preliminary findings on the cytotoxic properties of several rhenium (tricarbonyl) diimine complexes against breast, prostate, and lung cancer cell lines (Orsa, et. al., 2008). Here we present the cytotoxic properties of the following perrhenato (**1-5**), chloro (**6-9**) , and pentyloxo (**10-12**) complexes against prostate cancer cell lines.

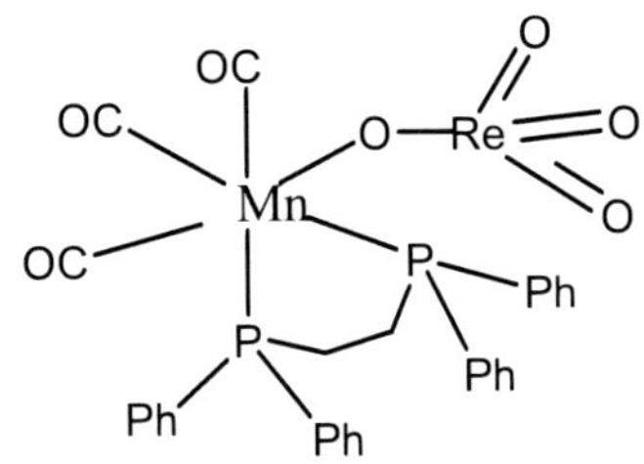

1

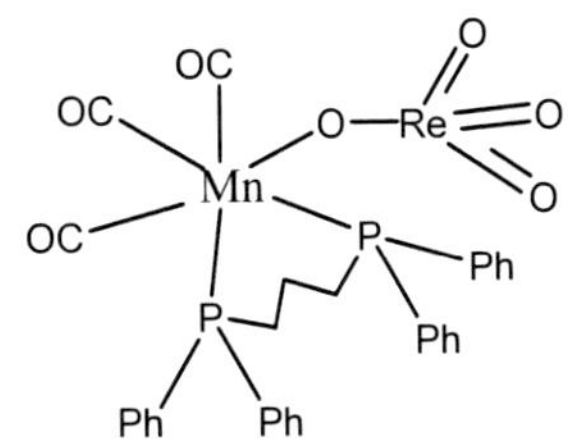

2

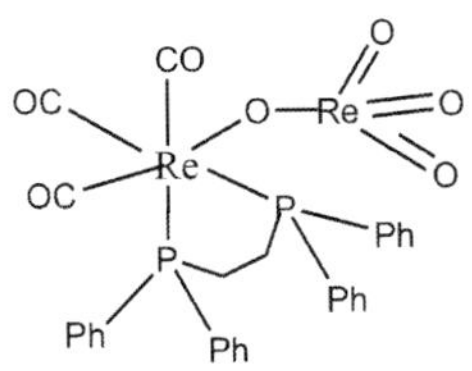

3

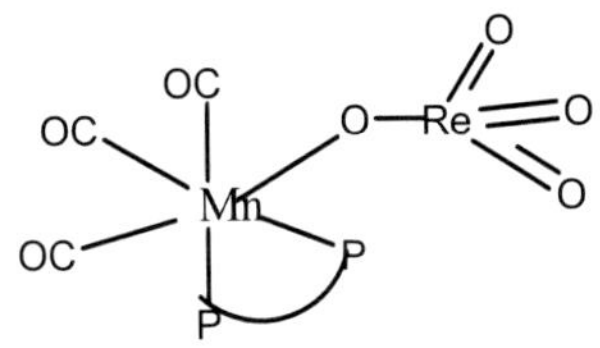

4

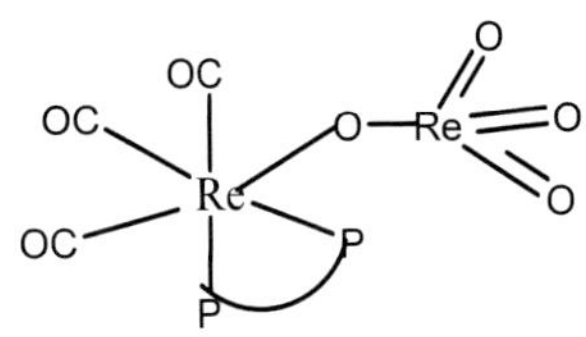

5

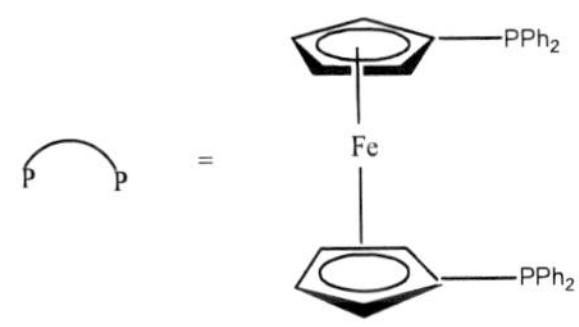

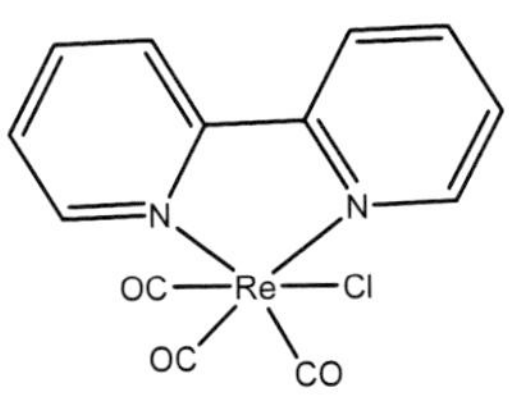

6

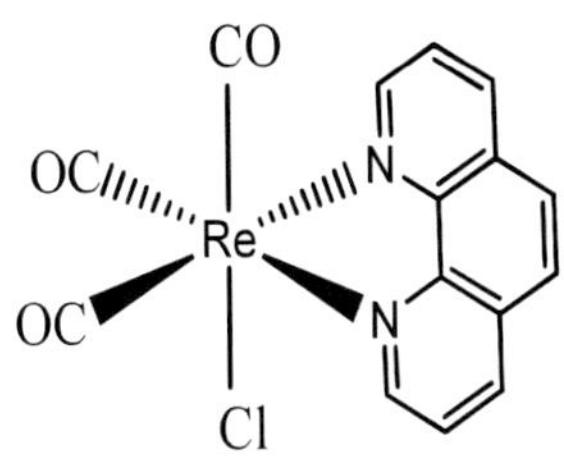

7

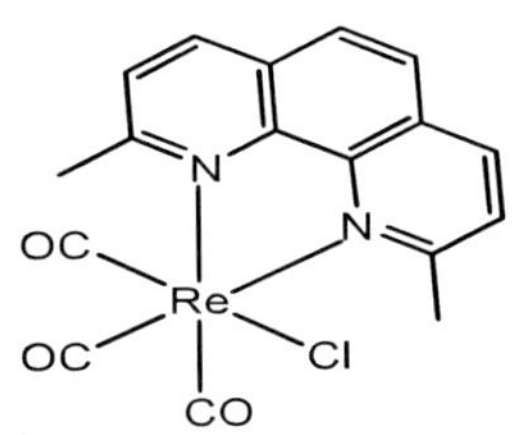

8

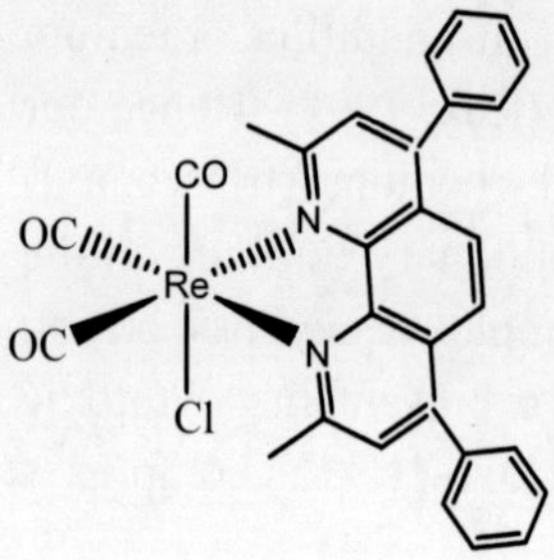

9

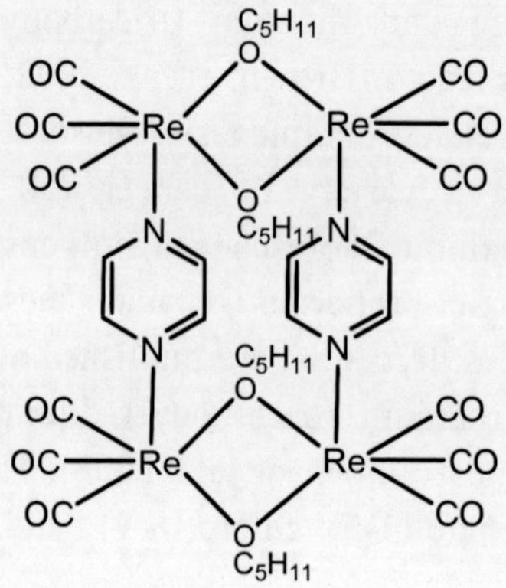

10

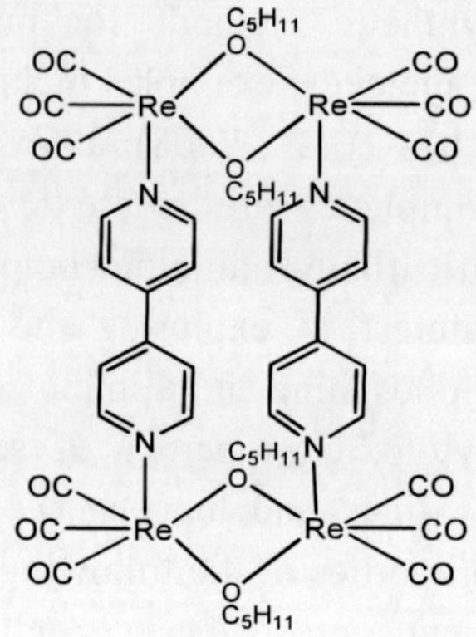

11

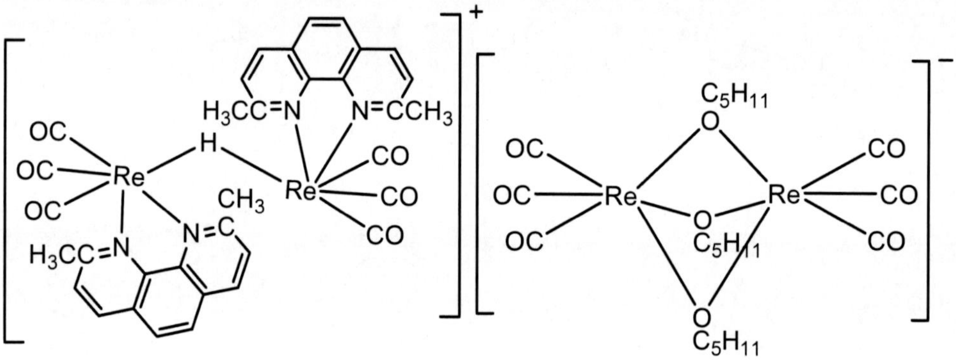

12

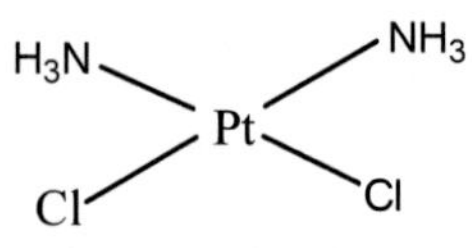

Cisplatin, **13**

Results and Discussion

*Synthesis and Characterization of Perrhenato Complexes, **1-5***

The perrhenato complexes were synthesized from the reactions of the corresponding hydrido complexes with Re_2O_7 according to the literature procedures (Brown, et. al., 2002; Kimari, et. al., 2006).The complexes were characterized previously through IR and NMR spectroscopic techniques and X-ray crystal structure determinations.

*Synthesis and Characterization of Chloro Complexes, **6-9***

The chloro complexes were synthesized from the reactions of $Re_2(CO)_{10}$ with the corresponding diimines in 2-chloroethanol according to literature procedures (Orsa, et. al., 2008). The complexes were characterized previously.

*Synthesis and characterization of the pentyloxo complexes (molecular rectangles), **10, 11***

The synthesis of **11** from the reaction of $Re_2(CO)_{10}$ and 4,4'-bipyridyl in 1-pentanol was reported previously (Orsa, et. al., 2007). The molecular rectangle **10** was synthesized in a similar manner from the reaction of $Re_2(CO)_{10}$ and pyrazine in 1-pentanol according to eq 1:

$$2\ Re_2(CO)_{10} + 2\ (\text{pyrazine}) + 4\ C_5H_{11}OH \\ \rightarrow [(CO)_3Re(\mu\text{-}OC_5H_{11})_2ReCO)_3]_2(\mu\text{-pyrazine})_2,\ \mathbf{10} + 8\ CO + 2\ H_2 \quad (1)$$

The spectral properties of **10** are very similar to the spectral properties of **11**. The structure of **10** was solved using X-ray crystallography and is shown in Figure 2. Crystal data for **10** were determined under the conditions summarized in Table 1. Atomic coordinates and equivalent isotropic displacement parameters are provided in Table 2. Bond distances and angles are provided in Table 3. Anisotropic displacement parameters and torsion angles are given in Tables 4 and 5. The compound crystallizes as 1.5 independent molecules in the crystal lattice. As expected, each rhenium atom is octahedrally coordinated to three terminal carbonyl groups in a facial arrangement, two bridging pentyloxo ligands, and one bridging pyrazine ligand. The structure is similar to the previously reported molecular rectangle **11** (see, Figure 3) (Orsa, et. al., 2007).

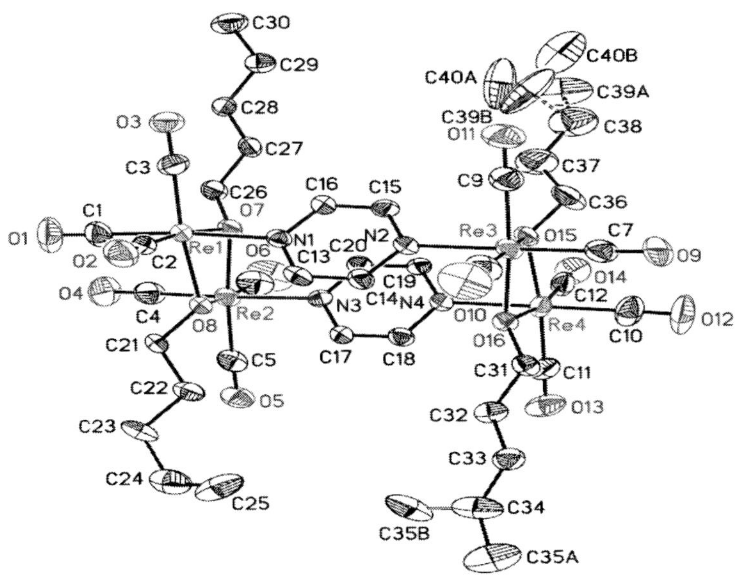

Figure 2. Molecular structure of molecular rectangle, **10**

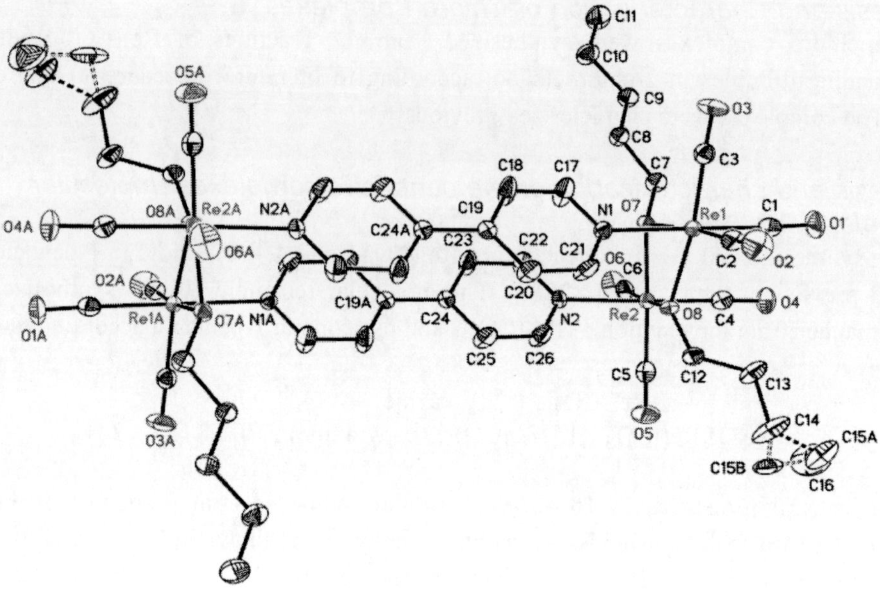

Figure 3. Molecular structure of the molecular rectangle, **11** (courtesy: Elsevier)

Table 1. Summary of crystal data for 10

Empirical formula	$C_{40}H_{52}N_4O_{16}Re_4$	
Formula weight	1589.66	
Temperature	150(2) K	
Wavelength	0.77490 Å	
Crystal system	Monoclinic	
Space group	$P2_1/c$	
Unit cell dimensions	a = 17.1427(9) Å	$\alpha = 90°$
	b = 17.5591(9) Å	$\beta = 93.106(1)°$
	c = 24.6637(13) Å	$\gamma = 90°$
Volume	7413.1(7) $Å^3$	
Z	6	
Density (calculated)	2.136 Mg/m^3	
Absorption coefficient	12.195 mm^{-1}	
F(000)	4488	
Crystal size	0.06 x 0.05 x 0.03 mm^3	
θ range for data collection	2.51 to 31.19°	
Index ranges	$-22 \leq h \leq 22$, $-23 \leq k \leq 23$, $-32 \leq l \leq 32$	
Reflections collected	94409	
Independent reflections	18490 [R_{int} = 0.0968]	
Completeness to θ = 31.19°	99.7 %	
Absorption correction	Multi-scan	
Max. and min. transmission	0.7111 and 0.5282	
Refinement method	Full-matrix least-squares on F^2	
Data / restraints / parameters	18490 / 9 / 901	
Goodness-of-fit on F^2	1.019	
Final R indices [$I>2\sigma(I)$]	R1 = 0.0401, wR2 = 0.0953	
R indices (all data)	R1 = 0.0535, wR2 = 0.1027	
Largest diff. peak and hole	1.571 and -1.299 $eÅ^{-3}$	

Table 2. Atomic coordinates [x10^4] and equivalent isotropic displacement parameters [Å^{2}x10^3] for 10. U(eq) is defined as one third of the trace of the orthogonalized U_{ij} tensor

	x	y	z	U(eq)
Re(1)	1038(1)	4492(1)	3293(1)	34(1)
Re(2)	2243(1)	3231(1)	3955(1)	34(1)
Re(3)	4320(1)	6943(1)	2898(1)	33(1)
Re(4)	5539(1)	5602(1)	3463(1)	31(1)
O(1)	-354(4)	3479(4)	3557(3)	81(2)
O(2)	61(3)	5875(3)	3600(2)	59(2)
O(3)	346(3)	4680(4)	2125(2)	65(2)
O(4)	893(4)	2145(4)	4140(3)	83(2)
O(5)	2918(3)	2968(3)	5117(2)	50(1)
O(6)	3148(5)	1797(3)	3653(3)	88(2)
O(7)	1864(3)	3590(2)	3154(2)	33(1)
O(8)	1643(2)	4299(2)	4066(2)	31(1)
O(9)	5659(4)	8044(3)	2686(2)	59(2)
O(10)	3380(4)	8294(3)	3302(3)	78(2)
O(11)	3634(4)	7297(4)	1758(2)	75(2)
O(12)	6958(4)	6620(4)	3291(3)	82(2)
O(13)	6256(3)	5169(3)	4579(2)	59(2)
O(14)	6407(3)	4248(3)	3000(2)	56(2)
O(15)	4909(3)	5902(2)	2716(2)	35(1)
O(16)	4763(2)	6512(2)	3671(2)	27(1)
N(1)	2050(3)	5232(3)	3180(2)	29(1)
N(2)	3332(3)	6176(3)	3034(2)	28(1)
N(3)	3233(3)	3973(3)	3788(2)	26(1)
N(4)	4515(3)	4890(3)	3596(2)	26(1)
C(1)	174(5)	3842(5)	3458(3)	56(2)
C(2)	425(4)	5342(5)	3483(3)	39(2)
C(3)	617(4)	4622(5)	2567(3)	46(2)
C(4)	1395(5)	2570(5)	4066(3)	55(2)
C(5)	2645(5)	3068(4)	4689(3)	41(2)
C(6)	2811(6)	2331(5)	3762(4)	62(2)
C(7)	5163(5)	7628(4)	2773(3)	44(2)
C(8)	3744(5)	7795(5)	3146(3)	51(2)
C(9)	3897(5)	7164(5)	2183(3)	54(2)
C(10)	6426(5)	6246(5)	3351(4)	57(2)
C(11)	5990(4)	5345(4)	4154(3)	41(2)
C(12)	6100(4)	4764(4)	3175(3)	42(2)
C(13)	2276(4)	5771(4)	3546(2)	35(1)
C(14)	2904(4)	6224(4)	3469(3)	35(1)
C(15)	3104(4)	5644(3)	2669(2)	32(1)
C(16)	2469(4)	5181(4)	2738(2)	31(1)
C(17)	3495(4)	4504(3)	4145(2)	30(1)
C(18)	4123(4)	4947(4)	4051(2)	32(1)
C(19)	4262(4)	4351(3)	3241(2)	30(1)

Table 2. (Continued)

	x	y	z	U(eq)
C(20)	3628(4)	3899(3)	3339(2)	31(1)
C(21)	1221(4)	4433(4)	4547(3)	37(2)
C(22)	1739(4)	4748(4)	5009(3)	44(2)
C(23)	1258(5)	4922(4)	5502(3)	45(2)
C(24)	1774(5)	5187(5)	5985(3)	63(2)
C(25)	2151(5)	5960(5)	5900(4)	73(3)
C(26)	1648(4)	3019(4)	2748(3)	41(2)
C(27)	1850(4)	3253(4)	2186(3)	35(1)
C(28)	1472(4)	2719(4)	1754(2)	33(1)
C(29)	1702(4)	2886(4)	1184(3)	40(2)
C(30)	1241(5)	2411(5)	761(3)	60(2)
C(31)	4977(4)	7055(3)	4084(2)	32(1)
C(32)	4881(4)	6743(4)	4651(3)	37(2)
C(33)	5027(4)	7374(4)	5076(3)	43(2)
C(34)	4971(6)	7097(5)	5651(3)	66(3)
C(35A)	5169(18)	7547(13)	6111(10)	79(9)
C(35B)	4271(9)	6749(7)	5780(5)	70(4)
C(36)	5302(5)	5859(5)	2214(3)	58(2)
C(37)	5000(6)	5204(8)	1857(4)	100(4)
C(38)	5330(7)	5122(7)	1335(4)	94(4)
C(39A)	5336(13)	4437(11)	970(10)	95(10)
C(40A)	4539(13)	4214(14)	999(16)	113(13)
C(39B)	4836(16)	4496(13)	1046(13)	122(17)
C(40B)	5158(15)	4357(14)	527(13)	114(11)
Re(5)	1424(1)	6703(1)	-378(1)	33(1)
Re(6)	2267(1)	5259(1)	350(1)	27(1)
O(17)	2973(4)	7571(3)	-420(2)	68(2)
O(18)	680(4)	8244(3)	-132(3)	78(2)
O(19)	1138(4)	7181(4)	-1557(3)	87(2)
O(20)	3823(3)	6128(3)	398(3)	70(2)
O(21)	2720(3)	4853(3)	1527(2)	46(1)
O(22)	3111(3)	3793(3)	54(2)	57(2)
O(23)	1863(2)	5561(2)	-462(2)	31(1)
O(24)	1562(3)	6267(2)	434(2)	32(1)
N(5)	304(3)	6107(3)	-373(2)	28(1)
N(6)	1123(3)	4673(3)	358(2)	26(1)
C(41)	2405(5)	7227(4)	-392(3)	49(2)
C(42)	969(4)	7664(4)	-222(3)	48(2)
C(43)	1256(5)	6987(4)	-1119(3)	48(2)
C(44)	3230(4)	5800(4)	363(3)	43(2)
C(45)	2534(4)	5015(4)	1085(3)	34(1)
C(46)	2800(4)	4346(4)	163(3)	38(2)
C(47)	-5(4)	5716(4)	-807(2)	32(1)
C(48)	-126(4)	6097(3)	65(2)	31(1)
C(49)	837(4)	4281(3)	-72(2)	30(1)

Table 2. (Continued)

	x	y	z	U(eq)
C(50)	702(4)	4661(4)	804(2)	31(1)
C(51)	2340(4)	5368(4)	-908(3)	42(2)
C(52)	1863(5)	5201(5)	-1419(3)	50(2)
C(53)	2354(5)	5004(5)	-1899(3)	50(2)
C(54)	1877(6)	4961(7)	-2444(3)	77(3)
C(55A)	1400(11)	4278(10)	-2463(7)	102(6)
C(55B)	1650(20)	5778(14)	-2585(10)	116(15)
C(56)	1739(5)	6783(4)	873(3)	45(2)
C(57)	1353(5)	6516(4)	1388(3)	54(2)
C(58)	1599(6)	6979(5)	1875(4)	70(3)
C(59)	1177(7)	6742(6)	2381(4)	81(3)
C(60)	342(6)	6974(7)	2383(4)	88(3)
H(13)	1992	5835	3863	42
H(14)	3044	6594	3738	42
H(15)	3390	5584	2353	38
H(16)	2324	4818	2466	37
H(17)	3235	4573	4472	36
H(18)	4291	5311	4317	38
H(19)	4526	4280	2916	36
H(20)	3467	3523	3079	37
H(21A)	986	3949	4663	45
H(21B)	791	4797	4460	45
H(22A)	1995	5221	4890	53
H(22B)	2153	4374	5112	53
H(23A)	868	5321	5404	54
H(23B)	972	4457	5604	54
H(24A)	2191	4805	6061	75
H(24B)	1457	5215	6308	75
H(25A)	2457	6107	6230	110
H(25B)	1744	6341	5819	110
H(25C)	2495	5929	5596	110
H(26A)	1079	2925	2750	50
H(26B)	1919	2536	2844	50
H(27A)	2425	3243	2161	42
H(27B)	1669	3780	2115	42
H(28A)	1619	2188	1847	40
H(28B)	897	2760	1765	40
H(29A)	2266	2781	1159	48
H(29B)	1614	3433	1105	48
H(30A)	1346	2594	397	91
H(30B)	681	2457	817	91
H(30C)	1398	1876	796	91
H(31A)	4648	7515	4031	38
H(31B)	5527	7207	4049	38
H(32A)	5256	6321	4724	45

Table 2. (Continued)

	x	y	z	U(eq)
H(32B)	4346	6539	4676	45
H(33A)	5554	7590	5035	51
H(33B)	4642	7786	5005	51
H(34A)	4423	6933	5685	79
H(34B)	5295	6631	5686	79
H(34C)	5400	6732	5737	79
H(34D)	5051	7538	5898	79
H(35A)	5077	7256	6441	119
H(35B)	4845	8007	6102	119
H(35C)	5721	7690	6110	119
H(35D)	4310	6582	6160	105
H(35E)	4171	6309	5543	105
H(35F)	3841	7115	5728	105
H(36A)	5870	5792	2296	70
H(36B)	5225	6343	2013	70
H(37A)	5092	4724	2061	120
H(37B)	4428	5264	1796	120
H(38A)	5886	5267	1396	113
H(38B)	5087	5532	1108	113
H(38C)	5874	4932	1375	113
H(38D)	5330	5606	1129	113
H(39A)	5464	4573	595	115
H(39B)	5700	4039	1114	115
H(40A)	4452	3731	807	170
H(40B)	4200	4607	830	170
H(40C)	4417	4152	1380	170
H(39C)	4852	4025	1266	146
H(39D)	4285	4663	994	146
H(40D)	4928	3892	367	170
H(40E)	5725	4297	576	170
H(40F)	5038	4788	284	170
H(47)	277	5703	-1128	38
H(48)	62	6359	383	37
H(49)	1126	4260	-389	36
H(50)	894	4919	1122	37
H(51A)	2663	4918	-808	50
H(51B)	2698	5797	-974	50
H(52A)	1536	5650	-1516	61
H(52B)	1509	4769	-1353	61
H(53A)	2612	4507	-1828	60
H(53B)	2768	5393	-1926	60
H(54A)	1537	5415	-2488	92
H(54B)	2234	4954	-2746	92
H(54C)	2192	4737	-2728	92
H(54D)	1402	4647	-2410	92

Table 2. (Continued)

	x	y	z	U(eq)
H(55A)	1098	4253	-2811	153
H(55B)	1043	4289	-2166	153
H(55C)	1739	3829	-2424	153
H(55D)	1346	5788	-2931	174
H(55E)	2129	6083	-2615	174
H(55F)	1344	5990	-2299	174
H(56A)	1548	7299	772	54
H(56B)	2312	6812	945	54
H(57A)	778	6547	1328	64
H(57B)	1491	5976	1459	64
H(58A)	1490	7523	1797	84
H(58B)	2169	6923	1949	84
H(59A)	1207	6181	2417	97
H(59B)	1459	6965	2705	97
H(60A)	149	6890	2745	133
H(60B)	34	6669	2116	133
H(60C)	294	7514	2289	133

Table 3. Bond distances [Å] and angles [°] for 10

Re(1)-C(2)	1.898(8)	Re(1)-C(3)	1.907(7)	Re(1)-C(1)	1.931(8)
Re(1)-O(8)	2.148(4)	Re(1)-O(7)	2.165(5)	Re(1)-N(1)	2.196(5)
Re(2)-C(4)	1.892(7)	Re(2)-C(5)	1.923(7)	Re(2)-C(6)	1.929(9)
Re(2)-O(7)	2.141(4)	Re(2)-O(8)	2.162(4)	Re(2)-N(3)	2.195(5)
Re(3)-C(9)	1.910(8)	Re(3)-C(8)	1.912(9)	Re(3)-C(7)	1.918(7)
Re(3)-O(15)	2.146(4)	Re(3)-O(16)	2.150(4)	Re(3)-N(2)	2.203(5)
Re(4)-C(11)	1.887(8)	Re(4)-C(12)	1.915(7)	Re(4)-C(10)	1.926(8)
Re(4)-O(15)	2.150(4)	Re(4)-O(16)	2.159(4)	Re(4)-N(4)	2.194(5)
O(1)-C(1)	1.145(9)	O(2)-C(2)	1.170(9)	O(3)-C(3)	1.167(8)
O(4)-C(4)	1.161(9)	O(5)-C(5)	1.146(8)	O(6)-C(6)	1.142(10)
O(7)-C(26)	1.451(7)	O(8)-C(21)	1.443(7)	O(9)-C(7)	1.150(8)
O(10)-C(8)	1.153(10)	O(11)-C(9)	1.142(9)	O(12)-C(10)	1.140(9)
O(13)-C(11)	1.163(9)	O(14)-C(12)	1.144(8)	O(15)-C(36)	1.445(8)
O(16)-C(31)	1.428(7)	N(1)-C(16)	1.340(7)	N(1)-C(13)	1.351(8)
N(2)-C(14)	1.336(8)	N(2)-C(15)	1.340(7)	N(3)-C(20)	1.337(7)
N(3)-C(17)	1.344(7)	N(4)-C(18)	1.342(7)	N(4)-C(19)	1.347(7)
C(13)-C(14)	1.359(9)	C(15)-C(16)	1.376(8)	C(17)-C(18)	1.357(8)
C(19)-C(20)	1.377(8)	C(21)-C(22)	1.511(9)	C(22)-C(23)	1.538(9)
C(23)-C(24)	1.518(11)	C(24)-C(25)	1.522(13)	C(26)-C(27)	1.504(9)
C(27)-C(28)	1.535(8)	C(28)-C(29)	1.509(9)	C(29)-C(30)	1.524(10)
C(31)-C(32)	1.518(9)	C(32)-C(33)	1.535(9)	C(33)-C(34)	1.509(10)
C(34)-C(35B)	1.398(13)	C(34)-C(35A)	1.408(15)	C(36)-C(37)	1.522(13)
C(37)-C(38)	1.441(13)	C(38)-C(39A)	1.503(16)	C(38)-C(39B)	1.540(16)
C(39A)-C(40A)	1.427(15)	C(39B)-C(40B)	1.442(16)	Re(5)-C(43)	1.900(8)
Re(5)-C(42)	1.907(7)	Re(5)-C(41)	1.918(7)	Re(5)-O(24)	2.145(4)
Re(5)-O(23)	2.157(4)	Re(5)-N(5)	2.189(5)	Re(6)-C(45)	1.894(7)
Re(6)-C(44)	1.904(7)	Re(6)-C(46)	1.915(7)	Re(6)-O(23)	2.151(4)
Re(6)-O(24)	2.159(4)	Re(6)-N(6)	2.215(5)	O(17)-C(41)	1.151(8)
O(18)-C(42)	1.159(9)	O(19)-C(43)	1.141(9)	O(20)-C(44)	1.167(8)

Table 3. (Continued)

O(21)-C(45)	1.154(8)	O(22)-C(46)	1.146(8)	O(23)-C(51)	1.446(7)
O(24)-C(56)	1.433(7)	N(5)-C(48)	1.340(8)	N(5)-C(47)	1.354(7)
N(6)-C(49)	1.334(7)	N(6)-C(50)	1.348(7)	C(47)-C(50)#1	1.367(8)
C(48)-C(49)#1	1.388(9)	C(51)-C(52)	1.494(10)	C(52)-C(53)	1.527(10)
C(53)-C(54)	1.538(11)	C(54)-C(55A)	1.451(14)	C(54)-C(55B)	1.520(18)
C(56)-C(57)	1.538(11)	C(57)-C(58)	1.492(11)	C(58)-C(59)	1.535(14)
C(59)-C(60)	1.488(13)				
C(2)-Re(1)-C(3)	87.2(3)	C(2)-Re(1)-C(1)	88.4(3)	C(3)-Re(1)-C(1)	90.7(4)
C(2)-Re(1)-O(8)	98.8(2)	C(3)-Re(1)-O(8)	172.8(2)	C(1)-Re(1)-O(8)	93.4(3)
C(2)-Re(1)-O(7)	171.9(2)	C(3)-Re(1)-O(7)	99.0(2)	C(1)-Re(1)-O(7)	96.7(3)
O(8)-Re(1)-O(7)	74.67(16)	C(2)-Re(1)-N(1)	91.0(2)	C(3)-Re(1)-N(1)	94.2(3)
C(1)-Re(1)-N(1)	175.0(3)	O(8)-Re(1)-N(1)	81.87(17)	O(7)-Re(1)-N(1)	83.38(17)
C(4)-Re(2)-C(5)	90.6(3)	C(4)-Re(2)-C(6)	86.3(4)	C(5)-Re(2)-C(6)	87.1(3)
C(4)-Re(2)-O(7)	96.7(3)	C(5)-Re(2)-O(7)	171.1(2)	C(6)-Re(2)-O(7)	98.3(3)
C(4)-Re(2)-O(8)	98.0(3)	C(5)-Re(2)-O(8)	99.2(2)	C(6)-Re(2)-O(8)	172.3(3)
O(7)-Re(2)-O(8)	74.87(16)	C(4)-Re(2)-N(3)	177.3(3)	C(5)-Re(2)-N(3)	91.3(2)
C(6)-Re(2)-N(3)	91.9(3)	O(7)-Re(2)-N(3)	81.46(16)	O(8)-Re(2)-N(3)	83.51(16)
C(9)-Re(3)-C(8)	87.6(4)	C(9)-Re(3)-C(7)	88.6(3)	C(8)-Re(3)-C(7)	88.1(3)
C(9)-Re(3)-O(15)	98.0(3)	C(8)-Re(3)-O(15)	172.0(2)	C(7)-Re(3)-O(15)	97.7(3)
C(9)-Re(3)-O(16)	171.1(3)	C(8)-Re(3)-O(16)	99.0(3)	C(7)-Re(3)-O(16)	97.5(2)
O(15)-Re(3)-O(16)	74.87(15)	C(9)-Re(3)-N(2)	90.5(3)	C(8)-Re(3)-N(2)	90.9(3)
C(7)-Re(3)-N(2)	178.6(3)	O(15)-Re(3)-N(2)	83.44(17)	O(16)-Re(3)-N(2)	83.58(16)
C(11)-Re(4)-C(12)	87.8(3)	C(11)-Re(4)-C(10)	89.0(4)	C(12)-Re(4)-C(10)	89.0(3)
C(11)-Re(4)-O(15)	173.9(2)	C(12)-Re(4)-O(15)	96.3(3)	C(10)-Re(4)-O(15)	95.6(3)
C(11)-Re(4)-O(16)	100.9(2)	C(12)-Re(4)-O(16)	170.1(2)	C(10)-Re(4)-O(16)	95.8(3)
O(15)-Re(4)-O(16)	74.60(16)	C(11)-Re(4)-N(4)	91.1(3)	C(12)-Re(4)-N(4)	92.3(2)
C(10)-Re(4)-N(4)	178.8(3)	O(15)-Re(4)-N(4)	84.22(17)	O(16)-Re(4)-N(4)	82.96(16)
C(26)-O(7)-Re(2)	119.2(4)	C(26)-O(7)-Re(1)	118.0(4)	Re(2)-O(7)-Re(1)	103.76(17)
C(21)-O(8)-Re(1)	117.7(4)	C(21)-O(8)-Re(2)	120.6(4)	Re(1)-O(8)-Re(2)	103.61(17)
C(36)-O(15)-Re(3)	118.1(4)	C(36)-O(15)-Re(4)	119.1(5)	Re(3)-O(15)-Re(4)	104.47(17)
C(31)-O(16)-Re(3)	117.6(3)	C(31)-O(16)-Re(4)	121.7(4)	Re(3)-O(16)-Re(4)	103.99(16)
C(16)-N(1)-C(13)	116.3(5)	C(16)-N(1)-Re(1)	121.7(4)	C(13)-N(1)-Re(1)	122.0(4)
C(14)-N(2)-C(15)	115.4(5)	C(14)-N(2)-Re(3)	123.0(4)	C(15)-N(2)-Re(3)	121.6(4)
C(20)-N(3)-C(17)	116.3(5)	C(20)-N(3)-Re(2)	122.0(4)	C(17)-N(3)-Re(2)	121.6(4)
C(18)-N(4)-C(19)	115.9(5)	C(18)-N(4)-Re(4)	121.7(4)	C(19)-N(4)-Re(4)	122.4(4)
O(1)-C(1)-Re(1)	177.6(8)	O(2)-C(2)-Re(1)	178.7(7)	O(3)-C(3)-Re(1)	177.9(8)
O(4)-C(4)-Re(2)	177.7(9)	O(5)-C(5)-Re(2)	176.9(7)	O(6)-C(6)-Re(2)	179.4(8)
O(9)-C(7)-Re(3)	178.3(7)	O(10)-C(8)-Re(3)	177.9(8)	O(11)-C(9)-Re(3)	179.0(8)
O(12)-C(10)-Re(4)	178.8(10)	O(13)-C(11)-Re(4)	178.2(7)	O(14)-C(12)-Re(4)	177.3(6)
N(1)-C(13)-C(14)	121.1(6)	N(2)-C(14)-C(13)	123.3(6)	N(2)-C(15)-C(16)	122.2(6)
N(1)-C(16)-C(15)	121.6(6)	N(3)-C(17)-C(18)	121.8(6)	N(4)-C(18)-C(17)	122.6(6)
N(4)-C(19)-C(20)	121.4(6)	N(3)-C(20)-C(19)	122.0(6)	O(8)-C(21)-C(22)	112.3(5)
C(21)-C(22)-C(23)	110.5(6)	C(24)-C(23)-C(22)	111.5(7)	C(23)-C(24)-C(25)	113.6(7)
O(7)-C(26)-C(27)	112.7(5)	C(26)-C(27)-C(28)	111.3(5)	C(29)-C(28)-C(27)	114.0(5)
C(28)-C(29)-C(30)	112.4(6)	O(16)-C(31)-C(32)	112.4(5)	C(31)-C(32)-C(33)	110.3(6)
C(34)-C(33)-C(32)	113.2(6)	C(35B)C(34)C(35A)	103.6(16)	C(35B)-C(34)-C(33)	117.0(9)
C(35A)-C(34)-C(3)	123.5(16)	O(15)-C(36)-C(37)	112.1(7)	C(38)-C(37)-C(36)	117.0(10)
C(37)C(38)C(39A)	129.0(15)	C(37)-C(38)-C(39B)	104.6(16)	C(39A)C(38)C(39)	34.1(14)
C(40A)C(39A)C(3)	98.7(18)	C(40B)C(39B)C(38)	108(2)	C(43)-Re(5)-C(42)	85.4(3)
C(43)-Re(5)-C(41)	86.8(3)	C(42)-Re(5)-C(41)	87.1(3)	C(43)-Re(5)-O(24)	173.9(2)
C(42)-Re(5)-O(24)	98.8(3)	C(41)-Re(5)-O(24)	97.8(2)	C(43)-Re(5)-O(23)	100.8(3)
C(42)-Re(5)-O(23)	172.3(3)	C(41)-Re(5)-O(23)	97.7(3)	O(24)-Re(5)-O(23)	74.70(15)
C(43)-Re(5)-N(5)	92.5(3)	C(42)-Re(5)-N(5)	93.0(3)	C(41)-Re(5)-N(5)	179.3(3)

Table 3. (Continued)

O(24)-Re(5)-N(5)	82.87(17)	O(23)-Re(5)-N(5)	82.33(17)	C(45)-Re(6)-C(44)	86.2(3)
C(45)-Re(6)-C(46)	87.0(3)	C(44)-Re(6)-C(46)	89.8(3)	C(45)-Re(6)-O(23)	175.0(2)
C(44)-Re(6)-O(23)	97.3(2)	C(46)-Re(6)-O(23)	96.6(2)	C(45)-Re(6)-O(24)	101.6(2)
C(44)-Re(6)-O(24)	94.5(2)	C(46)-Re(6)-O(24)	170.6(2)	O(23)-Re(6)-O(24)	74.55(16)
C(45)-Re(6)-N(6)	93.2(2)	C(44)-Re(6)-N(6)	177.3(3)	C(46)-Re(6)-N(6)	92.7(2)
O(23)-Re(6)-N(6)	83.20(16)	O(24)-Re(6)-N(6)	83.08(17)	C(51)-O(23)-Re(6)	119.0(4)
C(51)-O(23)-Re(5)	120.3(4)	Re(6)-O(23)-Re(5)	103.63(17)	C(56)-O(24)-Re(5)	119.2(4)
C(56)-O(24)-Re(6)	119.6(4)	Re(5)-O(24)-Re(6)	103.76(17)	C(48)-N(5)-C(47)	115.0(5)
C(48)-N(5)-Re(5)	122.1(4)	C(47)-N(5)-Re(5)	122.9(4)	C(49)-N(6)-C(50)	116.8(5)
C(49)-N(6)-Re(6)	121.4(4)	C(50)-N(6)-Re(6)	121.8(4)	O(17)-C(41)-Re(5)	176.2(7)
O(18)-C(42)-Re(5)	178.8(7)	O(19)-C(43)-Re(5)	177.3(8)	O(20)-C(44)-Re(6)	176.6(7)
O(21)-C(45)-Re(6)	177.6(6)	O(22)-C(46)-Re(6)	179.0(7)	N(5)-C(47)-C(50)#1	123.1(6)
N(5)-C(48)-C(49)#1	122.4(6)	N(6)-C(49)-C(48)#1	121.5(6)	N(6)-C(50)-C(47)#1	121.2(6)
O(23)-C(51)-C(52)	112.4(6)	C(51)-C(52)-C(53)	113.5(6)	C(52)-C(53)-C(54)	113.5(7)
C(55A)-C(54)-C(55B)	129.8(19)	C(55A)-C(54)-C(53)	109.8(10)	C(55B)-C(54)-C(53)	105.6(13)
O(24)-C(56)-C(57)	110.4(6)	C(58)-C(57)-C(56)	112.5(8)	C(57)-C(58)-C(59)	112.4(9)
C(60)-C(59)-C(58)	115.1(8)				

Symmetry transformations used to generate equivalent atoms: #1 -x,-y+1,-z

Table 4. Anisotropic displacement parameters [$Å^2 x10^3$] for 10. The anisotropic displacement factor exponent takes the form: $-2\pi^2[h^2a^{*2}U_{11}+ ... +2hka^*b^*U_{12}]$

	U_{11}	U_{22}	U_{33}	U_{23}	U_{13}	U_{12}
Re(1)	26(1)	50(1)	25(1)	-4(1)	4(1)	-6(1)
Re(2)	44(1)	30(1)	28(1)	2(1)	6(1)	-12(1)
Re(3)	42(1)	29(1)	29(1)	5(1)	5(1)	-7(1)
Re(4)	25(1)	31(1)	36(1)	-10(1)	5(1)	-3(1)
O(1)	52(4)	91(5)	103(5)	-41(4)	34(4)	-35(3)
O(2)	50(3)	76(4)	51(3)	3(3)	11(3)	14(3)
O(3)	45(3)	115(5)	34(3)	-13(3)	-1(2)	15(3)
O(4)	85(5)	87(5)	76(5)	24(4)	3(4)	-53(4)
O(5)	61(3)	53(3)	34(3)	14(2)	2(2)	8(3)
O(6)	146(7)	36(3)	84(5)	-8(3)	19(5)	22(4)
O(7)	38(2)	37(2)	25(2)	-6(2)	7(2)	-12(2)
O(8)	29(2)	42(3)	23(2)	-3(2)	6(2)	-6(2)
O(9)	73(4)	52(3)	54(3)	12(3)	12(3)	-31(3)
O(10)	108(6)	45(4)	82(5)	11(3)	8(4)	27(4)
O(11)	99(5)	82(5)	43(3)	27(3)	-12(3)	-21(4)
O(12)	52(4)	74(4)	123(6)	-32(4)	30(4)	-34(3)
O(13)	52(3)	72(4)	51(3)	-21(3)	-16(3)	26(3)
O(14)	52(3)	54(3)	63(4)	-21(3)	8(3)	22(3)
O(15)	42(3)	38(2)	26(2)	-6(2)	12(2)	-11(2)
O(16)	33(2)	23(2)	26(2)	-5(2)	4(2)	-1(2)
N(1)	31(3)	32(3)	25(2)	5(2)	3(2)	3(2)
N(2)	32(3)	27(3)	24(2)	3(2)	5(2)	0(2)
N(3)	29(3)	24(2)	25(2)	1(2)	-2(2)	2(2)

Table 4. (Continued)

	U_{11}	U_{22}	U_{33}	U_{23}	U_{13}	U_{12}
N(4)	28(3)	23(2)	26(2)	-1(2)	4(2)	3(2)
C(1)	51(5)	69(5)	48(5)	-24(4)	18(4)	-19(4)
C(2)	27(3)	66(5)	26(3)	3(3)	2(3)	-1(3)
C(3)	37(4)	65(5)	36(4)	-10(3)	-3(3)	10(3)
C(4)	72(6)	53(5)	42(4)	7(4)	3(4)	-30(4)
C(5)	57(5)	28(3)	39(4)	9(3)	8(3)	-2(3)
C(6)	91(7)	41(5)	54(5)	8(4)	4(5)	-6(4)
C(7)	58(5)	40(4)	34(4)	2(3)	3(3)	-11(3)
C(8)	66(5)	43(4)	43(4)	12(3)	5(4)	-3(4)
C(9)	65(5)	56(5)	41(4)	11(4)	-2(4)	-18(4)
C(10)	45(4)	51(5)	75(6)	-27(4)	12(4)	-12(4)
C(11)	36(4)	41(4)	46(4)	-17(3)	-3(3)	8(3)
C(12)	27(3)	50(4)	50(4)	-7(3)	6(3)	-4(3)
C(13)	41(4)	40(4)	24(3)	3(3)	10(3)	5(3)
C(14)	40(4)	36(3)	29(3)	2(3)	5(3)	-10(3)
C(15)	36(3)	37(3)	23(3)	-8(3)	3(3)	0(3)
C(16)	31(3)	35(3)	27(3)	-1(2)	4(2)	-1(3)
C(17)	31(3)	34(3)	26(3)	-1(2)	4(2)	-3(2)
C(18)	39(3)	31(3)	26(3)	-4(3)	5(3)	-4(3)
C(19)	33(3)	31(3)	25(3)	-5(2)	0(2)	-4(2)
C(20)	37(3)	27(3)	29(3)	-3(2)	4(3)	0(2)
C(21)	34(3)	53(4)	26(3)	-2(3)	10(3)	-10(3)
C(22)	53(4)	54(4)	27(3)	-4(3)	9(3)	-3(3)
C(23)	72(5)	40(4)	24(3)	2(3)	18(3)	9(4)
C(24)	67(6)	77(6)	44(5)	-4(4)	3(4)	36(5)
C(25)	70(6)	71(6)	75(6)	-36(5)	-31(5)	21(5)
C(26)	50(4)	42(4)	33(3)	-5(3)	7(3)	-22(3)
C(27)	35(3)	38(4)	32(3)	-4(3)	9(3)	-6(3)
C(28)	30(3)	38(3)	31(3)	-8(3)	4(3)	0(3)
C(29)	47(4)	40(4)	33(3)	-6(3)	1(3)	11(3)
C(30)	64(5)	76(6)	40(4)	-15(4)	-9(4)	6(5)
C(31)	40(3)	26(3)	30(3)	-6(2)	10(3)	0(3)
C(32)	42(4)	34(3)	35(4)	-7(3)	-3(3)	11(3)
C(33)	39(4)	52(4)	37(4)	-10(3)	1(3)	3(3)
C(34)	85(7)	75(6)	36(4)	-9(4)	-8(4)	44(5)
C(35A)	110(20)	39(13)	82(19)	-42(13)	-18(17)	26(14)
C(35B)	126(14)	47(8)	41(7)	-4(6)	35(8)	0(8)
C(36)	86(6)	55(5)	37(4)	-7(4)	29(4)	-17(4)
C(37)	77(7)	177(13)	48(6)	-44(7)	14(5)	0(8)
C(38)	89(8)	124(10)	69(7)	-31(7)	2(6)	29(7)
C(39A)	140(30)	72(15)	73(15)	-28(12)	-20(16)	62(17)
C(40A)	59(16)	52(16)	230(40)	16(18)	20(20)	-7(11)
C(39B)	90(20)	43(15)	220(40)	-42(19)	-90(30)	9(14)
C(40B)	82(18)	73(16)	180(30)	-41(19)	-30(19)	26(13)
Re(5)	37(1)	28(1)	34(1)	6(1)	-1(1)	-8(1)
Re(6)	30(1)	27(1)	25(1)	-1(1)	4(1)	-3(1)

Table 4. (Continued)

	U_{11}	U_{22}	U_{33}	U_{23}	U_{13}	U_{12}
O(17)	63(4)	69(4)	70(4)	22(3)	-10(3)	-34(3)
O(18)	89(5)	37(3)	101(5)	-13(3)	-44(4)	14(3)
O(19)	111(6)	99(5)	49(4)	36(4)	-10(4)	-21(4)
O(20)	49(3)	62(4)	95(5)	20(3)	-10(3)	-26(3)
O(21)	46(3)	63(3)	30(3)	0(2)	1(2)	-6(2)
O(22)	65(4)	39(3)	68(4)	-11(3)	9(3)	9(3)
O(23)	35(2)	36(2)	23(2)	1(2)	6(2)	-4(2)
O(24)	41(2)	24(2)	30(2)	-5(2)	1(2)	0(2)
N(5)	34(3)	22(2)	28(3)	3(2)	-3(2)	1(2)
N(6)	32(3)	26(2)	22(2)	-1(2)	0(2)	2(2)
C(41)	55(5)	42(4)	47(4)	15(3)	-9(4)	-19(4)
C(42)	54(5)	28(4)	58(5)	0(3)	-20(4)	-2(3)
C(43)	52(5)	44(4)	47(4)	17(3)	-3(4)	-12(3)
C(44)	49(4)	38(4)	42(4)	13(3)	-2(3)	-4(3)
C(45)	30(3)	39(4)	34(3)	-3(3)	13(3)	-4(3)
C(46)	30(3)	38(4)	47(4)	0(3)	1(3)	-2(3)
C(47)	33(3)	39(3)	23(3)	-1(3)	5(2)	-3(3)
C(48)	40(3)	28(3)	24(3)	-4(2)	-1(3)	6(3)
C(49)	39(3)	25(3)	26(3)	-1(2)	1(3)	0(2)
C(50)	34(3)	36(3)	24(3)	-1(2)	5(2)	-2(3)
C(51)	38(4)	60(5)	28(3)	-3(3)	7(3)	0(3)
C(52)	49(4)	67(5)	36(4)	-3(4)	7(3)	-10(4)
C(53)	51(4)	66(5)	34(4)	-9(4)	12(3)	-16(4)
C(54)	87(7)	108(8)	37(5)	-15(5)	10(5)	-24(6)
C(55A)	117(16)	122(16)	65(11)	-27(11)	-1(10)	-12(13)
C(55B)	130(30)	190(40)	26(13)	33(18)	7(15)	80(30)
C(56)	57(5)	34(4)	42(4)	-15(3)	-5(3)	1(3)
C(57)	81(6)	44(4)	35(4)	-12(3)	-1(4)	15(4)
C(58)	80(7)	71(6)	59(6)	-20(5)	-17(5)	33(5)
C(59)	116(9)	83(7)	43(5)	-9(5)	1(5)	48(6)
C(60)	88(8)	121(10)	58(6)	17(6)	14(5)	36(7)

Table 5. Torsion angles [°] for 10.

C(4)-Re(2)-O(7)-C(26)	-50.6(5)	C(5)-Re(2)-O(7)-C(26)	164.0(16)
C(6)-Re(2)-O(7)-C(26)	36.6(5)	O(8)-Re(2)-O(7)-C(26)	-147.1(5)
N(3)-Re(2)-O(7)-C(26)	127.3(5)	C(4)-Re(2)-O(7)-Re(1)	83.1(3)
C(5)-Re(2)-O(7)-Re(1)	-62.3(17)	C(6)-Re(2)-O(7)-Re(1)	170.3(3)
O(8)-Re(2)-O(7)-Re(1)	-13.42(16)	N(3)-Re(2)-O(7)-Re(1)	-98.96(19)
C(2)-Re(1)-O(7)-C(26)	-175.5(16)	C(3)-Re(1)-O(7)-C(26)	-35.5(5)
C(1)-Re(1)-O(7)-C(26)	56.2(5)	O(8)-Re(1)-O(7)-C(26)	147.9(4)
N(1)-Re(1)-O(7)-C(26)	-128.7(4)	C(2)-Re(1)-O(7)-Re(2)	50.1(17)
C(3)-Re(1)-O(7)-Re(2)	-169.9(3)	C(1)-Re(1)-O(7)-Re(2)	-78.1(3)
O(8)-Re(1)-O(7)-Re(2)	13.53(16)	N(1)-Re(1)-O(7)-Re(2)	96.87(19)
C(2)-Re(1)-O(8)-C(21)	35.6(5)	C(3)-Re(1)-O(8)-C(21)	-178(2)
C(1)-Re(1)-O(8)-C(21)	-53.2(5)	O(7)-Re(1)-O(8)-C(21)	-149.2(4)

Table 5. (Continued)

N(1)-Re(1)-O(8)-C(21)	125.4(4)	C(2)-Re(1)-O(8)-Re(2)	171.5(2)
C(3)-Re(1)-O(8)-Re(2)	-42(2)	C(1)-Re(1)-O(8)-Re(2)	82.6(3)
O(7)-Re(1)-O(8)-Re(2)	-13.38(16)	N(1)-Re(1)-O(8)-Re(2)	-98.70(19)
C(4)-Re(2)-O(8)-C(21)	52.9(5)	C(5)-Re(2)-O(8)-C(21)	-39.0(5)
C(6)-Re(2)-O(8)-C(21)	176(2)	O(7)-Re(2)-O(8)-C(21)	147.8(4)
N(3)-Re(2)-O(8)-C(21)	-129.3(4)	C(4)-Re(2)-O(8)-Re(1)	-81.3(3)
C(5)-Re(2)-O(8)-Re(1)	-173.3(2)	C(6)-Re(2)-O(8)-Re(1)	42(2)
O(7)-Re(2)-O(8)-Re(1)	13.52(16)	N(3)-Re(2)-O(8)-Re(1)	96.40(18)
C(9)-Re(3)-O(15)-C(36)	39.3(5)	C(8)-Re(3)-O(15)-C(36)	173.6(19)
C(7)-Re(3)-O(15)-C(36)	-50.3(5)	O(16)-Re(3)-O(15)-C(36)	-146.1(5)
N(2)-Re(3)-O(15)-C(36)	128.8(5)	C(9)-Re(3)-O(15)-Re(4)	174.3(3)
C(8)-Re(3)-O(15)-Re(4)	-51(2)	C(7)-Re(3)-O(15)-Re(4)	84.7(3)
O(16)-Re(3)-O(15)-Re(4)	-11.08(16)	N(2)-Re(3)-O(15)-Re(4)	-96.18(19)
C(11)-Re(4)-O(15)-C(36)	-171(2)	C(12)-Re(4)-O(15)-C(36)	-38.6(5)
C(10)-Re(4)-O(15)-C(36)	51.0(5)	O(16)-Re(4)-O(15)-C(36)	145.5(5)
N(4)-Re(4)-O(15)-C(36)	-130.2(5)	C(11)-Re(4)-O(15)-Re(3)	55(2)
C(12)-Re(4)-O(15)-Re(3)	-173.0(2)	C(10)-Re(4)-O(15)-Re(3)	-83.4(3)
O(16)-Re(4)-O(15)-Re(3)	11.05(16)	N(4)-Re(4)-O(15)-Re(3)	95.32(19)
C(9)-Re(3)-O(16)-C(31)	-174.1(18)	C(8)-Re(3)-O(16)-C(31)	-36.4(5)
C(7)-Re(3)-O(16)-C(31)	52.9(4)	O(15)-Re(3)-O(16)-C(31)	148.9(4)
N(2)-Re(3)-O(16)-C(31)	-126.2(4)	C(9)-Re(3)-O(16)-Re(4)	48.1(19)
C(8)-Re(3)-O(16)-Re(4)	-174.2(3)	C(7)-Re(3)-O(16)-Re(4)	-84.9(3)
O(15)-Re(3)-O(16)-Re(4)	11.01(16)	N(2)-Re(3)-O(16)-Re(4)	95.93(19)
C(11)-Re(4)-O(16)-C(31)	37.7(5)	C(12)-Re(4)-O(16)-C(31)	-170.9(13)
C(10)-Re(4)-O(16)-C(31)	-52.4(5)	O(15)-Re(4)-O(16)-C(31)	-146.6(4)
N(4)-Re(4)-O(16)-C(31)	127.5(4)	C(11)-Re(4)-O(16)-Re(3)	173.3(3)
C(12)-Re(4)-O(16)-Re(3)	-35.3(15)	C(10)-Re(4)-O(16)-Re(3)	83.2(3)
O(15)-Re(4)-O(16)-Re(3)	-11.00(16)	N(4)-Re(4)-O(16)-Re(3)	-96.92(19)
C(2)-Re(1)-N(1)-C(16)	-132.2(5)	C(3)-Re(1)-N(1)-C(16)	-44.9(5)
C(1)-Re(1)-N(1)-C(16)	144(3)	O(8)-Re(1)-N(1)-C(16)	129.1(5)
O(7)-Re(1)-N(1)-C(16)	53.7(5)	C(2)-Re(1)-N(1)-C(13)	46.6(5)
C(3)-Re(1)-N(1)-C(13)	133.8(5)	C(1)-Re(1)-N(1)-C(13)	-37(4)
O(8)-Re(1)-N(1)-C(13)	-52.1(5)	O(7)-Re(1)-N(1)-C(13)	-127.5(5)
C(9)-Re(3)-N(2)-C(14)	-128.9(6)	C(8)-Re(3)-N(2)-C(14)	-41.3(5)
C(7)-Re(3)-N(2)-C(14)	-84(11)	O(15)-Re(3)-N(2)-C(14)	133.1(5)
O(16)-Re(3)-N(2)-C(14)	57.7(5)	C(9)-Re(3)-N(2)-C(15)	48.6(5)
C(8)-Re(3)-N(2)-C(15)	136.2(5)	C(7)-Re(3)-N(2)-C(15)	93(11)
O(15)-Re(3)-N(2)-C(15)	-49.5(5)	O(16)-Re(3)-N(2)-C(15)	-124.9(5)
C(4)-Re(2)-N(3)-C(20)	-3(6)	C(5)-Re(2)-N(3)-C(20)	133.3(5)
C(6)-Re(2)-N(3)-C(20)	46.1(5)	O(7)-Re(2)-N(3)-C(20)	-52.0(4)
O(8)-Re(2)-N(3)-C(20)	-127.6(5)	C(4)-Re(2)-N(3)-C(17)	180(100)
C(5)-Re(2)-N(3)-C(17)	-43.7(5)	C(6)-Re(2)-N(3)-C(17)	-130.9(5)
O(7)-Re(2)-N(3)-C(17)	131.0(5)	O(8)-Re(2)-N(3)-C(17)	55.3(4)
C(11)-Re(4)-N(4)-C(18)	49.9(5)	C(12)-Re(4)-N(4)-C(18)	137.8(5)
C(10)-Re(4)-N(4)-C(18)	-46(15)	O(15)-Re(4)-N(4)-C(18)	-126.1(5)
O(16)-Re(4)-N(4)-C(18)	-50.9(5)	C(11)-Re(4)-N(4)-C(19)	-127.2(5)

Table 5. (Continued)

C(12)-Re(4)-N(4)-C(19)	-39.3(5)	C(10)-Re(4)-N(4)-C(19)	137(15)
O(15)-Re(4)-N(4)-C(19)	56.8(5)	O(16)-Re(4)-N(4)-C(19)	131.9(5)
C(2)-Re(1)-C(1)-O(1)	9(19)	C(3)-Re(1)-C(1)-O(1)	-78(19)
O(8)-Re(1)-C(1)-O(1)	107(19)	O(7)-Re(1)-C(1)-O(1)	-178(100)
N(1)-Re(1)-C(1)-O(1)	92(19)	C(3)-Re(1)-C(2)-O(2)	-105(25)
C(1)-Re(1)-C(2)-O(2)	165(100)	O(8)-Re(1)-C(2)-O(2)	71(25)
O(7)-Re(1)-C(2)-O(2)	36(26)	N(1)-Re(1)-C(2)-O(2)	-10(25)
C(2)-Re(1)-C(3)-O(3)	-110(19)	C(1)-Re(1)-C(3)-O(3)	-22(19)
O(8)-Re(1)-C(3)-O(3)	103(19)	O(7)-Re(1)-C(3)-O(3)	75(19)
N(1)-Re(1)-C(3)-O(3)	159(19)	C(5)-Re(2)-C(4)-O(4)	-56(19)
C(6)-Re(2)-C(4)-O(4)	31(19)	O(7)-Re(2)-C(4)-O(4)	129(19)
O(8)-Re(2)-C(4)-O(4)	-156(19)	N(3)-Re(2)-C(4)-O(4)	81(20)
C(4)-Re(2)-C(5)-O(5)	136(11)	C(6)-Re(2)-C(5)-O(5)	49(11)
O(7)-Re(2)-C(5)-O(5)	-79(12)	O(8)-Re(2)-C(5)-O(5)	-126(11)
N(3)-Re(2)-C(5)-O(5)	-42(11)	C(4)-Re(2)-C(6)-O(6)	-79(80)
C(5)-Re(2)-C(6)-O(6)	11(80)	O(7)-Re(2)-C(6)-O(6)	-176(100)
O(8)-Re(2)-C(6)-O(6)	156(100)	N(3)-Re(2)-C(6)-O(6)	103(80)
C(9)-Re(3)-C(7)-O(9)	7(27)	C(8)-Re(3)-C(7)-O(9)	-80(27)
O(15)-Re(3)-C(7)-O(9)	105(27)	O(16)-Re(3)-C(7)-O(9)	-179(100)
N(2)-Re(3)-C(7)-O(9)	-37(34)	C(9)-Re(3)-C(8)-O(10)	106(22)
C(7)-Re(3)-C(8)-O(10)	-165(22)	O(15)-Re(3)-C(8)-O(10)	-29(24)
O(16)-Re(3)-C(8)-O(10)	-68(22)	N(2)-Re(3)-C(8)-O(10)	16(22)
C(8)-Re(3)-C(9)-O(11)	-50(44)	C(7)-Re(3)-C(9)-O(11)	-138(44)
O(15)-Re(3)-C(9)-O(11)	124(44)	O(16)-Re(3)-C(9)-O(11)	88(44)
N(2)-Re(3)-C(9)-O(11)	41(44)	C(11)-Re(4)-C(10)-O(12)	20(43)
C(12)-Re(4)-C(10)-O(12)	-67(43)	O(15)-Re(4)-C(10)-O(12)	-164(43)
O(16)-Re(4)-C(10)-O(12)	121(43)	N(4)-Re(4)-C(10)-O(12)	116(41)
C(12)-Re(4)-C(11)-O(13)	-67(21)	C(10)-Re(4)-C(11)-O(13)	-156(21)
O(15)-Re(4)-C(11)-O(13)	66(21)	O(16)-Re(4)-C(11)-O(13)	109(21)
N(4)-Re(4)-C(11)-O(13)	26(21)	C(11)-Re(4)-C(12)-O(14)	108(15)
C(10)-Re(4)-C(12)-O(14)	-163(15)	O(15)-Re(4)-C(12)-O(14)	-67(15)
O(16)-Re(4)-C(12)-O(14)	-44(16)	N(4)-Re(4)-C(12)-O(14)	17(15)
C(16)-N(1)-C(13)-C(14)	-1.1(9)	Re(1)-N(1)-C(13)-C(14)	-179.9(5)
C(15)-N(2)-C(14)-C(13)	0.0(9)	Re(3)-N(2)-C(14)-C(13)	177.6(5)
N(1)-C(13)-C(14)-N(2)	0.4(10)	C(14)-N(2)-C(15)-C(16)	0.3(9)
Re(3)-N(2)-C(15)-C(16)	-177.4(5)	C(13)-N(1)-C(16)-C(15)	1.4(9)
Re(1)-N(1)-C(16)-C(15)	-179.8(5)	N(2)-C(15)-C(16)-N(1)	-1.0(10)
C(20)-N(3)-C(17)-C(18)	0.7(9)	Re(2)-N(3)-C(17)-C(18)	177.8(5)
C(19)-N(4)-C(18)-C(17)	-1.7(9)	Re(4)-N(4)-C(18)-C(17)	-179.0(5)
N(3)-C(17)-C(18)-N(4)	0.8(10)	C(18)-N(4)-C(19)-C(20)	1.2(9)
Re(4)-N(4)-C(19)-C(20)	178.5(4)	C(17)-N(3)-C(20)-C(19)	-1.1(9)
Re(2)-N(3)-C(20)-C(19)	-178.2(4)	N(4)-C(19)-C(20)-N(3)	0.1(9)
Re(1)-O(8)-C(21)-C(22)	-145.0(5)	Re(2)-O(8)-C(21)-C(22)	86.8(6)
O(8)-C(21)-C(22)-C(23)	177.1(6)	C(21)-C(22)-C(23)-C(24)	176.1(6)
C(22)-C(23)-C(24)-C(25)	67.3(9)	Re(2)-O(7)-C(26)-C(27)	-146.2(5)
Re(1)-O(7)-C(26)-C(27)	86.5(6)	O(7)-C(26)-C(27)-C(28)	-168.6(6)

Table 5. (Continued)

C(26)-C(27)-C(28)-C(29)	-175.6(6)	C(27)-C(28)-C(29)-C(30)	-173.0(6)
Re(3)-O(16)-C(31)-C(32)	150.9(4)	Re(4)-O(16)-C(31)-C(32)	-79.0(6)
O(16)-C(31)-C(32)-C(33)	-173.6(5)	C(31)-C(32)-C(33)-C(34)	-178.0(6)
C(32)-C(33)-C(34)-C(35B)	-56.8(11)	C(32)-C(33)-C(34)-C(35A)	172.4(16)
Re(3)-O(15)-C(36)-C(37)	-122.4(7)	Re(4)-O(15)-C(36)-C(37)	109.2(8)
O(15)-C(36)-C(37)-C(38)	177.7(10)	C(36)-C(37)-C(38)-C(39A)	160.5(14)
C(36)-C(37)-C(38)-C(39B)	-172.4(14)	C(37)-C(38)-C(39A)-C(40A)	46(3)
C(39B)-C(38)-C(39A)-C(40A)	-6(3)	C(37)-C(38)-C(39B)-C(40B)	-177.3(19)
C(39A)-C(38)-C(39B)-C(40B)	-36(3)	C(45)-Re(6)-O(23)-C(51)	-169(3)
C(44)-Re(6)-O(23)-C(51)	58.1(5)	C(46)-Re(6)-O(23)-C(51)	-32.6(5)
O(24)-Re(6)-O(23)-C(51)	150.8(5)	N(6)-Re(6)-O(23)-C(51)	-124.6(5)
(45)-Re(6)-O(23)-Re(5)	55(3)	C(44)-Re(6)-O(23)-Re(5)	-78.7(3)
C(46)-Re(6)-O(23)-Re(5)	-169.4(2)	O(24)-Re(6)-O(23)-Re(5)	13.98(16)
N(6)-Re(6)-O(23)-Re(5)	98.65(19)	C(43)-Re(5)-O(23)-C(51)	34.1(5)
C(42)-Re(5)-O(23)-C(51)	178.1(17)	C(41)-Re(5)-O(23)-C(51)	-54.1(5)
O(24)-Re(5)-O(23)-C(51)	-150.1(5)	N(5)-Re(5)-O(23)-C(51)	125.2(4)
C(43)-Re(5)-O(23)-Re(6)	170.2(3)	C(42)-Re(5)-O(23)-Re(6)	-45.8(18)
C(41)-Re(5)-O(23)-Re(6)	82.0(3)	O(24)-Re(5)-O(23)-Re(6)	-14.06(16)
N(5)-Re(5)-O(23)-Re(6)	-98.72(19)	C(43)-Re(5)-O(24)-C(56)	-167(3)
C(42)-Re(5)-O(24)-C(56)	-34.0(5)	C(41)-Re(5)-O(24)-C(56)	54.2(5)
O(23)-Re(5)-O(24)-C(56)	150.1(5)	N(5)-Re(5)-O(24)-C(56)	-126.0(5)
C(43)-Re(5)-O(24)-Re(6)	57(3)	C(42)-Re(5)-O(24)-Re(6)	-170.0(3)
C(41)-Re(5)-O(24)-Re(6)	-81.8(3)	O(23)-Re(5)-O(24)-Re(6)	14.02(16)
N(5)-Re(5)-O(24)-Re(6)	97.98(19)	C(45)-Re(6)-O(24)-C(56)	33.4(5)
C(44)-Re(6)-O(24)-C(56)	-53.6(5)	C(46)-Re(6)-O(24)-C(56)	-170.6(14)
O(23)-Re(6)-O(24)-C(56)	-149.9(5)	N(6)-Re(6)-O(24)-C(56)	125.3(5)
C(45)-Re(6)-O(24)-Re(5)	169.2(2)	C(44)-Re(6)-O(24)-Re(5)	82.2(3)
C(46)-Re(6)-O(24)-Re(5)	-34.8(15)	O(23)-Re(6)-O(24)-Re(5)	-14.07(16)
N(6)-Re(6)-O(24)-Re(5)	-98.90(19)	C(43)-Re(5)-N(5)-C(48)	-145.0(5)
C(42)-Re(5)-N(5)-C(48)	-59.5(5)	C(41)-Re(5)-N(5)-C(48)	-156(23)
O(24)-Re(5)-N(5)-C(48)	39.0(4)	O(23)-Re(5)-N(5)-C(48)	114.4(4)
C(43)-Re(5)-N(5)-C(47)	35.5(5)	C(42)-Re(5)-N(5)-C(47)	121.1(5)
C(41)-Re(5)-N(5)-C(47)	25(24)	O(24)-Re(5)-N(5)-C(47)	-140.4(5)
O(23)-Re(5)-N(5)-C(47)	-65.0(5)		
C(45)-Re(6)-N(6)-C(49)	-142.1(5)	C(44)-Re(6)-N(6)-C(49)	142(5)
C(46)-Re(6)-N(6)-C(49)	-55.0(5)	O(23)-Re(6)-N(6)-C(49)	41.3(4)
O(24)-Re(6)-N(6)-C(49)	116.5(5)	C(45)-Re(6)-N(6)-C(50)	34.2(5)
C(44)-Re(6)-N(6)-C(50)	-42(5)	C(46)-Re(6)-N(6)-C(50)	121.4(5)
O(23)-Re(6)-N(6)-C(50)	-142.3(5)	O(24)-Re(6)-N(6)-C(50)	-67.1(5)
C(43)-Re(5)-C(41)-O(17)	36(13)	C(42)-Re(5)-C(41)-O(17)	-50(13)
O(24)-Re(5)-C(41)-O(17)	-148(12)	O(23)-Re(5)-C(41)-O(17)	136(12)
N(5)-Re(5)-C(41)-O(17)	46(32)	C(43)-Re(5)-C(42)-O(18)	58(42)
C(41)-Re(5)-C(42)-O(18)	145(42)	O(24)-Re(5)-C(42)-O(18)	-117(42)
O(23)-Re(5)-C(42)-O(18)	-86(42)	N(5)-Re(5)-C(42)-O(18)	-34(42)

Table 5. (Continued)

C(42)-Re(5)-C(43)-O(19)	-11(16)	C(41)-Re(5)-C(43)-O(19)	-98(16)
O(24)-Re(5)-C(43)-O(19)	123(15)	O(23)-Re(5)-C(43)-O(19)	165(16)
N(5)-Re(5)-C(43)-O(19)	82(16)	C(45)-Re(6)-C(44)-O(20)	-12(12)
C(46)-Re(6)-C(44)-O(20)	-99(12)	O(23)-Re(6)-C(44)-O(20)	165(12)
O(24)-Re(6)-C(44)-O(20)	90(12)	N(6)-Re(6)-C(44)-O(20)	64(14)
C(44)-Re(6)-C(45)-O(21)	-63(15)	C(46)-Re(6)-C(45)-O(21)	28(15)
O(23)-Re(6)-C(45)-O(21)	164(13)	O(24)-Re(6)-C(45)-O(21)	-156(15)
N(6)-Re(6)-C(45)-O(21)	120(15)	C(45)-Re(6)-C(46)-O(22)	67(44)
C(44)-Re(6)-C(46)-O(22)	153(44)	O(23)-Re(6)-C(46)-O(22)	-110(44)
O(24)-Re(6)-C(46)-O(22)	-90(44)	N(6)-Re(6)-C(46)-O(22)	-27(44)
C(48)-N(5)-C(47)-C(50)#1	0.2(9)	Re(5)-N(5)-C(47)-C(50)#1	179.6(5)
C(47)-N(5)-C(48)-C(49)#1	-0.4(8)	Re(5)-N(5)-C(48)-C(49)#1	-179.9(4)
C(50)-N(6)-C(49)-C(48)#1	2.3(8)	Re(6)-N(6)-C(49)-C(48)#1	178.9(4)
C(49)-N(6)-C(50)-C(47)#1	-2.1(9)	Re(6)-N(6)-C(50)-C(47)#1	-178.6(5)
Re(6)-O(23)-C(51)-C(52)	148.1(5)	Re(5)-O(23)-C(51)-C(52)	-82.4(7)
O(23)-C(51)-C(52)-C(53)	179.4(6)	C(51)-C(52)-C(53)-C(54)	-170.5(8)
C(52)-C(53)-C(54)-C(55A)	-72.7(13)	C(52)-C(53)-C(54)-C(55B)	71.1(16)
Re(5)-O(24)-C(56)-C(57)	146.2(5)	Re(6)-O(24)-C(56)-C(57)	-84.7(6)
O(24)-C(56)-C(57)-C(58)	172.9(6)	C(56)-C(57)-C(58)-C(59)	176.6(7)
C(57)-C(58)-C(59)-C(60)	-73.3(11)		

Symmetry transformations used to generate equivalent atoms: #1 -x,-y+1,-z

*Synthesis and characterization of the bridging hydride salt, **12***

12 was synthesized from the reaction of $Re_2(CO)_{10}$ and 2,9-dimethyl-1,10-phenanthroline (abbreviated as, 2,9-Me_2-Phen) in 1-pentanol according to the eq. 2:

$$3\ Re_2(CO)_{10} + 4\ (2,9\text{-}Me_2\text{-Phen}) + 5\ C_5H_9OH \rightarrow \mathbf{12}\ (50\%) + 2\ \textit{Fac}\text{-}(CO)_3(2,9\text{-}Me_2\text{-Phen})ReOC_5H_{11}\ (50\%) + 2\ H_2 + 12\ CO \quad (2)$$

Molecular structure of **12** was solved using X-ray crystallography and is shown in Figure 6. Least-squares planes and crystal data for **12** are summarized in Tables 6 and 7, respectively. Atomic coordinates and equivalent isotropic displacement parameters are provided in Table 8. Bond distances and angles are provided in Table 9. Anisotropic displacement parameters, hydrogen coordinates, and torsion angles are given in Tables 10-12 respectively. As expected, each rhenium atom of the cation is octahedrally coordinated to three terminal carbonyl groups in a facial arrangement, two nitrogen atoms of the diimine ligand, and the bridging hydride ligand; each rhenium atom of the anion is coordinated to three terminal carbonyls in a facial arrangement and three bridging pentyloxo ligands. In our complex the Re1…Re2 distance is 3.4194(4) Å with a R1...H…Re2 angle of 137(5)°. A similar Re...Re distance of 3.49 Å was observed for the cation of $[Re_2(bipy)_2(CO)_6(\mu\text{-}H)]Cl$ (Guilhem et. al., 1989). In our complex Re3…Re4 distance is 3.1097(5) Å. A similar distance of 3.0919(8) Å was observed for the anion of DABCO-H$[Re_2(CO)_6(\mu\text{-}OMe)_3]$ (Kim and Verkade, 2003).

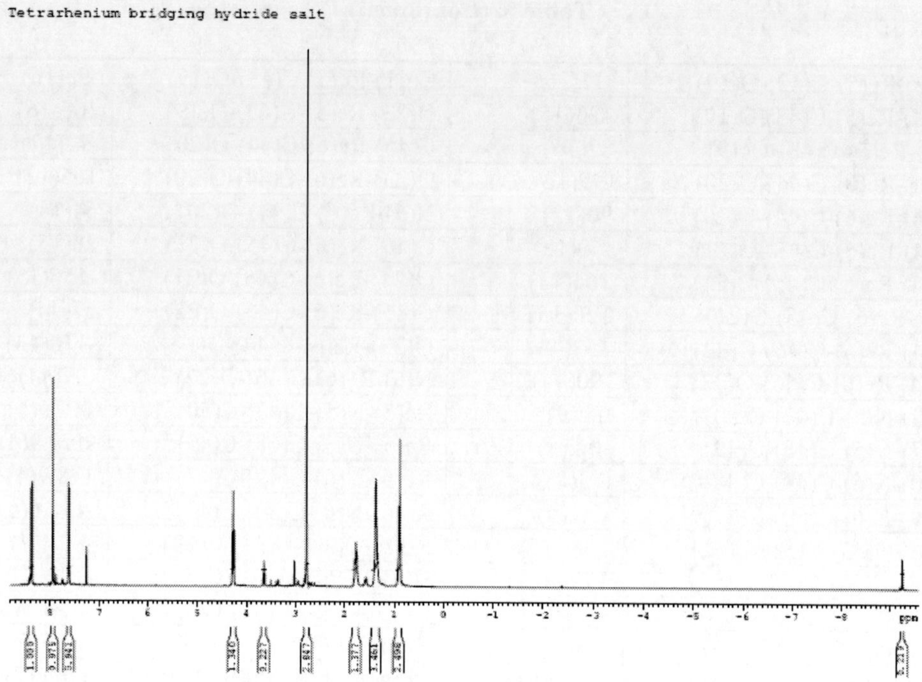

Figure 4. Proton NMR spectrum of **12** in $CDCl_3$.

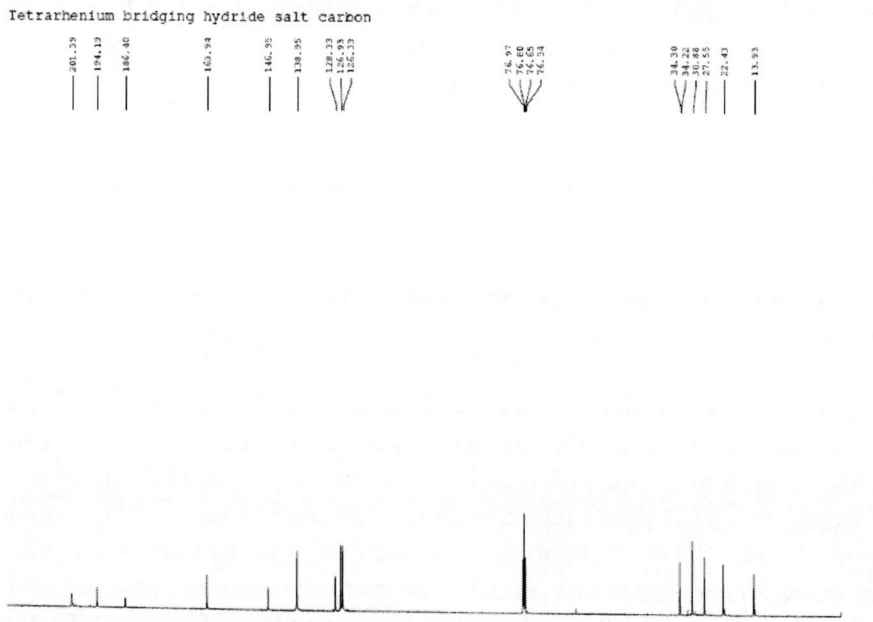

Figure 5. ^{13}C NMR spectrum of **12** in $CDCl_3$.

12 was characterized spectroscopically. The proton NMR spectrum exhibits the characteristic high field resonance at δ-9.24 attributed to the hydride ligand (see, Figure 4).

The ^{13}C NMR spectrum shows three downfield resonances due to three facial terminal carbonyls attached to each rhenium atom (see, Figure 5).

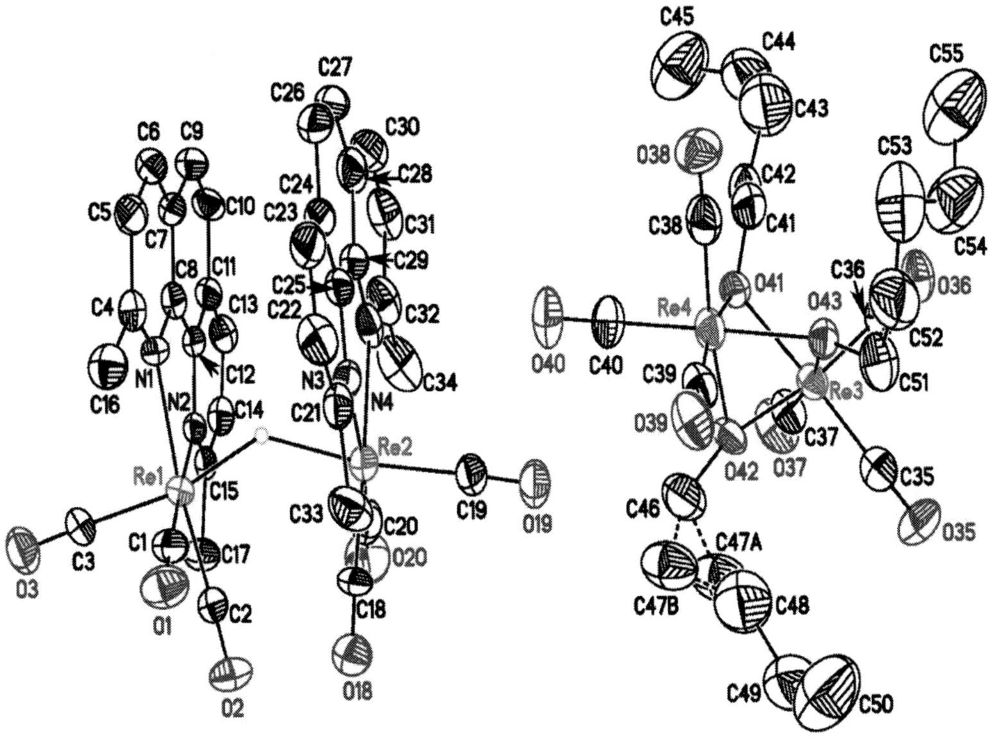

Figure 6. Molecular structure of **12**.

Table 6. Least Squares Planes for 12.

Least Squares Planes	Deviation from Planarity (Å)	Dihedral Angle (°)
N1-C4-C5-C6-C7-C8	0.0082	1.2(3)
C7-C8-C9-C10-C11-C12	0.0126	2.8(3)
N2-C11-C12-C13-C14-C15	0.0178	
N3-C21-C22-C23-C24-C25	0.0100	4.6(3)
C24-C25-C26-C27-C28-C29	0.0128	2.3(3)
N4-C28-C29-C30-C31-C32	0.0094	
N1-N2-phen carbons	0.0361	7.3(1)
N3-N4-phen carbons	0.0559	

Table 7. Crystal data and structure refinement for 12.

Empirical formula	$[C_{34}H_{25}O_6Re_2][C_{21}H_{33}O_9Re_2]$	
Formula weight	1759.85	
Temperature	150(2) K	
Wavelength	1.54178 Å	
Crystal system	Monoclinic	
Space group	$P2_1/c$	
Unit cell dimensions:	a = 18.5412(3) Å	α = 90°
	b = 19.5174(4) Å	β = 117.436(1)°
	c = 17.7521(3) Å	γ = 90°
Volume, Z	5701.52(18) $Å^3$, 4	
Density (calculated)	2.050 Mg/m^3	
Absorption coefficient	16.770 mm^{-1}	
F(000)	3344	
Crystal size	0.15 x 0.07 x 0.04 mm	
θ range for data collection	2.69 to 65.08°	
Limiting indices	$-21 < h < 21$, $-22 < k < 22$, $-20 < l < 20$	
Reflections collected	45889	
Independent reflections	9663 (R_{int} = 0.0532)	
Completeness to θ = 65.08°	99.5 %	
Absorption correction	Multi-scan	
Max. and min. transmission	0.5535 and 0.1876	
Refinement method	Full-matrix least-squares on F^2	
Data / restraints / parameters	9663 / 5 / 685	
Goodness-of-fit on F^2	1.033	
Final R indices [I>2σ(I)]	R1 = 0.0390, wR2 = 0.1007	
R indices (all data)	R1 = 0.0491, wR2 = 0.1062	
Largest diff. peak and hole	2.446 and -1.164 $eÅ^{-3}$	

Table 8. Atomic coordinates [$x10^4$] and equivalent isotropic displacement parameters [$Å^2x10^3$] for 12. U(eq) is defined as one third of the trace of the orthogonalized Uij tensor.

	x	y	z	U(eq)
Re(1)	451(1)	3978(1)	8375(1)	27(1)
Re(2)	1777(1)	3843(1)	7546(1)	31(1)
O(1)	-697(4)	4717(3)	6734(4)	51(2)
O(2)	1117(4)	5442(3)	8733(4)	48(1)

Table 8. (Continued)

	x	y	z	U(eq)
O(3)	-660(4)	4379(3)	9157(4)	51(2)
O(18)	1204(4)	5305(3)	6956(4)	57(2)
O(19)	3089(4)	4067(3)	6969(4)	53(2)
O(20)	2881(4)	4573(4)	9208(4)	68(2)
N(1)	86(3)	2893(3)	8096(4)	27(1)
N(2)	1320(3)	3445(3)	9542(4)	24(1)
N(3)	914(3)	3242(3)	6434(4)	29(1)
N(4)	2123(4)	2743(3)	7902(4)	37(2)
C(1)	-282(5)	4414(4)	7329(5)	38(2)
C(2)	902(5)	4886(4)	8639(5)	33(2)
C(3)	-256(5)	4216(4)	8865(5)	35(2)
C(4)	-527(4)	2628(4)	7400(5)	33(2)
C(5)	-592(5)	1913(4)	7259(5)	39(2)
C(6)	-42(5)	1480(4)	7842(5)	37(2)
C(7)	588(4)	1740(3)	8583(5)	30(2)
C(8)	640(4)	2459(4)	8691(5)	29(2)
C(9)	1180(5)	1307(4)	9212(5)	38(2)
C(10)	1784(5)	1564(4)	9939(5)	38(2)
C(11)	1837(4)	2302(4)	10075(5)	32(2)
C(12)	1288(4)	2740(3)	9453(5)	27(2)
C(13)	2440(5)	2591(4)	10823(5)	37(2)
C(14)	2443(5)	3293(4)	10920(5)	36(2)
C(15)	1867(4)	3710(4)	10278(5)	30(2)
C(16)	-1144(5)	3083(4)	6753(5)	43(2)
C(17)	1891(5)	4473(4)	10438(5)	42(2)
C(18)	1392(5)	4753(4)	7159(5)	39(2)
C(19)	2595(5)	3973(4)	7167(6)	41(2)
C(20)	2486(5)	4285(5)	8602(6)	48(2)
C(21)	326(5)	3475(4)	5679(5)	39(2)
C(22)	-231(5)	3028(5)	5078(6)	45(2)
C(23)	-201(5)	2345(5)	5216(6)	46(2)
C(24)	400(5)	2076(4)	5958(5)	39(2)
C(25)	957(5)	2543(4)	6566(5)	35(2)
C(26)	497(6)	1358(4)	6125(6)	46(2)
C(27)	1119(6)	1114(4)	6822(6)	47(2)
C(28)	1699(6)	1560(4)	7438(6)	45(2)
C(29)	1606(5)	2287(4)	7319(5)	34(2)
C(30)	2352(7)	1324(5)	8155(6)	55(2)
C(31)	2884(6)	1776(5)	8734(6)	55(3)
C(32)	2746(6)	2497(5)	8596(5)	48(2)
C(33)	267(6)	4224(4)	5470(6)	50(2)
C(34)	3332(7)	2975(6)	9251(6)	69(3)
Re(3)	6991(1)	3282(1)	8034(1)	37(1)
Re(4)	5163(1)	2877(1)	7168(1)	40(1)
O(35)	7730(4)	4504(4)	7548(5)	67(2)

Table 8. (Continued)

	x	y	z	U(eq)
O(36)	8405(4)	2395(4)	8173(4)	57(2)
O(37)	8093(5)	3578(4)	9906(5)	77(2)
O(38)	4380(4)	1547(3)	6230(5)	62(2)
O(39)	3794(4)	3666(4)	5722(4)	58(2)
O(40)	4037(4)	2704(4)	7985(5)	69(2)
O(41)	6267(3)	2486(3)	8174(4)	39(1)
O(42)	5878(3)	3720(3)	7859(4)	43(1)
O(43)	6105(3)	2999(3)	6798(3)	37(1)
C(35)	7459(5)	4045(5)	7743(6)	48(2)
C(36)	7876(5)	2737(5)	8106(5)	42(2)
C(37)	7670(5)	3481(5)	9208(6)	50(2)
C(38)	4674(5)	2052(5)	6570(6)	52(2)
C(39)	4312(5)	3377(5)	6263(6)	46(2)
C(40)	4470(5)	2768(5)	7697(6)	55(2)
C(41)	6475(7)	1780(5)	8186(7)	64(3)
C(42)	7129(7)	1561(6)	9029(8)	72(3)
C(43)	7373(13)	817(9)	8975(13)	143(5)
C(44)	7744(14)	465(8)	9687(13)	141(5)
C(45)	7358(13)	307(10)	10211(13)	155(4)
C(46)	5709(7)	4198(6)	8329(9)	83(4)
C(47A)	5919(13)	4884(10)	8141(15)	90(6)
C(47B)	5145(12)	4761(11)	7853(15)	90(6)
C(48)	5450(13)	5077(9)	7245(13)	143(5)
C(49)	5658(14)	5735(8)	6896(13)	141(5)
C(50)	5253(13)	5794(10)	5956(13)	155(4)
C(51)	6045(6)	3239(6)	6029(6)	63(3)
C(52)	5608(9)	2801(6)	5300(9)	90(4)
C(53)	5880(9)	2031(8)	5474(11)	115(5)
C(54)	6747(9)	1986(8)	5786(12)	115(6)
C(55)	6982(13)	1217(10)	5984(13)	155(4)

Table 9. Bond distances [Å] and angles [°] for 12.

Re(1)-C(2)	1.922(8)	Re(1)-C(1)	1.922(9)
Re(1)-C(3)	1.933(7)	Re(1)-N(1)	2.210(5)
Re(1)-N(2)	2.211(6)	Re(1)-H(1)	1.76(8)
Re(2)-C(18)	1.920(7)	Re(2)-C(20)	1.928(10)
Re(2)-C(19)	1.937(8)	Re(2)-N(3)	2.218(6)
Re(2)-N(4)	2.246(6)	Re(2)-H(1)	1.89(8)
O(1)-C(1)	1.144(10)	O(2)-C(2)	1.142(9)
O(3)-C(3)	1.136(9)	O(18)-C(18)	1.138(9)
O(19)-C(19)	1.138(10)	O(20)-C(20)	1.133(10)
N(1)-C(4)	1.339(9)	N(1)-C(8)	1.373(9)
N(2)-C(15)	1.335(9)	N(2)-C(12)	1.384(8)

Table 9. (Continued)

N(3)-C(21)	1.359(10)	N(3)-C(25)	1.380(9)
N(4)-C(32)	1.330(10)	N(4)-C(29)	1.367(10)
C(4)-C(5)	1.414(11)	C(4)-C(16)	1.484(11)
C(5)-C(6)	1.361(12)	C(6)-C(7)	1.391(11)
C(7)-C(8)	1.416(10)	C(7)-C(9)	1.426(11)
C(8)-C(12)	1.442(10)	C(9)-C(10)	1.356(11)
C(10)-C(11)	1.456(10)	C(11)-C(12)	1.395(10)
C(11)-C(13)	1.402(11)	C(13)-C(14)	1.381(11)
C(14)-C(15)	1.407(11)	C(15)-C(17)	1.513(10)
C(21)-C(22)	1.397(12)	C(21)-C(33)	1.498(11)
C(22)-C(23)	1.352(12)	C(23)-C(24)	1.379(12)
C(24)-C(25)	1.426(11)	C(24)-C(26)	1.427(12)
C(25)-C(29)	1.416(11)	C(26)-C(27)	1.330(13)
C(27)-C(28)	1.424(13)	C(28)-C(30)	1.371(13)
C(28)-C(29)	1.432(11)	C(30)-C(31)	1.369(14)
C(31)-C(32)	1.432(13)	C(32)-C(34)	1.495(14)
Re(3)…Re(4)	3.1097(5)	Re(3)-C(35)	1.910(10)
Re(3)-C(36)	1.910(9)	Re(3)-C(37)	1.913(9)
Re(3)-O(42)	2.119(5)	Re(3)-O(43)	2.119(5)
Re(3)-O(41)	2.141(5)	Re(4)-C(38)	1.913(11)
Re(4)-C(39)	1.920(10)	Re(4)-C(40)	1.921(10)
Re(4)-O(42)	2.114(6)	Re(4)-O(41)	2.142(5)
Re(4)-O(43)	2.144(5)	O(35)-C(35)	1.156(11)
O(36)-C(36)	1.146(10)	O(37)-C(37)	1.137(10)
O(38)-C(38)	1.153(11)	O(39)-C(39)	1.146(10)
O(40)-C(40)	1.140(11)	O(41)-C(41)	1.427(11)
O(42)-C(46)	1.381(12)	O(43)-C(51)	1.398(11)
C(41)-C(42)	1.490(15)	C(42)-C(43)	1.538(18)
C(43)-C(44)	1.32(2)	C(44)-C(45)	1.44(2)
C(46)-C(47A)	1.475(16)	C(46)-C(47B)	1.483(16)
C(47A)-C(48)	1.468(17)	C(47B)-C(48)	1.557(17)
C(48)-C(49)	1.55(2)	C(49)-C(50)	1.49(2)
C(51)-C(52)	1.449(15)	C(52)-C(53)	1.570(15)
C(53)-C(54)	1.440(19)	C(54)-C(55)	1.56(2)
C(2)-Re(1)-C(1)	82.1(3)	C(2)-Re(1)-C(3)	88.8(3)
C(1)-Re(1)-C(3)	90.2(3)	C(2)-Re(1)-N(1)	172.7(3)
C(1)-Re(1)-N(1)	101.5(3)	C(3)-Re(1)-N(1)	97.4(3)
C(2)-Re(1)-N(2)	99.1(3)	C(1)-Re(1)-N(2)	177.0(3)
C(3)-Re(1)-N(2)	92.5(3)	N(1)-Re(1)-N(2)	77.0(2)
C(2)-Re(1)-H(1)	104(3)	C(1)-Re(1)-H(1)	91(3)
C(3)-Re(1)-H(1)	167(3)	N(1)-Re(1)-H(1)	70(3)

Table 9. (Continued)

N(2)-Re(1)-H(1)	86(3)	C(18)-Re(2)-C(20)	84.7(4)
C(18)-Re(2)-C(19)	88.6(3)	C(20)-Re(2)-C(19)	89.2(4)
C(18)-Re(2)-N(3)	100.0(3)	C(20)-Re(2)-N(3)	172.3(3)
C(19)-Re(2)-N(3)	96.9(3)	C(18)-Re(2)-N(4)	174.9(3)
C(20)-Re(2)-N(4)	100.1(3)	C(19)-Re(2)-N(4)	93.2(3)
N(3)-Re(2)-N(4)	75.0(2)	C(18)-Re(2)-H(1)	98(3)
C(20)-Re(2)-H(1)	95(3)	C(19)-Re(2)-H(1)	172(3)
N(3)-Re(2)-H(1)	78(3)	N(4)-Re(2)-H(1)	80(3)
Re(1)-H(1)-Re(2)	137(5)	C(4)-N(1)-C(8)	119.2(6)
C(4)-N(1)-Re(1)	128.3(5)	C(8)-N(1)-Re(1)	12.0(4)
C(15)-N(2)-C(12)	118.0(6)	C(15)-N(2)-Re(1)	129.2(5)
C(12)-N(2)-Re(1)	112.8(4)	C(21)-N(3)-C(25)	117.1(6)
C(21)-N(3)-Re(2)	128.5(5)	C(25)-N(3)-Re(2)	114.2(5)
C(32)-N(4)-C(29)	118.3(7)	C(32)-N(4)-Re(2)	128.2(6)
C(29)-N(4)-Re(2)	113.5(5)	O(1)-C(1)-Re(1)	175.1(7)
O(2)-C(2)-Re(1)	174.3(7)	O(3)-C(3)-Re(1)	177.5(7)
N(1)-C(4)-C(5)	120.8(7)	N(1)-C(4)-C(16)	120.5(7)
C(5)-C(4)-C(16)	118.8(7)	C(6)-C(5)-C(4)	120.3(8)
C(5)-C(6)-C(7)	120.1(7)	C(6)-C(7)-C(8)	117.7(7)
C(6)-C(7)-C(9)	122.2(7)	C(8)-C(7)-C(9)	120.0(7)
N(1)-C(8)-C(7)	121.8(6)	N(1)-C(8)-C(12)	119.5(6)
C(7)-C(8)-C(12)	118.7(6)	C(10)-C(9)-C(7)	121.8(7)
C(9)-C(10)-C(11)	119.1(7)	C(12)-C(11)-C(13)	118.4(7)
C(12)-C(11)-C(10)	120.4(7)	C(13)-C(11)-C(10)	121.2(7)
N(2)-C(12)-C(11)	122.8(6)	N(2)-C(12)-C(8)	117.2(6)
C(11)-C(12)-C(8)	119.9(6)	C(14)-C(13)-C(11)	118.2(7)
C(13)-C(14)-C(15)	121.0(7)	N(2)-C(15)-C(14)	121.3(7)
N(2)-C(15)-C(17)	120.5(7)	C(14)-C(15)-C(17)	118.1(7)
O(18)-C(18)-Re(2)	176.4(8)	O(19)-C(19)-Re(2)	177.4(8)
O(20)-C(20)-Re(2)	176.7(9)	N(3)-C(21)-C(22)	121.3(8)
N(3)-C(21)-C(33)	120.3(7)	C(22)-C(21)-C(33)	118.4(8)
C(23)-C(22)-C(21)	121.5(8)	C(22)-C(23)-C(24)	119.8(8)
C(23)-C(24)-C(25)	117.6(7)	C(23)-C(24)-C(26)	122.9(8)
C(25)-C(24)-C(26)	119.5(8)	N(3)-C(25)-C(29)	117.8(7)
N(3)-C(25)-C(24)	122.6(7)	C(29)-C(25)-C(24)	119.5(7)
C(27)-C(26)-C(24)	121.2(8)	C(26)-C(27)-C(28)	121.3(8)
C(30)-C(28)-C(27)	122.6(8)	C(30)-C(28)-C(29)	117.7(9)
C(27)-C(28)-C(29)	119.7(8)	N(4)-C(29)-C(25)	118.7(7)
N(4)-C(29)-C(28)	122.6(7)	C(25)-C(29)-C(28)	118.8(7)
C(31)-C(30)-C(28)	120.3(8)	C(30)-C(31)-C(32)	119.6(9)
N(4)-C(32)-C(31)	121.5(9)	N(4)-C(32)-C(34)	120.3(8)

Table 9. (Continued)

C(31)-C(32)-C(34)	118.1(8)		
C(35)-Re(3)-C(36)	88.2(4)	C(35)-Re(3)-C(37)	89.0(4)
C(36)-Re(3)-C(37)	85.8(4)	C(35)-Re(3)-O(42)	100.3(3)
C(36)-Re(3)-O(42)	169.3(3)	C(37)-Re(3)-O(42)	100.7(3)
C(35)-Re(3)-O(43)	98.9(3)	C(36)-Re(3)-O(43)	99.0(3)
C(37)-Re(3)-O(43)	170.8(3)	O(42)-Re(3)-O(43)	73.5(2)
C(35)-Re(3)-O(41)	169.8(3)	C(36)-Re(3)-O(41)	98.8(3)
C(37)-Re(3)-O(41)	98.9(3)	O(42)-Re(3)-O(41)	71.9(2)
O(43)-Re(3)-O(41)	72.7(2)	C(35)-Re(3)-Re(4)	126.3(3)
C(36)-Re(3)-Re(4)	126.8(3)	C(37)-Re(3)-Re(4)	127.6(3)
O(42)-Re(3)…Re(4)	42.67(15)	O(43)-Re(3)…Re(4)	43.48(14)
O(41)-Re(3)…Re(4)	43.47(14)	C(38)-Re(4)-C(39)	88.1(4)
C(38)-Re(4)-C(40)	85.8(4)	C(39)-Re(4)-C(40)	89.4(4)
C(38)-Re(4)-O(42)	171.0(3)	C(39)-Re(4)-O(42)	98.3(3)
C(40)-Re(4)-O(42)	100.7(3)	C(38)-Re(4)-O(41)	100.9(3)
C(39)-Re(4)-O(41)	168.3(3)	C(40)-Re(4)-O(41)	98.7(3)
O(42)-Re(4)-O(41)	72.0(2)	C(38)-Re(4)-O(43)	99.7(3)
C(39)-Re(4)-O(43)	99.1(3)	C(40)-Re(4)-O(43)	170.0(3)
O(42)-Re(4)-O(43)	73.1(2)	O(41)-Re(4)-O(43)	72.2(2)
C(38)-Re(4)…Re(3)	128.2(3)	C(39)-Re(4)…Re(3)	124.9(3)
C(40)-Re(4)…Re(3)	127.5(3)	O(42)-Re(4)…Re(3)	42.78(14)
O(41)-Re(4)…Re(3)	43.44(14)	O(43)-Re(4)…Re(3)	42.85(14)
C(41)-O(41)-Re(3)	121.5(5)	C(41)-O(41)-Re(4)	120.3(6)
Re(3)-O(41)-Re(4)	93.1(2)	C(46)-O(42)-Re(4)	128.3(6)
C(46)-O(42)-Re(3)	131.4(6)	Re(4)-O(42)-Re(3)	94.5(2)
C(51)-O(43)-Re(3)	126.8(5)	C(51)-O(43)-Re(4)	128.8(5)
Re(3)-O(43)-Re(4)	93.7(2)	O(35)-C(35)-Re(3)	178.4(9)
O(36)-C(36)-Re(3)	177.3(8)	O(37)-C(37)-Re(3)	177.2(9)
O(38)-C(38)-Re(4)	177.9(9)	O(39)-C(39)-Re(4)	178.6(8)
O(40)-C(40)-Re(4)	177.7(9)	O(41)-C(41)-C(42)	112.4(9)
C(41)-C(42)-C(43)	109.6(12)	C(44)-C(43)-C(42)	118.1(17)
C(43)-C(44)-C(45)	122(2)	O(42)-C(46)-C(47A)	108.6(14)
O(42)-C(46)-C(47B)	116.9(14)	C(48)-C(47A)-C(46)	113.1(16)
C(46)-C(47B)-C(48)	107.7(15)	C(47A)-C(48)-C(49)	120.9(19)
C(49)-C(48)-C(47B)	147.4(17)	C(50)-C(49)-C(48)	115.1(18)
O(43)-C(51)-C(52)	115.8(10)	C(51)-C(52)-C(53)	113.1(12)
C(54)-C(53)-C(52)	109.5(14)	C(53)-C(54)-C(55)	107.0(14)

Table 10. Anisotropic displacement parameters [$Å^2x10^3$] for 12. The anisotropic displacement factor exponent takes the form: $-2\pi^2[(ha^*)^2U_{11} + ... + 2hka^*b^*U_{12}]$

	U11	U22	U33	U23	U13	U12
Re(1)	30(1)	21(1)	34(1)	0(1)	17(1)	1(1)
Re(2)	33(1)	30(1)	36(1)	-2(1)	20(1)	-2(1)
O(1)	53(4)	39(3)	46(4)	12(3)	12(3)	4(3)
O(2)	63(4)	28(3)	47(3)	-4(3)	22(3)	-8(3)
O(3)	52(3)	55(4)	58(4)	6(3)	36(3)	12(3)
O(18)	78(5)	39(4)	73(5)	-3(3)	52(4)	-1(3)
O(19)	54(4)	56(4)	67(4)	-13(3)	43(4)	-12(3)
O(20)	61(4)	75(5)	58(4)	-29(4)	19(4)	-19(4)
N(1)	28(3)	19(3)	35(3)	-3(2)	16(3)	-2(2)
N(2)	26(3)	25(3)	27(3)	3(2)	17(3)	3(2)
N(3)	27(3)	33(3)	31(3)	-3(3)	16(3)	-4(3)
N(4)	37(3)	46(4)	37(4)	17(3)	25(3)	20(3)
C(1)	46(5)	25(4)	46(5)	-1(4)	24(4)	1(4)
C(2)	36(4)	31(4)	36(4)	0(3)	19(3)	0(3)
C(3)	33(4)	35(4)	41(4)	-1(3)	21(4)	6(3)
C(4)	25(4)	37(4)	37(4)	0(3)	16(3)	1(3)
C(5)	48(5)	39(4)	40(5)	-11(4)	29(4)	-7(4)
C(6)	46(4)	29(4)	46(5)	-1(4)	31(4)	1(4)
C(7)	36(4)	26(4)	36(4)	0(3)	23(3)	-2(3)
C(8)	28(3)	33(4)	30(4)	0(3)	17(3)	2(3)
C(9)	47(5)	26(4)	47(5)	0(3)	26(4)	1(3)
C(10)	50(5)	26(4)	44(5)	8(3)	25(4)	11(4)
C(11)	35(4)	28(4)	38(4)	5(3)	21(4)	5(3)
C(12)	30(4)	24(3)	36(4)	-3(3)	23(3)	-2(3)
C(13)	32(4)	41(4)	37(4)	2(4)	16(4)	5(3)
C(14)	33(4)	44(5)	33(4)	-2(4)	17(3)	-2(3)
C(15)	29(4)	31(4)	38(4)	-4(3)	23(3)	-2(3)
C(16)	37(4)	43(5)	42(5)	-2(4)	11(4)	-2(4)
C(17)	48(5)	30(4)	43(5)	-7(4)	16(4)	-2(4)
C(18)	54(5)	18(4)	54(5)	-3(3)	34(4)	1(3)
C(19)	42(5)	41(4)	47(5)	-9(4)	28(4)	-5(4)
C(20)	43(5)	53(5)	57(6)	-8(5)	31(5)	-10(4)
C(21)	39(4)	45(5)	44(5)	2(4)	28(4)	1(4)
C(22)	36(4)	53(5)	39(5)	-11(4)	12(4)	0(4)
C(23)	43(5)	47(5)	57(6)	-14(4)	30(5)	-10(4)
C(24)	46(5)	35(4)	50(5)	-13(4)	34(4)	-8(4)
C(25)	44(4)	30(4)	43(4)	4(3)	31(4)	5(3)
C(26)	59(6)	37(4)	56(6)	-12(4)	39(5)	-12(4)
C(27)	71(6)	28(4)	63(6)	-3(4)	50(6)	-3(4)
C(28)	56(5)	45(5)	52(5)	14(4)	40(5)	11(4)
C(29)	43(4)	30(4)	42(4)	1(3)	29(4)	0(3)
C(30)	86(7)	40(5)	61(6)	13(5)	52(6)	12(5)
C(31)	64(6)	65(6)	48(5)	28(5)	34(5)	34(5)

Table 10. (Continued)

	U11	U22	U33	U23	U13	U12
C(32)	53(5)	61(6)	36(5)	9(4)	25(4)	19(5)
C(33)	66(6)	41(5)	49(5)	9(4)	31(5)	18(4)
C(34)	72(7)	86(8)	41(5)	9(5)	20(5)	32(6)
Re(3)	31(1)	38(1)	38(1)	-5(1)	14(1)	1(1)
Re(4)	32(1)	47(1)	45(1)	-5(1)	20(1)	0(1)
O(35)	64(4)	55(4)	81(5)	8(4)	33(4)	-17(4)
O(36)	41(3)	78(5)	55(4)	7(3)	24(3)	15(3)
O(37)	67(5)	98(6)	51(4)	-21(4)	15(4)	5(4)
O(38)	50(4)	52(4)	75(5)	-14(4)	20(4)	-14(3)
O(39)	40(3)	77(5)	49(4)	-3(3)	14(3)	13(3)
O(40)	61(4)	96(6)	70(5)	10(4)	47(4)	6(4)
O(41)	34(3)	36(3)	48(3)	-1(3)	20(3)	1(2)
O(42)	39(3)	37(3)	56(4)	-10(3)	23(3)	6(2)
O(43)	35(3)	42(3)	37(3)	-4(2)	19(2)	-5(2)
C(35)	32(4)	55(5)	48(5)	-10(4)	11(4)	-3(4)
C(36)	37(4)	53(5)	34(4)	4(4)	15(4)	2(4)
C(37)	39(5)	61(6)	43(5)	-14(4)	13(4)	1(4)
C(38)	39(5)	61(6)	59(6)	3(5)	24(5)	7(5)
C(39)	35(4)	55(5)	47(5)	-18(4)	18(4)	-5(4)
C(40)	37(5)	69(6)	61(6)	0(5)	25(5)	-2(5)
C(41)	65(6)	57(6)	87(8)	8(6)	49(6)	7(5)
C(42)	71(7)	75(7)	91(8)	33(7)	56(7)	31(6)
C(43)	186(14)	101(9)	185(14)	41(9)	122(12)	38(9)
C(44)	234(16)	80(7)	172(13)	31(8)	147(13)	33(9)
C(45)	176(11)	134(9)	167(11)	36(8)	89(10)	-13(8)
C(46)	71(7)	60(7)	128(11)	-25(7)	55(8)	-4(6)
C(47A)	56(9)	63(10)	123(16)	-10(10)	17(10)	6(8)
C(47B)	56(9)	63(10)	123(16)	-10(10)	17(10)	6(8)
C(48)	186(14)	101(9)	185(14)	41(9)	122(12)	38(9)
C(49)	234(16)	80(7)	172(13)	31(8)	147(13)	33(9)
C(50)	176(11)	134(9)	167(11)	36(8)	89(10)	-13(8)
C(51)	50(5)	85(8)	53(6)	0(5)	24(5)	6(5)
C(52)	95(10)	91(10)	94(10)	-27(8)	52(8)	-16(8)
C(53)	96(11)	156(16)	104(12)	-3(11)	57(10)	-2(11)
C(54)	85(10)	103(11)	152(16)	23(11)	49(11)	-4(9)
C(55)	176(11)	134(9)	167(11)	36(8)	89(10)	-13(8)

Table 11. Hydrogen coordinates (x10^4) and isotropic displacement parameters (Å^{2}x10^3) for 12.

	x	y	z	U(eq)
H(1)	980(50)	3590(50)	7880(50)	50
H(5)	-1022	1733	6754	47
H(6)	-89	1000	7743	44
H(9)	1151	827	9119	46
H(10)	2167	1267	10354	46
H(13)	2836	2311	11253	44
H(14)	2840	3498	11428	43
H(16A)	-1598	2806	6348	65
H(16B)	-897	3333	6452	65
H(16C)	-1344	3409	7033	65
H(17A)	2216	4699	10205	63
H(17B)	2134	4558	11050	63
H(17C)	1336	4657	10163	63
H(22)	-641	3208	4560	54
H(23)	-593	2052	4803	56
H(26)	112	1048	5732	55
H(27)	1177	633	6910	56
H(30)	2436	845	8250	66
H(31)	3343	1611	9226	66
H(33A)	75	4473	5823	75
H(33B)	-116	4291	4870	75
H(33C)	804	4396	5582	75
H(34A)	3498	3330	8973	103
H(34B)	3811	2718	9648	103
H(34C)	3070	3189	9562	103
H(41A)	6657	1700	7749	77
H(41B)	5985	1497	8039	77
H(42A)	6934	1597	9461	86
H(42B)	7607	1865	9202	86
H(43A)	6874	564	8594	172
H(43B)	7731	821	8700	172
H(44A)	8251	715	10052	169
H(44B)	7909	23	9540	169
H(45A)	7747	73	10728	233
H(45B)	6889	8	9896	233
H(45C)	7176	732	10363	233
H(46A)	6020	4109	8923	99
H(46B)	5137	4171	8190	99
H(46C)	6201	4398	8736	99
H(46D)	5467	3959	8636	99
H(47A)	5822	5226	8496	108
H(47B)	6505	4894	8298	108
H(47C)	5139	5113	8250	108
H(47D)	4587	4581	7522	108

Table 11. (Continued)

	x	y	z	U(eq)
H(48A)	4878	5122	7103	172
H(48B)	5466	4708	6887	172
H(48C)	5925	4820	7384	172
H(48D)	5054	4882	6696	172
H(49A)	5506	6138	7131	169
H(49B)	6253	5752	7105	169
H(50A)	5438	6213	5792	233
H(50B)	4663	5814	5743	233
H(50C)	5392	5396	5713	233
H(51A)	6602	3309	6099	75
H(51B)	5775	3692	5909	75
H(52A)	5694	2970	4821	108
H(52B)	5020	2830	5127	108
H(53A)	5731	1837	5899	138
H(53B)	5600	1762	4944	138
H(54A)	7032	2266	6306	138
H(54B)	6899	2156	5353	138
H(55A)	7569	1164	6191	233
H(55B)	6689	945	5467	233
H(55C)	6837	1058	6421	233

Table 12. Torsion angles [°] for 12.

C(2)-Re(1)-N(1)-C(4)	-124(2)	C(1)-Re(1)-N(1)-C(4)	-4.5(7)
C(3)-Re(1)-N(1)-C(4)	87.3(6)	N(2)-Re(1)-N(1)-C(4)	178.2(6)
C(2)-Re(1)-N(1)-C(8)	48(2)	C(1)-Re(1)-N(1)-C(8)	167.2(5)
C(3)-Re(1)-N(1)-C(8)	-101.0(5)	N(2)-Re(1)-N(1)-C(8)	-10.1(4)
C(2)-Re(1)-N(2)-C(15)	13.6(6)	C(1)-Re(1)-N(2)-C(15)	126(5)
C(3)-Re(1)-N(2)-C(15)	-75.6(6)	N(1)-Re(1)-N(2)-C(15)	-172.6(6)
C(2)-Re(1)-N(2)-C(12)	-163.8(5)	C(1)-Re(1)-N(2)-C(12)	-51(5)
C(3)-Re(1)-N(2)-C(12)	107.1(5)	N(1)-Re(1)-N(2)-C(12)	10.0(4)
C(18)-Re(2)-N(3)-C(21)	4.4(6)	C(20)-Re(2)-N(3)-C(21)	132(2)
C(19)-Re(2)-N(3)-C(21)	-85.4(6)	N(4)-Re(2)-N(3)-C(21)	-176.9(6)
C(18)-Re(2)-N(3)-C(25)	-170.9(5)	C(20)-Re(2)-N(3)-C(25)	-43(2)
C(19)-Re(2)-N(3)-C(25)	99.3(5)	N(4)-Re(2)-N(3)-C(25)	7.8(5)
C(18)-Re(2)-N(4)-C(32)	-170(3)	C(20)-Re(2)-N(4)-C(32)	-10.8(7)
C(19)-Re(2)-N(4)-C(32)	79.0(7)	N(3)-Re(2)-N(4)-C(32)	175.3(7)
C(18)-Re(2)-N(4)-C(29)	9(3)	C(20)-Re(2)-N(4)-C(29)	168.3(5)
C(19)-Re(2)-N(4)-C(29)	-101.9(5)	N(3)-Re(2)-N(4)-C(29)	-5.6(5)
C(2)-Re(1)-C(1)-O(1)	-13(8)	C(3)-Re(1)-C(1)-O(1)	76(8)
N(1)-Re(1)-C(1)-O(1)	174(8)	N(2)-Re(1)-C(1)-O(1)	-126(8)
C(1)-Re(1)-C(2)-O(2)	6(7)	C(3)-Re(1)-C(2)-O(2)	-85(7)
N(1)-Re(1)-C(2)-O(2)	126(6)	N(2)-Re(1)-C(2)-O(2)	-177(100)
C(2)-Re(1)-C(3)-O(3)	2(16)	C(1)-Re(1)-C(3)-O(3)	-80(16)
N(1)-Re(1)-C(3)-O(3)	178(100)	N(2)-Re(1)-C(3)-O(3)	101(16)

Table 12. (Continued)

C(8)-N(1)-C(4)-C(5)	-2.3(10)	Re(1)-N(1)-C(4)-C(5)	168.9(5)
C(8)-N(1)-C(4)-C(16)	178.9(7)	Re(1)-N(1)-C(4)-C(16)	-10.0(10)
N(1)-C(4)-C(5)-C(6)	1.8(11)	C(16)-C(4)-C(5)-C(6)	-179.3(7)
C(4)-C(5)-C(6)-C(7)	0.1(11)	C(5)-C(6)-C(7)-C(8)	-1.4(11)
C(5)-C(6)-C(7)-C(9)	-179.9(7)	C(4)-N(1)-C(8)-C(7)	1.0(10)
Re(1)-N(1)-C(8)-C(7)	-171.5(5)	C(4)-N(1)-C(8)-C(12)	-178.2(6)
Re(1)-N(1)-C(8)-C(12)	9.2(8)	C(6)-C(7)-C(8)-N(1)	0.9(10)
C(9)-C(7)-C(8)-N(1)	179.4(7)	C(6)-C(7)-C(8)-C(12)	-179.9(6)
C(9)-C(7)-C(8)-C(12)	-1.4(10)	C(6)-C(7)-C(9)-C(10)	-178.9(7)
C(8)-C(7)-C(9)-C(10)	2.7(11)	C(7)-C(9)-C(10)-C(11)	-1.0(12)
C(9)-C(10)-C(11)-C(12)	-1.9(11)	C(9)-C(10)-C(11)-C(13)	178.7(7)
C(15)-N(2)-C(12)-C(11)	-4.5(10)	Re(1)-N(2)-C(12)-C(11)	173.2(5)
C(15)-N(2)-C(12)-C(8)	173.7(6)	Re(1)-N(2)-C(12)-C(8)	-8.6(7)
C(13)-C(11)-C(12)-N(2)	0.7(10)	C(10)-C(11)-C(12)-N(2)	-178.8(6)
C(13)-C(11)-C(12)-C(8)	-177.4(7)	C(10)-C(11)-C(12)-C(8)	3.1(10)
N(1)-C(8)-C(12)-N(2)	-0.4(9)	C(7)-C(8)-C(12)-N(2)	-179.7(6)
N(1)-C(8)-C(12)-C(11)	177.8(6)	C(7)-C(8)-C(12)-C(11)	-1.5(10)
C(12)-C(11)-C(13)-C(14)	2.1(11)	C(10)-C(11)-C(13)-C(14)	-178.4(7)
C(11)-C(13)-C(14)-C(15)	-1.3(11)	C(12)-N(2)-C(15)-C(14)	5.4(10)
Re(1)-N(2)-C(15)-C(14)	171.9(5)	C(12)-N(2)-C(15)-C(17)	-175.4(6)
Re(1)-N(2)-C(15)-C(17)	7.4(9)	C(13)-C(14)-C(15)-N(2)	-2.6(11)
C(13)-C(14)-C(15)-C(17)	178.1(7)	C(20)-Re(2)-C(18)-O(18)	34(11)
C(19)-Re(2)-C(18)-O(18)	-55(11)	N(3)-Re(2)-C(18)-O(18)	-152(11)
N(4)-Re(2)-C(18)-O(18)	-166(9)	C(18)-Re(2)-C(19)-O(19)	71(18)
C(20)-Re(2)-C(19)-O(19)	-14(18)	N(3)-Re(2)-C(19)-O(19)	170(18)
N(4)-Re(2)-C(19)-O(19)	-114(18)	C(18)-Re(2)-C(20)-O(20)	14(14)
C(19)-Re(2)-C(20)-O(20)	103(14)	N(3)-Re(2)-C(20)-O(20)	-114(14)
N(4)-Re(2)-C(20)-O(20)	-164(14)	C(25)-N(3)-C(21)-C(22)	2.6(10)
Re(2)-N(3)-C(21)-C(22)	-172.5(6)	C(25)-N(3)-C(21)-C(33)	-176.8(7)
Re(2)-N(3)-C(21)-C(33)	8.0(10)	N(3)-C(21)-C(22)-C(23)	-1.1(13)
C(33)-C(21)-C(22)-C(23)	178.4(8)	C(21)-C(22)-C(23)-C(24)	-1.2(13)
C(22)-C(23)-C(24)-C(25)	1.8(12)	C(22)-C(23)-C(24)-C(26)	-176.3(8)
C(21)-N(3)-C(25)-C(29)	174.9(6)	Re(2)-N(3)-C(25)-C(29)	-9.2(8)
C(21)-N(3)-C(25)-C(24)	-2.0(10)	Re(2)-N(3)-C(25)-C(24)	173.8(6)
C(23)-C(24)-C(25)-N(3)	-0.2(11)	C(26)-C(24)-C(25)-N(3)	177.9(7)
C(23)-C(24)-C(25)-C(29)	-177.1(7)	C(26)-C(24)-C(25)-C(29)	1.1(11)
C(23)-C(24)-C(26)-C(27)	175.2(8)	C(25)-C(24)-C(26)-C(27)	-2.8(12)
C(24)-C(26)-C(27)-C(28)	1.6(13)	C(26)-C(27)-C(28)-C(30)	-178.9(8)
C(26)-C(27)-C(28)-C(29)	1.3(12)	C(32)-N(4)-C(29)-C(25)	-178.0(7)
Re(2)-N(4)-C(29)-C(25)	2.8(8)	C(32)-N(4)-C(29)-C(28)	1.6(11)
Re(2)-N(4)-C(29)-C(28)	-177.6(6)	N(3)-C(25)-C(29)-N(4)	4.3(10)
C(24)-C(25)-C(29)-N(4)	-178.6(7)	N(3)-C(25)-C(29)-C(28)	-175.3(6)
C(24)-C(25)-C(29)-C(28)	1.8(10)	C(30)-C(28)-C(29)-N(4)	-2.3(11)
C(27)-C(28)-C(29)-N(4)	177.4(7)	C(30)-C(28)-C(29)-C(25)	177.2(7)
C(27)-C(28)-C(29)-C(25)	-3.0(11)	C(27)-C(28)-C(30)-C(31)	-178.9(8)
C(29)-C(28)-C(30)-C(31)	0.9(13)	C(28)-N(30)-C(31)-C(32)	1.2(14)

Table 12. (Continued)

C(29)-N(4)-C(32)-C(31)	0.6(12)	Re(2)-N(4)-C(32)-C(31)	179.7(6)
C(29)-N(4)-C(32)-C(34)	179.7(8)	Re(2)-N(4)-C(32)-C(34)	-1.2(12)
C(30)-C(31)-C(32)-N(4)	-2.0(13)	C(30)-C(31)-C(32)-C(34)	178.9(9)
C(35)-Re(3)…Re(4)-C(38)	118.1(5)	C(36)-Re(3)…Re(4)-C(38)	-2.0(5)
C(37)-Re(3)…Re(4)-C(38)	-119.3(5)	O(42)-Re(3)…Re(4)-C(38)	-179.9(4)
O(43)-Re(3)…Re(4)-C(38)	57.8(4)	O(41)-Re(3)…Re(4)-C(38)	61.3(4)
C(35)-Re(3)…Re(4)-C(39)	-1.7(5)	C(36)-Re(3)…Re(4)-C(39)	-121.8(4)
C(37)-Re(3)…Re(4)-C(39)	120.9(5)	O(42)-Re(3)…Re(4)-C(39)	60.3(4)
O(43)-Re(3)…Re(4)-C(39)	-62.1(4)	O(41)-Re(3)…Re(4)-C(39)	178.9(4)
C(35)-Re(3)…Re(4)-C(40)	-122.9(5)	C(36)-Re(3)…Re(4)-C(40)	117.0(5)
C(37)-Re(3)…Re(4)-C(40)	-0.3(5)	O(42)-Re(3)…Re(4)-C(40)	-60.9(5)
O(43)-Re(3)…Re(4)-C(40)	176.7(4)	O(41)-Re(3)…Re(4)-C(40)	57.7(4)
C(35)-Re(3)…Re(4)-O(42)	-62.0(4)	C(36)-Re(3)…Re(4)-O(42)	178.0(4)
C(37)-Re(3)…Re(4)-O(42)	60.6(4)	O(43)-Re(3)…Re(4)-O(42)	-122.3(3)
O(41)-Re(3)…Re(4)-O(42)	118.6(3)	C(35)-Re(3)…Re(4)-O(41)	179.4(4)
C(36)-Re(3)…Re(4)-O(41)	59.3(4)	C(37)-Re(3)…Re(4)-O(41)	-58.0(4)
O(42)-Re(3)…Re(4)-O(41)	-118.6(3)	O(43)-Re(3)…Re(4)-O(41)	119.0(3)
C(35)-Re(3)…Re(4)-O(43)	60.3(4)	C(36)-Re(3)…Re(4)-O(43)	-59.7(4)
C(37)-Re(3)…Re(4)-O(43)	-177.1(4)	O(42)-Re(3)…Re(4)-O(43)	122.3(3)
O(41)-Re(3)…Re(4)-O(43)	-119.0(3)	C(35)-Re(3)-O(41)-C(41)	125.8(17)
C(36)-Re(3)-O(41)-C(41)	-7.1(7)	C(37)-Re(3)-O(41)-C(41)	-94.1(7)
O(42)-Re(3)-O(41)-C(41)	167.4(7)	O(43)-Re(3)-O(41)-C(41)	89.6(7)
Re(4)…Re(3)-O(41)-C(41)	128.7(7)	C(35)-Re(3)-O(41)-Re(4)	-2.9(18)
C(36)-Re(3)-O(41)-Re(4)	-135.8(3)	C(37)-Re(3)-O(41)-Re(4)	137.1(3)
O(42)-Re(3)-O(41)-Re(4)	38.7(2)	O(43)-Re(3)-O(41)-Re(4)	-39.06(19)
C(38)-Re(4)-O(41)-C(41)	5.8(7)	C(39)-Re(4)-O(41)-C(41)	-134.0(14)
C(40)-Re(4)-O(41)-C(41)	93.1(7)	O(42)-Re(4)-O(41)-C(41)	-168.4(6)
O(43)-Re(4)-O(41)-C(41)	-91.0(6)	Re(3)…Re(4)-O(41)-C(41)	-129.6(7)
C(38)-Re(4)-O(41)-Re(3)	135.4(3)	C(39)-Re(4)-O(41)-Re(3)	-4.4(15)
C(40)-Re(4)-O(41)-Re(3)	-137.3(3)	O(42)-Re(4)-O(41)-Re(3)	-38.8(2)
O(43)-Re(4)-O(41)-Re(3)	38.65(19)	C(38)-Re(4)-O(42)-C(46)	-155(2)
C(39)-Re(4)-O(42)-C(46)	70.8(9)	C(40)-Re(4)-O(42)-C(46)	-20.1(9)
O(41)-Re(4)-O(42)-C(46)	-115.8(9)	O(43)-Re(4)-O(42)-C(46)	167.9(9)
Re(3)…Re(4)-O(42)-C(46)	-155.2(10)	C(38)-Re(4)-O(42)-Re(3)	0(2)
C(39)-Re(4)-O(42)-Re(3)	-134.0(3)	C(40)-Re(4)-O(42)-Re(3)	135.1(3)
O(41)-Re(4)-O(42)-Re(3)	39.4(2)	O(43)-Re(4)-O(42)-Re(3)	-36.9(2)
C(35)-Re(3)-O(42)-C(46)	-72.4(9)	C(36)-Re(3)-O(42)-C(46)	145.1(16)
C(37)-Re(3)-O(42)-C(46)	18.6(9)	O(43)-Re(3)-O(42)-C(46)	-168.7(9)
O(41)-Re(3)-O(42)-C(46)	114.5(9)	Re(4)…Re(3)-O(42)-C(46)	154.0(10)
C(35)-Re(3)-O(42)-Re(4)	133.7(3)	C(36)-Re(3)-O(42)-Re(4)	-8.9(17)
C(37)-Re(3)-O(42)-Re(4)	-135.4(3)	O(43)-Re(3)-O(42)-Re(4)	37.3(2)
O(41)-Re(3)-O(42)-Re(4)	-39.4(2)	C(35)-Re(3)-O(43)-C(51)	11.5(7)
C(36)-Re(3)-O(43)-C(51)	-78.1(7)	C(37)-Re(3)-O(43)-C(51)	161.1(19)
O(42)-Re(3)-O(43)-C(51)	109.7(7)	O(41)-Re(3)-O(43)-C(51)	-174.6(7)
Re(4)…Re(3)-O(43)-C(51)	146.3(8)	C(35)-Re(3)-O(43)-Re(4)	-134.9(3)
C(36)-Re(3)-O(43)-Re(4)	135.6(3)	C(37)-Re(3)-O(43)-Re(4)	15(2)

Table 12. (Continued)

O(42)-Re(3)-O(43)-Re(4)	-36.7(2)	O(41)-Re(3)-O(43)-Re(4)	39.05(19)
C(38)-Re(4)-O(43)-C(51)	77.2(8)	C(39)-Re(4)-O(43)-C(51)	-12.4(8)
C(40)-Re(4)-O(43)-C(51)	-160.4(19)	O(42)-Re(4)-O(43)-C(51)	-108.4(7)
O(41)-Re(4)-O(43)-C(51)	175.6(7)	Re(3)...Re(4)-O(43)-C(51)	-145.2(8)
C(38)-Re(4)-O(43)-Re(3)	-137.6(3)	C(39)-Re(4)-O(43)-Re(3)	132.8(3)
C(40)-Re(4)-O(43)-Re(3)	-15(2)	O(42)-Re(4)-O(43)-Re(3)	36.9(2)
O(41)-Re(4)-O(43)-Re(3)	-39.16(19)	C(36)-Re(3)-C(35)-O(35)	99(30)
C(37)-Re(3)-C(35)-O(35)	-176(100)	O(42)-Re(3)-C(35)-O(35)	-75(30)
O(43)-Re(3)-C(35)-O(35)	0(30)	O(41)-Re(3)-C(35)-O(35)	-35(32)
Re(4)...Re(3)-C(35)-O(35)	-37(31)	C(35)-Re(3)-C(36)-O(36)	149(18)
C(37)-Re(3)-C(36)-O(36)	60(18)	O(42)-Re(3)-C(36)-O(36)	-68(18)
O(43)-Re(3)-C(36)-O(36)	-112(18)	O(41)-Re(3)-C(36)-O(36)	-39(18)
Re(4)...Re(3)-C(36)-O(36)	-75(18)	C(35)-Re(3)-C(37)-O(37)	-102(21)
C(36)-Re(3)-C(37)-O(37)	-14(21)	O(42)-Re(3)-C(37)-O(37)	157(21)
O(43)-Re(3)-C(37)-O(37)	108(20)	O(41)-Re(3)-C(37)-O(37)	84(21)
Re(4)...Re(3)-C(37)-O(37)	120(21)	C(39)-Re(4)-C(38)-O(38)	-128(23)
C(40)-Re(4)-C(38)-O(38)	-38(23)	O(42)-Re(4)-C(38)-O(38)	97(23)
O(41)-Re(4)-C(38)-O(38)	60(23)	O(43)-Re(4)-C(38)-O(38)	133(23)
Re(3)...Re(4)-C(38)-O(38)	98(23)	C(38)-Re(4)-C(39)-O(39)	37(35)
C(40)-Re(4)-C(39)-O(39)	-49(35)	O(42)-Re(4)-C(39)-O(39)	-150(35)
O(41)-Re(4)-C(39)-O(39)	177(100)	O(43)-Re(4)-C(39)-O(39)	136(35)
Re(3)...Re(4)-C(39)-O(39)	174(100)	C(38)-Re(4)-C(40)-O(40)	-60(23)
C(39)-Re(4)-C(40)-O(40)	29(23)	O(42)-Re(4)-C(40)-O(40)	127(23)
O(41)-Re(4)-C(40)-O(40)	-160(23)	O(43)-Re(4)-C(40)-O(40)	177(100)
Re(3)...Re(4)-C(40)-O(40)	164(23)	Re(3)-O(41)-C(41)-C(42)	79.9(9)
Re(4)-O(41)-C(41)-C(42)	-164.6(6)	O(41)-C(41)-C(42)-C(43)	-175.2(10)
C(41)-C(42)-C(43)-C(44)	-159.2(17)	C(42)-C(43)-C(44)-C(45)	63(3)
Re(4)-O(42)-C(46)-C(47A)	-136.3(10)	Re(3)-O(42)-C(46)-C(47A)	77.6(14)
Re(4)-O(42)-C(46)-C(47B)	-79.8(14)	Re(3)-O(42)-C(46)-C(47B)	134.0(12)
O(42)-C(46)-C(47A)-C(48)	61(2)	C(47B)-C(46)-C(47A)-C(48)	-48.8(17)
O(42)-C(46)-C(47B)-C(48)	-50(2)	C(47A)-C(46)-C(47B)-C(48)	43.3(15)
C(46)-C(47A)-C(48)-C(49)	-171.9(16)	C(46)-C(47A)-C(48)-C(47B)	46.9(16)
C(46)-C(47B)-C(48)-C(47A)	-44.6(15)	C(46)-C(47B)-C(48)-C(49)	-132(3)
C(47A)-C(48)-C(49)-C(50)	168(2)	C(47B)-C(48)-C(49)-C(50)	-127(3)
Re(3)...O(43)-C(51)-C(52)	157.4(8)	Re(4)...O(43)-C(51)-C(52)	-67.9(12)
O(43)-C(51)-C(52)-C(53)	-47.7(15)	C(51)-C(52)-C(53)-C(54)	-54.4(18)
C(52)-C(53)-C(54)-C(55)	177.5(14)		

Fluorescence Studies of 6-11

The chlorotricarbonyl complexes, **6-9** and the molecular rectangles, **10** and **11** exhibit fluorescence in DMSO and other solvents. A typical fluorescence spectrum of **11** in DMSO is shown in Figure 7. Table 13 summarizes the photophysical parameters of the choro complexes **6-9** (Orsa, et. al., 2008). The quantum yields of the molecular rectangles, **10** and **11**, are less than 0.01 (Orsa et. al., 2007).

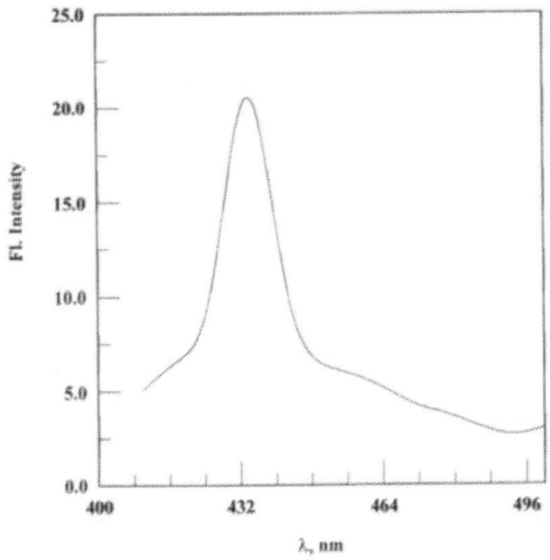

Figure 7. Fluorescence spectrum of 2.5 x 10^{-7} M DMSO solution of **11** (Courtesy: Elsevier).

Table 13. Photophysical parameters of 6-9 (Courtesy: Elsevier).

Compound	λ_{ex} (nm)	λ_{em} (nm)	ε (x 10^3/M-cm)	Φ	τ_o (ns)
6	319	576.47	2.897	0.261	21.0
7	319	577.19	4.375	0.603	32.0
8	319	577.98	3.259	0.483	38.0
9	319	578.31	5.036	0.459	29.0
Ref. Sample	319	630.21	-	0.56	-

*Cytotoxicity studies of **1-12** against prostate cancer cell lines (PC-3)*

The cytotoxicity studies in the concentration range of 0.5-4.0 μg/mL indicate that 1μg/mL of **1-12** are active against prostate cancer cell lines (see, Figure 8). Among the five perrhenato complexes four of them are very active. **6-11** are fluorescent and many of them are highly active against prostate cancer cell lines. Also, the molecular rectangle, **10** exhibits strongest cytotoxicity of all. Surprisingly, the molecular rectangle, **11** is not very active against PC-3 cell lines. Similar effects of **11** were also observed when LnCap cell lines were treated (Orsa, et. al., 2007). We included cisplatin in this study for comparison purposes. It is evident that **1,2,9,10** and **12** are better drugs than cisplatin. However, **9** and **10** are preferred drugs because they exhibit fluorescence. Strong activities of **9** and **10** may be due to their molecular structures possibly suggesting a DNA intercalation mechanism.

Chemotherapeutic agents have varying mechanisms of action with some agents utilizing multiple mechanisms and sites of action. Based on preliminary data, our experimental compounds exhibit anti-neoplastic activity against in vitro prostate cancer cell lines. While the exact mechanism(s) of action have not been determined, it is assumed that the flat topography of some of the compounds may work by way of DNA intercalation possibly resulting in single- and doubled stranded DNA cleavage. While this assumption signifies one possible mechanism of action, it does not preclude the plausibility of other mechanism(s) of action. The effect of the rhenium and/or other inorganic component(s) of our experimental compounds in eliciting anti-neoplastic activity have not been determined. The exact interaction between these compounds and DNA has not been established. The duration and stability of anti-neoplastic activity has not been analyzed. It should not be presumed without further study that any of our experimental compounds exploits known mechanisms in prostate cancer pathophysiology.

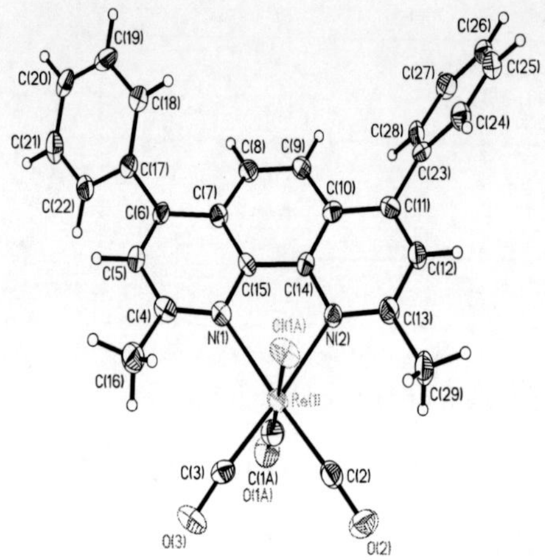

9 (Courtesy: Elsevier)

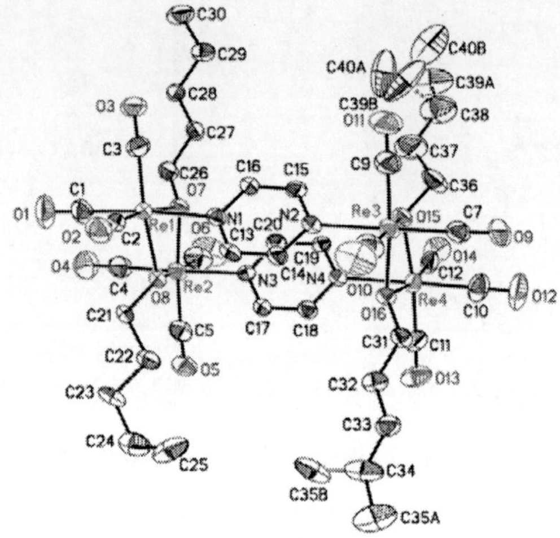

10

Detailed cytotoxicity studies using a wide range of concentrations of **1-12** and the mechanism of actions of **1-12** are in progress.

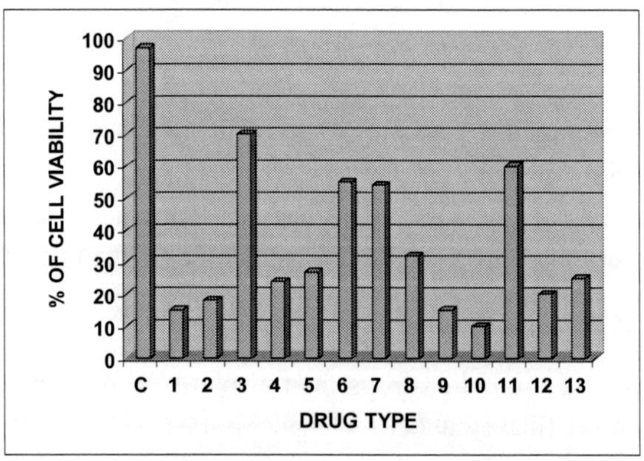

Figure 8. Plot of drug type vs. % of viable cells. C represents control – cells exposed to DMSO only. Prostate cancer cell lines (PC-3) were exposed to 1ug/ml of **1-13** for 72 hrs.

Conclusion

We have synthesized numerous manganese and rhenium complexes. Their synthetic methods are simple and convenient. We foresee that these complexes may be commercially available inexpensively. Many of these complexes exhibit powerful anticancer activity against prostate cancer cell lines. In the near future, we plan to do animal model studies with several of the complexes in the hope of establishing compounds that may facilitate strengthening of the therapeutic armamentarium against prostate cancer.

Acknowledgments

This research was supported in part by an appointment to the U.S. Nuclear Regulatory Commission HBCU Faculty Research Participation Program administered by the Oak Ridge Institute for Science and Education. Crystallographic data for **9** were collected through the SCrALS (Service Crystallography at Advanced Light Source) program at the Small-Crystal Crystallography Beamline 11.3.1 at the Advanced Light Source (ALS). The ALS is supported by the U.S. Department of Energy, Office of Energy Sciences Materials Sciences Division, under contract DE-AC02-05CH11231 at Lawrence Berkeley National Laboratory. Crystallographic data for **12** were collected on a Smart6000 diffractometer, funded through NSF-MRI grant CHE-0215950.

Experimental

All manipulations were carried out under a nitrogen atmosphere and solvents were dried prior to use. Reagent grade chemicals were used without further purification and were obtained from commercial sources. The prostate cancer cell lines were purchased from ATCC. The starting materials were synthesized through published procedures.

IR spectra were recorded on a Perkin-Elmer 1600 series FTIR instrument. NMR spectra were recorded on a Bruker 400 MHz spectrometer.

*Synthesis and characterizations of perrhenato complexes, **1-5***

The perrhenato complexes were synthesized from the treatment of the corresponding hydido complexes with Re_2O_7 according to published procedures (Brown, et. al., 2002; Kimari, et. al., 2006). The characterization of these complexes through IR and NMR techniques and X-ray crystallography was reported recently.

*Synthesis and characterizations of chlorotricarbonyl complexes, **6-9***

The chlorotricarbonyl complexes were synthesized from one-pot reaction of $Re_2(CO)_{10}$ and the corresponding diimines in refluxed 2-chloroethanol according to published procedures (Orsa, et. al., 2008). The complexes were characterized previously.

*Synthesis and characterizations of the bridging pentyloxo complexes (molecular rectangles) complexes, **10, 11***

The synthesis and characterization of the molecular rectangle, **11** was reported a few years ago (Orsa, et. al., 2007). The molecular rectangle, **10** was synthesized similarly from the reaction of an equimolar mixture of $Re_2(CO)_{10}$ and pyrazine in refluxed 1-pentanol for 24 hrs. After cooling to room temperature for several hours, purple-colored microcrystals of **10** were obtained by filtration. The filtrate was evaporated to dryness under reduced pressure. The residue and the micocrystals were combined together and recrystallized in a mixture of dichloromethane and hexane at -5° C. Purple-colored microcrystalline **10** were obtained after filtration. The yield of **10** was typically greater than 83%.

*Synthesis and characterization of the bridging hydrido salt, **12***

12 was synthesized from the reaction of one equivalent of $Re_2(CO)_{10}$ and two equivalents of 2,9-dimethyl-1,10-phenanthroline (abbreviated as, 2,9-Me_2-Phen) in refluxed 1-pentanol for 24 hours. On cooling to room temperature for several hours, brownish-red crystals of **12** were obtained. The crystals were separated by filtration. The filtrate was converted to the alkylcarbonato complex, *fac*-$(CO)_3$(2,9-Me_2-Phen)ReOC(O)OC_5H_{11} which will be reported elsewhere. The yield of **12** was typically less than 50%.

*Fluorescence studies of **6-11***

Steady-state fluorescence experiments were conducted on **6-11** using the Perkin Elmer Luminescence Spectrophotometer, model LS 50B in a quartz cuvet of 1.0 cm in radiation pathlength. DMSO was used as a solvent.

Cell culture and maintenance studies

Human bone metastasis of grade IV prostatic adenocarcinoma cell line, PC3 (CRL 1435) from Caucasian 62 years old was obtained from American Type Culture Collection (ATCC, VA). Cells were cultured in monolayer using RPMI-1640 (Invitrogen) medium with the supplements of 2.0 g/L glucose, 0.3 g/L L-glutamine, and 2.0 g/L $NaHCO_3$ plus 10% fetal bovine serum (FBS) and 100 U/ml penicillin-streptomycin sulfate. Incubation of the cell was carried out at 37° C with a continuous supply of 5% CO_2 gas and 95% air. For regular maintenance, the medium was renewed 2-3 times per week.

Sub-culturing of cell lines

For sub-culturing, the medium was removed at 70% confluences of the growth and then briefly rinsed with 1xPBS solution (Phosphate Buffer Saline, Invitrogen) to remove all traces of serum which contains trypsin inhibitors. To that, 1 ml of Trypsin (0.25%)-EDTA (0.53 mM) solution was added per 25 mm flask. Cells were observed under an inverted microscope until the monolayer was dispersed which usually takes 2-10 minutes (incubation at 37° C enhances disruption). To the flask, the complete RPMI- medium was added when the individual cell was completely detached without clumping. A sub-cultivation ratio 1:4 was followed for subsequent maintenance and the medium was renewed 2-3 times/week. This cell lines loose the adherent capacity after 29 passages.

Freezer stock

After trypsinization, the cells were centrifuged at 4k /4° C for 10 minutes and the pellet was collected. To that pellet, 90% FBS and 10% DMSO (Dimethy sulfoxide) was added and vortexes to get a uniform suspension of the cell. Cells were stored at ultra low freezer.

Cytotoxic studies and trypan blue staining

For cytotoxic studies, cells were grown in 12-well plates. When the cells growth reached to the 60% confluences, the medium was replaced with serum starved medium (2.0 g/L glucose, 0.3 g/L L-glutamine, and 2.0 g/L $NaHCO_3$ plus 0.1% FBS and 100 U/ml penicillin-streptomycin sulfate) and incubated for over night in order to get a synchronize culture. Following day, the old serum free- medium was replaced with a fresh medium containing 1 ug/ml of a rhenium-complex which was previously dissolved in DMSO. Cells were incubated at 37°C in 5% CO_2 environment for various time periods.

Trypan blue staining - cell viability

The medium was removed from each well and then cells were washed with PBS (Phosphate buffer saline) for two times. Cells were trypsinzied and centrifuged as described previously. The pellet was suspended in 200 ul Trypam Blue Stain (0.4%). A 10 ul suspension was spread out on a haemo-cytometer and then dead cells were counted under a microscope. The reactivity of Trypan Blue is based on negatively charged chromophore groups, which react on damaged membrane i.e., dead cells and viable cells will exclude the dye.

References

Balk, P., Ko,Y. J. & Bubley, G. J. (2003). Biology of Prostate-Specific Antigen. *J. Clinic. Oncol.*, *28 (2)*, 383–91.

Brown, D. A., Kimari, D. M., Dusz-Moore, A. M., Budzichowski, T. A., Ho, D. M. & Mandal, S. K. (2002). Reactions of dirhenium heptoxide with manganese(I) and rhenium(I) hydrido, alkoxo, methylcarbonato, carbonato-bridged, and methoxymethyl complexes. The X-ray structures of *fac*-(CO)3(dppp)$MnOReO_3$ and *fac*-$(CO)_3$(dppp) $ReOReO_3$. *J. Organomet. Chem.*, *658*, 88-93.

Callery, P. & Gannett, P. (2002). Cancer and Cancer Chemotherapy. *Foye's Principles of Medicinal Chemistry*. 5th ed. D. A. Williams, T. L. Lemke, (Eds.), Lippincott Williams and Wilkins: Baltimore, 924-51.

Guilhem, J., Pascard, C., Lehn, J. M. & Ziessel, R. (1989). Crystal and molecular structures of μ-Hydrido-bis[fac-(2,2'-Bipyridine)tricarbonylrhenium(I)]chloride, fac-(2,2'Bipyridine)-tricarbonylformatorhenium(I), and fac-(2,2'-Bipyridine)tricarbonyl(cyanotrihydroborato) -rhenium(I). *J. Chem. Soc. Dalton Trans.*, 1449-1454.

Hall, M. D., Dolman, R. C. & Hambley, T. W. (2004). Platinum(IV) Anticancer Complexes, *Metal Ions in Biological Systems* A. Sigel, & H. Sigel (Eds.), Marcel Dekker: New York, *vol. 42*, 297-317.

Johnson, K., Frazier, T., Becker, T. M., Miller, K., Ho, D. M., Krause-Bauer, J. & Mandal, S. K. (2001). One-pot synthesis of manganese(I) and rhenium(I) alkylcarbonato complexes, *fac*-$(CO)_3$(dppp)MOC(O)OR. Possible trapping of intermediate diphosphine dimers, $[(CO)_3(dppp)M]_2$. *Inorg. Chem. Commun.*, *4*, 602-605.

Kim, Y. & Verkade, J. G. (2003). Organic-organometallic crystal engineering: novel formation of a honeycomb supramolecular architecture of $(Re_2(\mu\text{-}OMe)_3(CO)_6]^-$ anions encapsulating a linear H-bonded chain of [DABCO-H]$^+$ cations. *Inorg. Chem., 42*, 4262-4264.

Kimari, D. M., Mbagu, M. K. Nettles, C. R.,Dusz-Moore, A. M., Budzichowski, T. A., Greco, G. E., Krause, J. A., Ho, D. M. & Mandal, S. K. (2006). Reactions of dirhenium heptoxide with organometallic compounds: syntheses, characterizations and structures of manganese(I) and rhenium(I) tricarbonyl O-rhenato complexes, *mer,trans*-$(CO)_3(PPh_3)_2MnOReO_3$ and *fac*-$(CO)_3$(L-L)$MOReO_3$, where, L-L are chelated diphosphines and diimines, *Frontiers in Organometallic Chemistry*. M. A. Cato (Ed.), Nova Science Publishers: New York, 191-228.

Orsa, D. K., Haynes, G. K., Pramanik, S. J., Iwunze, M. O., Greco, G. E., Ho, D. M., Krause, J. A., Hill, D. A., Williams, R. J. & Mandal, S. K. (2008). The one-pot synthesis and the fluorescence and cytotoxicity studies of chlorotricarbonyl(α-diimine)rhenium(I), *fac*-$(CO)_3$(α- diimine)ReCl, complexes. *Inorg. Chem. Commun., 11*, 1054-1056.

Orsa, D. K., Haynes, G. K., Pramanik, S. K., Iwunze, M. O., Greco, G. E., Krause, J. A., Ho, D. M., Williams, A.L., Hill, D.A., Mandal, S.K. (2007). Synthesis, characterization, and fluorescence and cytotoxicity studies of a tetrarhenium molecular rectangle. *Inorg. Chem. Commun., 10*, 821-824.

Pil, P. & Lippard, S. J. (1992). Cisplatin as an Anticancer Drug *Encyclopedia of Cancer*. J. R. Bertino (Ed.), Academic Press: *San Diego, vol. 1*, 392-410.

Rennie, J. (1996). *Fact Sheet* -Twelve Major Cancers, *Scientific American, 275*, 126-132

Roth, B. J. (1996). New therapeutic agents for hormone-refractory prostate cancer. *Semin Oncol., 23*, 49-55.

Stufkens, D. J. & Vlček, A. (1998). Ligand-dependent excited state behaviour of Re (I) and Ru (II) carbonyl–diimine complexes. *Coord. Chem. Rev., 177*, 127-179.

Wang, W., Yan, Y. K., Hor, T. S. A., Vittal, J. J., Wheaton, J. R. & Hall, I. H. (2002). Synthesis, X-ray structures, and cytotoxicity of rhenium(I) carbonyl2-(dimethylamino) ethoxide complexes. *Polyhedron, 21(20)*, 1991-1999.

Weiss, R. B. & Christian, M. C. (1993). New cisplatin analogues in development. A review. *Drugs, 46(3)*, 360-377.

Yan, Y. K., Cho, S. E., Shaffer, K. A., Rowell, J. E., Barnes, B. J. & Hall, I. H. (2000). Cytotoxicity of rhenium(I) alkoxo and hydroxo carbonyl complexes in murine and human tumor cells. *Pharmazie, 55(4)*, 307-313.

Zhang, J., Vittal, J. J., Henderson, W., Wheaton, J. R., Hall, I. H., Hor, T. S. A. & Yan, Y. K. (2002). Tricarbonylrhenium(I) complexes of phosphine-derivatized amines, amino acids, and a model peptide: structures, solution behavior, and cytotoxicity. *J. Organomet. Chem., 650*, 123-132.

In: Handbook of Prostate Cancer Cell Research...
Editor: Alan T. Meridith
ISBN: 978-1-60741-954-9

Chapter 10

The Role of COX-2 in Prostate Carcinoma: Therapeutical Implications

S. Díaz Prado[1,2], E. Grande Pulido [3], J. Cassinello Espinosa[4] and L.M. Antón Aparicio [1,5*]

[1]Medicine Department. Faculty of Healthy Sciences. University of A Coruña, Spain
[2]Research Unit. CHU A Coruña, Spain
[3]Medical Oncology Service. Ramón y Cajal Hospital. Madrid, Spain
[4]Medical Oncology Service. Guadalajara University Hospital. Guadalajara, Spain
[5]Medical Oncology Service. CHU A Coruña, Spain

Abstract

Eicosanoids have received considerable research attention as mediating agents in inflammatory and physiological processes, and there is some evidence for a role in neoplasia.

Eicosanoids′ well-recognised ability to modulate DNA/RNA synthesis, cell membrane activity, cell communication and immune response may mediate neoplastic processes including: tumor cell proliferation, tissue invasiveness, the spread of metastasis, immune surveillance and hormone responsiveness.

COX-2 induction is rapid but transient and there can be rapid induction in response to a variety of stimuli occurring over a similar time course to other immediate-early genes, like c-fos and c-myc.

Experimental studies have suggested that biosynthesis of arachidonic acid-derived eicosanoids such as prostaglandin E_2 via the cyclooxygenase pathway may play a significant role in supporting cell proliferation in human prostate cancer.

However, the exact role of COX-2 expression and eicosanoid production in prostate carcinogenesis and cancer progression remains unclear since precise mechanisms at the cellular and molecular level have not been established.

* Medical Oncology Service, CH Universitario A Coruña, Materno Infantil Hospital, C/ As Xubias, s/n, 15.006 A Coruña, Spain, E-mail: lantapa@canalejo.org.

Keywords: Prostate cancer, cyclooxygenase-2, prostaglandins, tumorigenesis, COX2 mRNA expression.

Introduction

COX (also referred as prostaglandin synthase or prostaglandin synthetase) is the rate-limiting enzyme for the production of prostaglandins (PGs) and thromboxanes from free archidonic acid [1-3] (Figure 1a). The enzyme is bifunctional, with fatty acid COX (producing PGG_2 from arachidonic acid) and PG hydroperoxidase activities (converting PGG_2 to PGH_2) (Figure 1b). Two forms of COX have now been described: a constitutively expressed enzyme COX-1 present in most cells and tissues, and an inducible isoenzyme COX-2 expressed in response to cytokines, growth factors and other stimuli. Despite the similar enzymatic activities of COX-1 and COX-2, the COX-1 and COX-2 genes have distinct properties, and differing expression patterns (Table 1).

Table 1. Properties of COX-1 and COX-2.

	COX-1	COX-2
Expression	Constitutive	Inducible
Localization	Endoplasmic reticulum Nuclear envelope	Endoplasmic reticulum Nuclear envelope
Expression pattern	Most tissues	Activated macrophages Stromal cells Inflammatory cells Malignant epithelial cells
Stimulate expression		Oncogenes Tumor promoters Growth factors Cytokines

The COX-1 enzyme is constitutively expressed at low levels in a majority of tissues and produces prostaglandins for normal physiological functions. By contrast, COX-2 is not normally present in most cells, but tight regulation allows it to be rapidly expressed in response to growth-related signals resulting in increased prostaglandin synthesis associated with inflammation and carcinogenesis.

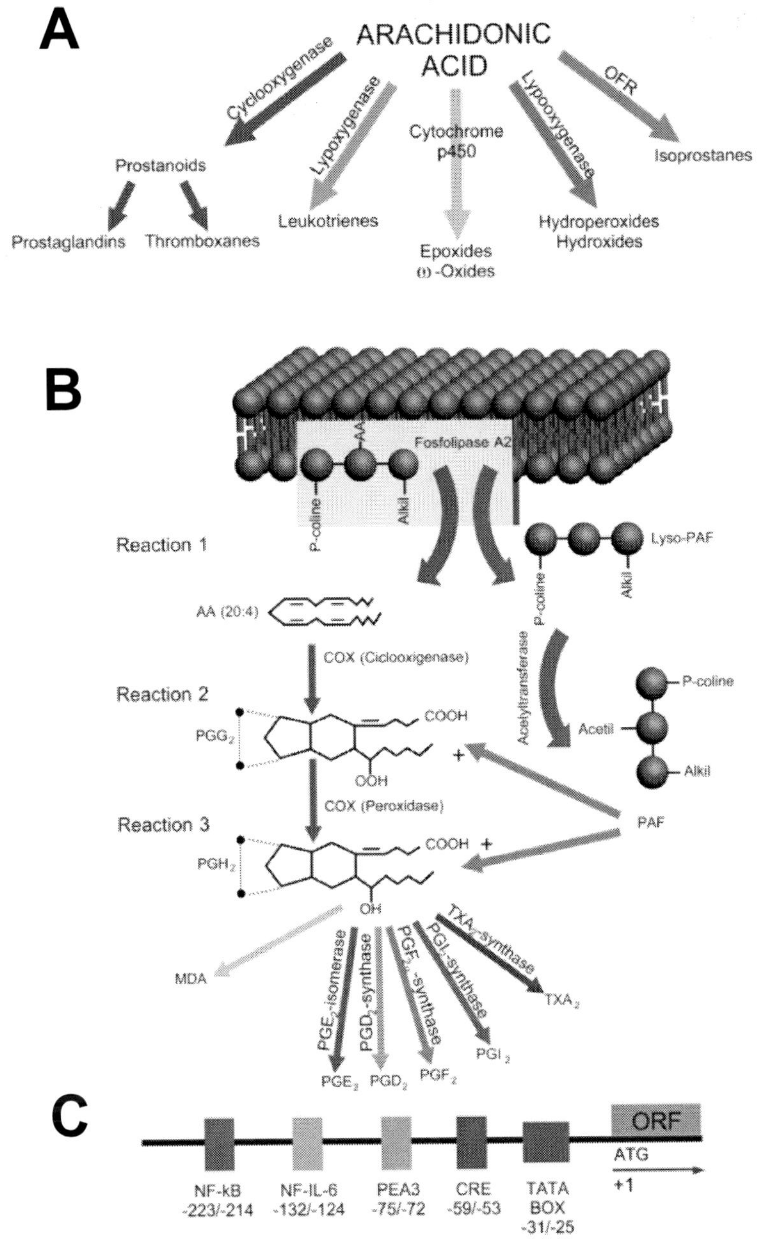

Figure 1. A. Cellular mechanisms of oxidation of arachidonic acid (AA) and their products. AA can be oxidized in the cells by different pathways giving rise different compound groups. Among them, the best well-known ones and with more important biological activities are prostanoids, cyclooxygenase products and Leukotrienes, a 5-lypooxigenase product. OFR: oxygen free radicals. B. Cyclooxygenase enzymatic pathway and their final products. Prostanoids synthesis requires three enzymes, phospolipase, which releases the AA of the membrane phospholipids, cyclooxygenase, which performs two activities: cyclooxygenation and peroxidation, and finally PG synthetases or isomerases that generate specific compounds. Phospholipids can be used as precursors for PAF (Platelet-activating factor) synthesis. **C.** Proximal region of human COX-2 promoter. Transcription start site (+1) is indicated with an arrow. TATA box, located in -31/-25, and well-characterized transcription factors binding sites are also shown.

Mechanisms of COX-2 Upregulation

There is evidence that COX-2 is upregulated in both neoplastic and stromal cells within tumors. Multiple mechanisms are likely to account for overexpression of COX-2 in these different cell types. COX-2 expression is normally regulated at both transcriptional and post-transcriptional levels, and can also be regulated by the rate of protein synthesis and /or degradation. The human COX-2 promoter contains multiple transcription factor binding sites, including a cAMP response element (CRE), and potential binding sites for Myb, nuclear factor interleukin-6 (NF-IL6), nuclear factor κB (NF-κB), and Ets factors. Of these, the sites proximal to the transcription start site have been shown to be differentially responsive to various stimuli (Figure 1c).

COX-2 can Contribute to Cancer

Since COX-2 is a PG synthase, the most obvious consequence of COX-2 overexpression is the increase of PG production, and indeed high PG levels have been detected in many cancers. Enhanced PG synthesis may contribute to carcinogenesis in several ways, including direct stimulation of cell growth. $PGE_{2\alpha}$ and $PGF_{2\alpha}$ can both stimulate mitogenesis in synergy with epidermal growth factor (EGF) [4]. Thus, inappropriate stimulation of cellular proliferation by PGs may contribute to tumorigenesis. However, PGs do not act as mitogens for all cell types, and in fact, they may restrict proliferation of some cells, particularly those of the immune system [5].

The potential contributions of PG overproduction to tumorigenesis, including increased cellular proliferation and diminished immune surveillance, have been discussed. In particular, COX-2 overexpression may result in increased production of mutagens: malondialdehyde (MDA) can be produced by isomerisation of PGH_2 both enzymatically and non-enzymatically. Additional carcinogens can be formed by oxidation of aromatic amines, heterocyclin amines, and dihydrodiol derivatives of polycyclic hydrocarbons [6]. Thus, COX-2 overexpression may lead to DNA damage, thereby contributing to carcinogenesis.

Overexpressing COX-2 shows several altered characteristics, including increased adhesion to extracellular matrix, and a delayed transit through the G1 phase of the cell cycle [7, 8].

Much interest has focused on the ability of COX-2 to suppress apoptosis. Diminished apoptosis is thought to favour carcinogenesis by permitting survival of cells that have acquired mutations, and thus is viewed as one of the central mechanisms of tumorigenesis.

Different hypotheses have been postulated to account for suppression of apoptosis in response to COX-2 overexpression: the ability of PGE_2 to inhibit apoptosis caused by a selective COX-2 inhibitor and concomitantly to induce Bcl-2, suggest that PG-mediated upregulation of Bcl-2 may suppress apoptosis [9]. Alternatively, since arachidonic acid stimulates apoptosis, enhanced COX-2 expression could inhibit apoptosis by increasing the conversion of arachidonic acid to PG [10, 11].

On the other hand, it has become apparent that cyclooxygenases are involved in angiogenesis [12]: this is a crucial facet of tumorigenesis since neovascularisation is required for tumors to grow. COX-2 apparently contributes to the production of pro-angiogenic factors, including vascular endothelial growth factor (VEGF), basic fibroblast growth factor (bFGF), transforming growth factor-1, (TGF), platelet-derived growth factor (PDGF), and endothelin-1.

Conclusive evidence linking cyclooxygenases to tumorigenesis was first provided by studies using mice with targeted disruptions of the COX-1 or COX-2 genes [13, 14]. Additionally, COX-2 is responsive to several oncogenes, including v-src, v-Ha-ras, HER-2/neu and Wnt genes [15-18]. COX-2 has been shown to be a downstream target of the Wnt signalling pathway, and high expression of COX-2 has been proposed to be correlated to poor survival in malignancy.

Regulation of COX-2 Expression by Signal Transduction Pathways

The regulation of COX-2 expression is a complex process involving multiple signal transduction pathways. This level of complexity underscores the requirement for tight regulation of the enzymatic action of COX-2, which has pathogenic effects if its expression is left unchecked. As a result of genetic mutations commonly detected in a majority of human tumors, cellular defects in these signalling pathways promote the expression of COX-2 through the loss of transcriptional regulation, rapid mRNA decay, and translational inhibition (Figure 2).

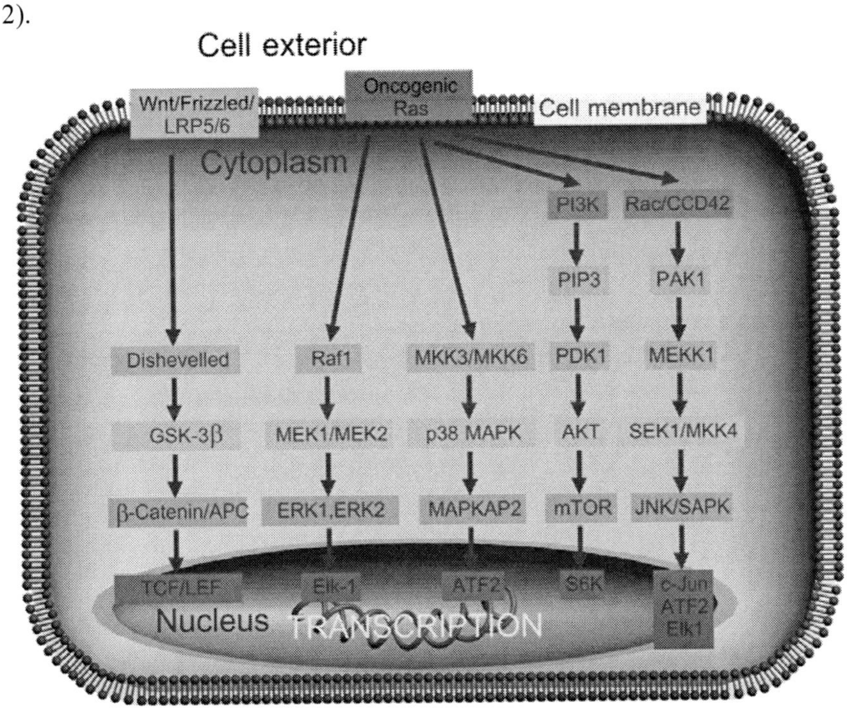

Figure 2. Signaling pathways regulating COX-2 expression. Schematic diagram outlining the major signal transduction pathways participating in regulation of COX-2 gene expression.

These signalling pathways and their involvement in regulating COX-2 expression in human cancer are discussed below:

1. Wnt/APC

Wnt-1 is a mammary oncogen that codes for a secreted signalling factor and its expression stimulates mammary carcinoma formation. A later modulator of the Wnt signalling is the APC tumor suppressor gene. As a consequence of the altered Wnt expression or APC mutations, citoplasmic stabilization and a nuclear accumulation of the β-catenin transcriptional activator occurs. In murine mammary epithelial cells transformed with Wnt-1, elevated COX-2 concentrations were detected due to the transcriptional activation. In human cancer cells it was found an association between the increased concentrations of nuclear β-catenin and COX-2 expression, which is related to APC defects, indicating that β-catenin may have some function in COX-2 transcription stimulation. Curiously, it seems that APC function in COX-2 regulation occurs in several phases.

COX-2 appears to be a common downstream target of Wnt expression and APC mutations in transformed murine mammary epithelial cells [17]. Association between increased nuclear β-catenin levels and COX-2 expression is seen in human cancer cells and is related to defects in APC [19, 20], suggesting a possible role for β-catenin in stimulating COX-2 transcription [21].

2. Ras

Ras proteins are membrane molecules regulated by GDP/GTP that transfer signals from cellular membrane to the nucleus. It is well known that Ras mutations exist in a great variety of human neoplasias. Ras oncogenic mutations cause a GTPase constitutive activation that lead to the activation of multiple signals of the MAPK translate cascade.

In the regulation of COX-2 expression, both oncogenic K-ras and H-ras stimulate COX-2 transcription through activation of the Rho, JNK, ERK, and p38 MAPK pathways , whereas COX-2 mRNA stability is modulated through activation of signalling pathway involving ERK, p38 MAPK, and Akt/PKB [22-27]. Oncogenic Ras can also modulate COX-2 translation because it has been observed that stimulates phosphorilation of eIF-4e translational initiation factor and an increase in protein synthesis.

Ras signalling has also been shown to participate with growth factors to enhance the expression of COX-2. There is increasing evidence that TGF-β can enhance malignant transformation and tumor progression of several different epithelial tumors. Ras mutations are found to occur in a wide variety of human malignancies and H-ras is frequently mutated in cancers of the urinary tract and bladder [28]. Oncogenic cellular transformation of cells by Ras is dependent of the Rho family of proteins [31]. Rho-dependent activation of COX-2 transcription has been demonstrated through cis-acting elements present in the proximal region of the COX-2 promoter [23, 29, 30].

3. JNK

Stimulation of the Ras/Rac1/MEKK1/JNK signal transduction pathway activates the transcription of COX-2. Ras activation of the JNK kinase cascade leads to phosphorylation of c-Jun, resulting in transcriptional activation of COX-2 via the CRE element [32, 33]. Increased COX-2 promoter activity, owing in part to JNK activation, has been observed in cancer cells [23, 24]. Some promoters of tumors, such as biliary acids and ceramide, also activate JNK pathway by causing COX-2 transcription activation, while COX-2 mRNA stability seems not to be regulated by this pathway.

4. ERK

Activation of the Ras/Raf1/MEK/ERK signal transduction pathway was originally shown to modulate COX-2 transcriptional regulation in fibroblasts [32]. This MAPK pathway is equally important in constitutive activation of COX-2 expression in cancer. An ERK pathway constitutive activation in colorectal cancer cell lines with high concentrations of COX-2 has been observed and that inhibition of this pathway decreases COX-2 promoter activity and mRNA stability. In the same way, the ERK pathway is essential for COX-2 promoter induction mediated by Ras and for mRNA stabilization in epithelial cells transformed with Ras. Also, in HER2/neu-transformed mammary epithelial cells, HER2/neu induced COX-2 promoter activity via ERK activation [24].

5. p38 MAPK

Different studies have demonstrated the ability of proinflammatory signals to induce COX-2 expression in a variety of cell types through p38 MAPK activation [27, 34, 35]. Activation of the p38 MAPK pathway is primarily associated with cellular stress and pro-inflammatory stimuli. Considerable evidence suggests that p38 MAPK promotes stabilization of COX-2 and other ARE-containing mRNAs (ARE, AU rich element) as well as COX-2 transcription.

Curiously, it has been observed that the capacity of dexametasona, an anti-inflammatory glucocorticoid, to inhibit COX-2 expression promotes a rapid COX-2 mRNA destabilization by p38 MAPK pathway inhibition. The capacity of p38 MAPK to contribute to COX-2 mRNA stabilization was detected in intestinal and mammary carcinoma cells. This is of great importance as it has been found that p38 MAPK is activated in neoplasic tissue and that also stimulates COX-2 transcription.

6. Akt/PKB

The Akt/PKB pathway predominantly regulates COX-2 expression by modulating the stability of COX-2 mRNA [23]. On the other hand, as oncogenic Ras-mediated post-transcriptional stabilization of COX-2 mRNA occurs through activation of the Akt/PKB pathway [26], COX-2 may represent a downstream mediator of this oncogenic pathway.

The kinase Akt, or protein kinase B (Akt/PKB), is a downstream effector of phosphophatidylinositol 3-kinase (PI3-K). Activation of this pathway is required for Ras-mediated transformation of a variety of cell types, including intestinal epithelial cells, and Akt/PKB activation facilitates a number of cellular events associated with cellular transformation [36, 37].

7. PPARs

Peroxisome proliferator-activated receptors (PPARs) are a family of ligand-activated transcription factors that are members of the nuclear-hormone-receptor superfamily.

PPARδ belongs to the nuclear receptor superfamily, which includes the steroid hormone, thyroid hormone, retinoid, and PPAR subfamilies which includes (PPARα, PPARδ, and PPARγ) as well as a growing number of orphan receptors [38, 39].

Several studies have suggested that APC´s association with β-catenin [40, 41] might be critical to its tumor-suppressive effects [42, 43]. PPARδ has been identified as a target of APC. PPARδ expression was elevated in human cancer and repressed by APC. This repression was mediated by β-catenin/Tcf-4-responsive elements in the PPARδ promoter.

COX-2-Associated Tumorigenesis and Cancer Pathogenesis

Epidemiological studies provided the first evidence that COX may be involved in the pathogenesis of cancer. In the process of carcinogenesis and in the route of intracellular signalling, COX-2 expression may be a universal phenomenon during carcinogenesis [44]. COX-2 is overexpressed along the continuum of oncogenesis and is likely to be a key player in a number of biologic pathways leading to cancer. Several lines of evidence gathered from epidemiological, experimental models, and cellular studies indicate that unregulated COX-2 expression is an important step in tumorigenesis and indicate that deregulation occurs early in carcinogenesis. Although a number of epidemiological studies have reported associations between prostate cancer and diet, particularly dietary fat, the role of dietary risk factors remains unclear, and no plausible underlying mechanisms have been established. The pattern of eicosanoid biosynthesis is modified by the availability from diet of n-6 and n-3 polyunsaturated fatty acid substrates which compete for the COX pathway.

In general, COX-2 is up-regulated throughout the tumorigenic process, from early hyperplasia to metastatic disease (Figure 3). Elevated levels of COX-2 immunoreactivity were primarily detected in the neoplastic epithelium, inflammatory cells, and vasculature within and adjacent to tumor nests. COX-2-derived metabolites from infiltrating inflammatory cells undoubtedly contribute to the tumorigenic process as well. Excessive local synthesis of prostaglandins has also been shown to disrupt immune surveillance that may otherwise suppress tumor growth. Furthermore, the direct product of COX-2, PGH_2, can be isomerized by both enzymatic and non-enzymatic reactions to form the potent mutagen malondialdehyde, which can induce frame shifts and base pair substitutions [45]. Additional free radical damage may occur via the peroxidative activity of COX-2 which can efficiently oxidize aromatic and heterocyclic amines and dihydrodiol derivatives [46].

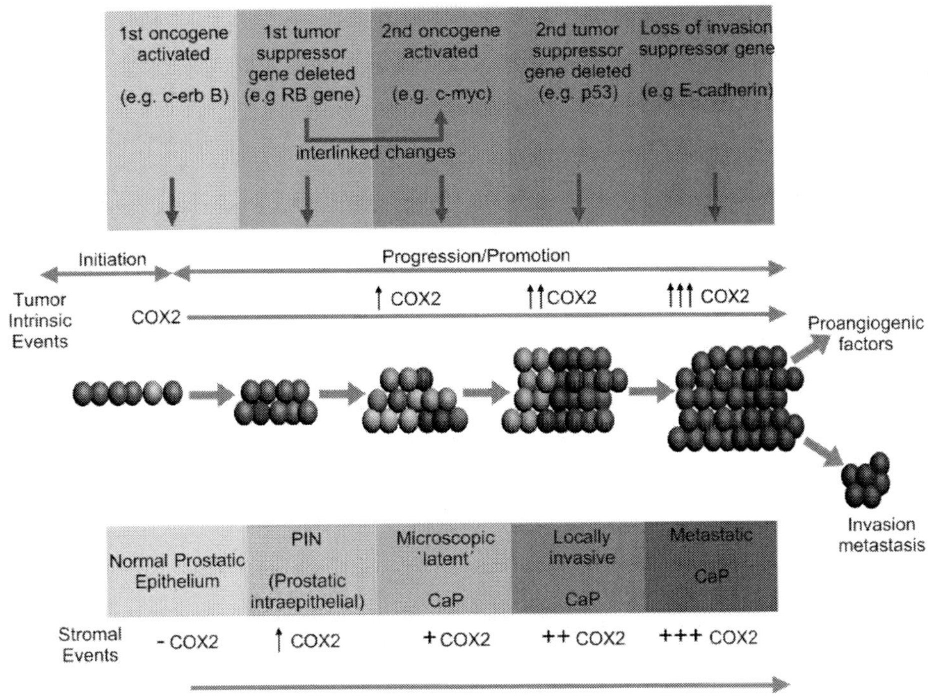

Figure 3. The development of prostate cancer is a multistep process involving the activation of oncogenes, such as *ras* and *c-erb B-2*, and the loss of tumor suppressor genes (TSG), such as p53 and the retinoblastoma (Rb) TSG. Metastatic capacity is then bestowed by the deletion of the cell adhesion molecule E-cadherin and the release of angiogenesis factors (VEGF, iNOS, IL-6, IL-8, TGFα, TGFβ, and FGF-2). In general COX-2 is up-regulated through the tumorigenic process from early hyperplasia to metastatic disease.

Current evidence indicates that COX-2 may promote tumor-specific angiogenesis [47-50], inhibits apoptosis [51-54], and induces proangiogenic factors such as VEGF [55-56], inducible nitrogen oxide synthetase promoter (iNOS) [57] and IL-6 [58].

COX-2 has been consistently overexpressed in premalignant lesions such as prostatic intraepithelial neoplasia, and carcinoma. The evolution of the proliferative inflammatory atrophy of the prostate to prostate carcinoma has been described [59], and as COX-2 has been found to be involved in the induction of inflammation and apoptosis, COX-2 may play an important role in the evolution of carcinogenesis in the prostate.

Finally both autocrine and paracrine mechanisms may be responsible for the development and/or progression androgen-independent prostate cancer either directly by stimulation of tumor cell proliferation or indirectly by enhanced prostaglandin-regulated local androgen synthesis.

COX-2 and Cell Growth

An expanding body of evidence indicates that the prostaglandin products of the COX-2 (or COX-1) pathway may enhance cell proliferation and growth in both normal and tumor cells [60, 61].

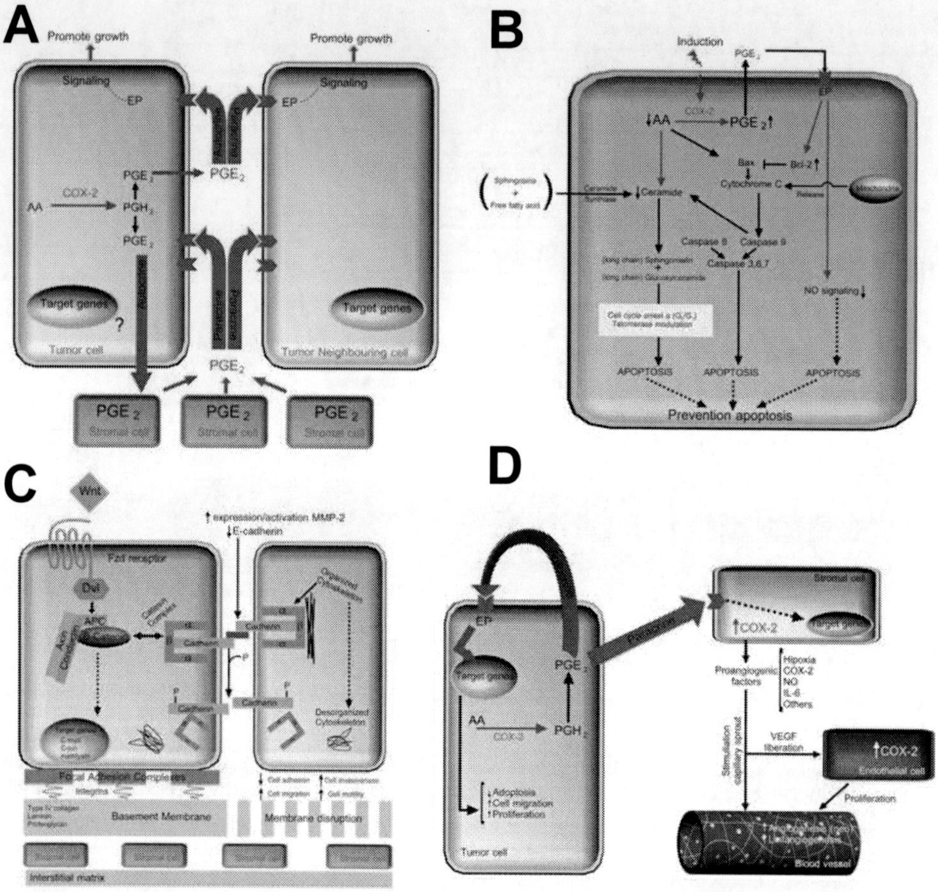

Figure 4. A: Cell growth promotion by PGE_2 implicated tumor itself and surrounding stromal cell. After PGE_2 is secreted into the extracellular space it binds either to membrane ephrin receptor (EP) on the same epithelial cell (autocrine) or a neighboring cell or the same or different type (paracrine). B: An increased level of PGE_2 activates the ephrin receptor (EP) leading to elevated expression of Bcl-2 and attenuation of nitric oxide (NO) signalling. Also reduction of the intracellular arachidonic acid (AA) level can lower the ceramide concentration and increase Bcl-2 expression. The consequence is prevention of apoptosis. C: Regulation of cadherin-mediated cellular adhesion. The catenin complex binds to the intracellular portion of the cadherin molecules and to the cytoskeleton, allowing cadherin molecules to interact intracellularly. Upon phosphorylation, through v-src, the cadherin-catenin interaction is dissociated and intercellular adhesion is abrogated. **D:** COX-2 promotes tumor-specific angiogenesis through induction of proangiogenic factors such as VEGF and iNOS. COX-2 has been localized to epithelial stromal and endothelial cells. In some models, COX-2 is primarily found in the stromal cells within a tumor.

COX-2 is upregulated in cancers. The excessive production of prostaglandins in cancerous tissues, mainly PGE_2, may enhance tumor cell growth. Several studies suggest that the COX-2 pathway may be involved in the signalling of other mitogens. It has been demonstrated that PGE_2 release may be a critical event in mediating the growth-promoting effect of growth factors and oncogenes.

Prostaglandins exert their effects locally in both autocrine and paracrine patterns. PGE_2 effects are mediated by a family of G-protein-coupled receptors, namely, EP_1, EP_2, EP_3, and EP_4 (Figure 4). In some cell types, PPAR nuclear receptors are also involved in mediating the prostaglandin effects.

COX-2 and apoptosis

Another important action of COX-2 on cellular processes is inhibition of apoptosis, which in many cases constitutes another mechanism to promote tumor cell growth. It is documented that the COX product PGE_2 contributes to the preventive action of COX-2 in apoptosis.

The preventive role of COX-2 in apoptosis is attributable to two possible mechanisms: it could be mediated by generation of prostaglandin products or mediated by removal of the substrate, arachidomic acid, via COX-2 catalytic activity [62]. Today at least three pathways have been shown to be involved: the Bcl-2 mediated pathway, the nitric oxide pathway, and production of ceramide [63] (Figure 4).

COX-2 and Cell Motility and Adhesion

COX inhibitors have been found to reduce tumor cell migration, cell adhesion, and tumor invasiveness in *in vivo* and *in vitro* experimental systems [64]. The biochemical changes associated with these altered cellular dynamics include increased expression and activation of metalloproteinase-2 and reduced expression of E-cadherin [65].

Invasiveness promoted by COX-2 is mediated by the cell surface receptor for hyaluronate, CD44. Overexpression of COX-2 caused cell invasiveness and also increased CD44 expression [66].

Also the phosphatidylinositol-3-kinase/Akt kinase/protein kinase B pathway was recently found to mediate PGE_2-promoted cell motility [67].

COX-2 and Angiogenesis

In addition to a causal role of COX-2 in tumorigenesis, compelling evidence also indicates that it may play an important role in tumor angiogenesis, a process involved in later stages of tumor development and migration. The evidence comes from data: COX-2 significantly increased angiogenesis [68], and this action involved activation of MAPK [68, 69]; COX-2 expression significantly induced the production of vascular endothelial growth factor (VEGF), a proangiogenic growth factor; and genetic studies with double deficient mice ($COX\text{-}2^{-/-}$ $APC^{\Delta 716}$ and EP $receptor^{-/-}$ $APC^{\Delta 716}$) strongly suggest that polyposis stromal expression of COX-2 plays a key role in promoting tumor angiogenesis [70].

COX-2 and Prostate Carcinoma

The molecular pathogenesis of prostate cancer has been characterized by alterations of genes and proteins involved in proinflammatory pathways. Some studies have postulated that proliferative inflammatory atrophy (PIA) and prostatic intraepithelial neoplasia (PIN) not only share some molecular traits but also they promote further cancer growth. In this way, there exists a strong correlation between the levels of arachidonic acid (AA) metabolites and accumulation of proinflammatory prostaglandins in prostate carcinogenesis [71]. The blockade of the eicosanoid production from arachidonic acid cascade metabolism by using specific inhibitors of lipoxygenases and cyclooxygenases result in an arrest of the growth and massive apoptotic death of prostate cancer cells *in vitro* and *in vivo* as well as in a decrease in tumor angiogenesis.

In a variety of malignancies COX-2 has been reported to be constitutively overexpressed. Particularly its expression is frequently constitutively elevated in prostate carcinoma. Numerous studies have evaluated the expression of COX-2 in human prostate cancer with conflicting results. Whereas some investigators have postulated that COX-2 is commonly overexpressed in prostate cancer [72-5], others have found that levels of COX-2 are very low or absent in the majority of prostate cancers [76, 77].

The immunohistochemical analysis of human prostate carcinoma reveals that COX-2 expression is multi-focal and moderate to strong in the majority of cases. Staining for COX-2 is mild in benign tissue as compared to prostate carcinoma tissue. Expression of COX-2 in prostate carcinoma tissue is predominantly granular and mostly localised in tumor cells. Interestingly, COX-2 expression in prostate carcinoma tissues is greater in the surrounding stromal cells or infiltrating inflammatory cells than in tumor cells. This observation raises interesting possibilities about the localisation of COX-2 expressing cells in prostate carcinoma. Therefore, stromally-derived COX-2 may promote tumor growth by producing bioactive prostaglandins which affect carcinoma cells in a paracrine fashion. COX-2 may thus be acting as a "landscaping tumor promoter" according to the landscaping model [78]. There are conflicting data regarding whether COX-2 is increased in the epithelial or in the stromal component of tumors. COX-2 has been found to be increased in the epithelium of adenomas (i.e colonic epithelium). It has been found that COX-2 expression is increased in the stromal component of adenomas [79, 80].

The regulator of human prostate cancer, the 12-lipoxygenase (12-LOX) [81], is expressed in prostate cancer cell lines [82] and its expression was associated with poor differentiation and invasiveness of prostate cancer [83]. Also, its overexpression was correlated with advanced stage, poor differentiation and invasive potential. More importantly, elevation in the expression of this marker in prostate cancer patients may predict the disease aggressiveness in this subset of patients [84]. Platelet-type 12-LOX has been shown to regulate growth, metastasis, and angiogenesis of prostate cancer: the 12-LOX pathway is a critical regulator of prostate cancer progression and apoptosis, by affecting various proteins regulating these processes.

The expression of COX-1 is found in prostate tissues as well as in prostate carcinoma samples [85]. The expression of COX-2 was modulated by the degree of adenocarcinoma differentiation (Figure 5): in the low and middle groups of adenocarcinoma the expression was strong and a very strong expression level in the high group of adenocarcinomas. The expression of COX-2 also was high in the blood vessels and stromal tissues of prostate carcinoma samples. However, the expressions of COX-2 in the blood vessels and stromal tissues from benign prostatic hypertrophy, PIN, and normal prostate samples were at basic levels [85].

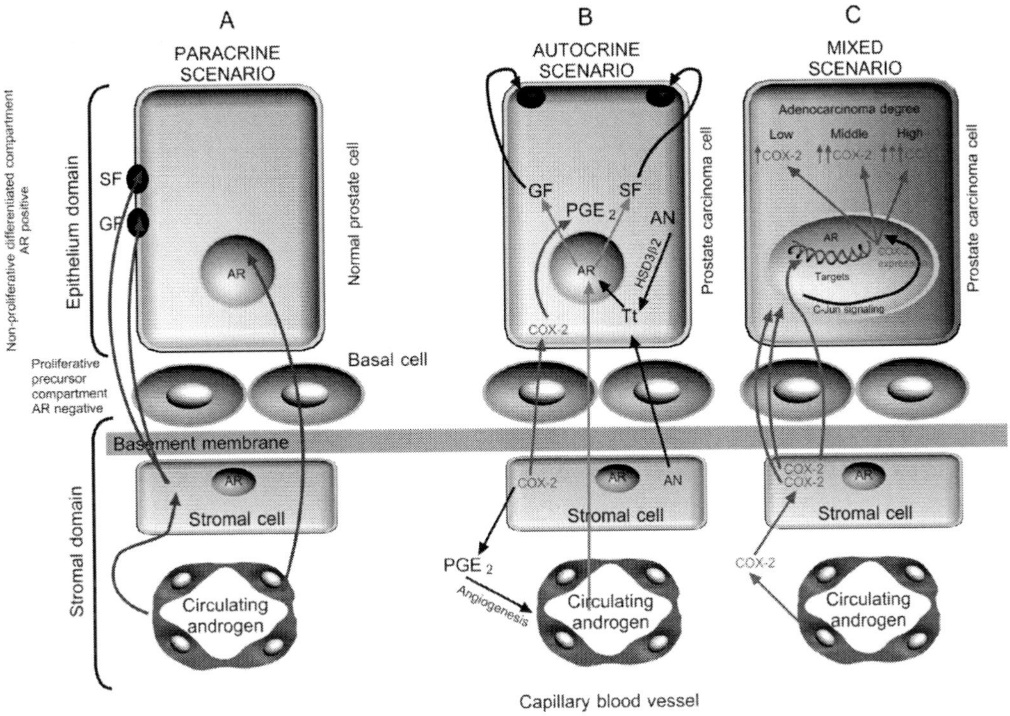

Figure 5. A. In normal prostate gland, growth and maintenance of prostate epithelium depends on growth factors (GF) and survival factors (SF) produced by stromal cells. GF and SF are produced as a result of androgen signalling through androgen receptor (AR). B. During transformation to cancer, the paracrine mechanism of AR action is replaced by an emergent autocrine mechanism, whereby cancer cells are less dependent on stromal cell factors. Hypothetical COX-2 (cyclooxygenase 2) derived PGE_2 (prostaglandin 2) potentiates androgen biosynthesis by the HSD3β2 (3β-hydroxysteroid dehydrogenase type 2) enzyme and promotes angiogenesis, cell proliferation survival and metastasis. AN (androstenedione), Tt (Testosterone). C. Both autocrine and paracrine mechanisms may be responsible for the development and/or progression androgen-independent prostate cancer by enhanced prostaglandin-regulated local androgen synthesis. COX-2 is a potent stimulator of androgen receptor (AR) function and the induction of c-Jun expression/phosphorylation plays an important role. C-Jun exhibits an inhibitory effect on AR protein activity; c-Jun can inhibit the transcription activities of the AR promoter.

Therapies with Cyclooxygenase Inhibitors

The molecular mechanisms by which non-steroidal anti-inflammatory drugs (NSAIDs) exhibit anticancer activities are not completely understood and it is plausible that COX-2 inhibitors-mediated apoptosis works through both COX-2 activity-dependent and independent pathways [71]. At least three distinct mechanisms have been proposed: first, NSAID treatment of cells resulted in an increase in arachidonic acid that, in turn, stimulated the production of ceramide, a known death signal, by activating neutral sphingomyelinase [86]; second, the COX-2 inhibitors have been reported to sensitize prostate cancer cells to apoptosis by down-regulating the anti-apoptotic protein Bcl-2 [87, 88]; third, a study shows the involvement of a COX-2-independent pathway during the apoptosis signalling [89].

Celecoxib induces apoptosis by blocking the activation of the antiapoptotic kinase Akt (also called protein kinase B). Among the selective COX-2 inhibitors, celecoxib and rifecoxib are also undergoing preclinical evaluation in human for the treatment of prostate cancer. Interestingly, since hormonal deprivation therapies seem to be accompanied by selective expression of COX-2, the inhibitors of this enzyme represent then potent adjuvant agents for a more effective treatment of recurrent prostate cancer. However there exist molecular targets, other than COX-2, in celecoxib-mediated *in vitro* antiproliferative effects in prostate [89-92]; the 3-phosphoinositide-dependent protein kinase-1 (PDK-1)/Akt signalling pathway represents a major COX-2-independent mechanism by which celecoxib exert *in vivo* antiproliferative effects against prostate tumors [93].

PDK-1/Akt signalling plays a central role in regulating cancer cell survival and proliferation by affecting the phosphorylation status of a plethora of downstream effectors that are either PDK-1 or Akt substrates [94]. Consequently, blockade of this signalling cascade leads to growth inhibition by cell cycle arrest and/or apoptosis in malignant cells.

Previous studies strongly postulated that NSAIDs are potent inhibitors of androgen receptor (AR) function and that the induction of c-Jun expression/phosphorylation [95] plays an important role in mediating their effects.

Clinical Prospects and Future Therapeutic Directions

Epidemiological studies have demonstrated that cyclooxygenase inhibition might be an effective approach to preventing cancer. This fact has stimulated clinical trials to investigate the efficacy of selective COX-2 inhibitors, as an attractive therapeutic or chemopreventive target, in individuals at risk for human cancer. Selective COX-2 inhibitors decrease the risks of secondary effects associated with the inhibition of COX-1 and they act preferably on metastatic prostate cancer cells that highly express this enzyme compared to normal and low-grade prostate cancer cells.

There are a lot of studies that support the use of selective COX-2 inhibitors in prostate cancer disease since they induce apoptosis of human prostate cancer cells *in vitro* and *in vivo*. Inhibition of cyclooxygenase by traditional NSAIDs and others selective inhibitors of the pro-inflammatory COX-2 isoform may also prove to have chemopreventive efficacy in prostate.

Treatment with COX-2 inhibitors has been shown to reduce the size and number of polyps in FAP (familial adenomatous polyposis) patients, and is currently being evaluated for efficacy in preventing sporadic colorectal adenomas [96, 97]. Moreover, aberrant COX-2 expression is also associated with breast cancer: selective COX-2 inhibitors have proved effective in preventing experimental breast cancer [98].

One study suggests that men who regularly took NSAIDs may have a reduced relative risk of prostate cancer, particularly for advanced cases of the disease [99]. Limited epidemiological studies suggest that NSAID use is associated with reduced risk of prostate cancer [99-101]. Two NSAID derivatives which have lost COX inhibitory but retained pro-apoptotic activity are also promising chemopreventive agents in prostate: the sulfone metabolite of sulindac, most probably stimulates apoptosis via inhibition of cyclic GMP phosphodiesterase. This drug was found to inhibit the rise of (prostate-specific antigen) PSA in patients with recurring prostate cancer after radical prostatectomy (RRP) [102].

NSAIDs have been to inhibit NF-κB signalling (Figure 6) by directly blocking the activity of IκB kinase β [103, 104]. Additionally, indomethacin, a non-selective COX inhibitor, can act as a direct ligand for peroxisome proliferator activating receptor (PPAR) α and γ, and thus may directly regulate gene transcription [105]. The ability of PPARs to bind eicosanoids suggested that PPARδ might be a target of chemopreventive non-steroidal anti-inflammatory drugs [106-8]. The findings suggest that PPARδ represent a β-catenin/Tcf-4 target with particular importance for chemoprevention. Whereas APC or β-catenin mutations can result in increased PPARδ activity, NSAIDs can compensate this defect by suppressing PPARδ activity and promoting apoptosis. This suppression of PPARδ is mediated in part by the ability of some NSAIDs to directly inhibit the DNA binding activity of PPARδ (Figure 6 and 7).

The currents trials advance prior preclinical works that implicate synergistic effects of combining COX inhibitors with cytotoxic chemotherapy and /or pharmacologic agents. Recently, there is observed evidence of modest biologic activity for celecoxib in men with rising PSA levels after RRP and/or radiation therapy for prostate cancer. In this study celecoxib, compared with placebo, significantly decreased mean PSA velocity and tended to increase the proportion of men who doubled their PSA doubling time (PSADT) [109]. Also, celecoxib may have an effect on serum PSA levels in patients with biochemical progression after definitive X-ray therapy (XRT) or RRP, delaying or preventing disease progression and thereby helping extend the time until androgen deprivation therapy [110]. A recent study suggest that celecoxib with weekly docetaxel and estramustine is an effective and safe treatment for patients with hormone-refractory prostate cancer (HRPC), but it does not seem to add any benefit to docetaxel [111]. In the "Prostacox" multicentre phase II trial, docetaxel combined with celecoxib in HRPC patients showed, as preliminary results, a significant reduction of chemotherapy haematological toxicity [112]. In other study, the selective COX-2 inhibitor etodolac exhibited an antitumor effect against androgen-dependent and independent prostate cancer cell lines *in vitro* and *in vivo*. In this way, etodolac induced downregulation of Bcl-2 and upregulation of E-cadherin *in vitro* and induced apoptosis *in vivo*, suggesting that it might be useful for the treatment of hormone-resistant prostate cancer [113].

A recent study investigated, *in vitro*, the effects of celecoxib treatment on expression levels of genes relevant to prostate cancer and drug resistance. Celecoxib resulted in highly significant increases in the mRNA expression of the smooth muscle component desmin, the detoxification enzyme glutathione S-transferase π (GSTpi), and non-steroidal anti-inflammatory response gene (NAG-1). Thus, indicating a role for celecoxib as both an inducer of apoptosis and an agent that can correct inappropriate gene expression levels associated with malignant progression in human prostate cancer. This report provided evidence to support the use of celecoxib in the management of prostate cancer in both the adjuvant and chemoprevention settings [114].

The first study in the literature using a COX-2 inhibitor during the off-phases of intermittent androgen deprivation (IAD) therapy in patients with biochemical progression after RRP was developed by Di Silverio and co-workers [115]. They used the selective COX-2 inhibitor etoricoxib to improve the whole response to AID. In particular to prolong the time of the off-treatment phases, to reduce the highest level of PSA reached during off-phases and to increase the percentage of time of the off-phases in three cycles of IAD, compared with the use of IAD alone.

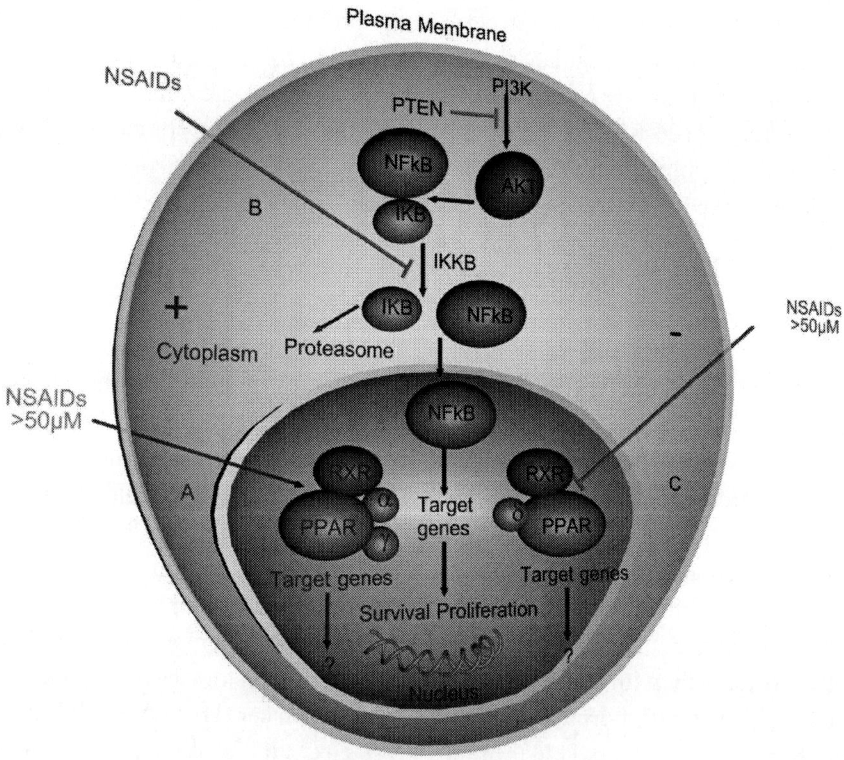

Figure 6. A. There is evidence that PGs can modulate cellular pathways by acting directly within the nucleus. Certain NSAIDs can activate peroxisome proliferator activated receptor subtypes (PPARα and PPARγ). B. NSAIDs can inhibit the activity of IkB kinase β (IKKβ); this leads to antagonism of the NF-kB signalling pathway by blocking the destruction of the NF-kB inhibitor IkB, preventing NF-kB from translocating to the nucleus and regulating key target genes. C. Other NSAIDs (sulindac) have been shown to inhibit the DNA-binding activity of peroxisome proliferator activated receptor subtype δ. Overexpression of PPARδ and can partially rescue sulindac-induced apoptosis.

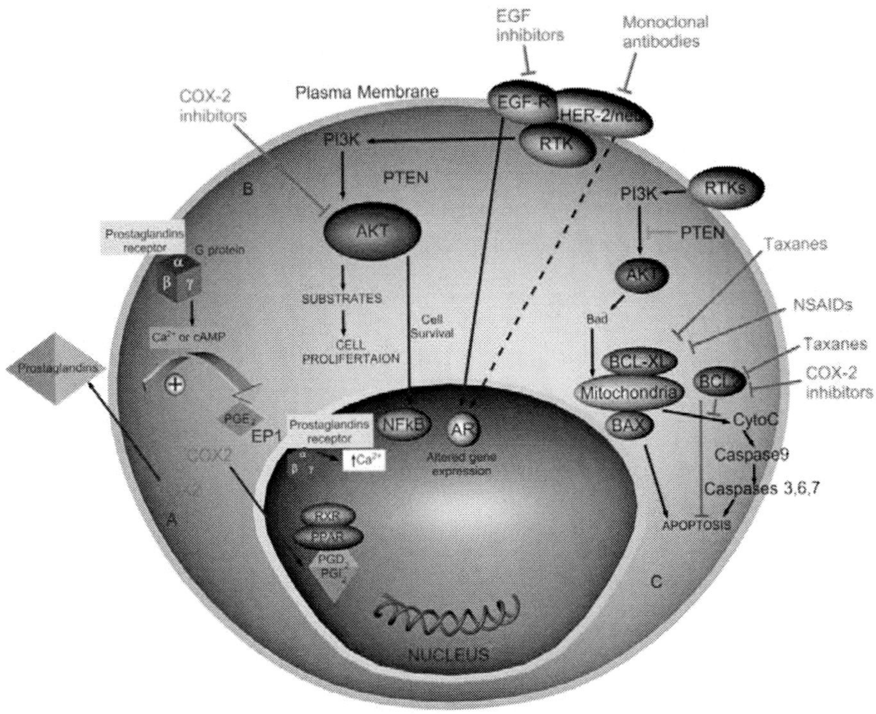

Figure 7. A. The different classes of PGs exert their effects by binding to a G-protein-coupled cell-surface receptor leading to changes in the cellular levels of cAMP and Ca^{2+}, or by acting directly within the nucleus. PGE_2 works by their receptor EP1 (localized to the nuclear envelope) and activation of the receptor leads to change in nuclear levels of Ca^{2+}, retinoid X receptor (RXR). Metabolites of PGD_2 can transactivate members of the peroxisome proliferator activated receptor (PPAR) family of nuclear hormone receptor. B. Epidermal growth factor receptor (EGFR), aberrantly expressed in prostate carcinoma can activated cell growth through RTK/PI3K/AKT/ pathway. COX-2 inhibitors induce apoptosis by blocking AKT activation in prostate cancer cells independently of Bcl-2. Otherwise, overexpression of Her2/neu has been implicated in the activation of the androgen receptor (AR) and in inducing androgen-resistant prostate cancer cell growth. C. NSAIDs can decrease the levels of the anti-apoptotic gene $BCL\text{-}_{xL}$, thereby increasing the cellular ratio of BAX/ $BCL\text{-}_{xL}$. There is also evidence that microtubule-interfering agents stimulate COX-2 transcription; otherwise, microtubule-interfering agents inhibit Bcl-2 activity by inducing Bcl-2 phosphorylation.

Recently, it was shown that Nexrutine^R, a *Phellodendrom amurense* bark extract, suppresses proliferation of prostate cancer cell lines and tumor development in the transgenic adenocarcinoma of mouse prostate (TRAMP) model. The anti-proliferative effects of this non-toxic complex mixture are mediated in part by Akt and Cyclic AMP response element binding protein (CREB). Since COX-2 is a CREB target that induces PGE_2 and suppresses apoptosis, the identification of a CREB-mediated activation of COX-2 as a signalling pathway in prostate cancer has been suggested the use of cost-effective dietary supplement such as Nexrutine^R for prostate cancer management. Furthermore, being a complex mixture, Nexrutine^R may target multiple targets, including CREB, NFkB, COX-2, and cyclin D1, that play a critical role in the carcinogenesis process [116]. In this way, the development of compounds, or a combination of compounds, targeting multiple targets has been proposed as an ideal approach for successful cancer management [117].

Taken together, these findings clearly indicate that chronic elevation of COX-2 is pathological and suggest that inhibition of COX-2 via pharmacological means or regulation of its expression can limit the development or progression of human prostate cancer. The investigators should be commended for designing and conducting trials which are excellent examples of translational research, by addressing whether NSAIDs can be administered in combination with chemotherapy. Undoubtedly the potential use of selective COX-2 inhibitors for the treatment of prostate cancer will also be investigated.

Conclusion

The COX-1 enzyme is constitutively expressed at low levels in a majority of tissues and makes prostaglandins for normal physiological functions.

COX-2 expression is normally regulated at both transcriptional and post-transcriptional levels, and can also be regulated by the rate of protein synthesis and / or degradation. The regulation of COX-2 expression is a complex process involving multiple signal transduction pathways. This level of complexity underscores the requirement for tight regulation of the enzymatic action of COX-2, which has pathogenic effects if its expression is left unchecked. As a result of genetic mutations commonly detected in the majority of human tumors, cellular defects in these signalling pathways promote the expression of COX-2 through the loss of transcriptional regulation, rapid mRNA decay, and translational inhibition.

Epidemiological studies provided the first evidence that COX may be involved in the pathogenesis of cancer. In the process of carcinogenesis and in the route of intracellular signalling, COX-2 expression may be a universal phenomenon during carcinogenesis. COX-2 is overexpressed along the continuum of oncogenesis and is likely to be a key player in a number of biologic pathways leading to cancer. In general, COX-2 is up-regulated throughout the tumorigenic process, from early hyperplasia to metastatic disease. Elevated levels of COX-2 immunoreactivity were primarily detected in the neoplastic epithelium, inflammatory cells, and vasculature within and adjacent to tumor nests. COX-2-derived prostaglandins may promote the development of cancer by various mechanisms, including stimulation of tumor cell growth and neovascularization. In contrast, COX-2 is induced in response to growth factors, ie., epidermal growth factor (EGF), vascular endothelial growth factor (VEGF), fibroblast growth factor (FGF-2), cytokines [eg., tumor necrosis factor-α (TNF-α), interleukin-α (IL-α), interleukin-1β (IL-1β)], and tumor promoters (eg, v-src, v-Ha-ras, HER-2/neu, and Wnt).

An increase in COX-2 expression also results in resistance to apoptosis. Clearly, the suppression of apoptosis, associated with COX-2 overexpression, could be an important factor in tumorigenesis, although the precise mechanistic basis remains uncertain. It is documented that the COX product PGE_2 contributes to the preventive action of COX-2 in apoptosis. Different hypotheses have been postulated to account for suppression of apoptosis in response to COX-2 overexpression: the ability of PGE_2 to inhibit apoptosis caused by a selective COX-2 inhibitor and concomitantly to induce Bcl-2, suggest that PG-mediated upregulation of Bcl-2 may suppress apoptosis.

In addition to a causal role of COX-2 in tumorigenesis, compelling evidence also indicates that it may play an important role in tumor angiogenesis, a process involved in later stages of tumor development and migration. Current evidence indicates that COX-2 promotes tumor-specific angiogenesis, inhibits apoptosis, and induces proangiogenic factor such as VEGF, and inducible nitrogen oxide synthetetase promoter (iNOS).

References

[1] Herschman, H. (1996). Review: Prostaglandin synthase 2. *Biochim Biophys Acta*, 1299: 125-140.

[2] DuBois, R. N., Abramson, S. B., Crofford, L., et al. (1998). Cyclooxygenase in biology and disease. *Faseb J.*, 12: 1063-1073.

[3] Smith, W. L., DeWitt, D. L. & Garavito, R. M. (2000). Cyclooxygenases: Structural, cellular and molecular biology. *Annu Rev Biochem.*, 69: 145-182.

[4] Nolan, R. D., Danilowicz, R. M. & Eling, T. E. (1988). Role of arachidonic acid metabolism in the mitogenic response of BALB/c 3T3 fibroblasts to epidermal growth factor. *Mol Pharmacol.*, 33: 650-656.

[5] Marnett, L. J. (1992). Aspirin and the potential role of prostaglandins in colon cancer. *Cancer Res.*, 52: 5575-5589.

[6] Wiese, F. W., Thompson, P. A. & Kadlubar, E. F. (2001). Carcinogen substrate specificity of human COX-1 and COX-2. *Carcinogenesis.*, 21: 5-10.

[7] Tsujii, M. & DuBois, R. N. (1995). Alterations in cellular adhesion and apoptosis in epithelial cells overexpressing prostaglandin endoperoxide synthase 2. *Cell*, 83: 498-501.

[8] DuBois, R. N., Shao, J., Tsujii, M., Sheng, H. & Beauchamp, R. D. (1996). G_1 delay in cells overexpressing prostaglandin endoperoxide synthase-2. *Cancer Res.*, 56: 733-737.

[9] Sheng, H., Shao, J., Morrow, J. D., Beauchamp, R. D. & DuBois, R. N. (1998). Modulation of apoptosis and Bcl/2 expression by prostaglandin E_2 in human colon cancer cells. *Cancer Res.*, 58: 362-366.

[10] Chan, T. A., Morin, P. J., Vogelstein, B. & Kinzler, K. W. (1998). Mechanisms underlying nonsteroidal anti-inflammatory drug-mediated apoptosis. *PNAS*, 95: 681-686.

[11] Cao, T., Pearman, A. T., Zimmerman, G. A., McIntyre, T. M. & Prescott, S. M. (2000). Intracellular unesterified arachidonic acid signals apoptosis. *Proc Natl Acad Sci.*,; 97: 11280-11285.

[12] Gately, S. (2000). The contributions of cyclooxygenase-2 to tumor angiogenesis. *Cancer and Metastasis Rev.*, 19: 19-27.

[13] Oshima, M., Dinchuk, J. E., Kargaman, S. L., et al. (1996). Suppression of intestinal polyposis in $Apc^{\Delta 716}$ knockout mice by inhibition of cyclooxygenase 2 (COX-2). *Cell*, 87: 803-809.

[14] Tsujii, M., Kawano, S., Tsuji, S., Sawaoka, H., Hori, H. & DuBois, R. N. (1998). Clycooxygenase regulates angiogenesis induced by colon cancer cells. *Cell*, 93: 705 - 716.

[15] Xie, W. & Herchman, H. R. (1995). v-src induces prostaglandin synthase 2 gene expression by activation of the c-Jun N-terminal kinase and the c-Jun transcription factor. *J Biol Chem.*, 270: 27622-27628.

[16] Subbaranmaiah, K., Norton, I., Gerald, W. & Dannenberg, A. J. (1999). Increased expression of cyclooxygenase-2 in HER-2/neu overexpressing human breast cancer cells. *National Cancer Institute 7th Spore Investigators Workshop*, Rockville, Md.

[17] Howe, L. R., Subbaramaiah, K., Chung, W. J., et al. (1999). Transcriptional activation of cyclooxygenase-2 in Wnt-1 transformed mouse mammary epithelial cells. *Cancer Res.*, 59: 1572-1577.

[18] Vadlamudi, R., Mandal, M., Adam, I., Steinbach, G., Mendelsohn, J. & Kumar, R. (1999). Regulation of the cyclooygenase-2 pathway by HER2 receptor. *Oncogene*; 18: 305-314.

[19] Mei, J. M., Hord, N. G., Winterstein, D. F., Donald, S. P. & Phang, J. M. (1999). Differential expression of prostaglandin endoperoxide H synthase-2 and formation of activated β-catenin-LEF-1 transcription complex in mouse colonic epithelial cells contrasting in APC. *Carcinogenesis*, 20: 737-740.

[20] Dimberg, J., Hugander, A., Sirsjo, A. & Soderkvist, P. (2001). Enhanced expression of cyclooxygenase-2 and nuclear β-catenin are related to mutations in the APC gene in human colorectal cancer. *Anticancer Res.*, 21: 911-915.

[21] Howe, L. R., Crawford, H. C., Subbaramaiah, D., Hassell, J. A., Dannenberg, A. J. & Brown, A. M. (2001). PEA3 is up-regulated in response to Wnt1 and activates the expression of cyclooxygenase-2. *J Biol Chem.*, 276: 20108-20115.

[22] Subbaramaiah, D., Chung, W. J. & Dannenberg, A. J. (1998). Ceramide regulates the transcription of cyclooxygenase-2. Evidence for involvement of extracellular signal-regulated kinase/c-Jun N-terminal kinase and p38 mitogen-activated protein kinase pathways. *J Biol Chem.*, 273: 32943-32949.

[23] Shao, J., Sheng, H., Inoue, H., Morrow, J. D. & Dubois, R. N. (2000). Regulation of constitutive cyclooxygenase-2 expression in colon carcinoma cells. *J Biol Chem.*, 43: 33951-33956.

[24] Subbaramaiah, D., Norton, L., Gerald, W. & Dannenberg, A. J. (2002). Cyclooxygenase-2 is overexpressed in HER-2/neu-positive breast cancer. Evidence for involvement of AP-1 and PEA3. *J Biol Chem.*, 277: 18649-18657.

[25] Sheng, H., Williams, C. S., Shao, J., Liang, P., DuBois, R. N. & Beauchamp, R. D. (1998). Induction of cyclooxygenase-2 by activated Ha-ras oncogene in Rat-1 fibroblasts and the role of mitogen-activated protein kinase pathway. *J Biol Chem.*, 273: 22120-22127.

[26] Sheng, H., Shao, J. & DuBois, R. N. (2001). K-Ras-mediated increase in cyclooxygenase-2 mRNA stability involves activation of the protein kinase B. *Cancer Res.*, 61: 2670-2675.

[27] Guan, Z., Buchman, S. Y., Pentland, A. P., Tampleton, D. J. & Morrison, A. R. (1998). Induction of cyclooxygenase-2 by the activated MEKK1 → SK1/MKK4 → p38 mitogen-activated protein kinase pathway. *J Biol Chem.*, 273:12901-12908.

[28] Adjei, A. A. (2001). Ras signaling pathway proteins as therapeutic targets. *Curr Pharm Des.*, 7: 1581-1594.

[29] Slice, L. W., Bui, L., Mak, C. & Walsh, J. H. (2000). Differential regulation of COX-2 transcription by Ras- and Rho-family of GTPases. *Biochem Biophys Res Commun.*, 276: 405-410.

[30] Subbaramiah, K., Hart, J. C., Norton, L. & Dannenberg, A. J. (2000). Microtubule-interfering agents stimulate the transcription of cyclooxygenase-2. Evidence for involvement of ERK1/2 and p38 mitogen-activated protein kinase pathways. *J Biol Chem.*, 275: 14848-14845.

[31] Zohn, I. M., Campbell, S. L., Khosravi-Far, R., Rossman, K. L. & Del, C. J. (1998). Rho family proteins and Ras transformation: The Rhoad less travelled gets congested. *Oncogene*, 17: 1415-1438.

[32] Xie, W. & Herschman, H. R. (1996). Transcriptional regulation of prostaglandin synthase 2 gene expression by platelet-derived growth factor and serum. *J Biol Chem.*, 271: 31742-31748.

[33] Reddy, S. T., Wadleigh, D. J. & Herschman, H. R. (2000). Transcriptional regulation of the cyclooxygenase-2 gene in activated mast cells. *J Biol Chem.*, 275: 3107-3113.

[34] Ridley, S. H., Dean, J. L., Sarsfield, S. J., Brook, M., Clark, A. R. & Saklatvala, J. (1998). A p38 MAP kinase inhibitor regulates stability of interleukin-1-induced cyclooxygenase-2 mRNA. *FEBS Lett.*, 439: 75-80.

[35] Dean, J. L., Brook, M., Clark, A. R. & Saklatvala, J. (1999). p38 mitogen-activated protein kinase regulates cyclooxygenase-2 mRNA stability and transcription in lipopolysaccharide-treated human monocytes. *J Biol Chem.*, 274: 264-269.

[36] Rodriguez-Viciana, P., Warne, P. H., Khwaja, A., et al. (1997). Role of phosphoinositide 3-OH kinase in cell transformation and control of the actin cytoskeleton by Ras. *Cell*, 89: 457-467.

[37] Sheng, H., Shao, J. & DuBois, R. N. (2001). Akt/PKB activity is required for Ha-Ras-mediated transformation of intestinal epithelial cells. *J Biol Chem.*, 276: 14498-14504.

[38] Mangelsdorf, D. J., Thummel, C., Beato, M., et al. (1995). The nuclear receptor superfamily: the second decade. *Cell*, 83: 835-839.

[39] Lemberger, T., Desvergne, B. & Wahli, W. (1996). Peroxisome proliferators-activated receptors: a nuclear receptor signaling pathway in lipid physiology. *Annu Rev Cell Dev Biol.*, 12: 335-363.

[40] Rubinfeld, B., Souza, B. & Albert, I. (1993). Associated of the APC gene product with β-catenin. *Science*, 262: 1731-1734.

[41] Su, L. K., Vogelstein, B. & Kinzler, K. W. (1993). Association of the APC tumor suppressor protein with catenins. *Science*, 262: 1734-1737.

[42] Behrens, J., von Kries, J. P., Kuhl, M., et al. (1996). Functional interaction of β-catenin with the transcriptional factor LEF-1. *Nature*, 382: 638-642.

[43] Molenaar, M., van de Wetering, M., Oosterwegel, M., et al. (1996). XTcf-3 transcription factor mediates β-catenin-induced axis formation in Xenopus embryos. *Cell*, 86: 391-399.

[44] Prescott, S. M. & Fitzpatrick, F. A. (2000). Cyclooxygenase-2 and carcinogenesis. Biochim Biophys *Acta.*, 1470: M69-M78.

[45] Wiese, F. W., Thompson, P. A. & Kadlubar, F. F. (2001). Carcinogen substrate specificity of human COX-1 and COX-2. *Carcinogenesis*, 21: 5-10.

[46] Eling, T. E., Thompson, D. C., Foureman, G. L., Curtis, J. F. & Hughes, M. F. (1990). Prostaglandin H synthase and xenobiticoxidation. Annu Rev. *Pharmacol Toxicol.*, 30: 1-45.

[47] Leahy, D. M., Ornberg, R. L., Want, Y., Zweifel, B. S., Koki, A. T. & Masferrer, J. L. (2002). Cyclooxygenase-2 inhibition by celecoxib reduces proliferation and induces apoptosis in angiogenic endothelial cells in vivo. *Cancer Res.*, 62: 625-631.

[48] Masferrer, J. L., Leahy, K. M., Koki, A. T., et al. (2000). Antiangiogenic and antitumor activities of cyclooxygenase-2 inhibitors. *Cancer Res.*, 60: 1306-1311.

[49] Leahy, K. M., Koki, A. T. & Masferrer, J. L. (2000). Role of cyclooxygenases in angiogenesis. *Curr Med Chem.*, 7: 1163-1170.

[50] Liu, X. H., Kirschenbaum, A., Yao, S., Lee, R., Holland, J. F. & Levine, A. C. (2000). Inhibition of cyclooxygenase-2 suppresses angiogenesis and the growth of prostate cancer in vivo. *J Urol.*, 164: 820-825.

[51] Souza, R. F., Shewmake, K., Beer, D. G., Cryer, B. & Spechler, S. J. (2000). Selective inhibition of cyclooxygenase-2 suppresses growth and induces apoptosis in human esophageal adenocarcinoma cells. *Cancer Res.*, 60: 5767-5772.

[52] Tsujii, M. & Dubois, R. N. (1995). Alterations in cellular adhesion and apoptosis in epithelial cells overexpressing prostaglandin endoperoxide synthase-2. *Cell*, 83: 493-501.

[53] Ding, X. Z., Tong, W. G. & Adrian, T. E. (2000). Blockade of cyclooxygenase-2 inhibits proliferation and induces apoptosis in human pancreatic cancer cells. *Anticancer Res.*, 20: 2625-2631.

[54] Hsu, A. L., Ching, T. T., Wang, D. S., et al. (2000). The cyclooxygenase-2 inhibitor celecoxib induces apoptosis by blocking Akt activation in human prostate cancer cells independently of Bcl-2. *J Biol Chem.*, 275: 11379-11403.

[55] Shweiki, D., Itin, A., Soffer, D. & Keshet, E. (1992). Vascular endothelial growth factor induced by hypoxia may mediate hypoxia-initiated angiogenesis. *Nature*, 359: 843-845.

[56] Liu, Y., Cox, S. R., Morita, T. & Kourembanas, S. (1995). Hypoxia regulates vascular endothelial growth factor gene expression in endothelial cells. *Circ Res.*, 77: 638-647.

[57] Mellilo, G., Musso, T., Sica, A., Taylor, L. S., Cox, G. W. & Valesio, L. (1995). A hypoxia-responsive element mediates a novel pathway of activation of the inducible nitric oxide synthase promoter. *J Exp Med.*, 182: 1683-1693.

[58] Yan, S. F., Tritto, I., Pinsky, D., et al. (1995). Induction of interleukin 6 (IL-6) by hypoxia in vascular cells. Central role of the binding site for nuclear factor IL-6. *J Biol Chem.*, 270: 11463-11471.

[59] De Mazro, A. M., Marchi, V. L., Epstein, J. I. & Nelson, W. G. (1999). Proliferation inflammatory atrophy of the prostate implications for prostatic carcinogenesis. *Am J Pathol.*, 155: 1985-92.

[60] Kimura, M., Osumi, S. & Ogihara, M. (2002). Stimulation of DNA synthesis and proliferation by prostaglandins in primary cultures of adult rat hepatocytes. *Eur J Pharmacol.*, 404: 259-271.

[61] Liu, C. H., Chang, S. H., Narko, K. et al. (2001). Overexpression of cyclooxygenase-2 is sufficient to induce tumorigenesis in transgenic mice. *J Biol Chem.*, 276: 18563-18569.

[62] Surette, M. E., Winkler, J. D., Fonteh, A. N. & Chilton, F. H. (1996). Relationship between arachidonate-phospholipid remodelling and apoptosis. *Biochemistry*, 35: 9187-9196.

[63] Condorelly, F., Canonico, P. L. & Fortino, M. A. (1999). Distinct effects of ceramide-generating pathways in prostate adenocarcinoma. *Br J Pharmacol.*, 127: 75-84.

[64] Attiga, F. A., Fernandez, P. M., Weeraratna, A. T., Manyak, M. J. & Pitierno, S. R. (2000). Inhibitors of prostaglandin synthesis inhibits human prostate tumor cell invasiveness and reduces the release of matrix metalloproteinases. *Cancer Res.*, 60: 4629-4637.

[65] Tsujii, M. & Dubois, R. N. (1995). Alterations in cellular adhesion and apoptosis in epithelial cells overexpressing prostaglandin endoperoxide synthase 2. *Cell*, 83: 493-501.

[66] Dohadwala, M., Luo, J. & Zhu, L. (2001). Non-small cell lung cancer cyclooxygenase-2-dependent invasion is mediated by CD44. *J Biol Chem.*, 276: 20809-20812.

[67] Sheng, H., Shao, J., Washington, M. K. & DuBois, R. N. (2001). Prostaglandin E2 increases growth and motility of colorectal carcinoma cells. *J Biol Chem.*, 176: 18075-18081.

[68] Jones, M. K., Wang, H., Peskar, B. M., et al. (1999). Inhibition of angiogenesis by nonsteroidal anti-inflammatory drugs: Insight into mechanisms and implications for cancer growth and ulcer healing. *Nat Med.*, 5: 1418-1423.

[69] Masferrer, J. L., Leahy, K. M. & Koi, A. T. (2000). Antiangiogenic and antitumor activities of cyclooxygenase-2 inhibitors. *Cancer Res.*, 60: 1306-1311.

[70] Sonoshita, M., Takaku, K., Sasaki, N., et al. (2001). Acceleration of intestinal polyposis through prostaglandin receptor EP2 in ApcDelta 716 knockout mice. *Nat Med.*, 7: 1048-1051.

[71] Aparicio Gallego, G., Díaz Prado, S., Jiménez Fonseca, P., García Campelo, R., Cassinello Espinosa, J. & Antón Aparicio, L. M. (2007). Cyclooxygenase-2 (COX-2): a molecular target in prostate cancer. *Clin Transl Oncol.*, 9: 694-702.

[72] Gupta, S., Srivastava, M., Ahmad, N., Bostwick, D. G. & Mukhtar, H. (2000). Overexpression of cyclooxygenase-2 in human prostate adenocarcinoma. *Prostate*, 42: 73-9.

[73] Kirschenbaum, A., Klausner, A. P., Lee, R., et al. (2000). Expression of cyclooxygenase-1 and cyclooxygenase-2 in the human prostate. *Urology*, 56: 671-6.

[74] Madaan, S., Abel, P. D., Chaudhary, D. S., et al. (2000). Cytoplasmic induction and over-expression of cyclooxygenase-2 in human prostate cancer: implications for prevention and treatment. *BJU Int.*, 86: 736-41.

[75] Tanji, N., Kikugawa, T. & Yokoyama, M. (2000). Immunohistochemical study of cyclooxygenases in prostatic adenocarcinoma; relationship to apoptosis and Bcl-2 protein expression. *Anticancer Res.*, 20: 2313-9.

[76] Zha, S., Gage, W. R., Sauvageot, J., et al. (2001). Cyclooxygenase-2 is upregulated in proliferative inflammatory atrophy of the prostate, but not in prostate, but not in prostate carcinoma. *Cancer Res.*, 61: 8617-23.

[77] Shappell, S. B., Manning, S., Boeglin, W. E., et al. (2001). Alterations in lipoxygenase and cyclooxygenase-2 catalytic and mRNA expression in prostate carcinoma. *Neoplasia*, 3: 287-303.

[78] Kinzler, K. W. & Vogelstein, B. (1996). Lessons from hereditary colorectal cancer. *Cell*, 87: 159-170.

[79] Eberhart, C. E., Coffey, R. J., Radhika, A., Giardiello, F. M., Ferressbach, S. & DuBois, R. N. (1994). Up-regulation of cyclooxygenase 2 gene expression in human colorectal adenomas and adenocarcinomas. *Gastroenterology*, 107: 1183-1188.

[80] Sano, H., Kawahito, Y., Wilder, R. L., et al. (1995). Expression of cyclooxygenase -1 and -2 in human colorectal cancer . *Cancer Res.*, 55: 3785-3789.

[81] Nie, D., Hillman, G. G., Geddes, T., et al. (1998). Platelet-type 12-lipoxyenase in a human prostate carcinoma stimulates angiogenesis and tumor growth. *Cancer Res.*, 58: 4047-4051.

[82] Timar, J., Raso, E., Dome, B., et al. (2000). Expression, subcellular localization and putative function of platelet-type 12-lipoxygenase in human prostate cancer cell lines of different metastatic potential. *Int. J Cancer*, 87: 37-43.

[83] Nolan, R. D., Danilowicz, R. M. & Eling, T. E. (1988). Role of arachidonic acid metabolism in the mitogenic response of BALB/c 3T3 fibroblasts to epidermal growth factor. *Mol. Pharmacol.*, 33: 650-656.

[84] Gao, X., Grignon, D. J., Chbihi, T., et al. (1995). Elevated 12-lipoxygenase mRNA expression correlates with advanced stage and poor differentiation of human prostate cancer. *Urology*, 46: 227-237.

[85] Yoshimura, R., Sano, H., Masuda, C., et al. (2000). Expression of cyclooxygenase-2 in prostate carcinoma. *Cancer*, 89: 589-596.

[86] Chan, T. A., Morin, P. J., Vogelstein, B. & Kinzler, K. W. (1998). Mechanisms underlying nonsteroidal antiinflammatory drug-mediated apoptosis. *Proc Natl Acad Sci USA*, 95: 681-686.

[87] Sheng, H., Shao, J., Morrow, J. D., Beauchamp, R. D. & DuBois, R. N. (1998). Modulation of apoptosis and Bcl-2 expression by prostaglandin E2 in human colon cancer cells *Cancer Res.*, 58: 362-366.

[88] Liu, X. H., Yao, S., Kirschenbaum, A. & Levine, A. C. (1998). NS398, a selective cyclooxygenase-2 inhibitor, induces apoptosis and down-regulates bcl-2 expression in LNCaP cells. *Cancer Res.*, 58: 4245-4249.

[89] Hsu, A. L., Ching, T. T., Wang, D. S., Song, X., Rangnekar, V. M. & Chen, C. S. (2000). The cyclooxygenase-2 inhibitor celecoxib induces apoptosis by blocking Akt activation in human prostate cancer cells independently of Bcl-2. *J Biol Chem.*, 275: 11397-11403.

[90] Johnson, A. J., Song, X., Hsu, A. & Chen, C. (2001). Apoptosis signalling pathways mediated by cyclooxygenase-2 inhibitors in prostate cancer cell. *Adv Enzyme Regul.*, 41: 221-235.

[91] Song, X., Lin, H. P., Johnson, A. J., et al. (2002). Cyclooxygenase-2 player or spectator in cyclooxygenase-2 inhibitor-induced apoptosis in prostate cancer cells. *J Natl Cancer Inst* (Bethesda), 94: 585-591.

[92] Zhu, J., Song, X., Lin, H. P., et al. (2002). Using cyclooxygenase-2 inhibitors as molecular platforms to develop a new class of apoptosis-inducing agents. *J Natl Cancer Inst.,* (Bethesda), 1745-1757.

[93] Kulp, S. K., Yang, Y. T., Hung, C. C., et al. (2004). 3-Phosphoinositide-dependent protein kinase-1 / Akt signalling represents a major cyclooxygenase-2-independent target for celecoxib in prostate cancer cells. *Cancer Res.*, 64: 1444-1451.

[94] Patel, M. I., Subbaramaiah, K., Du, B., et al. (2005). Celecoxib inhibits prostate cancer growth: evidence of a cyclooxygenase-2-independent mechanism. *Clin Cancer Res.*, 11: 1999-2001.

[95] Pan, Y., Zhang, J. S., Gazi, M. H. & Young, C. Y. F. (2003). The cyclooxygenase 2-specific nonsteroidal anti-inflammatory drugs celecoxib and nimesulide inhibit androgen receptor activity via induction of c-Jun in prostate cancer cells. *Cancer Epidemiol Biomarkers Prev.*, 12: 769-774.

[96] Wagner, E. F. (2001). AP-1: introductory remarks. *Oncogene*, 20: 2334-2335.

[97] Steinbach, G., Lynch, P. M., Phillips, R. K., et al. (2000). The effect of celecoxib a cycloxygenase-2 inhibitor, in familial adenomatous polyposis. *N Engl J Med.*, 342: 1946-1952.

[98] Howe, L. R., Subbaramaiah, K., Brown, A. M. & Dannenberg, A. J. (2001). Cyclooxygenase-2: a target for the prevention and treatment of breast cancer. *Endocr Relat Cancer*, 8: 97-114.

[99] Norrish, A. E., Jackson, R. T. & McRae, C. U. (1998). Non-steroidal anti-inflammatory drugs and prostate cancer progression. *Int J Cancer.*, 77: 1322-1327.

[100] Paganini-Hill, A., Chao, A., Ross, R. K. & Henderson, B. E. (1989). Aspirin use and chronic diseases: a cohort study of the elderly. *Br Med J.*, 299: 1247-1250.

[101] Thun, M. J., Namboodiri, M. M., Calle, E. E., Flanders, W. D. & Heath, Jr. C. W. (1993). Aspirin use and risk of fatal cancer. *Cancer Res.*, 53: 1322-1327.

[102] Goluboff, E. T., Prager, D., Rukstalis, D., et al. (2001). Safety and efficacy of exisulind for treatment of recurrent prostate cancer after radical prostatectomy. *J Urol.*, 166: 882-886.

[103] Yamamoto, K., Arakawa, T., Ueda, N. & Yamamoto, S. (1995). Transcriptional roles of nuclear factor κB and nuclear factor-interleukin-6 in the tumor necrosis factor α-dependent induction of cyclooxygenase-2 in MC3T3-E1 cells. *J Biol Chem.*, 270: 31315-31320.

[104] Yamamoto, Y., Yin, M. J., Lin, K. M. & Gaynor, R. B. (1999). Sulindac inhibits activation of the NF-κB pathway. *J Biol Chem.*, 274: 27307-27314.

[105] Lehmann, J. M., Lenhard, J. M., Oliver, B. B., Ringold, G. M. & Kliewer, S. A. (1997). Peroxisome proliferator-activated receptors alpha and gamma are activated by indomethacin and other non-steroidal anti-inflammatory drugs. *J Biol Chem.*, 272: 3406-3410.

[106] Keller, H., Dreyer, C., Mdin, J., Mahfoudi, A., Ozato, K. & Wahli, W. (1993). Fatty acids and retinoids control lipid metabolism through activation of peroxisome proliferators-activated receptor-retinoid X receptor heterodimers. *Proc Natl Acad Sci USA*, 90: 2160-2164.

[107] Forman, B. M., Chen, J. & Evans, R. M. (1997). Hypolipidemic drugs, polyunsaturated fatty acids, and eicosanoids are ligand for peroxisome proliferators-activated receptors α and δ. *Proc Natl Acd Sci USA*, 94: 4312-4317.

[108] Kliewer, S. A., Sundseth, S. S., Jones, S. A., et al. (1997). Fatty acids and eicosanoids regulate gene expression through direct interactions with peroxisome proliferator-activated receptors α and γ. *Proc Natl Acad Sci USA*, 94: 4318-4323.

[109] Smith, M. R., Manola, J., Kaufman, D. S., Oh, W. K. & Bubley, G. J. (2008). Celecoxib versus placebo for men with prostate cancer and a rising serum prostate-specific antigen after radical prostatectomy and/or radiation therapy. *J Clin Oncol.*, 24: 2723-8.

[110] Pruthi, R. S., Derksen, J. E., Moore, D., Carson, C. C., Grigson, G., Watkins, C. & Wallen, E. (2006). Phase II trial of celecoxib in prostate-specific antigen recurrent prostate cancer after definitive radiation therapy or radical prostatectomy. *Clin Cancer Res.*, 12: 2172-7.

[111] Carles, J., Font, A., Mellado, B., Domenech, M., Gallardo, E., González-Larriba, J. L., et al. (2007). Weekly administration of docetaxel in combination with estramustine and celecoxib in patients with advanced hormone-refractory prostate cancer: final results from a phase II study. *Br J Cancer*, 97: 1206-10.

[112] Albouy, B., Tourani, J. M., Allain, P., Rolland, F., Staerman, F., Eschwege, P. & Pfister, C. (2007). *BJU Int.*, 100: 770-4.

[113] Shigemura, K., Shirakawa, T., Wada, Y., Kamidono, S., Fujisawa, M. & Gotoh, A. (2005). Antitumor effects of etodolac, a selective cyclooxygenase-II inhibitor, against human prostate cancer cell lines in vitro and in vivo. *Urology*, 66: 1239-44.

[114] Mehar, A., Macanas-Pirard, P., Mizokami, A., Takahashi, Y., Kass, G. E. N. & Coley, H. M. (2008). The effects of cyclooxygenase-2 expression in prostate cancer cells: modulation of response to cytotoxic agents. *JPET.*, 324: 1181-7.

[115] Di Silverio, F., Sciarra, A. & Gentile, V. (2008). Etoricoxib and intermittent androgen deprivation therapy in patients with biochemical progression after radical prostatectomy. *Urology.*, 71: 947-51.

[116] Ghosh, R., Garcia, G. E., Crosby, K., Inoue, H., Thompson, I. M., Troyer, D. A. & Kumar, A. P. (2007). Regulation of Cox-2 by cyclic AMP response element binding protein in prostate cancer: potential role for Nexrutine. *Neoplasia*, 9: 893-9.

[117] McCarty, M. F. (2004). Targeting multiple signalling pathways as a strategy for management of prostate cancer: multifocal signal modulation therapy. *Integr Cancer Ther.*, 3: 349-80.

In: Handbook of Prostate Cancer Cell Research...
Editor: Alan T. Meridith
ISBN: 978-1-60741-954-9

Chapter 11

Cell Cycle Signaling Networks in Prostate Cancer: E2F Target Genes Play a Predominant Role in Radiation Response and are Potentially Co-Regulated by p53 and WT1

Kamini Singh[1], Kurtis Eisermann[3], Helen Piontkivska[3], Gail Fraizer[3] and Alexandru Almasan[1,2,*]

[1]Department of Cancer Biology, Lerner Research Institute, Cleveland, OH 44195, USA
[2] Department of Radiation Oncology, Cleveland Clinic, Cleveland, OH 44195, USA
[3]School of Biomedical Sciences and Department of Biological Sciences,
Kent State University, Kent, OH 44242, USA

Abstract

Delineating signaling pathways active in prostate epithelial cells will lead to a better understanding of those processes essential for prostate cancer progression, such as uncontrolled proliferation, survival as a consequence of resistance to cell death, invasion and metastasis, and for development of resistance to therapies such as androgen ablation and radiation therapy. Additionally, once mechanistic information is acquired, identified gene products can be used as determinants of high-grade tumor development leading to acquisition of prognostic biomarkers or therapeutic targets for intervention. Our approach involves identifying gene expression pathways that regulate these essential functions, especially focusing on those genes related to proliferation and genotoxic stress response, examining interactions of transcription factors (TFs) such as androgen receptor (AR), p53, retinoblastoma (Rb), E2F, and WT1. Although many gene targets regulated by these TFs are known, there is relatively little information available about common regulatory pathways shared between these genes. Thus, we have sought evidence of overlapping

[*] Correspondence to: Tel.: (216) 444-9970; Fax: (216) 445-6269; E-mail: almasaa@ccf.org

target genes common between these TFs. A mechanistic understanding of how these key TFs regulate their gene targets, such as identification of essential cofactors and other cooperating or competing TFs will lead to additional therapeutic and diagnostic applications. Here, we describe our global gene expression studies applied to prostate cancer epithelial cells using LNCaP derivative C4-2 cells, with altered androgen responsiveness. Irradiation used as a prototypic genotoxic stress response model system has revealed many induced and repressed genes. For those genes repressed by genotoxic stress that are candidate radiation sensitive genes we seek a better understanding of their mechanisms of coordinate expression. A list of genes repressed at two-fold level or higher among irradiated C4-2 cells was identified, and functional groups of genes were designated using Gene Ontology annotations. Additionally, promoter sequences of genes from the top two functional clusters were examined to identify the presence of TF binding sites (TFBS) of interest. Genes with detected TFBS were further considered as candidates for functional studies as potential target genes. Here we report the co-occurrence of predicted TFBS for E2F, p53, and WT1 within the promoters of radiation suppressed mitotic genes. Examining relationships between TFs and their potential target genes will provide important insight into pathways essential for prostate cancer progression that could lead to the development of the next generation of diagnostics and therapeutics.

Keywords: E2F, WT1, p53, prostate cancer, cell cycle, mitosis, genotoxic stress.

Introduction

In response to DNA damage, such as that caused by ionizing radiation, the most prominent arrest is at the G_2/M phase transition (Iliakis, G. et al. 2003), which occurs particularly in tumor cells that almost invariably lack an effective G_1 arrest. In normal, non-transformed cells that have a functional p53 pathway, both a G_1 and a G_2 arrest can take place in response to DNA damage (Crosby, M. E. et al. 2007; Wahl, G. M. and A. M. Carr 2001). It has been established that the G_2 checkpoint initiation is primarily regulated through the control of Cdk1 (Cdc2) activity, which is regulated at multiple levels, including periodic association with the B-type cyclins, reversible phosphorylation, and intracellular compartmentalization (Atherton-Fessler, S. et al. 1993). Of the over 1,000 proteins that are phosphorylated in mitosis most contain motifs that are suggesting that these targets are Cdk1 substrates (Dephoure, N. et al. 2008). Cyclin B1, the prototypical Cyclin B (B2 and B3 are less abundant) reaches maximum levels in G_2, when it enters into the nucleus to form a complex with Cdk1 in a phosphorylation dependent manner (Hagting, A. et al. 1998). The complex is activated by phosphorylation of Cdk1 on threonine 161 (T161) by Cdk-activating kinase (CAK), but is kept inactive by its phosphorylation on threonine 14 (T14) and tyrosine 15 (Y15) by the Myt1 and Wee1 kinases, respectively (Parker, L. L. and H. Piwnica-Worms 1992).

p130, a retinoblastoma (Rb)-family protein has been recently implicated in the mechanism of G_2 arrest following treatment with DNA-damaging chemotherapeutics (Jackson, M. W. et al. 2005; Polager, S. and D. Ginsberg 2003; Crosby, M. E. et al. 2007). The Rb family of pocket proteins, comprised of Rb, p130, and p107, associate with a wide variety of transcription factors and chromatin remodeling enzymes forming transcriptional repressor complexes that control gene expression (Macaluso, M. et al. 2006). Rb, a tumor suppressor critical for the control of cell cycle progression, regulates E2F activity. It has been known for some time that Rb-deficient cells have increased levels of mRNA and protein levels of genes that are under the transcriptional control of activating E2Fs (Almasan, A. et al. 1995; Herrera, R. E. et al. 1996). Recently, it has become clear that the inactivation of Rb-family members that are associated with these activating E2Fs reduces the cellular ability to also undergo G_2 arrest in response to DNA damage (Polager, S. and D. Ginsberg 2003). A small molecule based on the Rb2/p130 spacer domain as well as a specific E2F4 inhibitor proved to be effective in inhibiting cell cycle arrest and tumor growth *in vivo* (Bagella, L. et al. 2007; Cheung, I. Y. et al. 2008).

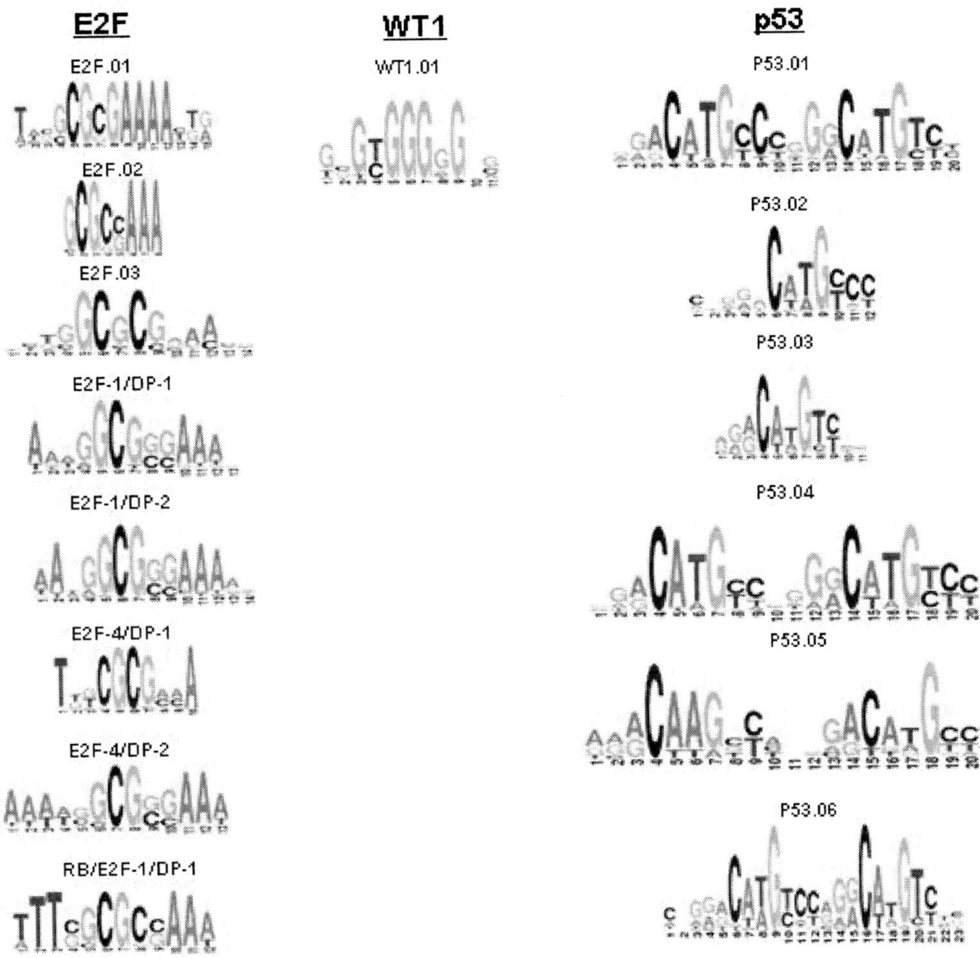

Figure 1. Consensus sequence logos used by MatInspector to predict binding sites for E2F, p53, and WT1. Transcription factor binding sites of E2F, WT1 and p53 were predicted for each gene using MatInspector (http://www.genomatix.de) that utilizes the TRANSFAC libraries of transcription factor binding motifs (Cartharius, K. et al. 2005). Matrices used were updated June 2008 and include several iterations of the p53 site, the E2F monomeric, heterodimeric, and heterotrimeric sites. Note the similarity of the E2F-1 and E2F-4 motifs.

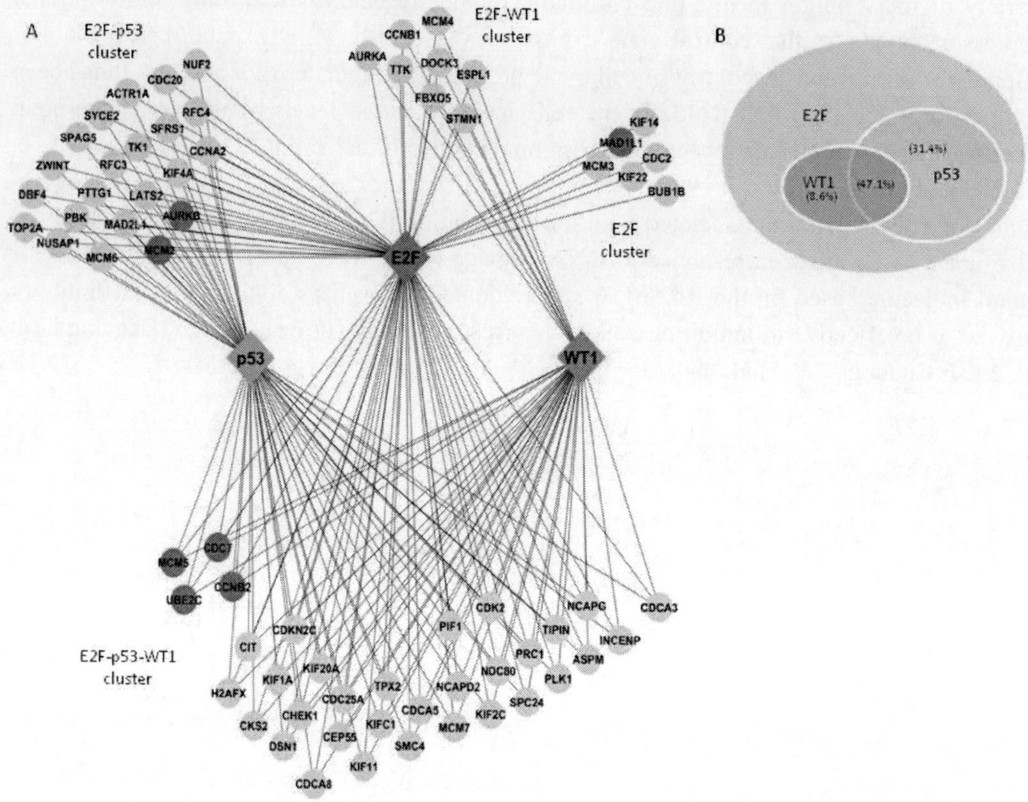

Figure 2. Distribution of predicted E2F, p53, and WT1 binding sites in promoters of 70 radiation suppressed genes.

A. The promoters of 70 genes involved in cell cycle, spindle checkpoint, and mitosis contain predicted E2F transcription factor binding sites (TFBS). Clusters show genes whose promoters contain predicted E2F binding sites, either alone or in combination with predicted p53 or WT1 sites, or predicted E2F, p53, and WT1 sites. TF interactions map was generated using Cytoscape version 2.5.1. Interaction of E2F, p53 and WT1 with the respective promoters is based on the consensus site analysis for these transcription factors. Genes in red circles indicate the presence of only heterodimeric E2F sites predicted to bind a complex of E2F and DP-1, E2F and DP-2, or a trimeric site predicted to bind E2F, Rb, and DP-1.

B. Distribution of p53 and WT1 binding sites in promoters of mitotic genes that also contain E2F consensus binding sites. The Venn diagram shows that 78% of genes whose promoters have predicted E2F binding sites also have predicted p53 sites. In contrast, 56% of these genes have predicted E2F and WT1 sites and 47% have all three sites present.

p53 can contribute to a sustained G_2 arrest primarily by inducing the transcription of *p21* and *14-3-3sigma* (Wahl, G. M. and A. M. Carr 2001). *14-3-3sigma* sequesters Cdk1/Cyclin B1 and Cdc25C in the cytoplasm (Chan, T. A. et al. 1999; Hermeking, H. et al. 1997), while p21 primarily inhibits the phosphorylation of p130 and p107 by Cdks, allowing them to repress the transcription of a large number of genes required for transit through G_2 (Crosby, M. E. et al. 2007; Taylor, W. R. and G. R. Stark 2001). Similar to p53, p130 can repress transcription of *cyclin B1* and *cdk1* and therefore, enforce the G_2 arrest. p130, just like p53 can have additional targets that do not affect Cdk1 but also contribute to the G_2 arrest. For example, Plk1 is phosphorylated by Wee1 and may degrade it in a proteasome-mediated manner. Chk1 and Plk1 have been also suggested to be important for the checkpoint adaptation, a phenomenon that allows cells to escape from a G_2 checkpoint into mitosis with damaged DNA (Syljuasen, R. G. et al. 2006). Promoter analyses combined with gene expression profiles indicate that in fact both transcription factors may cooperate in regulating target genes. Their role in co-regulating other gene targets is less clear.

Analysis of gene expression profiles based on microarrays of prostate tissue samples have led to a better understanding of the gene regulatory pathways important in prostate cancer progression. Previous studies using laser capture microscopy and microarray identified over 200 genes predominantly expressed in prostate cancer epithelial cells, as compared to adjacent normal stromal tissue (Brown, K. 2006). One of these genes expressed in prostate cancer epithelial cells is the zinc finger transcription factor, WT1 (Brown, K. 2006; Devilard, E. et al. 2006; Reese, J. L. et al. 2008). Importantly, several other genes coordinately expressed with WT1 have been demonstrated to contain evolutionarily conserved WT1 TFBS within their promoters (Eisermann, K. et al. 2008). These novel gene targets were demonstrated to bind WT1 *in vivo* by chromatin immunoprecipitation (ChIP). These androgen responsive genes are now being studied to determine how growth control genes in prostate cancer cells may be regulated by zinc finger transcription factors, such as WT1.

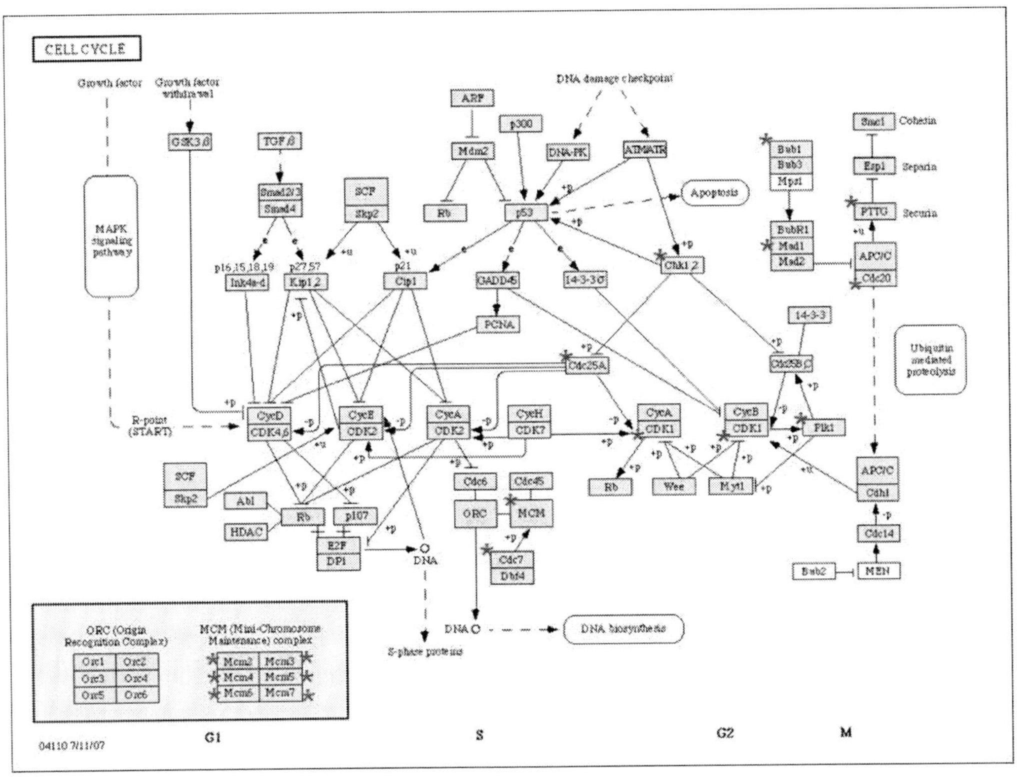

Figure 3. Cell Cycle signaling network.
Cell cycle signaling network as defined by the KEGG pathway (www.genome.ad.jp/kegg/pathway.html - 48k) includes 16 genes suppressed >2-fold in irradiated C4-2 cells. These 16 genes all have predicted E2F binding sites (*), including the MCM family of genes and several CDC genes.

Table 1. Gene Ontology terms that describe top two functional gene clusters among 747 radiation suppressed genes.

GO terms		Gene count#
Annotation Cluster 1, enrichment score@: 23.68		
GOTERM_BP_ALL	cell cycle phase	61
GOTERM_BP_ALL	M phase	55
GOTERM_BP_ALL	mitotic cell cycle	57
GOTERM_BP_ALL	mitosis	46
GOTERM_BP_ALL	M phase of mitotic cell cycle	46
Annotation Cluster 2, enrichment score: 4.51		
GOTERM_MF_ALL	ATP binding	84
GOTERM_MF_ALL	adenyl ribonucleotide binding	84
GOTERM_MF_ALL	adenyl nucleotide binding	86
GOTERM_MF_ALL	ribonucleotide binding	95
GOTERM_MF_ALL	purine ribonucleotide binding	95
GOTERM_MF_ALL	nucleotide binding	109
GOTERM_MF_ALL	purine nucleotide binding	97
Total gene count##		153

Notes: @ enrichment scores were calculated using Functional Annotation Clustering tool in DAVID (Huang da et al. 2007a; Huang da et al. 2007b); # clusters 1 and 2 contained 65 and 110 genes, respectively; differences in gene count reflect some genes lacking the particular GO annotation terms; ## some genes belong to both clusters.

In hematopoietic cells, WT1 regulates growth suppression and apoptosis, as well as differentiation (Ellisen, L. W. et al. 2001; Kudoh, T. et al. 1995; Murata, Y. et al. 1997). WT1 mediates G1 arrest and the induction of apoptosis *in vitro* and *in vivo* (Englert, C. et al. 1995; Fraizer, G. et al. 2004; Kudoh, T. et al. 1995). WT1 inhibition of cell cycle is correlated with induction of *p21* and transcriptional repression of *cyclin E* (Englert, C. et al. 1997; Loeb, D. M. and S. Sukumar 2002). In some cell types growth arrest leads to both p53 dependent and p53 independent apoptosis (Englert, C. et al. 1995; Menke, A. L. et al. 1997). Other apoptotic genes regulated by WT1 include *bcl-2* and *bak*. In some cell types WT1 promotes apoptosis by upregulating *bak* and repressing *bcl-2* (Heckman, C. et al. 1997; Hewitt, S. M. et al. 1995; Morrison, D. J. et al. 2005), while in others it promotes cell survival by, instead, activating *bcl-2* (Mayo, M. W. et al. 1999). Similarly, WT1 has been shown to inhibit p53-mediated apoptosis by binding and stabilizing p53 in kidney cells (Maheswaran, S. et al. 1995). In addition to apoptotic genes, WT1 regulates genes associated with cell proliferation, such as *myc* and *myb* (Hewitt, S. M. et al. 1995; Han, Y. et al. 2004; McCann, S. et al. 1995; Udtha, M. et al. 2003), and cell cycle regultory genes such as *cyclin E* and *p21* (although WT1 regulation of *p21* may be indirect) (Englert, C. et al. 1997; Loeb, D. M. and S. Sukumar 2002).

Table 2. Radiation suppressed genes with mitotic function that are likely E2F regulated.

Gene Symbol	Description	References
ASPM	Abnormal Spindle	(Fujimori, A. et al. 2008)
CCNB2	Cyclin B2	(Stav, D. et al. 2007)
CDC25A	Cell Division Cycle 25A	(Ray, D. and H. Kiyokawa 2007)
CDC7	Cell Division Cycle 7	(Tenca, P. et al. 2007)
CDCA3	Cell Division Cycle Associated 3	(Lim, H. H. and U. Surana 2003)
CDCA5	Cell Division Cycle Associated 5	(Schmitz, J. et al. 2007)
CDCA8	Cell Division Cycle Associated 8	(Date, D. A. et al. 2007)
CDK2	Cyclin-Dependent Kinase 2	(De Boer, L. et al. 2008)
CDKN2C	Cyclin-Dependent Kinase Inhibitor 2c	(Sanchez-Aguilera, A. et al. 2004)
CEP55	Centrosomal Protein 55kDa	(Martinez-Garay, I. et al. 2006)
CHEK1	Chk1 Checkpoint Homolog (S. pombe)	(Enders, G. H. 2008)
CIT	Citron (Rho-Interacting, Serine/Threonine Kinase 21)	(Liu, H. et al. 2003)
CKS2	Cdc28 Protein Kinase Regulatory Subunit 2	(Lan, Y. et al. 2008)
DSN1	Mind Kinetochore Complex Component	(Kline, S. L. et al. 2006)
H2AFX	H2a Histone Family, Member X	(Kinner, A. et al. 2008)
INCENP	Inner Centromere Protein Antigens 135/155kDa	(Ruchaud, S. et al. 2007)
KIF11	Kinesin Family Member 11	(Hayashi, N. et al. 2008)
KIF1A	Kinesin Family Member 1a	(Cheung, I. Y. et al. 2008)
KIF20A	Kinesin Family Member 20a	(Gruneberg, U. et al. 2004)
KIFC2	Kinesin Family Member C2	(Zhu, C. et al. 2005)
KIFC1	Kinesin Family Member C1	(Zhu, C. et al. 2005)
MCM5	Minichromosome Maintenance Deficient 5	(Boskovic, J. et al. 2007)
MCM7	Minichromosome Maintenance Deficient 7	(Ren, B. et al. 2006)
NCAPD2	Chromosome Condensation-Related Smc-Associated Protein 1	(Ball, A. R., Jr. et al. 2002)
NCAPG	Non-SMC condensing I complex, Subunit G	(Ryu, B. et al. 2007)
NDC80	Kinetochore Associated 2/Hec	(Ciferri, C. et al. 2007)
PIF1	5'-To-3' DNA Helicase homolog (S. cerevisiae)	(Futami, K. et al. 2007)
PLK1	Polo-Like Kinase 1	(Petronczki, M. et al. 2008)
PRC1	Protein Regulator of Cytokinesis 1	(Kurasawa, Y. et al. 2004)
SMC4	Structural Maintenance of Chromosomes 4-Like 1	(Losada, A. and T. Hirano 2005)
SPC24	Spindle Pole Body Component 24 Homolog (S. cerevisiae)	(Bharadwaj, R. et al. 2004)
TIPIN	Timeless-Interacting Protein	(Unsal-Kacmaz, K. et al. 2007)
TPX2	Tpx2, Microtubule-Associated, Homolog (Xenopus laevis)	(Gruss, O. J. and I. Vernos 2004)
UBE2C	Ubiquitin-Conjugating Enzyme E2c/ E2	(Summers, M. K. et al. 2008)

Table 3. Numbers of predicted binding sites for three transcription factors (E2F, p53 and WT1) in the promoters of radiation suppressed genes.

GenBank accession no.	Gene symbol	Gene name	Predicted binding sites[a].		
			E2F[b]	p53	WT1
NM_018136	ASPM	Hypothetical Protein Flj10517	2(2)	1	3
NM_004701	CCNB2	Cyclin B2	2(0)	2	5
NM_001789	CDC25A	Cell Division Cycle 25a	7(2)	1	2
NM_003503	CDC7	Cdc7 Cell Division Cycle 7 (S. cerevisiae)	6(0)	1	1
NM_031299	CDCA3	Cell Division Cycle Associated 3	8(5)	7	5
NM_080668	CDCA5	Cell Division Cycle Associated 5	8(2)	5	1
NM_018101	CDCA8	Cell Division Cycle Associated 8	7(3)	3	3
NM_052827	CDK2	Cyclin-Dependent Kinase 2	4(2)	5	1
NM_001262	CDKN2 C	Cyclin-Dependent Kinase Inhibitor 2c (P18, Inhibits Cdk4)	8(5)	2	2
NM_018131	CEP55	Centrosomal Protein 55kDa	6(4)	5	6
NM_001274	CHEK1	Chk1 Checkpoint Homolog (S. pombe)	11(6)	3	3
NM_007174	CIT	Citron (Rho-Interacting, Serine/Threonine Kinase 21)	5(5)	3	1
NM_001827	CKS2	Cdc28 Protein Kinase Regulatory Subunit 2	4(0)	2	3
NM_024918	DSN1	Chromosome 20 Open Reading Frame 172	9(4)	4	1
NM_002105	H2AFX	H2a Histone Family, Member X	5(3)	3	1
NM_020238	INCENP	Inner Centromere Protein Antigens 135/155kDa	8(2)	1	2
NM_004523	KIF11	Kinesin Family Member 11	9(6)	3	1
NM_004321	KIF1A	Kinesin Family Member 1a	10(6)	3	10
NM_005733	KIF20A	Kinesin Family Member 20a	5(3)	3	1
NM_006845	KIF2C	Kinesin Family Member 2c	3(2)	5	8
NM_002263	KIFC1	Kinesin Family Member C1	15(6)	1	1
NM_006739	MCM5	Mcm5 Minichromosome Maintenance Deficient 5, Cell Division Cycle 46 (S. cerevisiae)	5(0)	2	1
NM_182776	MCM7	Mcm7 Minichromosome Maintenance Deficient 7 (S. cerevisiae)	18(9)	4	5
NM_014865	NCAPD 2	Chromosome Condensation-Related Smc-Associated Protein 1	9(5)	3	1
NM_022346	NCAPG	Chromosome Condensation Protein G	9(5)	1	1
NM_006101	NDC80	Kinetochore Associated 2	6(6)	1	2
NM_025049	PIF1	Chromosome 15 Open Reading Frame 20	3(2)	3	3
NM_005030	PLK1	Polo-Like Kinase 1 (Drosophila)	3(2)	2	1
NM_199413	PRC1	Protein Regulator of Cytokinesis 1	7(3)	3	2
NM_0010028 00	SMC4	Smc4 Structural Maintenance of Chromosomes 4-Like 1 (Yeast)	9(6)	3	1
NM_182513	SPC24	Spindle Pole Body Component 24 Homolog (S. cerevisiae)	5(3)	3	1
NM_017858	TIPIN	Timeless-Interacting Protein	14(2)	2	1
NM_012112	TPX2	Tpx2, Microtubule-Associated, Homolog (Xenopus laevis)	3(2)	4	2
NM_181801	UBE2C	Ubiquitin-Conjugating Enzyme E2C	4(0)	4	2

a. E2F, WT1 and p53 binding sites were predicted for each gene using MatInspector (http://www.genomatix.de) as described in text.

b. Total number of E2F sites identified by MatInspector includes heterodimeric and trimeric sites, eg.E2F/DP and Rb/E2F/DP. In parenthesis are number of non-heteromeric E2F sites predicted.

Since many growth factor and growth factor receptors are transcriptionally regulated by WT1, it is possible that within the context of prostate epithelial cells, WT1 may play a role in mitogenic signaling via MAPK pathways. Indeed, WT1 has been demonstrated to regulate MAPK signaling in other cell types (Morrison, D. J. et al. 2008; Gross, I. et al. 2003). Importantly, for prostate cancer cells, WT1 transcriptionally regulates androgen receptor (AR) expression (Kohler, B. et al. 2007; Shimamura, R. et al. 1997) and AR signaling (Hanson, J. et al. 2007; Zaia, A. et al. 2001). Interestingly, the effect of WT1 and E2F on *AR* expression may be opposite, as E2F appears to repress the androgen response (Balk, S. P. and K. E. Knudsen 2008). Since androgen ablation therapy initially induces apoptosis and regression of the prostate epithelia, repression of *AR* expression by E2F may be viewed as another mode of inducing cell death, while upregulation of *AR* by WT1 may maintain proliferation of prostate cancer epithelial cells. Here, we found that WT1 is expressed constitutively in LNCaP C4-2 cells, independent of irradiation. This suggests that WT1 pro-survival signals may be part of the basal gene expression program and not responsive to radiation and other genotoxic agents.

Initially, prostate cancer usually responds to androgen ablation therapy, however most cancers eventually relapse with a hormone-refractory, or androgen-depletion-independent (ADI) phenotype. No treatment currently exists for this fatal stage of the disease. C4-2 cells have been used as models for ADI prostate cancer. They were derived from LNCaP cells that had been serially xenografted through castrated mice (Thalmann, G. N. et al. 1994). While they can grow in the absence of androgen, they are nevertheless hypersensitive to it since they respond robustly to very low concentrations of androgen that do not have any affect on the parental LNCaP, quite similar to many human tumors (Raclaw, K. A. et al. 2008). Understanding their gene expression signature could provide important therapeutic targets or markers for monitoring clinical response.

Results and Discussion

Microarray Analysis of Irradiated LNCaP C4-2 Cells

Prostate carcinoma cells are quite resistant to radiation treatment (Crosby, M. E. et al. 2007; DuPree, E. L. et al. 2004). To determine the nature of this resistance, the gene expression of LNCaP C4-2 cells was examined in response to irradiation. The response of C4-2 cells to ionizing radiation was measured by global gene expression studies using the HG-U95Av2 Affymetrix Gene Chip arrays. Gene expression in irradiated cells was determined as an average between three experiments consisting in C4-2 cells treated with 6 and 24 h of 10 Gy irradiation and C4-2 cells treated with fractionated low dose (2.5 Gy every 24 h for 4 days), which was compared to untreated C4-2 cells to determine those genes with more than 2-fold change in expression. There were 747 radiation suppressed genes that were identified to be down-regulated at least 2-fold in irradiated C4-2 cells (GSE762 record). Functional analysis of these repressed genes was conducted using the Gene Ontology tool DAVID (http://david.abcc.ncifcrf.gov/). The results showed that many of these genes could be classified as pertaining to the biological functions involved in mitosis and cell cycle. One of the most highly enriched functional categories was composed of mitotic genes, showing 5-6 fold enrichment, with p values of 10^{-22} to 10^{-26}. The second most enriched category included genes related to nucleotide binding, which showed a 1.4-1.6 fold enrichment, with p values of 10^{-4} to 10^{-6} (Table I). These two functional categories altogether included 151 genes. These genes were then further analyzed based on their relationship with the E2F transcription factor family, which is known to regulate cell cycle arrest in response to genotoxic stress (Crosby, M. E. et al. 2007; Plesca, D. et al. 2007). Genes suppressed in response to radiation included 34 known E2F targets and 36 additional genes with known mitotic functions (Table II).

Potential Target Genes

In silico analyses of promoter regions of the 34 known E2F target genes and the 36 additional mitotic genes was carried out to identify E2F binding sites. As location of the E2F binding sites in promoters of some of the previously published "target genes" had not been identified, all 70 genes were analyzed. Promoter sequences of the 70 potential E2F target genes (2 kb upstream of the transcriptional start site) were collected from the USCS Human Genome Browser (http://genome.ucsc.edu). Consensus binding sites for E2F, p53, and WT1 used by MatInspector are shown in Figure 1. To determine whether other transcription factors associated with cell cycle regulation might also be predicted within these regulatory regions, binding sites of E2F, WT1, and p53 were identified using MatInspector (http://www.genomatix.de) (Figure 2A and Table III)

As p53 is known to regulate many of the same gene targets as E2F, overlap in the list of potential target genes was expected and 78% of genes with predicted E2F sites also had predicted p53 binding sites. Surprisingly, 56% of E2F target gene promoters also contained predicted WT1 transcription factor binding sites (Figure 2B). Indeed, 34 of these genes contain E2F, p53, and WT1 predicted binding sites, suggesting the potential for either cooperative or competitive regulation by all three transcription factors (Table III). When genomic locations of predicted binding sites were compared, it was discovered that many of the E2F sites were associated with either p53 and/or WT1 binding sites (Figure 2B and Table III).

Cell Cycle Signaling Network

Not surprisingly, the genes suppressed by more than 2-fold in irradiated LNCaP C4-2 cells included 16 genes that have previously been described as part of the cell cycle signaling network (Figure 3). Promoter sequences of these 16 genes all contain potential E2F binding sites, and many have been previously described as E2F regulated (Table II). Five of these genes have predicted TFBS for all three factors: E2F, WT1, and p53. Although E2F and p53-regulated gene targets within the cell cycle network are well studied, the role of WT1 is not as clear (see above). However, WT1 has been shown to interact with and to stabilize p53 (Maheswaran, S. et al. 1995), and like p53, WT1 overexpression is associated with upregulation of p21 (Englert, C. et al. 1997), a key CDK inhibitor. Additionally, WT1 is known to regulate several growth factors and their cognate receptors and these mitogenic signals, as well as MAPK regulators, MKP3 and Sprouty (Morrison, D. J. et al. 2008; Gross I, et al. 2003) impact cell cycle. WT1 has also been implicated in regulation of the AR signaling pathway that is intact within the C4-2 cells (Kohler, B. et al. 2007; Shimamura, R. et al. 1997; Zaia, A. et al. 2001), so it is possible that WT1 may impact cell cycle regulation at several levels, via growth factor and hormone signaling pathways, or directly via regulation of the basal expression of newly identified mitotic genes with predicted WT1 binding sites.

E2F Family Proteins: From DNA Damage to a Stable G_2 Arrest

As the Rb family members interact with multiple binding partners and can control a wide range of signaling pathways (Khidr, L. and P. L. Chen 2006), the role of individual E2Fs has been of great interest. The E2F family of transcription factors controls the expression of ten genes encoded by eight independent loci that are divided, based on sequence homology, into three groups (Kong, L. J. et al. 2007). E2F1-3, which are highly regulated during cell growth and cell cycle generally act as transcriptional activators. E2F4-5, which are constitutively expressed, are transcriptional repressors when bound to p130 or Rb. E2F6-8 act also as transcriptional repressors but do so independently of Rb (Attwooll, C. et al. 2004). Various E2Fs have opposite biological effects but they can also complement each other (Crosby, M. E. and A. Almasan 2004). Of these, E2F4 and E2F5 are unique as they are regulated through their predominantly cytoplasmic cellular localization, which is facilitated by a nuclear export signal (Gaubatz, S. et al. 2001). However, they do not have a nuclear localization signal; therefore it is believed that nuclear translocation is dependent on binding to p130 (Chestukhin, A. et al. 2002) or the other pocket proteins, Rb or p107.

Deregulation of the Rb/E2F pathway in human fibroblasts following a genotoxic insult leads to E2F1-mediated apoptosis. This deregulation was first reported in Rb-deficient cells (Almasan, A. et al. 1995). E2F4 and E2F5, in complex with p130, are known to be present primarily in quiescent cells (Sardet, C. et al. 1995), with decreasing levels of E2F4 found associated with Rb and p107-Cdk2 during S-phase progression (Farkas, T. et al. 2002; Takahashi, Y. et al. 2000). Interestingly, we found that a typical genotoxic agent, ionizing radiation, induces complex formation of E2F4, but not E2F5, with the unphosphorylated form of p130 (DuPree, E. L. et al. 2004). Examining the subcellular localization indicated that E2F4 and p130 were present in the nucleus a short time following irradiation; their co-localization reached a peak at 24 h. In contrast, the protein levels of E2F1, which may counterbalance the activity of E2F4 (Crosby, M. E. and A. Almasan 2004), were dramatically down-regulated. This pattern of decreased E2F1 levels was sustained following irradiation or treatment with the topoisomerase II inhibitor, VP16.

Irradiation of C4-2 prostate carcinoma cells depleted of E2F4 by siRNA (siE2F4) markedly elevated DNA double strand break (DSB) levels leading to a predominant appearance of sub-G_2 DNA content indicating that E2F4 activity helps to sustain G_2 arrest and prevent cell death (Crosby, M. E. et al. 2007; Plesca, D. et al. 2007). Previously, Cyclin B1, Bub1B, CENP-E, Chk1, HEC, Mad2L1, and PTTG1, all known to be regulated by E2F4 and to have a critical function in mitosis, were found to be associated with E2F4 function. Cdc6, known to be essential for DNA replication could also play an important role in mitosis (Boronat, S. and J. L. Campbell 2007). Moreover, chromatin immunoprecipitation (ChIP) analyses demonstrated specific E2F4 recruitment to the promoter binding sites of *PTTG1* and *BUB3*. In contrast, E2F4 was bound constitutively to *Cdc6,* an observation similar to the constitutive chromatin binding by p53 to its target promoters (Crosby, M. E. and A. Almasan 2004), indicating the involvement of other factors or post-transcriptional regulation for their activation. PTTG1, BUB1, BUB3, Mad2, and CENP-E are mitotic checkpoint proteins. Stathmin (Polager, S. et al. 2002), HEC, and Ect-2, are involved in spindle formation (Eguchi, T. et al. 2007). The abundance of these targets and those described in this chapter and their individual role proposed in the DNA damage response is continuously expanding.

Conclusions

Several expression profiling experiments have documented that, similar to p130, E2F4 regulates many genes that are critical for mitotic entry. Our results are consistent with previous studies based on combined cross-linking DNA immunoprecipitation and microarray experiments (Cam, H. and B. D. Dynlacht 2003), which have identified some of the same targets we have uncovered as E2F4 regulated genes with a putative role in various phases of the cell cycle including G_2 and M (Chestukhin, A. et al. 2002). Similar targets were independently identified in studies examining specific p130 or Rb-regulated gene expression (Jackson, M. W. et al. 2005). At the same time, Rb-deficient U2OS cells showed an analogous dependence on E2F4-regulated genes (Eguchi, T. et al. 2007), suggesting a context-dependence for p130 or Rb requirement for regulating mitotic entry. From our results and other studies we conclude that in response to DNA damage, these tumor cells undergo cell cycle exit initiated in G_2 (Baus, F. et al. 2003; Crosby, M. E. et al. 2007; Jackson, M. W. et al. 2005). Independent profiling of the radiation response in human diploid fibroblasts has yielded similar results, with a prominent induction of additional markers of G_0 (Zhou, T. et al. 2006).

Interestingly, a number of E2F4-bound genes originally identified through array analyses were also bound by E2F1 (Chestukhin, A. et al. 2002). This observation was extended to the promoters of genes regulated during G_2 and M containing distinct E2F4 and E2F1 binding sites (Cam, H. and B. D. Dynlacht 2003; Zhu, W. et al. 2004). Binding sites for many E2Fs, including E2F1 and E2F4 are similar and could have interchangeable roles (Unsal-Kacmaz, K. et al. 2007). Some of these promoters may display collaborative binding of several E2F family members to the same elements, such as in case of *ect-2* (Eguchi, T. et al. 2007), while others distinct positive and negative E2F-dependent regulatory elements (Zhu, W. et al. 2004). In addition, E2F-dependent transcriptional activity may require binding of other transcription factors, such NF-Y or Myb to regulate the transcription of Cdk1 (Zhu, W. et al. 2004). The balance between the E2Fs that occupy these promoters determines the physiological outcome, with more E2F4 bound resulting in cell cycle arrest (Crosby, M. E. and A. Almasan 2004; Crosby, M. E. et al. 2007). Here, we show that many E2F or WT1 promoters have p53 binding sites. While fewer E2F promoters have WT1 binding sites, this is a novel observation suggesting a potential for their interaction or even co-regulation of target promoters. Global analysis of E2F-binding sites suggests their widespread presence in the human genome (Bieda, M. et al. 2006). In addition to targets with consensus binding sites (Figure 2A), there are many targets, *in vivo*, that do not have such binding sites (Rabinovich, A. et al. 2008). Clearly many radiation suppressed genes are E2F target genes that may provide important therapeutic targets.

Materials and Methods

Expression Profiling with Oligonucleotide Arrays

Total RNA was extracted from cells by the Trizol® (Invitrogen) method and reverse-transcribed into double-stranded cDNA following isolation of Poly(A) RNA with RNAeasy (Qiagen, Valencia, CA). The double-stranded cDNA was used to prepare biotinylated cRNA (Enzo Bioarray High Yield Kit, Affymetrix, Santa Clara, CA). These cRNA probes were hybridized to the Affymetrix HG-U95Av2 oligonucleotide array. The GenePix 4000A laser scanner (Axon Instruments, Foster City, CA) and GenePix v.3.0 scanning software were used to analyze the data. Quantitative RT-PCR experiments were performed to validate the levels of expression in the repressed gene targets. Total RNA was extracted from cells by the Trizol® method and reverse-transcribed into cDNA using Taqman® Reverse Transcription Reagents (Roche for ABI, Branchburg, NJ). The gene expression data is available in the GEO database (GSE762 record).

Microarray Data Analysis

Raw expression data was imported into the BRB Array Tools version 3.6.1 (http://linus.nci.nih.gov/BRB-ArrayTools.html), and filtered to exclude spots based on channel intensity and flag values; each array was normalized using the median over the entire array. Lists of genes differentially expressed at 2-fold or higher levels between untreated and treated cells (the later value was based on the average between three treated arrays) were identified, and 747 genes down-regulated in treated cells (i.e., irradiation responsive) were used for further analyses. DAVID Bioinformatics Resources were used to provide functional interpretation of the irradiation sensitive gene list. In particular, the DAVID Functional Annotation Clustering Tool was used to cluster 747 genes into 23 functional annotation clusters with enrichment scores > 1 (Huang da, W. et al. 2007a; Huang da, W. et al. 2007b). Enrichment scores reflect enriched annotation of terms that belong to the particular gene group, and generally the higher the score of the functional cluster, the more important (enriched) the role of its member genes (Huang da, W. et al. 2007a), although clusters with low enrichment scores are also potentially interesting (Huang da, W. et al. 2007a).

Promoter Sequences and Transcription Factor Binding Sites Analysis

Promoter sequences of 70 potential E2F target genes (2 kb upstream of the transcriptional start site) were collected from USCS Human Genome Browser (http://genome.ucsc.edu). Transcription factor binding sites of E2F, WT1 and p53 were predicted for each gene using MatInspector (http://www.genomatix.de) (Figure 3) (Cartharius, K. et al. 2005). The default parameters of similarity thresholds were used for all examined genes, and they were as follows: core similarity > 0.75 and optimized matrix similarity thresholds (i.e., those that minimize false positives for each individual matrix as available in the library). Core similarity refers to the four most conserved consecutive nucleotides of the matrix, usually the most critical sites for protein binding, and reaches 1.0 only when there is a perfect match. The Matrix Family Library Version 7.1 (June 2008) that included 519 matrices of the vertebrate group was used.

Acknowledgment

Work described in this article was supported by a research grant from the US National Institutes of Health (CA81504).

References

Almasan, A., Yin, Y., Kelly, R. E., Lee, E. Y., Bradley, A., Li, W., Bertino, J. R. & Wahl, G. M. (1995). Deficiency of retinoblastoma protein leads to inappropriate S-phase entry, activation of E2F-responsive genes, and apoptosis. *Proc Natl Acad Sci U S A*, *92*, 5436-5440.

Atherton-Fessler, S., Parker, L. L., Geahlen, R. L. & Piwnica-Worms, H. (1993). Mechanisms of p34cdc2 regulation. *Mol Cell Biol*, *13*, 1675-1685.

Attwooll, C., Lazzerini Denchi, E. & Helin, K. (2004). The E2F family: specific functions and overlapping interests. *Embo J*, *23*, 4709-4716.

Bagella, L., Sun, A., Tonini, T., Abbadessa, G., Cottone, G., Paggi, M. G., De Luca, A., Claudio, P. P. & Giordano, A. (2007). A small molecule based on the pRb2/p130 spacer domain leads to inhibition of cdk2 activity, cell cycle arrest and tumor growth reduction in vivo. *Oncogene*, *26*, 1829-1839.

Balk, S. P. & Knudsen, K. E. (2008). AR, the cell cycle, and prostate cancer. *Nuclear Receptor Signaling*, *6*, 1-12.

Ball, A. R., Jr., Schmiesing, J. A., Zhou, C., Gregson, H. C., Okada, Y., Doi, T. & Yokomori, K. (2002). Identification of a chromosome-targeting domain in the human condensin subunit CNAP1/hCAP-D2/Eg7. *Mol Cell Biol*, *22*, 5769-5781.

Baus, F., Gire, V., Fisher, D., Piette, J. & Dulic, V. (2003). Permanent cell cycle exit in G2 phase after DNA damage in normal human fibroblasts. *Embo J*, *22*, 3992-4002.

Bharadwaj, R., Qi, W. & Yu, H. (2004). Identification of two novel components of the human NDC80 kinetochore complex. *J Biol Chem*, *279*, 13076-13085.

Bieda, M., Xu, X., Singer, M. A., Green, R. & Farnham, P. J. (2006). Unbiased location analysis of E2F1-binding sites suggests a widespread role for E2F1 in the human genome. *Genome Res*, 16, 595-605.

Boronat, S. & Campbell, J. L. (2007). Mitotic Cdc6 stabilizes anaphase-promoting complex substrates by a partially Cdc28-independent mechanism, and this stabilization is suppressed by deletion of Cdc55. *Mol Cell Biol*, *27*, 1158-1171.

Boskovic, J., Coloma, J., Aparicio, T., Zhou, M., Robinson, C. V., Mendez, J. & Montoya, G. (2007). Molecular architecture of the human GINS complex. *EMBO Rep*, *8*, 678-684.

Brown, K. (2006). *Differential gene expression patterns in prostate cancer epithelial and interstitial stromal cells*. MS Thesis, Kent State University.

Cam, H. & Dynlacht, B. D. (2003). Emerging roles for E2F: beyond the G1/S transition and DNA replication. *Cancer Cell*, *3*, 311-316.

Cartharius, K., Frech, K., Grote, K., Klocke, B., Haltmeier, M., Klingenhoff, A., Frisch, M., Bayerlein, M. & Werner, T. (2005). MatInspector and beyond: promoter analysis based on transcription factor binding sites. *Bioinformatics*, *21*, 2933-2942.

Chan, T. A., Hermeking, H., Lengauer, C., Kinzler, K. W. & Vogelstein, B. (1999). 14-3-3Sigma is required to prevent mitotic catastrophe after DNA damage. *Nature*, *401*, 616-620.

Chestukhin, A., Litovchick, L., Rudich, K. & DeCaprio, J. A. (2002). Nucleocytoplasmic shuttling of p130/RBL2: novel regulatory mechanism. *Mol Cell Biol*, *22*, 453-468.

Cheung, I. Y., Feng, Y., Gerald, W. & Cheung, N. K. (2008). Exploiting gene expression profiling to identify novel minimal residual disease markers of neuroblastoma. *Clin Cancer Res*, *14*, 7020-7027.

Ciferri, C., Musacchio, A. & Petrovic, A. (2007). The Ndc80 complex: hub of kinetochore activity. *FEBS Lett*, *581*, 2862-2869.

Crosby, M. E. & Almasan, A. (2004). Opposing roles of E2Fs in cell proliferation and death. *Cancer Biol Ther*, *3*, 1208-1211.

Crosby, M. E., Jacobberger, J., Gupta, D., Macklis, R. M. & Almasan, A. (2007). E2F4 regulates a stable G2 arrest response to genotoxic stress in prostate carcinoma. *Oncogene*, *26*, 1897-1909.

Date, D. A., Jacob, C. J., Bekier, M. E., Stiff, A. C., Jackson, M. W. & Taylor, W. R. (2007). Borealin is repressed in response to p53/Rb signaling. *Cell Biol Int*, *31*, 1470-1481.

De Boer, L., Oakes, V., Beamish, H., Giles, N., Stevens, F., Somodevilla-Torres, M., Desouza, C. & Gabrielli, B. (2008). Cyclin A/cdk2 coordinates centrosomal and nuclear mitotic events. *Oncogene*, *27*, 4261-4268.

Dephoure, N., Zhou, C., Villen, J., Beausoleil, S. A., Bakalarski, C. E., Elledge, S. J. & Gygi, S. P. (2008). A quantitative atlas of mitotic phosphorylation. *Proc Natl Acad Sci U S A*, *105*, 10762-10767.

Devilard, E., Bladou, F., Ramuz, O., Karsenty, G., Dales, J. P., Gravis, G., Nguyen, C., Bertucci, F., Xerri, L. & Birnbaum, D. (2006). FGFR1 and WT1 are markers of human prostate cancer progression. *BMC Cancer*, *6*, 272.

DuPree, E. L., Mazumder, S. & Almasan, A. (2004). Genotoxic stress induces expression of E2F4, leading to its association with p130 in prostate carcinoma cells. *Cancer Res*, *64*, 4390-4393.

Eguchi, T., Takaki, T., Itadani, H. & Kotani, H. (2007). RB silencing compromises the DNA damage-induced G2/M checkpoint and causes deregulated expression of the ECT2 oncogene. *Oncogene*, *26*, 509-520.

Eisermann, K., Tandon, S., Bazarov, A., Brett, A., Fraizer, G. & Piontkivska, H. (2008). Evolutionary conservation of zinc finger transcription factor binding sites in promoters of genes co-expressed with WT1 in prostate cancer. *BMC Genomics*, *9*, 337.

Ellisen, L. W., Carlesso, N., Cheng, T., Scadden, D. T. & Haber, D. A. (2001). The Wilms tumor suppressor WT1 directs stage-specific quiescence and differentiation of human hematopoietic progenitor cells. *Embo J*, *20*, 1897-1909.

Enders, G. H. (2008). Expanded roles for Chk1 in genome maintenance. *J Biol Chem*, *283*, 17749-17752.

Englert, C., Hou, X., Maheswaran, S., Bennett, P., Ngwu, C., Re, G. G., Garvin, A. J., Rosner, M. R. & Haber, D. A. (1995). WT1 suppresses synthesis of the epidermal growth factor receptor and induces apoptosis. *Embo J*, *14*, 4662.

Englert, C., Maheswaran, S., Garvin, A., Kreidberg, J. & Daniel A. Haber, D. A. (1997). Induction of p21 by the Wilms' tumor suppressor gene WT1. *Cancer Res*, *57*, 1429-1434.

Farkas, T., Hansen, K., Holm, K., Lukas, J. & Bartek, J. (2002). Distinct phosphorylation events regulate p130- and p107-mediated repression of E2F-4. *J Biol Chem.*, *277*, 26741-26752.

Fraizer, G., Leahy, R., Priyadarshini, S., Graham, K., Delacerda, J. & Diaz, M. (2004). Suppression of prostate tumor cell growth in vivo by WT 1, the Wilms' tumor suppressor gene. *Intl J Oncol*, *24*, 461-471.

Fujimori, A., Yaoi, T., Ogi, H., Wang, B., Suetomi, K., Sekine, E., Yu, D., Kato, T., Takahashi, S., Okayasu, R., Itoh, K. & Fushiki, S. (2008). Ionizing radiation downregulates ASPM, a gene responsible for microcephaly in humans. *Biochem Biophys Res Commun*, *369*, 953-957.

Futami, K., Shimamoto, A. & Furuichi, Y. (2007). Mitochondrial and nuclear localization of human Pif1 helicase. *Biol Pharm Bull*, *30*, 1685-1692.

Gaubatz, S., Lees, J. A., Lindeman, G. J. & Livingston, D. M. (2001). E2F4 is exported from the nucleus in a CRM1-dependent manner. *Mol Cell Biol*, *21*, 1384-1392.

Gross, I. M. D. H. D. P. G. K. E. M. A. M. M. H. S. & Sassoon, D. W. P. L. M. L. J. D. (2003). The receptor tyrosine kinase regulator sprouty1 is a target of the tumor suppressor WT1 and important for kidney development. *J Biol Chem*, *278*, 41420-41430.

Gross, I., Morrison, D. J., Hyink, D. P., Georgas, K., English, M. A., Mericskay, M., Hosono, S., Sassoon, D., Wilson, P. D., Little, M. & Licht, J. D. (2003). The receptor tyrosine kinase regulator Sprouty1 is a target of the tumor suppressor WT1 and important for kidney development. *J Biol Chem*, *278*, 41420-41430.

Gruneberg, U., Neef, R., Honda, R., Nigg, E. A. & Barr, F. A. (2004). Relocation of Aurora B from centromeres to the central spindle at the metaphase to anaphase transition requires MKlp2. *J Cell Biol*, *166*, 167-172.

Gruss, O. J. & Vernos, I. (2004). The mechanism of spindle assembly: functions of Ran and its target TPX2. *J Cell Biol*, *166*, 949-955.

Hagting, A., Karlsson, C., Clute, P., Jackman, M. & Pines, J. (1998). MPF localization is controlled by nuclear export. *Embo J*, *17*, 4127-4138.

Han, Y., San-Marina, S., Liu, J. & Minden, M. D. (2004). Transcriptional activation of c-myc proto-oncogene by WT1 protein. *Oncogene, 23*, 6933-6941.

Hanson, J., Gorman, J., Reese, J. & Fraizer, G. (2007). Regulation of vascular endothelial growth factor, VEGF, gene promoter by the tumor suppressor, WT1. *Front Biosci, 12*, 2279-2290.

Hayashi, N., Koller, E., Fazli, L. & Gleave, M. E. (2008). Effects of Eg5 knockdown on human prostate cancer xenograft growth and chemosensitivity. *Prostate, 68*, 1283-1295.

Heckman, C., Mochon, E., Arcinas, M. & Boxer, L. M. (1997). The WT1 protein is a negative regulator of the normal bcl-2 allele in t (14; 18) lymphomas. *J Biol Chem, 272*, 19609-19614.

Hermeking, H., Lengauer, C., Polyak, K., He, T. C., Zhang, L., Thiagalingam, S., Kinzler, K. W. & Vogelstein, B. (1997). 14-3-3 sigma is a p53-regulated inhibitor of G2/M progression. *Mol Cell, 1*, 3-11.

Herrera, R. E., Chen, F. & Weinberg, R. A. (1996). Increased histone H1 phosphorylation and relaxed chromatin structure in Rb-deficient fibroblasts. *Proc Natl Acad Sci U S A, 93*, 11510-11515.

Hewitt, S. M., Fraizer, G. C. & Saunders, G. F. (1995). Transcriptional silencer of the Wilms' Tumor gene WT1 contains an Alu repeat. *J Biol Chem, 270*, 17908-17912.

Huang da, W., Sherman, B. T., Tan, Q., Collins, J. R., Alvord, W. G., Roayaei, J., Stephens, R., Baseler, M. W., Lane, H. C. & Lempicki, R. A. (2007a). The DAVID Gene Functional Classification Tool: a novel biological module-centric algorithm to functionally analyze large gene lists. *Genome Biol, 8*, R183.

Huang da, W., Sherman, B. T., Tan, Q., Kir, J., Liu, D., Bryant, D., Guo, Y., Stephens, R., Baseler, M. W., Lane, H. C. & Lempicki, R. A. (2007b). DAVID Bioinformatics Resources: expanded annotation database and novel algorithms to better extract biology from large gene lists. *Nucleic Acids Res, 35*, W169-175.

Iliakis, G., Wang, Y., Guan, J. & Wang, H. (2003). DNA damage checkpoint control in cells exposed to ionizing radiation. *Oncogene, 22*, 5834-5847.

Jackson, M. W., Agarwal, M. K., Yang, J., Bruss, P., Uchiumi, T., Agarwal, M. L., Stark, G. R. & Taylor, W. R. (2005). p130/p107/p105Rb-dependent transcriptional repression during DNA-damage-induced cell-cycle exit at G2. *J Cell Sci, 118*, 1821-1832.

Khidr, L. & Chen, P. L. (2006). RB, the conductor that orchestrates life, death and differentiation. *Oncogene, 25*, 5210-5219.

Kinner, A., Wu, W., Staudt, C. & Iliakis, G. (2008). Gamma-H2AX in recognition and signaling of DNA double-strand breaks in the context of chromatin. *Nucleic Acids Res, 36*, 5678-5694.

Kline, S. L., Cheeseman, I. M., Hori, T., Fukagawa, T. & Desai, A. (2006). The human Mis12 complex is required for kinetochore assembly and proper chromosome segregation. *J Cell Biol, 173*, 9-17.

Kohler, B., Delezoide, A. L., Boizet-Bonhoure, B., McPhaul, M. J., Sultan, C. & Lumbroso, S. (2007). Coexpression of Wilms' tumor suppressor 1 (WT1) and androgen receptor (AR) in the genital tract of human male embryos and regulation of AR promoter activity by WT1. *J Mol Endocrinol, 38*, 547-554.

Kong, L. J., Chang, J. T., Bild, A. H. & Nevins, J. R. (2007). Compensation and specificity of function within the E2F family. *Oncogene*, *26*, 321-327.

Kudoh, T., Ishidate, T., Moriyama, M., Toyoshima, K. & Akiyama, T. (1995). G1/S phase arrest induced by Wilms Tumor protein WT1 is abrogated by Cyclin/CDK complexes. *Proc Natl Acad Sci U S A*, *92*, 4517-4521.

Kurasawa, Y., Earnshaw, W. C., Mochizuki, Y., Dohmae, N. & Todokoro, K. (2004). Essential roles of KIF4 and its binding partner PRC1 in organized central spindle midzone formation. *Embo J*, *23*, 3237-3248.

Lan, Y., Zhang, Y., Wang, J., Lin, C., Ittmann, M. M. & Wang, F. (2008). Aberrant expression of Cks1 and Cks2 contributes to prostate tumorigenesis by promoting proliferation and inhibiting programmed cell death. *Int J Cancer*, *123*, 543-551.

Lim, H. H. & Surana, U. (2003). Tome-1, wee1, and the onset of mitosis: coupled destruction for timely entry. *Mol Cell*, *11*, 845-846.

Liu, H., Di Cunto, F., Imarisio, S. & Reid, L. M. (2003). Citron kinase is a cell cycle-dependent, nuclear protein required for G2/M transition of hepatocytes. *J Biol Chem*, *278*, 2541-2548.

Loeb, D. M. & Sukumar, S. (2002). The Role of WT1 in Oncogenesis: Tumor Suppressor or Oncogene? *Int J Hemato*, *76*, 117-126.

Losada, A. & Hirano, T. (2005). Dynamic molecular linkers of the genome: the first decade of SMC proteins. *Genes Dev*, *19*, 1269-1287.

Macaluso, M., Montanari, M. & Giordano, A. (2006). Rb family proteins as modulators of gene expression and new aspects regarding the interaction with chromatin remodeling enzymes. *Oncogene*, *25*, 5263-5267.

Maheswaran, S., Englert, C., Bennett, P., Heinrich, G. & Haber, D. A. (1995). The WT1 gene product stabilizes p53 and inhibits p53-mediated apoptosis. *Genes Dev*, *9*, 2143-2156.

Martinez-Garay, I., Rustom, A., Gerdes, H. H. & Kutsche, K. (2006). The novel centrosomal associated protein CEP55 is present in the spindle midzone and the midbody. *Genomics*, *87*, 243-253.

Mayo, M. W., Wang, C. Y., Drouin, S. S., Madrid, L. V., Marshall, A. F., Reed, J. C., Weissman, B. E. & Baldwin, A. S. (1999). WT1 modulates apoptosis by transcriptionally upregulating the bcl-2 proto-oncogene. *Embo J*, *18*, 3990-4003.

McCann, S., Sullivan, J., Guerra, J., Arcinas, M. & Boxer, L. M. (1995). Repression of the c-myb gene by WT1 protein in T and B cell lines. *J Biol Chem*, *270*, 23785-23789.

Menke, A. L., Shvarts, A., Riteco, N., van Ham, R. C. A., van der Eb, A. J. & Jochemsen, A. G. (1997). Wilms'tumor 1-KTS isoforms induce p53-independent apoptosis that can be partially rescued by expression of the epidermal growth factor receptor or the insulin receptor. *Cancer Res*, *57*, 1353-1363.

Morrison, D. J., English, M. A. & Licht, J. D. (2005). WT1 induces apoptosis through transcriptional regulation of the proapoptotic Bcl-2 family member Bak. *Cancer Res*, *65*, 8174-8182.

Morrison, D. J., Kim, M. K. H., Berkofsky-Fessler, W. & Licht, J. D. (2008). WT1 Induction of Mitogen-Activated Protein Kinase Phosphatase 3 Represents a Novel Mechanism of Growth Suppression. *Mol Cancer Res*, *6*, 1225.

Murata, Y., Kudoh, T., Sugiyama, H., Toyoshima, K. & Akiyama, T. (1997). The Wilms tumor suppressor gene WT1 induces G1 arrest and apoptosis in myeloblastic leukemia M1 cells. *FEBS letters*, *409*, 41-45.

Parker, L. L. & Piwnica-Worms, H. (1992). Inactivation of the p34cdc2-cyclin B complex by the human WEE1 tyrosine kinase. *Science*, *257*, 1955-1957.

Petronczki, M., Lenart, P. & Peters, J. M. (2008). Polo on the rise-from mitotic entry to cytokinesis with Plk1. *Dev Cell*, *14*, 646-659.

Plesca, D., Crosby, M. E., Gupta, D. & Almasan, A. (2007). E2F4 Function in G2: Maintaining G2-arrest to prevent mitotic entry with damaged DNA. *Cell Cycle*, *6*, 1147-1152.

Polager, S. & Ginsberg, D. (2003). E2F mediates sustained G2 arrest and down-regulation of Stathmin and AIM-1 expression in response to genotoxic stress. *J Biol Chem*, *278*, 1443-1449.

Polager, S., Kalma, Y., Berkovich, E. & Ginsberg, D. (2002). E2Fs up-regulate expression of genes involved in DNA replication, DNA repair and mitosis. *Oncogene*, *21*, 437-446.

Rabinovich, A., Jin, V. X., Rabinovich, R., Xu, X. & Farnham, P. J. (2008). E2F in vivo binding specificity: Comparison of consensus versus nonconsensus binding sites. *Genome Res*, *18*, 1763-1777.

Raclaw, K. A., Heemers, H. V., Kidd, E. M., Dehn, S. M. & Tindall, D. J. (2008). Induction of FLIP expression by androgens protects cancer cells from TRAIL mediated apoptosis. *The Prostate*, *68*, 1696-1706.

Ray, D. & Kiyokawa, H. (2007). CDC25A levels determine the balance of proliferation and checkpoint response. *Cell Cycle*, *6*, 3039-3042.

Reese, J. L., Brown, K. E., Magi-Galluzzi, C. & Fraizer, G. C. (2008). The role of WT1 in prostate cancer progression. *Proc American Assoc Cancer Res*, Abstract #2796.

Ren, B., Yu, G., Tseng, G. C., Cieply, K., Gavel, T., Nelson, J., Michalopoulos, G., Yu, Y. P. & Luo, J. H. (2006). MCM7 amplification and overexpression are associated with prostate cancer progression. *Oncogene*, *25*, 1090-1098.

Ruchaud, S., Carmena, M. & Earnshaw, W. C. (2007). Chromosomal passengers: conducting cell division. *Nat Rev Mol Cell Biol*, *8*, 798-812.

Ryu, B., Kim, D. S., Deluca, A. M. & Alani, R. M. (2007). Comprehensive expression profiling of tumor cell lines identifies molecular signatures of melanoma progression. *PLoS ONE*, *2*, e594.

Sanchez-Aguilera, A., Delgado, J., Camacho, F. I., Sanchez-Beato, M., Sanchez, L., Montalban, C., Fresno, M. F., Martin, C., Piris, M. A. & Garcia, J. F. (2004). Silencing of the p18INK4c gene by promoter hypermethylation in Reed-Sternberg cells in Hodgkin lymphomas. *Blood*, *103*, 2351-2357.

Sardet, C., Vidal, M., Cobrinik, D., Geng, Y., Onufryk, C., Chen, A. & Weinberg, R. A. (1995). E2F-4 and E2F-5, two members of the E2F family, are expressed in the early phases of the cell cycle. *Proc Natl Acad Sci U S A*, *92*, 2403-2407.

Schmitz, J., Watrin, E., Lenart, P., Mechtler, K. & Peters, J. M. (2007). Sororin is required for stable binding of cohesin to chromatin and for sister chromatid cohesion in interphase. *Curr Biol*, *17*, 630-636.

Shimamura, R., Fraizer, G. C., Trapman, J., Lau, Y. & Saunders, G. F. (1997). The Wilms' tumor gene WT1 can regulate genes involved in sex determination and differentiation: SRY, Mullerian-inhibiting substance, and the androgen receptor. *Clin Cancer Res*, *3*, 2571-2580.

Stav, D., Bar, I. & Sandbank, J. (2007). Usefulness of CDK5RAP3, CCNB2, and RAGE genes for the diagnosis of lung adenocarcinoma. *Int J Biol Markers*, *22*, 108-113.

Summers, M. K., Pan, B., Mukhyala, K. & Jackson, P. K. (2008). The unique N terminus of the UbcH10 E2 enzyme controls the threshold for APC activation and enhances checkpoint regulation of the APC. *Mol Cell*, *31*, 544-556.

Syljuasen, R. G., Jensen, S., Bartek, J. & Lukas, J. (2006). Adaptation to the ionizing radiation-induced G2 checkpoint occurs in human cells and depends on checkpoint kinase 1 and Polo-like kinase 1 kinases. *Cancer Res*, *66*, 10253-10257.

Takahashi, Y., Rayman, J. B. & Dynlacht, B. D. (2000). Analysis of promoter binding by the E2F and pRB families in vivo: distinct E2F proteins mediate activation and repression. *Genes Dev*, *14*, 804-816.

Taylor, W. R. & Stark, G. R. (2001). Regulation of the G2/M transition by p53. *Oncogene*, *20*, 1803-1815.

Tenca, P., Brotherton, D., Montagnoli, A., Rainoldi, S., Albanese, C. & Santocanale, C. (2007). Cdc7 is an active kinase in human cancer cells undergoing replication stress. *J Biol Chem*, *282*, 208-215.

Thalmann, G. N., Anezinis, P. E., Chang, S. M., Zhau, H. E., Kim, E. E., Hopwood, V. L., Pathak, S., von Eschenbach, A. C. & Chung, L. W. (1994). Androgen-independent cancer progression and bone metastasis in the LNCaP model of human prostate cancer. *Cancer Res*, *54*, 2577-2581.

Udtha, M., Lee, S. J., Alam, R., Coombes, K. & Huff, V. (2003). Upregulation of c-MYC in WT1-mutant tumors: assessment of WT1 putative transcriptional targets using cDNA microarray expression profiling of genetically defined Wilms' tumors. *Oncogene*, *22*, 3821-3826.

Unsal-Kacmaz, K., Chastain, P. D., Qu, P. P., Minoo, P., Cordeiro-Stone, M., Sancar, A. & Kaufmann, W. K. (2007). The human Tim/Tipin complex coordinates an Intra-S checkpoint response to UV that slows replication fork displacement. *Mol Cell Biol*, *27*, 3131-3142.

Wahl, G. M. & Carr, A. M. (2001). The evolution of diverse biological responses to DNA damage: insights from yeast and p53. *Nat Cell Biol*, *3*, E277-286.

Zaia, A., Fraizer, G. C., Piantanelli, L. & Saunders, G. F. (2001). Transcriptional regulation of the androgen signaling pathway by the Wilms' tumor suppressor gene WT1. *Anticancer Res*, *21*, 1-10.

Zhou, T., Chou, J. W., Simpson, D. A., Zhou, Y., Mullen, T. E., Medeiros, M., Bushel, P. R., Paules, R. S., Yang, X., Hurban, P., Lobenhofer, E. K. & Kaufmann, W. K. (2006). Profiles of global gene expression in ionizing-radiation-damaged human diploid fibroblasts reveal synchronization behind the G1 checkpoint in a G0-like state of quiescence. *Environ Health Perspect*, *114*, 553-559.

Zhu, C., Zhao, J., Bibikova, M., Leverson, J. D., Bossy-Wetzel, E., Fan, J. B., Abraham, R. T. & Jiang, W. (2005). Functional analysis of human microtubule-based motor proteins, the kinesins and dyneins, in mitosis/cytokinesis using RNA interference. *Mol Biol Cell*, *16*, 3187-3199.

Zhu, W., Giangrande, P. H. & Nevins, J. R. (2004). E2Fs link the control of G1/S and G2/M transcription. *Embo J*, *23*, 4615-4626.

In: Handbook of Prostate Cancer Cell Research...
Editor: Alan T. Meridith

ISBN: 978-1-60741-954-9

Chapter 12

Molecular Genetic Alterations in Prostate Cancer Microenvironment

Tatiana V. Kekeeva [1,2]***, M.V. Nemtsova*** [1]***, Y.Y. Andreeva*** [2]***, G.A. Frank*** [2]***, I.G. Rusakov*** [2] ***and Dmitri V. Zaletayev*** [1]

[1] Institute of Molecular Medicine, Sechenov Moscow Medical Academy, Moscow, Russian Federation

[2] Herzen Moscow Research Institute of Oncology Diseases, Moscow, Russian Federation

Abstract

Due to intensive investigations for the last ten years it has become increasingly clear that tumor microenvironment plays a critical role in prostate carcinogenesis. Cancer progression depends on either molecular alterations of epithelial cells and tumor-associated stroma, which provides required changes for growth, invasion and distant metastasis of tumor epithelium. Tumor microenvironment undergoes significant modification such as protein expression alterations and various genetic changes of stromal cells. Accumulation of multiplicity genetic alterations such as tumor suppressor genes hypermethylation, allelic imbalance, loss of heterozygosity, mutations, telomere shortening is typical not only for cancer epithelial cells but for tumor-associated fibroblasts and endothelial cells as well. Furthermore, these stromal cell alterations are not tumor epithelium specific.

We investigated epigenetic changes and allelic imbalance (LOH/AI) in prostate cancer epithelia, tumor-associated stroma and prostate intraepithelial neoplasia (PIN) in prostatectomy specimens from patients with pT1–T4 stage prostate cancer by using laser capture microdissection. Microsatellite allelotyping was evaluated using 4 highly polymorphic markers for chromosomal regions 8p22, 16q23-24 and 13q14. The methylation status of four genes significant in prostate carcinogenesis *P16*, *HIC1*, *N33* and *GSTP1* were investigated in tumor epithelia, tumor-associated stroma, normal epithelia, foci of PIN and benign prostate hyperplasia, and stroma adjacent to tumor tissues.

Our results show that genetic and epigenetic aberrations of stromal cells are the frequent event present on all stages of prostate cancer development. First, frequencies of genetic aberrations are as high in stromal cells as in tumor cells. Second, molecular aberration spectrum in stroma and tumor are discordant. Third, stromal AI frequencies for initial prostate cancer stages demonstrate at least a 2-fold excess over epithelium ones. Fourth, some methylation in both hyperplastic epithelium and stromal cells were observed in normal-appearing tissues located adjacent to tumors.

Thus numerous stroma investigations testify that cancer is not an epithelium cell disease but one of the whole tumor microenvironment. "Which genetic alterations are primary?" That is the question for elucidation of prostate cancer's main mechanisms. Further research of the molecular mechanisms underlying tumor-stromal interaction may yield improved medical treatments for prostate cancer growth and metastasis and open potentially novel compartments for targeted therapy and prevention.

Introduction

In the past decade, intense studies of the tumor microenvironment have yielded ample convincing evidence for the critical role of the stroma in carcinogenesis. Tumor progression depends, not only on the molecular changes in epithelial cells, but also on many alterations in the adjacent stroma, which provides a necessary environment for proliferation, migration, and invasion of the tumor epithelium during neoplastic growth. Stromal alterations include changes in the expression profile of growth factors and cytokines and reorganizations of the fibroblast cytoskeleton and extracellular matrix (ECM). The stroma is capable of stimulating the epithelial tumor growth, and, on the other hand, it itself undergoes substantial modification, including changes in the protein expression profile and various genetic alterations. Molecular alterations of the stromal cell genome—hypermethylation of tumor suppressor genes, changes in telomere length, loss of heterozygosity (LOH), microsatellite instability or allelic instability (AI) on various chromosomes, and mutations—form a specific spectrum, differing from that of the tumor epithelium. However, data on epigenetic and genetic alterations in the stroma associated with the tumor epithelium are recent and require detailed analysis to identify new markers and therapeutic targets and to better understand the mechanisms of carcinogenesis.

The objectives of this work were to study the genetic and epigenetic alterations in the tumor and its microenvironment and to evaluate the diagnostic significance of these alterations as potential markers of prostate cancer (PC).

Mechanisms of Stroma–Tumor Interactions

It is clear that the interaction of tumor and stromal cells results from certain coordinate changes in transcription of their genes, but the mechanism sustaining the key steps of early PC pathogenesis is obscure. The question as to what genetic alterations arising in neoplastic and stromal cells eventually result in malignant transformation of the epithelium is still open.

The development of normal and neoplastic glandular structures in the prostate is strictly controlled and is regulated via mesenchymal–epithelial interactions [41]. The interplay of the epithelium and stroma is essential for all steps of prostate development and substantially changes in carcinogenesis. Tumor progression depends, not only on the molecular alterations that arise in epithelial cells, but also on numerous changes that arise in the adjacent stroma, which provides a necessary environment for proliferation, migration, and invasion of the tumor epithelium in the course of neoplastic growth. The microenvironment includes the ECM, growth factors and cytokines, and nonepithelial cells: endothelial cells, smooth muscle cells, pericytes, lymphocytes, macrophages, and mast cells. The alterations arising in the stroma include changes in the expression profile of growth factors and cytokines, reorganizations of the fibroblast cytoskeleton and ECM, and stimulation of neoangiogenesis [4].

The reciprocal effects of stromal and epithelial cells are mediated by many cytokines dissolved in the ECM and their receptors present on the surface of target cells, occurring via paracrine and autocrine mechanisms. As a result of expression the vascular endothelial growth factor (VEGF) by glandular and stromal cells, there is angiogenesis stimulation by affection on the endothelial component of the stroma. The ErbB (HER) family of signaling proteins obtains especially important weight for PC pathogenesis during the transition from the androgen-dependent to the androgen-independent stage. Members of the family—the epidermal growth factor receptor (EGFR), HER2/neu, ErbB3, and ErbB4—interact with a broad range of growth factors and trigger the signaling cascades involved in cell proliferation.

A large family of heparin-binding growth factors, which includes the basic fibroblast growth factor (bFGF), VEGF, HGF/SF, and the heparin-binding EGF-like growth factor, stimulates angiogenesis and proliferation via the co-receptors perlecan and heparan sulfate proteoglycan. The insulin-like growth factor (IGF) family plays a key role in PC progression and the response to anticancer treatment. The TGFβ family performs many functions in carcinogenesis. Factors of this family affect the epithelial–mesenchymal transition and angiogenesis, control the ECM composition, and play a role in immune processes [31]. Various functions in PC have been reported for the platelet-derived growth factor (PDGF) and Wnt families, chemokine and c-met receptors, and interleukin 6 (Table 1).

Table 1. Most important growth factors involved in the signaling pathways that regulate PC growth.

Growth factor	Origin	Receptor	Receptor location
VEGF	Epithelium, stroma	VEGFR-1, 2	Epithelium, stroma
FGF-2	Stroma	FGF-2R	Epithelium, stroma
HGF/SF	Stroma	c-met	Epithelium
TGFB	Epithelium	TGFB I-III рецепторы	Stroma
IGF-1	Stroma	IGF-IR	Epithelium, stroma
IL-6	Epithelium, stroma	IL-6R, sIL-6R	Epithelium, stroma
KGF	Stroma	KGF-R	Epithelium

Soluble growth factors are not the only factors mediating the reciprocal influences of the stroma and epithelium; their reciprocal coordination is due to direct interactions between ECM proteins and surface integrins of epithelial cells [29]. These interactions provide for tumor cell proliferation, adhesion, migration, and metastasis. The ability of tumor cells to invade the stroma and vessels depends on the expression of proteases that are capable of disrupting cell–cell and cell–stroma interactions. The ECM is substantially degraded and reorganized in the site of invasion of a tumor cell cluster. Interestingly, the most important enzymes responsible for degradation of ECM components are secreted by stromal, rather than tumor, cells, including endothelial cells, myofibroblasts, and macrophages. The expression of some proteases is associated with clinical progression of PC. Matrix metalloproteinases (MMPs) are secreted in an inactive form by the ECM and are activated upon their cleavage by extracellular proteinases. Elevated expression of MMP-2, MMP-9, and MMP-11 is associated with a poor prognosis in PC. TIMP family proteins act as natural inhibitors of MMPs, but the promoters of their genes are hypermethylated in PC. Epigenetic alterations have been observed for *PLAU*, which codes for another protease, belonging to the plasminogen activation system along with PAI-1, PAI-2, and uPAR. Tissue kallikrein family proteases, such as PSA, KLK2, hK4, and hK15, are thought to mediate tumor invasion and metastasis by degrading the basal membrane and ECM or activating growth factors and other inactive proteases. Kallikreins are utilized as serum markers in PC early diagnosis and prognosis. Serine protease TMPRSS2 is better known for the formation of chimeric transcripts with ETS family transcription factors.

Neoangiogenesis is an early and critical event in carcinogenesis. Secretion of proangiogenic cytokines and growth factors and ECM reorganization change the phenotype of endothelial cells. When the basal membrane of capillaries is degraded, endothelial cells invade the adjacent stroma and migrate towards the angiogenic stimulus. After proliferation, cells form three-dimensional structures and, eventually, capillaries [13]. The factors that stimulate angiogenesis include PDGF, VEGF, IGF, cytokines, and metalloproteinases. The physiological inhibitors of the process include interferones, angiostatin, endostatin, and TIMP [44].

The ECM and its degradation products interact with integrins and nonintegrin associated receptors regulate cell polarity, adhesion, secretion, motility, proliferation, and differentiation [9]. Cell adhesion proteins (integrins, cadherins, ICAM family proteins, CD44, selectins, and mucins) control the effects of metalloproteinases and serine proteases to promote cell migration and to generate secondary cell growth foci (Figure 1). CD44 belongs to the family of transmembrane glycoproteins responsible for cell–stroma and cell–cell interactions. The extracellular domain of CD44 binds a broad variety of molecules: hyaluronic acid, HGF, EGF, VEGF, and cytokines. *CD44* inactivation via hypermethylation of the promoter region is a frequent event in PC. The ICAM gene family includes the *TSLC1/IGSF4* tumor suppressor gene, which is inactivated in 30–60% of various tumors, including PC [44]. E-CAD is a calcium-dependent cell–cell adhesion glycoprotein whose extracellular domain binds small peptides, such as cell and ECM fragments. Loss of E-CAD function facilitates cell proliferation and invasion, thereby leading to tumor progression.

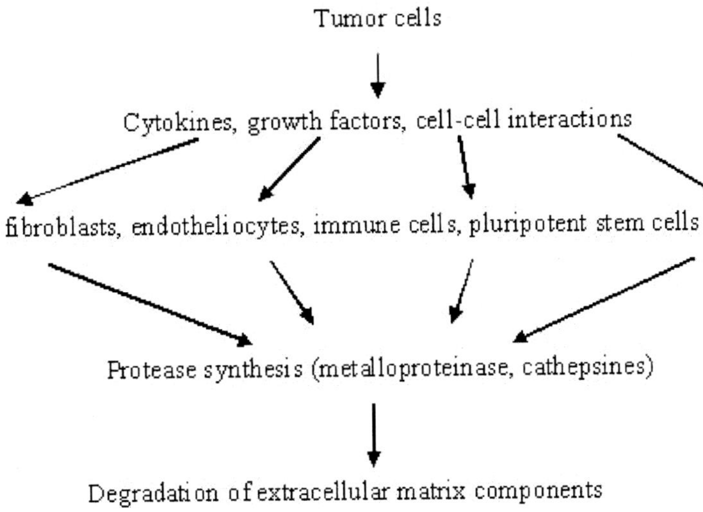

Figure 1. Interactions between tumor cells and stromal cells.

Another mechanism that increases the malignant potential of tumor cells is the epithelial–mesenchymal transition, during which epithelial cells increase the expression of genes characteristic of mesenchyme [8]. Epithelial cells acquiring a mesenchymal phenotype express smooth-muscle actin, type I collagen, vimentin, osteopontin, and N-cadherin with a simultaneous decrease in E-cadherin expression [11]. As a result, such cells can produce MMPs, become more sensitive to growth factors, and acquire properties essential for migration and invasion. PC cells that have acquired mesenchymal properties express RANKL, which stimulates the osteoclast function; such stimulation is probably a prerequisite to tumor cell metastasis into bones [53].

Thus, the microenvironment is capable of stimulating the epithelial tumor growth, while the stroma itself undergoes substantial modification, including changes in protein expression profile and various genetic alterations. Molecular alterations that occur in the stromal cell genome posess a specific profile that differ from that of the tumor epithelium. Such alterations have been observed in PC, breast cancer (BC), bladder cancer, colorectal cancer, and head-and-neck cancer (15, 16, 18, 25, 36, 39, 43]. A specific profile of allelic deletions in stromal cells has been described for BC and head-and-neck cancer [15]. For instance, LOH on various chromosomes and mutations of *TP53* and *PTEN* have been reported for BC-associated stromal cells. The tumor epithelium and stroma have separately been tested for LOH at 17p13, 10q23, and 16q24 and mutations of *TP53, PTEN*, and *WFDC1* [27]. *PTEN* and *WFDC1* mutations in the stromal component were more frequent than in the epithelial component. LOH at 2p, 3p, 11q, 16q, and 17q has been observed in both the stroma and epithelium in BC, but deletions of some loci were restricted to the stroma [36]. LOH analysis with 57 markers in BC has shown that deletions in the stroma are more diverse and form a broader spectrum than in the epithelium; the difference is so great that the genetic alterations have been assumed to arise independently in the stroma and in the tumor [16]. LOH/AI in the stroma has been reliably associated with tumor differentiation and metastasis; moreover, a greater number of positive correlations with clinical parameters have been found for the stroma than for the epithelium [17]. Methyl-specific direct karyotyping has revealed different epigenetic alterations for epithelial cells and stromal fibroblasts in both normal breast tissue and BC [20].

A testing of the tumor microenvironment with 366 microsatellite markers in head-and-neck cancer has revealed a high LOH/AI frequency in the stroma [51]. Correlations with tumor size and metastasis were observed for three chromosome regions 4q, 3p, and 19p in the stroma and LOH/AI of these regions was detected only in the stromal elements. Interestingly, LOH/AI hot spots in the stroma were more frequent than in the tumor epithelium. It is possible that fewer genetic alterations suffice to initiate tumorigenesis in the epithelium than in the stroma. In contrast, LOH/AI analysis of several tumor and stromal loci has revealed a significantly lower frequency of allelic alterations in the stroma than in the epithelium (24% vs. 65%) in oral squamous carcinoma [50]. The LOH/AI frequency in the stroma is high in early esophageal cancer; with disease progression, the frequency of genetic alterations gradually increases in the epithelium and remains the same in the microenvironment [46].

Interesting data have been obtained for *TGFBR1* polymorphism in stromal and epithelial cells in head-and-neck and colorectal cancers [5]. The population frequency of allele *TGFBR1*6A* is 14%, and its carriers have a higher risk of tumor diseases. The epithelium and adjacent fibroblasts acquire allele *TGFBR1*6A* during carcinogenesis, while the initial genotype is preserved in other cells of the body, including stromal lymphocytes. Cells that have acquired the pathological allele are detectable up to 2 cm away from the tumor.

Studies of bladder tumors have similarly revealed a high LOH frequency in stromal cells; nonrandom X inactivation has been observed in 50% of stromal cells [39]. Numerous genetic alterations (*MGMT* methylation, LOH/AI, and mutations of *K-RAS* and *TP53* in stromal cells have been found in colorectal cancer [45, 32]. AI of chromosome 3p in tumor-associated fibroblasts has been detected in ovarian cancer [49].

Aberrant methylation of *GSTP1, CD44, RARβ, HIC1, P16,* and *N33* has been observed in the PC microenvironment at a high frequency [19, 43]. Hypermethylation of *GSTP1* and *RARB* has been demonstrated for endothelial cells of tumor vessels in PC [18]. Changes in telomere size in the adjacent stroma of high-grade prostatic intraepithelial neoplasia (PIN) are associated with a high risk of progression to PC [24]. A high frequency of LOH/AI at 7q, 8p, 10q, 17p, 16q, 12q, 17q, 18p, 22q, and 13q has been observed in the PC stroma [26, 34].

Thus, there is multiple evidence that genetic alterations occur in stromal cells of the tumor, suggesting pathology, not only for the transformed epithelium, but also for its total microenvironment, including stromal cells. Genetic alterations in stromal cells, mostly fibroblasts, lead to abnormal protein expression and subsequent disintegration of signaling cascades, which eventually distorts the normal interactions between the stroma and tumor. It is of interest that molecular changes in the stroma provide significant prognostic markers, which is quite conceivable given the equal contributions of the epithelium and stroma to the tumor pathogenesis. Moreover, the same markers can be associated with different courses of the disease, depending on their stromal or tumor localization. Elevated expression of PDGF-beta and PDGFR-alpha in the tumor epithelium suggests a poor prognosis, while elevated expression of the same markers in the stroma correlates with a favorable prognosis [12].

It should be noted that some studies have not revealed any genetic alterations in the stroma. For instance, substantial changes in the expression profile have been reported for all cells of the tumor microenvironment and tumor epithelium in PC, while genetic alterations have only been found in tumor cells [1]. Microarray analysis has made it possible to assume that molecular genetic alterations in cancer-associated fibroblasts (CAFs) are a rare event in BC and ovarian cancer [40]. It was not until recently that epigenetic and genetic alterations were found in the stroma associated with the tumor epithelium. Hence, thorough studies are needed to identify new markers and therapeutic targets and to better understand the mechanisms of tumorigenesis.

We have employed laser microdissection to separate the epithelial and stromal components of PC. To study the methylation profile in various tumors, we have chosen the *GSTP1* xenobiotic metabolism gene, *P16* tumor suppressor gene, and *HIC1* and *N33* candidate tumor suppressor genes, which are involved in prostate carcinogenesis. To compare the frequency of genetic alterations between the tumor epithelium and adjacent stroma, we have examined chromosomes 8p22, 13q14, and 16q23, which are most commonly damaged in PC.

Analysis of Aberrant Methylation

Inactivation of tumor suppressor genes via methylation of CpG islands of their promoter regions is widespread in various tumors, including PC. More than 30 genes have been reported to undergo aberrant hypermethylation in PC [28]. This set includes both classical and potential tumor suppressor genes and genes involved in hormone-dependent signaling pathways, tumor invasion, cell cycle control, and DNA repair. Hypermethylation of cancer-related genes occurs in early tumorigenesis and provides a molecular marker for early diagnosis. Hence, one of the objectives of our study was to assess the aberrant methylation level in the tumor microenvironment.

Genes involved in prostate carcinogenesis— *P16, HIC1, N33*, and *GSTP1*—were tested for methylation in PC and benign prostatic hyperplasia (BPH) specimens. Tissue microdissection was used to isolate malignant acini, loci of PIN, hyperplasia regions, and connective tissue (stroma) surrounding glandular structures of either type on histological sections (Figure 2).

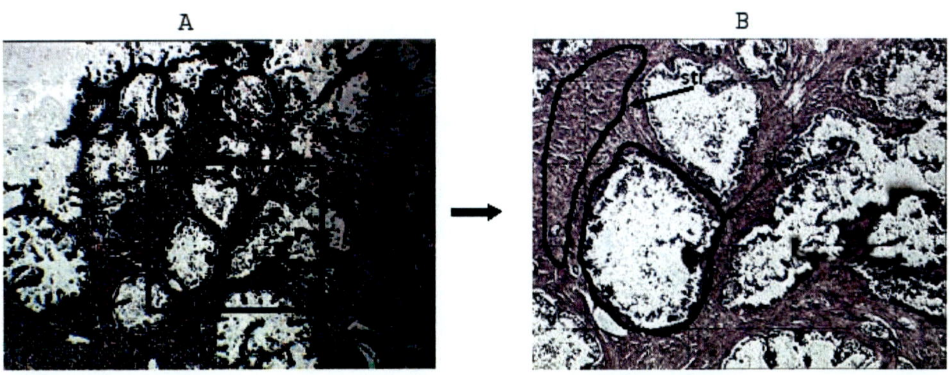

Figure 2. Laser microdissection of PC. Magnification (A) 4x10 or (B) 10x10. Arrows indicate the microdissected fragments: a hyperplastic acinus (gl) and adjacent stroma (str). Sections were prepared for laser microdissection and stained with hematoxylin and eosin.

Microdissection was performed with 30 histologically heterogeneous sections of surgery material obtained from adenocarcinoma patients. In total, we obtained 30 specimens of adenocarcinoma glandular structures, 15 PIN specimens of various grades, 8 hyperplasias, and 43 adjacent connective tissue specimens, of which 37 neighbored adenocarcinoma foci or high-grade PIN and 6 neighbored benign hyperplastic acini.

As a reference, we used two autopsy specimens of normal prostate tissue and two prostate tissue specimens of BPH patients. The specimens were microdissected to isolate normal and hyperplastic glandular epithelial structures and adjacent stromal fragments. Methylation of the target genes was not detected in both the epithelial and stromal components of the normal and hyperplastic specimens.

A high methylation frequency was observed in both malignant epithelium and adjacent stroma in PC (Table 2). The methylation frequency was high in adenocarcinoma: 77% (23/30) for *HIC1*, 50% (15/30) for *P16*, 93% (28/30) for *GSTP1*, and 27% (8/30) for *N33*, which agrees with published data. In the adjacent stroma, methylation affected all target genes but *N33* in almost all specimens: the frequency was 14/25 (56%) for *HIC1*, 9/25 (36%) for *P16*, 14/25 (56%) for *GSTP1,* and 2/25 (8%) for *N33*. A high methylation frequency of *HIC1* has been found in many investigations; *HIC1* is inactivated via methylation in 80-100% of cases in BC, PC, colorectal cancer, and kidney and brain tumors. The *N33* methylation frequency in adenocarcinomas was 33%. *N33* maps to chromosome 8p22, which is affected by deletions in 60-80% of cases in PC. Inactivation of *P16* via methylation is rather frequent in PC; the frequency was 50% in our sample.

Table 2. Methylation frequencies of *HIC1, P16, N33*, and *GSTP1* in microdissected prostate specimens.

	HIC1	*P16*	*GSTP1*	*N33*
norma	0/2	0/2	0/2	0/2
adjacent stroma	0/2	0/2	0/2	0/2
hyperplasia (BPH patients)	0/5	0/5	0/5	0/5
adjacent stroma	0/5	0/5	0/5	0/5
hyperplasia (PC patients)	2/8 (25%)	2/8 (25%)	1/8 (13%)	0/8
adjacent stroma	1/6	0/6	1/6	1/6
PIN (PC patients)	9/15 (60%)	8/15 (53%)	11/15 (73%)	5/15 (33%)
adjacent stroma	8/12 (66%)	7/12 (58%)	4/12 (33%)	0/12
adenocarcinoma	23/30 (77%)	15/30 (50%)	28/30 (93%)	8/30 (27%)
adjacent stroma	14/25 (56%)	9/25 (36%)	14/25 (56%)	2/25 (8%)

The methylation frequencies of the target genes in high-grade PIN were similar to those in adenocarcinoma: 60% (9/15) for *HIC1*, 53% (8/15) for *P16*, 73% (11/15) for *GSTP1*, and 33% (5/15) for *N33*. PIN is thought to directly precede invasive carcinoma in the prostate. It has been demonstrated that lesions are multifocal in PIN and that molecular genetic alterations are much the same as in early invasive PC, although less frequent. Such alterations include chromosomal deletions from 8p, 8q, 10q, and 16q; changes in telomere length; chimeric *TMPRSS2-ERG4*; and methylation of *HIC1, GSTP1, RARβ, RASSF1, ESR2*, etc. Like adenocarcinoma glands, PIN glands display a substantial genetic heterogeneity, detectable by LOH analysis, fluorescence *in situ* hybridization, comparative genome hybridization, and comparative gene expression profiling. All but one target gene (*N33*) proved to be methylated in the PIN-adjacent stroma of almost all specimens in our sample; the frequency was 8/12 for *HIC1,* 7/12 for *P16*, 4/12 for *GSTP1*, and 0/12 for *N33*. Molecular alterations in the adjacent stroma have already been observed in high-grade PIN. For instance, Joshua et al. have associated a significant risk of adenocarcinoma with short telomeres in the stroma adjacent to high-grade PIN [24]. Our finding that aberrant methylation occurs in the tumor stroma at a high frequency agrees with the results of recent studies focusing on gene methylation in the stromal microenvironment of prostate tumors and stromal fibroblasts of BC and colorectal cancer. Moreover, methylation of *GSTP1* and *RARβ2* has been observed in the endothelial component of the PC microenvironment [18]. Thus, histologically different structures from one section revealed distinct methylation profiles (Table 3).

Methylation of the target genes was not detected in hyperplastic glandular structures neighbor adenocarcinoma foci and in the adjacent stroma of our PC patients with two exceptions. In these two patients, each having early metastasis, *P16, HIC1,* and *GSTP1* proved to be methylated in hyperplastic foci and adjacent stroma.

Analysis of Gene Methylation in Biopsy and Surgery Material

Since aberrant methylation was observed both in cancer cells and the tumor microenvironment and proved to be widespread, we studied whether the cancer-related alterations could be detected in prostate specimens without microdissection. Methylation analysis was performed with biopsy specimens of patients with various disorders (BPH, PIN, and adenocarcinoma) and surgery material of prostate adenocarcinoma patients.

In total, 106 biopsy specimens were tested for *P16, HIC1, N33*, and *GSTP1* methylation. The prostate specimens were conventionally grouped by histological pattern. For instance, the PC group included all patients who had adenocarcinoma in at least one of 12 biopsy fragments. The patients who had at least one PIN focus in any of the 12 biopsy fragments were assigned to the PIN group; the same rule was applied to BPH patients.

The frequency of *HIC1* methylation was 33% in BPH and was significantly higher in PIN and adenocarcinoma (66 and 65%, respectively; $p = 0.03$; Table 4). Hypermethylation of the *P16* promoter region was observed in 18% of BPH cases, 48% of PIN cases, and 42% of adenocarcinoma cases. *GSTP1* displayed a similar methylation frequency (41%) in adenocarcinoma. While the *P16* methylation frequency in PIN was as high as in adenocarcinoma, the *GSTP1* methylation frequency significantly differed between PIN and adenocarcinoma ($p = 0.007$). *N33* methylation was not detected in the precancer conditions, while its frequency in adenocarcinoma was 19% ($p = 0.05$).

Table 3. Comparative methylation status of histologically different glandular structures from one section in various patients.

Patient	Tissue	*HIC1*	*P16*	*GSTP1*	*N33*
98	adenocarcinoma	dark gray	dark gray	dark gray	light gray
	high-grade PIN	dark gray	dark gray	dark gray	dark gray
	stroma	dark gray	dark gray	dark gray	light gray
94	adenocarcinoma	dark gray	dark gray	dark gray	light gray
	high-grade PIN	dark gray	light gray	dark gray	light gray
	stroma	dark gray	dark gray	light gray	light gray
86	high-grade PIN	dark gray	dark gray	dark gray	light gray
	stroma	dark gray	dark gray	light gray	light gray
149	high-grade PIN	dark gray	dark gray	dark gray	dark gray
	hyperplasia	light gray	light gray	light gray	light gray
58	low-grade PIN	dark gray	dark gray	dark gray	light gray
	stroma	light gray	light gray	dark gray	dark gray
55	hyperplasia	light gray	light gray	light gray	light gray
	stroma	light gray	light gray	light gray	light gray
137	hyperplasia	light gray	light gray	light gray	light gray
	stroma	light gray	light gray	light gray	light gray
77	hyperplasia	light gray	light gray	light gray	light gray
	stroma	dark gray	light gray	dark gray	light gray

Note: The presence (dark gray) or absence (light gray) of methylation is indicated. Patients 98, 94, 86, 149, 58, and 77 had adenocarcinoma; patients 55 and 137 had BPH.

Table 4. Methylation frequencies of *P16, HIC1, N33,* and *GSTP1* in biopsy and surgery prostate specimens.

Specimens	*HIC1*	*P16*	*GSTP1*	*N33*
BPH (biopsy material)	8/24 (33%)	2/11 (18%)	3/24 (13%)	0/12
PIN (biopsy material)	19/29 (66%)*	14/29 (48%)	4/29 (13%)	0/29
PC (biopsy material)	32/50 (65%)	19/45 (42%)	22/53 (41%)*	9/47 (19%)*
PC (surgery material)	20/28 (71%)	6/28 (22%)	9/28 (32%)	3/16 (20%)

* The difference was significant at $p < 0.05$ (Fisher's test).

In addition, *P16, HIC1, N33*, and *GSTP1* were tested for methylation in 28 PC specimens obtained by prostatectomy. As expected, the methylation frequencies of the target genes in these specimens did not significantly differ from the frequencies in adenocarcinoma biopsy material (Table 4).

The methylation frequencies of all genes in the PC surgery specimens examined without microdissection were substantially lower than in the microdissected adenocarcinoma specimens (71% vs. 77% for *HIC1*, 22% vs. 50% for *P16*, 32% vs. 93% for *GSTP1*, and 20% vs. 27% for *N33*). The difference can be explained by the admixture of nontumor cells present in the specimens as a result of a high heterogeneity of PC. The comparison confirmed again that microdissection is essential for molecular analysis in PC. Although aberrant methylation extends rather far away from the malignant focus according to our and published data, false negative results are possible when the heterogeneity of PC is neglected.

To study the clinical associations of the methylation frequencies of the target genes, the patients were grouped by histological picture (PC, PIN, and BPH) and the groups were compared with respect to methylation frequencies of individual genes by the χ^2 test. The methylation frequencies of *N33* and *GSTP1* significantly differed between PIN (0/29 (0%) and 4/29 (13%), respectively) and PC (9/47 (19%) and 22/53 (41%), respectively) (Table 4).

Loss of Heterozygosity and Allelic Instability in PC

Homozygous and heterozygous deletions are among the most common genetic alterations in epithelial tumors; their detection widely employs LOH and AI analyses. To assess the genetic alterations in the PC microenvironment, AI analysis of chromosome regions 8p22, 13q14, and 16q23 was performed with the epithelial and stromal components of PC specimens. Chromosome 8p22 harbors several tumor suppressor genes (*N33, DLC1,* and *LZTS1*) and is often affected by allelic deletions in epithelial tumors, such as PC, BC, and colorectal, bladder, lung, and head-and-neck cancers [2]. In PC, LOH at 8p is one of the most frequent somatic mutations, occurring in more than 60% of cases [33]. 16q23 deletions have been described in BC and hepatocellular carcinoma and are found in 50% of cases in PC [38]. Region 16q23 harbors the genes for heat shock proteins, cadherins, and antioxidation enzymes, which may be involved in carcinogenesis. Deletions from 13q14, which includes *RB1*, occur in 27-60% of cases in PC according to various estimates and are often found in most tumors [35]. It is thought that deletions from 8p22 and 13q14 are characteristic of early PC, while LOH at 16q23 is more common in advanced PC.

Microsatellite analysis was performed with 51 microdissected prostate specimens from patients with adenocarcinoma of the prostate. The LOH and AI frequencies were estimated for the tumor epithelium (51 specimens) and adjacent stroma (39 specimens). As an example, electrophoresis of specimens with LOH and AI is shown in Fig. 3. LOH/AI was revealed for at least one of the markers in 46 tumor epithelium specimens (91%). The frequency of allelic alterations in the epithelium was 49% for 8p22, 51% for *D16S534*, 57% for *D16S422*, 72% for 16q23 (as established with the two markers), and 37% for 13q14.

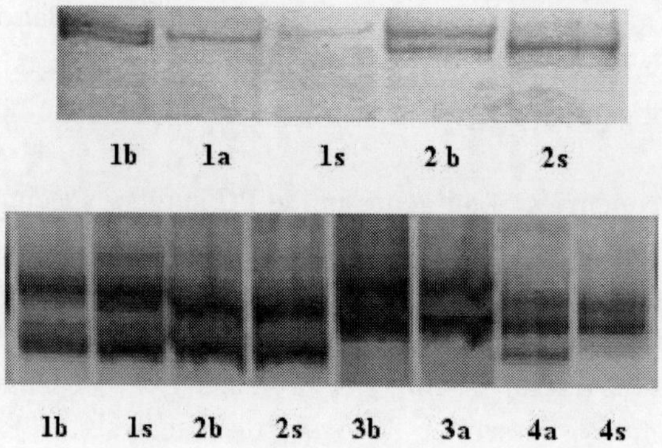

Figure 3. Microsatellite analysis of *D16S534*. Electrophoresis. Designations: b, blood; a, adenocarcinoma; s, stroma. (A) - LOH (lanes 1a, 1s, and 2s). (B) - AI (lanes 2s and 5a).

Table 5. Frequencies of allelic alterations in the neoplastic epithelium and adjacent stroma in patients with adenocarcinoma or PIN

	LOH/AI frequency			
Marker	adenocarcinoma	stroma	PIN	stroma
D8S1731	25/51 (49%)	22/39 (56%)	4/25 (16%)	1/19 (6%)
D16S534	26/51 (51%)	21/39 (54%)	7/25 (28%)	2/19 (11%)
D16S422	29/51 (57%)	20/39 (51%)	6/25 (24%)	2/20 (10%)
RB1int20	19/51 (37%)	23/39 (59%)	7/25 (28%)	4/19 (21%)

A somewhat higher frequency of LOH/AI at the target loci was observed for the stroma (Table 5). LOH/AI for a particular marker was observed exclusively in the epithelium or the stroma in some specimens, while some others displayed molecular alterations in both of the components but with different alleles. The highest LOH frequency was established for 16q23 in the tumor epithelium (72%) and for 13q14 in the stroma (59%) in our sample.

LOH and AI Analyses in PIN

Microsatellite analysis was performed with 25 biopsy specimens from patients with high- or low-grade PIN but without PC. The PIN epithelium and stroma adjacent to neoplastic acini were examined. Molecular alterations occurred at a low frequency in both the epithelial (16-28%) and stromal (6-21%) components of PIN (Table 5). The highest frequency of allelic alterations was observed for *RB1int20* in both the epithelium and stroma. In both the epithelium and stroma, the frequencies of allelic alterations in PIN were significantly lower than in PC in our sample. LOH and AI were less frequent in stromal fibroblasts than in the PIN epithelium, but the significance of this finding should be verified with a larger sample of patients.

Associations of LOH/AI Frequencies with Clinical Parameters

To correlate the LOH/AI frequencies with PC progression, we chose several clinical criteria: Gleason score, PC stage, and metastasis to regional lymph nodes. Significant associations with all of these clinical parameters were observed for *D16S534* alterations in the tumor epithelium. Positive associations with the disease stage were established for *RB1int20* and *D16S422* (Table 6).

The frequencies of allelic alterations of the target markers in the stromal component did not significantly correlate with PC progression. Positive associations with tumor progression have been reported for stromal alterations [15]. We did not observe such associations, possibly, because of the small sample size. A panel of prognostic LOH/AI markers can be used together with histological examination to improve the diagnosis and prognosis in PC.

Table 6. Associations of allelic alterations in the tumor epithelium with clinical parameters: Gleason score, PC stage, and metastasis by Fisher's exact test.

Marker	Correlation		
	Gleason score, p	metastasis, p	stage, p
D8S1731	NS	NS	NS
D16S534	0.01	0.04	0.02
D16S422	NS	NS	0.04
RB1int20	NS	NS	0.02

Note: p, probability; NS, nonsignificant ($p > 0,05$).

Comparison of the Frequencies of Genetic and Epigenetic Alterations for the Tumor Microenvironment and Tumor Epithelium in PC

The PC microenvironment plays a key role in carcinogenesis, actively inducing the tumor growth. Olumi et al. have demonstrated in their classical study that nontumor prostate cells undergo malignant transformation when cultured together with fibroblasts originating from the PC microenvironment, while normal fibroblasts do not induce malignant transformation [37]. A breast stroma exposed to radiation induces transformation of epithelial cells, while the normal stroma does not [3]. Thus, the microenvironment can stimulate the tumor growth of the epithelium. However, the stroma itself experience substantial modification, including changes in protein expression profile and various genetic alterations. We studied the epigenetic and genetic changes in the PC microenvironment.

Analysis of CpG methylation in the promoter regions of *P16, GSTP*, and *N33* in microdissected specimens obtained from adenocarcinoma patients revealed a high methylation frequency for both the malignant epithelium and adjacent stroma (Table 2). Our finding that aberrant methylation is frequent in the tumor stroma agrees with recent data on gene methylation in the stromal microenvironment of prostate tumors and stromal fibroblasts of BC and colorectal cancer [19]. Moreover, methylation of *GSTP1* and *RARβ2* has been observed in the endothelial component of the PC microenvironment [18]. Epigenetic alterations in the stroma are probably a part of the interactions between the tumor and its stromal microenvironment. It is still unknown how methylation arises in tumor cells and neighbor regions, including morphologically normal epithelium, stromal fibroblasts, and endothelial cells, where it has been observed in recent studies [28]. Several hypotheses have been advanced to explain gene methylation in the tumor microenvironment. According to one of the possible mechanisms, the changed expression pattern of cytokines and growth factors in the tumor epithelium affects neighbor tissues and causes a disregulation of the processes that control methylation in the cell. It remains unknown whether the microenvironment is involved in malignant transformation or its response is physiological and generates a marginal zone around the tumor, but genes do experience aberrant methylation in the tumor epithelium, adjacent stroma, and closely located PIN sites in PC (Table 3). Rodriguez-Canales et al. were the first to estimate how far methylation extends from an adenocarcinoma focus through its environment [43]. Methylation analysis with three-dimensional reconstruction of the prostate anatomy revealed aberrant methylation of *GSTP1* only in some cells of the tumor microenvironment, namely, in the local region that accounted for one-fourth of the adenocarcinoma focus and was closer to the apex of the prostate. Methylation was maximal in this region and gradually decreased from the apex to the base. The submicroenvironmental region consisted of the nonreactive tumor stroma, while reactive changes with the appearance of myofibroblasts were observed closer to the base. In any case, the presence of tumor-specific methylation in neighbor cells originating from different germ layers suggests a universal mechanism for methylation induction. Another hypothesis, which we think to be less plausible, explains methylation in the stroma by transdifferentiation of epithelial tumor cells into cells that express the mesenchymal phenotype but preserve the original methylation profile. Epithelial–mesenchymal transition (EMT) is a type of plasticity during which epithelial cells lose many of their epithelial characteristics and acquire properties that are typical of mesenchymal cells.

We detected a high LOH/AI frequency in the PC microenvironment for all loci examined (Table 5). It should be noted that LOH/AI for a particular marker was observed exclusively in the epithelium or the stroma in some specimens, while some others displayed molecular alterations in both of the components but with different alleles. In other words, the deletion profile was not the same in the epithelium and stroma. Such genetic differences between the tumor and stroma have often been observed in various tumors [20]. The highest LOH frequency was established for 16q23 in the tumor epithelium (72%) and for 13q14 in the stroma (59%) in our sample. An interesting feature was observed when patients were grouped by disease stage. The frequency of allelic alterations in the epithelium increased with PC progression, while that in the stroma remained the same or even decreased at advanced stages (Table 7). As mentioned above, such differences in behavior between the tumor and adjacent stroma have already been reported in the literature. Examination of esophageal tumors has shown that the LOH/AI frequency in the stroma is high at the early stage; with disease progression, the frequency of genetic alterations gradually increases in the epithelium and persists at the same high level in the microenvironment [46]. Ishiguro et al. found that epithelial genetic instability of colorectal tumors increases with progression of histological

lesions, whereas LOH and AI in stromal elements tend to be more constant in dysplasia, and invasive carcinomas. In line with the demonstration of conversion of colonic adenomas to adenocarcinomas in nude mice with inflammation associated stroma the results of this study suggest that early changes in the stroma may be associated with epithelial tumorigenesis [22]. The frequencies of stromal alterations in colorectal tumors were rather lower than in prostate cancer or esophageal tumors whereas tendency for constant stromal alterations and increasing epithelial alterations is common to different types of epithelial tumors.

Assuming that genetic alterations in the stroma arise earlier than in the epithelium, we performed LOH/AI analysis of PIN as a precancer condition and observed a low frequency of allelic deletions in both the stroma and tumor (Table 5). We did not observe a substantial prevalence of genetic alterations for any tumor component. To identify the component that is the first to acquire genetic alterations, it is required to examine in detail large samples of carcinoma *in situ* and its microenvironment.

Thus, we showed that epigenetic and genetic alterations of the stromal cell genome are frequent and occur at all stages of PC pathogenesis (Fig. 4). Alterations in the stroma are as frequent as in the tumor epithelium.

Table 7. Frequencies of allelic losses in epithelial and stromal elements at different stages of prostate cancer.

Stage	*D8S1731*		*D16S422*		*D16S534*		*RB1int20*	
	Epithelium,%	Stroma, %	Epithelium,%	Stroma,%	Epithelium,%	Stroma,%	Epithelium,%	Stroma,%
I-II	33	57	30	50	27	70	9	64
III-IV	58	55	78	50	68	50	56	57

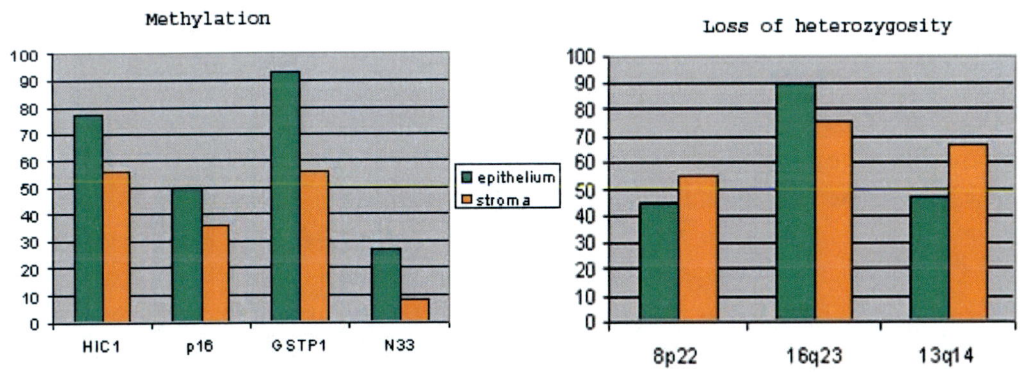

Figure 4. Comparison of the frequency of molecular genetic alterations for the tumor epithelium and stroma

Since epigenetic and genetic alterations occur in both the epithelium and stroma in carcinogenesis, it is possible to assume that stromal alterations arise as a consequence of the primary mutation process in the epithelium, or vice versa. The somatic mutation theory suggests that pathological proliferation results from an accumulation of mutations over a certain threshold in one epithelial cell of the gland. Since carcinogenic etiological factors probably affect the total organ rather than its epithelial component alone, identification of the cells most prone to mitotic defects may clarify the early steps of carcinogenesis. Maffini et al. have found that nitrosomethylurea (NMM) induces malignant transformation of mammary epithelial cells only when stromal cells are affected regardless of whether the epithelium itself is exposed to NMM; i.e., the stroma is a target of the mutagen [30].

Huang et al. have directly shown that ECM components affect angiogenesis and tumor development. Knockout in the MMP-9 gene decreases ECM alterations and angiogenesis in mice, thus suppressing the growth of ovarian tumors [21]. Johansson et al. have observed that androgen deprivation differently affects the tumor epithelium and stroma in the prostate [23]. Androgen blockage leads to involution of both epithelial and stromal cells in the normal prostate. In the tumor, androgen deprivation dramatically reduces the AR and VEGF expression in the epithelium, while the stroma remains intact. The expression of AR, growth factors, and their receptors in the stroma remains the same, providing a necessary microenvironment for proliferation of the tumor epithelium. Yang et al. have searched for the candidate genes that are expressed exclusively in stromal cells and stimulate tumor epithelial activity [52]. For instance, CTGF expression in the reactive stroma substantially increases the capillary density and stimulates tumor growth. *Scube 1*, another novel gene, is also expressed only in stromal cells of the prostate. Its expression is reduced in tumor-associated fibroblasts compared to normal fibroblasts, indicating that *Scube 1* is a potential tumor suppressor gene.

In 1953, Slaughter introduced the concept of field cancerization based on the analysis of the microenvironment in head-and-neck squamous carcinoma as a multifocal tumor with a high relapse rate. The concept has received ample support to date. In early carcinogenesis, a stem cell acquires genetic alterations and gives origin to a patch, a clone of changed daughter cells [6]. The alterations include somatic mutations, LOH, AI, and nonrandom X inactivation. At the next step, the patch develops into a vast field. Having an advantage in growth rate, the field gradually replaces normal tissue. The field has monoclonal origin and lacks invasive growth and aggressive behavior, which are characteristic of malignant cells. Such fields have been found up to 7 cm away from head-and-neck and esophageal tumor growth foci. Additional genetic alterations in the field yield various subclones, which gradually diverge. During selection, one or several clones reach the invasive carcinoma stage. Molecular evidence for the cancerization field has been obtained in many tumors, including neoplasms of the head and neck, esophagus, stomach, lung, cervix, vulva, bladder, colon, breast, ovaries, pancreas, and prostate [10]. *RARβ* and *GSTP1* are methylated, not only in the tumor epithelium and adjacent stroma, but also in the normal epithelium away from PC [19]. We have obtained similar results. In two out of eight adenocarcinoma patients, methylation of *HIC1, GSTP1*, and *N33* was observed in hyperplastic foci and adjacent connective tissue away from the adenocarcinoma focus (Table 3). Changes in telomere length have been detected in normal prostate glandular structures adjacent to the tumor focus [14]. It has been reported that the gene expression profile in normal prostate tissue adjacent to adenocarcinoma is similar to that in the tumor and differs from that in healthy subjects [7]. Mitochondrial DNA mutations have been found in both tumor prostate tissue and adjacent normal tissue.

The evidence that the stromal cell niche plays a key role in defining the stem cell niche and directing stem cell fate is described in review Risbridger [42]. Stromal-epithelial stem cell signaling is conserved across species, making it possible to use rodent stroma to direct the fate of human stem cells, and exemplified by the generation of human prostatic epithelia from pluripotent human embryonic stem cells. Secondly, malignant tumor stroma abrogates teratoma formation, but there is little evidence that tumor stroma initiates tumor formation, although it is known to promote tumor progression. Finally reprogramming of adult stem cells by alteration of the stromal cell niche resulting in transdifferentiation would provide novel evidence of the full potential of the stroma to direct and redirect stem cell fate. Is it possible that an early genetic event occurs in a single stem cell that acts as a precursor of different germ layers? The monoclonal character of the cancerization field is supported by molecular data obtained for tumors of the head and neck, bladder, and esophagus, which have extended epithelial regions. However, genetically different tumor clones have been observed in many tumors of epithelial glandular structures, such as PC, pancreatic cancer, and BC. The question arises as to whether a tumor can originate from polyclonal stem cells. To clarify this issue, it is necessary to identify the genes for constructing the evolutionary trees of tumor clones and to further study the mechanisms of carcinogenesis. In any case, the cancerization field is a well-documented fact of tumor transformation. Clinical morphological examination fails to yield a correct prognosis in the case of histologically normal tissue, even when this tissue has all signs of molecular pathology. It is clear that molecular genetic methods should supplement this gold standard of diagnosis.

The tissue organization field theory suggests that carcinogens change the stromal–epithelial interactions and that epithelial proliferation results from aberrant interactions of cells of the two types [47, 48]. Morphogenetic fields, or organization fields, are a set of cells whose interactions form the total organ. These fields maintain normal cell behavior and normal tissue microarchitecture. The theory is based on two main principles. First, proliferation is a consequence of aberrant interactions among cells of all types. Second, carcinogens disrupt the normal connections between the stroma and epithelium. Recent works on the tumor microenvironment better support this theory than the earlier theory of somatic mutations, at least in the case of epithelial tumors. These results highlight again the importance of studying various components of the microenvironment, including fibroblasts, endothelial cells, adipocytes, and ECM components. Further investigation of the molecular basis of stromal–epithelial interactions will probably identify the therapeutic targets for etiological treatment of PC in the tumor microenvironment.

References

[1] Allinen, M., Beroukhim, R., Cai, L. et al. (2004). Molecular characterization of the tumor microenvironment in breast cancer. *Cancer Cell.*, *6*, 17-32.

[2] Arbieva, Z., Banerjee, K., Kim, S. et al. (2000). High-resolution physical map and transcript identification of prostate cancer deletion interval on 8p22. *Genome Res.*, *10*, 244-257.

[3] Barcellos-Hoff, M. & Ravani, S. (2000). Irradiated mammary gland stroma promotes the expression of tumorigenic potential by unirradiated epithelial cells. *Cancer Res.*, *60*, 1254-1260.

[4] Bhowmick, N., Neilson, E. & Moses, H. (2004). Stromal fibroblasts in cancer initiation and progression. *Nature*, *432*, 332-337.

[5] Bian, Y., Knobloch, T., Sadim, M. et al. (2007). Somatic acquisition of TGFBR1*6A by epithelial and stromal cells during head and neck and colon cancer development. *Hum Mol Genet.*, *16*, 3128-3135.

[6] Braakhuis, B., Tabor, M., Kummer, A. et al. (2003). A genetic explanation of Slaughter's concept of field cancerization: evidence and clinical implications. *Cancer Res.*, *63*, 1727-1730.

[7] Chandran, U., Dhir, R., Michalopoulos, G. et al. (2005). Differences in gene expression in prostate cancer, normal appearing prostate tissue adjacent to cancer and prostate tissue from cancer free organ donors. *BMC Cancer*, *5*, 45.

[8] Christiansen, J. & Rajasekaran, A. (2006). Reassessing epithelial to mesenchymal transition as a prerequisite for carcinoma invasion and metastasis. *Cancer Res.*, *66*, 8319-8326.

[9] Chung, L., Baseman, A., Assikis, V. & Zhau, H. (2005). Molecular insights into prostate cancer progression: the missing link of tumor microenvironment. *J. Urol.*, *173*, 10-20.

[10] Dakubo, G., Jakupciak, J., Birch-Machin, M. & Parr, R. (2007). Clinical implications and utility of field cancerization. *Cancer Cell Int.*, *7*, 2.

[11] De Wever, O., Demetter, P., Mareel, M. & Bracke, M. (2008). Stromal myofibroblasts are drivers of invasive cancer growth. *Int J Cancer*, *123*, 2229-2238.

[12] Donnem, T., Al-Saad, S., Al-Shibli, K. et al. (2008). Prognostic impact of platelet-derived growth factors in non-small cell lung cancer tumor and stromal cells. *J Thorac Oncol.*, *3*, 963-970.

[13] Fidler, I. (2002). Critical determinants of metastasis. *Semin Cancer Biol.*, *12*, 89-96

[14] Fordyce, C., Heaphy, C., Joste, N. et al. (2005). Association between cancer-survival and telomere DNA content in prostate tumors. *J Urol.*, *173*, 610-614.

[15] Frank, W., Xu, M., Zhang, M. et al. (2007). Microenvironmental genomic alterations and clinicopathological behavior in head and neck squamous cell carcinoma targets for therapeutic and preventive intervention. *JAMA*, *297*, 187-195.

[16] Fukino, K., Shen, L., Matsumoto, S. et al. (2004). Combined total genome loss of heterozygosity scan of breast cancer stroma and epithelium reveals multiplicity of stromal targets. *Cancer Res.*, *64*, 7231-7236.

[17] Fukino, K., Shen, L., Patocs, A. et al. (2007). Genomic instability within tumor stroma and clinicopathological characteristics of sporadic primary invasive breast carcinoma. *JAMA.*, *297*, 2103-2111.

[18] Grover, A., Tangrea, M., Woodson, K. et al. (2006). Tumor-associated endothelial cells display GSTP1 and RARbeta2 promoter methylation in human prostate cancer. *J Transl Med.*, *2*, 4-13.

[19] Hanson, J., Gillespie, J., Grover, A. et al. (2006). Gene promoter methylation in prostate tumor – associated stromal cells. *J. Natl. Cancer Inst.*, *98*, 255-261.

[20] Hu, M., Yao, J., Cai, L. et al. (2005). Distinct epigenetic changes in the stromal cells of breast cancers. *Nature Genet.*, *37*, 899-905.

[21] Huang, S., Van Arsdall, M., Tedjarati, S. et al. (2002). Contribution of stromal metalloproteinase-9 to angiogenesis and growth of human ovarian carcinoma in mice. *J Natl Cancer Inst.*, *94*, 1134-1142.

[22] Ishiguro, K., Yoshida, T., Yagishita, H. et al. (2006). Epithelial and stromal genetic instability contributes to genesis of colorectal adenomas.*Gut.*, *55*, 695-702

[23] Johansson, A., Jones, J., Pietras, K., et al. (2007). A stroma targeted therapy enhances castration effects in a transplantable rat prostate cancer model. *Prostate*, *67*, 1664-1676

[24] Joshua, A., Vukovic, B., Braude, I. et al. (2007). Telomere attrittion in isolated high-grade prostatic intraepithelial neoplasia and surrounding stroma is predictive of prostate cancer. *Neoplasia*, *9*, 81-89.

[25] Kekeeva, T., Popova, O. & Shegai, P. (2008). Frequent allelic losses in tumor-associated stromal cells and tumor epitelium of prostate cancer. *Mol Biol* (Mosk). *42*, 96-101.

[26] Kekeeva, T., Popova, O., Shegai, P. et al. (2007). Abberant methylation of p16, HIC1, N33 and GSTP1 genes in tumor epitelium and tumor-associated stromal cells of prostate cancer. *Mol Biol* (Mosk). *41*, 79-85.

[27] Kurose, K., Gilley, K., Matsumoto, S. et al. (2002). Frequent somatic mutations in PTEN and TP53 are mutually exclusive in the stroma of breast carcinomas. *Nature Genet*, *32*, 355-357.

[28] Li, L., Carroll, P. & Dahiya, R. (2005). Epigenetic changes in prostate cancer: implication for diagnosis and treatment. *J. Natl. Cancer Inst.*, *97*, 103-115.

[29] Loffek, S., Zigrino, P. & Mauch, C. (2006). Tumor-stroma interactions: their role in the control of tumor cell invasion and metastasis. *JDDG*, *6*, 496-501.

[30] Maffini, M., Soto, A., Calabro, J., Ucci, A. & Sonnenschein, C. (2004). The stroma as a crucial target in rat mammary gland carcinogenesis. *J. Cell Sci.*, *117*, 1495-1502.

[31] Massague, J., Blain, S. & Roger, S. (2000). TGFB signaling in growth control, cancer and heritable disorders. *Cell*, *103*, 295-309.

[32] Matsumoto, N., Yoshida, T. & Okayasu, I. (2003). High epithelial and stromal genetic instability of chromosome 17 in ulcerative colitis-associated carcinogenesis. *Cancer Res.*, *63*, 6158-6161.

[33] Matsuyama, H., Pan, Y., Yoshihiro, S. et al. (2003). Clinical significance of chromosome 8p, 10q, and 16q deletions in prostate cancer. *Prostate*, *54*, 103-111.

[34] McCarthy, R., Zhang, S., Bostwick, D. et al. (2004). Molecular genetic evidence for different clonal origins of epithelial and stromal components of phyllodes tumor of the prostate. *Am J Pathol.*, *165*, 1395-1400.

[35] Melamed, J., Einhorn, J. & Ittmann, M. (1997). Allelic loss on chromosome 13q in human prostate carcinoma. *Clin. Cancer Res.*, *3*, 1867-1872.

[36] Moinfar, F., Man, Y., Arnould, L. et al. (2000). Concurrent and independent genetic alterations in the stromal and epithelial cells of mammary carcinoma: implications for tumorigenesism. *Cancer Res.*, *60*, 2562-2566.

[37] Olumi, A., Grossfeld, G., Hayward, S. et al. (1999). Carcinoma-associated fibroblasts direct tumor progression of initiated human prostate epithelium. *Cancer Res.*, *59*, 5002-5011.

[38] Paris, P., Witte, J., Kupelian, P. et al. (2000). Identification and fine mapping of a region showing a high frequency of allelic imbalance on chromosome 16q23.2 that corresponds to a prostate cancer susceptibility locus. *Cancer Res.*, *60*, 3645-3649.

[39] Paterson, R., Ulbright, T., MacLennan, G. et al. (2003). Molecular genetic alterations in the laser-capture-microdissected stroma adjacent to bladder carcinoma. *Cancer*, *98*, 1830-1836.

[40] Qiu, W., Hu, M., Sridhar, A. et al. (2008). No evidence of clonal somatic genetic alterations in cancer-associated fibroblasts from human breast and ovarian carcinomas. *Nat Genet.*, *40*, 650-655.

[41] Ricke, W., Wang, Y., Kurita, T. et al. (2005). Hormonal and Stromal Regulation of Normal and Neoplastic Prostatic Growth. *In: Progress in Molecular and Subcellular Biology. Developmental Biology of Neoplastic Growth.* W. Muller, P. Jeanteur, Y. Kuchino, A. Macieira-Coelho, & R. Rhoads (Eds.), Springer Berlin Heidelberg. 183-216.

[42] Risbridger, G. & Taylor, R. (2008). *Minireview: Regulation of Prostatic Stem Cells by Stromal Niche in Health and Disease Endocrinology*, *149*, 4303-4306.

[43] Rodriguez-Canales, J., Hanson, J., Tangrea, M. et al. (2007). Identification of a unique epigenetic sub-microenvironment in prostate cancer. *J Pathol.*, *211*, 410-419.

[44] Ross, J. & Foster, C. (2007). *Molecular Oncology of prostate cancer*. Jones and Bartlett publishers.

[45] Shen, L., Kondo, Y., Rosner, G. et al. (2005). MGMT promoter methylation and field defect in sporadic colorectal cancer. *J Natl Cancer Inst.*, *97*, 1330-1338.

[46] Shiraishi, H., Mikami, T. & Yoshida, T. (2006). Early genetic instability of both epithelial and stromal cells in esophageal squamous cell carcinomas, contrasted with Barrett's adenocarcinomas. *J Gastroenterol.*, *41*, 1186-1196.

[47] Sonnenschein, C. & Soto, A. (2008). Theories of carcinogenesis: An emerging perspective. *Seminars in Cancer Biology*, *18*, 372-377.

[48] Soto, A., Sonnenschein, C. et al. (2005). Emergentism as a default: cancer as a problem of tissue organization. *J Biosci.*, *30*, 103-118.

[49] Tuhkanen, H., Anttila, M. & Kosma, V. (2004). Genetic alterations in the peritumoral stromal cells of malignant and borderline epithelial ovarian tumors as indicated by allelic imbalance on chromosome 3p. *Int J Cancer*, *109*, 247-252.

[50] Wang, X., Fan, M., Sun, Z. et al. (2007). The TP53 and RPS6 alterations at the invasive tumor front, center and stroma of oral squamous cell carcinoma. *Zhonghua Kou Qiang Yi Xue Za Zhi.*, *42*, 140-143.

[51] Weber, F., Xu, Y., Zhang, L. et al. (2007). Microenvironmental genomic alterations and clinicopathological behavior in head and neck squamous cell carcinoma. *JAMA*, *297*, 187-195.

[52] Yang, F., Tuxhorn, J., Ressler, S. et al. (2005). Stromal expression of connective tissue growth factor promotes angiogenesis and prostate cancer tumorigenesis. *Cancer Res.*, *65*, 8887-8895.

[53] Zhau, H., Odero-Marah, V., Lue, H. et al. (2008). Epithelial to mesenchymal transition (EMT) in human prostate cancer: lessons learned from ARCaP model. *Clin Exp Metastasis.*, *25*, 601-610.

In: Handbook of Prostate Cancer Cell Research... ISBN: 978-1-60741-954-9
Editor: Alan T. Meridith

Chapter 13

Prostate Cancer Cell Growth and Death: Complex Roles of Pro- and Anti-Oncogenic Protein Signaling

Qun Lu[1, 2,*], *Jiao Zhang*[1, 3] *and Yan-Hua Chen*[1, 2]
[1]Department of Anatomy and Cell Biology.
[2]Department of Medicine, Leo Jenkins Cancer Center,
The Brody School of Medicine, East Carolina University,
Greenville, North Carolina, 27834 USA
[3]Department of Hematology and Oncology,
Southeast University Institute of Clinical Medicine,
Nanjing, 210009 P.R. China

Abstract

Prostate cancer pathogenesis is the result of loss of balance in a number of cellular functions including cell proliferation, growth, death and survival. It has been established that misregulation of cell cycle progression can lead to an accelerated cell division, loss of appropriate checkpoints, or increased genetic instability. Accumulating data showed that the number of signal transduction pathways that control these events leading to carcinogenesis is not unlimited in the cell. Therefore, interrupting these pathways should conceivably lead to the attenuation of cancer. However, the clinical resolution in prostate cancer cure is very limited. Despite the discovery of many oncogenes and tumor suppressors, their specific roles in prostate cancer development remain a considerable dichotomy. Researches have provided increasing evidences that oncogenes can promote apoptosis while the expression of tumor suppressors can be correlated with tumor progression. To understand the molecular and

* Qun Lu, Ph.D.Associate Professor, Department of Anatomy and Cell Biology, The Brody School of Medicine, East Carolina University, Greenville, North Carolina, 27834 , U.S.A, Tel: 252-744-2844, Fax: 252-744-2850, e-mail: luq@ecu.edu.

cellular mechanisms of prostate cancer, signaling pathways dictated by the anti- and pro-oncogenic pathways need to be integrated in the context-dependent manner, rather than as a generalization. This article will review some recent progresses and problems facing prostate cancer research. For future directions, defining the specific roles of pro- and anti-oncogenic elements in their respective signaling pathways at the different stages of prostate cancer growth and survival presents unique opportunities as well as challenges for drug designs and therapeutic intervention.

Introduction

Cancer, with its high prevalence and mortality rate, continues to rank among the world's deadliest and most costly diseases. According to the World Health Organization (WHO), over 25 million people live with cancer and nearly 11 million new cases are diagnosed every year. Among all human malignancies, prostate cancer is the second most common male cancer in the United States, and only lung cancer causes more deaths in American men [1]. The lifetime probability of developing prostate cancer is about 16%. But retrospective studies suggested that by age 80, more than half of all American men have some cancerous cells in their prostate glands [2]. Each year approximately 234,460 men will be diagnosed with prostate cancer. About 27,350 of these men will die from the disease.

Development of cancer, including prostate cancer, is the result of loss of balance in a number of cellular functions including cell growth, proliferation, death and survival. Numerous studies have established that misregulation of cell cycle progression can result in altered cell division, loss of appropriate checkpoints, or increased genetic instability. Many oncogenes and tumor suppressors have been discovered, and they are involved in controlling these complex signaling events. Because the accumulating evidences showed that the number of signal transduction pathways leading to carcinogenesis is not unlimited in the cell, so interrupting these pathways would lead to the attenuation of cancer.

However, the clinical resolution in prostate cancer cure is very limited. What is becoming more and more recognized is that oncogenes can promote apoptosis while the expression of tumor suppressors can be correlated with tumor progression. One classical example is the roles of transforming growth factor-β (TGF-β) in prostate cancer and metastasis. TGF-β inhibits tumor development during the early stage of tumorigenesis while it promotes tumor progression at the later stages [3-5]. To understand the molecular and cellular mechanisms of prostate cancer, signaling pathways dictated by the anti- and pro-oncogenic pathways need to be integrated in the context-dependent manner, rather than as a generalization. Due to space limitation and rapid accumulation of information, it is not possible to include all of the outstanding studies that help to reveal the dual functions of these signaling pathways. For example, oncogenic and tumor suppressive roles of TGF-β has been known for decades and reviewed extensively so readers are referred to the recent review by Jakowlew [4]. The current article will only serve to "*cast a brick to attract jade*", and to review some of the most recent progresses and problems facing prostate cancer research. For future directions, defining the specific roles of each pro- and anti-oncogenic element in their respective signaling pathways in different stages of prostate cancer progression presents unique opportunities as well as challenges for drug designs and therapeutic intervention.

Background on Prostate Cancer

The prostate gland is a complex branched ductal organ that surrounds the urinary urethra in men [6]. It secretes fluid, which mixes with sperm to make semen. The prostate gland undergoes many changes during the course of a man's life. At birth, the prostate is about the size of a pea. It grows only slightly until puberty. When a man reaches his early 20s, the prostate gland begins to enlarge rapidly to become the size of a walnut. The gland generally remains stable until men reach their mid-40s, when the prostate begins to enlarge again through cellular multiplication in most men. Prostate cancer is a malignant tumor that arises in the prostate gland. As with any cancer, if it is advanced or left untreated in early stages, it can eventually spread through the blood and lymphatic systems to other organs. While fortunately, prostate cancer tends to be slow growing compared to other cancers, once becoming metastatic and androgen independent, prostate cancer becomes incurable. So the question is what happens during the long period of time when the cancer is developing yet not a full-blown disease. While many factors contribute to the development of prostate cancer, one important factor appears to be the balance of cell growth and death.

Until the early 1990's, it was thought that the factors that regulate the entrance and progression through the proliferative cell cycle control the number of cells in the prostate exclusively. We now recognize that this assumption is naive because the number of cells in prostate tissue, whether benign or cancerous, depends largely on the rate of cell proliferation versus the rate of cell death. In prostate glands, this balance is regulated by both systemic hormones and local growth factors that control both the rate of cell proliferation and cell death [7, 8]. A fundamental characteristic of prostate cancer is that, unlike that of normal prostate tissue, the rate of proliferation for the cancer cells exceeds the rate of death, thus resulting in continuous net growth of cancer. This loss of balance can be simplistically viewed as the outcome of the war between the pro-oncogenic and anti-oncogenic signaling pathways.

Context Dependent Pro-oncogenic and Anti-oncogenic Regulations

Oncogenes and tumor suppressors

It is commonly believed that an oncogene is a protein encoding gene which can transform cells and induce cancer [9]. A proto-oncogene is a normal cellular gene that, when altered by DNA damaging agents or viral genomes, can acquire the ability to function as an oncogene [9]. Among these alterations, genetic mutations resulting in the activation of oncogenes increase the chance that a normal cell will transform into a tumor cell. Many proto-oncogenes code for proteins that help to regulate cell growth, proliferation, and differentiation, and are thus involved in the execution of mitogenic signals. Examples of proto-oncogenes include *ras, abl, wnt, myc, and erk*.

Oncogenes are thought to be in a perpetual tug-of-war with tumor suppressor genes which act to constrain cell proliferation or as surveillance instruments to weed out cells growing awry. Tumor suppressors are genes whose partial or complete inactivation, occurring in either germ line or the genome of a somatic cell, results in an increased likelihood of tumorigenesis [9]. There is much evidence to support the notion that loss of tumor suppressors or gain of oncogenes can lead to cancer. Examples of tumor suppressor genes include pRb, p53, PTEN, and APC.

Classical oncogenes

Many cells normally undergo an apoptosis program. In the presence of an activated oncogene, altered survival and proliferation push cells to deviate from their normal paths or fates. Most oncogenes require an additional step, such as mutations in another gene, or environmental factors such as viral infection or chemical exposure, to cause cancer. Since the 1970s, dozens of oncogenes have been identified in human cancer and shown to promote cell proliferation. However, there are also many examples of activation of oncogenic proteins leading to apoptosis and cell death in a context-dependent manner. Here we list many proteins that have been reported to display dual functions to promote both pro- and anti-oncogenc pathways (Table I) and select some to describe in more detail.

Myc gene expression is deregulated in 15 to 30 % of human cancers, resulting in elevated levels of Myc protein. Myc oncoprotein has mitogenic function as it perturbs the decision to phosphorylate tumor suppressor pRb and thereby alter cell cycle progression. However, it is known that *myc* oncogene also has pro-apoptotic effects on cells unless they have been protected from apoptosis by some other, previously acquired anti-apoptotic allele (e.g., a *ras* oncogene); in the presence of the anti-apoptotic mutation, the strong mitogenic effects of the myc oncogene then become apparent. Both Myc and Ras can activate E2F transcription factor, which can initiate both cell proliferation and death [10]. Further studies showed that myc activation can induce apoptosis by overactivating E2F, and also by overactivating the p19Arf tumor suppressor pathway [11]. This is consistent with an earlier report that one of the E2F family members, E2F1, has both oncogenic and tumor-suppressive properties [12]. In this study, E2F1 transgenic mice showed increases in spontaneous tumor formation in the skin, as well as the inhibition of tumor promotion by *O*-tetradecanoyl-phorbol-13-acetate (TPA).

Indeed, the oncogenic roles of E2F could be context-dependent. Homozygous deletion of Rb in a prostate tissue recombination model exhibits increased E2F activity, activation of E2F-target genes, and increased susceptibility to hormonal carcinogenesis [13]. A closer examination of E2F1 expression using tissue microarray found that E2F1 was low in benign and localized prostate cancer, modestly elevated in metastatic lymph nodes from hormone-naive patients, and significantly elevated in metastatic tissues from hormone-resistant prostate cancer patients [14]. In contrast, strong androgen receptor (AR) expression was detected in benign prostate, localized prostate cancer, and lymph node metastasis, but decreased in metastatic hormone-resistant prostate cancer. Since E2F1 can repress AR transcription, elevated E2F1 may contribute to the progression of hormone-refractory prostate cancer [15]. These studies suggest that in the early stages of prostate cancer development, E2F1 may need to be kept low to allow its oncogenic effects to be best accomplished with the cooperation of other oncogenic mutations, such as Ras, to outweigh its potential tumor suppressive effects.

Somatic mutations in the *K-ras* gene are involved in the development of several types of cancer and are common in pancreatic, lung, and colorectal cancers. While these mutations can lead to a K-Ras protein that is always active and can direct cells to grow and divide without control, some studies suggested *K-ras* as having a tumor suppressor role in lung cancer. For example, transgenic mice carrying an oncogenic mutation of *K-ras* gene (G12D) developed lung adenocarcinomas [16]. When *K-ras* G12D transgenic mice were crossed with integrin α1-null mice, K-RasG12D/α1-null mice have a decreased incidence of primary lung tumors and longer survival compared with K-RasG12D/α1 wild-type controls [17]. These data supported oncogenic roles of *K-ras* and further suggested that integrin $\alpha 1\beta 1$ cooperates with K-Ras to drive the growth of non-small cell lung cancer *in vivo*. Interestingly, Zhang et al., [18] reported that while the wild type *K-ras* allele is frequently lost in human lung adenocarcinimas, mice with a heterozygous K-ras deficiency were highly susceptible to the chemical induction of lung tumors when compared to wild type mice. These studies indicated that the wild type K-Ras has tumor suppressor activity. Although similar studies have not been reported for prostate cancer, recent evaluation of 206 prostatic adenocarcinomas found *K-ras* mutations in 7.3% of the patients. Their clinicopathologic features differ from those of prostate adenocarcinomas displaying other gene mutations such as *B-raf* [19]. Whether *K-ras* also has dual roles in prostate cancer remains to be seen.

Mitotic regulators

Aurora A is a serine/threonine protein kinase that functions in various stages of cell cycle. Accumulating evidence has demonstrated that gene amplification and overexpression of Aurora A are linked to tumorigenesis in many cancer types, including colon, breast, and prostate cancer [20], suggesting that Aurora A is an oncogene. In addition, Aurora A overexpression has been used as a negative prognostic marker, because it is associated with resistance to anti-mitotic agents commonly used for cancer therapy. In a recent study, Aurora A null mice died early during embryonic development and showed defects in mitosis, particularly in spindle assembly, supporting critical functions of Aurora A during mitotic transitions [21]. However, Aurora A heterozygosity resulted in a significantly increased tumor incidence in mice, suggesting that Aurora A may also act as a haploinsufficient tumor suppressor. Further analysis showed that embryonic fibroblast cells in Aurora A heterozygous mice displayed higher rates of aneuploidy. Interestingly, VX-680, an Aurora kinase inhibitor currently in phase II clinical trials for cancer treatment, could induce aneuploidy in the embryonic fibroblasts of wild type mice. These studies suggested that a balanced Aurora A level is critical for maintaining genomic stability, and we should be mindful of the potential complications involving the use of Aurora A-specific inhibitors as anti-cancer therapy.

Wnt signaling proteins

While the same genes can display dual functions as pro- and anti-oncogenic proteins, many proteins in the same family can display opposite roles in cancer. For example, β-catenin is the founding member of the armadillo/β-catenin superfamily that is characterized by the presence of a 42-amino acid repeating sequence (armadillo domain). This family contains several subfamilies including the p120^{ctn} subgroup that has 10 armadillo repeating sequences and β-catenin subgroup that has 13 armadillo repeating units [22]. One major function of armadillo domain containing proteins is to interact with cell adhesion receptors such as E-cadherin to regulate the strength of cell-cell adhesion. β-Catenin, for example, is a multifunctional protein that is also a key player in the Wnt/Wingless signaling pathways. Normally, β-catenin, when not associated with E-cadherin and free in the cytoplasm, is subject to ubiquitination and proteosome-mediated degradation. Binding of Wnt to frizzled receptors triggers a cascade of steps that inhibits glycogen synthase kinase 3β (GSK-3β) activity, which allows β-catenin to be stabilized and promote cell proliferation. In addition, when β-catenin is mutated, it can escape the recognition by GSK-3β and promote cell proliferation, thus qualifying for an oncogenic protein. Indeed, the mutations of β-catenin are linked to colorectal cancer development [23] and are also found in prostate cancer [24]; [25].

However, it has been reported that overexpression of β-catenin induces apoptosis in cultured cells [26]. In addition, multiple inhibitors of GSK-3 activity and genetic down-regulation of GSK3-α/β significantly inhibit glioma cell survival and clonogenicity [27]. It was found that the inhibition of GSK-3 activity results in c-Myc activation, leading to the induction of Bax, Bim, DR4/DR5, and tumor necrosis factor (TNF)-related apoptosis-inducing ligand expression and subsequent cytotoxicity. Therefore, inhibition of GSK-3, rather than the activation of GSK-3, may also be considered as an important therapeutic target for gliomas and other cancers.

In prostate cancer development, recent studies established the interaction of Wnt with AR. We now know that β-catenin is a ligand-dependent co-activator of AR, GSK-3β phosphorylates AR and modulates its activity, and LEF-1/TCF regulates expression of AR mRNA and protein [28]. Paradoxically, even though Wnt signaling upregulates AR mRNA levels, the expression of the human AR protein is actually downregulated through an AKT- and MDM2-mediated degradation process that is also activated by Wnt signaling in prostate cancer cells [29]. Nevertheless, the oncogenic roles of β-catenin are quite distinct since most recent studies demonstrated that activation of β-catenin promotes the androgen-independent prostate growth with high grade prostate intraepithelial neoplasia (HGPIN) [30].

Closely interacting with Wnt signaling pathway, adenomatosis polyposis coli (APC) has 6 armadillo domains and has established itself as a tumor suppressor. In fact, over 80% of the colon cancer patients show inactivating APC mutations with over 90 % of them displaying a premature protein termination mutation [31]. The role of APC as an antagonist of the Wnt signaling pathway is well known and it is widely accepted that inappropriate activation of this pathway through loss of APC function contributes to the progression of colon cancers [32]. Although stabilizing APC mutations are rare in primary prostatic tumors, recent data have now identified APC hypermethylation and inactivation during prostate cancer progression [33]. In colon cancer cells, studies also showed that APC mutations stimulate Rac-specific guanine nucleotide exchange factor ASEF-dependent cell migration and may be important for both adenoma formation and tumor progression to invasive malignancy [34]. However, APC cancer mutations recently were reported to compromise cellular proliferation [35] and dominantly inhibit cytokinesis to induce polyploid cells [36]. Although polyploidy is associated with prostate cancer progression, polyploid cell viability is clearly decreased and proliferation is reduced due to cell cycle interruption. This paradoxical feature of disruption in cell cycle and the link for cancer progression in vivo will need to be further addressed in the future.

Paradoxical expression of p120^{ctn} family proteins in vitro versus in vivo

In the p120^{ctn} subfamily, p120^{ctn} (*CTNND1*) is downregulated or redistributed in vivo in most epithelial cancers including prostate cancer [37, 38][39], consistent with its role as a tumor suppressor. However, in almost all tumor cell lines, p120^{ctn} maintains rather strong expression. In cultured tumor cell lines, p120^{ctn} expression stabilizes E-cadherin and cell-cell junction, whereas the loss of p120^{ctn} leads to the loosening of cell-cell junction and disruption of epithelial cell morphology [40]. It has been postulated that the tumor microenvironment *in vivo* may actively suppress p120^{ctn} expression, whereas this suppression could be lost when the cells are cultured *in vitro*. It is thus possible that the roles of p120^{ctn} during tumor progression actually differ, depending on the order in which p120^{ctn} or E-cadherin are downregulated. Some studies suggest that if p120^{ctn} is lost first, E-cadherin levels will follow and fall significantly, which is likely to be paralleled by reduced levels of α-, β-, and γ-catenins [41, 42]. Indeed, when all components of the E-cadherin/catenin complex were examined in prostate cancer specimens, it was found that p120^{ctn} loss was associated with general downregulation of all members of the complex including α-, β-, and γ-catenin [43]. Because E-cadherin is well established as a tumor suppressor, therefore, p120^{ctn} may function similarly as a tumor suppressor through its ability to stabilize and/or regulate E-cadherin [44, 45].

On the other hand, in the event that E-cadherin is lost first, p120^{ctn} may directly and actively promote metastasis. For example, once dissociated from cell-cell junction and released to the cytoplasm, p120^{ctn} can modulate the activities of Rho family small GTPases [46]. These effects are particularly pronounced when p120^{ctn} is overexpressed, a condition that results in increased levels of p120^{ctn} in the cytoplasm and significant changes in cell morphology and/or actin-based motility [44];[47]; [48]; [49]. Upon loss of E-cadherin, p120^{ctn} translocates from the plasma membrane to the cytoplasm [50], where it may actively promote some of the exaggerated effects observed in cells overexpressing p120^{ctn}. Taken together, these observations suggested that p120^{ctn} may behave as either tumor suppressor or metastasis promoter, depending on whether E-cadherin is downregulated first or not [45].

δ-Catenin/NPRAP/Neurojungin (*CTNND2*), a close member of the p120^{ctn} subgroup of the armadillo/β-catenin superfamily, shares over 30% amino acid sequence identity in the armadillo domains with p120^{ctn}, and binds to the same juxtamembrane region on E-cadherin as p120^{ctn}. Unlike p120^{ctn}, which is ubiquitously expressed in the body, δ-catenin is a neural specific protein that is primarily expressed in the central nervous system [51-54]. While p120^{ctn} is downregulated in prostate cancer, δ-catenin is overexpressed in over 80 % of prostatic adenocarcinomas [39, 55]. The increased δ-catenin expression is correlated with the increase in Gleason scores [39]. It is interesting and paradoxical that δ-catenin expression levels in prostate cancer cells in culture were generally moderate or minimal [56]. Its overexpression in prostate cancer cells altered tumor suppressor E-cadherin and p120^{ctn} distribution at the cell-cell junction [39], and disrupted normal monolayer in cell culture [53], thus acts as an oncogenic protein. However, in mammary epithelial cells, δ-catenin appears to act as a tumor suppressor to inhibit cell colony formation [57], raising intriguing possibilities that δ-catenin can exert context-dependent effects on normal epithelial cells versus on transformed cancer cells.

Raf/MEK/MAPK and PI3K/PTEN/AKT signaling

The Ras/Raf/MEK/ERK and PI3K/PTEN/AKT signaling cascades play critical roles in the transmission of signals from growth factor receptors to regulate gene expression involved in diverse cellular functions including cell growth, proliferation, differentiation, and survival.

It is well established that Ras/Raf/MEK/ERK pathway activation is associated with many cancers, and many clinical trials employing the inhibitors targeted at this pathway are currently underway. For example, second generation MEK inhibitors such as PD184352 (*a.k.a.,* CI1040) have been developed and are currently in clinical trials [58-61]. However, how the activation of Ras/Raf/MEK/ERK leads to a cellular decision to proliferate or differentiate appears to be highly context-dependent. A classical example is the neuronal differentiation of pheochromocytoma (PC12) cells. It is well documented that epidermal growth factor (EGF) mediated activation of the Ras/Raf/MEK/ERK axis promotes cell proliferation while nerve growth factor (NGF) activates the same pathway to induce differentiation and neurite outgrowth [62-64]. It was found that EGF stimulation led to a transient activation of ERK, while NGF elicits a sustained ERK activation, which leads to a neuronal differentiation program. However, this model may not be simply extended to other cancer cells under cytotoxic treatment. For example, the anti-cancer activity of ethyl-4-isothiocyanatobutanoate (E-4IB) led to the rapid activation of ERK signaling pathway coupled with delayed transition through the cell cycle and cell cycle arrest, which resulted in diminished mitochondrial membrane potential culminating in apoptosis in leukemic HL-60 cells [65].

The roles of Ras/Raf/MEK/ERK pathway in the development of specific cancer are also controversial [66]. Take Raf activation as an example, overexpression of activated Raf proteins is associated with cell growth, cell cycle arrest or even apoptosis [67-69]. It appears that the fate of the cells depends on the expression level and isoform of Raf kinase. Overexpression of Raf proteins is associated with proliferation in hematopoietic cells [67], erythroid progenitor cells [70], and A10 smooth muscle cells [71]. However, overexpression of activated Raf proteins is also associated with cell cycle arrest in Schwann cells, PC12 cells, HL-60 cells, small cell lung cancer cells and some hematopoietic cells [72-74]. Depending on the Raf isoform, overexpression of Raf can lead to cell proliferation (A-Raf or Raf-1) or cell growth arrest (B-Raf) in NIH-3T3 fibroblasts and FDC-P1 hematopoietic cells [67, 75, 76]. It is not clear why overexpression of the *Raf* gene can lead to such diverse and conflicting results. One possibility is that the opposite outcomes may be determined by the amount or activity of the particular Raf oncoprotein [76, 77]. Therefore, cancer therapies targeted at the inhibition of Raf kinase expression and/or activity should consider the potential off target effects in different types of cancer cells.

In the PI3K/PTEN/AKT pathway, PTEN (phosphatase and tensin homolog deleted on chromosome 10) is a PI3 kinase inhibitor, and its loss or mutation is common in human prostate cancer. 17-allyl-aminogeldanamycin (17-AAG) is a geldanamycin derivative currently employed in clinical trials [78]. In addition to downregulate Akt expression, 17-AAG also causes a rapid inhibition of Akt kinase activity prior to proteosomal degradation of

HER2 and Akt [79]. CCI-779 and 40-O-(2-Hydroxyethyl)-rapamycin are esterified rapamycin derivatives which improve solubility and oral bioavailability. Recent data suggests that tumor cell lines or murine tumors lacking PTEN are particularly sensitive to CCI-779 [80, 81]. Phase II trials are underway in the treatment of several cancers including prostate cancer. Responses have been seen in a small number of patients in Phase I trials. While we cannot conclude for the efficacy for the CCI-779 from Phase I trials, it would not be surprising if the outcomes were variable for different cancer types and cancers at the different stages.

The cross talks between Ras/Raf/MEK/ERK and PI3K/PTEN/AKT signaling pathways can be observed in some advanced prostate cancer cells. These cells express increased levels of activated Akt [82, 83], which may suppress Raf activation. Introduction of activated forms of Akt increased the drug resistance of advanced prostate cancer cells. However, introduction of activated forms of Raf did not increase the drug resistance of the prostate cancer cells. In contrast to the results observed in hematopoietic cells, Raf may normally promote differentiation in prostate cells, which is suppressed in advanced prostate cancer due to increased expression of activated Akt, resulted from PTEN mutation. Thus in advanced prostate cancer it may be advantageous to induce Raf expression to promote differentiation, while in hematopoietic cancers it may be better to inhibit Raf/MEK/ERK-induced proliferation to achieve optimal effects [66].

microRNAs and prostate cancer

miRNAs are a class of naturally occurring, small non-coding RNAs that function in the negative regulation of gene expression by degrading target mRNAs. Oncogenic miRNAs down-regulate the expression of apoptosis-related genes, and tumor suppressor miRNAs target the proliferation-related genes [84]. It is now clear that that aberrant expression of miRNAs occurs in many different types of human cancer and in different stages of disease progression [85, 86]. Many miRNAs have recently been found to act as oncogenes to promote cell survival and tumor growth [87, 88]. Moreover, oncogene expression, such as *myc*, can lead to a widespread regression of tumor suppressor miRNAs and promotes tumorigenesis[89]. Clearly, miRNAs may play a critical role in the pathogenesis of human cancer.

Recent studies also found aberrantly expressed miRNAs in prostate cancer cell lines, xenografts and clinical tissues. Although a number of prostate cancer-related miRNAs were discovered, to date, seven were characterized for their functionalities: three as oncogenes and four as tumor suppressors. *miR-20*, *miR-125b*, and *miR-221/222* promoted prostate cancer because they contributed to the reduction of cell death and increases in cell growth [84, 90, 91]. *miR126* and *miR146a*, on the other hand, led to the reduction of invasiveness of LNCaP cells and reduction of proliferation in PC-3 cells, respectively [92, 93]. There is evidence that *miR-16* and *miR-15* were deleted in prostate cancer and their knockdown in untransformed prostate cells promotes tumorigenesis [94, 95]. However, a recent study comparing invasive versus non-invasive prostate cancer specimens found *miR-15* to be 2 fold higher in the perineural invasion lesions [96]. If this finding is confirmed, *miR-15* may also be involved in the spread of cancers beyond the prostate. Furthermore, since *miR-20* forms an autoregulatory loop with E2F1-3, its function may be subject to the complexity of Myc-E2F pathway, although a mechanism through which c-Myc simultaneously activates E2F1 transcription and limits its translation can allow a tightly controlled proliferative signal [97]. Similarly, because AR regulation is a complex process, the regulation of *miR-125b* through androgen signaling could be complicated.

Table 1. Complex roles of oncogenes and tumor suppressors in carcinogenesis.

Name	Functions	Pro-oncogenic	Anti-oncogenic	Reference
C-myc	Cell proliferation	+	+	[11]
E2F1	Transcription factor	+	+	[10] [14, 15] [12] [13]
K-ras	Kirsten ras oncogene homolog	+	+	[16] [17] [18] [19]
Aurora A	A serine/threonine kinase in mitotic regulation	+	+	[20] [21]
β-catenin	Key player in cell-cell adhesion and also in the Wnt signaling pathways.	+	+	[102] [23] [24] [25] [26] [28] [29]. [30].
APC	Mitotic regulation	+	+	[31] [32, 33] [34, 35] [36].
$p120^{ctn}$	Regulate E-cadherin stability and Rho small GTPases	+	+	[38] [40] [41, 42, 44, 45] [46] [47] [48] [45, 49, 50].
δ-catenin	Neural gene misregulated in cancer; interact with actin in the synapse	+	+	[51-54] [39, 55] [39] [56] [53] [57]
Raf	Functions downstream of the Ras family of membrane associated GTPases	+	+	[67-69] [67] [70] [71] [72-74] [67, 75, 76] [68, 77].
microRNAs	Function in the negative regulation of gene expression	+	+	[84, 90, 91] [92, 93] [94, 95] [96] [97] [98-100] [100, 101]
EBNA1	Human herpesvirus	+	+	[103]
TGF-β	Growth factor initially growth inhibitory and then become tumor promoting	+	+	[3-5, 104, 105]
IRF-4	A hematopoietic cell-restricted transcription factor.	+	+	[106]
GSK-3	A key serine/threonine kinase in Wnt signaling pathway	+	+	[107]

Finally, some miRNAs display different expression patterns dependent on different cancer types. For example, *miR-18* is upregulated in hepatocelluar carcinomas while it is downregulated in brain cancer [98, 99]. *miR-21* expression is increased in brain cancer but its expression is reduced in breast cancer [100, 101]. At present, it is not clear whether differential expression of the same miRNA types also occurs in prostate cancer, but it should not be surprising if they do. Although the oncogenic or tumor-suppressive properties of prostate cancer-related miRNAs make them highly likely to be of clinical use as biomarkers as well as targets for prostate cancer treatment, the same scrutiny would be needed when therapeutic strategies are designed.

Conclusions

The discovery of oncogenes and tumor suppressors and the elucidation of their roles in specific signaling pathways have provided the outstanding mechanism-based treatment strategy of cancer. However, it is increasingly recognized that oncogenes can promote apoptosis while the expression of tumor suppressors can be correlated with tumor progression. We have presented some examples to demonstrate the complexity of actions of oncogenic protein and tumor suppressors in different cancer types including prostate cancer. To understand the molecular and cellular mechanisms of prostate cancer, signaling pathways dictated by the anti- and pro-oncogenic pathways need to be integrated in the context-dependent manner, rather than as a generalization. For future directions, defining the specific roles of each signaling event in prostate cancer growth and survival presents unique opportunities as well as challenges for drug designs and therapeutic intervention. Rationale drug design should consider the specific actions of the drug target at the certain stages of prostate cancer progression, with optimal efficacy/toxicity and right combinations of targeting different signaling pathways.

References

[1] Jemal, A., et al., (2008). Cancer statistics, 2008. *CA Cancer J Clin., 58(2)*, 71-96.

[2] Sakr, W. A., et al. (2000). Epidemiology and molecular biology of early prostatic neoplasia. *Mol Urol., 4(3)*, 109-13; discussion 115.

[3] Barrack, E. R. (1997). TGF beta in prostate cancer: a growth inhibitor that can enhance tumorigenicity. *Prostate, 31(1)*, 61-70.

[4] Jakowlew, S. B. (2006). Transforming growth factor-beta in cancer and metastasis. *Cancer Metastasis Rev., 25(3)*, 435-57.

[5] Bello-DeOcampo, D. & Tindall, D. J. (2003). TGF-betal/Smad signaling in prostate cancer. *Curr Drug Targets., 4(3)*, 197-207.

[6] Abate-Shen, C. & Shen, M. M. (2000). Molecular genetics of prostate cancer. *Genes Dev, 14(19)*, 2410-34.

[7] Isaacs, J. T. (1994). Advances and controversies in the study of programmed cell death/apoptosis in the development of and therapy for cancer. *Curr Opin Oncol., 6(1)*, 82-9.

[8] Berges, R. R. et al., (1995). Implication of cell kinetic changes during the progression of human prostatic cancer. *Clin Cancer Res., 1(5)*, 473-80.

[9] Weinberg, R. A. (2006). *The Biology of Cancer*. 2006 ed., London: Garland Science.

[10] Iaquinta, P. J. & Lees, J. A. (2007). Life and death decisions by the E2F transcription factors. *Curr Opin Cell Biol., 19(6)*, 649-57.

[11] Taghavi, P., et al. (2008). In vitro genetic screen identifies a cooperative role for LPA signaling and c-Myc in cell transformation. *Oncogene*.

[12] Pierce, A. M., et al. (1999). E2F1 has both oncogenic and tumor-suppressive properties in a transgenic model. *Mol Cell Biol., 19(9)*, 6408-14.

[13] Wang, D., Russell, J. L. & Johnson, D. G. (2000). E2F4 and E2F1 have similar proliferative properties but different apoptotic and oncogenic properties in vivo. *Mol Cell Biol., 20(10)*, 3417-24.

[14] Davis, J. N., et al. (2006). Elevated E2F1 inhibits transcription of the androgen receptor in metastatic hormone-resistant prostate cancer. *Cancer Res., 66(24)*, 11897-906.

[15] Kaseb, A. O., et al. (2007). Androgen receptor and E2F-1 targeted thymoquinone therapy for hormone-refractory prostate cancer. *Cancer Res., 67(16)*, 7782-8.

[16] Fisher, G. H., et al. (2001). Induction and apoptotic regression of lung adenocarcinomas by regulation of a K-Ras transgene in the presence and absence of tumor suppressor genes. *Genes Dev, 15(24)*, 3249-62.

[17] Macias-Perez, I., et al. (2008). Loss of integrin alpha1beta1 ameliorates Kras-induced lung cancer. *Cancer Res., 68(15)*, 6127-35.

[18] Zhang, Z., et al. (2001). Wildtype Kras2 can inhibit lung carcinogenesis in mice. *Nat Genet., 29(1)*, 25-33.

[19] Cho, N. Y., et al. (2006). BRAF and KRAS mutations in prostatic adenocarcinoma. *Int J Cancer, 119(8)*, 1858-62.

[20] Zhou, H., et al. (1998). Tumour amplified kinase STK15/BTAK induces centrosome amplification, aneuploidy and transformation. *Nat Genet, 20(2)*, 189-93.

[21] Lu, L. Y., et al. (2008). Aurora a is essential for early embryonic development and tumor suppression. *J Biol Chem., 283(46)*, 31785-90.

[22] Peifer, M., Berg, S. & Reynolds, A. B. (1994). A repeating amino acid motif shared by proteins with diverse cellular roles. *Cell, 76(5)*, 789-91.

[23] Polakis, P. (2000). Wnt signaling and cancer. *Genes Dev, 14(15)*, 1837-51.

[24] Voeller, H. J., Truica, C. I. & Gelmann, E. P. (1998). Beta-catenin mutations in human prostate cancer. *Cancer Res., 58(12)*, 2520-3.

[25] Chesire, D. R., et al. (2000). Detection and analysis of beta-catenin mutations in prostate cancer. *Prostate, 45(4)*, 323-34.

[26] Kim, K., et al. (2000). Overexpression of beta-catenin induces apoptosis independent of its transactivation function with LEF-1 or the involvement of major G1 cell cycle regulators. *Mol Biol Cell, 11(10)*, 3509-23.

[27] Kotliarova, S., et al. (2008). Glycogen synthase kinase-3 inhibition induces glioma cell death through c-MYC, nuclear factor-kappaB, and glucose regulation. *Cancer Res., 68(16)*, 6643-51.

[28] Terry, S., et al. (2006). Multifaceted interaction between the androgen and Wnt signaling pathways and the implication for prostate cancer. *J Cell Biochem., 99(2)*, 402-10.

[29] Yang, X., et al. (2006). Complex regulation of human androgen receptor expression by Wnt signaling in prostate cancer cells. *Oncogene, 25(24)*, 3436-44.

[30] Yu, X., et al. (2008). Activation of beta-Catenin in mouse prostate causes HGPIN and continuous prostate growth after castration. *Prostate*.

[31] Kinzler, K. W. & Vogelstein, B. (1996). Lessons from hereditary colorectal cancer. *Cell, 87(2)*, 159-70.

[32] Hanson, C. A. & Miller, J. R. (2005). Non-traditional roles for the Adenomatous Polyposis Coli (APC) tumor suppressor protein. *Gene, 361*, 1-12.

[33] Dobosy, J. R., et al. (2007). The expanding role of epigenetics in the development, diagnosis and treatment of prostate cancer and benign prostatic hyperplasia. *J Urol., 177(3)*, 822-31.

[34] Kawasaki, Y., Sato, R. & Akiyama, T. (2003). Mutated APC and Asef are involved in the migration of colorectal tumour cells. *Nat Cell Biol., 5(3)*, 211-5.

[35] Tighe, A., Johnson, V. L. & Taylor, S. S. (2004). Truncating APC mutations have dominant effects on proliferation, spindle checkpoint control, survival and chromosome stability. *J Cell Sci., 117(Pt 26)*, 6339-53.

[36] Caldwell, C. M., Green, R. A. & Kaplan, K. B. (2007). APC mutations lead to cytokinetic failures in vitro and tetraploid genotypes in Min mice. *J Cell Biol., 178(7)*, 1109-20.

[37] Kallakury, B. V., et al. (2001). Decreased expression of catenins (alpha and beta), p120 CTN, and E-cadherin cell adhesion proteins and E-cadherin gene promoter methylation in prostatic adenocarcinomas. *Cancer, 92(11)*, 2786-95.

[38] Kallakury, B. V., Sheehan, C. E. & Ross, J. S. (2001). Co-downregulation of cell adhesion proteins alpha- and beta-catenins, p120CTN, E-cadherin, and CD44 in prostatic adenocarcinomas. *Hum Pathol., 32(8)*, 849-55.

[39] Lu, Q., et al. (2005). Increased expression of delta-catenin/neural plakophilin-related armadillo protein is associated with the down-regulation and redistribution of E-cadherin and p120ctn in human prostate cancer. *Hum Pathol., 36(10)*, 1037-48.

[40] Ireton, R. C., et al. (2002). A novel role for p120 catenin in E-cadherin function. *J Cell Biol., 159(3)*, 465-76.

[41] Nagafuchi, A., Takeichi, M. & Tsukita, S. (1991). The 102 kd cadherin-associated protein: similarity to vinculin and posttranscriptional regulation of expression. *Cell, 65(5)*, 849-57.

[42] Papkoff, J. (1997). Regulation of complexed and free catenin pools by distinct mechanisms. Differential effects of Wnt-1 and v-Src. *J Biol Chem., 272(7)*, 4536-43.

[43] van Oort, I. M., et al. (2007). The prognostic value of E-cadherin and the cadherin-associated molecules alpha-, beta-, gamma-catenin and p120ctn in prostate cancer specific survival: a long-term follow-up study. *Prostate, 67(13)*, 1432-8.

[44] Anastasiadis, P. Z., et al. (2000). Inhibition of RhoA by p120 catenin. *Nat Cell Biol., 2(9)*, 637-44.

[45] Thoreson, M. A. & Reynolds, A. B. (2002). Altered expression of the catenin p120 in human cancer: implications for tumor progression. *Differentiation, 70(9-10)*, 583-9.

[46] Anastasiadis, P. Z. & Reynolds, A. B. (2001). Regulation of Rho GTPases by p120-catenin. *Curr Opin Cell Biol., 13(5)*, 604-10.

[47] Grosheva, I., et al. (2001). p120 catenin affects cell motility via modulation of activity of Rho-family GTPases: a link between cell-cell contact formation and regulation of cell locomotion. *J Cell Sci., 114(Pt 4)*, 695-707.

[48] Noren, N. K., et al. (2000). p120 catenin regulates the actin cytoskeleton via Rho family GTPases. *J Cell Biol., 150(3)*, 567-80.
[49] Reynolds, A. B., et al. (1996). The gene encoding p120cas, a novel catenin, localizes on human chromosome 11q11 (CTNND) and mouse chromosome 2 (Catns). *Genomics, 31(1)*, 127-9.
[50] Thoreson, M. A., et al. (2000). Selective uncoupling of p120(ctn) from E-cadherin disrupts strong adhesion. *J Cell Biol., 148(1)*, 189-202.
[51] Paffenholz, R. & Franke, W. W. (1997). Identification and localization of a neurally expressed member of the plakoglobin/armadillo multigene family. *Differentiation, 61(5)*, 293-304.
[52] Zhou, J., et al. (1997). Presenilin 1 interaction in the brain with a novel member of the Armadillo family. *Neuroreport, 8(6)*, 1489-94.
[53] Lu, Q., et al. (1999). delta-catenin, an adhesive junction-associated protein which promotes cell scattering. *J Cell Biol., 144(3)*, 519-32.
[54] Ho, C., et al. (2000). delta-catenin is a nervous system-specific adherens junction protein which undergoes dynamic relocalization during development. *J Comp Neurol., 420(2)*, 261-76.
[55] Burger, M. J., et al. (2002). Expression analysis of delta-catenin and prostate-specific membrane antigen: their potential as diagnostic markers for prostate cancer. *Int J Cancer, 100(2)*, 228-37.
[56] Kim, K., et al. (2008). Identification of E2F1 as a positive transcriptional regulator for delta-catenin. *Biochem Biophys Res Commun., 369(2)*, 414-20.
[57] Westbrook, T. F., et al. (2005). A genetic screen for candidate tumor suppressors identifies REST. *Cell, 121(6)*, 837-48.
[58] Yu, C., et al. (2002). Pharmacologic mitogen-activated protein/extracellular signal-regulated kinase kinase/mitogen-activated protein kinase inhibitors interact synergistically with STI571 to induce apoptosis in Bcr/Abl-expressing human leukemia cells. *Cancer Res., 62(1)*, 188-99.
[59] Allen, L. F., Sebolt-Leopold, J. & Meyer, M. B. (2003). CI-1040 (PD184352), a targeted signal transduction inhibitor of MEK (MAPKK). *Semin Oncol., 30(5 Suppl 16)*, 105-16.
[60] Milella, M., et al. (2001). Therapeutic targeting of the MEK/MAPK signal transduction module in acute myeloid leukemia. *J Clin Invest., 108(6)*, 851-9.
[61] Milella, M., et al. (2002). Synergistic induction of apoptosis by simultaneous disruption of the Bcl-2 and MEK/MAPK pathways in acute myelogenous leukemia. *Blood, 99(9)*, 3461-4.
[62] Greene, L. A. & Tischler, A. S. (1976). Establishment of a noradrenergic clonal line of rat adrenal pheochromocytoma cells which respond to nerve growth factor. *Proc Natl Acad Sci U S A, 73(7)*, 2424-8.
[63] Huff, K., End, D. & Guroff, G. (1981). Nerve growth factor-induced alteration in the response of PC12 pheochromocytoma cells to epidermal growth factor. *J Cell Biol., 88(1)*, 189-98.
[64] Marshall, C. J. (1995). Specificity of receptor tyrosine kinase signaling: transient versus sustained extracellular signal-regulated kinase activation. *Cell, 80(2)*, 179-85.

[65] Bodo, J., et al. (2007). Isothiocyanate E-4IB induces MAPK activation, delayed cell cycle transition and apoptosis. *Cell Prolif., 40(3)*, 316-26.

[66] McCubrey, J. A., et al. (2006). Roles of the RAF/MEK/ERK and PI3K/PTEN/AKT pathways in malignant transformation and drug resistance. *Adv Enzyme Regul., 46*, 249-79.

[67] Hoyle, P. E., et al. (2000). Differential abilities of the Raf family of protein kinases to abrogate cytokine dependency and prevent apoptosis in murine hematopoietic cells by a MEK1-dependent mechanism. *Leukemia, 14(4)*, 642-56.

[68] Chang, F. & McCubrey, J. A. (2001). P21(Cip1) induced by Raf is associated with increased Cdk4 activity in hematopoietic cells. *Oncogene, 20(32)*, 4354-64.

[69] Chang, F., Steelman, L. S. & McCubrey, J. A. (2002). Raf-induced cell cycle progression in human TF-1 hematopoietic cells. *Cell Cycle, 1(3)*, 220-6.

[70] Sanders, M. R., et al. (1998). The Raf-1 protein mediates insulin-like growth factor-induced proliferation of erythroid progenitor cells. *Stem Cells, 16(3)*, 200-7.

[71] Cioffi, C. L., et al. (1997). Selective inhibition of A-Raf and C-Raf mRNA expression by antisense oligodeoxynucleotides in rat vascular smooth muscle cells: role of A-Raf and C-Raf in serum-induced proliferation. *Mol Pharmacol., 51(3)*, 383-9.

[72] Yen, A., et al. (1994). Expression of activated RAF accelerates cell differentiation and RB protein down-regulation but not hypophosphorylation. *Eur J Cell Biol., 65(1)*, 103-13.

[73] Ravi, R. K., et al. (1998). Activated Raf-1 causes growth arrest in human small cell lung cancer cells. *J Clin Invest., 101(1)*, 153-9.

[74] Lloyd, A. C., et al. (1997). Cooperating oncogenes converge to regulate cyclin/cdk complexes. *Genes Dev, 11(5)*, 663-77.

[75] Woods, D., et al. (1997). Raf-induced proliferation or cell cycle arrest is determined by the level of Raf activity with arrest mediated by p21Cip1. *Mol Cell Biol., 17(9)*, 5598-611.

[76] Shelton, J. G., et al. (2004). B-raf and insulin synergistically prevent apoptosis and induce cell cycle progression in hematopoietic cells. *Cell Cycle, 3(2)*, 189-96.

[77] Shelton, J. G., et al. (2004). Synergy between PI3K/Akt and Raf/MEK/ERK pathways in IGF-1R mediated cell cycle progression and prevention of apoptosis in hematopoietic cells. *Cell Cycle, 3(3)*, 372-9.

[78] Chanan-Khan, A. & Porter, C. W. (2006). Immunomodulating drugs for chronic lymphocytic leukaemia. *Lancet Oncol., 7(6)*, 480-8.

[79] Basso, A. D., et al. (2002). Ansamycin antibiotics inhibit Akt activation and cyclin D expression in breast cancer cells that overexpress HER2. *Oncogene, 21(8)*, 1159-66.

[80] Neshat, M. S., et al. (2001). Enhanced sensitivity of PTEN-deficient tumors to inhibition of FRAP/mTOR. *Proc Natl Acad Sci U S A, 98(18)*, 10314-9.

[81] Podsypanina, K., et al. (2001). An inhibitor of mTOR reduces neoplasia and normalizes p70/S6 kinase activity in Pten+/- mice. *Proc Natl Acad Sci U S A, 98(18)*, 10320-5.

[82] Lee, J. T., Jr., Steelman, L. S. & McCubrey, J. A. (2005). Modulation of Raf/MEK/ERK kinase activity does not affect the chemoresistance profile of advanced prostate cancer cells. *Int J Oncol., 26(6)*, 1637-44.

[83] Lee, J. T., Jr. & McCubrey, J. A. (2002). The Raf/MEK/ERK signal transduction cascade as a target for chemotherapeutic intervention in leukemia. *Leukemia, 16(4)*, 486-507.

[84] Shi, X. B., Tepper, C. G. & White, R. W. (2008). microRNAs and prostate cancer. *J Cell Mol Med., 12(5A)*, 1456-65.

[85] Calin, G. A. & Croce, C. M. (2006). MicroRNA signatures in human cancers. *Nat Rev Cancer, 6(11)*, 857-66.

[86] Wu, W., et al. (2007). MicroRNA and cancer: Current status and prospective. *Int J Cancer, 120(5)*, 953-60.

[87] Si, M. L., et al. (2007). miR-21-mediated tumor growth. *Oncogene, 26(19)*, 2799-803.

[88] Voorhoeve, P. M., et al. (2006). A genetic screen implicates miRNA-372 and miRNA-373 as oncogenes in testicular germ cell tumors. *Cell, 124(6)*, 1169-81.

[89] Chang, T. C., et al. (2008). Widespread microRNA repression by Myc contributes to tumorigenesis. *Nat Genet, 40(1)*, 43-50.

[90] Sylvestre, Y., et al. (2007). An E2F/miR-20a autoregulatory feedback loop. *J Biol Chem., 282(4)*, 2135-43.

[91] Galardi, S., et al. (2007). miR-221 and miR-222 expression affects the proliferation potential of human prostate carcinoma cell lines by targeting p27Kip1. *J Biol Chem., 282(32)*, 23716-24.

[92] Musiyenko, A., Bitko, V. & Barik, S. (2008). Ectopic expression of miR-126*, an intronic product of the vascular endothelial EGF-like 7 gene, regulates prostein translation and invasiveness of prostate cancer LNCaP cells. *J Mol Med., 86(3)*, 313-22.

[93] Lin, S.L., et al. (2008). Loss of mir-146a function in hormone-refractory prostate cancer. *RNA, 14(3)*, 417-24.

[94] Yoon, S. & De Micheli, G. (2005). Prediction of regulatory modules comprising microRNAs and target genes. *Bioinformatics, 21 Suppl 2*, ii93-100.

[95] Bonci, D., et al. (2008). The miR-15a-miR-16-1 cluster controls prostate cancer by targeting multiple oncogenic activities. *Nat Med., 14(11)*, 1271-7.

[96] Prueitt, R. L., et al. (2008). Expression of microRNAs and protein-coding genes associated with perineural invasion in prostate cancer. *Prostate, 68(11)*, 1152-64.

[97] O'Donnell, K. A., et al. (2005). c-Myc-regulated microRNAs modulate E2F1 expression. Nature, *435(7043)*, 839-43.

[98] Murakami, Y., et al. (2006). Comprehensive analysis of microRNA expression patterns in hepatocellular carcinoma and non-tumorous tissues. *Oncogene, 25(17)*, 2537-45.

[99] Ciafre, S. A., et al. (2005). Extensive modulation of a set of microRNAs in primary glioblastoma. *Biochem Biophys Res Commun., 334(4)*, 1351-8.

[100] Chan, J. A., Krichevsky, A. M. & Kosik, K. S. (2005). MicroRNA-21 is an antiapoptotic factor in human glioblastoma cells. *Cancer Res., 65(14)*, 6029-33.

[101] Iorio, M. V., et al. (2005). MicroRNA gene expression deregulation in human breast cancer. *Cancer Res., 65(16)*, 7065-70.

[102] Weinberg, B. D., et al. (2007). Antitumor efficacy and local distribution of doxorubicin via intratumoral delivery from polymer millirods. *J Biomed Mater Res A., 81(1)*, 161-70.

[103] Liu, J. Y., et al. (2008). The N-terminal domain of EBNA1 acts as a suppressor of the HER2/neu oncogene. *Cancer Lett.*

[104] Criswell, T. L., et al. (2008). Knockdown of the transforming growth factor-beta type III receptor impairs motility and invasion of metastatic cancer cells. *Cancer Res.*, *68(18)*, 7304-12.

[105] Yang, J., et al. (2008). Rb/E2F4 and Smad2/3 link survivin to TGF-beta-induced apoptosis and tumor progression. *Oncogene*, *27(40)*, 5326-38.

[106] Acquaviva, J., Chen, X. & Ren, R. (2008). IRF-4 functions as a tumor suppressor in early B-cell development. *Blood.*

[107] Wang, Z., et al. (2008). Glycogen synthase kinase 3 in MLL leukaemia maintenance and targeted therapy. *Nature*, *455(7217)*, 1205-9.

In: Handbook of Prostate Cancer Cell Research...
Editor: Alan T. Meridith
ISBN: 978-1-60741-954-9

Chapter 14

Key Role of Bone Microenvironment on Prostate Cancer Bone Metastasis: The Tumor Sanctuary and Source of Prostate Cancer Cell Survival, Migration and Development in Bone

***K Mori*[1] *and D Heymann*[2, 3]**

[1] Department of Orthopaedic Surgery, Shiga University of Medical Science, Tsukinowa-cho, Seta, Otsu, Shiga, 520-2192, Japan

[2] INSERM, U957

[3] Nantes Atlantique Universités, Université de Nantes EA 3822; Physiopathologie de la Résorption Osseuse et Thérapie des Tumeurs Osseuses Primitives Faculté de Médecine, 1 rue Gaston Veil, 44035, Nantes cedex 1, France

Abstract

Nowadays, cancer is one of the major leading causes of death worldwide. Because of the improvement of primary cancer site control by significant advances in treatment, cancer bone metastases have become a major problem for patients with cancers. Indeed, prostate cancer metastases to bone are observed in around 80% of prostate cancer patients and represent the most critical complication of advanced prostate cancer, frequently resulting in significant morbidity and mortality.

Bone is a well-known target organ of cancer metastasis, as well as the lungs and the liver, and represents a good microenvironment for the colonization and progression of a tumor mass. The vicious cycle formed between cancer cells and bone cells in the bone microenvironment has been proposed to explain the high trend of tumor development in bone.

Recently, the biological mechanisms that govern prostate cancer metastases to bone at the molecular level were disclosed. Accumulating findings have strongly suggested the important involvement of the key regulators of bone metabolism, Receptor Activator of NF-κB ligand (RANKL)/RANK/Osteoprotegerin (OPG), in prostate cancer bone metastasis development, prostate cancer cell migration and its survival. It has been reported that blocking of RANKL-RANK binding significantly diminished prostate cancer progression in bone both in indirect and direct fashion.

The present chapter summarizes the current knowledge on the development of bone metastases derived from prostate cancer. Recent data on cancer cell migration and molecular pathways involved in cancer cell survival in the bone microenvironment identifying novel therapeutic targets of the prostate cancer bone metastases are described.

Introduction

Prostate cancer is the most common cancer diagnosed in men, currently the major leading cause of cancer death [1, 2]. Solid tumors such as prostate, breast and lung cancers preferentially metastasize to bone [3]. Indeed, around 350,000 patients with bone metastases die annually in the United States [1]. Bone is the most frequent site of metastases of prostate cancer, which are observed in approximately 80% of the patients with prostate cancer and represent the most serious complication of advanced prostate cancer, frequently causing significant morbidity and mortality [1, 2].

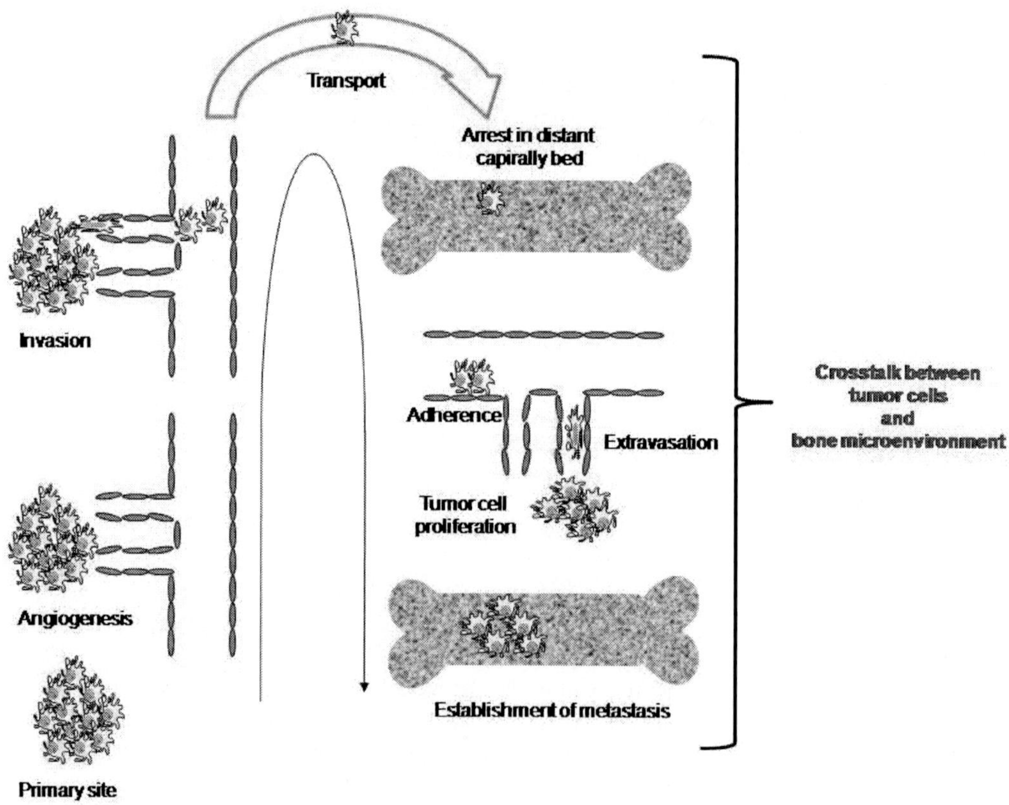

Figure 1. Multiple steps involved in tumor migration from primary site in prostate to bone. Tumor cells must escape from host's defense and survive. Each of these steps involves specific interactions between tumor cells and host cells and might be therapeutic targets.

In particular, bone is almost always the first and frequently the only site of metastases in patients with stage D3 prostate cancer [4]. Spinal cord compression and pathological fractures occur in approximately one-third of cases with stage D3, and anemia, cachexia, bone pain and sepsis are common complications in such prostate cancer patients [3, 5]. These complications markedly deteriorate the quality of life (QOL) and the activities of daily living (ADL) of the patients. Thus, the socio-economic burden of this disease is increasing steadily [6].

In addition, the number of skeletal metastatic foci is the most powerful independent prognostic factor of limited response to hormone ablation therapy and poor survival of patients with advanced prostate cancer [7]. It is therefore considered that one of the top targets of cancer treatment is the control of bone metastasis.

The precise mechanism how cancer cells metastasize to bone is not completely understood. However, it is generally hard and challenging for cancer cells to metastasize to distant tissues far from the original site, since cancer cells must escape from various hosts' defenses, survive and overcome biological processes before establishing metastatic lesions [1] (Figure 1). It is considered that one of the most critical steps in such a process is the crosstalk between the biological environment of bone and of cancer cells. As represented by Paget's "Seed and Soil" theory in the 1880s [8], bone is an innate fertile environment for tumor cell survival and development [9]. Furthermore, it is suggested that tumor cells migrated into bone modulate a tumoral bone microenvironment favorable for their own growth [10]. Recently, Kaplan *et al.* proposed an interesting "pre-metastatic niche" theory [11]. That is, tumor cells actively modulate the metastatic site microenvironment to make it convenient for their metastasis prior to their homing. Thus, we have to take into account the role of the bone microenvironment when we discuss prostate cancer bone metastasis.

In this context, we must understand the physiological working of bone. Bone is always and continuously remodeled and its homeostasis depends on the strictly balanced activities between bone formation by osteoblasts and bone resorption by osteoclasts [12]. The discovery of key factors involved in bone remodeling has moved bone research into a new era. Most notable of these factors belonging to the tumor necrosis factor (TNF)/TNF receptor family are the receptor activator of nuclear factor κB ligand (RANKL/TNFSF11), its cognate receptor RANK (TNFRSF11A) and decoy receptor for RANKL, osteoprotegerin (OPG/TNFRSF11B) [13-17]. Consequently, RANKL has been shown to both mediate osteoclastogenesis and activate mature osteoclasts, whereas OPG negatively regulates RANKL binding to RANK and reduces the half-life of membranous RANKL, therefore inhibiting bone resorption induced by osteoclasts [18]. Current findings have revealed that RANKL/RANK/OPG triad is the key regulator not only for physiological but also pathological bone metabolism [19-31]. Thus, the RANKL level is increased in osteolytic lesions associated with malignant tumors, whereas the serum OPG level is increased in osteoblastic lesions. *In fine*, local RANKL/OPG ratio determines local bone features.

Bone plays an important role in not only the skeleton but also the reservoir of minerals and growth factors. During the bone degradation by osteoclasts, several growth factors [insulin-like growth factor-I (IGF-I), transforming growth factor-β (TGF-β), etc.] stocked in the mineralized bone matrix are released and lead to tumor cell proliferation [32]. Furthermore, bone degradation provides the space for tumor cell homing and development in the hard bone tissue. In addition, it has been suggested that tumor cells produce soluble factors that activate osteoclast differentiation and maturation directly (RANKL) or indirectly *via* osteoblasts [parathyroid hormone-related peptide (PTH-rP), interleukin (IL)-8, monocyte chemotactic protein-1 (MCP-1) etc.] [33-37]. These activated osteoclasts induce further bone degradation and establish a vicious cycle between bone cells and prostate cancer cells in the tumoral bone microenvironment. This vicious cycle has been proposed to explain the mechanism of tumor development in bone sites [38] (Figure 2).

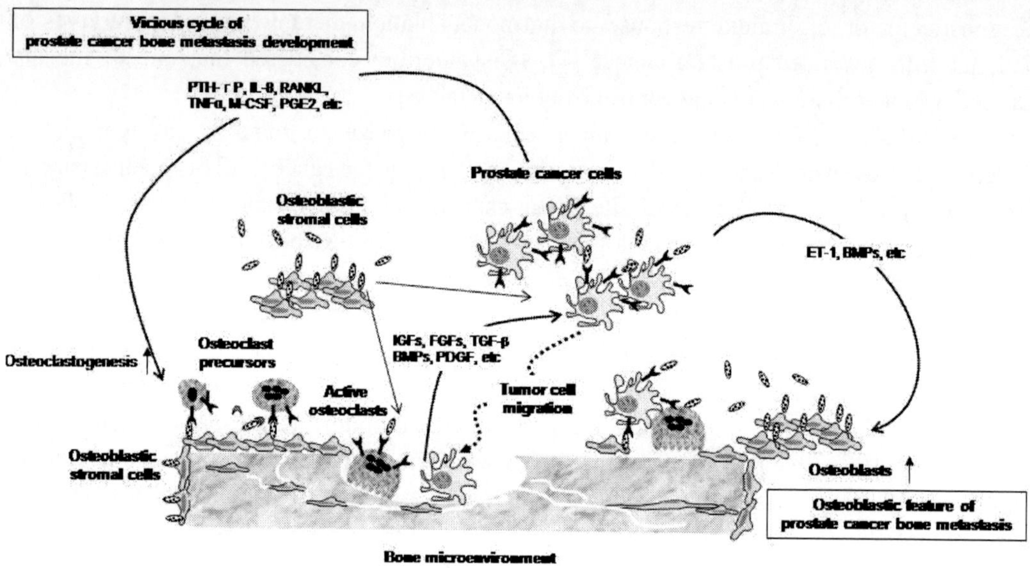

Figure 2. Schematic presentation of the interaction between prostate cancer cells and bone microenvironment. Significant crosstalk between prostate cancer cells and host cells in bone environment contribute to the development of prostate cancer bone metastasis and its osteoblastic feature. RANKL acts as a 'soil' factor that facilitates prostate cancer metastasis development in bone by activating both kinds of RANK-positive cells (osteoclasts/osteoclast precursors and prostate cancer cells).

: RANKL, : RANK, : OPG

Unlike other solid tumors such as breast cancer that are associated with osteolytic bone metastases, prostate cancer bone metastases stimulate an overall increase in both bone remodeling and bone volume [39]. In 1985, however, histological examination has already revealed that prostate cancer bone metastasis consists of both osteolytic and osteoblastic lesions [40]. Furthermore, recent findings strongly suggested the importance of osteoclast function in osteoblastic tumors as well as osteolytic tumors [41-46]. Thus, the role of osteoclasts is very important, even if prostate cancer represents osteoblastic bone metastases.

The present chapter summarizes the current knowledge on the development of bone metastases derived from prostate cancer. Recent data on cancer cell migration and molecular pathways involved in cancer cell survival in the bone microenvironment, identifying novel therapeutic targets of prostate cancer bone metastases, are described.

Inhibition of Bone Degradation

Blocking of the vicious cycle residing between bone microenvironment and prostate cancer cells by inhibition of bone degradation is one of the reasonable therapeutic approaches. In this context and together with the essential role of the RANKL/RANK/OPG triad in osteoclastogenesis, inhibition of osteoclastic activities by RANKL/RANK blocking and/or bisphosphonates (BPs) constitutes a potential novel therapeutic approach. Indeed, inactivation of osteoclastic function with BPs, soluble RANK (sRANK) or OPG has prevented the development of prostate cancer bone metastases [44-48]. These antitumor effects were uniquely revealed in bone, not in other tissues such as subcutaneous [43, 44]. Thus, it has been considered that the antitumor effect of RANKL/RANK blocking was the result of indirect effect by inhibition of osteoclastic activity [43, 44]. Furthermore, these findings revealed that osteoclasts, not tumor cells, are the effector cells of bone degradation in prostate cancer bone metastases. Thus, osteoclasts are a critical therapeutic target of prostate cancer bone metastases. Due to the inhibition of bone degradation, prostate cancer cells metastasizing to bone lose both the space and growth factors released from bone for their development. It is therefore rational that inhibition of bone degradation is a promising approach for prostate cancer bone metastases treatment, if BPs and RANKL/RANK blocking agents don't have a direct effect on tumor cells.

Functional RANK Expression on Prostate Cancer Cells

Alternatively, RANK expressed at the surface of osteoclasts/osteoclast precursors is the essential signaling receptor for osteoclast differentiation [14, 49-51]. However its expression is not restricted to this cell type, as it was also observed in other tissue including the mammary gland [52], lung [51], brain [51], kidney [51], thyroid gland [53] and hypertrophic chondrocytes [50].

Interestingly, prostate cancer bone metastases preferentially present the osteoblastic feature, however, prostate cancer cells express functional RANK [54-56]. We [54] and Jones *et al.* [55] demonstrated that human breast cancer cell lines (MDA-MB231, Hs578T, MCF7) and human prostate cancer cell lines (DU145, PC3, LNCaP) express functional RANK. Moreover, mouse melanoma cell lines (B16F10), originate another cell lineage also expressing functional RANK [55]. RANK expressed on these tumor cells does not change tumor cell proliferation and survival; however it clearly induces tumor cell migration [54-56].

These findings provide the evidence that RANKL can directly affect RANK-positive prostate cancer cells and induce their migration. To reveal visible RANKL-induced prostate cancer cell migration, it is necessary to provide enough space where they can migrate. In bone tissue, RANKL can activate osteoclasts/osteoclast precursors that initiate bone degradation, providing such space for tumor cell migration. Herein, we can be aware of the importance of bone microenvironment in bone metastases afresh.

Consequently, the previously reported OPG- and sRANK-Fc-induced tumor growth inhibition in bone might be due to the direct effect, at least in part, on the RANK-positive prostate cancer cells by blocking RANKL-induced tumor cell migration. In addition, in breast cancer, a positive correlation has been reported between constant expressions of RANK with decreased/absent expression of RANKL and a high metastatic phenotype [57]. Altogether, RANK-positive cells are starving RANKL and preferentially attracted by RANKL rich bone environment, where RANKL acts as a 'soil' factor that facilitates bone metastases development by activating both kinds of RANK-positive cells (osteoclasts/osteoclast precursors and prostate cancer cells). These findings are all consistent with a high trend of prostate cancer metastases to bone.

Therefore, a therapeutic approach using RANKL/RANK blocking (OPG, sRANK-Fc or denosumab: RANKL inhibitor) might be suggested even if they don't affect cell proliferation or survival, because these agents can inhibit tumor cell migration. This effect could be additional to the direct inhibition of bone resorption, leading to suppression of tumor growth. The therapeutic benefits of bi-therapy associating anti-bone resorption and antitumor agents in abrogating tumor recurrence have been reported in bone metastases and primary bone tumors [58-60]. RANK expressed on these tumor cells therefore appeared an attractive novel therapeutic target.

Intriguingly, osteosarcoma, the most frequent malignant primary bone tumor which presents osteoblastic profile also expresses functional RANK [61, 62]. RANK expressed on osteosarcoma cells were not implicated in RANKL-induced tumor cell migration, proliferation and survival, whereas cancer-related genes were significantly modulated by RANKL stimulation [61]. Thus, it is possible that RANK expressed on tumor cells plays an unknown role except for RANKL-induced tumor cell proliferation, we need further study to reveal the full potential of RANK expressed on tumor cells.

Prostate Cancer Cell-Derived Soluble Factors

It is well known that prostate epithelial cells produce several soluble factors which realize communications between tumor cells and between tumor and bone cells. Indeed, prostate cancer cells produce IGF-I, IGF-II, their receptors and IGF binding proteins (IGFBPs) [63]. Growth factors such as TGF-β, IGFs and FGF linked to glycoaminoglycans are stocked in a high amount in bone matrix [64]; however prostate cancer cells also produce these growth factors by themselves. The significant role of IGF-I as a growth factor in osteoblasts and prostate cancer cells is common knowledge.

The IGFBPs are molecules that have higher activity for IGF receptor type I (IGFR-I) than IGFs, thereby acting as controllers of extracellular IGF bioavailability [65]. One of the major IGFBPs produced by prostate cancer cells is IGFBP-2 and positive association between serum concentration of IGFBP-2 and disease stage is reported [66]. Furthermore, prostate-specific antigen (PSA), urokinase-type plasminogen activator (uPA) and cathepsin-D produced by prostate cancer cells act as a protease of IGFBPs and increase the effect of IGFs (see below).

In addition, prostate cancer cells produce several bone morphogenetic proteins (BMPs) (BMP-2, 3, 4 and 6). BMPs that belong to the TGF-β family protein are involved in bone formation [67]. Among them, significant attention has been paid to BMP-6 in prostate cancer bone metastasis, since it stimulated not only new bone formation, but also prostate cancer cell proliferation and invasion in bone [68]. In addition, significant association of BMPs with Wnt signals, critical signaling pathway in bone formation, was reported in prostate cancer bone metastases [69].

Moreover, they also produce Endothelin (ET)-1 and platelet-derived growth factor (PDGF). ET was formally recognised as a peptide produced by vascular endothelial cells induces strong vascular contraction [70]. Now, it is known that ET-1 was produced by several epithelial tumor cells including prostate cancer [71]. Among the association of malignant tumors with ET-1, prostate cancer is one of the most studied ones. In fact, the expression of ET A receptor (ETAR) is associated with progression and aggressive phenotype of prostate cancer [72]. ET-1 stimulated osteoblast proliferation as well as enhanced the effect of other osteoblastic factors [73]. Notably, over-expression of ET-1 induced new bone formation *in vivo* model animal study [74]. Furthermore, PTH-rP is well known as an osteoclast activator; however Guise *et al.* reported that N-terminal of PTH-rP activates ETAR and contributes to the osteoblastic profile of prostate cancer bone metastases [75].

To date, PDGF expression has been shown in a number of different solid tumors, from glioblastomas to prostate carcinomas and significantly associated with their metastatic behaviour [76, 77]. Among three isoforms of PDGF (PDGF-AA, AB, BB), significant association of PDGF-BB with osteosclerotic prostate cancer bone metastasis were reported in animal model [78].

In addition to these growth factors, prostate cancer cells directly act on key regulator of bone remodelling RANKL/RANK/OPG system. Prostate cancer cells produce OPG, decoy receptor of RANKL. Increased expression of OPG at the protein level was confirmed in bone metastatic site of prostate cancer cells [79].

These factors induce prostate cancer cell proliferation and/or bone formation by osteoblasts and contribute to osteoblastic phenotype of prostate cancer bone metastases (Figure 2). The precise mechanism why prostate cancer bone metastasis presents osteoblastic profile is much more complicated compared to that of osteolytic one, above factors might contribute a role, at least in part, in such condition.

Moreover, prostate cancer cells produce osteolytic soluble factors and should involve in vicious cycle of prostate cancer bone metastases. It has been reported that prostate cancer cell produce soluble RANKL and stimulate bone degradation by direct activation of osteoclasts/osteoclast precursors [80]. In addition, prostate cancer cells produce PTH-rP, which is a critical osteoclast activator other than RANKL and indirectly stimulate bone degradation [81]. Furthermore, prostate cancer cells produce IL-8 [82] and MCP-1 [83] and directly induce bone degradation by RANKL-independent pathway.

All these findings suggest that prostate cancer cells actively work on bone microenvironment and contribute to their growth and pathological features in bone metastases. Therapeutic strategy that targeting these both osteoblastic and osteolytic soluble factors produced by prostate cancer cells can therefore inhibit prostate cancer cell proliferation and normalize pathological bone remodelling and consists one of the treatment for prostate cancer bone metastases.

Growth Factors Released from Bone Matrix by Bone Degradation

The bone microenvironment is a rich milieu of growth factors/cytokines. These factors are released from bone during bone degradation by osteoclasts and lead to tumor cell proliferation migrated into the bone. Consequently, this step plays a part of vicious cycle factors of prostate cancer bone metastasis as a microenvironment factor.

Furthermore, organ sites frequently involved in metastatic diseases, such as bones, appear to confer to metastatic cell protection from anticancer drug-induced apoptosis. This protection is mediated by soluble growth factors/cytokines released by the normal cellular constitutes of the host tissue microenvironment [84]. A prime example of such observation is seen in prostate cancer bone metastases. Survival mechanism of prostate cancer cells are produced by the interactions of the tumor cells with the local microenvironment of the host tissue. Thus, explaining why the bone metastases are the almost exclusive site for the development of androgen ablation refractoriness in advanced prostate cancer [85].

It has been described that one of the most interesting anti-survival pathways appeared IGFR-1-induced activation of phosphatidylinositol 3-Kinase (PI3-K)/Akt-mediated phosphorylation of the androgen receptor (AR) and IGFR-1-mediated activation of nuclear factor kappa B (NF-κB) survival pathway. Apparently, growth factors, such as IGF-I, IL-6 and TGF-β, pathways can activate the transcriptional target of AR, even in the absence of ligand (androgen), thereby inhibiting directly and/or indirectly apoptosis of prostate cancer cells [86].

Among the effects of these growth factors/cytokines, the locally increased IGF-I bioavailability in the bone microenvironment of metastasis contributes as a major mediator of the rescue from anticancer drug apoptosis of metastatic prostate cancer. Prostate cancer cells metastasize to bone released uPA, which can bind to its receptor (uPA-R) on the cell surface of osteoblasts. This binding activates local increase of proteinolysis. Accordingly, uPA-mediated limited proteolysis of IGFBPs (mainly IGFBP-3) produces a local increase in IGF bioavailability which, in turn, stimulates the proliferation of osteoblasts and tumor cells, locally leading to the development of an osteoblastic reaction at the site of prostate cancer bone metastasis [65]. These self-amplifying molecular loops activate local IGF-I bioavailability and modulate biology of metastatic prostate cancer cells in terms of proliferation and/or antiapoptosis [86].

Notably, the management of bone metastases have traditionally been treated by therapeutic strategy that aims at directly inducing tumor cell death. However, accumulating evidence suggested major limitation of this conventional approach [87]. One of the causes of this limitation is that these conventional therapies lack concern for the interaction between tumors cell and bone microenvironment. Thus, we have to establish another strategy that takes into account both the tumor cells and the bone microenvironment to eradicate prostate cancer bone metastases. It appears that a new therapy based on such a strategy is a promising and unprecedented therapy for prostate cancer bone metastases.

Conclusion

In conclusion, the mechanism of the prostate cancer bone metastases development is the consequence of the significant interaction between tumor cells and bone microenvironment intricately. Bone microenvironment appears as a sanctuary, since it significantly contributes to prostate cancer bone metastasis development by supporting their proliferation and inhibition of their apoptosis. To eradicate prostate cancer bone metastases, we have to establish novel therapeutic approaches targeting both prostate cancer cells and bone microenvironment.

References

[1] Mundy, G. R. (2002). Metastasis to bone: causes, consequences and therapeutic opportunities. *Nat Rev Cancer*, *2*. 584-593.

[2] Landis, S. H., Murray, T., Bolden, S., Wingo, P. A. (1999). Cancer statistics. 1999, *CA. Cancer J Clin.*, *49*, 8-31.

[3] Coleman, R. E. (1997). Skeletal complications of malignancy. *Cancer*, *80*, 1588-1594.

[4] Koutsilieris, M. (1993). Osteoblastic metastasis in advanced prostate cancer. *Anticancer Res.*, *13*, 443-449.

[5] Small, E. J. & Harris, K. A. (2002). Secondary hormonal manipulation of prostate cancer. *Semin Urol Oncol.*, *20*, 24-30.

[6] Schulman, K. L. & Kohles, J. (2007). Economic burden of metastatic bone disease in the U.S. *Cancer*, *109*, 2334-2342.

[7] Crawford, E. D., Eisenberger, M. A., McLeod, D. G., Spaulding, J. T., Benson, R., Dorr, F. A., Blumenstein, B. A., Davis, M. A. & Goodman, P. J. (1989). A controlled trial of leuprolide with and without flutamide in prostatic carcinoma. *N Engl J Med*, (1989) *321*, 419-424. Erratum in: *N Engl J Med*, *16*, 321-1420.

[8] Paget, S. (1889). The distrubution of secondary growths in cancer of the breast. *Lancet*, *1*, 571-572.

[9] Hauschka, P. V., Mavrakos, A. E., Iafrati, M. D., Doleman, S. E. & Klagsbrun, M. (1986). Growth factors in bone matrix. Isolation of multiple types by affinity chromatography on heparin-Sepharose. *J Biol Chem.*, 261: 12665-12674.

[10] Yoneda, T. & Hiraga, T. (2005). Crosstalk between cancer cells and bone microenvironment in bone metastasis. *Biochem Biophys Res Commun.*, *328*, 679-687.

[11] Kaplan, R. N., Riba, R. D., Zacharoulis, S., Bramley, A. H., Vincent, L., Costa, C., MacDonald, D. D., Jin, D. K., Shido, K., Kerns, S. A., Zhu, Z., Hicklin, D., Wu, Y., Port, J. L., Altorki, N., Port, E. R., Ruggero, D., Shmelkov, S. V., Jensen, K. K., Rafii, S. & Lyden, D. (2005). VEGFR1-positive haematopoietic bone marrow progenitors initiate the pre-metastatic niche. *Nature*, *438*, 820-827.

[12] Théoleyre, S., Wittrant, Y., Tat, S. K., Fortun, Y., Rédini, F. & Heymann, D. (2004). The molecular triad OPG/RANK/RANKL: involvement in the orchestration of pathophysiological bone remodeling. *Cytokine Growth Factor Rev.*, *15*, 457-475.

[13] Yasuda, H., Shima, N., Nakagawa, N., Yamaguchi, K., Kinosaki, M., Mochizuki, S., Tomoyasu, A., Yano, K., et al. (1998). Osteoclast differentiation factor is a ligand for osteoprotegerin/osteoclastogenesis-inhibitory factor and is identical to TRANCE/ RANKL. *Proc Natl Acad Sci USA*, *95*, 3597-3602.

[14] Lacey, D. L., Timms, E., Tan, H. L., Kelley, M. J., Dunstan, C. R., Burgess, T., Elliott, R., Colombero, A, et al. (1998). Osteoprotegerin ligand is a cytokine that regulates osteoclast differentiation and activation. *Cell*, *93*, 165-176.

[15] Anderson, D. M., Maraskovsky, E., Billingsley, W. L., Dougall, W. C., Tometsko, M. E., Roux, E. R., Teepe, M. C., DuBose, R. F., et al. (1997). A homologue of the TNF receptor and its ligand enhance T-cell growth and dendritic-cell function. *Nature*, *390*, 175-179.

[16] Simonet, W. S., Lacey, D. L., Dunstan, C. R., Kelley, M., Chang, M. S., Luthy, R., Nguyen, H. Q., Wooden, S., et al. (1997). Osteoprotegerin: a novel secreted protein involved in the regulation of bone density. *Cell*, *89*, 309-319.

[17] Baud'huin, M., Lamoureux, F., Duplomb, L., Rédini, F. & Heymann, D. (2007). RANKL, RANK, osteoprotegerin: key partners of osteoimmunology and vascular diseases. *Cell Mol Life Sci.*, *64*, 2334-2350.

[18] Tat, S. K., Padrines, M., Theoleyre, S., Couillaud-Battaglia, S., Heymann, D., Rédini, F. & Fortun, Y. (2006). OPG/membranous-RANKL complex is internalized via the clathrin pathway before a lysosomal and a proteasomal degradation. *Bone*, *39*, 706-715.

[19] Goltzman, D. (2001). Osteolysis and cancer. *J Clin Invest.*, *107*, 1219-1220.

[20] Brown, J. M., Corey, E., Lee, Z. D., True, L. D., Yun, T. J., Tondravi, M. & Vessella, R. L. (2001). Osteoprotegerin and rank ligand expression in prostate cancer. *Urology*, *57*, 611-616.

[21] Granchi, D., Amato, I., Battistelli, L., Avnet, S., Capaccioli, S., Papucci, L., Donnini, M., Pellacani, A., et al. (2004). In vitro blockade of receptor activator of nuclear factor-kappaB ligand prevents osteoclastogenesis induced by neuroblastoma cells. *Int J Cancer*, *111*, 829-838.

[22] Farrugia, A. N., Atkins, G. J., To, L. B., Pan, B., Horvath, N., Kostakis, P., Findlay, D. M., Bardy, P., et al. (2003). Receptor activator of nuclear factorkappaB ligand expression by human myeloma cells mediates osteoclast formation in vitro and correlates with bone destruction in vivo. *Cancer Res.*, *63*, 5438-5445.

[23] Nosaka, K., Miyamoto, T., Sakai, T., Mitsuya, H., Suda, T. & Matsuoka, M. (2002). Mechanism of hypercalcemia in adult T-cell leukemia: Overexpression of receptor activator of nuclear factor kappaB ligand on adult T-cell leukemia cells. *Blood*, *99*, 634-640.

[24] Chen, G., Sircar, K., Aprikian, A., Potti, A., Goltzman, D. & Rabbani, S. A. (2006). Expression of RANKL/RANK/OPG in primary and metastatic human prostate cancer as markers of disease stage and functional regulation. *Cancer*, *107*, 289-298.

[25] Perez-Martinez, F. C., Alonso, V., Sarasa, J. L., Nam-Cha, S. G., Vela-Navarrete, R., Manzarbeitia, F., Callahora, F. J. & Esbrit, P. (2007). Immunohistochemical analysis of low-grade and high-grade prostate carcinoma: Relative changes of PTHrP and its PTH1 receptor, osteoprotegerin and receptor activator of nuclear factor kb ligand. *J Clin Pathol.*, *60*, 290-294.

[26] Michigami, T., Ihara-Watanabe, M., Yamazaki, M. & Ozono, K. (2001). Receptor activator of nuclear factor kappaB ligand (RANKL) is a key molecule of osteoclast formation for bone metastasis in a newly developed model of human neuroblastoma. *Cancer Res.*, *61*, 1637-1644.

[27] Lee, Y., Schwarz, E., Davies, M., Jo, M., Gates, J., Wu, J., Zhang, X. & Lieberman, J. R. (2003). Differences in the cytokine profiles associated with prostate cancer cell induced osteoblastic and osteolytic lesions in bone. *J Orthop Res.*, *21*, 62-72.

[28] Holen, I., Cross, S. S., Neville-Webbe, H. L., Cross, N. A., Balasubramanian, S. P., Croucher, P. I., Evans, C. A., Lippitt, J. M., et al. (2005). Osteoprotegerin (OPG) expression by breast cancer cells in vitro and breast tumours in vivo—a role in tumour cell survival? *Breast Cancer Res Treat*, *92*, 207-215.

[29] Holen, I., Croucher, P. I., Hamdy, F. C. & Eaton, C. L. (2002). Osteoprotegerin (OPG) is a survival factor for human prostate cancer cells. *Cancer Res.*, *62*, 1619-1623.

[30] Van Poznak, C., Cross, S. S., Saggese, M., Hudis, C., Panageas, K. S., Norton, L., Coleman, R. E. & Holen, I. (2006). Expression of osteoprotegerin (OPG), TNF related apoptosis inducing ligand (TRAIL), and receptor activator of nuclear factor {kappa}B ligand (RANKL) in human breast tumours. *J Clin Pathol.*, *59*, 56-63.

[31] Grimaud, E., Soubigou, L., Couillaud, S., Coipeau, P., Moreau, A., Passuti, N., Gouin, F., Redini, F. & Heymann, D. (2003). Receptor activator of nuclear factor kappaB ligand (RANKL)/osteoprotegerin (OPG) ratio is increased in severe osteolysis. *Am J Pathol.*, *163*, 2021-2031.

[32] Pfeilschifter, J. & Mundy, G. R. (1987). Modulation of type β transforming growth factor activity in bone cultures by osteotropic hormones. *Proc Natl Acad Sci USA*, *84*, 2024-2028.

[33] Grano, M., Mori, G., Minielli, V., Cantatore, F. P., Colucci, S. & Zallone, A. Z. (2000). Breast cancer cell line MDA-231 stimulates osteoclastogenesis and bone resorption in human osteoclasts. *Biochem Biophys Res Commun.*, *270*, 1097-1100.

[34] Wani, M. R., Fuller, K., Kim, N. S., Choi, Y. & Chambers, T. (1999). Prostaglandin E2 cooperates with TRANCE in osteoclast induction from hemopoietic precursors: synergistic activation of differentiation, cell spreading, and fusion. *Endocrinology*, *140*, 1927-1935.

[35] Guise, T. A., Yin, J. J., Taylor, S. D., Kumagai, Y., Dallas, M., Boyce, B. F., Yoneda, T. & Mundy, G. R. (1996). Evidence for a causal role of parathyroid hormone- related protein in the pathogenesis of human breast cancer-mediated osteolysis. *J Clin Invest.*, *98*, 1544-1549.

[36] Shen, X. & Falzon, M. (2003). PTH-related protein modulates PC-3 prostate cancer cell adhesion and integrin subunit profile. *Mol Cell Endocrinol.*, *199*, 165-177.

[37] Lu, Y., Cai, Z., Xiao, G., Keller, E. T., Mizokami, A., Yao, Z., Roodman, G. D. & Zhang, J. (2007). Monocyte chemotactic protein-1 mediates prostate cancer-induced bone resorption. *Cancer Res.*, *67*, 3646-3653.

[38] Halvorson, K. G., Sevcik, M. A., Ghilardi, J. R., Rosol, T. J. & Mantyh, P. W. (2006). Similarities and differences in tumor growth, skeletal remodeling and pain an osteolytic and osteoblastic model of bone cancer. *Clin J Pain*, *22*, 587-600.

[39] Kingsley, L. A., Fournier, P. G., Chirgwin, J. M. & Guise, T. A. (2007). Molecular biology of bone metastasis. *Mol Cancer Ther.*, *6*, 2609-2617.

[40] Urwin, G. H., Percival, R. C., Harris, S., Beneton, M. N., Williams, J. L. & Kanis, J. A. (1985). Generalised increase in bone resorption in carcinoma of the prostate. *Br J Urol.*, *57*, 721-723.

[41] Garnerog, P., Buchs, N., Zekri, J., Rizzoli, R., Coleman, R. E. & Delmas, P. D. (2000). Markers of bone turnover for the management of patients with bone metastases from prostate cancer. *Br J Cancer*, *82*, 858-864.

[42] Inoue, H., Nishimura, K., Oka, D., Nakai, Y., Shiba, M., Tokizane, T., Arai, Y., Nakayama, M., Shimizu, K., Takaha, N., Nonomura, N. & Okuyama, A. (2005). Prostate cancer mediates osteoclastogenesis through two different pathways. *Cancer Lett.*, *223*, 121-128.

[43] Zhang, J., Dai, J., Qi, Y., Lin, D. L., Smith, P., Strayhorn, C., Mizokami, A., Fu, Z., Westman, J. & Keller, E. T. (2001). Osteoprotegerin inhibits prostate cancer-induced osteoclastogenesis and prevents prostate tumor growth in the bone. *J Clin Invest.*, *107*, 1235-1244.

[44] Zhang, J., Dai, J., Yao, Z., Lu, Y., Dougall, W. & Keller, E. T. (2003). Soluble receptor activator of nuclear factor κB Fc diminishes prostate cancer progression in bone. *Cancer Res.*, *63*, 7883-7890.

[45] Corey, E., Brown, L. G., Kiefer, J. A., Quinn, J. E., Pitts, T. E., Blair, J. M. & Vessella, R. L. (2005). Osteoprotegerin in prostate cancer bone metastasis. *Cancer Res.*, *65*, 1710-1718.

[46] Whang, P. G., Schwarz, E. M., Gamradt, S. C., Dougall, W. C. & Lieberman, J. R. (2005). The effects of RANK blockade and osteoclast depletion in a model of pure osteoblastic prostate cancer metastasis in bone. *J Orthop Res.*, *23*, 1475-1483.

[47] Smith, M. R. (2008). Osteoclast targeted therapy for prostate cancer: bisphosphonates and beyond. *Urol Oncol.*, *26*, 420-425.

[48] Lamoureux, F., Ory, B., Battaglia, S., Pilet, P., Heymann, M. F., Gouin, F., Duteille, F., Heymann, D. & Rédini, F. (2008). Relevance of a new rat model of osteoblastic metastases from prostate carcinoma for preclinical studies using zoledronic acid. *Int J Cancer.*, *122*, 751-760.

[49] Kong, Y. Y., Yoshida, H., Sarosi, I., Tan, H. L., Timms, E., Capparelli, C., Morony, S., Oliveira-dos-Santos, A. J., Van, G., Itie, A., Khoo, W., Wakeham, A., Dunstan, C. R., Lacey, D. L., Mak, T. W., Boyle, W. J. & Penninger, J. M. (1999). OPGL is a key regulator of osteoclastogenesis, lymphocyte development and lymph-node organogenesis. *Nature*, *397*, 315-323.

[50] Hsu, H., Lacey, D. L., Dunstan, C. R., Solovyev, I., Colombero, A., Timms, E., Tan, H. L., Elliott, G., Kelley, M. J., Sarosi, I., Wang, L., Xia, X. Z., Elliott, R., Chiu, L., Black, T., Scully, S., Capparelli, C., Morony, S., Shimamoto, G., Bass, M. B. & Boyle, W. J. (1999). Tumor necrosis factor receptor family member RANK mediates osteoclast differentiation and activation induced by osteoprotegerin ligand. *Proc Natl Acad Sci USA*, *96*, 3540-3545.

[51] Nakagawa, N., Kinosaki, M., Yamaguchi, K., Shima, N., Yasuda, H., Yano, K., Morinaga, T. & Higashio, K. (1998). RANK is the essential signaling receptor for

osteoclast differentiation factor in osteoclastogenesis. *Biochem Biophys Res Commun.*, *253*, 395-400.

[52] Srivastava, S., Matsuda, M., Hou, Z., Bailey, J. P., Kitazawa, R., Herbst, M. P. & Horseman, N. D. (2003). Receptor activator of NF-kappaB ligand induction via Jak2 and Stat5a in mammary epithelial cells. *J Biol Chem.*; *278*, 46171-46178.

[53] Heymann, M. F., Riet, A., Le Goff, B., Battaglia, S., Paineau, J. & Heymann, D. (2008). OPG, RANK and RANK ligand expression in thyroid lesions. *Regul Pept.*, *148*, 46-53.

[54] Mori, K., Goff, B. L., Charrier, C., Battaglia, S., Heymann, D. & Rédini, F. (2007). DU145 human prostate cancer cells express functional receptor activator of NFκB: New insights in the prostate cancer bone metastasis process. *Bone*, *40*, 981-990.

[55] Jones, D. H., Nakashima, T., Sanchez, O. H., Kozieradzki, I., Komarova, S. V., Sarosi, I., Morony, S., Rubin, E., et al. (2006). Regulation of cancer cell migration and bone metastasis by RANKL. *Nature*, *440*, 692-696.

[56] Armstrong, A. P., Miller, R. E., Jones, J. C., Zhang, J., Keller, E. T. & Dougall, W. C. (2008). RANKL acts directly on RANK-expressing prostate tumor cells and mediates migration and expression of tumor metastasis genes. *Prostate*, *68*, 92-104.

[57] Bhatia, P., Sanders, M. M. & Hansen, M. F. (2005). Expression of receptor activator of nuclear factor-kappaB is inversely correlated with metastatic phenotype in breast carcinoma. *Clin Cancer Res.*, *11*, 162-165.

[58] Holen, I. & Shipman, C. M. (2006). Role of osteoprotegerin (OPG) in cancer. *Clin Sci.*, *110*, 279-291.

[59] Hiraga, T., Ueda, A., Tamura, D., Hata, K., Ikeda, F., Williams, P. J. & Yoneda, T. (2003). Effects of oral UFT combined with or without zoledronic acid on bone metastasis in the 4T1/luc mouse breast cancer. *Int J Cancer*, *106*, 973-979.

[60] Lamoureux, F., Richard, P., Wittrant, Y., Battaglia, S., Pilet, P., Trichet, V., Blanchard, F., Gouin, F., Pitard, B., Heymann, D. & Rédini, F. (2007). Therapeutic relevance of osteoprotegerin gene therapy in osteosarcoma: blockade of the vicious cycle between tumor cell proliferation and bone resorption. *Cancer Res.*, *67*, 7308-7318.

[61] Mori, K., Le Goff, B., Berreur, M., Riet, A., Moreau, A., Blanchard, F., Chevalier, C., Guisle-Marsollier, I., Léger, J., Guicheux, J., Masson, M., Gouin, F., Rédini, F. & Heymann, D. (2007). Human osteosarcoma cells express functional receptor activator of nuclear factor-kappa B. *J Pathol.*, *211*, 555-562.

[62] Wittrant, Y., Lamoureux, F., Mori, K., Riet, A., Kamijo, A., Heymann, D. & Redini, F. (2006). RANKL directly induces bone morphogenetic protein-2 expression in RANK-expressing POS-1 osteosarcoma cells. *Int J Oncol.*, *28*, 261-269.

[63] Peehl, D. M., Cohen, P. & Rosenfeld, R. G. (1995). The insulin-like growth factor system in the prostate. *World J Urol.*, *13*, 306-311.

[64] Lamoureux, F., Baud'huin, M., Duplomb, L., Heymann, D. & Rédini, F. (2007). Proteoglycans: key partners in bone cell biology. *Bioessays*, *29*, 758-771.

[65] Koutsilieris, M. & Polychronakos, C. (1992). Proteinolytic activity against IGF-binding proteins involved in the paracrine interactions between prostate adenocarcinoma cells and osteoblasts. *Anticancer Res.*, *12*, 905-910.

[66] Kehinde, E. O., Maghrebi, M. A. & Anim, J. T. (2008). The importance of determining the aggressiveness of prostate cancer using serum and tissue molecular markers. *Can J Urol., 15*, 3967-3974.

[67] Wozney, J. M., Rosen, V., Celeste, A. J., Mitsock, L. M., Whitters, M. J., Kriz, R. W., Hewick, R. M. & Wang, E. A. (1988). Novel regulators of bone formation: molecular clones and activities. *Science.* Dec 16, 242(4885), 1528-34.

[68] Dai, J., Keller, J., Zhang, J., Lu, Y., Yao, Z. & Keller, E. T. (2005). Bone morphogenetic protein-6 promotes osteoblastic prostate cancer bone metastases through a dual mechanism. *Cancer Res., 65*, 8274-8285.

[69] Dai, J., Hall, C. L., Escara-Wilke, J., Mizokami, A., Keller, J. M. & Keller, E. T. (2008). Prostate cancer induces bone metastasis through Wnt-induced bone morphogenetic protein-dependent and independent mechanisms. *Cancer Res., 68*, 5785-5794.

[70] Maguire, J. J. & Davenport, A. P. (2002). Is urotensin-II the new endothelin? *Br J Pharmacol. 137*, 579-588.

[71] Bagnato, A. & Rosanò, L. (2008). The endothelin axis in cancer. *Int J Biochem Cell Biol., 40*, 1443-1451.

[72] Godara, G., Cannon, G. W., Cannon, G. M. Jr, Bies, R. R., Nelson, J. B. & Pflug, B. R. (2005). Role of endothelin axis in progression to aggressive phenotype of prostate adenocarcinoma. *Prostate, 65*, 27-34.

[73] Guise, T. A., Yin, J. J. & Mohammad, K. S. (2003). Role of endothelin-1 in osteoblastic bone metastases. *Cancer, 97*, 779-784.

[74] Nelson, J. B., Nguyen, S. H., Wu-Wong, J. R., Opgenorth, T. J., Dixon, D. B., Chung, L. W. & Inoue, N. (1999). New bone formation in an osteoblastic tumor model is increased by endothelin-1 overexpression and decreased by endothelin A receptor blockade. *Urology, 53*, 1063-1069.

[75] Guise, T. A., Mohammad, K. S., Clines, G., Stebbins, E. G., Wong, D. H., Higgins, L. S., Vessella, R., Corey, E., Padalecki, S., Suva, L. & Chirgwin, J. M. (2006). Basic mechanisms responsible for osteolytic and osteoblastic bone metastases. *Clin Cancer Res., 12*, 6213s-6216s.

[76] Fudge, K., Bostwick, D. G. & Stearns, M. E. (1996). Platelet-derived growth factor A and B chains and the alpha and beta receptors in prostatic intraepithelial neoplasia. *Prostate, 29*, 282-286.

[77] George, D. (2001). Platelet-derived growth factor receptors: a therapeutic target in solid tumors. *Semin Oncol., 28*, 27-33.

[78] Yi, B., Williams, P. J., Niewolna, M., Wang, Y. & Yoneda, T. (2002). Tumor-derived platelet-derived growth factor-BB plays a critical role in osteosclerotic bone metastasis in an animal model of human breast cancer. *Cancer Res., 62*, 917-923.

[79] Brown, J. M., Corey, E., Lee, Z. D., True, L. D., Yun, T. J., Tondravi, M. & Vessella, R. L. (2001). Osteoprotegerin and rank ligand expression in prostate cancer. *Urology, 57*, 611-616.

[80] Chen, G., Sircar, K., Aprikian, A., Potti, A., Goltzman, D. & Rabbani, S. A. (2006). Expression of RANKL/RANK/OPG in primary and metastatic human prostate cancer as markers of disease stage and functional regulation. *Cancer, 107*, 289-298.

[81] Wu, G., Iwamura, M., di Sant'Agnese, P. A., Deftos, L. J., Cockett, A. T. & Gershagen, S. (1998). Characterization of the cell-specific expression of parathyroid hormone-related protein in normal and neoplastic prostate tissue. *Urology, 51*, 110-120.

[82] Bendre, M. S., Margulies, A. G., Walser, B., Akel, N. S., Bhattacharrya, S., Skinner, R. A., Swain, F., Ramani, V., Mohammad, K. S., Wessner, L. L., Martinez, A., Guise, T. A., Chirgwin, J. M., Gaddy, D. & Suva, L. J. (2005). Tumor-derived interleukin-8 stimulates osteolysis independent of the receptor activator of nuclear factor-kappaB ligand pathway. *Cancer Res., 65*, 11001-11009.

[83] Lu, Y., Cai, Z., Xiao, G., Keller, E. T., Mizokami, A., Yao, Z., Roodman, G. D. & Zhang, J. (2007). Monocyte chemotactic protein-1 mediates prostate cancer-induced bone resorption. *Cancer Res., 67*, 3646-3653.

[84] Koutsilieris, M., Mitsiades, C. & Sourla, A. (2000). Insulin-like growth factor I and urokinase-type plasminogen activator bioregulation system as a survival mechanism of prostate cancer cells in osteoblastic metastases: development of anti-survival factor therapy for hormone-refractory prostate cancer. *Mol Med., 6*, 251-267.

[85] Reyes-Moreno, C., Sourla, A., Choki, I., Doillon, C. & Koutsilieris, M. (1998). Osteoblast-derived survival factors protect PC-3 human prostate cancer cells from adriamycin apoptosis. *Urology, 52*, 341-347.

[86] Bogdanos, J., Karamanolakis, D., Tenta, R., Tsintavis, A., Milathianakis, C., Mitsiades, C. & Koutsilieris, M. (2003). Endocrine/paracrine/autocrine survival factor activity of bone microenvironment participates in the development of androgen ablation and chemotherapy refractoriness of prostate cancer metastasis in skeleton. *Endocr Relat Cancer, 10*, 279-289.

[87] Koutsilieris, M. (1995). Skeletal metastases in advanced prostate cancer: cell biology and therapy. *Crit Rev Oncol Hematol. Jan 18(1),* 51-64.

In: Handbook of Prostate Cancer Cell Research...
Editor: Alan T. Meridith
ISBN: 978-1-60741-954-9

Chapter 15

Prostate Cancer: Are We Close to Knowing the Disease?

***B Shannan**[*1,2], **D A Boothman**[3] and **G Akhras**[1]*

[1]College of Pharmacy, International University for Science and Technology, Damascus, Syria

[2]Dermatology Department, The Saarland University Hospital, 66421 Hombrug, Germany

[3]Laboratory of Molecular Stress Responses, Program in Cell Stress and Cancer Nanomedicine, Department of Oncology, University of Texas Southwestern Medical Center at Dallas, Dallas, TX, USA 75390-8807

Abstract

Prostate cancer has become the major public health issue for men in the past decade. Prostate cancer incidences have increased in the past few years, although rates vary considerably across the world. More advanced technology to increase early detection rates must also be taken into consideration and have somewhat countered the increased rates of prostate cancer. Nevertheless, prostate cancer is the second most leading cause of death after lung cancer in both the UK and USA. Risk factors, include eating habits (e.g., fats vs. vegetables and antioxidants, vitamin D deficiency), race, sexual habits, and inherented genetic susceptibilities all contribute to prostate cancer rates. The exact causes of prostate cancer are unclear and promise to be complex. As in all other cancers, the development of prostate cancer is a multi-step process involving genes that govern cell growth and differentiation and their interaction with environmental and micro-environmental changes during life. Major host factors affecting the progression of prostate cancer include hormonal levels, androgen receptor (AR) and polymorphisms in AR, intracellular heat shock proteins, the extracellular heat shock factor, clusterin, and cell cycle regulators (e.g., cyclin A1). A prognosis of prostate cancer depends greatly on patient responses to androgen withdrawal therapy. Treatment options have varied greatly in the past decade and have matched our knowledge of the disease. For example, antisense therapies against proposed cancer stem cell targets, bcl-2, clusterin, or telomerase have been proposed. Treatments targeting tumour angiogenesis and metastasis, as well as more conventional chemotherapeutic or radiotherapeutic methods, are described. We focus on prostate cancer risk factors, causes, and progression and shed more light into some of the newer treatment modalities available.

Introduction

Prostate cancer has become a public issue of interest during the past decade (Robertson *et al.*, 2008). Rates of prostate cancer vary across the world. Based on statistics by the American Cancer Society and the US National Cancer Institute, prostate cancer rates are lowest among Asian men and highest among Black men, with rates for White men in between (www.cancer.gov). Some studies have shown that the risk of prostate cancer has increased when Asian men immigrated to North America strongly suggesting the involvement of environmental and dietary factors in prostate cancer development (Shiminzu *et al.*, 1991, Whittemore *et al.*, 1995, Nelson *et al.*, 2003). The improvement of detection rates must also be taken into consideration. Prostate cancer is also documented to be the second most leading cause of death after lung cancer in both the UK and USA (Whittermore *et al.*, 1995).

A number of factors have been proposed and identified as risk factors, including fats that are documented to increase the risk of prostate cancer, while vegetables and antioxidants reduce the risk of prostate cancer (reviewed in Nelson *et al.*, 2003). The big difference in the incidence of prostate cancer between men in developed countries and Asian men is well established, and has been attributed to important differences in lifestyles. The importance of nutritional factors for development of prostate cancer was apparent in a Swedish study, in which both body-mass index and lean body mass were positively associated with the risk of prostate cancer development, and were strongly related, to death rather than to new cases of prostate cancer (Andersson *et al.*, 1997). Even though dietary fat seems to be an important environmental risk factor for prostate cancer identified so far, the evidence is open to doubt. Only three of 14 studies reported a statistically significant increase in risk ratio for total fat intake (Vainio *et al.*, 2002). The strong antioxidants lycopenes, found for example in tomatoes, have been extensively studied as possible protective agents for the development of prostate cancer. Moreover, some other micronutrients and vitamins—such as selenium, vitamin E, and vitamin D—have been associated with the risk of prostate cancer. Vitamin D deficiency has also been proposed a risk factor for prostate cancer development (Schwarts *et al.*, 1990, Shannan *et al.*, 2007). Other risk factors include age (risk increases for those over 50 years of age), race (as stated above), ejaculation rate (Giles *et al.*, 2003, BBC news/Health), and mutations of BRCA1 and BRCA2 (Struewing *et al.*, 1997). Other gene susceptibilities have been reported to be associated with prostate cancer development, for example, hereditary prostate cancer locus-1 (HPC1) gene.

As stated above, one of the possible risk factors for prostate cancer developement, in general, and for the aggression of this cancer specifically, is the inheritance of genes that cause prostate cancer, some of which show high prevalence, whereas other genes show polymorphism and low prevalence (Damber and Aus, 2008). One of these genes identified as potentially associated with prostate cancer is the HPC1, which may account for a low number of prostate cancer patients (Smith *et al.*, 1996). The genes most plausibly linked to familial prostate cancer are *EPAC2*, *RNASEL*, *MSR1*, *CHEK2*, *CAPZB*, vitamin D receptor, and *PON1* (Deutsch *et al.*, 2004, Porkka *et al.*, 2004, Dong *et al.*, 2006). A germline mutation in *BRCA2* can increase the risk of prostate cancer (Edwards *et al.*, 2004), and could cause about 5% of cases in men younger than 55 years. A genetic variant associated with prostate cancer in European and African populations has been identified on chromosome 8q24.1.

Development and Progression

Despite the many risk factors, the exact causes of prostate cancer are ambiguous. The development of prostate cancer is a multi-step process heavily dependent on genes that govern cell growth and differentiation (e.g. oncogenes and tumour suppressor genes). Male hormone levels may play a role in the pathogenesis of prostate cancer. Androgens are documented to be important for growth, development and survival of prostate cancer (Jenster *et al.*, 1999, Klotz *et al.*, 2000, Balk *et al.*, 2002, Culing *et al.*, 2006). Androgen carries out its biological function through binding to the Androgen receptor (AR) (Evans *et al.*, 1988, Gelmann *et al.*, 2006). AR is a member of the superfamily of nuclear receptors. The regulation of AR is complex and needs many protein-protein and protein-ligand binding.

Upon activation, AR dissociates and recognises the androgen response elements, located in the promoter regions of AR target genes, such as the prostate specific antigen (PCA) gene (Papatsoris *et al.*, 2001).

Normally, AR is kept inactive through the association with heat shock proteins (HSPs) and is prohibited from DNA binding (Gelmann *et al.*, 2002, Balk *et al.*, 2008). When a ligand binds, the inhibitory HSPs are released and bind DNA on androgen responsive elements within the regulatory regions of target genes.

Many theories have emerged about the involvement of AR in prostate cancer metastasis. However, the theory that AR is an oncogene was challenged when over-expression of AR in transgenic mice resulted in neoplastic growth, but not in metastatic cancer (Stanbrough *et al.*, 2001, Balakumaran *et al.*, 2006). Oncogenic activation seems to be required in addition to AR activity for the initiation and progression of prostate cancer. However, no single event has been isolated as the sole genetic cause. Some of the oncogenes activated in prostate cancer include: The phosphotidylinositol 3-phosphate kinase/phosphatase and tensin homologue AK pathway (Deocampo *et al.*, 2003, Sansal *et al.*, 2004), c-myc over-expression, loss of p53 and Rb (Theodorescu *et al.*, 1997).

Many signaling pathways affect AR signal transduction processes, including chaperons and coregulators (such as nuclear receptor cosuppressor) that interacts directly with AR and represses AR transcriptional activity, (Scher *et al.*, 2005). Smad3, a downstream effector of TGF-β1 signaling, specifically inhibits transcriptional activation mediated by AR, and loss of TGF-β1 enhances AR-mediated activation (Hayes *et al.*, 1999, 2001). Androgen withdrawal will decrease cell proliferation, as a start, and increase apoptosis (Scher *et al.*, 2005).

Other than androgens, growth factors, neuropeptides and cytokines can also induce AR activation (reviewed in Papatsoris *et al.*, 2001). During prostate cancer progression, androgen-dependent cells are transformed to androgen-independent, resulting in the failure of androgen ablation therapy (Papatsoris *et al.*, 2001). This transformation may happen by bypassing or adapting the AR pathway (Jenster *et al.*, 1999). A shift from paracrine to autocrine stimulation, by expressing growth factors and their receptors, has been reported for several growth factors. This may happen by AR mutation and/or deletions, or by AR gene activation via DNA methylation, resulting in loss of AR expression with cancer progression (Russell *et al.*, 1998, Jarrard *et al.*, 1998, Kinoshita *et al.*, 2000). As prostate cancer progresses, AR mutations are more frequent, cells become stroma-independent and therefore they can migrate.

Apoptosis, or cellular mechanisms blocking the process, is involved in the initiation, progression and metastasis of prostate cancer, as demonstrated by Isaacs *et al.*, where androgen ablation triggered apoptosis in most prostate glandular epithelial and androgen-dependent cancer cells. Although androgen-independent cells do not undergo apoptosis with androgen deprivation, they retain the capacity to do so with chemotherapeutics, radiation and TNF-alpha (Sinha *et al.*, 1995).

AR polymorphic repeats have been described (Edwards *et al.*, 1992). Various studies suggested that shorter polymorphic repeats may be associated with increased AR transcriptional transactivation and thus increased risk of prostate cancer (Chamberlain *et al.*, 1994, Irvine *et al.*, 2000). It remains to be said that other studies did not find any correlation between the polymorphisms and the risk of prostate cancer (Hakimi *et al.*, 1997, Hsing *et al.*, 2000, Bennett *et al.*, 2002). Inhibition of AR activity is still the major therapeutic goal for management of metastatic prostate cancer (Balk *et al.*, 2002, 2008).

HSP are ubiquitous proteins and have been characterized as cytoprotective molecular chaperones. Chaperones typically function to assist a protein to attain its functional. conformation (So *et al.*, 2007). HSP are important in signalling and their need increases upon injury or stress to the cell, as they allow cells to survive (So *et al.*, 2007).

Clusterin (CLU) is another stress protein that is involved in numerous physiological processes important for carcinogenesis and tumour growth, including apoptotic cell death, cell cycle regulation, DNA repair, cell adhesion, tissue remodelling, lipid transportation, membrane recycling, and immune system regulation. Increased CLU mRNA and protein levels have been consistently detected in various tissues undergoing stress, including heart, brain, liver, kidney, breast, and retinal tissues both in vivo and in vitro. Early studies seemed to establish that CLU gene expression was a marker of apoptotic cell loss. Importantly, CLU upregulation has been reported in various human malignancies, including bladder, kidney, prostate, colon, breast, and lung tumours. CLU was also shown to exert cytoprotective effects on various tissues and cell line, including prostate cancer cells. Moreover, silencing CLU expression using ASO in prostate cancer enhanced the apoptotic response. In prostate cancer, various studies support the theory that CLU expression is linked with androgen-independent progression (Gleave *et al.*, 2005, Shannan *et al.*, 2007). As stated above, androgen withdrawal is the best treatment option. However, more often than not, the cancer turns into androgen-independent progression which is followed by death (Gleave *et al.*, 2005). Introduction of sCLU (an isoform of CLU called secretory CLU) in LNCaP prostate cancer cells increases resistance to apoptosis and is associated with a higher Gleason score and cancer progression (Miyake *et al.*, 2000). Some studies (Shannan *et al.*, 2007) have shown that sCLU over-expression protects LNCaP cancer cells from the antiproliferative effects of $1,25(OH)_2D_3$.

The cell cycle is regulated by many factors, and any failure in this regulation might have important roles in cancer development and/or metastasis (Wegiel *et al.*, 2008). Cyclin A1, a cell cycle regulator factor is required for the G_2/M phase in mitotic division of male germ cells (Liu *et al.*, 1998, Wegiel *et al.*, 2008). Elevated levels of cyclin A1 were reported in testicular, ovarian and breast cancer (Liao *et al.*, 2004, Rivera *et al.*, 2006). Moreover, wegiel et al reported that cyclin A1 levels were elevated in primary and metastastic prostate cancer specimens used compared with benign specimens (Liu *et al.*, 1998, Wegiel *et al.*, 2008).

In a study done by Wegiel *et al.*, it was found that cyclin A1 plays an important role in prostate cancer invasion and metastasis. It was demonstrated that cyclin A1 expression was elevated in most of the primary and metastatic prostate cancer specimens examined compared with adjacent benign specimens. Furthermore, there was a statistically significant correlation between cyclin A1 and VEGF, MMP2, and MMP9 protein expression in tumour specimens from prostate cancer patients. Over-expression of cyclin A1 in PC3 cells by transient transfection or by adenoviral transduction promoted tumour cell invasion in vitro, and the cells became more invasive in vivo and metastatic. Silencing of cyclin A1 by shRNA inhibited migration and invasion in PC3 cells. Another study showed that silencing cyclin A1 inhibited migration and invasion of PC3 prostate cancer cells, while its over-expression promoted tumour cell invasion in vitro in the above mentioned cell line (Wegiel *et al.*, 2008).

Further, this study showed that inhibition of MMP2 or uPA by siRNA-mediated knockdown could partially inhibit the increased invasiveness of PC3 cells that over-expressed cyclin A1. This data suggested that cyclin A1 may promote tumour cell invasion in part by increasing the expression and activity of MMPs, uPA, and VEGF. What is more, the induced AR expression enhanced nuclear cyclin A1 expression. Finally, it was observed that cyclin A1 interacted with AR and enhanced AR-mediated activation of AR target genes, suggesting that cyclin A1 and AR may act as cofactors, possibly to promote prostate tumour progression and invasion. The importance of cyclin A1 in tumour cell invasion and metastasis is suggested by its high level of nuclear expression in aggressive prostate cancer and its ability to mediate the activities of MMPs and VEGF in cultured cells.

The expression of anti-apoptotic genes, for example bcl-2 increases during the progression of prostate cancer, whereas the expression of pro-apoptotic gene bax, remains constant (Krajewiska *et al.*, 1996). The bcl-2 protein, which dimerizes with bax, was found to be over-expressed as prostate cancer progresses from localized to metastatic androgen-dependent, to androgen-independent tumours (Dorai *et al.*, 1999). Also, bcl-2 over-expression is associated with resistance to several cytotoxic chemotherapeutics and radiotherapy (Van Brussel *et al.*, 2001).

The p53 tumour-suppressor gene is able to transcriptionally up-regulate the cell-death effector, bax, and potentially down-regulate the prosurvival factors, bcl-2 itself.

Diagnosis

At present, diagnosis is based on examination of histopathological or cytological specimens from the gland. The most common way to obtain the necessary tissue specimen is by several systematic transrectal core biopsies, with guidance by transrectal ultrasound. The most commonly used system for grading adenocarcinoma of the prostate is the Gleason score (Gleason *et al.*, 1974). The system describes a score between 2 and 10, with 2 being the least aggressive and 10 the most aggressive. This score is the sum of the two most common patterns (grades 1–5) of tumour growth. The risk that a prostate cancer will have metastasised is closely associated with a high serum PSA (an organ-specific marker with greater sensitivity and high specificity for prostate tissue) concentration, a locally advanced prostate cancer, or a poorly differentiated tumour. In practice, this association means that a staging bone scan is not routinely recommended to patients with a PSA concentration of less than 20 ng/mL in the absence of poorly differentiated cancer (Aus *et al.*, 2005). The first sign of failure after treatment with curative intent is generally a rising serum PSA concentration, occurring months to years before clinical symptoms or radiographic signs of recurrent disease. Thus, the choice of treatment for these men with rising PSA concentration after curative therapy is generally made without the contribution of any other objective signs of where the disease has recurred (i.e., local or systemic failure). Generally, after either surgery or radiotherapy, local failure is characterised by a late PSA increase (more than 12 months after primary treatment), a long PSA-doubling time (i.e., a slow rise in PSA over time), and a not too aggressive disease at diagnosis (no Gleason 8–10 score, or no invasion of seminal vesicle or lymph nodes) (Aus *et al.*, 2005).

Prognosis

Other factors affecting the prognosis of patients with prostate cancer that may be useful in making therapeutic decisions include histologic grade of the tumour, patient's age, other medical illnesses, and level of PSA (Paulson *et al.*, 1990, Gittes *et al.*, 1991, Matzkin *et al.*, 1992, Pisansky *et al.*, 1993, Chodak *et al.*, 1994, Gohagan *et al.*, 2001). Poorly differentiated tumours are more likely to have already metastasized by the time of diagnosis and are associated with a poorer prognosis. For patients treated with radiation therapy, the combination of clinical tumour (T) stage, Gleason score, and pre-treatment PSA level can be used to more accurately estimate the risk of relapse. Elevations of serum acid phosphatase are associated with poor prognosis in both localized and disseminated disease. PSA is often used as a tumour marker.

Treatment

The median age at diagnosis is 72 years, many patients, especially those with localized tumour, may die of other illnesses without ever having suffered significant disability from their cancer. The approach to treatment is influenced by age and coexisting medical problems. Side-effects of various forms of treatment should be considered in selecting appropriate management (Garnick *et al.*, 1993).

Primary therapy for localized prostate cancer typically includes radical prostatectomy, external beam radiotherapy. Brachytherapy, or active surveillance, but about third of the patients will develop metastatic disease (Dillioglugil *et al.*, 1997). Androgen deprivation is the therapy of choice; however, almost all patients progress to androgen-independent cancer (Sharifi *et al.*, 2005). Metastatic androgen-independent prostate cancer (AIPC) then develops and the response to hormonal therapy or chemotherapy decreases (Aragon-Ching *et al.*, 2007).

The use of single chemotherapeutic agents has been disappointing (Yagoda *et al.*, 1993). However, chemotherapy has recently been considered active in the treatment of prostate cancer (Aragon-Ching *et al.*, 2007). Many chemotherapeutic regimens are considered promising agents in AIPC treatment, and some of them are: mitoxantrone, which has a moderate response; satraplatin and other platinum agents, which are active treatment options for AIPC patients; and epothilones.

Novel Strategies

Many factors have contributed to the continuous introduction of new therapies for various diseases, and to prostate cancer. Some of these factors are the increased knowledge of the signlaing pathway of androgen, prostate cancer resistance, and the discovery of new drug regimens.

Strategies targeting cell-cycle and apoptosis, angiogenesis, and stem cells are currently under development. For example, replacement of a defective p53 is a promising treatment agent (Papatsoris *et al.*, 2001). Also, a tumour suppressor gene called C-CAMI is a cell-adhesion molecule which was found to be down-regulated in early stages of prostate cancer (Kleinerman *et al.*, 1995).

What is more, gene therapies and strategies aiming at inducing passive immunity have been greatly tested and evaluated in prostate cancer with promising results (reviewed in Papatsoris *et al.*, 2001).

Antisense therapy has also been proposed as a treatment for prostate cancer, where bcl-2 expression can be inhibited to induce apoptosis, and improve response to therapy. The inhibition is done using single-stranded oligonucleotides of 10-50 bases (unlike gene therapy which targets the gene itself). Cancer cells, in general, are characterized by immortality, partly due to the presence of telomerase. This suggests that telomerase may also be a target for anti-cancer therapy. The most promising telomerase inhibitors are the anti-sense oligonucleotides 2'-O-methyl-RNA molecules and peptide nucleic acids (Eastham *et al.*, 2000). Other methods include targeting tumour angiogenesis to stop metastasis.

It is hypothesized that inside the cancer cell mass, there should be a number of 'cancer stem cells' with tumour-initiating and sustaining capacity. This was well shown in the haematologic systems (Greaves *et al.*, 2007). Under certain conditions, these tumour-initiating cells could provide a reservoir for recurrent disease after therapy (Maitland *et al.*, 2008). Because the stem cell is likely to be the cell in the prostate with the most chances to accumulate mutations, it is suggested that this stem cell would share antigen expression with its tumour initiating cells. This stem cell was named a cancer stem cell (CSC), as it had a longer life span and a higher cell output capacity than its normal equivalents. The origin of these prostate CSCs remains controversial. They share the normal expression of a number of normal cell-surface antigens. The gene expression patterns of the CSCs and amplifying cells can also provide some clues to mechanisms of therapy resistance. In prostate cancer, the AR-negative, basal phenotype of the CSCs provides a ready explanation for minimal residual disease after castration therapy. If these CSCs could be isolated, then second-line therapy against the resistant phenotype could be assigned (Maitland *et al.*, 2008).

A new breakthrough study was recently completed by Boëda *et al.,* where Mena protein was found to be increased in tumours in general. Mena was previously reported to aid in cancer metastasis. Tes, is a tumour suppressor gene that normally bind and blocks Mena. Tes was found absent in many tumours. Thus if a drug could be designed to mimic tes's actions in blocking Mena, a novel therapy would be made to prevent metastasis.

Despite all of the mechanisms and therapies mentionedabove, there is no optimal cure for advanced prostate cancer. Novel therapeutic approaches are shedding more light for the development of better therapies. However, more basic research, and funding of that research, is still needed to win the battle with prostate cancer.

In conclusion, understanding the mechanisms underlying the invasion and metastasis of prostate cancer cells may be essential to developing effective therapeutics, and thus, aid in the treatment of the second deadliest disease in men.

Acknowledgements

This work was supported, in part by DOE grant DE-FG02-06ER64186-19 to DAB. This is CSCN #050.

References

[1] Andersson, S. O., Wolk, A., Bergström, R., Adami, H. O., Engholm, G., Englund, A. & Nyrén, O. (1997). Body size and prostate cancer: a 20-year follow-up study among 135006 Swedish construction workers. *J Natl Cancer Inst.*, *5*, 89(5):385-9.

[2] Aragon-Ching, J. B. & Dahut, W. L. (2007). Chemotherapy in Androgen-Independent Prostate Cancer (AIPC): What's next after taxane progression? *Cancer Ther.*, *5A*, 151-160.

[3] Aus, G., Damber, J. E., Khatami, A., Lilja, H., Stranne, J. & Hugosson, J. (2005). Individualized screening interval for prostate cancer based on prostate-specific antigen level: results of a prospective, randomized, population-based study. *Arch Intern Med.*, *165(16)*, 1857-61.

[4] Balakumaran, B. S. & Febbo, P. G. (2006). New insights into prostate cancer biology. *Hematol Oncol Clin North Am.*, *20(4)*, 773-96.

[5] Balk, S. P. & Knudsen, K. E. (2008). AR, the cell cycle, and prostate cancer. *Nucl Recept Signal*, *6*, e001.

[6] Balk, S. P. (2002). Androgen receptor as a target in androgen-independent prostate cancer. *Urology*, *60(3 Suppl 1)*, 132-8; discussion 138-9.

[7] Bennett, C. L., Price, D. K., Kim, S., Liu, D., Jovanovic, B. D., Nathan, D., Johnson, M. E., Montgomery, J. S., Cude, K., Brockbank, J. C., Sartor, O. & Figg, W. D. (2002). Racial variation in CAG repeat lengths within the androgen receptor gene among prostate cancer patients of lower socioeconomic status. *J Clin Oncol.*, *20(17)*, 3599-604.

[8] Boëda, B., Briggs, D. C., Higgins, T., Garvalov, B. K., Fadden, A. J., McDonald, N. Q. & Way, M. (2007). Tes, a specific Mena interacting partner, breaks the rules for EVH1 binding. *Mol Cell.*, *28(6)*, 1071-82.

[9] Chamberlain, R. M., Winter, K. A., Vijayakumar, S., Porter, A. T., Roach, M. 3rd, Streeter, O., Cox, J. D. & Bondy, M. L. (1998). Sociodemographic analysis of patients in radiation therapy oncology group clinical trials. *Int J Radiat Oncol Biol Phys.*, *40(1)*, 9-15.

[10] Chodak, G. W. (1994). The role of watchful waiting in the management of localized prostate cancer. *J Urol.*, *152(5 Pt 2)*, 1766-8.

[11] Culig, Z. & Bartsch, G. (2006). Androgen axis in prostate cancer. *J Cell Biochem.*, *99(2)*, 373-81.

[12] Damber, J. E. & Aus, G. (2008). Prostate cancer. *Lancet*, *371(9625)*, 1710-21.

[13] Deocampo, N. D., Huang, H. & Tindall, D. J. (2003). The role of PTEN in the progression and survival of prostate cancer. *Minerva Endocrinol.*, *28(2)*, 145-53.

[14] Deutsch, E., Maggiorella, L., Eschwege, P., Bourhis, J., Soria, J. C. & Abdulkarim, B. (2004). Environmental, genetic, and molecular features of prostate cancer. *Lancet Oncol.*, *5(5)*, 303-13.

[15] Dillioglugil, O., Leibman, B. D., Kattan, M. W., Seale-Hawkins, C., Wheeler, T. M. & Scardino, P. T. (1997). Hazard rates for progression after radical prostatectomy for clinically localized prostate cancer. *Urology*, *50(1)*, 93-9.

[16] Dong, X. Y., Chen, C., Sun, X., Guo, P., Vessella, R. L., Wang, R. X., Chung, L. W., Zhou, W. & Dong, J. T. (2006). FOXO1A is a candidate for the 13q14 tumor suppressor gene inhibiting androgen receptor signaling in prostate cancer. *Cancer Res.*, *66(14)*, 6998-7006.

[17] Dorai, T., Perlman, H., Walsh, K., Shabsigh, A., Goluboff, E. T., Olsson, C. A. & Buttyan, R. (1999). A recombinant defective adenoviral agent expressing anti-bcl-2 ribozyme promotes apoptosis of bcl-2-expressing human prostate cancer cells. Int J *Cancer*, *82(6)*, 846-52.

[18] Eastham, J. A., Grafton, W., Martin, C. M. & Williams, B. J. (2000). Suppression of primary tumor growth and the progression to metastasis with p53 adenovirus in human prostate cancer. *J Urol.*, *164(3 Pt 1)*, 814-9.

[19] Edwards, J., Mukherjee, R., Munro, A. F., Wells, A. C., Almushatat, A. & Bartlett, J. M. (2004). HER2 and COX2 expression in human prostate cancer. *Eur J Cancer.*, *40(1)*, 50-5.

[20] Evans, G. S. & Chandler, J. A. (1987). Cell proliferation studies in the rat prostate: II. The effects of castration and androgen-induced regeneration upon basal and secretory cell proliferation. *Prostate*, *11(4)*, 339-51.

[21] Garnick, M. B. (1994). Prostate cancer: screening, diagnosis, and management. *Ann Intern Med.* 1993, *118(10)*, 804-18. Erratum in: *Ann Intern Med.*, *120(8)*, 698.

[22] Gelmann, E. P., Chia, D., Pinsky, P. F., Andriole, G. L., Crawford, E. D., Reding, D., Hayes, R. B., Kramer, B. S., Woodrum, D. L., Gohagan, J. K. & Levin, D. L. (2001). Relationship of demographic and clinical factors to free and total prostate-specific antigen. PLCO Screening Trial Investigators. *Urology*, *58(4)*, 561-6.

[23] Giles, G. G., Severi, G., English, D. R., McCredie, M. R., Borland, R., Boyle, P. & Hopper, J. L. (2003). Sexual factors and prostate cancer. *BJU Int.*, *92(3)*, 211-6.

[24] Gittes, R. F. (1991). Carcinoma of the prostate. *N Engl J Med.*, *324(4)*, 236-45.

[25] Gleason, D. F. & Mellinger, G. T. (2002). Veterans Administration Cooperative Urological Research Group. Prediction of prognosis for prostatic adenocarcinoma by combined histological grading and clinical staging. 1974. *J Urol.*, *167(2 Pt 2)*, 953-8; discussion 959.

[26] Gleave, M. & Miyake, H. (2005). Use of antisense oligonucleotides targeting the cytoprotective gene, clusterin, to enhance androgen- and chemo-sensitivity in prostate cancer. *World J Urol.*, *23(1)*, 38-46.

[27] Gohagan, J. K., Prorok, P. C., Hayes, R. B. & Kramer, B. S. (2000). Prostate, Lung, Colorectal and Ovarian Cancer Screening Trial Project Team. The Prostate, Lung, Colorectal and Ovarian (PLCO) Cancer Screening Trial of the National Cancer Institute: history, organization, and status. *Control Clin Trials.*, *21(6 Suppl)*, 251S-272S.

[28] Greaves, W. W., Rom, W. N., Lyon, J. L., Varley, G., Wright, D. D. & Chiu, G. (1981). Relationship between lung cancer and distance of residence from nonferrous smelter stack effluent. *Am J Ind Med.*, *2(1)*, 15-23.

[29] Hakimi, J. M., Schoenberg, M. P., Rondinelli, R. H., Piantadosi, S. & Barrack, E. R. (1997). Androgen receptor variants with short glutamine or glycine repeats may identify unique subpopulations of men with prostate cancer. *Clin Cancer Res.*; *3(9)*, 1599-608.

[30] Hayes, R. B., Ziegler, R. G., Gridley, G., Swanson, C., Greenberg, R. S., Swanson, G. M., Schoenberg, J. B., Silverman, D. T., Brown, L. M., Pottern, L. M., Liff, J., Schwartz, A. G., Fraumeni, J. F. Jr & Hoover, R. N. (1999). Dietary factors and risks for prostate cancer among blacks and whites in the United States. *Cancer Epidemiol Biomarkers Prev.*, *8(1)*, 25-34.

[31] Hayes, S. A., Zarnegar, M., Sharma, M., Yang, F., Peehl, D. M., ten Dijke, P. & Sun, Z. (2001). SMAD3 represses androgen receptor-mediated transcription. *Cancer Res.*, *61(5)*, 2112-8.

[32] Hsing, A. W., Gao, Y. T., Wu, G., Wang, X., Deng, J., Chen, Y. L., Sesterhenn, I. A., Mostofi, F. K., Benichou, J. & Chang, C. (2000). Polymorphic CAG and GGN repeat lengths in the androgen receptor gene and prostate cancer risk: a population-based case-control study in China. *Cancer Res.*, *60(18)*, 5111-6.

[33] Irvine, R. A., Ma, H., Yu, M. C., Ross, R. K., Stallcup, M. R. & Coetzee, G. A. (2000). Inhibition of p160-mediated coactivation with increasing androgen receptor polyglutamine length. *Hum Mol Genet.*, *9(2)*, 267-74.

[34] Isaacs, J. T., Lundmo, P. I., Berges, R., Martikainen, P., Kyprianou, N. & English, H. F. (1992). Androgen regulation of programmed death of normal and malignant prostatic cells. *J Androl.*, *13(6)*, 457-64.

[35] Jarrard, D. F., Kinoshita, H., Shi, Y., Sandefur, C., Hoff, D., Meisner, L. F., Chang, C., Herman, J. G., Isaacs, W. B. & Nassif, N. (1998). Methylation of the androgen receptor promoter CpG island is associated with loss of androgen receptor expression in prostate cancer cells. *Cancer Res.*, *58(23)*, 5310-4.

[36] Jenster, G. (1999). The role of the androgen receptor in the development and progression of prostate cancer. *Semin Oncol.*, *(4)*, 407-21.

[37] Jenster, G. (1999). The role of the androgen receptor in the development and progression of prostate cancer. *Semin Oncol.*, *26(4)*, 407-21.

[38] Kinoshita, H., Shi, Y., Sandefur, C., Meisner, L. F., Chang, C., Choon, A., Reznikoff, C. R., Bova, G. S., Friedl, A. & Jarrard, D. F. (2000). Methylation of the androgen receptor minimal promoter silences transcription in human prostate cancer. *Cancer Res.*, *60(13)*, 3623-30.

[39] Kleinerman, D. I., Zhang, W. W., Lin, S. H., Nguyen, T. V., von Eschenbach, A. C. & Hsieh, J. T. (1995). Application of a tumor suppressor (C-CAM1)-expressing recombinant adenovirus in androgen-independent human prostate cancer therapy: a preclinical study. *Cancer Res.*, *55(13)*, 2831-6.

[40] Klotz, L. H. (2000). The role of nutrition in prostate cancer prevention. *Can J Urol.*, *7(1)*, 925.

[41] Krajewska, M., Krajewski, S., Epstein, J. I., Shabaik, A., Sauvageot, J., Song, K., Kitada, S. & Reed, J. C. (1996). Immunohistochemical analysis of bcl-2, bax, bcl-X, and mcl-1 expression in prostate cancers. *Am J Pathol.*, *148(5)*, 1567-76.

[42] Liao, C., Li, S. Q., Wang, X., Muhlrad, S., Bjartell, A. & Wolgemuth, D. J. (2004). Elevated levels and distinct patterns of expression of A-type cyclins and their associated cyclin-dependent kinases in male germ cell tumors. *Int J Cancer.*, *108(5)*, 654-64.

[43] Liu, D., Matzuk, M. M., Sung, W. K., Guo, Q., Wang, P. & Wolgemuth, D. J. (1998). Cyclin A1 is required for meiosis in the male mouse. *Nat Genet.*, *20(4)*, 377-80.

[44] Maitland, N. J. & Collins, A. T. (2008). Prostate cancer stem cells: a new target for therapy. *J Clin Oncol.*, *26(17)*, 2862-70.

[45] Matzkin, H., Eber, P., Todd, B., van der Zwaag, R. & Soloway, M. S. (1992). Prognostic significance of changes in prostate-specific markers after endocrine treatment of stage D2 prostatic cancer. *Cancer*, *70(9)*, 2302-9.

[46] Miyake, H., Nelson, C., Rennie, P. S. & Gleave, M. E. (2000). Acquisition of chemoresistant phenotype by overexpression of the antiapoptotic gene testosterone-repressed prostate message-2 in prostate cancer xenograft models. *Cancer Res.*; *60(9)*, 2547-54.

[47] Nazareth, L. V. & Weigel, N. L. (1996). Activation of the human androgen receptor through a protein kinase A signaling pathway. *J Biol Chem.*, *271(33)*, 19900-7.

[48] Nelson, P. S. & Montgomery, B. (2003). Unconventional therapy for prostate cancer: good, bad or questionable? *Nat Rev Cancer, 3(11)*, 845-58.

[49] Papatsoris, A. G. & Papavassiliou, A. G. (2001). Prostate cancer: horizons in the development of novel anti-cancer strategies. *Curr Med Chem Anticancer Agents, 1(1)*, 47-70.

[50] Paulson, D. F., Moul, J. W., Robertson, J. E. & Walther, P. J. (1990). Postoperative radiotherapy of the prostate for patients undergoing radical prostatectomy with positive margins, seminal vesicle involvement and/or penetration through the capsule. *J Urol., 143(6)*, 1178-82.

[51] Pisansky, T. M., Cha, S. S., Earle, J. D., Durr, E. D., Kozelsky, T. F., Wieand, H. S. & Oesterling, J. E. (1993). Prostate-specific antigen as a pretherapy prognostic factor in patients treated with radiation therapy for clinically localized prostate cancer. *J Clin Oncol., 11(11)*, 2158-66.

[52] Porkka, K. P. & Visakorpi, T. (2004). Molecular mechanisms of prostate cancer. *Eur Urol., 45(6)*, 683-91.

[53] Rivera, A., Mavila, A., Bayless, K. J., Davis, G. E. & Maxwell, S. A. (2006). Cyclin A1 is a p53-induced gene that mediates apoptosis, G2/M arrest, and mitotic catastrophe in renal, ovarian, and lung carcinoma cells. *Cell Mol Life Sci., 63(12)*, 1425-39.

[54] Robertson, L. M., Douglas, F., Ludbrook, A., Reid, G. & van Teijlingen, E. (2008). What works with men? A systematic review of health promoting interventions targeting men. *BMC Health Serv Res., 3*, 8:141.

[55] Russell, P. J., Bennett, S. & Stricker, P. (1998). Growth factor involvement in progression of prostate cancer. *Clin Chem., 44(4)*, 705-23.

[56] Sansal, I. & Sellers, W. R. (2004). The biology and clinical relevance of the PTEN tumor suppressor pathway. *J Clin Oncol., 22(14)*, 2954-63.

[57] Scher, H. I. & Sawyers, C. L. (2005). Biology of progressive, castration-resistant prostate cancer: directed therapies targeting the androgen-receptor signaling axis. *J Clin Oncol., 23(32)*, 8253-61.

[58] Schwartz, G. G. (2005). Vitamin D and the epidemiology of prostate cancer. *Semin Dial., 18(4)*, 276-89.

[59] Shannan, B., Seifert, M., Boothman, D. A., Tilgen, W. & Reichrath, J. (2006). Clusterin and DNA repair: a new function in cancer for a key player in apoptosis and cell cycle control. *J Mol Histol., 37(5-7)*, 183-8.

[60] Sharifi, N., Gulley, J. L. & Dahut, W. L. (2005). Androgen deprivation therapy for prostate cancer. *JAMA, 294(2)*, 238-44.

[61] Shimizu, H., Ross, R. K., Bernstein, L., Yatani, R., Henderson, B. E. & Mack, T. M. (1991). Cancers of the prostate and breast among Japanese and white immigrants in Los Angeles County. *Br J Cancer., 63(6)*, 963-6.

[62] Sinha, B. K., Yamazaki, H., Eliot, H. M., Schneider, E., Borner, M. M. & O'Connor, P. M. (1995). Relationships between proto-oncogene expression and apoptosis induced by anticancer drugs in human prostate tumor cells. *Biochim Biophys Acta, 1270(1)*, 12-8.

[63] Smith, J. R., Freije, D., Carpten, J. D., Grönberg, H., Xu, J., Isaacs, S. D., Brownstein, M. J., Bova, G. S., Guo, H., Bujnovszky, P., Nusskern, D. R., Damber, J. E., Bergh, A., Emanuelsson, M., Kallioniemi, O. P., Walker-Daniels, J., Bailey-Wilson, J. E., Beaty,

T. H., Meyers, D. A., Walsh, P. C., Collins, F. S., Trent, J. M. & Isaacs, W. B. (1996). Major susceptibility locus for prostate cancer on chromosome 1 suggested by a genome-wide search. *Science, 274(5291)*, 1371-4.

[64] So, A., Hadaschik, B., Sowery, R. & Gleave, M. (2007). The role of stress proteins in prostate cancer. *Curr Genomics., 8(4)*, 252-61.

[65] Stanbrough, M., Leav, I., Kwan, P. W., Bubley, G. J. & Balk, S. P. (2001). Prostatic intraepithelial neoplasia in mice expressing an androgen receptor transgene in prostate epithelium. *Proc Natl Acad Sci U S A, 98(19)*, 10823-8.

[66] Struewing, J. P., Hartge, P., Wacholder, S., Baker, S. M., Berlin, M., McAdams, M., Timmerman, M. M., Brody, L. C. & Tucker, M. A. (1997). The risk of cancer associated with specific mutations of BRCA1 and BRCA2 among Ashkenazi Jews. *N Engl J Med., 336(20)*, 1401-8.

[67] Theodorescu, D., Broder, S. R., Boyd, J. C., Mills, S. E. & Frierson, H. F. Jr. (1997). p53, bcl-2 and retinoblastoma proteins as long-term prognostic markers in localized carcinoma of the prostate. *J Urol., 158(1)*, 131-7.

[68] Vainio, H., Kaaks, R. & Bianchini, F. (2002). Weight control and physical activity in cancer prevention: international evaluation of the evidence. *Eur J Cancer Prev., Suppl 2*, S94-100.

[69] Van Brussel, J. P., Jan Van Steenbrugge, G., Van Krimpen, C., Bogdanowicz, J. F., Van Der Kwast, T. H., Schröder, F. H. & Mickisch, G. H. (2001). Expression of multidrug resistance related proteins and proliferative activity is increased in advanced clinical prostate cancer. *J Urol., 165(1)*, 130-5.

[70] Wegiel, B., Bjartell, A., Culig, Z. & Persson, J. L. (2008). Interleukin-6 activates PI3K/Akt pathway and regulates cyclin A1 to promote prostate cancer cell survival. *Int J Cancer., 122(7)*, 1521-9.

[71] Wegiel, B., Bjartell, A., Tuomela, J., Dizeyi, N., Tinzl, M., Helczynski, L., Nilsson, E., Otterbein, L. E., Härkönen, P. & Persson, J. L. (2008). Multiple cellular mechanisms related to cyclin A1 in prostate cancer invasion and metastasis. *J Natl Cancer Inst., 100(14)*, 1022-36.

[72] Whittemore, A. S., Kolonel, L. N., Wu, A. H., John, E. M., Gallagher, R. P., Howe, G. R., Burch, J. D., Hankin, J., Dreon, D. M., West, D. W., et al. (1995). Prostate cancer in relation to diet, physical activity, and body size in blacks, whites, and Asians in the United States and Canada. *J Natl Cancer Inst., 3*, 87(9):652-61.

[73] www.cancer.gov.

[74] Yagoda, A. & Petrylak, D. (1993). Cytotoxic chemotherapy for advanced hormone-resistant prostate cancer. *Cancer, 71(3 Suppl)*, 1098-109.

[75] Zhang, H., Kim, J. K., Edwards, C. A., Xu, Z., Taichman, R. & Wang, C. Y. (2005). Clusterin inhibits apoptosis by interacting with activated Bax. *Nat Cell Biol., 7(9)*, 909-15.

In: Handbook of Prostate Cancer Cell Research…
Editor: Alan T. Meridith

ISBN: 978-1-60741-954-9

Chapter 16

Implications of Lysophosphatidic Acid Signaling and Receptor-Mediated Mechanisms Contributing to the Development of Prostate Carcinoma

Mandi M. Murph
Department of Pharmacology and Biomedical Sciences,
The University of Georgia College of Pharmacy,
R.C. Wilson Pharmacy Building, Athens, Georgia 30602, USA

Abstract

In the United States, prostate cancer ranks as the second leading cause of cancer-related mortality for men, behind only lung cancer. Prostate cancer is also the most frequently diagnosed cancer overall, affecting almost 200,000 each year. While the 5-year survival rates are excellent, especially in comparison to all other cancers, metastasis remains a problem for clinicians and is a leading contributor to mortality. In addition, the molecular mechanisms governing the development of prostate disease need to be elucidated. Lysophosphatidic acid (LPA) signaling induces significant molecular pathways that may contribute and/or exacerbate cellular effects in prostate cancer. Changes in extracellular protein expression, motility, cross-communication with the epidermal growth factor signaling pathway and enhanced ERK signaling are effects mediated by LPA in prostate cell lines. In addition, intracellular enzymatic production of LPA may provide a survival signal in prostate. Herein we revisit and summarize these LPA-mediated molecular mechanisms and receptor signaling which contribute to prostate cancer cell differentiation, migration, survival and growth and discuss LPA signaling as an early inducer of abnormalities that occur in prostate disease. Future drug targeting of LPA signaling pathways may have clinical benefit against prostate cancer.

Introduction

One out of every four deaths in the U.S. is caused by cancer (1). While prostate cancer has a large number of men affected, nearly 200,000, the mortality is very low comparatively, 28,000 or 14% of the total estimated in 2008. In the 1970's the 5-year survival rate was 70% but today that number is a staggering 99%. Screening methods, better awareness and new treatment options helped decrease the mortality rate an additional 34% between 1990-2004. (1)

The positive outlook and progress made towards fighting prostate cancer deserves significant recognition and should serve as a model for other types of cancer. The regular assessment of a male's prostate specific antigen (PSA) during routine physician examinations can help guide medical decisions regarding disease management. Recently, this test has become controversial in terms of its therapeutic benefit, especially for older men. Although false positives occur, the existence of such PSA screening is a great achievement in the direction of cancer prevention. PSA is also a coveted example of an early-detection biomarker that functions as a goal for other cancers, in particular ovarian cancer where there is no early screening method and detection often occurs at a late stage of disease.

Since some men still succumb to prostate cancer, it is important to understand the outstanding challenges facing the patients, researchers and clinicians. For the patient, statistical analysis indicates several predispositions exist that are associated with poor outcome and these are race, age and education. Thus, the combination of these three risk factors yields the greatest non-clinical, predisposing risk influencing mortality: African-American race, over 70 years of age and less than 12 years of formal education. (1)

There are other outstanding factors and basic science questions that need to be addressed by researchers and clinicians in order to continue the progress towards fighting prostate disease. In terms of the clinic, the biggest hurdle that exists today for prostate cancer treatment is metastasis and this is the primary concern for patient outcome post radical prostatectomy. With most types of cancer, the tumor type, stage or grade presented at the time of clinical diagnosis correlates with prognosis. Advanced stage, metastatic disease coincides with poor prognosis. It is unknown whether the specific mechanisms that govern metastatic progression of prostate cancer are unique or similar to other cancers.

The underlying question for basic scientists to understand is the development of hyperplasia and the molecular etiology of prostate cancer. To achieve this end, research is underway to identify potential aspects of prostate cancer signaling that may drive tumor progression. Lysophosphatidic acid (LPA) signaling elicits proliferation, invasion and migration in PC-3 and DU-145 prostate cancer cells (2) and may contribute to tumor progression. Experimental compounds are under investigation as therapeutic inhibitors against LPA signaling in prostate cancer and these reduce motility in response to LPA (2).

LPA Receptor Expression in Prostate

To date there are nine reported receptors for LPA: LPA1 (3), LPA2 (4), LPA3 (5), LPA4/GPR23 (6), LPA5/GPR92/93 (7,8), GPR87 (putative LPA6) (9), P2Y5 (putative LPA7) (10), PPARγ (11) and P2Y10 (putative dual LPA and S1P receptor) (12), verified

LPA receptors are numbered sequentially in the order of their molecular cloning and identification. Not all of these receptors have been extensively investigated in prostate cancer, with most of the literature analyzing the function of LPA1 in this disease. Future research is needed to elucidate the role of LPA receptors in normal prostate and prostate cancer.

The most common lineages for studying LPA signaling in prostate cancer are DU-145, LNCaP and PC-3 cells. DU-145 cells are androgen-independent, epithelial metastatic prostate cells isolated from the brain in a 69 year-old male; LNCaP are androgen-dependent, epithelial metastatic prostate cells isolated from the left supraclavicular lymph node of a 50 year-old male; PC-3 cells are androgen-independent, epithelial metastatic prostate cells isolated from the bone of a 62 year-old male with grade IV prostatic adenocarcinoma. Although each lineage was isolated from a metastatic site, LPA induces motility in DU-145 and PC-3 cells in culture, yet LNCaP cells are unresponsive and do not migrate (data not shown). The divergent LPA-mediated responses are likely due to varying LPA receptor expression.

It is unknown whether all LPA receptors have been identified, especially among the recently recognized purinergic family. In terms of known or accepted LPA receptor expression, DU-145 cells express LPA1>LPA2 and little or no LPA3 and LPA4 (13); LNCaP cells express LPA3>LPA4>>LPA2 but no LPA1 (13); PC-3 cells express LPA1>LPA3>LPA2 and little or no LPA4 and LPA5 (13,14). To date there is little information in the literature regarding the expression of GPR87 (LPA6) or p2y5 (LPA7) in DU-145, LNCaP or PC-3 cell lines; however, RT-PCR analysis of mouse tissue indicates GPR87 (LPA6) has a minute level of expression in the prostate (9) and the dual LPA/S1P receptor, p2y10, shows strong expression in the prostate, along with uterus, lung, skeletal muscle and weaker expression in brain (12).

Motility and LPA Receptor Regulation

A report analyzing both PC-3 and LNCaP cells suggested the LPA1 receptor is the sole receptor responsible for LPA-mediated cell motility since LPA1 is expressed highly in PC-3 but not in LNCaP (15). Indeed, both cell lines express autotaxin (ATX) to a lesser extent (13) which would provide the stimulus necessary to induce autocrine activation and/or prime the cells for LPA signaling responses. The similarly highly motile, LPA-responsive DU-145 cells express abundant LPA1 receptors and moderate to low autotaxin which would be consistent with the hypothesis pinning motility exclusively on LPA1-mediated receptor functioning.

In contrast, a recent report suggests the LPA-mediated mechanism governing cell motility is more complicated than the idea that one receptor controls all decisions regarding LPA-mediated migration. Lee et al. demonstrates in multiple cell lines (mouse embryonic fibroblasts (MEFs), DLD1 colon cancer cells and B103 neuroblastoma cells) a functional antagonism between the LPA4 and LPA1 receptors regulating LPA-stimulated motility (16). Knocking out the LPA4 receptor in MEFs created a hypersensitive cell line responsive to LPA-mediated signaling and motility. Re-introducing the LPA4 receptor into MEFs attenuated the LPA-mediated hypersensitivity. Similar results were observed in DLD1 and B103 cells upon enforced expression of the LPA4 receptor, which significantly reduced LPA-mediated motility. (16) Taken together, this data suggests that LPA1 positively and LPA4 negatively regulates LPA-induced cell motility and it is the combination of signaling events that determine whether and how pronounced a cell will migrate in response to LPA stimulation.

Although we can only hypothetically extrapolate the mechanism in prostate cells, the observations are consistent. Both PC-3 and DU-145 cells which express LPA1 but little to no LPA4 are highly motile in response to LPA (2); whereas LNCaP cells do not migrate through a Boyden chamber (unpublished results) and they express LPA4 but not LPA1. Therefore, it is indeed likely that a combination of LPA1 and LPA4 signaling regulate LPA-mediated motility in a variety of cell lines. In most biological systems redundancy, feedback loops and complex layers of regulation for key functions are present. Extrapolation of the data from Lee et al. creates the hypothesis that prostate cells or carcinomas capable of silencing the LPA4 receptor and/or significantly enhancing the LPA1 receptor gene expression through mutational or epigenetic events will acquire a metastatic advantage in the presence of LPA, a prominent component in normal human blood circulation and increased in wound healing.

LPA Affects Androgen Receptor Signaling

It is unclear whether the effects of LPA on prostate cancers are dependent on androgen sensitivity or receptor selectivity. LPA elicits an increase in cell proliferation in PC-3 and DU-145 cells but LPA fails to elicit an increase in cell proliferation in androgen sensitive LNCaP cells (17,18). At first glance this would suggest androgens and LPA signaling are connected but not explain the mechanism. The androgen receptor is fundamental to prostate cancer development because it acts like a transcription factor and enhances the expression of prostate specific antigen (PSA). Additional work demonstrated LNCaP cells forced to overexpress the LPA1 receptor have enhanced LPA-mediated mitogenic outcomes, such as cell proliferation, viability and doubling time because LPA1 translocates the androgen receptor to the nucleus where it promotes cell cycle progression and proliferation. PC-3 cells which express the LPA1 receptor endogenously also show nuclear translocation of the androgen receptor when it is ectopically expressed. Furthermore, in LNCaP cells expressing the LPA1 receptor, LPA induced an androgen-dependent increase in the expression of PSA. (18) Taken together this data suggests a critical role for LPA1 in androgen receptor signaling in prostate cancer.

Enzymatic Regulation of LPA Governs Cell Survival

ATX levels in serum are not statistically different in patients with prostate cancer versus healthy controls (19) suggesting that enzymes other than ATX may be responsible for the increased LPA production capabilities seen in activated prostate cancer cells (20). In addition, gene expression analyses show most LPA receptors with reduced expression in metastatic prostate cancer (Figure 1). This enigmatic observation is coupled with multiple studies citing LPA as a potent mitogen for prostate cancer cells.

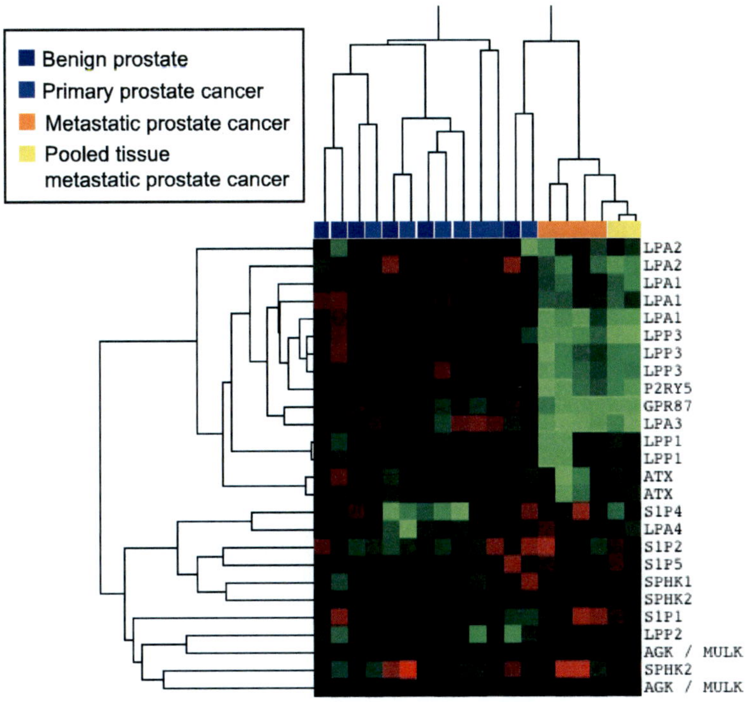

Figure 1.

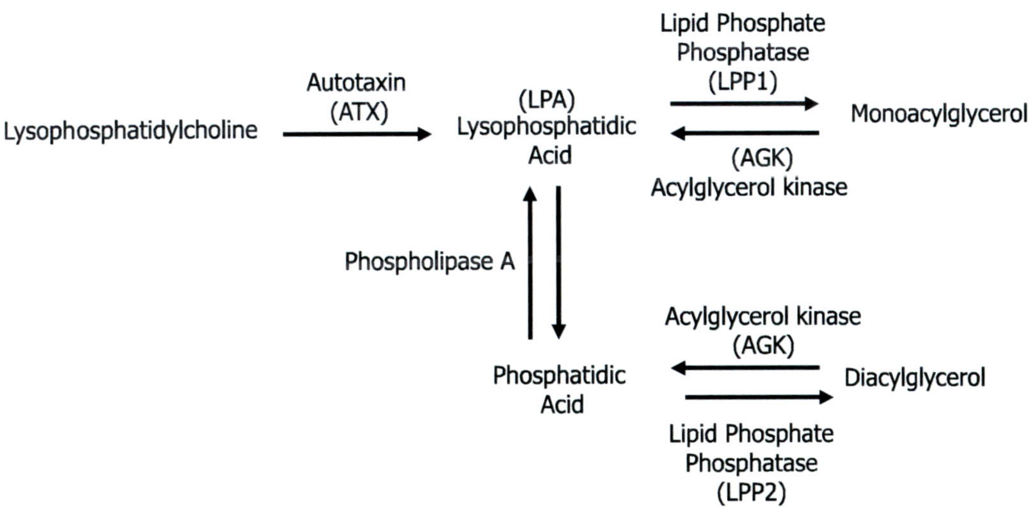

Figure 2.

The cloning of the lipid kinase, acylglycerol kinase (AGK) (21), may help reconcile the two contrasting observations, that LPA receptor expression and ATX are decreased in metastatic prostate cancer yet LPA is a potent mitogen for metastatic and primary prostate cancer cells in vitro. AGK catalyzes the phosphorylation of monoacylglycerol, yielding LPA and acting in opposition with lipid phosphate phosphatases (**Figure 2**). Unlike the LPA receptors and ATX, AGK is not decreased in metastatic prostate cancer in gene expression analysis of patient samples (Figure 1) suggesting this could be the major route of LPA production in prostate. Indeed, AGK phosphorylates and converts the pro-apoptotic 2-arachidonylglycerol into the anti-apoptotic LPA in PC-3 and DU-145 cells, thus AGK may regulate survival outcomes in prostate cancer (22,23). Further studies show that in LPA treatment of PC-3 cells 24 hours after serum starvation-induced autophagy reduces LC3 redistribution in cells, a marker of autophagy, and reduces the ratios of LC3-II to total LC3 (24). Taken together, these data suggest that LPA signals survival in prostate cancer.

Autocrine Signaling and CYR61

Interestingly both PC-3 and DU-145 cells produce and then secrete LPA after stimulation with bombesin (20). This data suggests that prostate agonists can facilitate the creation of an autocrine and/or paracrine LPA signaling environment in prostate carcinomas. Couple the ability to produce LPA with a change in LPA receptor expression and the cells could thus acquire a migratory capability. It is unclear what epigenetic or mutational event would alter LPA receptor expression that might lead to enhanced migration but research is underway to elucidate such an event.

LPA also increases the expression of CYR61, a heparin-binding extracellular matrix signaling protein that is related to the development and progression of benign prostatic hyperplasia (25). If true, then it implicates LPA as an early inducer and initiator of this disease. Coupled with work citing LPA as a mediator of androgen-dependent PSA increase (18) and that LPA signaling is decreased in metastatic prostate cancer patient profiles (**Figure 1**), LPA may indeed function in this capacity as an early, rather than late, inducer of abnormalities contributing to the etiology of prostate disease.

LPA-Mediated ERK Signaling

Mitogen-activated protein kinase or more specifically, extracellular signal-regulated kinase is activated by extracellular growth factors through a cascade of events that leads to its translocation into the nucleus where it phosphorylates transcription factors, increasing the transcription of genes associated with cell growth and proliferation. Interestingly, non-neoplastic prostate tissue has little ERK activity but prostate tumors have elevated levels of ERK activity that increases with tumor stage (26). The level of ERK expression is not overwhelmingly different in normal prostate tissue compared with prostate cancer, but the level of ERK activation can reach upwards of 15-fold increase in prostate cancer (27). In addition, both LPA and EGF activate ERK signaling in prostate cancer cells and together they cooperatively enhance the level of ERK activity through EGF receptor activation and transactivation by LPA through shedding of EGF-like peptides (28). Furthermore, cross-talk and interconnected signaling pathways that exist between LPA and EGF may have a strong influence over tumor development and progression in prostate (21,29).

Drug Targeting of LPA Signaling Pathways

We have recently demonstrated that experimental therapeutic compounds, called carba analogs of cyclic phosphatidic acid, have in vitro efficacy to inhibit LPA chemotactic migration (2). It will be interesting to further characterize the functionality of LPA receptor signaling in prostate and the outcome of inhibiting LPA signaling pathways and potentially cross-communication with the EGF signaling pathway that has been implicated in prostate disease. This future research will determine whether targeting this pathway is beneficial in vivo and therapeutically viable in terms of efficacy and toxicity.

Acknowledgments

Thanks to Harish Radhakrishna for critically reading this manuscript.

References

[1] Jemal, A., Siegel, R., Ward, E., Hao, Y., Xu, J., Murray, T. & Thun, M. J. (2008). *CA: a cancer journal for clinicians*, *58(2)*, 71-96.

[2] Hasegawa, Y., Murph, M., Shuangxing, Y., Tigyi, G. & Mills, G.B. (2008). *Molecular Oncology*, *2(1)*, 54-69.

[3] Hecht, J. H., Weiner, J. A., Post, S. R. & Chun, J. (1996). *The Journal of cell biology*, *135(4)*, 1071-1083.

[4] An, S., Bleu, T., Hallmark, O. G. & Goetzl, E. J. (1998). *The Journal of biological chemistry*, *273(14)*, 7906-7910.

[5] Bandoh, K., Aoki, J., Hosono, H., Kobayashi, S., Kobayashi, T., Murakami-Murofushi, K., Tsujimoto, M., Arai, H. & Inoue, K. (1999). *The Journal of biological chemistry*, *274(39)*, 27776-27785.

[6] Noguchi, K., Ishii, S. & Shimizu, T. (2003). *The Journal of biological chemistry*, *278(28)*, 25600-25606.

[7] Kotarsky, K., Boketoft, A., Bristulf, J., Nilsson, N. E., Norberg, A., Hansson, S., Owman, C., Sillard, R., Leeb-Lundberg, L. M. & Olde, B. (2006). *The Journal of pharmacology and experimental therapeutics*, *318(2)*, 619-628.

[8] Lee, C. W., Rivera, R., Gardell, S., Dubin, A. E. & Chun, J. (2006). *The Journal of biological chemistry*, *281(33)*, 23589-23597.

[9] Tabata, K., Baba, K., Shiraishi, A., Ito, M. & Fujita, N. (2007). *Biochemical and biophysical research communications*, *363(3)*, 861-866.

[10] Pasternack, S. M., von Kugelgen, I., Aboud, K. A., Lee, Y. A., Ruschendorf, F., Voss, K., Hillmer, A. M., Molderings, G. J., Franz, T., Ramirez, A., Nurnberg, P., Nothen, M. M. & Betz, R. C. (2008). *Nature genetics*, *40(3)*, 329-334.

[11] McIntyre, T. M., Pontsler, A. V., Silva, A. R., St Hilaire, A., Xu, Y., Hinshaw, J. C., Zimmerman, G. A., Hama, K., Aoki, J., Arai, H. & Prestwich, G. D. (2003). *Proceedings of the National Academy of Sciences of the United States of America*, *100(1)*, 131-136.

[12] Murakami, M., Shiraishi, A., Tabata, K. & Fujita, N. (2008). *Biochemical and biophysical research communications*, *371(4)*, 707-712.

[13] Kishi, Y., Okudaira, S., Tanaka, M., Hama, K., Shida, D., Kitayama, J., Yamori, T., Aoki, J., Fujimaki, T. & Arai, H. (2006). *The Journal of biological chemistry*, *281(25)*, 17492-17500.

[14] Valentine, W. J., Fujiwara, Y., Tsukahara, R. & Tigyi, G. (2008). *Biochimica et biophysica acta*, *1780(3)*, 597-605.

[15] Hama, K., Aoki, J., Fukaya, M., Kishi, Y., Sakai, T., Suzuki, R., Ohta, H., Yamori, T., Watanabe, M., Chun, J. & Arai, H. (2004). *The Journal of biological chemistry*, *279(17)*, 17634-17639.

[16] Lee, Z., Cheng, C. T., Zhang, H., Subler, M. A., Wu, J., Mukherjee, A., Windle, J. J., Chen, C. K. & Fang, X. (2008). *Molecular biology of the cell*.

[17] Guo, C., Luttrell, L. M. & Price, D. T. (2000). *The Journal of urology*, *163(3)*, 1027-1032.

[18] Guo, R., Kasbohm, E. A., Arora, P., Sample, C. J., Baban, B., Sud, N., Sivashanmugam, P., Moniri, N. H. & Daaka, Y. (2006). *Endocrinology*, *147(10)*, 4883-4892.

[19] Nakamura, K., Takeuchi, T., Ohkawa, R., Okubo, S., Yokota, H., Tozuka, M., Aoki, J., Arai, H., Ikeda, H., Ohshima, N., Kitamura, T. & Yatomi, Y. (2007). *Annals of clinical biochemistry*, *44(Pt 6)*, 549-556.

[20] Xie, Y., Gibbs, T. C., Mukhin, Y. V. & Meier, K. E. (2002). *The Journal of biological chemistry*, *277(36)*, 32516-32526.

[21] Bektas, M., Payne, S. G., Liu, H., Goparaju, S., Milstien, S. & Spiegel, S. (2005). *The Journal of cell biology*, *169(5)*, 801-811.

[22] Melck, D., De Petrocellis, L., Orlando, P., Bisogno, T., Laezza, C., Bifulco, M. & Di Marzo, V. (2000). *Endocrinology*, *141(1)*, 118-126.

[23] Spiegel, S. & Milstien, S. (2005). *Biochemical Society transactions*, *33(Pt 6)*, 1362-1365.

[24] Chang, C. L., Liao, J. J., Huang, W. P. & Lee, H. (2007). *Autophagy*, *3(3)*, 268-270.

[25] Sakamoto, S., Yokoyama, M., Zhang, X., Prakash, K., Nagao, K., Hatanaka, T., Getzenberg, R. H. & Kakehi, Y. (2004). *Endocrinology*, *145(6)*, 2929-2940.

[26] Gioeli, D., Mandell, J. W., Petroni, G. R., Frierson, H. F., Jr. & Weber, M. J. (1999). *Cancer research*, *59(2)*, 279-284.

[27] Price, D. T., Della Rocca, G., Guo, C., Ballo, M. S., Schwinn, D. A. & Luttrell, L. M. (1999). *The Journal of urology*, *162(4)*, 1537-1542.

[28] Kue, P. F., Taub, J. S., Harrington, L. B., Polakiewicz, R. D., Ullrich, A. & Daaka, Y. (2002). *International journal of cancer*, *102(6)*, 572-579.

[29] Prenzel, N., Zwick, E., Daub, H., Leserer, M., Abraham, R., Wallasch, C. & Ullrich, A. (1999). *Nature*, *402(6764)*, 884-888.

In: Handbook of Prostate Cancer Cell Research... ISBN: 978-1-60741-954-9
Editor: Alan T. Meridith

Chapter 17

Redox State and Prostate Cancer Behavior

Luksana Chaiswing and Terry D. Oberley
Pathology and Laboratory Medicine Service,
William S. Middleton Memorial Veterans Hospital and Department of Pathology and Laboratory Medicine, School of Medicine and Public Health, University of Wisconsin, Madison, Wisconsin, USA

Abstract

Cellular or tissue reduction-oxidation (redox) state is the result of the net balance of molecular reducing and oxidizing equivalents, and can be subdivided into intra-or extracellular redox state (inside or outside the cell), respectively. Under physiological conditions, redox state regulates cell metabolism, survival, and the cell cycle. Prostate cancer is the most common malignancy diagnosed in men and is the second leading cause of cancer deaths in males in the U.S.A. Many studies have documented that prostate cancer cell lines have altered levels of intracellular reactive oxygen species/reactive nitrogen species and/or antioxidant levels in relation to nonmalignant cells. This abnormal intracellular redox state may play an important role in prostate cancer cell cycle progression and cell survival through regulation of redox- modulated transcription factors. Our laboratory recently reported that prostate cancer cells in the cell cultures exhibited an increased glutathione/glutathione disulfide ratio in the media in comparison to immortalized but nonmalignant prostate epithelial cells, suggesting an alteration of extracellular redox state of prostate cancer cells. We also demonstrated that modulation of extracellular redox state by overexpression of extracellular superoxide dismutase resulted in inhibition of prostate cancer cell invasiveness through a mechanism responsive to superoxide radical and/or nitric oxide. We propose that both intra- and extracellular redox states play complementary role(s) in regulation of prostate cancer cell proliferation and/or invasion. This possible relationship between redox states may lead to the development of anti-cancer therapeutics utilizing redox-modulating compounds that are specifically designed to target prostate cancer cells and/or the prostate cancer microenvironment.

Introduction

Prostate cancer is the most common malignancy diagnosed in men and is the second leading cause of cancer deaths in males in the U.S.A. Androgens play a major role in the development of prostate cancer since males castrated during childhood do not develop this disease. Another factor known to be involved in prostate cancer development is age, since the majority of prostate cancer patients are aging men. More recently, inflammation has been postulated to be involved in the development of prostate cancer [Alberti, 2000]. Since androgen levels [Gurel et al., 2008], aging [Blank & Bellizzi, 2008], and inflammation may cause or be the result of oxidative stress, our laboratory has been interested in the possible role of cell redox state in prostate cancer.

Cellular or tissue redox state results from the net balance of molecular reducing and oxidizing equivalents, and can be subdivided into intra-or extracellular redox state (inside or outside the cell), respectively. The factors that determine the redox state of a cell include the levels of antioxidant enzymes (AEs) and other antioxidant proteins, reactive oxygen species (ROS)/reactive nitrogen species (RNS), pro-oxidant transition metal compounds, and molecules that can repair oxidative damage. Other factors shown to be involved in modulation of redox state are thiol/disulfide couples such as glutathione (GSH)/glutathione disulfide (GSSG), cysteine (Cys)/cystine (Cyss), and the thioredoxin system.

Evidence has accumulated that each subcellular compartment of each distinct cell type has a unique redox state designed to allow optimal physiological functions. Alteration in redox state in specific subcellular compartments and/or extracellular spaces will result in cell adaptation or cell dysfunction. In normal cells under physiological conditions, the intracellular redox state is more reduced than the extracellular space [Kemp, Go, & Jones, 2008], which may allow cells to modulate activities of extracellular proteins, examples being adhesion molecule expression, growth receptor binding, and secretion of inflammation-related proteins.

We hypothesize that cancer cells develop alterations in redox state in both intra-and extracellular spaces throughout cancer progression, with these changes postulated to promote abnormal behavior of cancer cells (Figure 1). Studies from our laboratory have demonstrated that the redox states of cancer cells are different in prostate cancer cells in comparison to normal cells [Bostwick et al., 2000; Oberley, Zhong, Szweda, & Oberley, 2000; Chaiswing, Zhong, Cullen, Oberley, & Oberley, 2008]. These results are in agreement with studies from other laboratories. By using EPR spectroscopy and imaging techniques with a nitroxide redox probe, Kuppusamy *et al.*, [1998] have shown that tumor tissues in a murine model were more highly reduced in comparison to normal tissues. In addition, using mercury orange staining, analysis of a human colon carcinoma demonstrated higher levels of non-protein thiols in neoplastic tissue than in the surrounding adjacent normal tissue counterpart [Rubartelli & Lotze, 2007]. More importantly, we have shown that modulation of redox state uniquely alters biochemical properties of prostate cancer cells with different degress of aggressiveness when compared to effects on normal prostate epithelial cells [Chaiswing, Bourdeau-Heller, Zhong, & Oberley, 2007; Chaiswing, Zhong, Cullen, Oberley, & Oberley, 2008]. Biochemical properties altered by redox state in cancer cells include alterations of protein conformation, transcriptional and translational processes, signal transduction, enzyme activities, protein-protein/DNA interactions, ligand binding to receptors, protein trafficking, and protein degradation. These biochemical changes are postulated to result in changes in cell behavior, including cell cycle progression, growth, and invasion. The aim of the current commentary is to define the role(s) of intra-/extracellular redox states on the behavior of prostate cancer cells.

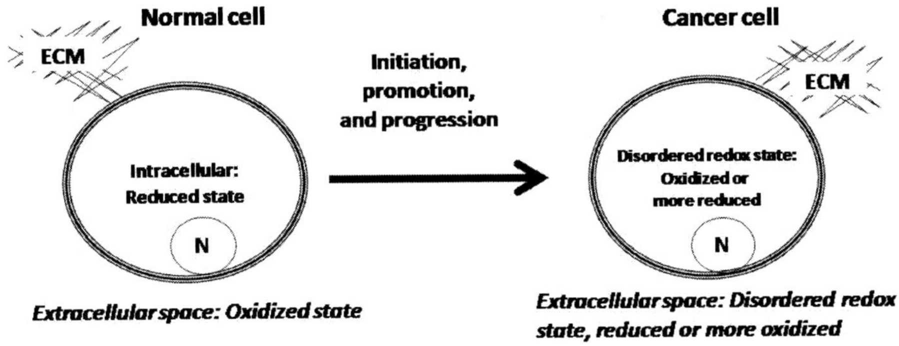

Figure 1. Redox state alterations during cancer progression. In normal cells, the outside of the cell is more oxidized than the inside, allowing cells to modulate activities of extracellular proteins such as adhesion molecule expression, growth receptor binding and/or secretion of inflammation-related proteins. During cancer progression, we hypothesize that redox states of cancer cells become disorganized both intra-and extracellularly, thus promoting abnormal behavior of cancer cells or neighboring cells. ECM=Extracellular matrix.

Intracellular Redox State and Prostate Cancer

Several studies have documented that alterations of redox state affect cancer initiation, promotion, and/or progression [Halliwell, 2007; Laurent et al., 2005]. It is possible that different types of cancer cells respond to redox alterations uniquely depending on their intrinsic redox states. The most commonly described change in prostate cancer cells is elevation of oxidative stress due to high levels of ROS/RNS and/or lowered levels of AEs [Laurent et al., 2005]. Our laboratory was the first to describe alterations in levels of oxidative/nitrative damage products in normal prostate tissues in comparison with prostate cancer tissues using immunoperoxidase and immmunogold labeling techniques; we demonstrated that metastatic human prostate cancer has altered subcellular distributions of specific oxidative/nitrative damage products (increased nuclear 8-hydroxy-2′-deoxyguanosine (8OHdG) and 3-nitrotyrosine (3NT) levels) in comparison with normal prostate epithelium or primary prostate cancer [Bostwick et al., 2000; Oberley et al., 2000; Chaiswing et al., 2007]. Using the same techniques, we also demonstrated lowered immunoreactive protein levels of the AE manganese superoxide dismutase (MnSOD) in prostatic intraepithelial neoplasia and primary prostatic adenocarcinomas in comparison with normal prostate epithelial tissues, whereas foci of cells at the infiltrating edge of invasive prostate adenocarcinomas showed increased MnSOD immunoreactive protein levels [Bostwick et al., 2000]. Of particular interest was the elevation of nuclear 8OHdG in metastatic human prostate cancer in comparison to primary prostate cancer or normal tissues.

This change was also observed in less aggressive vs. highly aggressive prostate cancer cell lines in cell culture, which demonstrated higher 8OHdG levels in PC3 cells than LNCaP cells [Chaiswing et al., 2007]. We also demonstrated that total ROS and RNS levels were higher in LNCaP cells than in PC3 cells during cell growth and during S and G_2/M phases of the cell cycle during logarithmic growth [Chaiswing et al., 2007]. The fact that PC3 cells had more oxidative DNA damage but less ROS/RNS production suggests a possible role for oxidative DNA repair enzymes in determining the levels of oxidative DNA damage observed. Since DNA repair enzymes may be decreased or inactivated in cancer, we suggest the possibility that the changes in 8OHdG observed were due to a decrease of oxidative damage DNA repair enzymes.

Alterations in several redox-modulated transcription factors have been described in prostate cancer. These include p53, nuclear factor-kappa B (NF-κB), and NF-E2-related factor 2 (Nrf2). These transcription factors have been shown to be regulated by oxidative stress and have pro-survival and pro-apoptotic functions often due to modulation of downstream genes that have redox functions [Frohlich, McCabe, Arnold, & Day, 2008; Lee et al., 2008; Royuela, Rodriguez-Berriguete, Fraile, & Paniagua, 2008]. NF-κB is a regulator of genes that control cell proliferation and cell survival genes such as TNF receptor-associated factor (TRAF) 1 and TRAF2. Nrf2 is a regulator of stress-responsive and cytoptotective enzymes which include NAD(P)H:quinone oxidoreductase-1, heme oxygenase-1, glutathione S-transferase, glutathione peroxidase, and thioredoxin. Finally, under normal physiologic conditions, p53 is responsible for transcribing pro-survival redox-modulating genes such as sestrins and GPx1, whereas under conditions of oxidative stress, p53 changes to a pro-apoptotic function, and downstream genes (e.g., PUMA and BAX) are upregulated to initiate apoptosis. In addition, other genes are upregulated that increase oxidative stress, and these modulate the apoptotic process. The functional status of these transcription factors and downstream protein levels has been analyzed in several prostate cancer cell lines under control conditions and following oxidative stress. Most of the prostate cancer cell lines analyzed demonstrated alterations of p53, NF-κB, and Nrf2 functional activities when compared with normal epithelial prostate cells [unpublished data from our laboratory]. We hypothesize that increase of these transcription factors may modulate expression of pro-survival and anti-apoptotic genes even in the presence of anti-cancer agents. Further studies should be directed at overexpression or knockdown of genes that are important in modulation of intracellular redox balance to directly determine effects on cancer behavior, such as cell growth and proliferation; such studies may provide insights into anti-cancer therapeutics based on intracellular redox state modulation.

Extracellular Redox State and Prostate Cancer

Extracellular redox state is one of the factors that serves as an important regulator of tumor cell–microenvironment interactions. Extracellular redox state is determined largely by several known variables: i) redox-modulating proteins that are located on the plasma membrane or outside of cells [including NADPH oxidase, extracellular superoxide dismutase (EC-SOD), thioredoxin-1, peroxiredoxin-IV, and extracellular glutathione peroxidase 3], ii) extracellular thiol/disulfide couples, and iii) ROS/ RNS that are diffusible and thus capable of traveling across plasma membranes into the extracellular space, possibly by free diffusion through lipid bilayers; however, recent data suggest that membrane channels (aquaporin) may also transport uncharged molecules including hydrogen peroxide and nitric oxide [Dynowski, Schaaf, Loque, Moran, & Ludewig, 2008]. ROS/RNS in the extracellular space originates both from within and outside the cell. Under physiological conditions, the extracellular space is known to have a relatively more oxidized redox state than the interior of the cell (Figure 1). In cancer, the extracellular redox state may be abnormal, which may cause particular proteins such as proteases, soluble factors or the extracellular matrix to have altered functions or activities. Many previous studies have focused on the role of the *intracellular* redox state in regulation of prostate cancer cell function. Recent studies from our laboratory have strongly supported an important relationship between the *extracellular* redox state and prostate cancer cell aggressiveness. We were the first to demonstrate that an aggressive prostate cancer cell line (WPE1-NB26) in cell culture exhibited an increased GSH/GSSG ratio in the media in comparison to RWPE1 cells (immortalized but nonmalignant prostate epithelial cells), indicating an alteration of extracellular redox state of prostate cancer cells (reduced state) [Chaiswing et al., 2008]. Significantly, we have shown that superoxide radical production at or near the plasma membrane partially regulated prostate cancer cell invasion. In addition, modulation of superoxide radical levels by overexpression of EC-SOD resulted in alteration of extracellular redox state in aggressive prostate cancer cells. The resulting changes inhibited prostate cancer cell invasion ability *in vitro*. This inhibition correlated with reduced metalloproteinase (MMP2/ MT1-MMP) activities and increased levels of extracellular nitrite. Our results strongly suggested a prominent role of extracellular redox status in regulation of cancer metastasis which may provide opportunities for therapeutic interventions (Figure 2).

Conclusions

Intra-and extracellular redox systems serve an important role(s) in mammalian cells. They function as buffers to maintain redox-regulated protein activities of transcription factors, signaling transduction proteins, receptors, and growth factors. These proteins are involved in several physiological functions. Extracellular redox state works in concert with intracellular redox state to control cell behavior in both normal and cancer cells. Thus, redox states of cancer cells are important parameters that may determine the response of cancers to certain chemotherapeutic agents, radiation, and redox-modulating small molecules. Based on these observations, it is tempting to speculate that treatment aimed at selective modulation of both intra-and extracellular redox states rather than intracellular redox state alone might offer the most effective strategy against prostate cancer. It should be emphasized that both intra- and extracellular redox states can be manipulated in several ways such as overexpression/inhibition of antioxidant proteins, alteration of thiol/disulfide couples levels, and/or modulation of ROS/RNS levels, thus offering several possible pathways for cancer therapeutic intervention. Although the cellular redox system is complex and intricate, continued research of its various components singly and in the context of the complete redox network will yield valuable information that may be translated into the clinical setting.

Understanding the role(s) of crosstalk between intracellular and extracellular redox states will help us further understand their impact on cancer cell progression. Therefore, a noninvasive means of determining the redox status of a tissue should be developed as a useful tool for design of novel cancer therapies.

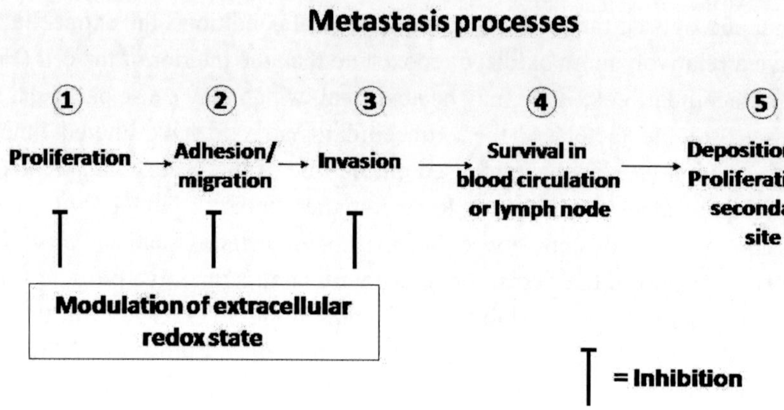

Figure 2. Manipulation and exploitation of tumor environment for cancer therapeutics. We propose new insights into the role of the extracellular redox state in regulation of cancer cell metastasis as indicated in the diagram. Metastases are initiated by cancer cell proliferation at the original site of the primary tumor (step 1). Cells then adhere and migrate to nearby extracellular matrix and/or vessel basement membrane (step 2). Proteolytic enzymes secreted by cancer cells subsequently degrade basement membrane molecules such as collagen type I to produce a pathway for invading cells (step 3). Invasion of cancer cells through extracellular matrix/or vessel basement membrane is the next step in the metastasis cascade, before cancer cells ultimately enter the blood circulation/lymphatic channels (step 3). Cancer cells must survive in the blood/lymphatic circulation and subsequently enter secondary sites and finally proliferate to form metastases (steps 4 and 5). Modulation of extracellular redox state could interfere with ligand-binding receptor processes such as growth factor binding to receptors, thereby inhibiting cell growth and proliferation (step 1). Additionally, the expression and/or conformation of cell adhesion molecules may be altered, thus interfering with the migration ability of cancer cells (step 2). Finally, alterations of extracellular redox state may result in inhibition of protease enzymes such as MMPs that are secreted into the extracellular space, resulting in degradation of the extracellular matrix (step 3). Inhibition of the first three steps may inhibit the ability of cancer cells to survive in blood circulation/lymphatic channels and relocate to secondary sites (steps 4 and 5).

References

Alberti, C. (2008). Genetic and microenvironmental implications in prostate cancer progression and metastasis. *Eur Rev Med Pharmacol Sci., 12*(3), 167-175.

Blank, T. O. & Bellizzi, K. M. (2008). A gerontologic perspective on cancer and aging. *Cancer, 112*(11 Suppl), 2569-2576.

Bostwick, D. G., Alexander, E. E., Singh, R., Shan, A., Qian, J., Santella, R. M., Oberley, L. W., Yan, T., Zhong, W., Jiang, X. & Oberley, T. D. (2000). Antioxidant enzyme expression and reactive oxygen species damage in prostatic intraepithelial neoplasia and cancer. *Cancer, 89*(1), 123-134.

Chaiswing, L., Bourdeau-Heller, J. M., Zhong, W. & Oberley, T. D. (2007). Characterization of redox state of two human prostate carcinoma cell lines with different degrees of aggressiveness. *Free Radic Biol Med., 43*(2), 202-215.

Chaiswing, L., Zhong, W., Cullen, J. J., Oberley, L. W. & Oberley, T. D. (2008). Extracellular redox state regulates features associated with prostate cancer cell invasion. *Cancer Res., 68*(14), 5820-5826.

Dynowski, M., Schaaf, G., Loque, D., Moran, O. & Ludewig, U. (2008). Plant plasma membrane water channels conduct the signalling molecule H2O2. *Biochem J, 414*(1), 53-61.

Frohlich, D. A., McCabe, M. T., Arnold, R. S. & Day, M. L. (2008). The role of Nrf2 in increased reactive oxygen species and DNA damage in prostate tumorigenesis. *Oncogene, 27*(31), 4353-4362.

Gurel, B., Iwata, T., Koh, C. M., Yegnasubramanian, S., Nelson, W. G., & De Marzo, A. M. (2008). Molecular alterations in prostate cancer as diagnostic, prognostic, and therapeutic targets. *Adv Anat Pathol., 15*(6), 319-331.

Halliwell, B. (2007). Oxidative stress and cancer: have we moved forward? *Biochem J, 401*(1), 1-11.

Kemp, M., Go, Y. M. & Jones, D. P. (2008). Nonequilibrium thermodynamics of thiol/disulfide redox systems: a perspective on redox systems biology. *Free Radic Biol Med., 44*(6), 921-937.

Kuppusamy, P., Wang, P., Shankar, R. A., Ma, L., Trimble, C. E., Hsia, C. J. & Zweier, J. L. (1998). In vivo topical EPR spectroscopy and imaging of nitroxide free radicals and polynitroxyl-albumin. *Magn Reson Med., 40*(6), 806-811.

Laurent, A., Nicco, C., Chereau, C., Goulvestre, C., Alexandre, J., Alves, A., Levy, E., Goldwasser, F., Panis, Y., Soubrane, O., Weill, B. & Batteux, F. (2005). Controlling tumor growth by modulating endogenous production of reactive oxygen species. *Cancer Res., 65*(3), 948-956.

Lee, J. T., Lehmann, B. D., Terrian, D. M., Chappell, W. H., Stivala, F., Libra, M., Martelli, A. M., Steelman, L. S. & McCubrey, J. A. (2008). Targeting prostate cancer based on signal transduction and cell cycle pathways. *Cell Cycle, 7*(12), 1745-1762.

Oberley, T. D., Zhong, W., Szweda, L. I. & Oberley, L. W. (2000). Localization of antioxidant enzymes and oxidative damage products in normal and malignant prostate epithelium. *Prostate, 44*(2), 144-155.

Royuela, M., Rodriguez-Berriguete, G., Fraile, B. & Paniagua, R. (2008). TNF-alpha/IL-1/NF-kappaB transduction pathway in human cancer prostate. *Histol Histopathol, 23*(10), 1279-1290.

Rubartelli, A. & Lotze, M. T. (2007). Inside, outside, upside down: damage-associated molecular-pattern molecules (DAMPs) and redox. *Trends Immunol, 28*(10), 429-436.

In: Handbook of Prostate Cancer Cell Research...
Editor: Alan T. Meridith

ISBN: 978-1-60741-954-9

Chapter 18

Telomerase mRNA Detection in Serum of Patients with Prostate Cancer

Luca Dalle Carbonare[*1], *Maria Teresa Valenti*[1], *Sandro Giannini*[2], *Giuseppe Realdi*[2], *and* [1]*Vincenzo Lo Cascio*

[1]Department of Biomedical and Surgical Sciences,
Clinic of Internal Medicine D, University of Verona, Italy
[2]Department of Medical and Surgical Sciences, 1st Medical Clinic,
University of Padova, Italy

Abstract

Purpose

To investigate human telomerase mRNA in the serum of prostate cancer patients.

Experimental Design

The aim of this study was to investigate the telomerase mRNA detection in the serum of patients with a prostate tumour by using Real-Time reverse transcription PCR. The results were compared with biological samples obtained by age-matched normal donors.

Results

Our data demonstrated that telomerase mRNA is detectable in the serum of patients with different prostate tumours whereas it is not amplifiable in normal donors.

* Correspondence: Department of Biomedical and Surgical Sciences, Clinic of Internal Medicine D, University of Verona, Piazzale Scuro, 10, 37134 Verona, Italy, E-mail: luca.dallecarbonare@univr.it, Tel. +39-045-8124684, Fax +39-045-8027496

Conclusions

This marker, revealed with the molecular method of quantitative PCR in serum, may be useful for diagnosing and monitoring prostate cancer patients.

Keywords: Prostate, RNA, serum, RT Real-time PCR

Introduction

Estimates indicate that in the United States 1 out of 10 men will develop prostate cancer in their lifetime. The incidence of prostate cancer increases with age. Since the advent of prostate-specific antigen (PSA) screening, prostate cancer is being detected and treated earlier; however, approximately 10–20% of newly diagnosed prostate cancer cases involve locally advanced disease. It is comparably less common because more early-stage cancer is currently being discovered (1).

Advanced prostate cancer results from any combination of lymphatic, blood, or contiguous local spread. Manifestations of metastatic and advanced prostate cancer may include anemia, bone marrow suppression, weight loss, pathologic fractures, spinal cord compression, pain, hematuria, ureteral and/or bladder outlet obstruction, urinary retention, chronic renal failure, urinary incontinence, and symptoms relating to bony or soft-tissue metastases. Treatment-related symptoms also should be kept in mind, such as rectal bleeding, gross hematuria, and urethrorectal fistula, which sometimes are associated with radiation therapy. The most important and established prognostic factors for prostate carcinoma include the Gleason score, the extent of tumor volume, and the presence of capsular penetration or margin positivity at the time of prostatectomy. The fact that high-grade cancer, particularly the percent presence of Gleason grades 4 and 5, is associated with adverse pathologic findings and disease progression is well recognized (2). Conversely, low-grade tumors also can be biologically aggressive. Differential diagnosis of advanced prostate cancer usually does not present any difficulty; however, certain caveats must be considered. Radiologic findings of bony metastases can mimic Paget disease of bone. Even though bony metastases are blastic in nature, lytic lesions can occur, resulting in pathologic fractures. Osteoporotic fractures due to prolonged luteinizing hormone-releasing hormone (LHRH) therapy must be distinguished from pathologic fractures. Neurological manifestations should be underscored, and elderly patients presenting in the emergency department with sudden onset of weakness of the legs with a history of prostate cancer should raise the suspicion of spinal cord compression, necessitating emergency treatment (spinal cord decompression). Similarly, even though brain metastases with associated neurological manifestations are rare, they do occur with enough frequency to deserve recognition. Lymphomas can present as pelvic masses and bone lesions. Coexistence of lymphomas with prostate cancer also has been reported. Transitional cell carcinoma and sarcoma of the prostate are more common in men with prior pelvic radiation for prostate cancer compared to men who have not received pelvic radiation therapy. Likewise, squamous cell carcinoma of the prostate can be observed in men treated with hormonal therapy. All of these can present as a large pelvic mass with or without metastases.

Hematological workup should include a CBC count and a chemistry profile, including serum creatinine, liver function tests, serum PSA, free-to-total PSA ratio, and acid and alkaline phosphatase (3). Further studies may be ordered if any of these test results demonstrate abnormalities. Urinalysis should be performed. If results are abnormal, i.e., indicating the presence of an infection, urinalysis should be followed by a urine culture, especially if the patient is symptomatic. Certain molecular markers, such as E-cadherin, p53 and p21, DNA ploidy analysis, human kallikrein 2, and microvessel density (histologic marker of tumor angiogenesis) also are being evaluated to help characterize disease progression. Note that elevated PSA levels are not always present in a relatively higher-grade prostate cancer, nor are they always observed in disease progression.

The idea that it should be possible to obtain circulating RNA from peripheral blood is intriguing because this is an easily available potential biomarker representing a non-invasive technique for diagnosing and monitoring tumours (4); in patients with breast, hepatocellular, colorectal and other cancers, circulating RNA has been detected at high levels; in addition, it has been demonstrated that circulating RNA is derived from tumour tissues (5).

The human telomerase, which is composed of two subunits including telomerase RNA template (hTR) and telomerase trabscriptase protein (hTERT), functions as a reverse transcriptase enzyme in the process of telomere synthesis (6). Telomerase activity was detected in 85–100% of cancer patients whereas in normal somatic cells, it was low or undetectable (7).

The aim of this study was to evaluate the telomerase mRNA detection in the serum of patients with prostate tumours by using Real-Time reverse transcription PCR. Our data demonstrated that telomerase mRNA is detectable in the serum of patients with different solid tumours, whereas it is not amplifiable in normal donors; however, in the light of this finding, this marker revealed with the molecular method of quantitative PCR in serum, may be useful for diagnosing and monitoring prostate tumours.

Material and Methods

Patients

We selected 26 patients, consecutively hospitalized in our Institute. Patients have previous diagnosis of prostate tumour or were diagnosed during the hospitalization. Results from these patients were compared with 29 normal age-matched donors, hospitalized in our Institution for cardiovascular or metabolic diseases. All the subjects had given written informed consent.

Serum Preparation

Centrifugation of collected blood and harvesting serum samples were done by using three steps of centrifugation (800 x *g* with 0.45 μm filtration, 1,000 x *g*, and 1,500 x *g*) to decrease lymphocyte to a minimum as previously described (8). The samples were then used for RNA extraction or stored at –80°C.

RNA extraction

Total RNA was extracted from 1 ml of serum using the QIAamp UltraSens Virus kit (Quiagen) with DNAse I treatment. The amount of extracted RNA was quantified by measuring the absorbance at 260 nm. The purity of the RNA was checked by measuring the ratio of the absorbance at 260 and 280 nm, where a ratio ranging from 1.8–2.0 was taken to be pure.

RT Real-Time Quantification of Telomerase mRNA.

cDNA was synthesized in a volume of 15 μl using (Gealthcare) containing random hexamer and 8 μl of RNA extracted from serum samples. The cycling conditions were 10 min at 65°C for RNA and 60 min at 37°C for the mix. The cDNA obtained was stored at -80°C for no more than 6 weeks. The reverse transcription product (5 μl) was used for each subsequent round of Real-Time PCR (50 cycles). A standard curve was constructed with pure and diluted cDNA synthesized with total RNA extracted from 10^7 prostate cancer cells (LNCaP). A new standard curve was prepared for each sequence of samples. The Real Time PCRs, TaqMan analysis, and subsequent calculations were performed using the ABI Prism 7300 Sequence Detection System (Applied Biosystems) which detects the signal from the fluorogenic probes during PCR. hTR, hTERT and B2M (used to verify the success of reverse transcription) cDNA were amplified using TaqMan Assay in a total volume of 50 μl containing 1x TaqMan Universal PCR master mix, no Amperase UNG, and 5 μl of cDNA; pre-designed, Gene-specific primers and probe sets for hTERT gene (Hs00162669-m1), B2M gene (Hs99999907) and hTR gene (PROBE: FAM-CGCGCTGTTTTTCTCGCTGACTTTC; RV: CTAGAATGAACGGTGGAAGGC; FW: GGTGGTGGCCATTTTTTGTC) were obtained from Applied Biosystems. The amount of telomerase cDNA in each sample, obtained as average of triplicates, was determined from the standard curve and the results are expressed as pg equivalent to hTR and hTERT RNA obtained from LNCaP and simplified as hTR/hTERT RNA pg/ml.

Calculation of Performance

Results of hTR and hTERT RT-Real-Time PCR in serum were plotted against disease status (present or absent) and sensitivity, specificity, positive predictive value (PPV) and negative predictive value (NPV) with 95% confidence interval were calculated using a standard formula.

Results

Standard Curve from LNCaP

In each quantitative assay, a strong linear relationship was demonstrated between pg hTR and hTERT RNA and PCR cycles using LNCaP RNA for concentration (r^2>0.99) and the efficiency of the reaction was always over 90%, as assessed by the slope of standard curve for the target gene.

hTR and hTERT RNA Quantification

We obtained good quality of RNA from the serum of patients and donors as evaluated by the ratio of the absorbance at 260 and 280 nm (ranging from 1.8–2.0). In addition, by RT Real Time PCR, beta-2-microglobulin was positive in all samples (ct<30).

To quantify the telomerase RNA in the serum of patients, we performed the Real Time PCR by quantitative relative standard curve, and we obtained the amplification corresponding to hTR and hTERT RNA in the serum from patients and donors.

The hTR RNA in serum was positive in 88.4% of patients (ct<50) whereas 2 normal donors presented low levels of hTR RNA amplifiable in serum (7.7%) (ct>50); on the contrary, in remaining donors, RNA was not amplifiable. The hTERT RNA in serum resulted lower sensible but more specific than hTR RNA as was positive in 46.1% of patients and negative in normal donors.

Performance of the Assay

Sensitivity, specificity positive predictive value and negative predictive value of this molecular test calculated from results obtained from patients and normal donors sera in are shown in Table 1. The calculated performance with 95% Confidence Interval is shown in Table 2.

Table 1A. hTR expression in cancer patient and donor serum.

N° Tumour Patients	Test results	RNA STEAP pg/ml
		Median 9.4
23	Positive	Mean 15.34
		Ranks range 0.2-49
3	Negative	not detectable

N° Donors	Test results	RNA STEAP pg/ml
		Median –
1	Positive	Mean 0.1
		Ranks range
28	Negative	not detectable

Table 1B. hTERT expression in cancer patient serum.

N° Tumour Patients	Test results	RNA STEAP pg/ml
		Median 1.3
12	Positive	Mean 3.3
		Ranks range 0.8-5.1
14	Negative	not detectable
26	Negative	not detectable

N° Donors	Test results	RNA STEAP pg/ml
0	Positive	not detectable
29	Negative	not detectable

Table 2. Performance of hTR RT-Real-Time PCR in serum of tumour patients.

Parameter	Value	95% Confidence Interval
Sensitivity	88 %	76.1-100
Specificity	96.5 %	89.9-100
PPV	95.8%	87.8-100
NPV	90.3 %	79.9-100

Parameter	Value	95% Confidence Interval
Sensitivity	46.1 %	26.9-65.3
Specificity	100 %	100-100
PPV	100%	100-100
NPV	67.4 %	53.4-81.4

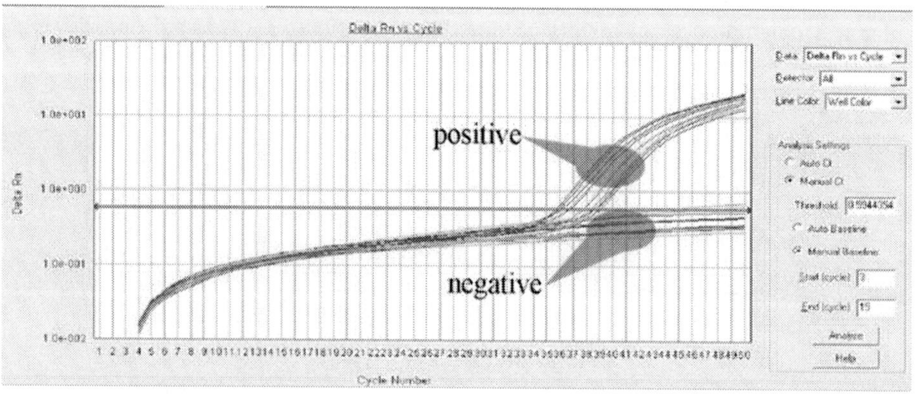

Figure 1. Telomerase mRNA amplification in serum from cancer patients and non malignant individuals by RT Real Time PCR.

Discussion

Eukaryotic chromosomal ends consist of repeating DNA sequences (TTAAGG) termed telomeres. An enzyme that adds telomeric repeats onto chromosomal ends is telomerase. This enzyme is composed of two subunits; hTR and hTERT (9). Accordingly, both hTR and hTERT are necessary for telomerase activity. Therefore, the detection of hTR and hTERT mRNA is a guarantee for the present of telomerase activity (10). In most somatic cells in which telomerase activity is undetectable, telomeric sequences are lost with each cell division because of the end-replication problem. Unlike healthy cells, most malignant human cells are capable of escaping senescence and sustaining infinite proliferation through the activation of telomerase to stabilize their telomere length (11). Human cancer cells, unlike their normal counterparts, have shed the molecular restraints to limited cell growth and are immortal. How cancer cells manage this at the molecular level is beginning to be understood. Human cells must overcome two barriers to cellular proliferation. The first barrier, referred to as senescence, minimally involves the p53 and Rb tumor-suppressor pathways. Inactivation of these pathways results in some extension of lifespan. However, inactivation of these pathways is insufficient for immortalization. As normal cells undergo repeated rounds of DNA replication, their telomeres shorten due to the inability of traditional DNA polymerases to completely replicate the end of the chromosomal DNA. This shortening continues until the cells reach a second proliferative block referred to as crisis, which is characterized by chromosomal instability, end-to-end fusions, and cell death. Stabilization of the telomeric DNA through either telomerase activation or the activation of the alternative mechanism of telomere maintenance (ALT) is essential if the cells have to survive and proliferate indefinitely. Conversely, loss of telomere stabilization by an already-immortalized cell results in loss of immortality and cell death. Together, these aspects indicate that telomere maintenance is a critical component of immortality (12). In fact, a clear link between telomere maintenance and the process of immortalization, tumorogenesis and chromosomal stability has been established (13). In addition, there is accumulating evidence that when only a few telomeres are short, they form end-associations, leading to a DNA damage signal resulting in replicative senescence (a cellular growth arrest, also called the M1 stage). In the absence of cell-cycle checkpoint pathways (e.g. p53 and or p16/Rb), cells bypass M1 senescence and telomeres continue to shorten eventually resulting in crisis (also called the M2 stage). M2 is characterized by many 'uncapped' chromosome ends, end-fusions, chromosome breakage fusion-bridge cycles, mitotic catastrophe and a high fraction of apoptotic cells. In a rare M2 cell, telomerase (a cellular reverse transcriptase) can be reactivated or up-regulated, resulting in indefinite cell proliferation. This cellular immortalization is a potentially rate-limiting step in carcinogenesis that is important for the continuing evolution of most advanced cancers. Thus, limiting or restricting telomerase function may be a novel working paradigm for the design of complementary strategies in cancer therapies. Recent progress has shed light on the hormonal control of telomerase by estrogens, progesterone, vitamin D3, and GH-RH(14).

Traditionally, telomerase activity has been assessed based on a biochemical primer extension assay, the inefficiency and low sensitivity of which, together with the low amounts of telomerase activity, greatly limit the application of the assay in primary human tumors

(15). Therefore, the detection of telomerase mRNA by real-time RT-PCR was used in this study. This assay determined the expression of telomerase by measuring the amount of the mRNA encoding the two subunits hTR and hTERT.

The presence of telomerase RNA in the serum of prostate cancer patients is probably the reflexion of this RNA in the tumour cells; in addition, a high percentage of sera from normal donors resulted negative and an important percentage of tumour patient sera were positive. The approach by RT Real-Time PCR allowed also a quantification of this marker in the serum of patients. On the basis of this finding, telomerase RNA should be utilized as a useful cancer marker and, as previous studies have shown that this protein is frequently expressed in different neoplastic diseases, the relative serum analysis may be applied to the detection of a broad range of cancers.

Further and larger studies are needed to confirm the role of this marker in detecting tumours. Anyway, we think that this non-invasive method can be useful for detecting prostate tumour diseases at a molecular level and can contribute to obtain an accurate diagnosis and follow-up in a broad range of cancer patients. In addition, this method may have implications also for minimal residual disease detection.

References

[1] Nelson, MA; Porterfield, BW; Jacobs, ET; Clark, LC. Selenium and prostate cancer prevention. *Semin Urol Oncol.*, May 1999, 17(2), 91-6.

[2] Allsbrook, WC; Mangold, KA; Yang, X; et al. The Gleason grading system: an overview. *J Urologic Path*, 1999, 10, 141-57.

[3] American Society for Therapeutic Radiology and Oncology Consensus Panel. Consensus statement: guidelines for PSA following radiation therapy. *Int J Radiat Oncol Biol Phys.*, Mar 15 1997, 37(5), 1035-41.

[4] Zhou, H; Xu, W; Qian, H; Yin, Q; Zhu, W; Yan, Y. Circulating RNA as a novel tumour marker: An in vitro study of the origins and characteristics of extracellular RNA. *Cancer Lett.*, 2008, 259(1), 50-60.

[5] Fleischhacker, M. Biology of circulating mRNA still more questions than answers, *Ann NY Acad Sci.*, 2006, 1075, 40-49.

[6] Bachand, F; Triki, I; Autexier, C. Human telomerase RNA-protein interactions *Nucleic Acids Res.*, 2001, Aug 15, 29(16), 3385-93.

[7] Yan, P; Coindre, JM; Benhattar, J; Bosman, FT; Guillou, L. Telomerase activity and human telomerase reverse transcriptase mRNA expression in soft tissue tumors: correlation with grade, histology, and proliferative activity. *Cancer Res.*, 1999, 59, 3166-3170.

[8] Miura, N; Shiota, G; Nakagawa, T; Maeda, Y; Sano, A, Marumoto, A; et al., Sensitive detection of hTERT mRNA in the serum of patients with hepatocellular carcinoma, *Oncology*, 2003, 64, 430-4.

[9] Bachand F, Triki I, Autexier C. Human telomerase RNAprotein interactions. *Nucleic Acids Res* 2001; 29, 3385-3393.

[10] Novakovic, S; Hocevar, M; Zgajnar, J; Besic, N; Stegel, V. Detection of telomerase RNA in the plasma of patients with breast cancer, malignant melanoma or thyroid cancer. *Oncol Rep.*, 2004, 11, 245-252.

[11] Li, YR; Wu, JM; Wang, L; Huang, X; Shi, J; Hu, LH. Human telomerase reverse transcriptase expression and its clinical significance in laryngeal squamous cell carcinoma. *Acta Otolaryngol*, 2005, 125, 409-414.

[12] Stewart, SA; Weinberg, RA. Telomerase and human tumorigenesis. *Semin Cancer Biol.*, 2000, 10, 399-406.

[13] Masutomi, K; Hahn, WC. Telomerase and tumorigenesis. *Cancer Lett.*, 2003, 194, 163-172.

[14] Orlando, C; Gelmini, S. Telomerase in endocrine and endocrine-dependent tumors. *J. Steroid Biochem Mol Biol.*, 2001, 78, 201-214.

[15] Mabruk, MJ; O'Flatharta, C. Telomerase: is it the future diagnostic and prognostic tool in human cancer? *Expert Rev Mol Diagn.*, 2005, 5, 907-916.

In: Handbook of Prostate Cancer Cell Research...
Editor: Alan T. Meridith
ISBN: 978-1-60741-954-9

Chapter 19

Piezoelectric Interaction with Prostatic Crystalloids May Explain the Relationship between Environmental Electromagnetic Field and Prostate Cancer

Kamyar Ghabili [1*]***, Paul S. Agutter*** [2] ***and Mohammadali M. Shoja*** [1]

[1]Tuberculosis and Lung Disease Research Center,
Tabriz University of Medical Sciences, Daneshgah St., Tabriz, Iran
[2]Theoretical Medicine and Biology Group, 26 Castle Hill, Glossop SK13 7RR, UK

Abstract

Electromagnetic fields (EMF) may contribute to the pathogenesis of prostate cancer, but the mechanism is obscure. Numerous calcium/phosphorus crystalloids with various morphologies have been found in the normal prostate gland and in both benign and malignant prostate lesions. Their distribution within the gland is heterogeneous and diffuse. We propose that an environmental EMF may exert a piezoelectric effect on these crystalloids, resulting in simultaneous, multidirectional and diffuse compression or expansion and consequently a slight mechanical distortion of the prostate. Such distortion is known to alter gene expression in many kinds of cells and therefore cell phenotype; such alterations can suppress apoptosis. In this paper we explore the scientific and clinical evidence favoring this hypothesis and present a mathematical model of the proposed cell-mechanical effect.

Keywords: Prostate cancer; Crystalloid; Electromagnetic field; Piezoelectric effect; Cell mechanics

---

* Correspondence: Kamyar Ghabili, No. 47, Near Aber Bank Sepah, Crossing Abresan and Chaikenar, Tabriz, Iran, Tel: +989144106136, Fax: +984113363846, E-mail: kghabili@gmail.com.

Prostate Cancer and Tissue Crystalloids

Intraluminal crystalloids were first demonstrated histologically by Holmes (1977) in 23% of 335 prostatic cancer cases. Christian et al. (2005) found them in needle biopsy specimens from 42% of such cancers. The crystalloids may be rectangular, hexagonal, triangular or rod-shaped; radiographic microanalysis has shown that they contain abundant sulfur, calcium and phosphorus (Bostwick et al., 2004). The association between these crystalloids and prostate cancer is not unequivocal (Henneberry et al., 1997; Hu et al., 1998); they also occur in normal prostate tissue (Hu et al., 1998). Nevertheless, light microscopic detection of crystalloids is considered helpful in the diagnosis of prostatic cancer, particularly when the tissue is artificially distorted or when the tumor is so well-differentiated that it can be confused with a benign lesion (Ro et al., 1986). Their potential role in malignant transformation of prostate tissue has not been explored.

Prostate cancer is the most commonly diagnosed cancer and the second leading cause of cancer deaths (Mills and Yang, 2003), but no single etiology has been established (Zeegers et al., 2004). Various different factors may be implicated. Exposure to electromagnetic fields (EMF) has been provisionally identified as an environmental risk factor (Charles et al., 2003), but it has not been causally or mechanistically linked to prostate cancer (Ahlbom et al., 2001). Nevertheless, causal connections can be envisaged: for instance, EMF are known to alter melatonin and testosterone secretion (Charles et al., 2003) or may induce cell injury directly by generating free radicals (Yokus et al., 2005).

Effects of Small Electrical Currents in Biological Systems

Small electrical currents have been detected in many biological systems, including developing embryos, and they may have at least three distinct effects on cells (McLeod, 1992): alteration of the plasma membrane potential; modification of ionic currents and ion distributions at the extracellular surface; and imposition of mechanical forces at the cell surface, and *ipso facto* throughout the cell via the cytoskeleton. Small electric currents also help to direct cell migration during wound healing (Nuccitelli, 1990, 1992; Reid et al., 2007).

Small applied electric fields cause phenotypic changes in vascular endothelial cells (Bouaziz et al., 1997) and regenerating nerve fibers (Valentini et al., 1992). Embryonic fibroblasts migrate towards the cathode in DC fields of 1-10 mV mm^{-1} and become orientated perpendicularly to DC fields of 50-150 mV mm^{-1}; at higher field strengths, they elongate (Erickson and Nuccitelli, 1984). This illustrates the fact that biological systems are highly sensitive to small currents and electric fields. The embryonic fibroblasts, for example, can respond to potential differences as low as 0.2 mV across a single-cell width.

Piezoelectric Effects in Biology

A piezoelectric effect occurs when a local electrical/electromagnetic current leads to the compression or expansion of crystals in a particular direction, or vice versa (Chakkalakal and Johnson, 1981). Piezoelectric transduction of mechanical stimuli produces small electrical signals in some biological systems and leads to physiological effects. This phenomenon is best established in two mammalian systems: the collagen matrix of bone, where it is essential for mineralization; and the cochlear outer hair cell, where it converts sound energy into nerve impulses. The outer hair cell of the cochlea shows a number of voltage-dependent properties: mechanical stiffness, capacitance and ability to generate force at high frequencies. Mechanical deformation of bone collagen produces piezoelectric dipoles that promote the electrochemical precipitation of hydroxyapatite. On the other hand, mineralization induces significant changes in the dielectric spectra of deformed collagen (Noris-Sua´rez et al., 2007).

There is a complicated and incompletely understood relationship between transmembrane voltages and electromechanical strains (see Spector et al., 2006, for review). Nevertheless, the relationship appears to be precise. For instance, piezoelectric resonance may serve to identify species-specific vocalizations in some mammals (Weitzel et al., 2003), and this may illustrate a widespread phenomenon in biology. A piezoelectric olfactory sensor has been described in rats (Ko and Park, 2005), and electromechanical coupling has been characterized in the purple membrane of Halobacterium salinarium (Kietis et al., 2001).

We suggest that piezoelectric transduction may account for the pathogenic link between EMF and prostate cancer (Ghabili et al., 2008).

Proposed Mechanistic Connection between EMF and Prostate Cancer

The prostate crystalloids resemble hydroxyapatite in composition, and their morphology, distribution and cell-like size suggest the possibility of an electromechanical response analogous to that of bone. We therefore hypothesize that a piezoelectric effect contributes to prostate cancer following exposure to EMF.

Assuming that the prostatic crystalloids are heterogeneously distributed within the prostate, an environmental EMF may effect a slight change in their dimensions in different directions simultaneously. Such multidirectional and diffuse compression or expansion of the crystalloids would induce mechanical distortion of the prostate gland and therefore of its component cells. Among other effects, this would significantly alter plasma membrane function; and mechanical stress operating via the cytoskeleton may alter gene expression (cf. Lee et al., 1982; Aikawa et al., 2002; Mayr et al., 2002; Martinez-Ramos et al., 2005).

In this context, Farge (2003) has demonstrated that mechanical pressure may increase the expression of *Twist*, which is essential for embryonic development in Drosophila. The human homologue of this gene is known to be involved in the development of certain cancers through the suppression of apoptosis (Farge, 2003). Moreover, mechanical stress enhances matrix metalloproteinase expression, which is up-regulated in several human cancers (Rundhaug, 2003). Other mechanisms may also connect mechanical stress with cell behavior and in particular with tissue morphogenesis and remodeling; for example, it promotes the generation of reactive oxygen species, which can cause oxidative DNA damage and mutations (Mayr et al., 2002).

Current and Prospective Evidence

Our proposal that prostate cancer is (at least in part) a consequence of mechanical distortion of the gland caused by piezoelectric interaction between EMF and prostatic crystalloids has some indirect support. First, the hypothesis obviously provides a mechanistic link between EMF and the increased risk of prostate cancer, for which there is evidence (Charles et al., 2003). Secondly, prostate cancer is mechanically stiffer than healthy tissue (Jalkanen et al., 2006), perhaps because it contains more crystalloids with altered morphology and distribution. Thirdly, clinical practice affords circumstantial support: low-intensity electric fields act synergistically with chemotherapy in treating metastatic prostate cancer (Plotnikov et al., 2006), and metastatic prostate cancer cells are strongly galvanotactic (Siwy et al., 2003; Mycielska and Djamgoz, 2004).

However, our proposal stands in need of direct experimental validation. The pressure differential formula (see below) can be tested in the laboratory. If the volume percentage occupied by the crystalloids and their volumetric coefficient of expansion are measured and compared with the maximal pressure differential with respect to a reference pressure and an external field of known strength, a quantitative relationship may be established between the strength of the applied EMF and the range of pressures to which the cells are subjected. This information will suffice for simulation of cell mechanical stress by applying external pressures within this range to cellular samples. The effects of the external pressure on cellular dynamics can then be noted.

Within the range of pressures calculated from the external field applied, we can determine the threshold value at which the field exerts observable effects on cellular dynamics. The absence of any such threshold value will necessitate rejection of our hypothesis.

Mathematical Model

Calculation of Cell Mechanical Stress

A fluctuating environmental EMF causes expansion/compression of the prostatic crystalloids (piezoelectric effect). Let v^0 be the volume of a generic crystalloid in the absence of external EMF, and let v (ΔE) be the volume of the same crystalloid in an external field ΔE. If ΔE is small we have

$$v\left(\overrightarrow{\Delta E}\right)= v^0 + \sum_{i=1}^{3} \frac{\partial v}{\partial E_i}\bigg|_{\overrightarrow{\Delta E}=0} \Delta E_i = v^0\left(1+\sum_{i=1}^{3}\epsilon_i \Delta E_i\right)$$

where

$$\epsilon_i = \frac{1}{v^0}\frac{\partial v}{\partial E_i}\bigg|_{\overrightarrow{\Delta E}=0}$$

is the linear expansion coefficient of the crystalloid. Let the external EMF have frequency ω and average amplitude $\langle E \rangle$ along any axis; if

$$\epsilon = \sum_{i=1}^{3} \epsilon_i$$

is the volumetric expansion coefficient of the crystalloid, we have

$$v\left(\overrightarrow{\Delta E}\right) = v^0\left(1 + \epsilon\langle E\rangle \sin(\omega t)\right)$$

Let V_0 be the volume of a sample having N_0 prostatic crystalloids in the absence of external field. The volume of the sample in the presence of an EMF will be:

$$V = V_0 + N_0 v^0 \epsilon\langle E\rangle \equiv V_0 + \Delta V_0$$

If the sample has temperature T_0 and pressure P_0 in the absence of EMF and if we assume that the crystalloids behave like an ideal gas at a fixed temperature T_0, then

$$P(V_0 + \Delta V_0) = N_0 k T_0$$

where P is the pressure due to the crystalloids and k is the Boltzmann constant. For $\Delta V_0 << V_0$

$$P = \frac{N_0 k T_0}{V_0 + \Delta V_0} \cong \frac{1}{V_0}\left(1 - \frac{\Delta V_0}{V_0}\right) N_0 k T_0 = P_0\left(1 - \frac{\Delta V_0}{V_0}\right)$$

Let ΔP be the pressure differential from equilibrium in the presence of an external field. We then have:

$$\frac{\Delta P}{P_0} = -\frac{N_0 v^0 \epsilon\langle E\rangle}{V_0}\sin(\omega t) = -\delta V_0 \epsilon\langle E\rangle \sin(\omega t)$$

where δV_0 is the portion of the volume filled with crystalloids with respect to the sample volume in absence of external field. Cells are subjected to a maximum pressure differential:

$$\frac{\Delta P_{Max}}{P_0} = \delta V_0 \varepsilon\langle E\rangle$$

depending on the amplitude of the external electric field $\langle E \rangle$.

References

Ahlbom, A., Cardis, E., Green, A., Linet, M., Savitz, D. & Swerdlow, A. (2001). Review of the epidemiologic literature on EMF and health. *Environ Health Perspect*, *109*(Suppl. 6), 911-933.

Aikawa, R., Nagai, T., Kudoh, S., Zou, Y., Tanaka, M., Tamura, M., Akazawa, H., Takano, H., Nagai, R. & Komuro, I. (2002). Integrins play a critical role in mechanical stress-induced p38 MAPK activation. *Hypertension*, *39*, 233-238.

Bostwick, D. G., Burke, H. B., Djakiew, D., Euling, S., Ho, S. M., Landolph, J., Morrison, H., Sonawane, B., Shifflett, T., Waters, D. J. & Timms, B. (2004). Human prostate cancer risk factors. *Cancer*, *101*(Suppl. 10), 2371-2390.

Bouaziz, A., Richert, A. & Caprani, A. (1997). Vascular endothelial cell responses to different electrically charged poly(vinylidene fluoride) supports under static and oscillating flow conditions. *Biomaterials*, *18*, 107-112.

Chakkalakal, D. A. & Johnson, M. W. (1981). Electrical properties of compact bone. *Clin Orthop.*, *161*, 133-145.

Charles, L. E., Loomis, D., Shy, C. M., Newman, B., Millikan, R., Nylander-French, L. A. & Couper, D. (2003). Electromagnetic fields, polychlorinated biphenyls, and prostate cancer mortality in electric utility workers. *Am J Epidemiol.*, *157*, 683-691.

Christian, J. D., Lamm, T. C., Morrow, J. F. & Bostwick, D. G. (2005). Corpora amylacea in adenocarcinoma of the prostate: incidence and histology within needle core biopsies. *Mod Pathol.*, *18*, 36-39.

Erickson, C. A. & Nuccitelli, R. (1984). Embryonic fibroblast motility and orientation can be influenced by physiological electric fields. *J Cell Biol.*, *98*, 296-307.

Farge, E. (2003). Mechanical induction of Twist in the Drosophila foregut/stomodeal primordium. *Curr Biol.*, *13*, 1365-1377.

Ghabili, K., Shoja, M. M. & Agutter, P. S. (2008). Piezoelectricity and prostate cancer: Proposed interaction between electromagnetic field and prostatic crystalloids. *Cell Biol Int.*, *32*, 688-691.

Henneberry, J. M., Kahane, H., Humphrey, P. A., Keetch, D. W. & Epstein, J. I. (1997). The significance of intraluminal crystalloids in benign prostatic glands on needle biopsy. *Am J Surg Pathol.*, *21*, 725-728.

Holmes, E. J. (1977). Crystalloids of prostatic carcinoma: relationship to Bence-Jones crystals. *Cancer*, *39*, 2073-2080.

Hu, J. C., Palapattu, G. S., Kattan, M. W., Scardino, P. T. & Wheeler, T. M. (1998). The association of selected pathological features with prostate cancer in a single-needle biopsy accession. *Human Pathol.*, *29*, 1536-1538.

Jalkanen, V., Andersson, B. M., Bergh, A., Ljungberg, B. & Lindahl, O. A. (2006). Resonance sensor measurements of stiffness variations in prostate tissue in vitro - a weighted tissue proportion model. *Physiol Meas.*, *27*, 593-603.

Kietis, P., Vengris, M. & Valkunas, L. (2001). Electrical-to-mechanical coupling in purple membranes: membrane as electrostrictive medium. *Biophys J.*, *80*, 1631-1640.

Ko, H. J. & Park, T. H. (2005). Piezoelectric olfactory biosensor: ligand specificity and dose-dependence of an olfactory receptor expressed in a heterologous cell system. *Biosens Bioelectron*, *20*, 1327-1332.

Lee, R. C., Rich, J. B., Kelley, K. M., Weiman, D. S. & Mathews, M. B. (1982). A comparison of in vitro cellular responses to mechanical and electrical stimulation. *Am Surg.*, *48*, 567-574.

Martinez-Ramos, I., Maya-Mendoza, A., Gariglio, P. & Aranda-Anzaldo, A. (2005). A global but stable change in HeLa cell morphology induces reorganization of DNA structural loop domains within the cell nucleus. *J Cell Biochem.*, *96*, 79-88.

Mayr, M., Hu, Y., Hainaut, P. & Xu, Q. (2002). Mechanical stress-induced DNA damage and racp38MAPK signal pathways mediate p53-dependent apoptosis in vascular smooth muscle cells. *FASEB J*, *16*, 1423-1425.

McLeod, K. J. (1992). Microelectrode measurements of low frequency electric field effects in cells and tissues. *Bioelectromagnetics*, (Suppl. 1), 161-178.

Mills, P. K. & Yang, R. (2003). Prostate cancer risk in California farm workers. *J Occup Environ Med.*, *45*, 249-258.

Mycielska, M. E. & Djamgoz, M. B. (2004). Cellular mechanisms of direct-current electric field effects: galvanotaxis and metastatic disease. *J Cell Sci.*, *117*, 1631-1639.

Noris-Suárez, K., Lira-Olivares, J., Ferreira, A. M., Feijoo, J. L., Suárez, N., Hernández, M. C. & Barrios, E. (2007). In vitro deposition of hydroxyapatite on cortical bone collagen stimulated by deformation-induced piezoelectricity. *Biomacromolecules*, *8*, 941-948.

Nuccitelli, R. (1990). Vibrating probe technique for studies of ion transport. In J. K. Foskett, & S. Grinstein, (Eds.), *Noninvasive techniques in cell biology* (1st edition, pp. 273-310). New York: Wiley-Liss.

Nuccitelli, R. (1992). Endogenous ionic currents and DC electric fields in multicellular animal tissues. *Bioelectromagnetics*, (Suppl. 1), 147-157.

Plotnikov, A., Niego, B., Ophir, R., Korenstein, R. & Keisari, Y. (2006). Effective treatment of mouse metastatic prostate cancer by low electric field enhanced chemotherapy. *Prostate*, *66*, 1620-1630.

Reid, B., Nuccitelli, R. & Zhao, M. (2007). Non-invasive measurement of bioelectric currents with a vibrating probe. *Nat Protoc.*, *2*, 661-669.

Ro, J. Y., Ayala, A. G., Ordonez, N. G., Cartwright Jr, J. & Mackay, B. (1986). Intraluminal crystalloids in prostatic adenocarcinoma immunohistochemical, electron microscopic, and X-ray microanalytic studies. *Cancer*, *57*, 2397-2407.

Rundhaug, J. E. (2003). Matrix metalloproteinases, angiogenesis, and cancer. *Clin Cancer Res.*, *9*, 551-554.

Siwy, Z., Mycielska, M. E. & Djamgoz, M. B. (2003). Statistical and fractal analyses of rat prostate cancer cell motility in a direct current electric field: comparison of strongly and weakly metastatic cells. *Eur Biophys J*, *32*, 12-21.

Spector, A. A., Deo, N., Grosh, K., Ratnanather, J. T. & Raphael, R. M. (2006). Electromechanical models of the outer hair cell composite membrane. *J Membr Biol.*, *209*, 135-152.

Valentini, R. F., Vargo, T. G., Gardella Jr, J. A. & Aebischer, P. (1992). Electrically charged polymeric substrates enhance nerve fibre outgrowth in vitro. *Biomaterials*, *13*, 183-190.

Weitzel, E. K., Tasker, R. & Brownell, W. E. (2003). Outer hair cell piezoelectricity: frequency response enhancement and resonance behavior. *J Acoust Soc Am.*, *114*, 1462-1466.

Yokus, B., Cakir, D. U., Akdag, M. Z., Sert, C. & Mete, N. (2005). Oxidative DNA damage in rats exposed to extremely low frequency electromagnetic fields. *Free Radic Res.*, *39*, 317-323.

Zeegers, M. P., Friesema, I. H., Goldbohm, R. A. & van den Brandt, P. A. (2004). A prospective study of occupation and prostate cancer risk. *J Occup Environ Med.*, *46*, 271-279.

In: Handbook of Prostate Cancer Cell Research... ISBN: 978-1-60741-954-9
Editor: Alan T. Meridith

Chapter 20

Developments, Directions and Problems in the Management of Prostate Cancer

Oluwakemi Obajimi
Department of Medicine, University of Medicine and Dentistry of New Jersey, Robert Wood Johnson Medical School, 401 Haddon Avenue, Camden, NJ 08103, USA and Cooper Cancer Institute, Cooper University Hospital, Camden, NJ 08103, USA

In men, prostate cancer (PCa) is the most common malignancy and a leading cause of cancer-related deaths in the western world. So far, measurements of prostate specific antigen (PSA) levels in the blood has proved fruitful diagnostically, increasing the number of prostate cancer patients diagnosed with the disease at its early-stage, when it is still confined to the prostate gland. This allows for early-detection and treatment of the disease at the early stages resulting in a significant decline in mortality rates [1]. However, current treatments have limited efficacies. Following androgen ablation therapy, there is the onset of the hormone-refractory or androgen-independent prostate cancer. In general, a major problem faced in the management of prostate cancer is the lack of a reliable prognostic biomarker that will predict the clinical course of the disease.

The limited therapeutic options available to patients with refractory prostate cancer underscore the need for the development of novel therapies. Furthermore, the identification and characterization of pathways and molecular targets driving prostate cancer cell growth is crucial for future drug discovery and development. With PCa progression, already dysfunctional cells evolve and develop new characteristic features enabling their survival in an androgen-ablated environment. Importantly, most androgen-independent prostate cancers express the androgen receptor (AR) [2] and despite the depleted androgen environment resulting from anti-androgen treatments, AR signaling remains functional [3]. Activation of AR regulates the transcriptional activity of a diverse array of AR target genes involved in many cellular processes including proliferation, differentiation and apoptosis. Since PCa cells have the ability to bypass the effects of androgen ablation therapy by developing growth mechanisms which promote survival at depleted androgen levels [4], an important role exists for novel strategies targeting AR signaling and its converging points of crosstalk with other major signaling pathways driving cell proliferation and concomitantly capable of inhibiting apoptosis. With the continuous extension of knowledge, through research, regarding the genes, proteins and multilinked pathways that significantly contribute to prostate cancer progression from androgen-dependent early-stage PCa to hormone-refractory metastatic late-stage disease, new and optimized therapies may emerge.

Presently, PSA does not predict survival and therefore is unsuitable for the assessment of drug response. However, the rate at which PSA levels increase was recently correlated with the period of survival (time-to-death) in treated patients with hormone-refractory metastatic PCa [5]. With the advent of cutting-edge technologies, such as proteomic and genomic tools, and our increasing understanding of the pathways that drive androgen-independent prostate cancer, it is anticipated that with the coming decade, scientific research will identify reliable prognostic markers able to predict the outcome of the disease especially in patients with early-stage PCa.

References

[1] Ward, E., et al. (2008). Association of Insurance with Cancer Care Utilization and Outcomes. *CA Cancer J Clin.*, *58*(1), 9-31.

[2] van Bokhoven, A., et al. (2003). Molecular characterization of human prostate carcinoma cell lines. *Prostate*, *57*(3), 205-225.

[3] Tararova, N. D., et al. (2007). Prostate cancer cells tolerate a narrow range of androgen receptor expression and activity. *The Prostate*, *67*(16), 1801-1815.

[4] Suzuki, H., et al. (2003). Androgen receptor involvement in the progression of prostate cancer. *Endocrine-Related Cancer*, *10*(2), 209-216.

[5] Rozhansky, F., et al. (2006). Prostate-specific antigen velocity and survival for patients with hormone-refractory metastatic prostate carcinoma. *Cancer*, *106*(1), 63-67.

Index

#

A

B

C

E

F

G

H

I

J

K

L

M

N

O

P

Q

R

S

S

T

U

V